Mathematics
in Our World

Mathematics
in Our World

Allan G. Bluman

Professor Emeritus,
Community College of Allegheny County

Boston   Burr Ridge, IL   Dubuque, IA   Madison, WI   New York   San Francisco   St. Louis
Bangkok   Bogotá   Caracas   Kuala Lumpur   Lisbon   London   Madrid   Mexico City
Milan   Montreal   New Delhi   Santiago   Seoul   Singapore   Sydney   Taipei   Toronto

Higher Education

MATHEMATICS IN OUR WORLD

Published by McGraw-Hill, a business unit of The McGraw-Hill Companies, Inc., 1221 Avenue of the Americas, New York, NY 10020. Copyright © 2005 by The McGraw-Hill Companies, Inc. All rights reserved. No part of this publication may be reproduced or distributed in any form or by any means, or stored in a database or retrieval system, without the prior written consent of The McGraw-Hill Companies, Inc., including, but not limited to, in any network or other electronic storage or transmission, or broadcast for distance learning.

Some ancillaries, including electronic and print components, may not be available to customers outside the United States.

 This book is printed on recycled, acid-free paper containing 10% postconsumer waste.

2 3 4 5 6 7 8 9 0 CTP/CTP 0 9 8 7 6
1 2 3 4 5 6 7 8 9 0 VNH/VNH 0 9 8 7 6 5 4

ISBN-13: 978-0-07-245107-8 / ISBN-10: 0-07-245107-6 (Student Edition)
ISBN-13: 978-0-07-294063-3 / ISBN-10: 0-07-294063-8 (Instructor's Edition)

Publisher: *William K. Barter*
Executive editor: *Robert E. Ross*
Director of development: *David Dietz*
Developmental editor: *Peter Galuardi*
Executive marketing manager: *Marianne C. P. Rutter*
Senior project manager: *Vicki Krug*
Lead production supervisor: *Sandy Ludovissy*
Media project manager: *Sandra M. Schnee*
Senior media technology producer: *Jeff Huettman*
Senior coordinator of freelance design: *Michelle D. Whitaker*
Cover/interior designer: *Elise Lansdon*
Cover images: Flower close-up © *Peter Holst/Getty Images;* Frost pattern © *Phil Degginger/Getty Images;*
Panther Chameleon © *Gail Shumway/Getty Images;* Soap bubbles © *Martin Rogers/Getty Images*
Senior photo research coordinator: *Lori Hancock*
Photo research: *Toni Michaels/PhotoFind, LLC*
Supplement producer: *Brenda A. Ernzen*
Compositor: *Interactive Composition Corporation*
Typeface: *10/12 Times Roman*
Printer: *CTPS*

The credits section for this book begins on page 760 and is considered an extension of the copyright page.

Library of Congress Cataloging-in-Publication Data

Bluman, Allan G.
    Mathematics in our world / Allan G. Bluman. — 1st ed.
      p.   cm.
    Includes index.
     ISBN 0–07–245107–6 (acid-free paper)
     1. Mathematics. I. Title.

    QA39.3.B597   2005
    510—dc22
                          2003061412
                          CIP

www.mhhe.com

To Betty Bluman, Earl McPeek,
and Dr. G. Bradley Seager, Jr.

Contents

Chapter Five

The Real Number System 159

Chapter Six

Other Mathematical Systems 239

Chapter Seven

Topics in Algebra 267

Chapter Thirteen

Voting Methods 689

Mathematics in Our World: Academy Awards 690

Preface

Purpose

After many years of teaching a course in mathematics for liberal arts majors, I decided to write a textbook to better meet the needs of my students. My primary goal in writing this textbook is to provide college students with a mathematics book that they can read and understand. In addition, I have tried to make my book interesting and even enjoyable by including a variety of topics and contemporary real-life applications. In part, I hope to answer the age-old question, "When will I ever use this stuff?"

My ultimate goal in writing this book, as well as my other books, is to develop each topic as I would teach it in the classroom. Because students reading a textbook, unlike those in the classroom, cannot ask the instructor questions, I have tried to anticipate students' needs for clarification of difficult concepts.

To show students that mathematics is a human endeavor, I included vignettes called "Sidelights." Some of these sidelights include minibiographies of famous mathematicians of past and present. Others show interesting applications of mathematics to areas such as weather, photography, music, health, money, etc.

In order to facilitate learning, I wrote the book so that students with a basic algebra background can experience success. For students without an adequate mathematical background, the first three sections of Chapter 5 present arithmetic, and Chapters 7 and 8 explain the basic concepts of algebra. In addition, I included "Math Notes" in the margin. These notes clarify concepts, emphasize important points, and provide helpful suggestions for learning the principles. When the solutions to example problems require several steps, I have used a numbered, step-by-step procedure to facilitate solving the examples.

After each topic, a "Try This One" box that consists of one or more sample exercises gives students an opportunity to confirm that they have learned the material before proceeding to the next topic. The topics are chosen because of their usefulness and inherent interest. The topics are presented in an order that provides the instructor flexibility in developing a syllabus.

Chapters are largely independent of each other. With the exception of Chapters 7 and 8, topics can be covered in nearly any order the instructor wishes. Chapter 8 should be covered after Chapter 7 as it builds on topics presented in Chapter 7. Depending on the mathematical background of the students, instructors may want to cover Chapter 5 prior to Chapter 7.

The book is designed for liberal arts students taking mathematics survey courses. The number of topics covered in the text enable it to be used in a two-term course. However, the flexibility of the presentation enable it to be used in one-term courses as well.

Supplements for the Instructor

Instructor's Edition

This ancillary contains answers to problems and exercises in the text, including answers to all section exercises, all *Review Exercises,* and *Chapter Tests*. These answers are printed in an appendix at the end of the text.

Instructor's Testing and Resource CD-ROM

This cross-platform CD-ROM provides a wealth of resources for the instructor. Supplements featured on this CD-ROM include a computerized test bank utilizing Brownstone Diploma® testing software to quickly create customized exams. This user-friendly program enables instructors to search for questions by topic, format, or difficulty level; edit existing questions or add new ones; and scramble questions and answer keys for multiple versions of the same test.

Instructor's Solutions Manual

This supplement contains detailed solutions to all the exercises in the text. The methods used to solve the problems in the manual are the same as those used to solve the examples in the textbook.

 **www.mathzone.com***

Free, Easy, Has it all...
*web-based product also available on CD-ROM
McGraw-Hill's MathZone is a complete, online tutorial and course management system for mathematics and statistics, designed for greater ease of use than any other system available. Free upon adoption of a McGraw-Hill title, instructors can create and share courses and assignments with colleagues and adjuncts in a matter of a few clicks of the mouse. All assignments, questions, e-Professors, online tutoring and video lectures are directly tied to text-specific materials in *Mathematics in Our World.* MathZone courses are customized to your textbook, but you can edit questions and algorithms, import your own content, create announcements and due dates for assignments. MathZone has automatic grading and reporting of easy-to-assign algorithmically generated homework, quizzing and testing. All student activity within MathZone is automatically recorded and available to you through a fully integrated grade book that can be downloaded to Excel.

PageOut

PageOut is McGraw-Hill's unique point-and-click course website tool, enabling you to create a full-featured, professional-quality course website without knowing HTML coding. With PageOut you can post your course syllabus, assign McGraw-Hill Online Learning Center content, add links to important off-site resources, and maintain student results in the online grade book. You can send class announcements, copy your course site to share with colleagues, and upload original files. PageOut is free for every McGraw-Hill user, and if you're short on time, we even have a team ready to help you create your site!

Supplements for the Student

Student's Solutions Manual

The Student's Solutions Manual contains complete worked-out solutions to all the odd-numbered exercises in the text. The procedures followed in the solutions in the manual match exactly those shown in worked examples in the text.

 **www.mathzone.com***

*web-based product also available on CD-ROM

McGraw-Hill's MathZone is a powerful new online tutorial for homework, quizzing, testing and interactive applications. There are an unlimited number of exercises to allow for as much practice as needed. MathZone offers videos that feature classroom instructors giving a lecture and **e-Professor** takes you through animated, step-by-step instruction for solving problems in the book allowing you to digest each step at your own pace. **NetTutor** offers live, personalized tutoring via the Internet. Every assignment, question, e-Professor, and video lecture is derived directly from *Mathematics in Our World.*

Video Series

The video series is composed of 13 videocassettes (one for each chapter of the text). An on-screen instructor introduces topics and works through examples using the methods presented in the text. The video series is also available on video CD-ROMs.

NetTutor

NetTutor is a revolutionary system that enables students to interact with a live tutor over the World Wide Web. Students can receive instruction from live tutors using NetTutor's Web-based, graphical chat capabilities. They can also submit questions and receive answers, browse previously answered questions, and view previous live chat sessions.

Acknowledgments

I would like to thank all the dedicated people who have given their time and energy to make this book possible:

To Eugene Mastroianni and Connie Hobbs for their help on the initial manuscript.
To Joyce Smith for her interesting technology applications.

To the following reviewers for their suggestions and recommendations:

Marwan Abu-Sawwa, *Florida Community College at Jacksonville*
Fusun Akman, *Coastal Carolina University*
Margaret M. Balachowski, *Michigan Technology University*
Elaine Barber, *Germanna Community College*
Larry Boyd, *Iona Central Community College*
Dr. Linda Kay Buchanan, *Howard College*
Ed Curtis, *Wilkes Community College*
Dr. Laura Dyer, *Southwestern Illinois College*
Barbara H. Glass, *Sussex County Community College*
Russell K. Gusack, *Suffolk County Community College*
Helen Harris, *Blinn College*
Neal Hart, *Macmurray College*
Marcia Hovinga, *Ellsworth Community College*
Harriet Kiser, *Floyd College*
Thomas J. Lankston, *Ivy Tech State College*
James Lapp, *Adams State College*
Kathryn Lavelle, *Westchester Community College*
Richard Leedy, *Polk Community College*

Winifred A. Mallam, *Texas Women's University*
Diane Metzger, *Rend Lake College*
Daniel Munton, *Santa Rosa Junior College*
Bill Naegele, *South Suburban College*
Matthew Pascal, *Northern Virginia Community College*
Cyril Petras, *Lord Fairfax Community College*
Kathy Lyn Pinchback, *University of Memphis*
Mary Lee Seitz, *Erie Community College–City Campus*
Mike Shirazi, *Germanna Community College*
Deirdre Longacher Smeltzer, *Eastern Mennonite University*
Hortensia Soto-Johnson, *University of Southern Colorado*
Kristin Stoley, *Blinn College*
Tom Tredon, *Lord Fairfax Community College*
Tammy Voepel, *Southern Illinois University–Edwardsville*
Alice Williamson, *Sussex County Community College*
Rob Wylie, *Carl Albert State College*

To the McGraw-Hill Company for permission to use material in Chapters 11 and 12 from another of his books, *Elementary Statistics, A Step by Step Approach,* 4th edition.

To my wife Betty Claire for her help and support in all aspects of this textbook. Finally, I would like to thank David Dietz, Peter Galuardi, Erin Brown, and Vicki Krug at McGraw-Hill for their efforts and support.

Allan G. Bluman

Walk-Through

The learning system found in *Mathematics in Our World* provides students with a useful framework in which to learn and apply concepts.

Chapter Outline and Objectives

Each chapter begins with an outline and a list of learning objectives. These help students focus on the key concepts presented in the chapter.

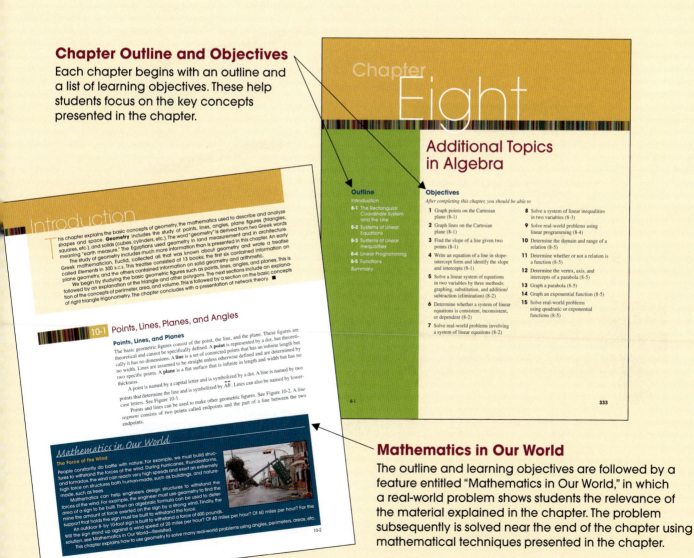

Chapter Eight

Additional Topics in Algebra

Outline

Introduction

8-1 The Rectangular Coordinate System and the Line

8-2 Systems of Linear Equations

8-3 Systems of Linear Inequalities

8-4 Linear Programming

8-5 Functions

Summary

Objectives

After completing this chapter, you should be able to

1 Graph points on the Cartesian plane (8-1)

2 Graph lines on the Cartesian plane (8-1)

3 Find the slope of a line given two points (8-1)

4 Write an equation of a line in slope-intercept form and identify the slope and intercepts (8-1)

5 Solve a linear system of equations in two variables by three methods: graphing, substitution, and addition/subtraction (elimination) (8-2)

6 Determine whether a system of linear equations is consistent, inconsistent, or dependent (8-2)

7 Solve real-world problems involving a system of linear equations (8-2)

8 Solve a system of linear inequalities in two variables (8-3)

9 Solve real-world problems using linear programming (8-4)

10 Determine the domain and range of a relation (8-5)

11 Determine whether or not a relation is a function (8-5)

12 Determine the vertex, axis, and intercepts of a parabola (8-5)

13 Graph a parabola (8-5)

14 Graph an exponential function (8-5)

15 Solve real-world problems using quadratic or exponential functions (8-5)

8-1

333

Introduction

This chapter explains the basic concepts of geometry, the mathematics used to describe and analyze shapes and space. **Geometry** includes the study of points, lines, angles, plane figures (triangles, squares, etc.), and solids (cubes, cylinders, etc.). The word "geometry" is derived from two Greek words meaning "earth measure." The Egyptians used geometry in land measurement and in architecture.

The study of geometry includes much more information than is presented in this chapter. An early Greek mathematician, Euclid, collected all that was known about geometry and wrote a treatise called *Elements* in 300 B.C.E. This treatise consisted of 13 books; the first six contained information on plane geometry, and the others contained information on solid geometry and arithmetic.

We begin by studying the basic geometric figures such as points, lines, angles, and planes. This is followed by an explanation of the triangle and other polygons. The next sections include an explanation of the concepts of perimeter, area, and volume. This is followed by a section on the basic concepts of right triangle trigonometry. The chapter concludes with a presentation of network theory. ∎

10-1 Points, Lines, Planes, and Angles

Points, Lines, and Planes

The basic geometric figures consist of the point, the line, and the plane. These figures are theoretical and cannot be specifically defined. A **point** is represented by a dot, but theoretically it has no dimensions. A **line** is a set of connected points that has an infinite length but no width. Lines are assumed to be straight unless otherwise defined and are determined by two specific points. A **plane** is a flat surface that is infinite in length and width but has no thickness.

A point is named by a capital letter and is symbolized by a dot. A line is named by two points that determine the line and is symbolized by $\overleftrightarrow{AB}$. Lines can also be named by lower-case letters. See Figure 10-1.

Points and lines can be used to make other geometric figures. See Figure 10-2. A *line segment* consists of two points called endpoints and the part of a line between the two endpoints.

Mathematics in Our World

The Force of the Wind

People constantly do battle with nature. For example, we must build structures to withstand the forces of the wind. During hurricanes, thunderstorms, and tornados, the wind can reach very high speeds and exert an extremely high force on structures both human-made, such as buildings, and nature-made, such as trees.

Mathematics can help engineers design structures to withstand the forces of the wind. For example, the engineer must use geometry to find the area of a sign to be built. Then an algebraic formula can be used to determine the amount of force exerted on the sign by a strong wind. Finally, the support that holds the sign must be built to withstand the force.

An outdoor 8- by 10-foot sign is built to withstand a force of 600 pounds. Will the sign stand up against a wind speed of 20 miles per hour? Of 40 miles per hour? Of 60 miles per hour? For the solution, see Mathematics in Our World—Revisited.

This chapter explains how to use geometry to solve many real-world problems using angles, perimeters, areas, etc.

10-2

458

Mathematics in Our World

The outline and learning objectives are followed by a feature entitled "Mathematics in Our World," in which a real-world problem shows students the relevance of the material explained in the chapter. The problem subsequently is solved near the end of the chapter using mathematical techniques presented in the chapter.

Worked Examples

There are over 400 example problems with detailed solutions to help students master the concepts. Where solutions consist of more than one step, a numbered, step-by-step procedure is used.

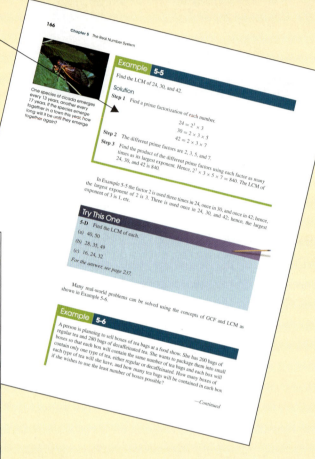

Try This One

After each topic, a "Try This One" box provides one or more problems for the student to try. This enables the students to determine whether or not they have learned the concepts before proceeding. There are over 200 of these boxes in *Mathematics in Our World*. The answers to these exercises appear at the end of each chapter.

Sidelights

Throughout the book, there are over 60 features called "Sidelights." Sidelights help display the connection of mathematics to other disciplines and to the world around us. Through these interesting real-world applications, historical perspectives, and biographies students will see how mathematics is a human endeavor.

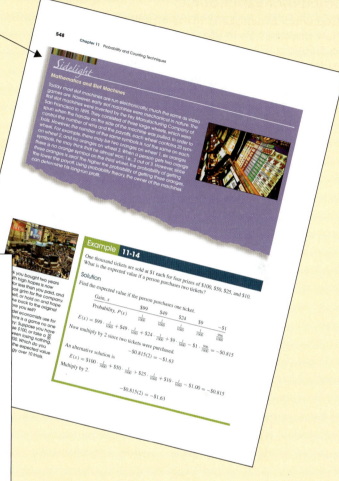

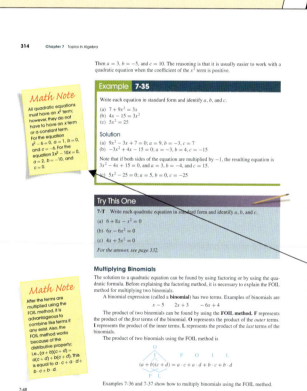

Math Notes

"Math Notes" are included in the margins to help clarify concepts, emphasize important points, and provide helpful suggestions for learning.

Calculator Explorations

"Calculator Explorations" in the margins show students how to use a graphing calculator to solve problems and further explore concepts. Usage of a calculator with this text is entirely optional.

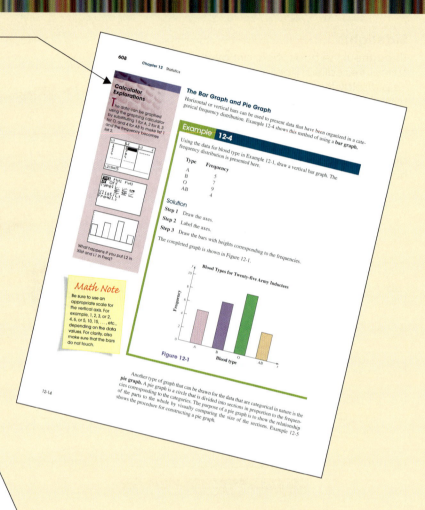

Ample Exercise Sets

There are over 4000 exercises in *Mathematics in Our World*. They are divided into four categories: Computational Exercises, Real World Applications, Writing Exercises, and Critical Thinking Exercises.

Chapter Summaries

Chapter Summaries include important terms and the important concepts presented in each section. They are a useful tool for test preparation.

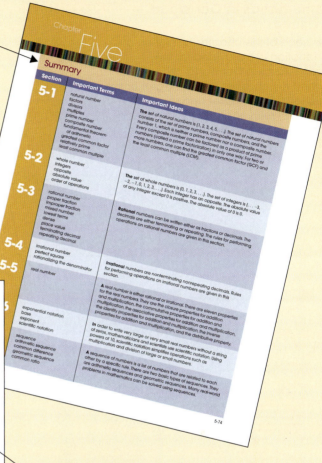

Chapter Five

Summary

Section	Important Terms	Important Ideas
5-1	natural number factors divisors multiples prime number composite number fundamental theorem of arithmetic greatest common factor relatively prime least common multiple	**The** set of natural numbers is {1, 2, 3, 4, 5, . . .}. The set of prime numbers consists of the set of prime numbers, composite numbers, and the number 1, which is neither a prime number nor a composite number. Every composite number can be factored as a product of prime numbers (called a prime factorization) in only one way. For two or more numbers, one can find the greatest common factor (GCF) and the least common multiple (LCM).
5-2	whole number integers opposite absolute value order of operations	**The** set of whole numbers is {0, 1, 2, 3, . . .}. The set of integers is {. . . , −3, −2, −1, 0, 1, 2, 3, . . .}. Each integer has an opposite. The absolute value of any integer except 0 is positive. The absolute value of 0 is 0.
5-3	rational number proper fraction improper fraction mixed number lowest terms dense place value terminating decimal repeating decimal	**Rational** numbers can be written either as fractions or decimals. The decimals are either terminating or repeating. The rules for performing operations on rational numbers are given in this section.
5-4	irrational number perfect square rationalizing the denominator	**Irrational** numbers are nonterminating nonrepeating decimals. Rules for performing operations on irrational numbers are given in this section.
5-5	real number	**A** real number is either rational or irrational. There are eleven properties for the real numbers. They are the closure properties for addition and multiplication, the commutative properties for addition and multiplication, the associative properties for addition and multiplication, the identity properties for addition and multiplication, the inverse properties for addition and multiplication, and the distributive property.
	exponential notation base exponent scientific notation	**In** order to write very large or very small real numbers without a string of zeros, mathematicians and scientists use scientific notation. Using powers of 10, scientific notation simplifies operations such as multiplication and division of large or small numbers.
	sequence arithmetic sequence common difference geometric sequence common ratio	**A** sequence of numbers is a list of numbers that are related to each other by a specific rule. There are two basic types of sequences. They are arithmetic sequences and geometric sequences. Many real-world problems in mathematics can be solved using sequences.

5-74

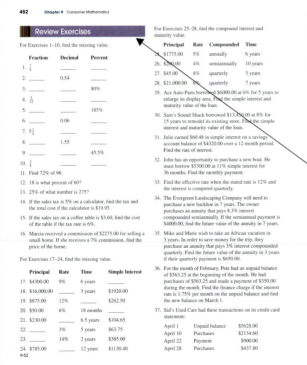

452   Chapter 9   Consumer Mathematics

Review Exercises

For Exercises 1–10, find the missing value.

	Fraction	Decimal	Percent
1.	$\frac{7}{8}$		
2.		0.54	
3.			80%
4.	$\frac{X}{12}$		
5.			185%
6.		0.06	
7.	$5\frac{1}{4}$		
8.		1.55	
9.			45.5%
10.	$\frac{3}{8}$		

11. Find 72% of 96.

12. 18 is what percent of 60?

13. 25% of what number is 275?

14. If the sales tax is 5% on a calculator, find the tax and the total cost if the calculator is $19.95.

15. If the sales tax on a coffee table is $3.60, find the cost of the table if the tax rate is 6%.

16. Marcia received a commission of $2275.00 for selling a small home. If she receives a 7% commission, find the price of the home.

For Exercises 17–24, find the missing value.

	Principal	Rate	Time	Simple Interest
17.	$4300.00	9%	6 years	
18.	$16,000.00		3 years	$1920.00
19.	$875.00	12%		$262.50
20.	$50.00	6%	18 months	
21.	$230.00		6.5 years	$104.65
22.		3%	5 years	$63.75
23.		14%	2 years	$385.00
24.	$785.00		12 years	$1130.40

9-52

For Exercises 25–28, find the compound interest and maturity value.

	Principal	Rate	Compounded	Time
25.	$1775.00	5%	annually	6 years
26.	$260.00	4%	semiannually	10 years
27.	$45.00	8%	quarterly	3 years
28.	$21,000.00	6%	quarterly	7 years

29. Ace Auto Parts borrowed $6000.00 at 6% for 5 years to enlarge its display area. Find the simple interest and maturity value of the loan.

30. Sam's Sound Shack borrowed $13,450.00 at 8% for 15 years to remodel its existing store. Find the simple interest and maturity value of the loan.

31. Julie earned $60.48 in simple interest on a savings account balance of $4320.00 over a 12-month period. Find the rate of interest.

32. John has an opportunity to purchase a new boat. He must borrow $5300.00 at 11% simple interest for 36 months. Find the monthly payment.

33. Find the effective rate when the stated rate is 12% and the interest is computed quarterly.

34. The Evergreen Landscaping Company will need to purchase a new backhoe in 7 years. The owner purchases an annuity that pays 8.3% interest compounded semiannually. If the semiannual payment is $4000.00, find the future value of the annuity in 7 years.

35. Mike and Marie wish to take an African vacation in 3 years. In order to save money for the trip, they purchase an annuity that pays 3% interest compounded quarterly. Find the future value of the annuity in 3 years if their quarterly payment is $650.00.

36. For the month of February, Pete had an unpaid balance of $563.25 at the beginning of the month. He had purchases of $563.25 and made a payment of $350.00 during the month. Find the finance charge if the interest rate is 1.75% per month on the unpaid balance and find the new balance on March 1.

37. Sid's Used Cars had these transactions on its credit card statement:

April 1	Unpaid balance	$5628.00
April 10	Purchases	$2134.60
April 22	Payment	$900.00
April 28	Purchases	$437.80

Review Exercises

The review exercises at the end of each chapter give students an opportunity to "put it all together."

Mathematics in Our World Revisited

At the end of the chapter, a solution and explanation are provided for the "Mathematics in Our World" feature that opened the chapter. This ties together the material in the chapter and reminds students of relevant ways in which such techniques can be applied in the real world.

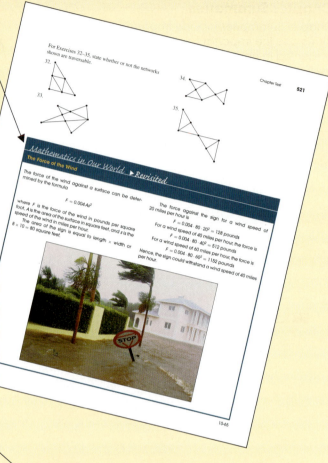

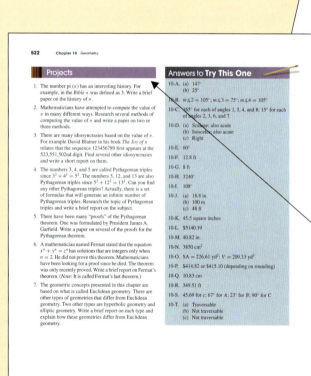

Projects

"Projects" can be completed by individuals or students working in groups. These projects extend the concepts presented in the chapter and most require more in-depth thought or require the student to do some research.

Chapter One

Problem Solving

Objectives

After completing this chapter, you should be able to

1 Explain the two types of mathematics (1-1)

2 Explain the difference between inductive and deductive reasoning (1-1)

3 State the steps for the basic problem-solving procedure (1-2)

4 Solve problems using the basic problem-solving procedure (1-2)

5 Solve problems using estimation (1-3)

Introduction

very day we face a multitude of problems that must be solved. These problems occur in our jobs, in school, and in our personal lives. For example, which computer should we buy? What should we do when our car breaks down? How can we decide on a topic for our research paper? The purpose of this chapter is to introduce you to some of the techniques of problem solving. The techniques you will learn in this chapter may prove to be useful tools that you can apply to situations you face outside of the classroom as much as to those you face inside the classroom. ∎

Mathematics in Our World

Rhind *Papyrus*

People have been intrigued with mathematical problems since ancient times. An Egyptian mathematical textbook called the Rhind *Papyrus,* written about 1650 B.C.E., consisted of 85 mathematical problems.

The *Papyrus* contained many real-life problems such as how to determine the strength of beer, how to mix feed for cattle and other domestic animals, and how to store grain. In addition, it contained geometric problems and problems that were theoretical in nature.

The text was translated from an earlier work by a scribe named Ahmes. It was purchased in 1858 by Egyptologist Henry Rhind and later willed to the British Museum where it remains today.

Although most of the problems were easily deciphered and interpreted, one problem confused translators even though its solution was given in the text. In 1907, a famous mathematician, Moritz Cantor, suggested a problem that would yield the same solution as found in the *Papyrus*. Here is his conjective for problem number 79.

An estate consisted of seven houses.

Each house had seven cats.

Each cat ate seven mice.

Each mouse ate seven heads of wheat.

Each head of wheat was capable of yielding seven hekat measures of grain.

Houses, cats, mice, heads of wheat, and hekat measures of grain—how many of these in all were in the estate?

Can you solve this ancient problem?

The solution is given on page 32.

Rhind Papyrus

Source: *An Introduction to the History of Mathematics,* Third Edition, by Howard Eves. Holt, Reinhart, and Winston.

1-1   The Nature of Mathematical Reasoning

In everyday life, we use two types of reasoning: *inductive reasoning* or *induction* and *deductive reasoning* or *deduction*.

> **Inductive reasoning** is the process of reasoning that arrives at a general conclusion or conjecture based on the observation of specific examples.

For example, a student might observe that his or her instructor gives a surprise quiz every Friday. Based on this event happening every Friday for the first 5 weeks of the semester, the student makes a conjecture or guess and concludes that he or she better study before next Friday's class.

This is an example of inductive reasoning. By observing the events for five specific Fridays, the student arrives at a general conclusion.

Inductive reasoning can also be used as a problem-solving tool in mathematics, as shown in Example 1-1.

Sports writers predict how teams and individual players will do based on their past performance. This is an example of inductive reasoning.

Example   1-1

Use inductive reasoning to find a pattern, and then find the next three numbers using that pattern.

$$1, 2, 4, 5, 7, 8, 10, 11, 13, \_\_, \_\_, \_\_.$$

Solution

In order to find the pattern, look at the first number and see how to obtain the second number. Then look at the second number and see how to obtain the third number, etc.

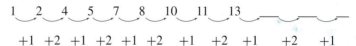

$$+1 \quad +2 \quad +1 \quad +2 \quad +1 \quad +2 \quad +1 \quad +2 \quad +1 \quad +2 \quad +1$$

The pattern seems to be to add 1, then add 2, then add 1, then add 2, etc. Hence a reasonable conjecture for the next three numbers is 14, 16, and 17.

Example   1-2

Make a reasonable conjecture for the next figure in the sequence.

Solution

The flat part of the figure is up, right, down, and then left. There is a solid circle ● in each figure. The sequence then repeats with an open circle ○ in each figure. Hence we could reasonably expect the next figure to be ⬡○.

Sidelight

Mathematics and Weather

Weather involves numerous variables, such as temperatures, wind velocity and direction, type and amount of precipitation, cloud cover, and atmospheric pressures, to name just a few. All of these variables use numbers.

In order to create a weather forecast, the National Weather Service collects information from various sources such as satellites, weather balloons, weather stations, and human observers around the world. The information is then input to a computer and a grid with over 200,000 points is drawn up by the computer showing the weather conditions at each point at the time the data were collected.

From this grid, the computer is able to predict the movement of weather fronts at 10-minute intervals into the future. Thus a forecast of tomorrow's weather can be generated.

Short-term forecasts are generally accurate, but due to variations in atmospheric conditions, long-range forecasts are at best guesses based on current conditions. Sometimes, however, atmospheric conditions change so rapidly that even short-term forecasts are inaccurate.

Recognizing, describing, and creating patterns is important in many fields. Many types of patterns are used in music such as following an established pattern, altering an established pattern, and producing variations on a familiar pattern.

Try This One

1-A Use inductive reasoning to find a pattern and make a reasonable conjecture for the next three numbers using that pattern.

$$1 \quad 4 \quad 2 \quad 5 \quad 3 \quad 6 \quad 4 \quad 7 \quad 5 \quad \_\_ \quad \_\_ \quad \_\_$$

For the answer, see page 32.

Induction is a very powerful method of reasoning. It is used often in our lives. Many discoveries in mathematics and science have resulted from inductive reasoning. The problem with inductive reasoning is that we can never be 100% sure that the conclusion or conjecture is true since it is not possible to verify the conclusion for every specific case. For example, the student can never be 100% sure that his or her professor will give a quiz every Friday for the rest of the semester. If there is one case where the result does not occur, then the conjecture is false.

In order to disprove a conjecture, a *counterexample* must be found. A **counterexample** is a specific example that shows the conjecture is false.

For example, one may think that a number is divisible by 3 if the last two digits are divisible by 3. Try a few numbers.

$$15\underline{27} \div 3 = 509$$

$$11,7\underline{45} \div 3 = 3915$$

In both cases, the last two digits of the number form a number divisible by 3 (i.e., $27 \div 3 = 9$ and $45 \div 3 = 15$).

Sidelight

Hypatia (370–415)

The First Female Mathematician

Hypatia lived in the turbulent times of the 4th century in the Hellenic Age of Greece. Her father was Theon of Alexandria. She was educated in mathematics, science, medicine, philosophy, and the arts. But mathematics was her first love.

She studied the algebra of Diophantus, a famous 3rd-century Greek mathematician who developed symbols to write and solve equations. Most of Hypatia's work was lost when the library at Alexandria was destroyed by fire in the 7th century, although it is known that she wrote about conic sections and the work of Euclid as well as philosophy and astronomy.

During her lifetime, she traveled extensively. She never married but claimed she was "married to the truth." Because her strong beliefs conflicted with the doctrine of the time, she was dragged from her chariot by religious zealots and killed.

However, if you keep trying, you will soon find a number that shows the conjecture is false. For example, the number 1136 provides a counterexample to this conjecture since it is not divisible by 3 even though 36 is!

The other method of reasoning is called *deductive reasoning* or *deduction*.

> **Deductive reasoning** is the process of reasoning that arrives at a specific conclusion based on previously accepted general statements.

For example, many states require that automobiles be inspected at least once a year. In this case, a general statement such as "All automobiles that pass a proper inspection are safe to drive" is accepted or assumed to be true. Then when a second statement such as "My automobile has been inspected" is paired with the first statement, a specific statement such as "My automobile is safe to drive" can be concluded to be true.

Deductive reasoning can also be used as a problem-solving tool in mathematics. Both types of reasoning are used in this example. In the first case, inductive reasoning will be used. Several specific numbers will be selected to use as examples, and then the general conclusion will be stated. In the second case, deductive reasoning will be used. A statement using any number, x, will be proved; then it can be concluded that the statement is true for all specific numbers.

Think of a number:
Multiply it by 2:
Add 6:
Divide by 2:
Subtract the original number:
What number will you get?
Approach: Induction

Calculator Explorations

A calculator can be used to test the conjecture, "A number is divisible by 3 if the last two digits are divisible by 3."

```
1527/3
              509
11745/3
             3915
1136/3
     378.6666667
```

Since 1136 is not evenly divisible by 3, but 36 (the last two digits of 1136) is, the conjecture is false.

What is another counterexample that shows the conjecture is false?

Calculator Explorations

Some calculators have a feature that enables them to generate tables that show input and output values based on an expression entered.

If 4 is entered as a value for X into the equation shown, what value do you think the calculator will return for Y_1?

Looking at the values shown in the X and Y_1 columns of the table, what conjecture can you form?

Why might a calculator be useful in forming a conjecture in this example?

Let's try several different numbers and see what we get.

Number:	12	5	43
Multiply by 2:	$2 \times 12 = 24$	$2 \times 5 = 10$	$2 \times 43 = 86$
Add 6:	$24 + 6 = 30$	$10 + 6 = 16$	$86 + 6 = 92$
Divide by 2:	$30 \div 2 = 15$	$16 \div 2 = 8$	$92 \div 2 = 46$
Subtract the original number:	$15 - 12 = 3$	$8 - 5 = 3$	$46 - 43 = 3$
Result:	3	3	3

Here we have tried several specific examples, and it seems that the answer will always be 3. But we cannot be 100% sure since we have not tried every single number.

Approach: Deduction

Rather than using specific numbers, select a general number called x and use some algebra to see what we get.

Think of a number:	x
Multiply by 2:	$2x$
Add 6:	$2x + 6$
Divide by 2:	$\dfrac{2x + 6}{2} = x + 3$
Subtract the original number:	$x + 3 - x = 3$

Hence, the result will always be 3.

In the first case, several *specific* numbers, 12, 5, and 43, were selected, and then a *general* conclusion was stated. In the second case, a *general* number (x) was selected and principles of arithmetic and algebra were used to prove that the answer will always be 3.

Example   1-3

Use inductive reasoning to arrive at a general conclusion, and then prove your conclusion is true using deductive reasoning.

Select a number:
Add 50:
Multiply by 2:
Subtract the original number:
Result:

Solution

Approach: Induction

Use several different numbers and make a conjecture.

Original number:	12	50
Add 50:	$12 + 50 = 62$	$50 + 50 = 100$
Multiply by 2:	$62 \times 2 = 124$	$100 \times 2 = 200$
Subtract the original number:	$124 - 12 = 112$	$200 - 50 = 150$
Result:	112	150

—Continued

Example 1-3 *Continued—*

The conjecture is that the final answer is 100 more than the original number.

Approach: Deduction
Select a number: x
Add 50: $x + 50$
Multiply by 2: $2(x + 50) = 2x + 100$
Subtract the original number: $2x + 100 - x$
Result: $x + 100$

Hence the final answer is 100 more than the original number.

Try This One

1-B Arrive at a conclusion using inductive reasoning, then try to prove your conclusion by using deductive reasoning.

Select a number:

Add 16:

Multiply by 3:

Add 2:

Subtract twice the original number:

Subtract 50:

Result:

For the answer, see page 32.

Calculator Explorations

Two expressions were entered on a calculator to generate the table of values shown.

```
Plot1 Plot2 Plot3
\Y1∎2(X+50)-X
\Y2∎X+100
\Y3=∎
\Y4=
\Y5=
\Y6=
\Y7=
```

X	Y₁	Y₂
-3	97	97
-2	98	98
-1	99	99
0	100	100
1	101	101
2	102	102
3	103	103

X= -3

What do you notice about the values in columns Y_1 and Y_2?

How could you prove this observation or conjecture is true for all values of X?

Would a calculator be useful in helping you do this? Why or why not?

Exercise Set 1-1

Computational Exercises

For Exercises 1–8, use inductive reasoning to find a pattern, and then make a reasonable conjecture for the next number or item in the sequence.

1. 1   2   4   7   11   16   22   29   __

2. 5   15   45   135   405   1215   __

3. 10   20   11   18   12   16   13   14   14   12   15   __

4. 1   4   9   16   19   24   31   34   39   46   __

5. 100   99   97   94   90   85   79   __

6. 9   12   11   14   13   16   15   18   __

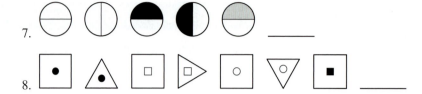

7. _____

8. _____

For Exercises 9–12, find a counterexample to show each statement is false.

9. The sum of any three odd numbers is even.

10. When an even number is added to the product of two odd numbers, the result will be even.

11. When an odd number is squared and divided by 2, the result will be a whole number.

12. When any number is multiplied by 6 and the digits of the answer are added, the sum will be divisible by 6.

For Exercises 13–16, use inductive reasoning to conjecture the rule that relates the number you selected to the final answer. Try to prove your conjecture using deductive reasoning.

13. Select a number:

    Double it:

    Subtract 20 from the answer:

    Divide by 2:

    Subtract the original number:

    Result:

14. Select a number:

    Multiply it by 9:

    Add 21:

    Divide by 3:

    Subtract three times the original number:

    Result:

15. Select a number:

    Add 50:

    Multiply by 2:

    Subtract 60:

    Divide by 2:

    Subtract the original number:

    Result:

16. Select a number:

    Multiply by 10:

    Subtract 25:

    Divide by 5:

    Subtract the original number:

    Result:

Writing Exercises

17. Give a real-life example of how to use inductive reasoning.

18. Explain the difference between inductive reasoning and deductive reasoning.

19. Explain why you can never be sure that the conclusion you arrived at using the inductive reasoning process is true.

20. What is a "counterexample" and when does one use it?

Critical Thinking

21. Can you draw a map on a flat surface consisting of five countries and color each country with a different color so that you need at least five different colors?

22. Find the next three numbers in the sequence.

    (a) 1, 1, 2, 3, 5, 8, 13, 21, . . .

    (b) 1, 4, 9, 16, 25, 36, . . .

    (c) 1, 2, 4, 8, 16, 32, . . .

23. Find the next numbers in the table.

```
                        1
                  1           1
               1     2     1
            1     3        3     1
         1     4     6     4     1
      1     5    10      10     5     1
   1     __    __    __    __    __         1
```

   ## Problem Solving

Problem solving is inherent to the nature of mathematics. A great deal has been written about this topic. A Hungarian mathematician, **George Polya,** did much research and writing on the nature of problem solving. His book, entitled *How to Solve It,* has been translated into at least 17 languages, and it sets forth the basic steps of problem solving. These steps are explained next.

Polya's Four-Step Problem-Solving Procedure

Step 1 *Understand the problem.* First, read the problem carefully several times and underline or write down the basic information given in the problem. Second, write down what you are being asked to find.

Step 2   *Devise a plan to solve the problem.* There are many ways to solve problems. It may be helpful to draw a picture. You could make an organized list. You may be able to solve it by trial and error. You might be able to find a similar problem that has already been solved and apply the technique to the problem that you are trying to solve. You may be able to solve the problem by using the arithmetic operations of addition, subtraction, multiplication, or division. You may be able to solve the problem using algebraic equations or geometric formulas.

Step 3   *Carry out the plan to solve the problem.* After you have devised a plan, try it out. If you can't get the answer, try a different strategy. Also be advised that sometimes there are several ways to solve a problem, so different students can arrive at the correct answer by different methods.

Step 4   *Check the answer.* Some problems can be checked by using mathematical methods. Sometimes you may have to ask yourself if the answer is reasonable, and sometimes the problem can be checked by using a different method. Finally, you may use estimation to approximate the answer. (See Section 1-3.)

Examples 1-4 through 1-7 show how to apply Polya's four-step procedure.

Example 1-4

Eight clothespins are placed on a clothesline at 2-foot intervals. How far is it from the first clothespin to the last one?

Solution

Step 1   *Understand the problem.* In this case, the information given is that there are eight clothespins on a clothesline, and there is a distance of 2 feet between them. You are being asked to find the total distance from the first one to the last one.

Step 2   *Devise a plan to solve the problem.* In this case, drawing a picture may be useful.

Step 3   *Carry out the plan to solve the problem.* The figure would look like this.

After seeing the picture, it is obvious that the distance from the first one is found by adding $2 + 2 + 2 + 2 + 2 + 2 + 2 = 14$ feet.

Step 4   *Check the answer.* There are eight clothespins and seven spaces of 2 feet between them. Hence, $7 \times 2 = 14$ feet.

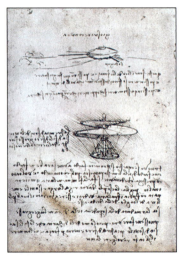

Scientists and inventors sometimes use sketches to organize their thoughts like Leonardo da Vinci. He wrote backwards in Latin to protect his work.

Try This One

1-C   Mike wishes to cut a log into 10 pieces. How many cuts are necessary?
For the answer, see page 32.

Sidelight

George Polya (1887–1985)

Born in Hungary, George Polya received his Ph.D. from the University of Budapest in 1912. He studied law and literature before turning his interests to mathematics. He came to the United States in 1940 and taught at Brown University. In 1942, he moved to Stanford University. His most famous book entitled *How to Solve It* was published in 1945 by Princeton University Press.

This book explained his method of problem solving, which consisted of four major steps. They are

1. Understand the problem
2. Devise a plan
3. Carry out the plan
4. Look back

One million copies of the book have been sold, and it is still being published today.

His four-step process is used as a blueprint for solving problems not only in mathematics but in other areas of study as well. The book has been translated into at least 17 languages and is sold worldwide.

In addition to his famous book, Polya has written over 200 articles, nine other books, and numerous other papers.

Example 1-5

A person has 10 coins consisting of quarters and dimes. If the person has a total of $1.90, find the number of coins of each denomination he has.

Solution

Step 1 *Understand the problem.* The person has 10 coins consisting of quarters and dimes. A quarter is worth $0.25 and a dime is worth $0.10. The total is $1.90. You are being asked to find how many quarters and how many dimes added together will give $1.90.

Step 2 *Devise a plan to solve the problem.* One strategy that can be used is to make an organized list of possible combinations of quarters and dimes and see if the sum is $1.90. For example, you may try one quarter and nine dimes. This gives $1 \times \$0.25 + 9 \times \$0.10 = \$1.15$.

Quarters	Dimes	Amount
1	9	$1.15

Step 3 *Carry out the plan.* Since one quarter and nine dimes is incorrect, try two quarters and eight dimes. This is also incorrect.

—Continued

Example **1-5** *Continued—*

Quarters	Dimes	Amount
2	8	$1.30

Continue until you get the correct answer.

Quarters	Dimes	Amount
1	9	$1.15
2	8	$1.30
3	7	$1.45
4	6	$1.60
5	5	$1.75
6	4	$1.90 ← Correct

Answer: six quarters and four dimes

Step 4 *Check the answer.* In this case, the mathematics can be checked by working out the problem again. $6 \times \$0.25 + 4 \times \$0.10 = \$1.90$.

Try This One

1-D Michelle bought seven stamps. If the total cost was $2.98 and some stamps were $0.37 each and some were $0.50 each, how many stamps of each denomination did she buy?
For the answer, see page 32.

Example **1-6**

One person earns $7.20 per hour and works 40 hours per week for 50 weeks and gets a 2-week paid vacation. Another person earns $15,000 per year. Which person has a higher yearly income?

Solution

Step 1 *Understand the problem.* The information given is that one person earns $7.20 per hour and works 40 hours per week while another person earns $15,000. You are being asked to compare the two salaries and determine which person makes the most money.

Step 2 *Devise a plan to solve the problem.* It is necessary to determine how much the first person can earn in 1 year given the conditions stated in the problem and then to see if it is more than, equal to, or less than $15,000.

—Continued

Example   1-6   Continued—

Step 3   *Carry out the plan to solve the problem.* Multiply the hourly wage by 40 hours and then by 52 weeks to get the first person's yearly salary: $7.20 × 40 × 52 = $14,976. Hence, the first person earns less than the person who earns $15,000 per year.

Step 4   *Check the answer.* In this case, we can find the hourly salary of the person who earns $15,000 per year. Divide $15,000 by 52 to find the weekly salary. It is $288.46. Then divide $288.46 by 40 hours to get $7.21 per hour. This is $0.01 more per hour than the first person earns.

Sidelight

Archimedes (287 B.C.E.–212 B.C.E.)

Archimedes is considered by some to be the greatest mathematician of antiquity. He was born in the Greek city of Syracuse in 287 B.C.E. He was the son of an astronomer and was a good friend to and was possibly related to King Hieron of Syracuse. He studied mathematics at the University of Alexandria in Egypt. After his studies, he returned to Syracuse to live. He wrote many papers or treatises in the areas of arithmetic, geometry, and the rudiments of calculus. He is credited with devising the classical method of computing the value of π and for developing the mathematical theory for hydrostatics of fluids.

Archimedes was a man of strong mental concentration. He solved problems by drawing figures in fireplace ashes or in the oil he placed on his body after his baths. There are many stories of Archimedes' feats. Here are several of them.

King Hieron had ordered a solid gold crown from his goldsmith. When the king received the crown, he suspected that it was not solid gold, but a mixture of gold and silver. The king did not want to have the crown melted down, so he asked Archimedes if he could figure out a way to determine if the crown was pure gold without destroying it.

After some serious thought, Archimedes decided to relax by taking a bath. While sitting in the bathtub, Archimedes saw that the water level rose in proportion to his weight. Suddenly he realized that he had

ARCHIMEDES PHILOSOPHE
Grec. Chap. 23.

discovered the solution to the problem. By placing the crown in a full tub of water, he could tell how much of it was gold by weighing the amount of water that overflowed. It is rumored that Archimedes became so excited that he forgot to clothe himself and ran down the streets of Syracuse naked, shouting, "Eureka, I have found it," right to the king's palace to tell him of the discovery. Needless to say, the king's suspicions were confirmed. The crown was not solid gold, and the goldsmith was placed in jail.

—Continued

Continued—

At another time, Archimedes boasted to the king, "Give me a lever long enough, and I will move the earth." King Hieron challenged the now-famous man to prove his boast. So one day, on the beach, Archimedes set up a complicated set of levers and pulleys and sat down on a chair. With little effort, he proceeded single-handedly to pull a large ship, fully loaded with people and cargo, out of the water. Heretofore, this task required many men and a great deal of effort.

King Hieron was so amazed with this feat that he immediately placed Archimedes in charge of the defense of Syracuse, which was being attacked by the Romans commanded by General Marcellus. First, the great mathematician devised huge catapults with adjustable ranges that were capable of hurling huge stones and fireballs at the enemy. Then he devised long poles that would drop large boulders on the Roman ships that sailed too close to the wall of Syracuse. Archimedes also created a device that resembled a huge grappling hook that was capable of lifting a Roman ship out of the water and shaking it violently until it fell apart or until the frightened sailors jumped overboard.

Some historians believe that Archimedes created a huge magnifying glass that was capable of focusing the sun's rays on the Roman ships and burning them up. It is said that the Romans became so fearful of the noble man that one day when they saw a rope hanging over a wall, they dared not approach, fearful that some dreadful contraption was on the other end.

Because of Archimedes' defenses, the soldiers of Syracuse held out for 3 years; however, during a festival celebration, they relaxed their defenses, and the Romans overran the city. At that time, Archimedes, 75, was engrossed in working out a mathematical problem in the sand. A Roman soldier approached the great man and ordered him to move to the center of the city. The old mathematician told him to wait until he finished the problem. The soldier became angry and killed him with his sword. The man who once ran naked through the streets has been called the "Father of Mathematics" by many.

Example 1-7

A 150-pound person walking briskly for 1 mile can burn about 100 calories. How many miles per day would the person have to walk to lose 1 pound in 1 week? It is necessary to burn 3500 calories to lose 1 pound.

Solution

Step 1 *Understand the problem.* A 150-pound person burns 100 calories per 1 mile and needs to burn 3500 calories in 7 days to lose 1 pound. The problem asks how many miles per day does the person have to walk to lose 1 pound in 1 week.

Step 2 *Devise a plan to solve the problem.* It is necessary to see how many calories need to be burned per day and then divide by 100 to see how many miles need to be walked.

Step 3 *Carry out the plan.* Since 3500 calories need to be expended in 7 days, divide 3500 by 7 to get 500 calories per day. Then divide 500 by 100 to get 5 miles. Hence, it is necessary to walk briskly 5 miles per day to lose 1 pound in a week.

Step 4 *Check the answer.* Multiply 5 miles $\times$ 100 calories $\times$ 7 days to get 3500 calories.

Try This One

1-E   If it costs $35.00 per day and $0.25 per mile to rent an automobile, how much will it cost a person to rent an automobile for a 2-day, 300-mile round trip?
For the answer, see page 32.

Exercise Set   1-2

Real World Applications

Select an appropriate strategy and solve each problem.

1. A pile of coins is worth $4.25. It consists of quarters and half dollars. If there are two more quarters than half dollars, how many coins of each are there?

2. A girl has twice as many dimes as nickels. If she has a total of $0.75, how many of each of the coins does she have?

3. A backyard fair charged $1.00 admission for adults and $0.50 for children. The fair made $25.00 and sold 30 tickets. How many adult tickets were sold?

4. One number is 6 more than another number, and their sum is 22. Find the numbers.

5. In a barnyard, there is an assortment of chickens and cows. Counting heads, one gets 12; counting legs, one gets 38. How many of each are there?

6. While sitting on the boardwalk, I counted nine people riding a bicycle or a tricycle, which made a total of 21 wheels. How many bicycles and how many tricycles passed by me?

7. The sum of the digits of a two-digit number is 7. If 9 is subtracted from the number, the answer will be a number with the digits reversed. Find the number.

8. A mother is four times as old as her daughter. In 18 years, the mother will be twice as old as her daughter. Find their present ages.

9. One number is 7 more than another number. Their sum is 23. Find the numbers.

10. A piece of rope is 48 inches long and is cut so that one piece is twice as long as the other. Find the lengths of each piece.

11. In 28 years, Mark will be five times as old as he is now. Find his present age.

12. A person has 10 stamps of two denominations, $0.25 and $0.15. How many of each does the person have if the stamps total $1.80?

13. Pete is twice as old as Lashanna. In 5 years, the sum of their ages will be 37. Find their present ages.

14. May receives $87 for working one 8-hour day. One day she had to stop after working 5 hours. How much was she paid for that day?

15. A property owner pays $60 to have his grass cut. It takes approximately 2 hours. If Sam works $1\frac{1}{2}$ hours and Pete works $\frac{1}{2}$ hour cutting the grass, how much does each receive? (Assume each receives the same hourly wage.)

16. A recipe calls for $\frac{1}{3}$ cup of sugar. If the chef wants to make $\frac{1}{2}$ of the recipe, how much sugar should be used?

17. A bag of apples weighs 8 pounds and a bag of pears weighs 10 pounds. Find the total weight of three bags of apples and six bags of pears.

18. A small beverage company has 832 bottles of water to ship. If there are six bottles per case, how many cases are needed and how many bottles will be left over?

19. A person's monthly budget includes $256 for food, $125 for gasoline, and $150 for utilities. If a person earns $1624 per month after taxes, how much money is left for other expenses?

20. One cup of Quaker Oat Bran cereal contains 3 grams of soluble fiber. If a person wanted to eat 300 grams of soluble fiber per month, how many cups of Oat Bran would the person need to consume?

21. A bag of pretzels contains 1650 calories. If a person purchases 26 bags of pretzels for a picnic, how many calories would be consumed if all the bags were eaten?

22. An automobile travels 527 miles on 12.8 gallons of gasoline. How many miles per gallon did the car get?

23. If a family borrows $12,381 for an addition to their home, and the loan is to be paid off in monthly payments over a period of 5 years, how much should each payment be? (Interest has been included in the total amount borrowed.)

24. Assuming that you could average driving 55 miles per hour, how long would it take you to drive from one city to another if the distance was 327 miles?

25. A landscaper sells maple saplings for $32, oak saplings for $25, and birch saplings for $20. If a person purchases five maple saplings, eight birch saplings, and three oak saplings, what is the total cost of all the trees?

26. Four persons decide to rent an apartment. Because each will be using it for different lengths of time, Mary will pay $\frac{1}{2}$ of the monthly rent, Jean will pay $\frac{1}{4}$ of the monthly rent, Claire will pay $\frac{1}{8}$ of the monthly rent, and Margie will pay the rest. If the monthly rent is $2375, how much will each person pay?

27. A business will reimburse its employees 26.5¢ per mile for using their own cars on business-related trips. How much should a person be reimbursed if a business trip was 864 miles?

28. Harry fills up his Jeep with gasoline and notes that the odometer reading is 23,568.7 miles. After a trip, he fills up his Jeep again and pays for 12.6 gallons of gasoline. He notes his odometer reading is 23,706.3 miles. How many miles per gallon did he get?

29. A clerk earns $9.50 per hour and is paid time and a half for any hours worked over 40. Find the clerk's pay if he worked 46 hours during a specific week.

30. The peak rate of a phone company is 25¢ per minute, and the off-peak rate is 15¢ per minute. Find the savings for a 32-minute call if it was made during off-peak time as opposed to peak time.

Writing Exercise

31. List and explain the steps in problem solving.

Critical Thinking

32. With a power mower, Phil can cut a large lawn in 3 hours. His younger brother can cut the same lawn in 6 hours. If both worked together using two power mowers, how long would it take them to cut the lawn?

33. A king decided to pay a knight one piece of gold for each day's protection on a 6-day trip. The king took a gold bar 6 inches long and paid the knight at the end of each day; however, he made only two cuts. How did he do this?

34. What is the smallest number of band members that can march in a straight line so that two marchers are in front of a member, two marchers are behind a member, and one marcher is between two members?

35. A new automobile depreciated 20% of its value after 1 year. If the automobile is priced at $18,000 after 1 year, what was its cost when it was new?

36. A college student with a part-time job budgets $\frac{1}{5}$ of her income for clothes, $\frac{2}{5}$ for living expenses (including food and rent), $\frac{1}{10}$ for entertainment, and $\frac{1}{2}$ of the remainder for savings. If she saves $1200 per year, what is her yearly income?

37. In order to purchase a computer for the Student Activities office, the freshman class decides to raise $\frac{1}{3}$ of the money, and the sophomore class decides to raise $\frac{1}{2}$ of the money. The Student Government Association agrees to contribute the rest, which amounted to $400. What was the cost of the computer?

1-3 Estimation

The problems in Section 1-2 required an exact answer. There are times when getting an exact answer is not feasible. In this case, we use estimation. **Estimation** is the process of finding an approximate answer to a mathematical problem. For example, if a contractor is building a game room for a customer, the contractor may give the customer an approximate cost for the job before starting. This is called an *estimate*.

Estimation has many uses. For example, you may be shopping for groceries, and you have only $20.00. It would be helpful to find the approximate cost of the groceries

Sidelight

Julia Robinson (1919–1985)

Julia Robinson became interested in mathematics at an early age, and her interest continued in high school where she was the only female student in her physics class. She studied mathematics at the University of California at Berkeley and later received her doctorate at Berkeley. Because of ill health, she never taught full time, but devoted her life to mathematics. She was able to solve problems similar to the ones studied by Hypatia in 400 c.e.

Her talents were widely respected during her lifetime, and she was the first female mathematician elected to the National Academy of Sciences and the first female president of the American Mathematical Society.

you have in your shopping cart so that you do not try to purchase more than you can pay for. Another area where estimation is helpful is in purchasing carpet for your home. First, you would have to estimate the area of the rooms in square yards; then you would need to know the approximate cost of the carpet, padding, and installation; and then you could find the approximate cost of the entire job.

Estimation can also be used to check the answers to mathematics problems. Although you cannot be 100% sure your exact answer is correct by estimating it, you can be somewhat sure if your estimated answer is reasonably close to your exact answer. For example, if a service station owner bought 24 cases of windshield washer fluid at $3.95 per case, his bill would be $24 \times \$3.95 = \$94.80$. You would know this is reasonable since 24 is about 25 and $3.95 is about $4.00; hence, $25 \times \$4.00 = \$100.00$.

Since the process of estimating uses rounding, a brief review of rounding is given here. Rounding uses the concept of place value. The place value of a digit in a number tells the value of the digit in terms of ones, tens, hundreds, etc. For example, in the number 325, the 3 means 3 hundreds or 300 since its place value is hundreds. A place value chart is shown here entitled "Rounding Numbers."

Rounding Numbers

In order to round numbers, use these rules:

1. Locate the place-value digit of the number that is being rounded. Here is the place value chart for whole numbers and decimals:

8	9	8	5,	7	3	0,	2	6	1	.	2	3	5	6	7	5
billions	hundred-millions	ten-millions	millions	hundred-thousands	ten-thousands	thousands	hundreds	tens	ones		tenths	hundredths	thousandths	ten-thousandths	hundred-thousandths	millionths

2a. If the digit to the right of the place-value digit is 0 through 4, then do not change the place-value digit.

2b. If the digit to the right of the place-value digit is 5 through 9, add one to the place-value digit.

Note: When rounding whole numbers, replace all digits to the right of the digit being rounded with zeros. When rounding decimal numbers, drop all digits to the right of the digit that is being rounded.

Example 1-8

Round each number to the place value given.

(a) 7328 (hundreds)
(b) 15,683 (thousands)
(c) 32.4817 (tenths)
(d) 0.047812 (ten-thousandths)

—Continued

Example **1-8** *Continued—*

Solution

(a) In the number 7328, the 3 is the digit being rounded. Since the digit to the right is 2, the digit 3 remains the same, and the 2 and 8 are replaced by zeros. The rounded number is 7300.

(b) In the number 15,683, the 5 is the digit to be rounded. Since the digit to the right is 6, 1 is added to the 5 and the digits 6, 8, and 3 are replaced by zeros. The rounded number is 16,000.

(c) In the number 32.4817, the 4 is the digit to be rounded. Since the digit to the right of the 4 is 8, 1 is added to the 4 to get 5 and all digits to the right of the 4 are dropped. The rounded number is 32.5.

(d) In the number 0.047812, the 8 is the digit to be rounded. Since the digit to the right of 8 is 1, the 8 remains the same. The digits 1 and 2 are dropped. The rounded number is 0.0478.

Math Note

One question students ask is, "How do I know what digit to round to?" There is no exact answer. Use the most convenient one. For example, in Example 1-9, the cost of the refrigerator, $579.99, could have been rounded to $580.00. Then the estimated answer would be $580.00 × 6 = $3,480.00. This is a better estimate than the one given in the example since the estimated cost has been rounded to the nearest $1.00, whereas in the example, the cost was rounded to the nearest $100.00.
    Since there is no exact rule to tell you what place value the number should be rounded to, it is important to use common sense. When estimating, there is no one correct answer.

Try This One

1-F   Round each number to the place value given.

(a) 372,651 (hundreds)

(b) 32.971 (ones)

(c) 0.37056 (thousandths)

(d) 1465.983 (hundredths)

For the answer, see page 32.

Estimation

When estimating answers for numerical calculations, two steps are required.

1. Round the numbers being used.
2. Perform the operation or operations involved.

Example **1-9**

An apartment owner wishes to purchase six refrigerators for her apartments. The cost of each refrigerator is $579.99. Estimate the cost of the refrigerators.

Solution

Step 1   Round the cost of the refrigerators to $600.00.

Step 2   Multiply $600.00 × 6 = $3600.00

Hence, it will cost approximately $3600.00 for the refrigerators.

Sidelight

The Largest Number

It seems that large numbers are a part of our advancing civilization. The cavemen probably relied on their fingers and toes for counting, while large numbers today tend to overwhelm us.

Do you know what the largest number is? A million? A billion? A trillion? Do you know how large a million is? A billion? A trillion?

A million is no longer considered a large number. Many times, we hear the word billion being used. Government budgets are in the billions of dollars, the world population is approaching six billion, etc. Just how large are these numbers?

If you counted to a million, one number per second with no time off to eat or sleep, it would take you approximately $11\frac{1}{2}$ days. A million pennies would make a stack almost a mile high and a million one dollar bills would weigh about a ton.

A million is written as a one followed by six zeros (1,000,000). Millions are used for measuring the distance from the earth to the sun, about 93 million miles.

The number one followed by nine zeros is called a billion (1,000,000,000). It is equal to 1000 millions.

It would take you 32 years to count to a billion with no rest. A billion dollars in one dollar bills would weigh over a thousand tons. The age of the earth is estimated to be almost 4.5 billion years. There are more than 100 billion stars in our galaxy. A billion pennies would make a stack almost 1000 miles high.

The number one followed by 12 zeros is a trillion. It is written as 1,000,000,000,000. Counting to a trillion would take 32 thousand years. The nearest star is 27 trillion miles away.

A quadrillion, a quintillion, and a sextillion are written with 15, 18, and 21 zeros, respectively. The weight of the earth is 6 sextillion and 570 quintillion tons (6.6 sextillion tons). Adding three zeros brings another -illion until a vigintillion is reached. That number has a total of 63 zeros.

Is this the largest number? No. A nine-year-old child invented a number with 100 zeros. It is called a googol. It looks like this:

10,000,000,000,000,000,000,000,000,000,000,000,000,000,
000,000,000,000,000,000,000,000,000,000,000,000,000,000,
000,000,000,000,000,000

This number is said to be more than the total number of protons in the universe. Yet an even larger number is the googolplex. It is a one followed by a googol of zeros. If you tried to write it, it would take a piece of paper larger than the distance from here to the moon.

Naturally, any number you can think of can be made larger by adding 1. A wise man once defined a myriad to be a numberless number, although many unwise consider it to be 10 thousand.

Example 1-10

A person wishes to purchase a digital cell phone at a cost of $179.00 and a monthly charge of $29.99. Estimate the cost of the phone for 1 year assuming that there are no additional charges.

Solution

The cost of the phone can be rounded to $200.00, and the cost per month to $30.00. Hence,

$$\$30.00 \times 12 = \begin{array}{r} \$360.00 \\ + 200.00 \\ \hline \$560.00 \end{array}$$

The estimated cost of the phone is $560.00.

—Continued

Example 1-10 Continued—

Note: If the cost of the phone would have been rounded to $180.00, then the estimate would be $360.00 + $180.00 = $540.00. This is a somewhat better estimate since it is closer to the exact answer of $538.88.

Try This One

1-G Estimate the total cost of five suits at $89.95 each and three pairs of shoes at $39.95 each.
For the answer, see page 32.

Sometimes you will need to overestimate or underestimate an answer. For example, if you were buying groceries and had only $20.00, you would want to overestimate the cost of your groceries so that you would have enough money to pay for them. If you would need to earn money to pay the cost of your college tuition for the next semester, you would want to underestimate how much you could earn each week in order to determine the number of weeks you need to work in order to pay your tuition.

At times you will need to make estimations from information displayed in various graphs, such as bar graphs, pie graphs, or time series graphs. Examples 1-11, 1-12, and 1-13 illustrate how this can be done.

A bar graph is used to compare amounts or percents using either vertical or horizontal bars of various heights which correspond to the amounts or percents.

Example 1-11

The graph shown represents the percent of electricity generated by nuclear energy in five countries. Find the approximate percent of electricity generated by nuclear energy in France.

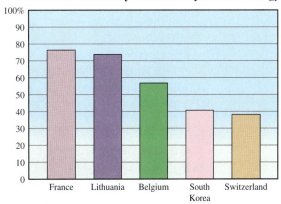

Percent of Electricity Generated by Nuclear Energy

Source: The World Almanac and Book of Facts

—Continued

Example   1-11   *Continued—*

Solution

Locate the bar representing the amount of nuclear energy generated in France and read across to the vertical axis. Since the top of the bar extends to about halfway between 70% and 80%, we would estimate the answer to be about 75%.

Try This One

1-H   Using the graph shown in Example 1-11, find the approximate percent of electricity generated by nuclear energy for Switzerland.
For the answer, see page 32.

A pie graph, also called a circle graph, is constructed by drawing a circle and dividing it into parts called sectors, according to the size of the percents of the parts in relationship to the whole.

Example   1-12

The pie graph shown represents the number of fatal occupational injuries in the United States for a selected year. If the total number of fatal injuries was 5915 for the year, estimate how many resulted from assaults and violent acts.

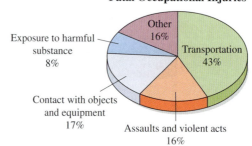

Fatal Occupational Injuries

Other 16%

Transportation 43%

Assaults and violent acts 16%

Contact with objects and equipment 17%

Exposure to harmful substance 8%

Source: The World Almanac and Book of Facts

Solution

The sector labeled "Assaults and Violent Acts" indicates that 16% of the total fatal injuries resulted from assaults and violent acts. In order to estimate the number of fatal injuries for that year, multiply 16% × 5915 or 0.16 × 5915 = 946.4 or about 946 when rounding. (*Note:* To change a percent to a decimal, move the point two places to the left and drop the percent sign. 16% = 0.16.)

Try This One

1-I   Using the graph shown in Example 1-12, find the approximate number of fatal occupational injuries that resulted from transportation accidents.
For the answer, see page 32.

A time series graph or line graph represents how something varies or changes over a specific time period.

Example   1-13

The graph shown indicates the number of cable systems in the United States from 1960 to 2000. Find the approximate number of cable systems in 1970.

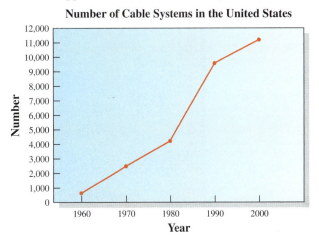

Number of Cable Systems in the United States

Source: The New York Times Almanac

Solution

Locate the year 1970 on the horizontal axis and move up to the line on the graph. At this point, move horizontally to the point on the vertical axis as shown.

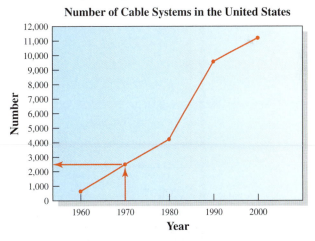

Number of Cable Systems in the United States

There were approximately 2400 cable systems in the United States in 1970.

Try This One

1-J   Using the graph shown in Example 1-13, find the approximate number of cable systems in the United States for the year 1985.
For the answer, see page 32.

Distances between two cities can be estimated by using a map and its scale. The procedure is shown in Example 1-14.

Example 1-14

Estimate the distance between two cities that are $5\frac{3}{4}$ inches apart on a map if the scale is $\frac{1}{2}$ inch = 25 miles.

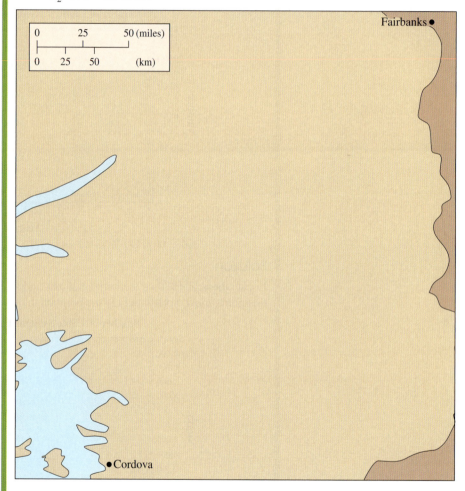

Solution

First it is necessary to find how many half-inches there are in $5\frac{3}{4}$ inches. Divide $5\frac{3}{4}$ by $\frac{1}{2}$ to get $11\frac{1}{2}$. Now, since each half-inch is equivalent to 25 miles, multiply 25 by $11\frac{1}{2}$ to get 287.5 miles. Hence, the approximate distance is 287.5 miles.

Try This One

1-K Estimate the distance between two cities that are $2\frac{1}{8}$ inches apart on a map if the scale is $\frac{1}{8}$ inch = 50 miles.

For the answer, see page 32.

Exercise Set 1-3

Computational Exercises

For Exercises 1–20, round the number to the place value given.

1. 2861 (hundreds)
2. 732.6498 (thousandths)
3. 3,261,437 (ten-thousands)
4. 9347 (tens)
5. 62.67 (ones)
6. 45,371,999 (millions)
7. 218,763 (hundred-thousands)
8. 923 (hundreds)
9. 3.671 (hundredths)
10. 56.3 (ones)
11. 327.146 (tenths)
12. 83,261,000 (millions)
13. 5,462,371 (ten-thousands)
14. 7.123 (hundredths)
15. 272,341 (hundred-thousands)
16. 63.715 (tenths)
17. 264.97348 (ten-thousandths)
18. 1,655,432 (millions)
19. 563.271 (hundredths)
20. 426.861356 (hundred-thousandths)

Real World Applications

For Exercises 21–36, estimate the answer.

21. Estimate the total cost of eight tires on sale for $56.99 each.
22. Estimate the total cost of five electric grills which cost $39.95 each.
23. Estimate the time it would take you to drive 237 miles at 37 miles per hour.
24. Estimate the distance you can travel in 3 hours and 25 minutes if you drive on average 42 miles per hour.
25. Estimate the sale price of a set of luggage that originally cost $178.99 and is now on sale for 60% off.
26. Estimate the sale price of an electric iron that costs $32.99, on sale for 15% off.
27. Estimate the total cost of the following meal at a fast-food restaurant:

Deluxe Hamburger	$1.89
Giant Fries	1.29
Soda	0.89

28. Estimate the total cost of the following:

Desk	$159.95
Chair	69.99
File Cabinet	29.95
Light	19.95

29. Estimate the time it would take you to bicycle 86 miles at 8 miles per hour.
30. If a person earns $48,300.00 per year, estimate how much the person earns per hour. Assume a person works 40 hours per week and 50 weeks per year.
31. If a person earns $8.75 per hour, estimate how much the person would earn per year. Assume a person works 40 hours per week and 50 weeks per year.

32. If a salesperson earns a commission of 12%, estimate how much money the salesperson would earn on an item that sold for $529.85.

33. Estimate the cost of fencing in a yard that is 24 feet long and 18 feet wide if the fence costs $5.95 per foot.

34. Estimate the cost of painting the walls of a room that is 12 feet by 16 feet by 9 feet if 1 gallon of paint costs $11.99 and covers 40 square feet.

35. Estimate the cost of planting grass if one box of grass seed will cover 24 square feet and the yard is 16 feet by 16 feet. The cost of the grass seed is $5.95 per box.

36. Estimate your cost to live in an apartment for 1 year if the rent is $365.00 per month and utilities are $62.00 per month.

Use the information shown in the graph for Exercises 37–40. The graph shown gives the areas of the Great Lakes in square miles.

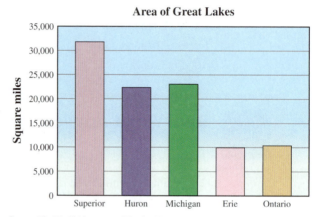

Area of Great Lakes

Source: The World Almanac and Book of Facts

37. Estimate the area of Lake Huron.

38. Estimate the area of Lake Erie.

39. Estimate the difference in areas of the largest and smallest of the Great Lakes.

40. Estimate the areas of each of the two lakes which are approximately the same size.

Use the information shown in the graph for Exercises 41 and 42. The graph represents a survey of 1385 office workers and shows the percents of people who indicated what time of day they feel is the most productive for them.

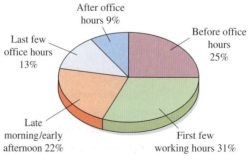

Most Productive Time of Day

After office hours 9%

Before office hours 25%

Last few office hours 13%

Late morning/early afternoon 22%

First few working hours 31%

Source: USA Today

41. Approximately how many people in the survey feel that late morning and early afternoon are the most productive times of day?

42. Approximately how many people in the survey feel that the time after office hours is the most productive time of the day?

Use the information shown in the graph for Exercises 43 and 44. The graph shows the cigarette consumption (in billions) in the United States for the years 1900 to 2000.

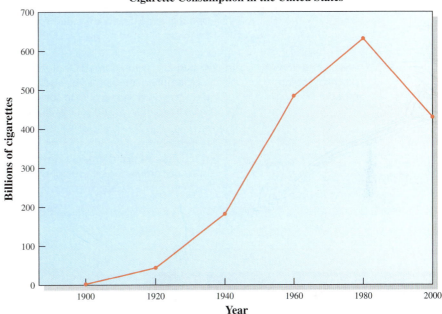

Cigarette Consumption in the United States

Source: The New York Times Almanac

43. Estimate the number of cigarettes smoked in 1950.

44. Estimate the number of cigarettes smoked in 1985.

45. Estimate the distance between two cities that are $3\frac{1}{2}$ inches apart on a map if the scale is $\frac{1}{4}$ inch = 40 miles.

Writing Exercises

46. What is estimation?

47. Suggest three areas where estimation can be used.

48. Explain why the exact answer to a mathematical problem is not always necessary.

49. How can estimation be used to "check" the exact answer to a mathematical problem?

50. Select an advertisement and estimate the cost of a home entertainment center. It should consist of a 27″ television, a VCR, a CD player, and a cabinet.

51. Select an advertisement and estimate the cost of a computer center. It should consist of a computer desk, a computer, a printer, and a 17″ color monitor.

Summary

Section	Important Terms	Important Ideas
1-1	inductive reasoning counterexample deductive reasoning	**In** mathematics, two types of reasoning can be used. They are inductive and deductive reasoning. Inductive reasoning is the process of arriving at a general conclusion based on the observation of specific examples. Deductive reasoning is the process of reasoning that arrives at a specific conclusion based on previously accepted statements.
1-2	Polya's four-step problem solving procedure	**A** mathematician named George Polya devised a procedure to solve mathematical problems. The steps of his procedure are (1) understand the problem, (2) devise a plan to solve the problem, (3) carry out the plan to solve the problem, and (4) check the answer.
1-3	estimation	**In** many cases, it is not necessary to find the exact answer to a problem. When only an approximate answer is needed, you can use estimation. This is accomplished by rounding the numbers used in the problem, then performing the necessary operation or operations.

Review Exercises

For Exercises 1–4, make reasonable conjectures for the next three numbers or letters in the sequence.

1. 3   4   6   7   9   10   12   13   15
   16   __18__   __19__   __21__

2. 2   7   4   9   6   11   8   13   ___
   ___   ___

3. 4   z   16   w   64   t   256   ___
   ___   ___

4. 20   A   18   B   16   C   ___   ___
   ___

For Exercises 5 and 6, make a reasonable conjecture and draw the next figure.

5.  _____

6.  _____

For Exercises 7 and 8, find a counterexample to show that each statement is false.

7. The product of three odd numbers will be even.

8. The sum of three multiples of 5 will always end in a 5.

For Exercises 9 and 10, use inductive reasoning to find a rule that relates the number selected to the final answer, and then try to prove your conjecture using deductive reasoning.

9. Select an even number:

   Add 6:
   Divide the answer by 2:
   Add 10:
   Result:

10. Select a number:

    Multiply it by 9:
    Add 18 to the number:
    Divide by 3:
    Subtract 6:
    Result:

For Exercises 11–25, use an appropriate strategy to solve each problem.

11. Cindy had 32 fish. She gave away all but nine. How many did she have left?

12. A tennis team played 40 games. The team won 20 more games than they lost. How many games did the team lose?

13. If a person weighs 110 pounds when standing on one foot on a scale, how much would that person weigh when he stands on a scale with both feet?

14. A cup of coffee and a Danish cost $1.40. If the Danish costs $0.40 more than the coffee, how much did each cost?

15. Harry is 10 years old, and his brother Bill is twice as old. When Harry is 20, how old will Bill be?

16. Mary had $30.00. She spent $3.00 on bus fare and $\frac{1}{3}$ of the remainder on lunch. How much did she have left?

17. A baseball glove and a baseball together cost $20.00. If the glove costs twice as much as the baseball, how much did the baseball cost?

18. Fill in the squares with digits to complete the problem. There are several correct answers.

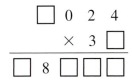

19. In 10 years, Harriet will be three times as old as she was 10 years ago. How old is she now?

20. In Alaska, the day is 18 hours longer than the night. How long is each?

21. Babs is 9 years older than Debbie, and Jack is 7 years younger than Babs. The sum of their ages is 20. How old is Debbie?

22. Using $+$, $-$, and $\times$ make a true equation. Do not change the position of any digits.

$$2 \quad 9 \quad 6 \quad 7 \quad = \quad 17$$

23. Harry says to Bill, "If you give me one baseball card, then we will have an equal number of cards." Bill replied, "If you give me one card then I will have double the number you have!" How many cards does each have?

24. Can you divide a pie into 11 pieces with four straight cuts? The cuts must go from rim to rim but not necessarily through the center. The pieces need not be identical.

25. How many triangles are in the figure shown here?

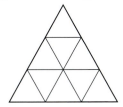

26. In 20 years, Charles will be three times as old as he is now. How old is he now?

27. The sum of two numbers is 20 and the difference is 5. Find the numbers.

28. Grass seed A costs $2.00 per pound, and grass seed B costs $4.00 per pound. If a mixture of 10 pounds costs $24.00, find the amount of each type that was mixed.

29. Mr. Taylor had $1000 to invest. He invested part of it at 8% and part of it at 6%. If his total simple interest was $76.00, find how much he invested for 1 year at each rate.

For Exercises 30–34, round each number to the place value given.

30. 132,356 (thousands)

31. 186.75 (ones)

32. 14.63157 (ten-thousandths)

33. 0.6314 (tenths)

34. 3725.63 (tens)

35. Estimate the cost of four lawnmowers if each one costs $329.95.

36. Estimate the cost of five textbooks if they cost $115.60, $89.95, $29.95, $62.50, and $43.10.

Use the information shown in the graph for Exercises 37 and 38. The graph shows the percents of the pollutants released into the environment for a specific year.

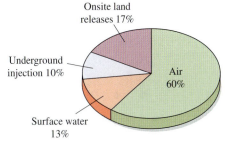

Pollutant Releases in the United States

Onsite land releases 17%

Underground injection 10%

Air 60%

Surface water 13%

Source: The World Almanac and Book of Facts

37. If the total amount of pollutants released is 1953 million pounds, find the amount of pollutant released in the air.

38. If the total amount of pollutants released is 1953 million pounds, find the amount of pollutants released in surface water.

Use the information shown in the graph for Exercises 39 and 40. The graph shows the average weekly salary (in dollars) for United States workers from 1970 to 2000.

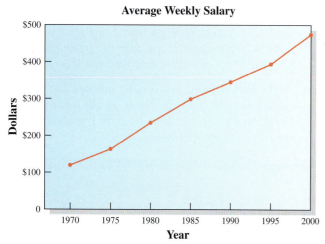

Average Weekly Salary

Source: World Almanac and Book of Facts

39. Estimate the weekly salary in 1988.

40. Estimate the weekly salary in 1972.

41. Estimate the distance between two cities that are 8 inches apart on a map if the scale is 2 inches = 15 miles.

Chapter Test

For Exercises 1 and 2, make reasonable conjectures for the next three numbers in the sequence.

1. 2   4   3   6   5   9   8   __12__   __12__   __18__

2. 5   10   20   40   80   ___   ___   ___

3. Determine the next figure in the sequence. (The numbers associated with the figures are called triangular numbers.)

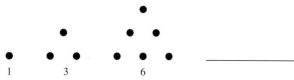

4. Determine the next figure in the sequence. (The numbers associated with the figures are called square numbers.)

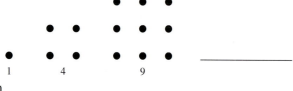

5. Use inductive reasoning to find a rule that relates the number you selected to the final answer and try to prove your conjecture.

   Select a number:
   Add 10 to the number:
   Multiply the answer by 5:
   Add 15 to the answer:
   Divide the answer by 5:
   Result:

For Exercises 6–21, use an appropriate strategy to solve each problem.

6. There were 12 birds sitting on a fence. All but 2 flew away. How many birds were left sitting on the fence?

7. For a job, a worker receives a salary that doubles each day. After 10 days of work, the job is finished and the person receives $100.00 for the final day's work. On which day did the worker receive $25.00?

8. What are the next two letters in the sequence O, T, T, F, F, S, S, . . .? (*Hint:* It has something to do with numbers.)

9. Move one line to make a correct equation:

$$| \, | \; - \; | \, | \, | \; = \; | \, |$$

10. Write the number eleven thousand, eleven hundred eleven using digits.

11. The problem here was written by the famous mathematician Diophantus. Can you find the solution?
    The boyhood of a man lasted $\frac{1}{6}$ of his life; his beard grew after $\frac{1}{12}$ more; after $\frac{1}{7}$ more he married; 5 years later his son was born; the son lived to half the father's age; and the father died 4 years after the son. How old was the father when he died?

12. A number divided by 3 less than itself gives a quotient of $\frac{8}{5}$. Find the number.

13. The sum of $\frac{1}{2}$ of a number and $\frac{1}{3}$ of the same number is 10. Find the number.

14. Add five lines to the square to make three squares and two triangles.

15. One person works for 3 hours and another person works for 2 hours. They are given a total of $60.00. How should it be divided up so that each person receives a fair share?

16. The sum of the reciprocals of two numbers is $\frac{5}{6}$ and the difference is $\frac{1}{6}$. Find the numbers. (*Hint:* The reciprocal of a number n is $\frac{1}{n}$.)

17. If Sam scored 87% on his first exam, what score would he need on his second exam to bring his average up to 90%?

18. Mt. McKinley is about 20,300 feet above sea level, and Death Valley is 280 feet below sea level. Find the vertical distance from the top of Mt. McKinley to the bottom of Death Valley.

19. The depth of the ocean near the island of Mindanao is about 36,400 feet. The height of Mt. Everest is about 29,000 feet. Find the vertical distance from the top of Mt. Everest to the floor of the ocean near the island of Mindanao.

20. A person is drawing a map and using a scale of 2 inches = 500 miles. How far apart on the map would two cities be located if they are 1800 miles apart?

21. Mark's mother is 32 years older than Mark. The sum of their ages is 66 years. How old is each?

22. Round 1,674,253 to the nearest hundred thousand.

23. Round 1.3752 to the nearest hundredth.

24. Estimate the cost of building a backyard playground if a swing set costs $69.95, a sandbox and sand cost $32.54, and a children's pool costs $42.99.

25. Using the graph shown here, estimate the average number of hours per week Americans worked in 1980.

Average Hours Worked

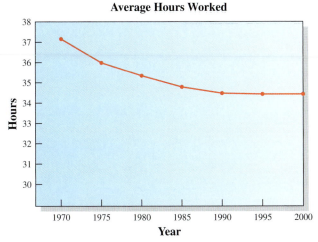

Source: World Almanac and Book of Facts

26. Using the graph shown here, estimate the number of working women surveyed who say they work for the extra money. The total number of women surveyed was 1250.

Reasons for Working Outside the Home

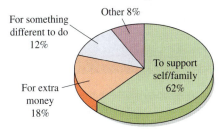

Source: USA Today

27. For a specific year, the number of homicides reported for selected cities is shown here. Estimate the number of homicides for Baltimore.

Number of Homicides

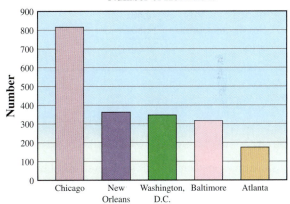

Source: USA Today

28. Estimate the distance between two cities that are 4 inches apart if the scale on a map is $\frac{1}{2}$ inch = 8 miles.

Mathematics in Our World
▶*Revisited*

Rhind Papyrus

The solution to problem 79 of the Rhind *Papyrus* is found by using 7 as a multiplier. Since there are seven houses and each house has seven cats, there are 7 × 7 or 49 cats. Since there are 49 cats and each cat eats seven mice, there are 49 × 7 or 343 mice, etc. Hence the estate consists of 19,607 items, as shown.

Houses	7
Cats	49
Mice	343
Heads of wheat	2,401
Hekat measures	16,807
Total	19,607

Answers to **Try This One**

1-A. Pattern: Alternate entries increase by 1. Hence 8, 6, 9.

1-B. Your final answer will be the original number.

1-C. Nine

1-D. She purchased four $0.37 stamps and three $0.50 stamps.

1-E. $145.00

1-F. (a) 372,700 (b) 33 (c) 0.371 (d) 1465.98

1-G. $570

1-H. Approximately 39%

1-I. Approximately 2543

1-J. Approximately 6800

1-K. Approximately 850 miles

Projects

1. Sometimes mathematicians study systems of equations where the number of unknowns is greater than the number of equations. In this case, there may be more than one solution. Can you solve the next problem?

   A person paid $100 for 100 animals (horses, cows, and chickens). Each horse costs $3.00, each cow costs $1.00, and each chicken costs $0.50. How many of each animal did the person buy?

2. There are many books on problem solving and there are critical thinking problems in other textbooks. Find a book that has a unique problem. Copy the problem and write a short paper explaining how you solved it using the four steps suggested by Polya.

Chapter

Two

Sets

Objectives

After completing this chapter, you should be able to

1 Define set, element, and null set (2-1)

2 Designate sets in three different ways (2-1)

3 Classify sets as finite or infinite (2-1)

4 Identify equal sets, equivalent sets (2-1)

5 Find subsets and proper subsets of a set (2-2)

6 Find the union and intersection of two sets (2-2)

7 Find the complement of a set (2-2)

8 Draw Venn diagrams for set operations (2-3)

9 Solve survey and classification problems using sets (2-4)

10 Find the general term of an infinite set (2-5)

11 Determine whether or not a set is infinite (2-5)

Introduction

The idea of a set or collection of objects has been used by people since time began. Primitive societies had a set of family members, a set of weapons, a set of tools, etc. Today people have collections of coins, stamps, automobiles, and so on.

The notion of a set has been used since people began studying mathematics. For example, the Romans used the set of symbols, I, V, X, L, C, D, and M to write their numbers. In their study of geometry, the Greeks used sets of points, lines, and planes. However, it wasn't until the late 1800s that the theory of sets was studied as a branch of mathematics. Much of this theory was developed by the German mathematician Georg Cantor. Other mathematicians who contributed to the early development of set theory were Leonard Euler and John Venn.

This chapter explains the basic concepts of sets, subsets, elements, basic set operations, and the pictorial representation of sets called Venn diagrams. The last section of this chapter shows how many real-life problems can be solved by using the principles of set theory. ∎

Georg Cantor

Mathematics in Our World

Survey: Blood Types

Human blood can be classified by the type of antigens it contains. There are three types of antigens. They are the A antigen, the B antigen, and the Rh antigen. If blood contains both the A antigen and the B antigen, then it is called type AB. If blood contains neither the A antigen nor the B antigen, it is called type O. Therefore, there are four blood types: A, B, AB, and O. If any of the four types contain the Rh antigen, then it is classified as positive (i.e., A+, B+, AB+, O+). If any of the blood types does not contain the Rh factor, then it is classified as negative.

In a certain hospital, the records of the patients show that

23 have type A antigen

32 have type B antigen

11 have both the A and B antigens

10 have the A antigen and the Rh antigen

13 have the B antigen and the Rh antigen

32 have the Rh antigen

3 have all three antigens

4 have none of the antigens

Based on the results of this survey, can you answer the following questions?

(a) How many patients are A positive?

(b) How many patients are O negative?

(c) How many patients are AB+?

(d) How many patients have exactly two antigens?

(e) How many patients are in the hospital?

The answers are found on page 71.

Health professionals need to know the blood type of both donor and recipient for a safe transfusion.

2-1 The Nature of Sets

Basic Concepts

Although a set is an abstract concept, we can define it as follows:

> A **set** is a well-defined collection of objects.

One way to designate a set is to use braces and a capital letter to name the set. This is called the *list* or **roster method.** Each object of a set is called an **element** or *member* of the set. The members of a set are separated by commas. This is called the list or roster method of showing a set.

A set is said to be **well defined** when there is no misunderstanding as to whether or not an element belongs to a set. For example, the set of the "letters of the English alphabet" is a well-defined set since it consists of the 26 symbols we use to make our alphabet, whereas the set of "tall students attending your school" is relative, which can be interpreted differently by different people.

The laws about who can vote determine a well-defined set of people. Other laws might affect sets (e.g., businesses of a certain size) that must be well defined so that the law will stand. If the set is not well defined, it will be hard to enforce the law.

> **Math Note**
>
> The commas make it clear that it is the words not the letters that are the elements of the set.

> **Math Note**
>
> Redundancy of elements is not a problem in sets; however, it is a convention that we chose not to repeat listing the elements of a set. The set of letters in the word "minimum" is written as {m, i, n, u}.

Example 2-1

Write the set of months of the year that begin with "*M*."

Solution

The months that begin with *M* are March and May. Hence, the answer can be written in set notation as

$$M = \{\text{March, May}\}$$

Each element in the set is separated by a comma.

Example 2-2

Write the set of Great Lakes.

Solution

The set of Great Lakes is {Ontario, Erie, Huron, Michigan, Superior}.

In mathematics, the set of *counting* or **natural numbers** is defined as $N = \{1, 2, 3, 4, \ldots\}$. The three dots or ellipsis mean that the numbers go on indefinitely in the same pattern.

Example 2-3

Write the set of natural numbers less than 6.

Solution

$\{1, 2, 3, 4, 5\}$

Example 2-4

Write the set of natural numbers greater than 4.

Solution

$\{5, 6, 7, 8, \ldots\}$

Try This One

2-A Write each in set notation using braces.

(a) The set of even natural numbers from 80 to 90.

(b) The set of days of the week that begin with the letter "S."

For the answer, see page 72.

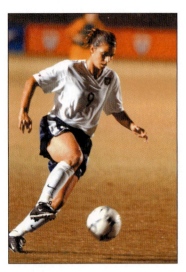

The U.S. Women's Soccer Team is a set of athletes. Mia Hamm is an element of the set.

The symbol $\in$ is used to show that an object is a member or element of a set. For example, let set $A = \{2, 3, 5, 7, 11\}$. Since 2 is a member of set A, it can be written as

$$2 \in \{2, 3, 5, 7, 11\} \quad \text{or} \quad 2 \in A$$

likewise,

$$5 \in \{2, 3, 5, 7, 11\} \quad \text{or} \quad 5 \in A$$

When an element is not a member of a set, the symbol $\notin$ is used. Because 4 is not an element of set A, this fact is written as

$$4 \notin \{2, 3, 5, 7, 11\} \quad \text{or} \quad 4 \notin A$$

Math Note

Be sure to use correct symbols when showing membership of a set. For example, the notation $\{6\} \in \{2, 4, 6\}$ is incorrect since a set is not a member of this set.

Try This One

2-B State whether each of these is true or false:

(a) $16 \in \{1, 3, 5, 7, \ldots\}$

(b) $a \notin \{m, a, p\}$

(c) $15 \in \{1, 2, 3, \ldots, 19, 20\}$

For the answer, see page 72.

There are three ways to designate sets. They are

1. The *list* or *roster* method.
2. The *descriptive* method.
3. *Set-builder* notation.

Recall that in designating sets using the list or roster method, the elements of the set are listed in braces and are separated by commas, as shown in Examples 2-1 through 2-4 and in Example 2-5.

Example 2-5

Write the set of natural numbers less than 8.

Solution

$\{1, 2, 3, 4, 5, 6, 7\}$

The **descriptive method** uses a short statement to describe the set.

Example 2-6

Use the descriptive method to describe the set containing 2, 4, 6, 8, ...

Solution

Since the elements in the set are called the even natural numbers, the answer is

$$E = \text{even natural numbers}$$

The third method of designating a set is **set-builder notation,** and this method uses *variables*.

A **variable** is a symbol (usually a letter) that can represent different elements of a set.

Example 2-7

Use set-builder notation to designate each set.

(a) The set containing the elements 2, 4, and 6.
(b) The set containing the elements red, yellow, and blue.

Solution

(a) $\{x \mid x \in E \text{ and } x < 7\}$
(b) $\{x \mid x \text{ is a primary color}\}$

In part a of Example 2-7, the | is read as "such that," and the entire expression is read, "The set of all x such that x is an even natural number and x is less than seven."

Example 2-8

Designate the set 32, 33, 34, 35, . . . using

(a) The roster method.
(b) The descriptive method.
(c) Set-builder notation.

—Continued

Example   2-8   *Continued—*

Solution

(a) {32, 33, 34, 35, . . .}
(b) Natural numbers greater than 31
(c) {$x \mid x \in N$ and $x > 31$}

Try This One

2-C   Designate the set of planets in our solar system using

(a) The roster method.

(b) The descriptive method.

(c) Set-builder notation.

For the answer, see page 72.

If a set contains many elements, three dots can be used to represent the missing elements. For example, the set {1, 2, 3, . . . , 99, 100} includes all the natural numbers from 1 to 100. Likewise, the set {a, b, c, . . . , x, y, z} includes all the letters of the alphabet.

Example   2-9

Using the roster method, write the set containing all even natural numbers between, but not including, 99 and 201.

Solution

{100, 102, 104, . . . , 198, 200}

Example   2-10

Write the set of odd natural numbers from 50 to 500.

Solution

{51, 53, 55, . . . , 497, 499}

Finite and Infinite Sets

Sets can be classified as *finite* or *infinite*. A set is said to be a **finite set** if the number of elements contained in the set is either 0 or a natural number; otherwise, it is said to be an **infinite set.**

The set {p, q, r, s} is considered a finite set since it has four members, namely, p, q, r, and s, whereas the set {10, 20, 30, . . .} is considered infinite since it has an unlimited number of elements, namely, natural numbers that are the multiples of 10.

Example 2-11

Classify each set as finite or infinite.

(a) $\{x \mid x \in N$ and $x < 100\}$
(b) The set of letters used to make Roman numerals
(c) $\{100, 102, 104, 106, \dots\}$
(d) The set of members of your immediate family
(e) The set of songs that can be written

Solution

(a) Finite
(b) Finite
(c) Infinite
(d) Finite
(e) Infinite

Math Note

The null set is represented by either { } or ∅, but not {∅} since the set {∅} has one element, namely, ∅.

Try This One

2-D Classify each set as finite or infinite.

(a) {the natural numbers that are multiples of 6}

(b) $\{x \mid x$ is a member of the U.S. Senate$\}$

(c) $\{3, 6, 9, \dots, 24\}$

For the answer, see page 72.

There are some situations in which it is necessary to define a set with no elements. For example, the set of female presidents of the United States would contain no people; hence, it would have no elements.

A set with no elements is called an *empty set* or **null set.** The symbols used to represent the null set are { } or ∅.

Calculator Explorations

Use your calculator to enter 3 divided by 0.

3/0

When you press [ENTER], what information does the calculator indicate?

What does it mean? (To remove the message from the HOMESCREEN press [CLEAR])

Write the set of answers to 3 divided by 0.

Example 2-12

Write the set of numbers that represents the spots on a six-sided die (singular for dice) that are greater than 6.

Solution

Since each side of a die contains only 1 through 6 spots, the set would be empty; hence, it is written as { } or ∅.

Math Note

The order in which the elements of equal sets are written does not matter. For example, {a, b, c} = {c, a, b}.

Math Note

All equal sets are equivalent since both sets will have the same number of members, but not all equivalent sets are equal. For example, the sets {x, y, z} and {10, 20, 30} have three members, but in this case, the members of the sets are not identical; hence, they are equivalent but not equal.

Calculator Explorations

Use your calculator to enter 0 divided by 3.

```
0/3
```

Write the set of answers to 0 divided by 3.
   Is this set equal to the set of answers for 3 divided by 0?

Try This One

2-E   Which of the following sets are empty sets?

(a) $\{x \mid x$ is a natural number divisible by seven$\}$

(b) $\{x \mid x$ is a human being living on Mars$\}$

(c) $\{+, -, \times, \div\}$

(d)   {living people on Earth who are over 200 years old}

For the answer, see page 72.

Equal and Equivalent Sets

In set theory, it is important to understand the concepts of *equal* sets and *equivalent* sets.

> Two sets, A and B, are **equal** (written $A = B$) if they have exactly the same members or elements.

   For example, the two sets {a, b, c} and {c, b, a} are equal since they have exactly the same members, a, b, and c.
   Two finite sets, A and B, are said to be **equivalent** (written $A \cong B$) if they have the same number of elements. For example, the set $C = \{x, y, z\}$ is equivalent to the set $D = \{10, \ 20, \ 30\}$ since both sets have three elements (i.e., $C \cong D$).

Example   **2-13**

State whether each pair of sets is equal, equivalent, or neither.

(a)   {p, q, r, s}; {a, b, c, d}
(b)   {8, 10, 12}; {12, 8, 10}
(c)   {213}; {2, 1, 3}
(d)   {1, 2, 10, 20}; {2, 1, 20, 11}
(e)   {even natural numbers less than 10}; {2, 4, 6, 8}

Solution

(a)   Equivalent
(b)   Equal and equivalent
(c)   Neither
(d)   Equivalent
(e)   Equal and equivalent

   The elements of two equivalent sets can be paired in such a way that they are said to have a *one-to-one correspondence* between them.

> Two sets have a **one-to-one correspondence** if and only if it is possible to pair the elements of one set with the elements of the other set in such a way that for each element in the first set there exists one and only one element in the second set.

   Sets with the same number of elements have a one-to-one correspondence.

Two sets of basketball teams on the court have a one-to-one correspondence.

Example 2-14

Show that (a) the sets {8, 16, 24, 32} and {s, t, u, v} have a one-to-one correspondence and (b) the sets {x, y, z} and {5, 10} do not have a one-to-one correspondence.

Solution

(a) It must be demonstrated that each element of one set can be paired with one and only one element of the second set. One possible way to show a one-to-one correspondence is

$$\{8, \ 16, \ 24, \ 32\}$$
$$\updownarrow \quad \updownarrow \quad \updownarrow \quad \updownarrow$$
$$\{s, \quad t, \quad u, \quad v\}$$

(b) The elements of the sets {x, y, z} and {5, 10} cannot be shown to have a one-to-one correspondence since there are three elements in the first set and two elements in the second set.

$$\{x, \quad y, \quad z\}$$
$$\updownarrow \quad \updownarrow$$
$$\{5, \quad 10\}$$

Try This One

2-F   State whether each pair of sets is equal, equivalent, or neither.

(a) {d, o, g}; {c, a, t}

(b) {run}; {r, u, n}

(c) {t, o, p}; {p, o, t}

(d) {10, 20, 30}; {1, 3, 5}

For the answer, see page 72.

As shown in this section, a set is a well-defined collection of objects. Sets are either finite or infinite. Two sets can be equal, equivalent, or neither, depending on the relationship of their elements. In Section 2-2, subsets and set operations will be explained.

Exercise Set 2-1

Computational Exercises

For Exercises 1–14, write each set using roster notation.

1. The set of letters in the word "stress".

2. The set of letters in the word "ALABAMA"

3. The set of natural numbers between 50 and 60

4. The set of even natural numbers between 10 and 40

5. The set of odd natural numbers less than 15

6. The set of even natural numbers less than 8

7. $\{x \mid x \in N \text{ and } x > 10\}$

8. The set of natural numbers greater than 100

9. The set of natural numbers between 2000 and 3000

10. $\{x \mid x \in N \text{ and } 500 < x < 6000\}$

11. {the days in the week}   12. {the colors in the spectrum}

13. The set of suits in a deck of cards   14. The set of face cards in a deck of cards

For Exercises 15–22, write each set using the descriptive method.

15. $\{2, 4, 6, 8, \ldots\}$ 16. $\{1, 3, 5, 7, \ldots\}$

17. $\{9, 18, 27, 36\}$ 18. $\{5, 10, 15, 20\}$

19. $\{m, a, r, y\}$ 20. $\{t, h, o, m, a, s\}$

21. $\{100, 101, 102, \ldots, 199\}$ 22. $\{21, 22, 23, \ldots, 29, 30\}$

For Exercises 23–28, write each set using set-builder notation.

23. $\{10, 20, 30, 40, \ldots\}$ 24. $\{55, 65, 75, 85\}$

25. {natural numbers greater than 20}   26. {even natural numbers less than 12}

27. $\{1, 3, 5, 7, 9\}$ 28. $\{18, 21, 24, 27, 30\}$

For Exercises 29–34, list the elements in each set.

29. {natural numbers less than 0}   30. $\{x \mid x \in N \text{ and } 70 < x < 80\}$

31. $\{7, 14, 21, \ldots, 63\}$ 32. $\{5, 12, 19, \ldots, 40\}$

33. $\{x \mid x \text{ is an even natural number greater than } 100\}$

34. $\{x \mid x \text{ is an odd natural number less than } 100\}$

For Exercises 35–42, state whether each set is well defined or not well defined.

35. {seasons of the year}   36. $\{x \mid x \in N\}$

37. $\{1, 2, 3, \ldots, 100\}$ 38. $\{x \mid x \text{ is a good professional golfer}\}$

39. $\{x \mid x \text{ is an excellent instructor}\}$ 40. $\{10, 15, 20\}$

41. $\{100, \ldots\}$ 42. {days of the week}

For Exercises 43–48, state whether each is true or false.

Let $A = \{\text{Saturday, Sunday}\}$

   $B = \{1, 2, 3, 4, 5\}$

   $C = \{p, q, r, s, t\}$

43. $3 \in B$ 44. $a \in C$

45. Wednesday $\notin A$ 46. $7 \notin B$

47. $r \in C$ 48. $q \in B$

For Exercises 49–56, state whether each set is infinite or finite.

49. $\{x \mid x \in N \text{ and } x \text{ is even}\}$

50. $\{1, 2, 3, \ldots, 999, 1000\}$

51. {the letters of the English alphabet}

52. {the years in which the past presidents of the United States were born}

53. $\{3, 6, 9, 12, \ldots\}$ 54. $\varnothing$

55. $\{x \mid x \text{ is a current television program}\}$ 56. $\{x \mid x \text{ is a fraction}\}$

For Exercises 57–64, state whether each pair of sets is equal, equivalent, or neither.

57. {6, 12, 18, 20} and {20, 12, 6, 18}   58. {p, q, r, s, t} and {5, 3, 4, 2, 1}

59. {2, 3, 7, 8} and {1, 4, 5}   60. {2, 4, 6, 8} and {2, 4, 6, 8, . . .}

61. {1, 2, 3, . . . , 99, 100} and {1001, 1002, 1003, . . . , 1100}

62. {the letters in the word "stop"} and {the letters in the word "pots"}

63. {three-digit numbers} and {1, 2, 3, . . . , 100}

64. {January, June, July} and {months beginning with the letter "J"}

For Exercises 65–68, show that each pair of sets is equivalent by using a one-to-one correspondence.

65. {10, 20, 30, 40} and {40, 10, 20, 30}

66. {w, x, y, z} and {1, 2, 3, 4}

67. {1, 2, 3, . . . , 25, 26} and {a, b, c, . . . , x, y, z}

68. {odd natural numbers less than 11} and {even natural numbers less than 12}

Real World Applications

The list shows the number of ships and berths for the leading cruise lines. Use the information for Exercises 69–72.

Cruise Line	Ships	Berths
Carnival	11	20,330
Royal Caribbean	11	19,770
Princess	9	12,250
Holland America	8	10,061
Norwegian	8	9,556
Costa	7	7,180
Celebrity	5	7,454
Cunard	6	3,131
Seabourn	3	612
Windstar	3	444

Source: USA Today, June 18, 1997

69. List the set of cruise lines that have eight or more ships.

70. List the set of cruise lines that have fewer than 8000 berths.

71. Find {x | x is a cruise line with 3 ships}.

72. Find {x | x is a cruise line with 10 ships}.

The graph shown represents the number of kidnappings for ransom worldwide. Use the information for Exercises 73–76.

73. Find the set of years when the number of kidnappings for ransom per year was less than 1000.

74. Find the set of years when the number of kidnappings per year was between 1000 and 1600.

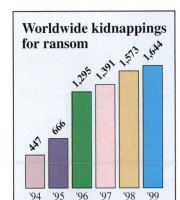

Worldwide kidnappings for ransom

'94 447
'95 666
'96 1,295
'97 1,391
'98 1,573
'99 1,644

Most kidnappings by country, 1999

1.	Colombia	972
2.	Mexico	402
3.	Former Soviet Union	250
4.	Brazil	51
5.	Nigeria	24
6.	Philippines	39
7.	India	17
8.	Ecuador	12
9.	Venezuela	12
10.	South Africa	10

Source: Hiscox Group (based on cases which Hiscox has obtained reasonably reliable information).

By Sam Ward, USA TODAY

Source: Copyright 2000, *USA Today.* Reprinted with permission.

75. Find $\{x \mid x$ is a year that the number of kidnappings was less than 400$\}$.

76. Find $\{x \mid x$ is a year that the number of kidnappings was greater than 1500$\}$.

Writing Exercises

77. Explain what a set is.

78. List three ways to write sets.

79. What is the difference between equal and equivalent sets?

80. Explain the difference between a finite and an infinite set.

81. What is meant by "one-to-one correspondence between two sets"?

82. Define the empty set and give two examples of an empty set.

Critical Thinking

83. If $A \cong B$ and $A \cong C$, is $B \cong C$? Explain your answer.

84. Is $\varnothing$ equivalent to $\{0\}$? Explain your answer.

85. Give an example of a set that is not well defined and change the definition so it is well defined.

86. How many different ways can you show a one-to-one correspondence between the two sets $\{a, b, c\}$ and $\{1, 3, 5\}$?

87. Are empty sets equivalent?

 **2-2** # Subsets and Set Operations

Subsets

When all, some, or none of the elements of one set are used in another set, the second set is called a subset of the original set. Formally defined,

> If every element of set A is also an element of set B, then set A is called a **subset** of set B. The symbol $\subseteq$ is used to designate a subset, and the relationship is written $A \subseteq B$.

Two important things should be noted about subsets. First, *every subset is a subset of itself.* This is true by definition. All elements of a set A would be contained in set A; hence, $A \subseteq A$. Second, *the empty set is a subset of every set.* Since the empty set contains no elements, it is impossible to find an element in the empty set that is not in any other set.

For example, eight subsets can be made from the elements of {x, y, z} as shown.

Number of elements taken at a time	Subsets
3	{x, y, z}
2	{x, y}, {x, z}, {y, z}
1	{x}, {y}, {z}
0	$\varnothing$

In this example, eight subsets can be made from the original set by taking three elements at a time, two elements at a time, one element at a time, and no elements at a time.

Example 2-15

Find all subsets of {bacon, egg}.

Solution

The subsets are

{bacon, egg}

{bacon}

{egg}

$\varnothing$

To indicate that a set is not a subset of another set, the symbol $\not\subseteq$ is used. For example, {1, 3} $\not\subseteq$ {0, 3, 5, 7} since 1 $\notin$ {0, 3, 5, 7}.

There is a rule to determine the number of subsets for a given set. It is based on the number of elements in the set. For example, there is one subset of the empty set, and that is the empty set. If a set has one element, there are two subsets, the set containing the element and the empty set. If a set has two elements, there are four subsets as shown in Example 2-15. If a set has three elements, there are eight subsets, as shown in the example at the beginning of this section. To generalize: if a set has n elements, there are 2^n subsets. For instance, if a set has eight elements, there will be 2^8 or $2 \times 2 \times 2 \times 2 \times 2 \times 2 \times 2 \times 2$, or 256 subsets.

It is important to distinguish subsets from *proper subsets*.

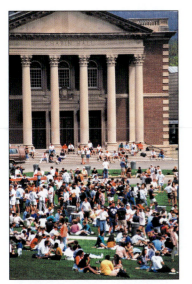

There are many subsets of this group: the subset of women students, the subset of sophomores, the subset of students who are taking history, etc.

If a subset of a given set is not equal to the original set, then the subset is called a **proper subset** of the original set. The symbol $\subset$ is used to indicate a proper subset.

For example, the set {1, 3} is a subset of the set {1, 3, 5}, and it is also a proper subset of {1, 3, 5} since it does not equal {1, 3, 5}; that is, {1, 3, 5} contains an element, namely, 5, that is not in {1, 3}.

The symbol $\not\subset$ is used to indicate that the set is not a proper subset. For example, {1, 3} $\subset$ {1, 3, 5}, but {1, 3, 5} $\not\subset$ {1, 3, 5}.

For finite sets, the number of proper subsets will always be one less than the total number of subsets; hence, the number of proper subsets of a set is $2^n - 1$, where n is the number of elements in the set.

Example 2-16

Find all proper subsets of {x, y, z}.

Solution

{x, y}   {x, z}   {y, z}

{x}   {y}   {z}

∅

Try This One

2-G For the set { ♦, ♥, ♠, ♣ }

(a) Find all subsets.

(b) Find all proper subsets.

For the answer, see page 72.

Example 2-17

State whether each statement is true or false.

(a) {1, 3, 5} ⊆ {1, 3, 5, 7}
(b) {a, b} ⊂ {a, b}
(c) {x | x ∈ N and x > 10} ⊂ N
(d) {2, 10} ⊄ {2, 4, 6, 8, 10}
(e) {r, s, t} ⊄ {t, s, r}
(f) {Erie, Huron} ⊂ Great Lakes

Solution

(a) True
(b) False, since a proper subset cannot equal the original set
(c) True
(d) False, since it is a subset
(e) True
(f) True

Example 2-18

State whether each statement is true or false.

(a) ∅ ⊂ {5, 10, 15}
(b) {u, v, w, x} ⊆ {x, w, u}
(c) ∅ ⊆ ∅
(d) {0} ⊆ ∅
(e) ∅ ⊂ ∅

—Continued

Math Note

It is important not to confuse the concept of subsets with the concept of elements. For example, the statement $6 \in \{2, 4, 6\}$ is true since 6 is an element of the set $\{2, 4, 6\}$, but the statement $\{6\} \in \{2, 4, 6\}$ is false since it states that the set containing the element 6 is an element of the set containing 2, 4, and 6. To show that an element does not belong to a set, recall that the symbol ∉ is used. Hence $8 \notin \{2, 4, 6\}$.

Example 2-18 *Continued—*

Solution

(a) True
(b) False; it is not a subset
(c) True, since every set is a subset of itself
(d) False, since the set {0} has one element and the empty set has no elements
(e) False, since a proper subset cannot be equal to itself

Example 2-19

State whether each statement is true or false.

(a) $\{8\} \in \{x \mid x$ is an even natural number$\}$
(b) $6 \in \{1, 3, 5, 7, \ldots\}$
(c) $\{2, 3\} \in \{x \mid x \in N\}$
(d) $\{a, b, c\} \subset \{$letters of the alphabet$\}$
(e) $\varnothing \in \{x, y, z\}$

Solution

(a) False, $\subset$ or $\subseteq$ should be used with subsets.
(b) False, $6 \notin \{1, 3, 5, 7, \ldots\}$ since it is an even number
(c) False, $\subset$ or $\subseteq$ should be used with subsets.
(d) True
(e) False since $\varnothing$ is a subset or proper subset of any set. It is not an element of $\{x, y, z\}$.

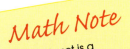

Math Note

The empty set is a subset of itself, but it is not a proper subset of itself.

Union and Intersection of Sets

There are two basic set operations. They are union and intersection.

> The **union** of two sets A and B (symbolized by $A \cup B$) consists of the elements of set A or set B, or both sets.

For example, if set $A = \{10, 12, 14, 15\}$ and $B = \{13, 14, 15, 16, 17\}$, $A \cup B = \{10, 12, 13, 14, 15, 16, 17\}$. Even though 14 and 15 are elements in both sets, they are not written twice in the union of two sets.

Also, the union of three or more sets consists of the set of all the elements in the three or more sets.

Note that the word "or" is sometimes used to indicate union; i.e., $A \cup B$ means the elements in set A or set B.

Example 2-20

If $A = \{0, 1, 2, 3, 4, 5\}$, $B = \{2, 4, 6, 8, 10\}$, and $C = \{1, 3, 5, 7\}$, find each.

(a) $A \cup B$
(b) $A \cup C$
(c) $B \cup C$
(d) $A \cup B \cup C$

Solution

(a) $A \cup B = \{0, 1, 2, 3, 4, 5, 6, 8, 10\}$
(b) $A \cup C = \{0, 1, 2, 3, 4, 5, 7\}$
(c) $B \cup C = \{1, 2, 3, 4, 5, 6, 7, 8, 10\}$
(d) $A \cup B \cup C = \{0, 1, 2, 3, 4, 5, 6, 7, 8, 10\}$

The other operation is called *intersection*.

> The **intersection** of two sets, A and B, symbolized by $A \cap B$, is the set of elements that are common to both sets.

For example, if $A = \{10, 12, 14, 15\}$ and $B = \{13, 14, 15, 16, 17\}$, then $A \cap B = \{14, 15\}$ since 14 and 15 are contained in both sets.

The intersection of three or more sets consists of the set of elements that all sets have in common.

Note that the word "and" is sometimes used to indicate the intersection; i.e., $A \cap B$ means the set of elements in set A and set B.

The advertisement is looking for people in the *intersection* of three sets of potential employees.

Example 2-21

Let $A = \{9, 10, 11, 12\}$, $B = \{10, 12, 14, 16\}$, and $C = \{12, 14, 16, 17, 18\}$. Find each.

(a) $A \cap B$
(b) $A \cap C$
(c) $B \cap C$
(d) $A \cap B \cap C$

Solution

(a) $A \cap B = \{10, 12\}$
(b) $A \cap C = \{12\}$
(c) $B \cap C = \{12, 14, 16\}$
(d) $A \cap B \cap C = \{12\}$

When the intersection of two sets is the empty set, the sets are said to be *disjoint*. For example, if $P = \{a, b, c\}$ and $Q = \{x, y, z\}$, then $P \cap Q = \varnothing$ since they have no elements in common. P and Q are said to be disjoint sets.

Unions and intersections of three or more sets can be performed together by completing the operations in parentheses first. This is shown in Example 2-22.

Example   2-22

Let $A = \{l, m, n, o, p\}$, $B = \{o, p, q, r\}$, and $C = \{r, s, t, u\}$. Find each.

(a) $(A \cup B) \cap C$
(b) $A \cap (B \cup C)$
(c) $(A \cap B) \cup C$

Solution

(a) First find $A \cup B$: $A \cup B = \{l, m, n, o, p, q, r\}$. Then intersect this set with set C to get $\{r\}$.
(b) First find $B \cup C$: $B \cup C = \{o, p, q, r, s, t, u\}$. Then intersect this set with set A to get $\{o, p\}$.
(c) First find $A \cap B$: $A \cap B = \{o, p\}$. Then find the union of this set with set C to get $\{o, p, r, s, t, u\}$.

Try This One

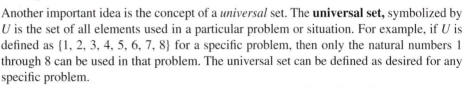

2-H   Let $R = \{\star, \triangle, !, \$, \bigcirc\}$ $S = \{\star, !, \bigcirc\}$ $T = \{\star, \triangle, \$\}$

Find.

(a) $S \cup T$

(b) $R \cap S$

(c) $(R \cup S) \cap T$

For the answer, see page 72.

There are many instances where unions and intersections of sets are used in real life. For example, when mixing paint, the intersection of yellow paint and blue paint makes green paint; yellow paint and red paint make orange, etc.

Complement of a Set

Another important idea is the concept of a *universal* set. The **universal set,** symbolized by U is the set of all elements used in a particular problem or situation. For example, if U is defined as $\{1, 2, 3, 4, 5, 6, 7, 8\}$ for a specific problem, then only the natural numbers 1 through 8 can be used in that problem. The universal set can be defined as desired for any specific problem.

Related to the universal set is the *complement* of a set. Formally defined:

The **complement** of set A, denoted by $\overline{A}$, is the set of elements contained in the universal set that are not contained in A.

For example, if $U = \{1, 2, 3, 4, 5, 6, 7, 8\}$ and set $A = \{2, 4, 6, 8\}$, the complement of A, denoted by $\overline{A}$ is the set $\{1, 3, 5, 7\}$. Sometimes the complement is denoted by A'; however, $\overline{A}$ will be used throughout this textbook.

The bar over a group of letters means perform the operations under the bar first, and then do the complement. Also, perform the set operations in parentheses before doing the operations outside the parentheses.

Example   2-23

If $U = \{10, 20, 30, 40, 50, 60, 70, 80\}$, $A = \{10, 30, 50, 70\}$, $B = \{40, 50, 60, 70\}$, and $C = \{20, 40, 60\}$, find each.

(a) $\overline{A}$
(b) $\overline{B}$
(c) $\overline{C}$
(d) $\overline{A \cup C}$
(e) $\overline{B} \cap \overline{C}$

Solution

(a) $\overline{A} = \{20, 40, 60, 80\}$
(b) $\overline{B} = \{10, 20, 30, 80\}$
(c) $\overline{C} = \{10, 30, 50, 70, 80\}$
(d) First find $A \cup C$, then take the complement: $A \cup C = \{10, 20, 30, 40, 50, 60, 70\}$ and $\overline{A \cup C} = \{80\}$.
(e) First find $\overline{B}$, then $\overline{C}$, and then find the intersection of the two complements: $\overline{B} = \{10, 20, 30, 80\}$ and $\overline{C} = \{10, 30, 50, 70, 80\}$ so $\overline{B} \cap \overline{C} = \{10, 30, 80\}$.

Try This One

2-I   Let $U = \{1, 2, 3, 4, 5, 6, 7, 8\}$, $A = \{1, 3, 5, 7\}$, $B = \{2, 4, 6, 8\}$, and $C = \{2, 3, 5, 7\}$. Find each.

(a) $\overline{C}$

(b) $\overline{A \cup B}$

(c) $\overline{A} \cap \overline{C}$

(d) $(A \cup B) \cap \overline{C}$

For the answer, see page 72.

Math Note

The complement of the universal set is the empty set, i.e., $\overline{U} = \varnothing$; and the complement of the empty set is the universal set, i.e., $\overline{\varnothing} = U$.

Two basic set operations are union and intersection. Once a universal set is defined, only those elements contained in the universal set can be used in the problem. Every set also has a complement with respect to the universal set.

Exercise Set   2-2

Computational Exercises

For Exercises 1–8, find all subsets of each set.

1. $\{r, s, t\}$ 2. $\{2, 5, 7\}$

3. $\{1, 3\}$ 4. $\{p, q\}$

5. $\{\ \}$ 6. $\varnothing$

7. $\{5, 12, 13, 14\}$ 8. $\{m, o, r, e\}$

For Exercises 9–14, find all proper subsets of each set.

9. {1, 10, 20}

10. {March, April, May}

11. {6}

12. {t}

13. ∅

14. { }

For Exercises 15–24, state whether each is true or false.

15. {3} ⊆ {1, 3, 5}

16. {a, b, c} ⊂ {c, b, a}

17. {1, 2, 3} ⊆ {123}

18. ∅ ⊂ ∅

19. ∅ ∈ { }

20. {Mars, Venus, Sun} ⊂ {planets in our solar system}

21. {3} ∈ {1, 3, 5, 7, …}

22. $\{x \mid x \in N \text{ and } x > 10\} \subseteq \{x \mid x \in N \text{ and } x \geq 10\}$

23. ∅ ⊂ {a, b, c}

24. ∅ ∈ {r, s, t, u}

For Exercises 25–30, find the number of subsets each set has. Do not list the subsets.

25. {25, 50, 75}

26. {1, 2, 3, 4, 5, 6, 7, 8, 9, 10}

27. ∅

28. {0}

29. {x, y}

30. {a, b, c, d, e}

For Exercises 31–40, let

$U = (10, 20, 30, 40, 50, 60, 70, 80, 90, 100\}$

$A = \{10, 30, 50, 70, 90\}$

$B = \{20, 40, 60, 80, 100\}$

$C = \{30, 40, 50, 60\}$

Find each.

31. $A \cup C$

32. $A \cap B$

33. $\overline{A}$

34. $(A \cap B) \cup C$

35. $\overline{A} \cap (B \cup C)$

36. $(A \cap B) \cap C$

37. $\overline{(A \cup B)} \cap C$

38. $A \cap \overline{B}$

39. $(B \cup C) \cap \overline{A}$

40. $(\overline{A} \cup \overline{B}) \cup \overline{C}$

For Exercises 41–50, let

$U = \{a, b, c, d, e, f, g, h\}$

$P = \{b, d, f, g\}$

$Q = \{a, b, c, d\}$

$R = \{e, f, g\}$

Find each.

41. $P \cap Q$

42. $Q \cup R$

43. $\overline{P}$

44. $\overline{Q}$

45. $\overline{R} \cap \overline{P}$

46. $P \cup (Q \cap R)$

47. $\overline{(Q \cup P)} \cap R$

48. $P \cap (Q \cap R)$

49. $(P \cup Q) \cap (P \cup R)$

50. $\overline{Q} \cup \overline{R}$

For Exercises 51–60, let

$U = \{1, 2, 3, 4, 5, 6, 7, 8, 9, 10, 11, 12\}$
$W = \{2, 4, 6, 8, 10, 12\}$
$X = \{1, 3, 5, 7, 9, 11\}$
$Y = \{1, 2, 3, 4, 5, 6\}$
$Z = \{2, 5, 6, 8, 10, 11, 12\}$

Find each.

51. $W \cap Y$

52. $X \cup Z$

53. $W \cup X$

54. $X \cap Y \cap Z$

55. $W \cap X$

56. $\overline{Y \cup Z}$

57. $(X \cup Y) \cap Z$

58. $(Z \cap Y) \cup W$

59. $\overline{W} \cap \overline{X}$

60. $(\overline{Z \cup X}) \cap Y$

For Exercises 61–64, let

$U = \{1, 2, 3, \ldots\}$
$A = \{3, 6, 9, 12, \ldots\}$
$B = \{9, 18, 27, 36, \ldots\}$
$C = \{2, 4, 6, 8, \ldots\}$

Find.

61. $A \cap B$

62. $\overline{A} \cap C$

63. $A \cap (B \cup \overline{C})$

64. $A \cup B$

Real World Applications

65. A household can have a telephone, television, and Internet service. List the different communication options a household can select considering all, some, or none of the services.

66. If a person is dealt five cards and has a chance of discarding any number including 0, how many choices would the person have?

67. A landscaper can select any, some, or all of these tools for a job: lawnmower, weedwacker, tiller, trimmer, chain saw, hedge trimmer, and leaf blower. How many different possibilities does the landscaper have in his or her selection? *Note:* If the landscaper does not select any tool, he or she is just giving a potential customer an estimate.

68. For a picnic, a person can select none, any, some, or all of these food or drink items: hamburgers, hot dogs, potatoes, vegetables, buns, and beverage. How many different selections can be made?

69. A person wishes to buy a computer and can select none, some, or all of these options: digital camera, printer, scanner, and fax machine. How many different selections are possible?

70. In order to integrate aerobics into her exercise program, Claire can select one, some, or all of these machines: treadmill, cycle, and stair stepper. List all possibilities for her aerobics selection.

Writing Exercises

71. Define a subset.

72. Explain the difference between a subset and a proper subset.

73. If a set has *n* elements, how many subsets of the original set can be made?

74. If a set has *n* elements, how many proper subsets of the original set can be made?

75. Explain the difference between a subset and an element of a set.

76. Explain why the empty set is a subset, but not a proper subset, of itself.

77. Explain the difference between the union and intersection of two sets.

78. When are two sets said to be disjoint?

79. What is a universal set?

80. Define the complement of a set.

Critical Thinking

81. Can you find two sets whose union and intersection are the same set?

82. Write a short paragraph listing five characteristics of George Washington and five characteristics of Abraham Lincoln. Find the union of the two sets. Find the intersection of the two sets.

83. Select two medications and write a paragraph listing the possible side effects of each. Find the intersection of the sets.

84. Select two breakfast cereals and write a short paragraph listing the ingredients of each. What ingredients do they have in common? If a person mixed them together, what ingredients would that person be eating?

2-3 # Venn Diagrams

Sets can be represented pictorially by using **Venn diagrams.** The universal set is shown as a rectangle, and various sets, such as *A*, *B*, and *C*, are drawn as circles inside the rectangle to represent various relationships among these sets. See Figure 2-1.

The Venn diagram for the intersection of two sets ($A \cap B$) is shown in Figure 2-2(a). The shaded area represents the elements set *A* and set *B* have in common.

The Venn diagram for the union of two sets ($A \cup B$) is shown in Figure 2-2(b). The shaded area represents the elements in set *A* or set *B* or both.

The Venn diagram for disjoint sets is shown in Figure 2-2(c). Since disjoint sets have no elements in common, the circles do not overlap.

The Venn diagram for the complement of a set, ($\overline{A}$), is shown in Figure 2-2(d). The shaded area represents the elements in the universal set that are not in set *A*.

Using the basic Venn diagrams shown in Figure 2-2, set operations can be illustrated pictorially. The shaded areas represent the regions for the set operations.

A helpful procedure for identifying the areas is to number each area, as shown in Figure 2-3 of Example 2-24, and then perform the set operations using the numbers. Finally, shade the appropriate area for the answer. Examples 2-24 and 2-25 illustrate this procedure.

Sidelight

John Venn (1834–1923)

The Venn diagrams were named after the mathematician John Venn (1834–1923) who was also an ordained clergyman. Venn studied and taught symbolic logic after leaving the clergy. He wrote a book entitled *Symbolic Logic* in 1881.

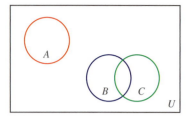

Figure 2-1

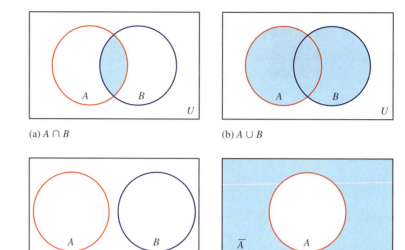

(a) $A \cap B$ (b) $A \cup B$

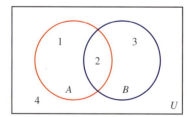

(c) Disjoint sets   (d) $\overline{A}$

Figure 2-2

Example   2-24

Draw a Venn diagram to show $\overline{A \cup B}$.

Solution

Step 1   Draw the set diagram and label each area, as shown in Figure 2-3.

Figure 2-3

Step 2   From the diagram, list the elements in each set.

$$U = \{1,\ 2,\ 3,\ 4\}$$
$$A = \{1,\ 2\}$$
$$B = \{2,\ 3\}$$

Step 3   Using the sets in step 2, find $\overline{A \cup B}$.

$$A \cup B = \{1,\ 2,\ 3\}$$
$$\overline{A \cup B} = \{4\}$$

—Continued

Example 2-24 *Continued—*

Step 4 Shade area 4, as shown in Figure 2-4.

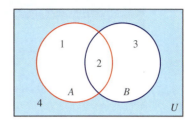

Figure 2-4

Hence, $\overline{A \cup B}$ would be the area outside $A \cup B$.

Example 2-25

Draw a Venn diagram to show $A \cap \overline{B}$.

Solution

Step 1 Draw the set diagram and label each area, as shown in Figure 2-5.

Step 2 From the diagram, list the elements in each set.

$$U = \{1, 2, 3, 4\}$$
$$A = \{1, 2\}$$
$$B = \{2, 3\}$$

Step 3 Using the sets in step 2, find $A \cap \overline{B}$.

$$\overline{B} = \{1, 4\}$$
$$A \cap \overline{B} = \{1\}$$

Step 4 Shade area 1, as shown in Figure 2-5.

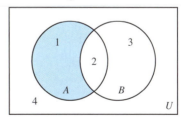

Figure 2-5

The shaded area represents $A \cap \overline{B}$.

Try This One

2-J Draw a Venn diagram to show $\overline{A} \cup B$.
For the answer, see page 72.

This procedure also works when drawing Venn diagrams for three sets, as shown in Examples 2-26 and 2-27.

Example   2-26

Draw a Venn diagram and show $A \cap (\overline{B \cap C})$.

Solution

Step 1   Draw and label the diagram, as shown in Figure 2-6.

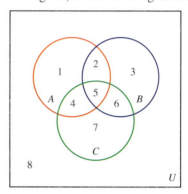

Figure 2-6

Step 2   From the diagram, list the elements in each set.

$$U = \{1, 2, 3, 4, 5, 6, 7, 8\}$$
$$A = \{1, 2, 4, 5\}$$
$$B = \{2, 3, 5, 6\}$$
$$C = \{4, 5, 6, 7\}$$

Step 3   Find the solution to $A \cap (\overline{B \cap C})$.

$$B \cap C = \{5, 6\}$$
$$\overline{B \cap C} = \{1, 2, 3, 4, 7, 8\}$$
$$A \cap (\overline{B \cap C}) = \{1, 2, 4\}$$

Step 4   Shade the sections, as shown in Figure 2-7.

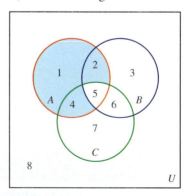

Figure 2-7

Try This One

2-K Draw a Venn diagram and show $(A \cap \bar{B}) \cup C$.
For the answer, see page 72.

Two expressions about sets are mathematically equivalent if they have the same Venn diagrams.

Example 2-27

Show that the two expressions are equivalent by using Venn diagrams: $(A \cup B) \cap C$ and $(A \cap C) \cup (B \cap C)$.

Solution

The Venn diagram for $(A \cup B) \cap C$ is shown in Figure 2-8(a) and the Venn diagram for $(A \cap C) \cup (B \cap C)$ is shown in Figure 2-8(b).

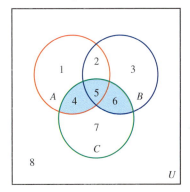

(a) $(A \cup B) \cap C$

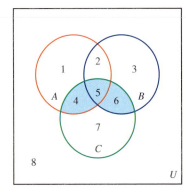

(b) $(A \cap C) \cup (B \cap C)$

Figure 2-8

Since the shaded areas are the same, the expressions are mathematically equivalent.

Venn diagrams, then, can be used to represent sets pictorially. Equivalent expressions using set operations will have the same Venn diagram.

Sets and Venn diagrams can be used to solve problems in surveys, logic, and in the area of probability. Section 2-4 shows how Venn diagrams can be used to solve problems involving surveys.

Exercise Set **2-3**

Computational Exercises

For Exercises 1–26, draw a Venn diagram and shade the sections representing each set.

1. $\overline{A} \cap B$	2. $A \cup \overline{B}$
3. $\overline{A \cap B}$	4. $\overline{A \cup B}$
5. $\overline{A} \cup \overline{B}$	6. $\overline{A} \cup B$
7. $\overline{A} \cap \overline{B}$	8. $A \cap \overline{B}$
9. $A \cup (B \cap C)$	10. $A \cap (B \cup C)$
11. $(A \cup B) \cup (A \cap C)$	12. $(A \cup B) \cap C$
13. $(A \cup B) \cap (A \cup C)$	14. $(A \cap B) \cup C$
15. $\overline{(A \cap B)} \cup C$	16. $(A \cup B) \cup \overline{C}$
17. $A \cap (\overline{B \cup C})$	18. $\overline{A} \cap (\overline{B} \cup \overline{C})$
19. $(\overline{A} \cup \overline{B}) \cap C$	20. $A \cap (\overline{B \cap C})$
21. $(\overline{A \cup B}) \cap (A \cup C)$	22. $(B \cup C) \cup \overline{C}$
23. $\overline{A} \cap (\overline{B} \cap \overline{C})$	24. $(\overline{A \cup B}) \cap C$
25. $\overline{A} \cap (\overline{B \cup C})$	26. $(A \cup B) \cup (A \cap C)$

For Exercises 27–33, determine whether the two expressions are mathematically equivalent by using Venn diagrams.

27. $\overline{A \cap B}$ and $\overline{A} \cup \overline{B}$	28. $\overline{A \cup B}$ and $\overline{A} \cup \overline{B}$
29. $(A \cup B) \cup C$ and $A \cup (B \cup C)$	30. $A \cap (B \cup C)$ and $(A \cap B) \cup (A \cap C)$
31. $\overline{A} \cup (B \cap \overline{C})$ and $(\overline{A} \cup B) \cap \overline{C}$	32. $(A \cap B) \cup \overline{C}$ and $(A \cap B) \cup (B \cap \overline{C})$
33. $\overline{(A \cap B)} \cup C$ and $(\overline{A} \cup \overline{B}) \cap C$	

Writing Exercises

34. Explain why a Venn diagram for the intersection of two sets is drawn as shown in Figure 2-2(a).

35. Explain why a Venn diagram for the union of two sets is drawn as shown in Figure 2-2(b).

36. Explain why the Venn diagram for a complement is drawn as shown in Figure 2-2(d).

37. Explain why the Venn diagram for disjoint sets is not drawn with intersecting circles.

Critical Thinking

Use this information for Exercises 38–43.
For two sets A and B, their difference A − B, is defined as $A \cap \overline{B}$. Using
$U = \{1, 2, 3, 4, 5, 6, 7, 8\}$, $A = \{1, 2, 4, 5\}$, $B = \{2, 3, 5, 6\}$, *and* $C = \{4, 5, 6, 7\}$,
find each.

38. $A - B$

39. $B - C$

40. $(A - B) - C$

41. $A - C$

42. $(A - B) - (B - C)$

43. Draw and shade a Venn diagram for the area representing $A - B$.

2-4 Using Sets to Solve Problems

Many problems involving classifications can be solved using sets and Venn diagrams. Examples 2-28 and 2-29 illustrate the techniques.

Example 2-28

In a restaurant survey, it was found that out of 65 people, 25 ordered steak, 15 ordered lobster, and 5 ordered both steak and lobster. Draw a Venn diagram to represent the survey results, and find how many people ordered steak only, lobster only, or neither steak nor lobster.

Solution

Step 1 Draw a Venn diagram, as shown in Figure 2-9(a) where

U = universal set

S = number of people who ordered steak

L = number of people who ordered lobster

Step 2 Since 5 customers ordered both steak and lobster, place "5" in the intersection, as shown in Figure 2-9(b).

Step 3 Since 25 customers ordered steak and 5 ordered both steak and lobster, subtract $25 - 5 = 20$ to get the number of customers who ordered steak only.

In a similar manner, 15 customers ordered lobster and 5 ordered steak and lobster; subtract $15 - 5 = 10$ to get the number of customers who ordered lobster only.

Place the numbers 20 and 10 in the appropriate sections of the Venn diagram, as shown in Figure 2-9(c).

—Continued

Example 2-28 *Continued—*

Step 4 To find how many customers ordered neither, add $20 + 5 + 10 = 35$ and subtract this value from 65.

$$65 - 35 = 30$$

Hence, 30 customers ordered neither steak nor lobster. Place this value in the appropriate section of the Venn diagram, as shown in Figure 2-9(d).

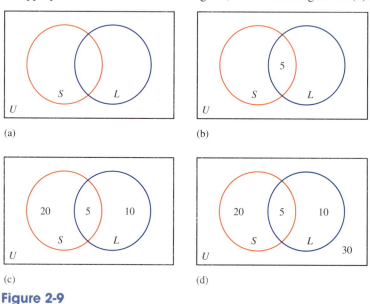

(a)   (b)

(c)   (d)

Figure 2-9

Try This One

2-L   At a newsstand, out of 46 customers, 27 bought the *Daily News,* 18 bought the *Tribune,* and 6 bought both papers. Draw a Venn diagram to represent the situation and determine how many customers bought only one paper and how many customers bought something other than either of the two newspapers.
For the answer, see page 72.

Example 2-29 illustrates the use of sets and Venn diagrams to solve a problem that involves three classifications.

Example 2-29

A survey of 100 students found that 20 students are taking English, 26 students are taking psychology, 25 students are taking statistics, 5 students are taking both English and psychology, 9 students are taking both psychology and statistics,

—Continued

> **Example** **2-29** *Continued—*

13 students are taking English and statistics, and 3 students are taking all three subjects. Draw a Venn diagram to represent the survey and find

(a) The number of students who are taking only English.
(b) The number of students who are taking psychology and statistics but not English.
(c) The number of students who are taking English or statistics.
(d) The number of students who are not taking any of the three subjects.

Solution

Step 1 Draw a Venn diagram, as shown in Figure 2-10(a).

Step 2 Since three students are taking all three subjects, place 3 where all three sets intersect, as shown in Figure 2-10(b).

Step 3 Find the number of students who are taking English and psychology but not statistics. The answer can be found by subtracting the number of students who are taking all three subjects (3) from the number of students who are taking English and psychology (5), $5 - 3 = 2$. Place 2 in the appropriate section of the Venn diagram. See Figure 2-10(c).

Find the number of students who are taking psychology and statistics but not English by subtracting $9 - 3 = 6$. Place 6 in the appropriate section of the Venn diagram.

In the same manner, find the number of students who are taking English and statistics but not psychology: $13 - 3 = 10$. Place 10 in the appropriate section of the Venn diagram. See Figure 2-10(d).

Step 4 By subtracting, find the number of students who are taking only English; $20 - (10 + 3 + 2) = 5$.

By subtracting, find the number of students who are taking only psychology; $26 - (2 + 3 + 6) = 15$.

By subtracting, find the number of students who are taking only statistics; $25 - (10 + 3 + 6) = 6$.

Place these numbers in the appropriate section of the Venn diagram. See Figure 2-10(e).

Step 5 Find the number of students who are not taking any of the three subjects by adding all the numbers, $5 + 2 + 15 + 10 + 6 + 6 + 3 = 47$, and subtracting that number from the total number of students, 100.

$$100 - 47 = 53$$

Place 53 in the appropriate section of the Venn diagram.

The completed Venn diagram is shown in Figure 2-10(f). Once the Venn diagram has been completed, the answers to the specific questions can be found by looking at the various sections of the diagram.

(a) The number of students taking only English is 5.
(b) The number of students who are taking psychology and statistics but not English is 6.

—Continued

Example **2-29** *Continued—*

(c) The number of students taking English or statistics is $5 + 2 + 10 + 3 + 6 + 6 = 32$. This includes the students who are also taking psychology.

(d) The number of students who are not taking any of the three subjects is 53.

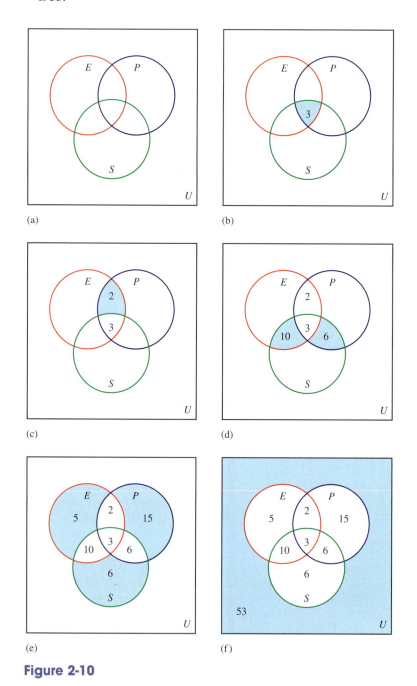

Figure 2-10

Try This One

2-M   A survey of 50 readers found that 27 like to read novels, 32 like to read biographies, 16 like to read self-help books, 14 like to read novels and biographies, 12 like to read biographies and self-help books, and 8 like to read novels and self-help books. Finally, 5 like to read all three types of books. Draw a Venn diagram to represent the results of the survey and answer these questions:

(a)   How many readers like to read only novels?

(b)   How many readers like to read biographies and self-help books but not novels?

(c)   How many readers do not like to read any of the three types of books?

(d)   How many readers do not like to read self-help books?

For the answer, see page 72.

Exercise Set   2-4

Real World Applications

1. In a survey of 40 musicians, it was found that 20 people played guitar, 25 played piano, and 7 played both instruments.

    (a)   How many musicians played guitar only?

    (b)   How many musicians played piano only?

    (c)   How many musicians played neither instrument?

2. In a class of 25 students, 18 were mathematics majors, 12 were computer science majors, and 7 were dual majors in mathematics and computer science.

    (a)   How many students were majoring in mathematics only?

    (b)   How many students were not majoring in computer science?

    (c)   How many students were not mathematics or computer science majors?

3. In a parking lot, there was a total of 16 automobiles, consisting of 8 red automobiles, 5 white automobiles, and 3 two-toned (red and white) automobiles.

    (a)   How many automobiles were red only?

    (b)   How many were red or white but not two-toned?

    (c)   How many were neither red nor white?

4. In the cafeteria, 25 students were seated at tables. Fifteen were enrolled in psychology, 9 were enrolled in physics, and 4 were enrolled in both psychology and physics.

    (a)   How many students were enrolled in psychology only?

    (b)   How many students were not enrolled in physics?

    (c)   How many students were enrolled in at least one of these courses?

5. Seventy women were surveyed and asked what activities they participate in. The results were that 16 played golf, 24 went fishing, and 20 played racquetball. Nine played golf and fished, 11 fished and played racquetball, and 7 played golf and racquetball. Two women participated in all three activities.

(a) How many women only played golf?

(b) How many women fished and played racquetball but not golf?

(c) How many women did not participate in any of the three activities?

6. In a college dormitory, 47 students were asked if they had a television set, VCR, or radio in their rooms. The following results were obtained: 17 had a television set, 13 had a VCR, 12 had a radio, 9 had a television and a VCR, 3 had a VCR and radio, and 5 had a television set and a radio, and 2 had all three.

(a) How many students had a radio and television but no VCR?

(b) How many students had exactly two of these appliances?

(c) How many students had a television set, but not a VCR?

7. A garage that does state automobile inspections found that 14 cars failed because of bad tires, 23 cars failed because of bad brakes, 15 cars failed because of faulty exhaust systems, 9 cars failed because of bad tires and brakes, 8 cars failed because of bad brakes and exhaust systems, 5 cars failed because of bad tires and exhaust systems, and 2 cars failed because of all three conditions.

(a) How many cars failed because of at least two of these problems?

(b) How many cars failed because of tires and brakes but not exhaust systems?

(c) How many cars failed inspection because of only one item?

(d) How many failed because of tires or brakes, but not exhaust systems?

8. At a local restaurant, there were 109 hamburgers served over a three-day period. Thirty-two customers ordered a hamburger with cheese, 40 ordered a hamburger with pickles, 18 ordered a hamburger with ketchup, 13 people ordered a hamburger with cheese and pickles, 10 with pickles and ketchup, 9 with cheese and ketchup, and 7 customers ordered a hamburger with all three items.

(a) How many customers ordered a hamburger with cheese and pickles but not ketchup?

(b) How many people ordered a hamburger with ketchup or pickles?

(c) How many people ordered a plain hamburger?

9. At a speciality sales show, there were 70 vendors: 19 sold cards, 21 sold picture frames, and 19 sold sunglasses. Nine sold both cards and picture frames, 8 sold picture frames and sunglasses, 5 sold cards and sunglasses, and 3 sold all three items.

(a) How many vendors sold at most two of these items?

(b) How many vendors sold cards and sunglasses but not picture frames?

(c) How many vendors sold neither cards nor sunglasses?

10. A survey of 96 train commuters showed that 29 read the *Press* newspaper, 24 read the *Daily News,* and 20 read the *Times.* Eight read the *Press* and *Daily News* but not the *Times,* while 4 read the *Daily News* and *Times,* and 7 read the *Press* and *Times.* One person reads all three newspapers.

(a) How many read the *Times* or *Daily News* but not both?

(b) How many read the *Times* and *Daily News* but not the *Press*?

(c) How many read the *Times* or the *Press*? (*Note:* "or" means one or the other or both)

11. When examining 61 briefcases, 20 contained paper clips, 26 contained a pencil, and 20 contained a pen. Nine contained both paper clips and pencils, 15 contained both pencils and pens, 6 contained pens and paper clips, and 3 contained all three items.

(a) How many briefcases contained none of the three articles?

(b) How many contained exactly two of the three items?

(c) How many contained exactly one of the three items?

12. At a collector's flea market, a survey of 200 people found that 128 collected coins, 131 collected stamps, 114 collected baseball cards, 75 collected coins and stamps, 59 collected stamps and baseball cards, 81 collected coins and baseball cards, and 33 people collected all three items.

(a) How many people collected exactly two items?

(b) How many people collected exactly one item?

(c) How many people did not collect stamps?

13. In a survey of 121 visitors to Atlantic City, 39 visited the casinos, 51 went shopping, and 26 went swimming. Furthermore, 25 went shopping and visited the casinos, 14 went shopping and swimming, 18 went swimming and to the casinos, and 10 people participated in all three activities.

(a) How many people did exactly one thing?

(b) How many people went shopping, but not swimming?

(c) How many people went neither shopping nor swimming?

14. In a survey of 20 homes, 15 had a family room, 13 had a deck, and 2 had neither a deck nor a family room. How many houses had both a deck and a family room?

15. In a survey of 34 fruit growers, 18 grew apples, 20 grew pears, and 2 grew neither apples nor pears. How many growers in the survey grew both apples and pears?

Critical Thinking

16. Explain why the following researcher was fired when the results of his survey were published. Forty people were asked if they would purchase Brand X, Brand Y, or Brand Z. Their responses follow:

23 said they would purchase Brand X.

18 said they would purchase Brand Y.

19 said they would purchase Brand Z.

12 said they would purchase Brands X and Y.

6 said they would purchase Brands Y and Z.

7 said they would purchase Brands X and Z.

2 said they would purchase all three brands.

2 said they would not purchase any of the brands.

2-5 Infinite Sets

Recall from Section 2-1 that a set is considered to be finite if the number of elements is either 0 or a natural number; otherwise, it is considered to be an infinite set. For example, the set {a, b, c, d} is said to be finite since it has four elements. The set {10, 20, 30, . . .} is said to be infinite since it has an unlimited number of elements.

Georg Ferdinand Ludwig Philipp Cantor (1845–1918)

The Father of Set Theory

Georg Cantor formulated the concepts of set theory. He was born in St. Petersburg, Russia, and in 1856, at age 11, he and his family moved to Germany, where he spent the rest of his life. Georg's father wanted him to pursue a career in engineering. Instead, he studied philosophy, physics, and mathematics. His main interest was number theory. In 1869, he received a doctorate at age 22. He spent most of his academic career at the University of Halle (1869–1905). His goal was to become a professor at the University of Berlin, but his views of mathematics met with opposition from a professor at Berlin, Leopold Kronecker, who opposed his appointment. He suffered a nervous breakdown and died in a mental hospital in 1918.

His work gained little recognition during his lifetime, but later, his work was praised as an "astonishing product of mathematical thought, one of the most beautiful realizations of human activity."

The German mathematician Georg Cantor studied infinite sets. His definition of an infinite set is given next.

> An **infinite set** is a set that can be placed in a one-to-one correspondence with a proper subset of itself.

For example, the set of natural numbers is an infinite set since there exists a one-to-one correspondence between the natural numbers and the even natural numbers. The set of even numbers is a proper subset of the natural numbers, as shown.

$$\{1, 2, 3, 4, \ldots, n, \ldots\}$$
$$\updownarrow \;\updownarrow\; \updownarrow\; \updownarrow \qquad\quad \updownarrow$$
$$\{2, 4, 6, 8, \ldots, 2n, \ldots\}$$

In this case, n represents the general term of the set of natural numbers, and $2n$ represents the general term of the set of even natural numbers. When $n = 6$, it is paired with $2n$ or 12 of the set of even natural numbers.

Different sets have different general terms. A rule of thumb is that a general term of a set should be written in terms of n such that when 1 is substituted for n, you get the first term of the set; when 2 is substituted for n, you get the second term of the set. Consider the set $\{4, 7, 10, 13, 16, \ldots\}$; the general term is $3n + 1$ since each element of the set is obtained by multiplying by 3 and adding 1.

Math Note

It should be pointed out that the solution to finding the general term is not always so obvious. It is sometimes necessary to use trial and error to see if you can find the pattern.

Example 2-30

Find the general term of the set $\{1, 3, 5, 7, \ldots\}$.

Solution

In this case, it is necessary to determine what is being done to 1 to get 1, 2 to get 3, 3 to get 5, etc. You can use inductive reasoning and see that by multiplying by 2 and subtracting 1, you get the desired elements of the set, as shown here.

$$2(1) - 1 = 1$$
$$2(2) - 1 = 3$$
$$2(3) - 1 = 5$$
$$2(4) - 1 = 7$$
$$\text{etc.}$$

Hence, the general term is $2n - 1$.

Try This One

2-N Find the general term for the set $\{6, 11, 16, 21, 26, \ldots\}$.
For the answer, see page 72.

Recall that a set is said to be infinite if it can be placed into a one-to-one correspondence with a proper subset of itself. Example 2-31 illustrates how to do this.

Example 2-31

Show that the set $\{5, 10, 15, 20, \ldots, 5n, \ldots\}$ is an infinite set.

Solution

Select a proper subset of the given set; then show a one-to-one correspondence between the two sets. In this case, use the set $\{10, 20, 30, \ldots, 10n, \ldots\}$.

$$\{5, 10, 15, 20, \ldots, 5n, \ldots\}$$
$$\updownarrow \ \updownarrow \ \updownarrow \ \updownarrow \qquad \updownarrow$$
$$\{10, 20, 30, 40, \ldots, 10n, \ldots\}$$

Hence, the set $\{5, 10, 15, 20, \ldots, 5n, \ldots\}$ is an infinite set.

Try This One

2-O Show that the set $\{-1, -2, -3, \ldots, -n, \ldots\}$ is an infinite set.
For the answer, see page 72.

Cantor defined a set as **countable** if it is finite or if there is a one-to-one correspondence between the members of the set and the natural numbers. Cantor assigned the number a cardinal number $\aleph_0$ (aleph-null) to the set of natural numbers. Hence any infinite set that can be placed into a one-to-one correspondence with the set of natural numbers is said to have a cardinality of $\aleph_0$.

Exercise Set 2-5

Computational Exercises

For Exercises 1–10, find the general term of the set.

1. $\{7, 14, 21, 28, 35, \ldots\}$
2. $\{1, 8, 27, 64, 125, \ldots\}$
3. $\{4, 16, 64, 256, 1024, \ldots\}$
4. $\{1, 4, 9, 16, 25, \ldots\}$
5. $\{-3, -6, -9, -12, -15, \ldots\}$
6. $\{22, 44, 66, 88, 110, \ldots\}$
7. $\{\frac{1}{2}, \frac{1}{3}, \frac{1}{4}, \frac{1}{5}, \frac{1}{6}, \ldots\}$
8. $\{\frac{1}{3}, \frac{2}{3}, \frac{3}{3}, \frac{4}{3}, \frac{5}{3}, \ldots\}$
9. $\{2, 6, 10, 14, 18, \ldots\}$
10. $\{1, 4, 7, 10, 13, \ldots\}$

For Exercises 11–20, show each set is an infinite set.

11. $\{3, 6, 9, 12, 15, \ldots\}$
12. $\{10, 15, 20, 25, 30, \ldots\}$
13. $\{9, 18, 27, 36, 45, \ldots\}$
14. $\{4, 10, 16, 22, 28, \ldots\}$
15. $\{2, 5, 8, 11, 14, \ldots\}$
16. $\{20, 24, 28, 32, 36, \ldots\}$
17. $\{10, 100, 1000, 10,000, \ldots\}$
18. $\{100, 200, 300, 400, 500, \ldots\}$
19. $\{\frac{5}{1}, \frac{5}{2}, \frac{5}{3}, \frac{5}{4}, \frac{5}{5}, \ldots\}$
20. $\{\frac{1}{2}, \frac{1}{4}, \frac{1}{8}, \frac{1}{16}, \ldots\}$

Writing Exercises

21. Define an infinite set.
22. What is meant by the general term of an infinite set?
23. Explain in your own words what is meant by a countable set.

Critical Thinking

Find the answer to each.

24. $\aleph_0 + 1 = $ _____
25. $5 \times \aleph_0 = $ _____
26. $\aleph_0 + \aleph_0 = $ _____

Summary

Section	Important Terms	Important Ideas
2-1	set roster method element well defined natural numbers descriptive method set-builder notation variable finite set infinite set null set equal sets equivalent sets one-to-one correspondence	**A** set is a well-defined collection of objects. Each object is called an element or member of the set. There are three ways to identify sets. They are the roster method, the descriptive method, and set-builder notation. Sets can be finite or infinite. A finite set contains a specific number of elements, while an infinite set contains an unlimited number of elements. If a set has no elements, it is called an empty set or a null set. Two sets are equal if they have the same elements, and two finite sets are equivalent if they have the same number of elements. Two sets are said to have a one-to-one correspondence if it is possible to pair the elements of one set with the elements of the other set in such a way that for each element in the first set there exists one and only one element in the second set.
2-2	subset proper subset union intersection universal set complement	**If** every element of a set is also an element of another set, then the first set is said to be a subset of the second set. A subset is a proper subset of another set if it is not equal to the original set. The union of two sets is a set containing all elements of one set or the other set, while the intersection of two sets is a set that contains the elements that both sets have in common. The universal set is a specifically defined set that contains all elements used in a specific problem or situation. The complement of a specific set is a set that consists of all elements in the universal set that are not in this specific set.
2-3	Venn diagram	**A** mathematician named John Venn devised a way to represent sets pictorially. His method uses overlapping circles to represent the sets.
2-4		**Venn** diagrams can be used to solve real-world problems involving surveys and classifications.
2-5	countable sets	**A** mathematician named Georg Cantor studied infinite sets. An infinite set can be placed in a one-to-one correspondence with a proper subset of itself. Cantor defined a set as countable if it is finite or if there is a one-to-one correspondence between the set and the set of natural numbers.

Review Exercises

For Exercises 1–8, write each set in roster notation.

1. The set of even numbers between 50 and 60

2. The set of odd numbers between 3 and 40

3. {letters in the word "letter"}

4. The set of letters in the word A R K A N S A S

5. $\{x \mid x \in N \text{ and } x > 500\}$

6. The set of counting numbers between 5 and 12

7. The set of months in the year that begin with the letter "P"

8. {days in the week that end with the letter "e"}

For Exercises 9–12, write each set using set-builder notation.

9. {18, 20, 22, 24}   10. {5, 10, 15, 20}

11. {101, 103, 105, 107, ...}   12. {8, 16, 24, ..., 72}

For Exercises 13–18, state whether the set is finite or infinite.

13. $\{x \mid x \in N \text{ and } x \geq 9\}$ 14. {4, 8, 12, 16, ...}

15. {x, y, z}   16. {3, 7, 9, 12}

17. ∅

18. {people who have naturally green hair}

19. Find all subsets of {r, s, t}.

20. Find all subsets of {m, n, o}.

21. How many subsets and proper subsets will the set {p, q, r, s, t} have?

22. How many subsets and proper subsets will the set {a, e, i, o, u, y} have?

For Exercises 23–32, let U = {p, q, r, s, t, u, v, w, x, y, z}, A = {p, r, t, u, v}, B = {t, u, v, x, y}, and C = {s, w, z}. Find each.

23. $A \cap B$ 24. $B \cup C$

25. $(A \cap B) \cap C$ 26. $\overline{B}$

27. $\overline{(A \cup B)} \cap C$ 28. $\overline{B} \cap \overline{C}$

29. $(B \cup C) \cap \overline{A}$ 30. $(A \cup B) \cap \overline{C}$

31. $(\overline{B} \cap \overline{C}) \cup \overline{A}$ 32. $(\overline{A} \cap B) \cup C$

For Exercises 33–36, draw a Venn diagram and shade the appropriate area for each.

33. $\overline{A} \cap B$ 34. $\overline{A \cup B}$

35. $(\overline{A} \cap \overline{B}) \cup C$ 36. $A \cap (\overline{B \cup C})$

37. In a recent survey of 25 students, 10 had cereal for breakfast, 5 had toast, and 2 had both cereal and toast.
    (a) How many did not eat cereal or toast?
    (b) How many had toast only?

38. In a mathematics department, it was found that 9 courses were taught via the computer, 18 were taught via lecture, and 3 were taught using the computer and lecture.
    (a) How many courses were taught via computer only?
    (b) How many courses were not taught using the computer? (Be careful on this one!)

39. Fifty-three eighth-grade students were asked what type of music they listened to. Twenty-two listened to the radio, 18 listened to tapes, 33 listened to CDs, 8 listened to the radio and tapes, 13 listened to tapes and CDs, 11 listened to the radio and CDs, and 6 listened to all three.
    (a) How many listened to tapes only?
    (b) How many listened to the radio and CDs but not tapes?
    (c) How many did not listen to any of the three types?

40. A survey of 41 artists showed that 15 used chalk, 16 used oils, and 20 used watercolors. Furthermore, 4 used both chalk and oils, 8 used oils and watercolors, and 8 used chalk and watercolors. One used all three.
    (a) How many artists surveyed did not use any of the three types of materials?
    (b) How many artists used only oils?
    (c) How many artists used chalk and watercolors but not oils?

For Exercises 41–52, state whether each is true or false.

41. {a, b, c} is equal to {x, y, z}.

42. {1, 2, 3, 4} is equivalent to {p, q, r, s}.

43. {80, 100, 120, ...} ⊆ {40, 80, 120, ...}

44. {6} ⊂ {6, 12, 18, ...}

45. 4 ∈ {even natural numbers}

46. 9 ∉ {2, 4, 5, 6, 10}

47. {5, 6, 7} ⊆ {5, 7}

48. {12} ∈ {12, 24, 36, ...}

49. For any set, $\overline{\varnothing} = U$

50. 0 ∈ ∅

51. For any set, $A \cap \varnothing = U$

52. For any set, $A \cup B = B \cup A$

53. Find the general term of the set {−5, −7, −9, −12, −15, ...}

54. Show that the set {12, 24, 36, 48, 60, ...} is an infinite set.

Chapter Test

For Exercises 1–8, write each set in roster notation.

1. The set of even numbers between 90 and 100

2. The set of odd numbers between 40 and 50

3. The set of letters in the word "envelope"

4. The set of letters in the word "Washington"

5. $\{x \mid x \in N \text{ and } x < 80\}$

6. $\{x \mid x \in N \text{ and } 16 < x < 25\}$

7. The set of months in the year that begins with the letter "J"

8. The set of days in the week that end with the letter "a"

For Exercises 9–12, write each set using set-builder notation.

9. $\{12, 14, 16, 18\}$ 10. $\{30, 35, 40, 45\}$

11. $\{201, 203, 205, 207, \ldots\}$ 12. $\{4, 8, 16, \ldots, 128\}$

For Exercises 13–17, state whether the set is finite or infinite.

13. $\{15, 16, 17, 18, \ldots\}$

14. $\{x \mid x \in N \text{ and } x \text{ is a multiple of } 6\}$

15. $\{a, b, c\}$

16. $\{4, 7, 8, 10\}$

17. {people who have three eyes}

18. Find all subsets of $\{d, e, f\}$.

19. Find all subsets of $\{p, q, r\}$.

20. How many subsets and proper subsets will the set $\{a, b, c, d, e\}$ have?

For Exercises 21–25, let $U = \{a, b, c, d, e, f, g, h, i, j, k\}$, $A = \{a, b, d, e, f\}$, $B = \{a, g, i, j, k\}$, and $C = \{e, h, j\}$. Find each.

21. $A \cap B$ 22. $B \cup C$

23. $\overline{B}$ 24. $\overline{(A \cup B)}$

25. $(A \cap \overline{B}) \cup \overline{C}$

For Exercises 26–29, draw a Venn diagram for each.

26. $\overline{A} \cap B$ 27. $\overline{\overline{A} \cap B}$

28. $(\overline{A} \cup \overline{B}) \cap \overline{C}$ 29. $A \cap \overline{(B \cup C)}$

30. In a recent survey of 24 secretaries, 8 had a salad for lunch, 9 had a sandwich, and 4 had both a salad and a sandwich.
    (a) How many did not eat a salad or a sandwich?
    (b) How many had a salad only?

31. Find the general term of the set $\{15, 30, 45, 60, 75, \ldots\}$.

32. Show that the set $\{1, -1, 2, -2, 3, -3, \ldots\}$ is an infinite set.

Mathematics in Our World ▶Revisited

Survey: Blood Types

The problem can be solved by using a Venn diagram, as shown in Section 2-4.

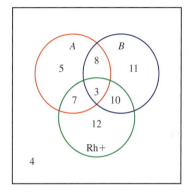

Then each question can be answered by looking at the diagram.

(a) 7

(b) 4

(c) 3

(d) 25

(e) 60

▌Projects

1. Have the students in your class fill out this questionnaire:

    A. Gender: Male _____   Female _____

    B. Age: under 21 _____   21 or older _____

    C. Work: Yes _____   No _____

Draw a Venn Diagram, and from the information answer these questions:

    (a) How many students are female?
    (b) How many students are under 21?
    (c) How many students work?
    (d) How many students are under 21 and work?
    (e) How many students are males and do not work?
    (f) How many students are 21 or older and work?
    (g) How many students are female, work, and are under 21?

2. Select five of these set properties. Explain why they are true for all sets.

   (a) $\overline{\overline{A}} = A$
   (b) $\overline{\varnothing} = U$ and $\overline{U} = \varnothing$
   (c) $A \cup \varnothing = A$ and $A \cap \varnothing = \varnothing$
   (d) $A \cup U = U$ and $A \cap U = A$
   (e) $A \cup \overline{A} = U$ and $A \cap \overline{A} = \varnothing$

   (f) $A \cup A = A$ and $A \cap A = A$
   (g) $A \cup B = B \cup A$ and $A \cap B = B \cap A$
   (h) $(A \cup B) \cup C = A \cup (B \cup C)$ and $(A \cap B) \cap C = A \cap (B \cap C)$
   (i) $A \cup (B \cap C) = (A \cup B) \cap (A \cup C)$ and $A \cap (B \cup C) = (A \cap B) \cup (A \cap C)$
   (j) $\overline{A \cup B} = \overline{A} \cap \overline{B}$ and $\overline{A \cap B} = \overline{A} \cup \overline{B}$

Answers to **Try This One**

2-A. (a) {80, 82, 84, 86, 88, 90}
   (b) {Saturday, Sunday}

2-B. (a) False; (b) False; (c) True

2-C. (a) {Earth, Jupiter, Mars, Mercury, Neptune, Pluto, Saturn, Uranus, Venus}
   (b) P = planets in our solar system
   (c) $P = \{x \mid x$ is a planet in our solar system$\}$

2-D. (a) Infinite; (b) Finite; (c) Finite

2-E. b; d

2-F. (a) Equivalent; (b) Neither;
   (c) Equal and equivalent; (d) Equivalent

2-G. (a) 1. No members: $\varnothing$

   2. Single-member subsets:

   $\{\blacklozenge\}; \{\blacktriangledown\}; \{\spadesuit\}; \{\clubsuit\}$

   3. Double-member subsets:

   $\{\blacklozenge, \blacktriangledown\}; \{\blacklozenge, \spadesuit\}; \{\blacklozenge, \clubsuit\}; \{\blacktriangledown, \spadesuit\}; \{\blacktriangledown, \clubsuit\}; \{\spadesuit, \clubsuit\}$

   4. Triple-member subsets:

   $\{\blacklozenge, \blacktriangledown, \spadesuit\}; \{\blacklozenge, \spadesuit, \clubsuit\}$
   $\{\blacklozenge, \blacktriangledown, \clubsuit\}; \{\blacktriangledown, \spadesuit, \clubsuit\}$

   5. Quadruple-member subset:

   $\{\blacklozenge, \blacktriangledown, \spadesuit, \clubsuit\}$

   (b) Same as part a, but with answer 5 deleted.

2-H. (a) $\{\star, !, \bigcirc, \triangle, \$\}$

   (b) $\{\star, !, \bigcirc\}$

   (c) $\{\star, \triangle, \$\}$

2-I. (a) {1, 4, 6, 8}; (b) $\varnothing$; (c) {4, 6, 8}; (d) {1, 4, 6, 8}

2-J. $\overline{A} \cup B = \{2, 3, 4\}$

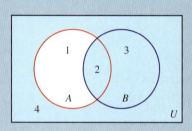

2-K. $(A \cap \overline{B}) \cup C = \{1, 3, 4, 5, 6\}$

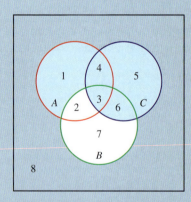

2-L. 21 bought the *Daily News* only, 12 bought the *Tribune* only; 7 bought something else.

2-M. (a) 10; (b) 7; (c) 4; (d) 34

2-N. $5n + 1$

2-O. $\{-1, -2, -3, \ldots, -n, \ldots\}$
   $\updownarrow \quad \updownarrow \quad \updownarrow \qquad \updownarrow$
   $\{-2, -4, -6, \ldots, -2n, \ldots\}$

Chapter Three

Logic

Outline

Objectives

After completing this chapter, you should be able to

1 Determine whether or not a sentence is a statement (3-1)

2 Classify statements as simple or compound (3-1)

3 Write compound statements in symbols using the four basic connectives (3-1)

4 Write symbolic statements in words (3-1)

5 Construct truth tables for statements (3-2)

6 Identify the type of statement according to the hierarchy of connectives (3-2)

7 Determine whether or not two statements are logically equivalent (3-3)

8 Determine if one statement is the negation of another statement (3-3)

9 Write the converse, inverse, or contrapositive of a statement (3-3)

10 Determine the validity of arguments by using truth tables (3-4)

11 Determine whether or not an argument is valid using Euler circles (3-5)

Introduction

ogic is sometimes defined as correct thinking or correct reasoning. It is important that we be able to reason correctly since we are bombarded every day by advertisements, contracts, product or service warranties, political debates, and news commentaries. People often have problems with these statements because of misinterpretation, misunderstanding, and faulty logic. In addition to everyday life, logic is used in fields such as science, psychology, law, philosophy, and mathematics.

Logic has been studied formally since the time of Aristotle (384–322 B.C.E.). In the 1600s and 1700s, a mathematician named Gottfried Wilhelm Leibniz used symbols to represent words and statements in logic. Another mathematician, Leonhard Euler, used circles to prove or disprove arguments. Other mathematicians and philosophers have made contributions to the field of logic. In fact, even Lewis Carroll, who is most noted for his book *Alice's Adventures in Wonderland*, wrote a book on logic entitled *The Game of Logic*. See Mathematics in Our World.

This chapter introduces the basic concepts of symbolic logic and shows how to determine whether or not arguments are valid or invalid by using truth tables. It concludes by explaining how Euler circles can be used to determine the validity of arguments called syllogisms. ■

3-1 Statements

Logic uses words and symbols. Words are put together to make sentences. In the English language, there are several different types of sentences, such as declarative sentences, interrogative sentences, etc. Logic uses sentences that can be determined to be either true or false. These sentences are called *statements*.

> A **statement** is a sentence that can be determined to be true or false but not both at the same time.

Mathematics in Our World

"Alice in Logicland"

A mathematics professor at Oxford University named Charles Lutwidge Dodgson (1832–1898) wrote two famous children's books entitled *Alice's Adventures in Wonderland* and *Through the Looking Glass*. He used the pseudonym Lewis Carroll.

Alice In Wonderland

It was rumored that Queen Victoria enjoyed *Alice's Adventures in Wonderland* so much that she requested a copy of Dodgson's next book as soon as it was published. To her dismay, he sent her a copy of his book entitled *The Elements of Determinants* (with their applications to *Simultaneous Linear Equations in Algebraic Geometry*).

In addition to his children's classics, Dodgson wrote several books and pamphlets on mathematics.

Dodgson's major interest was the logical foundation of mathematics, and he used logic throughout his children's books. He liked to invent things and wrote letters that had to be read backwards, that is, from the last word to the first word.

Here is an excerpt from *Alice's Adventures in Wonderland*. Alice is having a conversation with the March Hare, Mad Hatter, and Doormouse at the Mad Tea Party.

The March Hare questions Alice's response to a riddle the Hatter posed.

"Then you should say what you mean," the March Hare went on.

"I do," Alice hastily replied, "at least—at least I mean what I say—that's the same thing, you know."

Is Alice correct in her assertion?

The answer is found on page 115.

For example, sentences like

"It is raining."

"Harrisburg is the capital of Pennsylvania."

"$2 + 2 = 4$"

"$10 - 5 = 4$"

are called statements since they are either true or false. Notice the last one is false; nevertheless, it is called a statement.

Sentences like

"Bring me a glass of milk."

"How old are you?"

"Help me!"

are not statements since it is not possible to determine whether they are true or false.

A famous wrong statement is "Don't worry lieutenant, they can't possibly hit us at this dis—"

Example 3-1

Decide which of the following are statements and which are not.

(a) "Pike's Peak is in Colorado."
(b) "Is that your car?"
(c) "Hello!"
(d) "$8 - 2 = 6$"
(e) "This book has a blue cover."

Solution

a, d, and e are statements since they can be judged as being true or false.

Try This One

3-A   Which of the following are statements?

(a) Help!

(b) $12 - 8 = 5$

(c) David Letterman is a nighttime talk show host.

(d) Harry can lift 80 pounds.

(e) What time is it?

For the answer, see page 116.

Statements can be classified as *simple* or *compound*. A **simple statement** contains only one idea. Each of these statements is an example of a simple statement.

"This blouse is yellow."

"My car has four doors."

"This book has over 200 pages."

Sidelight

History of Logic

The basic concepts of logic can be attributed to Aristotle, who lived in the fourth century B.C.E. He used words, sentences, and deduction to prove arguments using syllogisms. Not much was done with logic until the 19th century, when it was taught in schools in the way Aristotle had developed it. In addition, in 300 B.C.E. Euclid formalized geometry using deductive proofs. Both subjects were considered to be the "inevitable truths" of the universe revealed to rational people.

In the 19th century, people began to reject the idea of "inevitable truths" and realized that a deductive system like Euclidean geometry is only true based on the original assumption. When the original assumption is changed, a new deductive system can be created. This is what happens with geometry. (See the Sidelight entitled "Non-Euclidean Geometry" in Chapter 10.)

In addition, several people developed the use of symbols rather than words and sentences in logic. One such person was George Boole (1815–1864). Boole created the symbols used in this chapter and developed the theory of symbolic logic. He also used

symbolic logic in mathematics. His manuscript, entitled "An Investigation into the Laws of Thought, on Which Are Founded the Mathematical Theories of Logic and Probabilities," was published when he was 39 in 1854. Boole was a friend of Augustus DeMorgan, who formulated DeMorgan's laws. In addition, Leonhard Euler (1707–1783) used circles to represent logical statements and proofs. The idea was refined by John Venn (1834–1923).

A simple statement can be negated by using or omitting "not." For example, the statement "It is sunny" can be negated by saying, "It is not sunny." A statement such as "The exam is not difficult" can be negated by saying, "The exam is difficult," or "It is not true that the exam is not difficult."

A statement such as "I will buy a quart of milk, and I will make a milk shake" is called a **compound statement** since it consists of two simple statements.

Compound statements are formed by joining two simple statements with what is called a *connective*.

> The basic **connectives** are "*and*," "*or*," "*if . . . then*," and "*if and only if*."

Table 3-1 shows the basic connectives, their symbols, and their common names. Each of these statements is an example of a compound statement.

"His name is John, and her name is Mary." (**conjunction**)

"She will run in the race, or she will play in the tennis tournament." (**disjunction**)

"If it snows, then I will go skiing." (**conditional**)

Table 3-1

Symbols for the Connectives

Connective	Symbol	Name
and	$\wedge$	Conjunction
or	$\vee$	Disjunction
if . . . then	$\rightarrow$	Conditional
if and only if	$\leftrightarrow$	Biconditional

Note that in this type of statement, the word "then" is often deleted; i.e., "If it snows, I will go skiing."

"A triangle is called an equilateral triangle if and only if it has three congruent sides." **(biconditional)**

In logic, statements are represented symbolically using the letters of the alphabet and the symbols for the connectives shown in Table 3-1.

Example 3-2

Classify each statement as simple or compound.

(a) The automobile is red.
(b) If you buy one, then you will get one free.
(c) Tomorrow is Christmas.
(d) I will buy a pencil or I will buy a pen.

Solution

(a) Simple
(b) Compound
(c) Simple
(d) Compound

Try This One

3-B   Classify each statement as simple or compound.

(a) It is hot, and it is humid.

(b) This is a history book.

(c) If it does not rain, then I will go fishing.

(d) I will buy a Mustang, or I will buy a Jeep.

(e) Yesterday was Tuesday.

For the answer, see page 116.

Simple statements are represented by lowercase letters. For example, if p represents the statement "It is cold," and q represents the statement "I will go to the mall," then the compound statement "If it is cold, then I will go to the mall" can be written in symbols as $p \rightarrow q$.

The symbol $\sim$ (tilde) represents a **negation.** For example, if p represents the statement "It is cold," the statement "It is not cold" is written symbolically as $\sim p$. The negation of the statement p can also be written as "It is not true that it is cold."

It is necessary to explain the difference between the compound statement $\sim p \wedge q$ and the compound statement $\sim (p \wedge q)$. The statement $\sim p \wedge q$ means to negate the statement p first, then use the negation of p in conjunction with the statement q. For example, if p is the statement "Fido is a dog," and q is the statement "Pumpkin is a cat," then $\sim p \wedge q$ would read "Fido is not a dog and Pumpkin is a cat." The statement $\sim p \wedge q$ could also be written as $(\sim p) \wedge q$. The statement $\sim (p \wedge q)$ means to negate the conjunction of the statement p and the statement q. Using the same statements for p and q as before, the statement $\sim (p \wedge q)$ would be written as "It is not the case that Fido is a dog and Pumpkin is a cat."

The same reasoning applies when the negation is used with other connectives. For example, $\sim p \rightarrow q$ means $(\sim p) \rightarrow q$.

Example 3-3 illustrates in more detail how to write statements symbolically.

Example 3-3

Let $p =$ "It is hot." Let $q =$ "I will play tennis." Write each statement in symbols.

(a) I will not play tennis.
(b) It is hot, and I will play tennis.
(c) If it is hot, then I will not play tennis.
(d) I will play tennis if and only if it is not hot.

Solution

(a) $\sim q$
(b) $p \wedge q$
(c) $p \rightarrow \sim q$
(d) $q \leftrightarrow \sim p$

Try This One

3-C Let $p =$ "I will buy a loaf of bread." Let $q =$ "I will buy some luncheon meat." Write each statement in symbols.

(a) I will buy a loaf of bread, and I will buy some luncheon meat.

(b) I will not buy a loaf of bread.

(c) If I buy some luncheon meat, then I will buy a loaf of bread.

(d) I will not buy a loaf of bread, and I will buy some luncheon meat.

For the answer, see page 116.

As we have said it is not necessary to repeat the subject and verb in a compound statement using "and" or "or." For example, the statement "It is cold, and it is snowing" can be written "It is cold and snowing." The statement "I will go to a movie, or I will go to a play" can be written "I will go to a movie or a play." Also the words "but" and "although" can be used in place of "and." For example, the statement "I will not buy a television set, and I will buy a CD player" can also be written as "I will not buy a television set, but I will buy a CD player."

Statements written in symbols can also be written in words, as shown in Example 3-4.

If this is your dog, then a, b, and c could all describe it.

Example 3-4

Write each statement in words. Let p = "My dog is a poodle." Let q = "My dog is brown."

(a) $\sim p$
(b) $p \vee q$
(c) $\sim p \rightarrow q$
(d) $q \leftrightarrow p$
(e) $q \wedge p$

Solution

(a) My dog is not a poodle.
(b) My dog is a poodle or my dog is brown.
(c) If my dog is not a poodle, then my dog is brown.
(d) My dog is brown if and only if my dog is a poodle.
(e) My dog is brown, and my dog is a poodle. (*Note:* This statement can also be written as "My dog is a brown poodle.")

Try This One

3-D Write each statement in words. Let p = "Math is easy." Let q = "I will study."

(a) $\sim q$

(b) $p \rightarrow \sim q$

(c) $q \wedge p$

(d) $\sim p \vee q$

(e) $p \leftrightarrow q$

For the answer, see page 116.

As shown previously, statements are declarative sentences that can be either true or false. A compound statement consists of two or more simple statements. The simple statements are joined by connectives. There are four basic connectives. They are *and* ($\wedge$), *or* ($\vee$), *if . . . then* ($\rightarrow$), and *if and only if* ($\leftrightarrow$). The symbol tilde $\sim$ represents the word *not*.

Exercise Set 3-1

Real World Applications

For Exercises 1–10, state whether the sentence is a statement or not.

1. Please do not whistle.
2. $5 + 9 = 14$
3. Bob is a dentist.
4. $9 - 3 = 2$
5. Who will be elected mayor of Springfield?
6. Neither Sam nor Mary attend art class.
7. Philadelphia is the capital of Pennsylvania.
8. Abraham Lincoln was the 16th president of the United States.
9. Go home.
10. It is not windy.

For Exercises 11–20, decide if each statement is simple or compound.

11. He likes to fish and swim.
12. Bob bought a new tennis racquet.
13. Mary will buy a hot dog or hamburger.
14. Pool is fun if and only if you win.
15. March is a month of the year.
16. Diane is a chemistry major.
17. If you win the contest, you will get a prize.
18. He cut the grass and trimmed the bushes.
19. $8 + 9 = 12$
20. Neither Bill nor Sally will go on the trip.

For Exercises 21–26, write the negation of each.

21. The sky is blue.
22. It is not true that the mail is late.
23. The blanket is not red.
24. The flower is not yellow.
25. It is not true that Harry failed statistics.
26. She has red hair.

For Exercises 27–34, identify each statement as a conjunction, disjunction, conditional, or biconditional.

27. Bob and Ron like sailing.
28. Either he passes the test, or he fails that course.
29. A number is even if and only if it is divisible by two.
30. His boat is red, and it has an outboard motor.
31. I will go camping, or I will go sightseeing.
32. If a number is divisible by 3, then it is an odd number.
33. A triangle is equiangular if and only if three angles are congruent.
34. If your battery is dead, then your automobile will not start.

For Exercises 35–44, write each statement in symbols. Let p = "Spot is a beagle." Let q = "Rover is a collie."

35. Spot is a beagle, and Rover is a collie.
36. Spot is not a beagle.
37. If Rover is not a collie, then Spot is a beagle.

38. It is not true that Rover is a collie or Spot is a beagle.

39. It is false that Rover is not a collie.

40. It is not true that Spot is a beagle.

41. Rover is a collie, or Spot is not a beagle.

42. Rover is not a collie, and Spot is a beagle.

43. Rover is a collie if and only if Spot is a beagle.

44. If Spot is a beagle, then Rover is not a collie.

For Exercises 45–54, write each statement in symbols. Let p = "Pete is happy." Let q = "Sue is happy." Let "sad" mean "not happy."

45. Sue is not happy.   46. Both Sue and Pete are sad.

47. If Sue is happy, then Pete is happy.   48. It is not true that Pete is happy.

49. Either Sue is sad, or Pete is happy.

50. It is not true that Sue is sad and Pete is happy.

51. Sue is happy if and only if Pete is happy.

52. Neither Pete nor Sue is happy.

53. If Sue is sad, then Pete is happy.

54. Pete is sad if and only if Sue is not happy.

For Exercises 55–64, write each statement in words. Let p = "The plane is on time." Let q = "The sky is clear."

55. $p \wedge q$ 56. $\sim p \vee q$ 57. $q \rightarrow p$
58. $q \rightarrow \sim p$ 59. $\sim p \wedge \sim q$ 60. $q \leftrightarrow p$
61. $p \vee \sim q$ 62. $\sim p \leftrightarrow \sim q$ 63. $q \rightarrow (p \vee \sim p)$
64. $(p \rightarrow q) \vee \sim p$

For Exercises 65–74, write each statement in words. Let p = "Mark is handsome." Let q = "Trudy is attractive."

65. $\sim q$ 66. $p \rightarrow q$ 67. $p \vee \sim q$
68. $q \leftrightarrow p$ 69. $\sim p \rightarrow \sim q$ 70. $\sim p$
71. $p \vee q$ 72. $(\sim p \vee q) \vee \sim q$ 73. $q \vee p$
74. $(p \vee q) \rightarrow \sim(\sim q)$

Writing Exercises

75. Define a statement.

76. Explain the difference between a simple and a compound statement.

77. What symbols are used for the four connectives?

78. Explain why the sentence "As a rule, all rules have exceptions" is not a statement.

79. Explain why the sentence "This statement is false" is not a statement.

80. A magician claims that he can create anything and do all things. A disbeliever says, "Okay, create a rock so large that you cannot move it." Why has the magician been bested?

3-2 Truth Tables

In Section 3-1, the various types of logical connectives were introduced. A more detailed explanation of each connective is given in this section. Each connective can be analyzed by a *truth table*. A **truth table** is a table that is used to show when a compound statement is true or false given the truth values of the simple statements that make up the compound statement.

The Negation, Conjunction, and Disjunction

A statement p is either true or false but not both. Let p denote the statement "It is snowing." If it is actually snowing outside, then p is true, and its negation ($\sim p$) "It is not snowing" is false. Now if I say, "It is snowing" and see no snow, the statement p is false, but its negation ($\sim p$) "It is not snowing" is true. The truth table for the negation looks like this:

p	$\sim p$
T	F
F	T

In summary, then, when p is true, its negation is false, and when p is false, its negation must be true.

When a compound statement involves two letters, namely, p and q, there are four possibilities that need to be analyzed when considering whether or not the entire statement is true or false. These possibilities are

1. Both p and q are true.
2. p is true, and q is false.
3. p is false, and q is true.
4. p and q are both false.

Now let's analyze the conjunction. Suppose someone says to you, "I bought a new home and a new car." This statement can symbolically be represented by $p \wedge q$, where $p =$ "I bought a new home" and $q =$ "I bought a new car." In analyzing this statement or any other compound statement consisting of two simple statements, there are four possibilities. They are

Possibilities	**Symbolic value of each**	
	p	q
1. p and q are both true.	T	T
2. p is true, and q is false.	T	F
3. p is false, and q is true.	F	T
4. p and q are both false.	F	F

The truth table for a compound statement is set up as

p	q	$p \wedge q$
T	T	
T	F	
F	T	
F	F	

When would the conjunctive statement $p \wedge q$ be true? If indeed the person purchased both items—the home and the car—the compound statement "I bought a new home and a new car" would be true.

Now suppose the person bought only one item, either the home or the car. Then the compound statement "I bought a new home and a new car" would be false. Finally, suppose the person really did not buy either item but was just trying to impress you; then the compound statement would be false. Hence, the completed truth table for the conjunction would look like this:

p	q	$p \wedge q$
T	T	T
T	F	F
F	T	F
F	F	F

In summary, the conjunction $p \wedge q$ is true only when both p and q are true.

The analysis of the disjunction or the "or" statements is slightly different from the conjunction. Assume that someone tells you, "I bought a new home or a new car." If the person indeed bought only one of the items, the disjunction would be true, and if the person purchased neither item, the disjunction would be false. Hence, the truth table would partially look like this:

p	q	$p \vee q$
T	T	
T	F	T
F	T	T
F	F	F

But what if the person really did purchase both items? You might be inclined at first to say the first statement, "I bought a new home or a new car," is false. Logically, however, the statement is true. It is important to distinguish between the two meanings of the word "or." The *inclusive* "or" means one or the other or both. In logic, we always strive to give the benefit of the doubt. So, if it is possible for p and q to be true at the same time, then the statement $p \vee q$ will be true. The completed truth table would be constructed as follows.

p	q	$p \vee q$
T	T	T
T	F	T
F	T	T
F	F	F

Now consider the statement "All entrees come with soup or salad." In this case, you cannot order both soup *and* salad at the same meal. This type of "or" usage is called the *exclusive* "or," or *exclusive* disjunction, and is true only when p is true and q is false or when p is false and q is true.

In summary, the inclusive disjunction is true when either p or q or both are true. The exclusive disjunction is true only when one statement is true.

The Conditional and the Biconditional Statements

The conditional statement, sometimes called the *implication,* is used extensively in mathematics as well as logic. For example, you might recall from high school geometry a statement such as "If two sides and the included angle of one triangle are congruent to two sides and the included angle of another triangle, then the triangles are congruent."

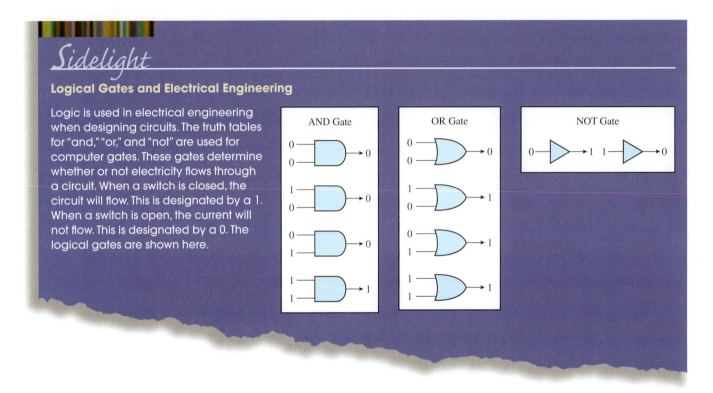

Sidelight

Logical Gates and Electrical Engineering

Logic is used in electrical engineering when designing circuits. The truth tables for "and," "or," and "not" are used for computer gates. These gates determine whether or not electricity flows through a circuit. When a switch is closed, the circuit will flow. This is designated by a 1. When a switch is open, the current will not flow. This is designated by a 0. The logical gates are shown here.

The truth table for the implication looks like this:

p	q	$p \to q$
T	T	T
T	F	F
F	T	T
F	F	T

Let's analyze the truth value of the statement $p \to q$. Suppose I say to you, "If it rains, I will go to a movie." Naturally, if it rains, and I attend a movie, my conditional statement is true, but if it rains and I go somewhere else, my original statement is false.

Now suppose it's sunny and hot, and you see me at the movie theater. What do you conclude about my original statement? You might be inclined to say it's false, but you would be incorrect since I did not say what I would do if it did *not* rain. Hence, when p is false and q is true, you cannot infer that I was lying, and the conditional statement $p \to q$ is true. Finally, when p and q are both false (i.e., it's not raining, and I did not go to a movie), the conditional statement $p \to q$ is also true.

In summary, the conditional statement $p \to q$ is false only when p is true and q is false.

The biconditional statement "I will go to a movie *if and only if* it rains" is the conjunction of two conditional statements: "If it is raining, then I will go to a movie" and "If I go to a movie, then it is raining." Therefore, it is true when both p and q are true and when both p and q are false. In other words, if it is raining and I go to a movie, the biconditional statement is true, and if it's not raining and I don't go to a movie, the biconditional statement is also true. When one of the statements is true and the other is false,

Table 3-2

Truth Tables for the Connectives

Conjunction "and"

p	q	$p \wedge q$
T	T	T
T	F	F
F	T	F
F	F	F

Disjunction "or" (inclusive)

p	q	$p \vee q$
T	T	T
T	F	T
F	T	T
F	F	F

Conditional "if . . . then"

p	q	$p \rightarrow q$
T	T	T
T	F	F
F	T	T
F	F	T

Biconditional "if and only if"

p	q	$p \leftrightarrow q$
T	T	T
T	F	F
F	T	F
F	F	T

Negation "not"

p	$\sim p$
T	F
F	T

A technician who designed an automated irrigation system would need to decide whether the system should turn on *if* the water in the soil falls below a certain level or *if and only if* the water in the soil falls below a certain level. In the first instance, other inputs could also turn on the system.

then the biconditional statement is false. The truth table for the biconditional statement looks like this:

p	q	$p \leftrightarrow q$
T	T	T
T	F	F
F	T	F
F	F	T

In summary, then, the biconditional statement $p \leftrightarrow q$ is true when both p and q are true or when both p and q are false.

Table 3-2 shows a summary of the truth values of the basic compound statements.

Truth Tables for Other Statements

Once the truth values of the basic connections are known, the truth value of any logical statement can be found by constructing a truth table. The procedure shown next may look complicated, but with a little practice it can be mastered.

Example 3-5

Construct a truth table for the statement $\sim p \vee q$.

Solution

Step 1 Set up a table as shown.

p	q	$\sim p \vee q$
T	T	
T	F	
F	T	
F	F	

—Continued

Example 3-5 Continued—

Note that the order in which you list the T's and F's in a truth table does not matter as long as you cover all the possibilities. Also note that once the T's and F's are listed for p and q, they remain in the same order as they move through the table. For consistency in this book, we will use the order TTFF for p and TFTF for q when there are only two letters in the logical statement.

Step 2 Write the truth values for p and q underneath the respective letters in the statement as shown, and label the columns as column 1 and column 2.

p	q	$\sim p$	$\vee$	q
T	T	T		T
T	F	T		F
F	T	F		T
F	F	F		F
		①		②

Step 3 Find the negation of p and place the truth values in column 3. (Remember when p is true, $\sim p$ is false, and when p is false, p is true.) Draw a line through the values in column 1, since they will not be used again.

p	q	$\sim$	p	$\vee$	q
T	T	F	T		T
T	F	F	T		F
F	T	T	F		T
F	F	T	F		F
		③	①		②

Step 4 Find the truth values for the disjunction ($\vee$) using the T and F values in column 3 and column 2 and the disjunction truth table.

For row 1, F $\vee$ T is T, i.e., False or True is True
For row 2, F $\vee$ F is F, i.e., False or False is False
For row 3, T $\vee$ T is T, i.e., True or True is True
For row 4, T $\vee$ F is T, i.e., True or False is True

Place these values in column 4. Draw a line through the truth values in column 3. The completed truth table is shown next.

"My leg isn't better, or I'm taking a break" is an example of a statement that can be written as $\sim p \vee q$.

p	q	$\sim$	p	$\vee$	q
T	T	F	T	T	T
T	F	F	T	F	F
F	T	T	F	T	T
F	F	T	F	T	F
		③	①	④	②

The truth value of $\sim p \vee q$ is found in column 4.

In summary then, $\sim p \vee q$ is true when p and q are true, when p is false and q is true, and when both p and q are false.

When the statement contains parentheses, find the truth values for the statement contained in the parentheses first, as shown in Example 3-6. This is similar to the order of operations used in algebra.

Example 3-6

Construct a truth table for $\sim(p \rightarrow \sim q)$.

Solution

Step 1   Set up the table, as shown.

p	q	$\sim(p \rightarrow \sim q)$
T	T	
T	F	
F	T	
F	F	

Step 2   Write the truth values for p and q underneath the respective letters in the statement as shown and label the columns as 1 and 2.

p	q	$\sim(p$	$\rightarrow$	$\sim q)$
T	T	T		T
T	F	T		F
F	T	F		T
F	F	F		F
		①		②

Step 3   Find the negation of the q since it is inside the parentheses, and place the truth values in column 3. Draw a line through the truth values in column 2 since they will not be used again.

p	q	$\sim(p$	$\rightarrow$	$\sim$	$q)$
T	T	T		F	T
T	F	T		T	F
F	T	F		F	T
F	F	F		T	F
		①		③	②

Step 4   Find the truth values for the conditional ($\rightarrow$) using the T and F values in columns 1 and 3 and the conditional truth table.

$$T \rightarrow F \text{ is } F$$
$$T \rightarrow T \text{ is } T$$
$$F \rightarrow F \text{ is } T$$
$$F \rightarrow T \text{ is } T$$

—Continued

"It is not true that if it rains, then we can't go out," is an example of a statement that can be written as $\sim(p \to \sim q)$.

Example   **3-6**   *Continued—*

Place these values in column 4 and draw a line through the T and F values in columns 1 and 3, as shown.

p	q	$\sim$	$(p$	$\to$	$\sim$	$q)$
T	T		T	F	F	T
T	F		T	T	T	F
F	T		F	T	F	T
F	F		F	T	T	F
			①	④	③	②

Step 5 Find the negations of the truth values in column 4 (since the negation sign is outside the parentheses).

p	q	$\sim$	$(p$	$\to$	$\sim$	$q)$
T	T	T	T	F	F	T
T	F	F	T	T	T	F
F	T	F	F	T	F	T
F	F	F	F	T	T	F
		⑤	①	④	③	②

Hence the truth value of $\sim(p \to \sim q)$ is found in column 5. In summary, the statement $\sim(p \to \sim q)$ is true only when p and q are true.

Try This One

3-E   Construct a truth table for the statement $p \leftrightarrow \sim q$.
For the answer, see page 116.

Truth tables can be constructed for compound statements with three or more variables. For a compound statement with three variables, there are eight possible combinations of T's and F's to consider. The truth table is set up as shown in step 1 of Example 3-7.

Example   **3-7**

Construct a truth table for $p \lor (q \to r)$.

Solution

Step 1 Set up the table as shown.

p	q	r	$p \lor (q \to r)$
T	T	T	
T	T	F	
T	F	T	
T	F	F	
F	T	T	
F	T	F	
F	F	T	
F	F	F	

—Continued

"I'll do my math assignment, or if I think of a good topic, then I'll start my English essay" is an example of a statement that can be written $p \vee (q \rightarrow r)$.

Example 3-7 Continued—

Again the order of the T's and F's is not important as long as all possibilities are covered. When there are three letters in the statement, the order just shown will be used for consistency throughout the textbook.

Step 2 Transpose the values of p, q, and r under their respective letters in the statement as shown.

p	q	r	p	$\vee$	$(q$	$\rightarrow$	$r)$
T	T	T	T		T		T
T	T	F	T		T		F
T	F	T	T		F		T
T	F	F	T		F		F
F	T	T	F		T		T
F	T	F	F		T		F
F	F	T	F		F		T
F	F	F	F		F		F
					①	②	③

Step 3 Using the truth values in columns 2 and 3 and the truth table for the conditional ($\rightarrow$), find the values inside the parentheses for the conditional.

p	q	r	p	$\vee$	$(q$	$\rightarrow$	$r)$
T	T	T	T		T	T	T
T	T	F	T		T	F	F
T	F	T	T		F	T	T
T	F	F	T		F	T	F
F	T	T	F		T	T	T
F	T	F	F		T	F	F
F	F	T	F		F	T	T
F	F	F	F		F	T	F
			①		②	④	③

Step 4 Complete the truth table using the truth values in columns 1 and 4 and the table for the disjunction ($\vee$), as shown.

p	q	r	p	$\vee$	$(q$	$\rightarrow$	$r)$
T	T	T	T	T	T	T	T
T	T	F	T	T	T	F	F
T	F	T	T	T	F	T	T
T	F	F	T	T	F	T	F
F	T	T	F	T	T	T	T
F	T	F	F	F	T	F	F
F	F	T	F	T	F	T	T
F	F	F	F	T	F	T	F
			①	⑤	②	④	③

Hence, the truth value for $p \vee (q \rightarrow r)$ is found in column 5.

In summary, the statement $p \vee (q \rightarrow r)$ is true in all cases except when p and r are false and q is true.

Try This One

3-F Construct a truth table for the statement $(p \land q) \lor \sim r$.
For the answer, see page 116.

Some suggestions for constructing a truth table are

1. Set up all truth tables as shown.

p	q	Statement
T	T	
T	F	
F	T	
F	F	

2. Transpose the truth values for the letters p and q underneath the corresponding letters in the statement.
3. Do any negations within parentheses.
4. Do the connectives within the parentheses.
5. Find the truth values for any connectives outside the parentheses using the truth values for the statement contained inside the parentheses.

When constructing truth tables, the truth value of the part of the statement that is inside the parentheses is found first. However, parentheses are not always necessary since there is a hierarchy of connectives. It is as follows.

1. Biconditional $\leftrightarrow$
2. Conditional $\rightarrow$
3. Conjunction $\land$, disjunction $\lor$
4. Negation $\sim$

When finding the truth value for a compound statement without parentheses, the truth value of a lower order connective is found first. For example, $p \lor q \rightarrow r$ is a conditional statement since the conditional ($\rightarrow$) is of a higher order than the disjunction ($\lor$). If you were constructing a truth table for the statement, you would find the truth value for $\lor$ first. The statement $p \leftrightarrow q \land r$ is a biconditional statement since the biconditional ($\leftrightarrow$) is of a higher order than the order of the conjunction ($\land$). When constructing a truth table for the statement, the truth value for the conjunction ($\land$) would be found first. The conjunction and disjunction are of the same order. The statement $p \land q \lor r$ cannot be identified unless parentheses are used. Hence $(p \land q) \lor r$ is a disjunction and $p \land (q \lor r)$ is a conjunction.

Example   3-8

For each, identify the type of statement using the hierarchy of connectives.

(a) $\sim p \lor \sim q$
(b) $p \rightarrow \sim q \land r$
(c) $p \lor q \leftrightarrow q \lor r$
(d) $p \rightarrow q \leftrightarrow r$

—Continued

Example 3-8   *Continued—*

Solution

(a) The ∨ is higher than the ∼; hence, the statement is a disjunction.
(b) The → is higher than the ∧; hence, the statement is a conditional statement.
(c) The ↔ is higher than ∨; hence, the statement is a biconditional statement.
(d) The ↔ is higher than →; hence, the statement is a biconditional statement.

Try This One

3-G   For each, identify the type of statement using the hierarchy of connectives.

(a) $\sim p \vee q$

(b) $p \vee \sim q \rightarrow r$

(c) $p \vee q \leftrightarrow \sim p \vee \sim q$

(d) $p \wedge \sim q$

(e) $p \leftrightarrow q \rightarrow r$

For the answer, see page 116.

Truth tables can be constructed for any logical statement, as shown in this section. These truth tables will be used in Section 3-4 to determine the validity of logical arguments.

Exercise Set 3-2

Computational Exercises

For Exercises 1–30, construct a truth table for each.

1. $\sim(p \vee q)$

2. $q \rightarrow p$

3. $\sim p \wedge q$

4. $\sim q \rightarrow \sim p$

5. $\sim p \leftrightarrow q$

6. $(p \vee q) \rightarrow \sim p$

7. $\sim(p \wedge q) \rightarrow p$

8. $(p \vee q) \wedge (q \wedge p)$

9. $(\sim q \wedge p) \rightarrow \sim p$

10. $q \wedge \sim p$

11. $(p \wedge q) \leftrightarrow (q \vee \sim p)$

12. $p \rightarrow (q \vee \sim p)$

13. $(p \wedge q) \vee p$

14. $(q \rightarrow p) \vee \sim r$

15. $(r \wedge q) \vee (p \wedge q)$

16. $(r \rightarrow q) \vee (p \rightarrow r)$

17. $\sim(p \vee q) \rightarrow \sim(p \wedge r)$

18. $(\sim p \vee \sim q) \rightarrow \sim r$

19. $(\sim p \vee q) \wedge r$

20. $p \wedge (q \vee \sim r)$

21. $(p \wedge q) \leftrightarrow (\sim r \vee q)$

22. $\sim(p \wedge r) \rightarrow (q \wedge r)$

23. $r \rightarrow \sim(p \vee q)$

24. $(p \vee q) \vee (\sim p \vee \sim r)$

25. $p \rightarrow (\sim q \wedge \sim r)$

26. $(q \vee \sim r) \leftrightarrow (p \wedge \sim q)$

27. $\sim(q \rightarrow p) \wedge r$

28. $q \rightarrow (p \wedge r)$

29. $(r \vee q) \wedge (r \wedge p)$

30. $(p \wedge q) \leftrightarrow \sim r$

Writing Exercises

31. Explain the purpose of a truth table.

32. Explain in words the difference between the conjunction and disjunction.

33. Explain in words the difference between the inclusive and exclusive disjunctions.

34. Explain in words the difference between the conditional and biconditional statements.

35. What is the hierarchy for the basic connectives?

   **3-3**   ## Types of Statements

Tautologies and Self-Contradictions

When statements are translated into words, one should think about when they are true or when they are false.

Some compound statements are true for all cases, and some statements are false for all cases. Consider these two logical statements $p \vee (q \vee \sim p)$ and $(q \wedge \sim p) \wedge p$.

The truth values for each are

p	q	$p \vee (q \vee \sim p)$	$(q \wedge \sim p) \wedge p$
T	T	T T T T F T	T F F T F T
T	F	T T F F F T	F F F T F T
F	T	F T T T T F	T T T F F F
F	F	F T F T T F	F F T F F F

The first statement is true in every case, and the second statement is false in every case. The first statement is called a *tautology,* and the second statement is called a *self-contradiction.* Formally defined,

> When a compound statement is always true, it is called a **tautology.**
> When a compound statement is always false, it is called a **self-contradiction.**

In order to determine if a statement is a tautology or a self-contradiction, a truth table must be constructed. If the truth value of this statement consists of all T's, then the statement is a tautology. If the truth value for the statement consists of all F's, then the statement is a self-contradiction.

Example   3-9

Determine if each statement is a tautology, self-contradiction, or neither.

(a) $(p \wedge q) \rightarrow p$
(b) $(p \wedge q) \wedge (\sim p \wedge \sim q)$
(c) $(p \vee q) \rightarrow q$

—Continued

In Example 3-9, let $p =$ "it is sunny" and $q =$ "I am wearing red." Translate each logic statement into a word statement. Can you predict which statements will be tautologies, self-contradictions, and neither?

Example 3-9 *Continued—*

Solution

(a) The truth table for statement a is

p	q	$(p$	$\wedge$	$q)$	$\rightarrow$	p
T	T	T	T	T	T	T
T	F	T	F	F	T	T
F	T	F	F	T	T	F
F	F	F	F	F	T	F

Since the truth table value consists of all T's, the statement is a tautology.

(b) The truth table for statement b is

p	q	$(p$	$\wedge$	$q)$	$\wedge$	$(\sim$	p	$\wedge$	$\sim$	$q)$
T	T	T	T	T	F	F	T	F	F	T
T	F	T	F	F	F	F	T	F	T	F
F	T	F	F	T	F	T	F	F	F	T
F	F	F	F	F	F	T	F	T	T	F

Since the truth value consists of all F's, the statement is a self-contradiction.

(c) The truth table for statement c is

p	q	$(p$	$\vee$	$q)$	$\rightarrow$	q
T	T	T	T	T	T	T
T	F	T	T	F	F	F
F	T	F	T	T	T	T
F	F	F	F	F	T	F

Since the statement can be true in some cases and false in some cases, it is neither a tautology nor a self-contradiction.

Try This One

3-H Determine if each statement is a tautology, self-contradiction, or neither.

(a) $(p \vee q) \wedge (\sim p \rightarrow q)$

(b) $(p \wedge \sim q) \wedge \sim p$

(c) $(p \rightarrow q) \wedge \sim p$

For the answer, see page 116.

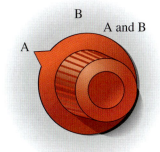

The statements "If the red dial is set to A, then use only speaker A" and "The red dial is not set to speaker A, or only speaker A is used" can be modeled with the logic statements $p \rightarrow q$ and $\sim p \vee q$.

Logically Equivalent Statements

Next, consider the two logical statements $p \rightarrow q$ and $\sim p \vee q$. The truth tables for the two statements are

p	q	p	$\rightarrow$	q	$\sim$	p	$\vee$	q
T	T	T	T	T	F	T	T	T
T	F	T	F	F	F	T	F	F
F	T	F	T	T	T	F	T	T
F	F	F	T	F	T	F	T	F

Notice that the truth values for both statements are *identical;* that is, TFTT. When this occurs, the statements are said to be *logically equivalent;* that is, both compositions of the same simple statements have the same meaning. For example, the statement "If it snows, I will go skiing" is logically equivalent to saying "It is not snowing or I will go skiing." Formally defined,

> Two compound statements are **logically equivalent** if and only if they have the same truth table values. The symbol for logically equivalent statements is ≡.

Example 3-10

Determine if the two statements, $p \rightarrow q$ and $\sim q \rightarrow \sim p$, are logically equivalent.

Solution

The truth tables for the statements are

p	q	$p \rightarrow q$	$\sim$	q	$\rightarrow$	$\sim$	p
T	T	T	F	T	T	F	T
T	F	F	T	F	F	F	T
F	T	T	F	T	T	T	F
F	F	T	T	F	T	T	F

Since both statements have the same truth values, they are logically equivalent.

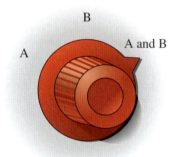

In the red dial example, $\sim q \rightarrow \sim p$ would be "If speaker A is not the only one used, then the red dial is not set to A."

Try This One

3-I Determine which two statements are logically equivalent.

(a) $\sim(p \wedge \sim q)$

(b) $\sim p \wedge q$

(c) $\sim p \vee q$

For the answer, see page 116.

Negations of Compound Statements

Recall that the negation of a simple statement p is $\sim p$, and when p is true, its negation is false, or when p is false, its negation is true. Compound statements can also be negated. Like a simple statement, the negation of a compound statement has the opposite truth value of the truth value of the original compound statement. For example, consider the statement $p \wedge q$. You might think that the negation of $p \wedge q$ is $\sim p \wedge \sim q$. However, the truth tables for both statements show that this is not true.

p	q	$p \wedge q$	p	q	$\sim$	p	$\wedge$	$\sim$	q
T	T	T	T	T	F	T	F	F	T
T	F	F	T	F	F	T	F	T	F
F	T	F	F	T	T	F	F	F	T
F	F	F	F	F	T	F	T	T	F

Notice the only case where $p \wedge q$ is true is when p and q are both true. By definition, then, the negation of $p \wedge q$ should be true in all cases except when p and q are both true. The truth table for $\sim p \wedge \sim q$ shows that this is not the case.

The negation of the statement $p \wedge q$ is the statement $\sim p \vee \sim q$. This can be verified by the truth table for $\sim p \vee \sim q$.

Notice that the truth value of $\sim p \vee \sim q$ is exactly the opposite of the truth value for the statement $p \wedge q$.

p	q	$\sim$	p	$\vee$	$\sim$	q
T	T	F	T	F	F	T
T	F	F	T	T	T	F
F	T	T	F	T	F	T
F	F	T	F	T	T	F

A similar situation occurs with the disjunction $p \vee q$. Its negation is $\sim p \wedge \sim q$, as shown in Example 3-11.

The two laws

$$\sim(p \wedge q) \equiv \sim p \vee \sim q$$

$$\sim(p \vee q) \equiv \sim p \wedge \sim q$$

are called De Morgan's laws. These laws were named after the English mathematician Augustus De Morgan (1806–1871) who first expressed them using logical symbols.

Example   3-11

Show that the negation of $p \vee q$ is $\sim p \wedge \sim q$.

Solution

The truth tables for the two statements are

p	q	$p \vee q$	$\sim$	p	$\wedge$	$\sim$	q
T	T	T	F	T	F	F	T
T	F	T	F	T	F	T	F
F	T	T	T	F	F	F	T
F	F	F	T	F	T	T	F

Since the truth values of the two statements are exactly opposites, $\sim p \wedge \sim q$ is the negation of $p \vee q$.

Try This One

3-J   Show that the negation of the statement $p \rightarrow q$ is $p \wedge \sim q$ by using truth tables.
For the answer, see page 116.

Variations of the Conditional Statement

From the conditional statement, three other statements can be formed. They are called the **converse,** the **inverse,** and the **contrapositive.** They are shown here symbolically.

Statement	$p \rightarrow q$
Converse	$q \rightarrow p$
Inverse	$\sim p \rightarrow \sim q$
Contrapositive	$\sim q \rightarrow \sim p$

Using the statement "If Daisy is a cow, then Daisy is black" as the original statement, the converse of the statement is "If Daisy is black, then Daisy is a cow." The inverse of the statement is "If Daisy is not a cow, then Daisy is not black." The contrapositive of the statement is "If Daisy is not black, then Daisy is not a cow."

Example 3-12

Write the converse, the inverse, and the contrapositive for the statement, "If you live in Pittsburgh, then you live in Pennsylvania."

Solution

Converse: "If you live in Pennsylvania, then you live in Pittsburgh."
Inverse: "If you do not live in Pittsburgh, then you do not live in Pennsylvania."
Contrapositive: "If you do not live in Pennsylvania, then you do not live in Pittsburgh."

Try This One

3-K   Write the converse, the inverse, and the contrapositive for the statement "If you are a mathematician, then you are intelligent."
For the answer, see page 116.

> ### Math Note
>
> Many people using logic in real life assume that if a statement is true, the converse is automatically true. Consider the following statement: "If a person earns more than $100,000 per year, that person can buy a Corvette." The converse would be stated as, "If a person can buy a Corvette, then that person earns more than $100,000 per year." This may be far from the truth since a person may have to make very large payments, live in a tent, or work three jobs in order to afford the expensive car.

The relationships of the variations of the conditional statements can be determined by looking at the truth tables for each of the statements.

p	q	$p \rightarrow q$	$q \rightarrow p$	$\sim p \rightarrow \sim q$	$\sim q \rightarrow \sim p$
T	T	T	T	T	T
T	F	F	T	T	F
F	T	T	F	F	T
F	F	T	T	T	T

Since the original statement ($p \rightarrow q$) and the contrapositive statement ($\sim q \rightarrow \sim p$) have the same truth values, they are equivalent. Also note that the converse ($q \rightarrow p$) and the inverse ($\sim p \rightarrow \sim q$) have the same truth values; hence, they are equivalent. Finally, notice that the original statement is not equivalent to the converse or the inverse since the truth values of the converse and inverse differ from those of the original statement.

The conditional statement $p \rightarrow q$ is used quite often in logic as well as mathematics and a more detailed analysis is needed.

The conditional statement $p \rightarrow q$ is also called an *implication* and consists of two simple statements; the first is the *antecedent* (p) and the second is the *consequent* (q). For example, the statement "If I jump in the pool, then I will get wet" consists of the antecedent

p, "I jump in the pool," and the consequent q, "I will get wet" connected by the if . . . then connective.

The conditional can also be stated in these other ways:

"p implies q"

"q if p"

"p only if q"

"p is sufficient for q"

"q is necessary for p"

Example 3-13

Write each statement in symbols. Let $p =$ "A person is over $6'6''$." Let $q =$ "A person is tall."

(a) If a person is over $6'6''$, then the person is tall.
(b) Being tall is necessary for being over $6'6''$.
(c) A person is over $6'6''$ only if the person is tall.
(d) Being $6'6''$ is sufficient for being tall.
(e) A person is tall if the person is over $6'6''$.

Solution

(a) If p, then q, $p \rightarrow q$
(b) q is necessary for p, $p \rightarrow q$
(c) p only if q, $p \rightarrow q$
(d) p is sufficient for q, $p \rightarrow q$
(e) q if p, $p \rightarrow q$

Actually, these statements all say the same thing!

Try This One

3-L   Write each statement in symbols. Let $p =$ "A student does his homework." Let $q =$ "A student is successful."

(a) A student is successful if a student does his homework.

(b) Doing your homework is necessary for being successful.

(c) A student is successful only if a student does his homework.

(d) Doing homework is sufficient for being successful.

For the answer, see page 116.

In summary, then, if a statement has all T's for its truth table, it is called a tautology. If it has all F's, then it is called a self-contradiction. Two statements are logically equivalent if they have the same truth values. If they have opposite truth values, they are negations of each other.

The conditional statement can be written several different ways. Also from the conditional statement, three other statements—the inverse, the converse, and the contrapositive—can be formed. The statement and the contrapositive are logically equivalent as are the inverse and the converse.

Exercise Set   3-3

Computational Exercises

For Exercises 1–10, determine which statements are tautologies, self-contradictions, or neither.

1. $(p \vee q) \vee (\sim p \wedge \sim q)$

2. $(p \rightarrow q) \wedge (p \vee q)$

3. $(p \wedge q) \wedge (\sim p \vee \sim q)$

4. $\sim p \vee (p \rightarrow q)$

5. $(p \leftrightarrow q) \vee \sim(q \leftrightarrow p)$

6. $(p \wedge q) \leftrightarrow (p \rightarrow \sim q)$

7. $(p \vee q) \wedge (\sim p \vee \sim q)$

8. $(p \wedge q) \vee (p \vee q)$

9. $(p \leftrightarrow q) \wedge (\sim p \leftrightarrow \sim q)$

10. $(p \rightarrow q) \wedge (\sim p \vee q)$

For Exercises 11–20, determine if the two statements are logically equivalent statements, negations, or neither.

11. $\sim q \rightarrow p; \sim p \rightarrow q$

12. $p \wedge q; \sim q \vee \sim p$

13. $\sim(p \vee q); p \rightarrow \sim q$

14. $\sim(p \rightarrow q); \sim p \wedge q$

15. $q \rightarrow p; \sim(p \rightarrow q)$

16. $p \vee (\sim q \wedge r); (p \wedge \sim q) \vee (p \wedge r)$

17. $\sim(p \vee q); \sim(\sim p \wedge \sim q)$

18. $(p \vee q) \rightarrow r; \sim r \rightarrow \sim(p \vee q)$

19. $(p \wedge q) \vee r; p \wedge (q \vee r)$

20. $p \leftrightarrow \sim q; (p \wedge \sim q) \vee (\sim p \wedge q)$

For Exercises 21–26, write the converse, inverse, and contrapositive of each.

21. $p \rightarrow q$

22. $\sim p \rightarrow \sim q$

23. $\sim q \rightarrow p$

24. $\sim p \rightarrow q$

25. $p \rightarrow \sim q$

26. $q \rightarrow p$

Real World Applications

For Exercises 27–30, write the converse, inverse, and contrapositive of each.

27. If he graduated, then he will get a job.

28. If she does not earn $5000, then she cannot buy a car.

29. If it is my birthday, then I will have a party.

30. If my car won't start, then I will check the battery.

Writing Exercises

31. When are two statements logically equivalent?

32. What is a self-contradiction?

33. What three other statements can be made from the if . . . then statement?

34. How does one show that one compound statement is the negation of another compound statement?

Critical Thinking

35. The negation of the statement $p \wedge q$ is $\sim p \vee \sim q$, and the negation of the statement $p \vee q$ is $\sim p \wedge \sim q$. Write the negation of the statement $p \rightarrow q$ using the conjunction.

36. Write the negation of the statement $p \leftrightarrow q$ using only the conjunction and disjunction.

37. Write a statement using p and q that is a tautology.

38. Write a statement using p and q that is a self-contradiction.

3-4   Arguments

Truth tables can also be used to determine the validity of **arguments.** Consider the following example:

> If a figure has three sides, then it is a triangle.
>
> The figure is not a triangle.
>
> Therefore, the figure does not have three sides.

In the preceding argument, the first two statements are called *premises,* and the third statement is called a *conclusion.* When the premises are true, the question is "Does the conclusion follow from the premises?" If so, the argument is said to be *valid.* If not, the argument is said to be *invalid.*

The basic procedure for determining the validity of arguments is given next.

Procedure for Determining the Validity of Arguments

Step 1   Write the argument in symbols.

Step 2   Use a conjunction between the premises and the implication ($\Rightarrow$) for the conclusion. (*Note:* The $\Rightarrow$ is the same as $\rightarrow$ but will be used to designate an argument.)

Step 3   Set up a truth table as follows:

Symbols	Premise	$\wedge$	Premise	$\Rightarrow$	Conclusion

Step 4   Construct a truth table.

Step 5   If all truth values under $\Rightarrow$ are T's (i.e., a tautology), then the argument is valid; otherwise, it is invalid.

Example   3-14

Determine if the following argument is valid or invalid.

> If a figure has three sides, then it is a triangle.
>
> This figure is not a triangle.
>
> Therefore, this figure does not have three sides.

—Continued

Example 3-14   *Continued—*

Solution

Step 1   Translate the argument into symbols. Let p = "The figure has three sides."
Let q = "The figure is a triangle."
Translated into symbols:

$$p \rightarrow q \quad \text{(Premise)}$$
$$\underline{\sim q \qquad \text{(Premise)}}$$
$$\therefore \ \sim p \ \text{(Conclusion)}$$

A line is used to separate the premises from the conclusion and the three
dots $\therefore$ mean "therefore."

Step 2   Make a truth table by connecting the premises with a conjunction and
implying the conclusion as shown.

		Premise 1		Premise 2		Conclusion
p	q	$(p \rightarrow q)$	$\wedge$	$\sim q$	$\Rightarrow$	$\sim p$

Step 3   Construct a truth table as shown.

p	q	$(p$	$\rightarrow$	$q)$	$\wedge$	$\sim$	q	$\Rightarrow$	$\sim$	p
T	T	T	T	T	F	F	T	T	F	T
T	F	T	F	F	F	T	F	T	F	T
F	T	F	T	T	F	F	T	T	T	F
F	F	F	T	F	T	T	F	T	T	F
		①	④	②	⑤	③	②	⑦	⑥	①

Step 4   Determine the validity of the argument. Since all the values under the $\Rightarrow$
are true, the argument is valid.

Example 3-15

Determine the validity of the following argument. "If a professor is rich, then he will
buy an expensive automobile. The professor bought an expensive automobile.
Therefore, the professor is rich."

Solution

Step 1   Translate the argument into symbols. Let p = "The professor is rich." Let
q = "The professor buys an expensive automobile."

$$p \rightarrow q$$
$$\underline{q}$$
$$\therefore p$$

Step 2   Set up a truth table as shown.

p	q	$(p \rightarrow q) \wedge q$	$\Rightarrow$	p

—Continued

Example 3-15 Continued—

Step 3 Construct the truth table for the argument.

p	q	$(p$	$\rightarrow$	$q)$	$\wedge$	q	$\Rightarrow$	p
T	T	T	T	T	T	T	T	T
T	F	T	F	F	F	F	T	T
F	T	F	T	T	T	T	F	F
F	F	F	T	F	F	F	T	F
		①	③	②	④②		⑤	①

Step 4 Determine the validity of the argument. This argument is invalid since it is not a tautology. That is, when the values are not all T's, the argument is invalid. In this case, it cannot be concluded that the professor is rich.

In symbolic logic, it is not important whether or not the conclusion is true. The main concern is whether or not the conclusion follows from the premises.

Consider the following two arguments.

1. Either $2 + 2 \neq 4$ or $2 + 2 = 5$

$$\dfrac{2 + 2 = 4}{\therefore 2 + 2 = 5}$$

2. If $2 + 2 \neq 5$, then I passed the math quiz.

$$\dfrac{\text{I did not pass the quiz.}}{\therefore 2 + 2 \neq 5}$$

The truth tables show that the first argument is valid even though the conclusion is false, and the second argument is invalid even though the conclusion is true!

Let p be the statement "$2 + 2 = 4$" and q be the statement "$2 + 2 = 5$"

Then the first argument is written as

$$(\sim p \vee q)$$
$$\dfrac{p}{\therefore q}$$

Truth table for argument 1

p	q	$(\sim$	p	$\vee$	$q)$	$\wedge$	p	$\Rightarrow$	q
T	T	F	T	T	T	T	T	T	T
T	F	F	T	F	F	F	T	T	F
F	T	T	F	T	T	T	F	T	T
F	F	T	F	T	F	F	F	T	F

Let p be the statement "$2 + 2 \neq 5$" and let q be the statement "I passed the math quiz."

The second argument is written as

$$p \rightarrow q$$
$$\dfrac{\sim q}{\therefore p}$$

Truth table for argument 2

p	q	$(p$	$\rightarrow$	$q)$	$\wedge$	$\sim$	q	$\Rightarrow$	p
T	T	T	T	T	F	F	T	T	T
T	F	T	F	F	F	T	F	T	T
F	T	F	T	T	F	F	T	T	F
F	F	F	T	F	T	T	F	F	F

Try This One

3-M Translate the following argument into symbols; then use a truth table to determine whether the argument is valid or invalid.

If the whole number is not an odd number, then the whole number is divisible by 2.
The whole number is not divisible by 2.
_____
∴ The whole number is an odd number.

For the answer, see page 116.

The validity of arguments that have three variables can also be determined by truth tables, as shown in Example 3-16.

Example **3-16**

Determine the validity of the following argument.

$p \rightarrow r$

$q \wedge r$

p
_____
$\therefore \sim q \rightarrow p$

Solution

The truth table is as follows. When there are three premises, work the conjunctions left to right as shown:

p	q	r	$(p \rightarrow r)$	$\wedge$	$(q \wedge r)$	$\wedge\ p$	$\Rightarrow$	$\sim\ q \rightarrow p$
T	T	T	T T T	T	T T T	T T	T	F T T T
T	T	F	T F F	F	T F F	F T	T	F T T T
T	F	T	T T T	F	F F T	F T	T	T F T T
T	F	F	T F F	F	F F F	F T	T	T F T T
F	T	T	F T T	T	T T T	F F	T	F T T F
F	T	F	F T F	F	T F F	F F	T	F T T F
F	F	T	F T T	F	F F T	F F	T	T F F F
F	F	F	F T F	F	F F F	F F	T	T F F F
			①④③	⑤	②④③	⑥①	⑨	⑦②⑧①

Since the truth value for ⇒ is all T's, the argument is valid.

Try This One

3-N Determine whether the following argument is valid or invalid.

$p \vee q$

$p \vee \sim r$
_____
$\therefore q$

For the answer, see page 116.

An argument consisting of two or more premises and a conclusion can be determined to be valid or invalid by constructing a truth table. An argument is valid if its truth value consists of all T's; otherwise, it is invalid.

In logic and mathematics, as well as everyday life, we use arguments to arrive at conclusions that may or may not be valid. Some of the commonly used valid arguments are as follows.

1. Law of detachment

$$p \rightarrow q$$
$$\underline{p}$$
$$\therefore q$$

2. Law of composition

$$p \rightarrow q$$
$$\underline{\sim q}$$
$$\therefore \sim p$$

3. Law of syllogism or transitional law

$$p \rightarrow q$$
$$\underline{q \rightarrow r}$$
$$\therefore p \rightarrow r$$

4. Law of disjunctive syllogism

$$p \vee q$$
$$\underline{\sim p}$$
$$\therefore q$$

(These arguments can be shown to be valid by constructing truth tables. See Exercises 2 and 10 in this section.)

For example, the law of detachment or *Modus Ponens* (the Latin name) goes something like this:

If a geometric figure has three sides, then it is a triangle.

This geometric figure has three sides.

Therefore, it is a triangle.

The law of syllogism can be applied to this situation:

If I exceed the speed limit, then I will get a speeding ticket.

If I get a speeding ticket, then I will lose my license.

Therefore, if I exceed the speed limit, then I will lose my license.

There are also two commonly used arguments that are invalid. They are

Fallacy of the converse

$$p \rightarrow q$$
$$\underline{q}$$
$$\therefore p$$

Fallacy of the inverse

$$p \rightarrow q$$
$$\underline{\sim p}$$
$$\therefore \sim q$$

For example, if your instructor makes the following statement, "If it rains on Friday, then I will give a test," and you wake up on Friday and see the sun shining, can you conclude that you will not have a test?

The answer is *no*! This argument is an example of the fallacy of the inverse.

You can verify that the argument is invalid by constructing a truth table for it. (See Exercise 13 in the exercise set.) However, a little common sense tells you that it is invalid since your instructor did not say what he or she would do if it is not raining on Friday!

Exercise Set 3-4

Computational Exercises

Using truth tables, determine whether each argument is valid.

1. $p \rightarrow q$
   $p \wedge q$
   ___
   $\therefore p$

2. $p \rightarrow q$
   p
   ___
   $\therefore q$

3. $p \vee q$
   $\sim q$
   ___
   $\therefore p$

4. $\sim p \vee q$
   p
   ___
   $\therefore p \wedge \sim q$

5. $p \leftrightarrow \sim q$
   $p \wedge \sim q$
   ___
   $\therefore p \vee q$

6. $\sim q \vee p$
   q
   ___
   $\therefore \sim p$

7. $p \vee q$
   $\sim p \wedge \sim q$
   ___
   $\therefore p$

8. $p \leftrightarrow q$
   $\sim q$
   ___
   $\sim p$

9. $p \vee \sim q$
   $\sim q \rightarrow p$
   ___
   $\therefore p$

10. $p \rightarrow q$
    $q \rightarrow r$
    ___
    $\therefore p \rightarrow r$

11. $p \wedge \sim q$
    $\sim r \rightarrow q$
    ___
    $\therefore q$

12. $p \leftrightarrow q$
    $q \leftrightarrow r$
    ___
    $\therefore p \wedge q$

Real World Applications

Identify p, q, and if necessary, r; then translate each argument into symbols, then determine whether each argument is valid or invalid.

13. If it rains, then I will do my homework.
    It did not rain.
    ___
    $\therefore$ I did not do my homework.

14. Ted will buy a cat or a dog.
    Ted did not buy a dog.
    ___
    $\therefore$ Ted bought a cat.

15. If Sam gains 15 pounds, he will make the football team.
    Sam made the football team.
    ___
    $\therefore$ Sam gained 15 pounds.

16. If it snows, I will do my homework.

    It did not snow.

    ∴ I did not do my homework.

17. I will cut the grass if and only if it does not rain.

    It did not rain.

    ∴ I cut the grass.

18. If you are in doubt, then do not act.

    If you do not act, then you are in doubt.

    ∴ You are in doubt and you did not act.

19. Either I did not study or I passed the exam.

    I did not study.

    ∴ I failed the exam.

20. I will enter the race if and only if I get in shape.

    I will get in shape or I will not enter the race.

    ∴ If I entered the race, then I was in shape.

21. If you do 20 pushups, then you will pass physical education.

    If you pass physical education, then you will graduate.

    ∴ If you can do 20 pushups, then you will graduate.

22. If you smile, then you are happy.

    Either you are happy or you are poor.

    ∴ If you are poor, you won't smile.

Writing Exercises

23. Explain the structure of an argument.

24. Is it possible for an argument to be valid yet have a false conclusion? Explain your answer.

25. Is it possible for an argument to be invalid yet have a true conclusion? Explain your answer.

26. When setting up a truth table in order to determine the validity of an argument, what connective is used between the premises of an argument? What connective is used between the premises and the conclusion?

Critical Thinking

27. Oscar Wilde once said, "Few parents nowadays pay any regard to what their children say to them. The old-fashioned respect for the young is fast dying out." This statement can be translated into an argument, as shown next.

If parents respected their children, then they would listen to them.

Parents do not listen to their children.

∴ Parents do not respect their children.

Using a truth table, determine whether the argument is valid or invalid.

28. Winston Churchill once said, "If you have an important point to make, don't try to be subtle or clever. Use a pile driver. Hit the point once. Then come back and hit it again. Then a third time—a tremendous wack!" This statement can be translated into an argument as shown.

If you have an important point to make, then you are not subtle or clever.

You are not subtle or clever.

∴ You will make your point.

Using a truth table, determine whether the argument is valid or invalid.

3-5 Euler Circles

Another way to determine whether an argument is valid or invalid is to use circles called **Euler circles.**

This method was developed by Leonhard Euler (1707–1783) and later refined by John Venn in 1881. The circles are similar to the Venn diagrams shown in Chapter 2.

The circle method uses four basic types of statements. They are

Type	General form	Example
Universal affirmative	All A is B	All chickens have wings.
Universal negative	No A is B	No horses have wings.
Particular affirmative	Some A is B	Some horses are black.
Particular negative	Some A is not B	Some horses are not black.

Each statement can be represented by a specific diagram. The universal affirmative "All A is B" means that every member of set A is also a member of set B. For example, the statement "All chickens have wings" means that the set of all chickens is a subset of the set of animals that have wings. The diagram for the universal affirmative statement is shown in Figure 3-1(a).

The universal negative "No A is B" means that no member of set A is a member of set B. In other words, set A and set B are *disjoint sets*. For example, "No horses have wings," means that the set of all horses and the set of all animals with wings are disjoint (nonintersecting). The diagram for the universal negative is shown in Figure 3-1(b).

The particular affirmative "Some A is B" means that there is at least one member of set A that is also a member of set B. For example, the statement "Some horses are black" means that there is at least one horse that is a member of the set of black animals. The diagram for the particular affirmative statement is shown in Figure 3-1(c). The x means that there is at least one black horse.

The particular negative "Some A is not B" means that there is at least one member of set A that is not a member of set B. For example, the statement "Some horses are not black" means that there is at least one horse that does not belong to the set of black animals.

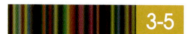

Math Note

If one states that "Some horses are black," it cannot be assumed that "Some horses are not black."

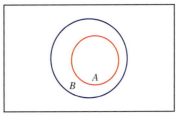

(a) Universal affirmative "All *A* is *B*."

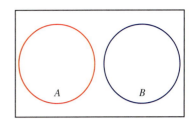

(b) Universal negative "No *A* is *B*."

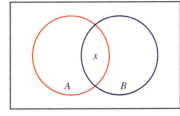

(c) Particular affirmative "Some *A* is *B*."

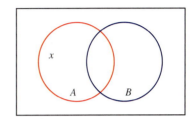

(d) Particular negative "Some *A* is not *B*."

Figure 3-1

The diagram for the particular negative is shown in Figure 3-1(d). The *x* is placed in circle *A* but not in circle *B*. The *x* in this example means that there exists at least one horse that is some color other than black.

Once the diagrams for the four basic types of statements are shown, they can be used to test the validity of an argument. These arguments are called *syllogisms,* which consist of two *premises* (i.e., the first two statements) and a *conclusion* (i.e., the third statement) as shown.

> **Math Note**
>
> The statement "Some horses are not black" does *not* imply that there are some horses that are black!

Premise	All cats have four legs.
Premise	Some cats are black.
Conclusion	Therefore, some four-legged animals are black.

As stated previously, we are not concerned with whether the conclusion is true or false, but only whether or not the conclusion logically follows or can logically descend from the premises. If yes, the argument is valid. If not, the argument is invalid.

Circle Method for Testing the Validity of an Argument

In order to determine whether or not an argument is valid, diagram both premises in the same figure, and if the conclusion is shown in the figure, the argument is valid.

Many times the premises can be diagrammed in several ways. If there is one way that the diagram contradicts the conclusion, the argument is *invalid* since the conclusion does not necessarily follow from the premises.

Examples 3-17, 3-18, and 3-19 show how to determine the validity of an argument using Euler circles.

Example 3-17

Using Euler circles, determine whether the argument is valid or invalid.

All cats have four legs.

Some cats are black.
_____

Therefore, some four-legged animals are black.

Solution

The first premise, "All cats have four legs," is the universal affirmative; hence, the set of cats diagrammed as a subset of four-legged animals is shown.

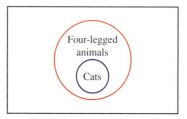

The second premise, "Some cats are black," is the particular affirmative and is shown by placing an x in the intersection of the cats circle and the black animals circle. The diagram for this premise is drawn on the diagram of the first premise and can be done in two ways, as shown.

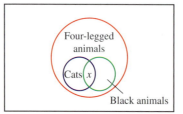 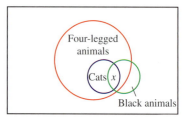

Now the conclusion states "Some four-legged animals are black," and the diagram for the statement must show an x in the four-legged animals circle and also in the black animal circle. Notice that _both_ diagrams have an x in the two circles; hence, the argument is valid. Since there is no other way to diagram the premises, the conclusion is shown "without a doubt."

It is not necessary to use actual subjects such as cats, four-legged animals, etc. in syllogisms. Arguments can use letters to represent the various sets, as shown in Example 3-18.

Example 3-18

Use Euler circles to determine whether the argument is valid or invalid.

Some A is not B.

All C is B.
_____

∴ Some A is C.

—Continued

Example 3-18   *Continued—*

Solution

The first premise, "Some A is not B," is diagrammed as shown.

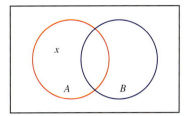

The second premise, "All C is B," is diagrammed by placing circle C inside circle B. This can be done in several ways, as shown.

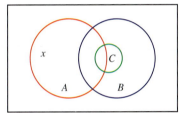
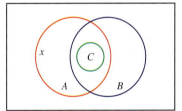

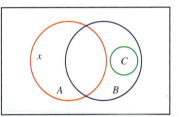

Notice that the first two ways show that it might be possible that "Some A is C." However, the third figure shows the two premises diagrammed in a way that contradicts the conclusion. Since we are not "forced" to accept the conclusion (as shown in Example 3-17), the argument is invalid.

Example 3-19

Determine whether the following argument is valid or invalid.

No A is B.
Some C is not A.
∴ Some B is C.

—Continued

Example 3-19 Continued—

Solution

The first premise, "No *A* is *B*," is diagrammed as shown.

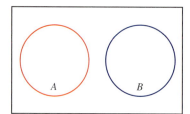

The second premise, "Some *C* is not *A*," can be diagrammed in two ways, as shown.

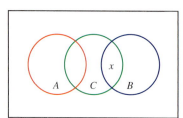 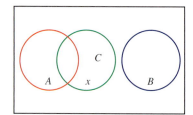

In the first case, the conclusion "Some *B* is *C*" is shown. But in the second case, it is not shown. Hence, we are not forced into the conclusion, and the argument is invalid.

> **Math Note**
>
> You do not need to draw all possible diagrams to disprove an argument, just determine if you can draw one diagram that disproves the argument.

Try This One

3-O Determine whether each argument is valid or invalid.

a. Some *A* is *B*.
   Some *A* is not *C*.
   ∴ Some *B* is not *C*.

b. All dogs bark.
   No barking animals are cats.
   Therefore, no dogs are cats.

For the answer, see page 116.

In summary, then, diagram both premises on the same figure. If the conclusion is shown without a doubt, the argument is valid. If you are not forced into the conclusion, the argument is invalid.

Exercise Set 3-5

Computational Exercises

For Exercises 1–10, draw a diagram for each statement.

1. All dogs are mammals.

2. No pigs are blue.

3. Some people do not go to college.

4. Some horses are brown.

5. No mathematics courses are easy.

6. Some books are not for children.

7. Some laws in Pennsylvania are laws in Ohio.

8. All members of Club Walk are healthy.

9. No policemen are dishonest.

10. Some politicians are dishonest.

For Exercises 11–30, determine whether each argument is valid or invalid.

11. All X is Y.
    Some Y is Z.
    ∴ Some X is Z.

12. Some A is not B.
    No B is C.
    ∴ Some A is not C.

13. Some P is Q.
    No Q is R.
    ∴ Some P is not R.

14. All S is T.
    No S is R.
    ∴ Some T is R.

15. No M is N.
    No N is O.
    ∴ Some M is not O.

16. Some U is V.
    Some U is not W.
    ∴ No W is U.

17. Some A is not B.
    No A is C.
    ∴ Some A is not C.

18. All P is Q.
    All Q is R.
    ∴ All P is R.

19. No S is T.
    No T is R.
    ∴ No S is R.

20. Some M is N.
    Some N is O.
    ∴ Some M is O.

Real World Applications

21. All fathers are men.
    Some men are wealthy.
    ∴ Some fathers are wealthy.

22. Some policemen are dishonest.
    No dishonest person is happy.
    ∴ Some policemen are not happy.

23. Some windstorms are violent.
    No snowstorms are violent.
    ∴ No snowstorms are windy.

24. Some golfers are angry.
    Some golfers are not pleasant.
    ∴ Some pleasant people are not angry.

25. Some nurses are patient.
    No patient people are rude.
    ∴ Some nurses are not rude.

26. Some lifeguards are brown-eyed.
    Some brown-eyed people are not intelligent.
    ∴ Some lifeguards are not intelligent.

27. Some gamblers are poor.
    No gamblers are intelligent.
    ∴ No poor people are intelligent.

28. Some generals are women.
    Some women are tall.
    ∴ Some generals are not tall.

29. Some machines have gears.
    Some vehicles have gears.
    ∴ No machines are vehicles.

30. Some cold days are sunny.
    No rainy days are sunny.
    ∴ No cold days are rainy.

Writing Exercises

31. Name and give an example of each of the four types of statements that can be diagrammed with Euler circles.

32. State the condition used to determine whether an argument is valid or invalid.

Critical Thinking

For Exercises 33–36, write a conclusion so that the argument is valid.

33. All A is B.
    All B is C.
    ∴

34. No M is P.
    All S is M.
    ∴

35. All birds can fly.
    No flying animals have teeth.
    ∴

36. Some cats are gray.
    All cats have claws.
    ∴

Summary

Section	Important Terms	Important Ideas
3-1	statement simple statement compound statement connective conjunction disjunction conditional biconditional negation	**Formal** logic uses statements. A statement is a sentence that can be determined to be true or false but not both at the same time. A simple statement contains only one idea. A compound statement is formed by joining two simple statements with connectives. The four basic connectives are the conjunction ($\wedge$), the disjunction ($\vee$), the conditional ($\rightarrow$), and the biconditional ($\leftrightarrow$). The symbol for negation is ($\sim$). Statements can be written using logical symbols and the letters of the alphabet.
3-2	truth table	**A** truth table can be used to determine when a compound statement is true or false. A truth table can be constructed for any logical statement.
3-3	tautology self-contradiction logically equivalent statements converse inverse contrapositive	**A** statement that is always true is called a tautology. A statement that is always false is called a self-contradiction. Two statements that have the same truth values are said to be logically equivalent. From the conditional statement, three other statements can be made. They are the converse, the inverse, and the contrapositive.
3-4	argument	**Truth** tables can be used to determine the validity of an argument. An argument consists of two or more statements called premises and a statement called the conclusion. An argument is valid if when the premises are true, the conclusion is true. Otherwise, the argument is invalid.
3-5	Euler circles universal affirmative universal negative particular affirmative particular negative	**A** mathematician named Leonhard Euler developed a method using circles to determine the validity of an argument. This method uses four types of statements. They are (1) the universal affirmative, (2) the universal negative, (3) the particular affirmative, and (4) the particular negative.

Review Exercises

For Exercises 1–10, let p = "It is cool." Let q = "It is cloudy." Write each in symbols.

1. It is cool and cloudy.

2. If it is cloudy, then it is cool.

3. It is cloudy if and only if it is cool.

4. It is cloudy or not cool.

5. If it is not cool, then it is not cloudy.

6. It is not true that it is cloudy and cool.

7. It is not true that if it is cool, then it is cloudy.

8. It is not cloudy if and only if it is not cool.

9. It is not true that it is not cloudy.

10. It is neither cool nor cloudy.

For Exercises 11–15, let p = "It is cool." Let q = "It is cloudy." Write each statement in words.

11. $p \vee \sim q$

12. $q \rightarrow p$

13. $p \leftrightarrow q$

14. $(p \vee q) \rightarrow p$

15. $\sim(\sim p \vee q)$

For Exercises 16–20, construct a truth table for each statement.

16. $p \leftrightarrow \sim q$ 17. $(p \rightarrow \sim q) \vee r$

18. $(p \vee \sim q) \wedge r$ 19. $\sim p \rightarrow (\sim q \vee p)$

20. $r \rightarrow (\sim p \vee q)$

For Exercises 21–25, determine which statement is a tautology, self-contradiction, or neither.

21. $p \rightarrow (p \vee q)$ 22. $(p \rightarrow q) \rightarrow (p \vee q)$

23. $(p \wedge \sim q) \leftrightarrow (q \wedge \sim p)$ 24. $q \rightarrow (p \vee \sim p)$

25. $(\sim q \vee p) \wedge q$

For Exercises 26–28, determine whether the two statements are logically equivalent.

26. $\sim(p \rightarrow q); \sim p \wedge \sim q$ 27. $\sim p \vee \sim q; \sim(p \leftrightarrow q)$

28. $(\sim p \wedge q) \vee r; (\sim p \vee r) \wedge (q \vee r)$

For Exercises 29–32, use truth tables to determine whether each argument is valid or invalid.

29. $\quad p \rightarrow \sim q$ 30. $\sim q \vee p$
$\quad \underline{\sim q \leftrightarrow \sim p}$ $\quad \underline{\quad p \wedge q \quad}$
$\quad \therefore p$ $\quad \therefore \sim q \leftrightarrow p$

31. $\sim p \vee q$ 32. $\sim r \rightarrow \sim p$
$\quad \underline{\quad q \vee \sim r \quad}$ $\quad \underline{\quad \sim q \vee \sim r \quad}$
$\quad \therefore q \rightarrow (\sim p \wedge \sim r)$ $\quad \therefore p \leftrightarrow q$

For Exercises 33–35, write the converse, inverse, and contrapositive of each.

33. If it is a cat, then it says, "Meow."

34. If the cover is good, then the book is good.

35. If the car is red, then it is not very fast.

For Exercises 36 and 37, use Euler circles to determine whether the argument is valid or invalid.

36. No A is B.   37. Some A is not C.
$\quad \underline{\text{Some } B \text{ is } C.}$ $\quad \underline{\text{Some } B \text{ is not } C.}$
$\quad \therefore \text{No } A \text{ is } C.$ $\quad \therefore \text{Some } A \text{ is not } B.$

Chapter Test

For Exercises 1–6, let $p =$ "It is warm." Let $q =$ "It is sunny." Write each statement in symbols.

1. It is warm and sunny.

2. If it is sunny, then it is warm.

3. It is warm if and only if it is sunny.

4. It is warm or sunny.

5. It is false that it is not warm and sunny.

6. It is not sunny, and it is not warm.

For Exercises 7–11, let $p =$ "It is sunny." Let $q =$ "It is warm." Write each in words.

7. $p \vee \sim q$ 8. $q \rightarrow p$

9. $p \leftrightarrow q$ 10. $(p \vee q) \rightarrow p$

11. $\sim(\sim p \vee q)$

For Exercises 12–16, construct a truth table for each statement.

12. $p \rightarrow \sim q$ 13. $(p \rightarrow \sim q) \wedge r$

14. $(p \wedge \sim q) \vee \sim r$ 15. $(\sim q \vee p) \wedge p$

16. $p \rightarrow (\sim q \vee r)$

For Exercises 17–21, determine whether each statement is a tautology, self-contradiction, or neither.

17. $(p \vee q) \wedge p$ 18. $(p \vee q) \rightarrow (p \rightarrow q)$

19. $(p \vee \sim q) \leftrightarrow (p \rightarrow \sim q)$ 20. $q \wedge (p \vee \sim p)$

21. $(p \wedge q) \wedge p$

For Exercises 22–25, use truth tables to determine the validity of each argument.

22. $\quad p \leftrightarrow \sim q$ 23. $\quad p \rightarrow q$
$\quad \underline{\sim q \rightarrow \sim p}$ $\quad \underline{\sim q \vee \sim r}$
$\quad \therefore p$ $\quad \therefore q \leftrightarrow (\sim p \wedge \sim r)$

24. $\sim q \vee p$ 25. $\sim p \rightarrow \sim r$
$\quad \underline{\quad p \vee q \quad}$ $\quad \underline{\sim r \vee \sim q}$
$\quad \therefore \sim q \rightarrow p$ $\quad \therefore q \leftrightarrow p$

For Exercises 26–27, determine if the two statements are logically equivalent.

26. $p; \sim(\sim p)$

27. $(p \vee q) \wedge r; (p \wedge r) \vee (q \wedge r)$

28. Write the converse, inverse, and contrapositive for the statement, "If I exercise regularly, then I will be healthy."

For Exercises 29 and 30, use Euler circles to determine whether the argument is valid or invalid.

29. No B is A.   30. Some C is not A.
$\quad \underline{\text{Some } A \text{ is } C.}$ $\quad \underline{\text{Some } B \text{ is not } A.}$
$\quad \therefore \text{No } B \text{ is } C.$ $\quad \therefore \text{Some } C \text{ is not } B.$

Mathematics in Our World
▶Revisited

"Alice in Logicland"

Actually, Dodgson answers the question in the next several paragraphs of the story.

"Not the same thing a bit!" said the Hatter. "Why, you might just as well say that 'I see what I eat' is the same thing as 'I eat what I see.'"

"You might as well say," added the March Hare, "that 'I like what I get' is the same thing as 'I get what I like.'"

As shown in this chapter, a statement and its converse are not necessarily equivalent.

Projects

1. Truth tables are related to Euler circles. Arguments in the form of Euler circles can be translated into statements using the basic connectives and the negation as follows:

    Let p be "The object belongs to set A." Let q be "the object belongs to set B."

    All A is B is equivalent to $p \rightarrow q$.

    No A is B is equivalent to $p \rightarrow \sim q$.

    Some A is B is equivalent to $p \wedge q$.

    Some A is not B is equivalent to $p \wedge \sim q$.

    Determine the validity of the next arguments by using Euler circles, then translate the statements into logical statements using the basic connectives, and using truth tables, determine the validity of the arguments. Compare your answers.

    (a) No A is B.
      Some C is A.
      ∴ Some C is not B.

    (b) All B is A.
      All C is A.
      ∴ All C is B.

2. The logical statements that contain the basic connectives and the negation are related to the statements containing union, intersection, and complements as follows:

    Let p be "The object belongs to set P." Let q be "The object belongs to set Q."

    $p \wedge q$ is equivalent to $P \cap Q$.

    $p \vee q$ is equivalent to $P \cup Q$.

    $p \rightarrow q$ is equivalent to $\overline{P} \cup Q$.

    $\sim p$ is equivalent to $\overline{P}$.

    In an argument, the premises and conclusion can be translated into equivalent set statements, and Venn diagrams can be drawn for the statements. The argument is valid if the Venn diagram for the conclusion is a subset of the intersection of the premises. Using Venn diagrams, determine whether or not these arguments are valid.

    (a) $p \rightarrow q$
      p
      ∴ q

    (b) $p \rightarrow q$
      $\sim p$
      ∴ q

3. Electrical circuits are designed using truth tables. A circuit consists of switches. Two switches wired in *series* can be represented as $p \wedge q$. Two switches wired in *parallel* can be represented as $p \vee q$.

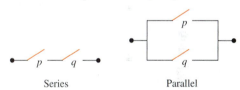

Series   Parallel

    In a series, circuit electricity will only flow when both switches, p and q, are closed. In a parallel circuit, electricity will flow when one or the other or both switches are closed. In a truth table, T represents a closed switch and F represents an open switch. Hence the truth table for $p \wedge q$ shows electricity flowing only when both switches are closed.

Truth table			**Circuit**		
p	q	$p \wedge q$	p	q	$p \wedge q$
T	T	T	closed	closed	current
T	F	F	closed	open	no current
F	T	F	open	closed	no current
F	F	F	open	open	no current

    Also, when switch p is closed, switch $\sim p$ will be open and vice versa, and p and $\sim p$ are different switches. Using this knowledge, design a circuit for a hall light that has switches at both ends of the hall such that the light can be turned on or off from either switch.

Answers to **Try This One**

3-A. b; c; d

3-B. (a) Compound
(b) Simple
(c) Compound
(d) Compound
(e) Simple

3-C. (a) $p \wedge q$
(b) $\sim p$
(c) $q \rightarrow p$
(d) $(\sim p) \wedge q$

3-D. (a) I will not study.
(b) If math is easy, then I will not study.
(c) I will study, and math is easy.
(d) Math is not easy, or I will study.
(e) Math is easy if and only if I will study.

3-E.

p	q	$p \leftrightarrow \sim q$		
T	T	T	F	F T
T	F	T	T	T F
F	T	F	T	F T
F	F	F	F	T F

①④③②

3-F.

p	q	r	$(p \wedge q) \vee \sim r$				
T	T	T	T	T	T	T F	T
T	T	F	T	T	T	T T	F
T	F	T	T	F	F	F F	T
T	F	F	T	F	F	T T	F
F	T	T	F	F	T	F F	T
F	T	F	F	F	T	T T	F
F	F	T	F	F	F	F F	T
F	F	F	F	F	F	T T	F

①④②⑥⑤③

3-G. (a) Disjunction
(b) Conditional
(c) Biconditional
(d) Conjunction
(e) Biconditional

3-H. (a) Neither
(b) Self-contradiction
(c) Neither

3-I. a and c

3-J.

p	q	$p \rightarrow q$	$p \wedge \sim q$		
T	T	T	T F	F	T
T	F	F	T T	T	F
F	T	T	F F	F	T
F	F	T	F F	T	F

① ②④③②

Columns 1 and 4 have opposite values.

3-K. Converse: "If you are intelligent, then you are a mathematician."

Inverse: "If you are not a mathematician, then you are not intelligent."

Contrapositive: "If you are not intelligent, then you are not a mathematician."

3-L. (a) $p \rightarrow q$
(b) $q \rightarrow p$
(c) $q \rightarrow p$
(d) $p \rightarrow q$

3-M. Let p = "The whole number is an odd number."
Let q = "The whole number is divisible by 2."

$$\sim p \rightarrow q$$
$$\underline{\sim q}$$
$$\therefore p$$

p	q	$[(\sim p \rightarrow q) \wedge \sim q]$							$\Rightarrow$	p
T	T	F	T	T	T	F F	T		T	T
T	F	F	T	T	F	T T	F		T	T
F	T	T	F	T	T	F F	T		T	F
F	F	T	F	F	F	F T	F		T	F

③①④②⑤③② ⑥ ①

Since column 6 has all T's, the argument is valid.

3-N. Invalid

3-O. (a) Invalid
(b) Valid

Chapter Four

Numeration Systems

Objectives

After completing this chapter, you should be able to

1 Define a numeration system (4-1)

2 Convert Egyptian numerals to Hindu-Arabic numerals and convert Hindu-Arabic numerals to Egyptian numerals (4-1)

3 Perform addition and subtraction using Egyptian numerals (4-1)

4 Convert Babylonian numerals to Hindu-Arabic numerals and Hindu-Arabic numerals to Babylonian numerals (4-1)

5 Convert Hindu-Arabic numerals to Roman numerals and Roman numerals to Hindu-Arabic numerals (4-1)

6 Determine the place value of a digit in a Hindu-Arabic numeral (4-1)

7 Write a Hindu-Arabic numeral in expanded notation (4-1)

8 Convert a base ten number to a base five number and vice versa (4-2)

9 Perform addition and subtraction using base five numbers (4-2)

10 Convert base ten numbers to numbers written in other bases, such as base two and base three, and vice versa (4-3)

Introduction

The concept of numbers originated long before any written record of humankind's history. People first used numbers to keep track of their possessions. Generations went by before the spoken number evolved into a written one. The first written notation system involved drawing pictures of objects. If a person wanted to show three arrows, he would draw pictures on the wall of his dwelling. These pictures represented the first step toward the abstraction of numbers.

With the increase in social living, personal property, and domesticated animals, people needed a better way to keep track of their possessions. Drawing pictures of their possessions became too cumbersome, so they developed a system of notation to represent the number of objects they owned by putting notches on a stick, pebbles in a bag, or dots on a clay tablet.

Next came a name or spoken word for each number. As the need for larger and larger numbers became evident, people developed various combinations of the symbols to write larger numbers.

This chapter begins by showing some of the numeration systems used by early civilizations, starting with the Egyptian system and concluding with the present system, called the Hindu-Arabic system.

The Hindu-Arabic numeration system was originated by the Hindus in India about 200–300 B.C.E. This system was brought to Spain by Arab traders and eventually replaced the Roman numeral system by 1500 C.E.

Our present system uses the base ten; however, numbers can be written in other bases, as shown in Sections 4-2 and 4-3. ∎

4-1 Early and Modern Numeration Systems

A **number** is an abstract idea or concept that is used to represent a quantity, whereas a **numeral** is a symbol that represents a number. For example, there is only one concept of the number "five"; however, there are many different ways or symbols that can be used to represent five. Some of these are 5, V, *cinq*, 〢〢, etc.

Since there is an infinitude of numbers, people have devised various ways to write numbers using symbols, thus giving rise to *numeration systems*.

> A **numeration system** consists of a set of symbols and various rules for combining the symbols to represent numbers.

Mathematics in Our World

Bar Codes

Shown here is a typical bar code. Almost all products sold today carry a bar code. These codes are also used to keep track of items in an organization for inventory purposes. They are used to lock and unlock entrances and exits to a building. They are used by supermarkets, department stores, and other businesses to identify an item and its cost. Just how does a bar code work? For the answer, see page 156. You may be surprised to learn that bar codes use the mathematics presented in this chapter.

780697 069795

Symbol	Number	Description
|	1	Vertical staff
∩	10	Heel bone
⑨	100	Scroll
⌇	1000	Lotus flower
⌀	10,000	Pointing finger
⋈	100,000	Burbot fish (or tadpole)
⚇	1,000,000	Astonished person

Figure 4-1

Some of the earliest numeration systems are the Egyptian, the Babylonian, and the Roman numeration systems.

The Egyptian Numeration System

One of the earliest numeration systems was developed by the Egyptians before 3000 B.C.E. It was a system of hieroglyphics using pictures to represent numbers. The symbols are shown in Figure 4-1.

The value of Egyptian hieroglyphic numerals can be found by multiplying each symbol by its corresponding numerical value and adding the answers, as shown in Example 4-1.

Example 4-1

Find the numerical value of each Egyptian numeral.

(a) ∩∩∩∩|||

(b) ⋈⋈⋈⌀⌀⌀⑨⑨∩∩∩||||||

(c) ⚇⌀⌀⌇⌇⑨∩|||

Solution

(a) The number has 4 heel bones or 4 tens and 3 vertical staffs or 3 ones; hence, it equals $4 \times 10 + 3 \times 1$ or 43.

(b) The number consists of 3 one hundred thousands, 3 ten thousands, 2 hundreds, 3 tens, and 6 ones; hence, it equals $3 \times 100,000 + 3 \times 10,000 + 2 \times 100 + 3 \times 10 + 6 \times 1 = 300,000 + 30,000 + 200 + 30 + 6 = 330,236$.

(c) The number consists of 1 million, 2 ten thousands, 2 thousands, 1 hundred, 1 ten, and 3 ones; hence, the number is
$1,000,000 + 20,000 + 2000 + 100 + 10 + 3 = 1,022,113$.

Try This One

4-A Find the numerical value of each Egyptian numeral.

(a) ᗆᗆᗆᗆᗆ∩∩∩∩∩∩|||||||

(b) 𓆎ᗆᗆ∩|

(c) 𓀀⌒𓏏𓏏𓆎∩∩∩∩∩|

For the answer, see page 157.

Sidelight

Mayan Mathematics

The Mayans lived in Southeastern Mexico and the Central American countries of Guatemala, Honduras, and El Salvador. Their civilization lasted from 2000 B.C.E. until 1700 C.E., although some archaeological finds date it to earlier times. They built many temples, most on the Yucatan Peninsula, similar to the pyramids of Egypt. They had a highly developed civilization that contained the elements of religion, trade, government, mathematics, and astronomy. They developed an advanced form of writing consisting of symbols similar to Egyptian hieroglyphics. They made paper from fig tree bark and wrote books that contained astronomical tables and religious ceremonies. They had two calendars. One consisted of 260 days and was used for religious purposes. The other consisted of 365 days and was based on the orbit of the earth about the sun. This calendar divided the year into 18 months with 20 days in each month and 5 days were added at the end of the year.

Their mathematical symbols consisted of three symbols. A dot represented a one. A horizontal line represented a five, and the symbol ⌓ was used for a zero. (This is one of the first cultures to have a specific symbol for zero.)

It is interesting to examine how these symbols were used to create numbers. For example, the number 3 was written as •••. The number 12 was written ‗••. The numbers were written vertically, except for the ones. Using the two symbols • and —, the Mayans could write all the numbers up through 19. However, once the number 20 was reached, a space would be needed between the digits. For example, the number

represents 14 ones plus 7 twenties or $14 \times 1 + 7 \times 20 = 154$.

The next place value after 20 is 360, which is 18×20. This value was used because the calendar had 360 + 5 days. The numeral

‗‗••• Three hundred sixties
‗‗ Twenties
•••• Ones

represents $9 \times 1 + 10 \times 20 + 13 \times 360 = 4889$.

The next higher place value is 18×20^2 or 7200. The symbol for zero was used to indicate that there were no digits in a place value position of a number. For example, the number

means $11 \times 1 + 0 \times 20 + 17 \times 360 = 6131$.

In order to write numbers using hieroglyphic symbols, simple *groupings* of ones, tens, hundreds, etc. are used. For example, 28 is equal to $10 + 10 + 8$ and is equal to

Thus, this system is called a **grouping system.**

Example 4-2

Write each number in the Egyptian notation.

(a) 42
(b) 137
(c) 5283
(d) 3,200,419

Solution

(a) Since 42 consists of 4 tens and 2 ones, it is written as

(b) Since 137 consists of 1 hundred, 3 tens, and 7 ones, it is written as

∽∩∩∩||||||

(c) Since 5283 consists of 5 thousands, 2 hundreds, 8 tens, and 3 ones, it is written as

(d) Since 3,200,419 consists of 3 millions, 2 hundred thousands, 4 one hundreds, 1 ten, and 9 ones, it is written as

Try This One

4-B Write each number using the Egyptian notation.

(a) 43

(b) 627

(c) 3286

For the answer, see page 157.

Addition and subtraction can be performed by grouping, as shown in Examples 4-3 and 4-4.

Example 4-3

Find the sum of

Solution

The answer is found by taking the total number of each symbol and converting the appropriate symbols. The total number of symbols is

For each 10 heel bones, replace them with a scroll and for each 10 vertical staffs, replace them with a heel bone. The final answer is

Example 4-4

Subtract

Solution

In this case, it is necessary to borrow since there are more vertical staffs and heel bones in the number being subtracted (i.e., the bottom number). In the top number, one heel bone (10) must be converted to 10 vertical staffs (1), and one scroll (100) must be converted to 10 heel bones (10). Once this is done, the symbols can be subtracted, and the answer is as shown.

Try This One

4-C Perform each operation.

(a)

(b)

For the answer, see page 157.

The Babylonian Numeration System

The Babylonians had a numerical system consisting of two symbols. They are ◀ and ▮. The ◀ represents the number of 10s, and ▮ represents the number of 1s.

Example   4-5

What number does ◀◀◀▮▮▮▮▮▮ represent?

Solution

Since there are 3 tens and 6 ones, the number represents 36.

You might think it would be cumbersome to write large numbers in this system; however, the Babylonians used a *positional* system. Numbers from 1 to 59 were written using the two symbols shown in Example 4-5, but after the number 60, a space is left between the groups of numbers. For example, the number 2538 was written as

◀◀◀◀▮▮   ◀▮▮▮▮▮▮▮▮

and means that there are 42 sixties and 18 ones. The space separates the 60s from the ones. The number is found as follows:

$$
\begin{aligned}
42 \times 60 &= 2520 \\
+\,18 \times 1 &= 18 \\
\hline
&2538
\end{aligned}
$$

A cuneiform tablet

Example 4-6

What number does ⟨⟨⟨⟨⟨⟨ ❙❙   ⟨⟨⟨ ❙❙❙❙ represent?

Solution

There are 52 sixties and 34 ones; hence,

$$52 \times 60 = 3120$$
$$\underline{+\ 34 \times\ \ 1 =\ \ \ \ 34}$$
$$3154$$

The symbols to the left of the first space represent the number of 3600s (i.e., 60×60); hence, the number ⟨❙❙   ⟨⟨⟨⟨⟨⟨❙   ⟨⟨❙❙❙ means that there are twelve 3600s, fifty-one 60s, and twenty-three 1s. It represents

$$12 \times 3600 = 43{,}200$$
$$51 \times\ \ \ \ 60 =\ \ \ 3060$$
$$\underline{23 \times\ \ \ \ \ 1 =\ \ \ \ \ \ 23}$$
$$46{,}283$$

Example 4-7

Write 5217 using the Babylonian system of notation.

Solution

Since the number is greater than 3600, it must be divided by 3600 to see how many of these are contained in the number.

$$5217 \div 3600 = 1 \text{ remainder } 1617$$

The remainder, 1617 is then divided by 60.

$$1617 \div 60 = 26 \text{ remainder } 57$$

Hence, the number 5217 consists of

$$1 \times 3600 = 3600 \quad ❙$$
$$26 \times\ \ \ \ 60 = 1560 \quad ⟨⟨❙❙❙❙❙❙$$
$$\underline{57 \times\ \ \ \ \ 1 =\ \ \ 57} \quad ⟨⟨⟨⟨⟨❙❙❙❙❙❙❙$$
$$\text{Total} = 5217$$

It can be written as

❙   ⟨⟨❙❙❙❙❙❙   ⟨⟨⟨⟨⟨❙❙❙❙❙❙❙

> **Math Note**
>
> The Babylonians had no symbol for zero. This complicated their writings. For example, how is the number 7200 distinguished from the number 72?

Try This One

4-D   Write the number represented.

(a) ⟨⟨⟨⟨⟨⟨❢❢

(b) ⟨⟨❢❢   ⟨⟨⟨⟨⟨⟨❢❢❢❢❢

(c) ⟨❢   ⟨⟨❢   ⟨❢❢❢

Write each number using the Babylonian symbols.

(d)   42

(e)   384

(f)   4278

For the answer, see page 157.

The Roman Numeration System

The Romans used letters to represent their numbers. They are

Symbol	Number
1	I
5	V
10	X
50	L
100	C
500	D
1000	M

In order to save space, the Romans also used the concept of subtraction. For example, 8 is written as VIII, but 9 is written as IX, meaning that 1 is subtracted from 10 to get 9. There are three rules for writing numbers in Roman numerals:

1. When a letter is repeated in sequence, its numerical value is added. For example, XXX represents $10 + 10 + 10$ or 30.
2. When smaller value letters follow larger value letters, the numerical values of each are added. For example, LXVI represents $50 + 10 + 5 + 1$ or 66.
3. When a smaller value letter precedes a larger value letter, the smaller value is subtracted from the larger value. For example, IV represents $5 - 1$ or four, and XC represents $100 - 10$ or 90.

In addition, I can only precede V or X, X can only precede L or C, and C can only precede D or M. Then 4 is written as IV, 9 is written as IX, 40 is written as XL, 90 is written XC, 400 is written as CD, and 900 is written as CM.

Example 4-8 shows how to convert Roman numerals to Hindu-Arabic numerals.

Roman numerals are still in use today. For example, many clocks and watches contain Roman numerals. Can you think of any other place Roman numerals are used?

Example   4-8

Find the value of each Roman numeral.

(a) LXVIII
(b) XCIV

—Continued

Calculator Explorations

On some calculators, it is possible to convert a Roman numeral into its Hindu-Arabic equivalent. (The values of L, X, V, and I were stored using the **STO▸** key.)

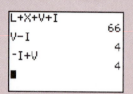

```
L+X+V+I
                66
V−I
                 4
−I+V
                 4
▪
```

Which expression, when entered on the calculator, correctly produces the Hindu-Arabic equivalent on the Roman numeral IV? What operation does the calculator perform when the expression IV is entered into the calculator?

How would you enter XL on the calculator if you wanted the calculator to correctly produce the Hindu-Arabic equivalent?

Math Note

For larger numbers, the Romans placed a bar over their symbols. The bar means to multiply the numerical value of the number under the bar by 1000. For example, $\overline{VII}$ means 7×1000 or 7000, and $\overline{XL}$ means 40,000.

Example   4-8   *Continued—*

(c) MCML
(d) CCCXLVI
(e) DCCCLV

Solution

(a) L = 50, X = 10, V = 5, and III = 3; hence, LXVIII = 68.
(b) XC = 90 and IV = 4; hence, XCIV = 94.
(c) M = 1000, CM = 900, L = 50; hence, MCML = 1950.
(d) CCC = 300, XL = 40, V = 5, and I = 1; hence, CCCXLVI = 346.
(e) D = 500, CCC = 300, L = 50, V = 5; hence, DCCCLV = 855.

Try This One

4-E   Convert each Roman numeral to a Hindu-Arabic numeral.

(a) XXXIX

(b) MCLXIV

(c) CCCXXXIII

For the answer, see page 157.

Numbers can be written using Roman numerals as shown in Example 4-9.

Example   4-9

Write each number using Roman numerals.

(a) 19
(b) 238
(c) 1999
(d) 840
(e) 72

Solution

(a) 19 is written as $10 + 9$ or XIX.
(b) 238 is written as $200 + 30 + 8$ or CCXXXVIII.
(c) 1999 is written as $1000 + 900 + 90 + 9$ or MCMXCIX.
(d) 840 is written as $500 + 300 + 40$ or DCCCXL.
(e) 72 is written as $50 + 20 + 2$ or LXXII.

Try This One

4-F   Write each number using Roman numerals.

(a) 67

(b) 192

(c) 202

(d) 960

For the answer, see page 157.

The Hindu-Arabic Numeration System

The Romans spread their system of numerals throughout the world as they conquered their enemies. The system was well entrenched in Europe until the 1500s when our present system, called the Hindu-Arabic system, became widely accepted.

The present system is thought to have been invented by the Hindus before 200 B.C.E. It was spread throughout Europe by the Arabs, who traded with the Europeans and traveled

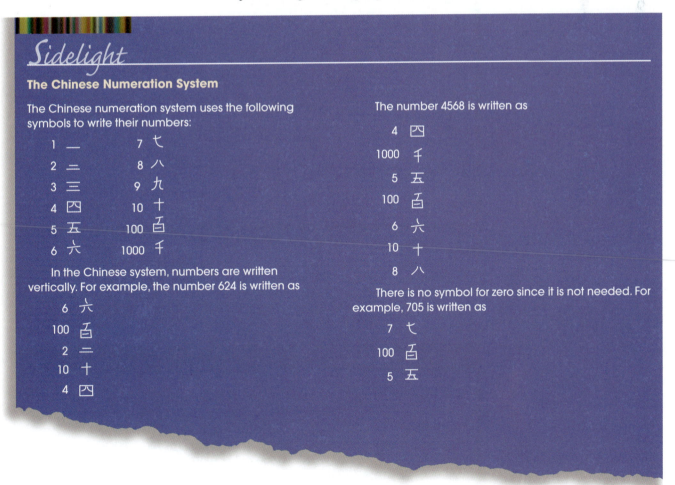

Sidelight

The Chinese Numeration System

The Chinese numeration system uses the following symbols to write their numbers:

1	一	7	七
2	二	8	八
3	三	9	九
4	四	10	十
5	五	100	百
6	六	1000	千

In the Chinese system, numbers are written vertically. For example, the number 624 is written as

6 六
100 百
2 二
10 十
4 四

The number 4568 is written as

4 四
1000 千
5 五
100 百
6 六
10 十
8 八

There is no symbol for zero since it is not needed. For example, 705 is written as

7 七
100 百
5 五

throughout the Mediterranean region. It is interesting to note that for about 400 years, the mathematicians of early Europe were divided into two groups—those favoring the use of the Roman system and those favoring the use of the Hindu-Arabic system. The Hindu-Arabic system eventually won out, although Roman numerals are still widely used today.

Our present system uses 10 symbols called **digits:** 0, 1, 2, 3, 4, 5, 6, 7, 8, 9, and powers of 10 to write larger numbers. For this reason, it is called a *base ten* system. More will be said about the number bases in Section 4-2.

The Hindu-Arabic system is a **place value** or a **positional system** since the position of each digit indicates a specific value. The place value of each number is given as

billion	hundred million	ten million	million	hundred thousand	ten thousand	thousand	hundred	ten	one
10^9	10^8	10^7	10^6	10^5	10^4	10^3	10^2	10^1	1

The number 82,653 means there are 8 ten thousands, 2 thousands, 6 hundreds, 5 tens, and 3 ones. We say that the place value of the 6 in this example is hundreds.

Example 4-10

In the number 153,946, what is the place value of each digit?

(a) 9
(b) 3
(c) 5
(d) 1
(e) 6

Solution

(a) hundreds
(b) thousands
(c) ten-thousands
(d) hundred thousands
(e) ones

Hindu-Arabic numbers can be written in **expanded notation.** For example, 32,569 can be written as $30,000 + 2000 + 500 + 60 + 9$, which can be written as $3 \times 10,000 + 2 \times 1000 + 5 \times 100 + 6 \times 10 + 9 \times 1$, which can be written as $3 \times 10^4 + 2 \times 10^3 + 5 \times 10^2 + 6 \times 10^1 + 9$.

Example 4-11

Write 9,034,761 in expanded notation.

Solution

9,034,761 can be written as $9,000,000 + 30,000 + 4000 + 700 + 60 + 1 = 9 \times 1,000,000 + 3 \times 10,000 + 4 \times 1000 + 7 \times 100 + 6 \times 10 + 1 = 9 \times 10^6 + 3 \times 10^4 + 4 \times 10^3 + 7 \times 10^2 + 6 \times 10^1 + 1$.

Try This One

4-G Write each number in expanded notation.

(a) 573

(b) 86,471

(c) 2,201,567

For the answer, see page 157.

Exercise Set 4-1

Computational Exercises

For Exercises 1–10, write each number in the Hindu-Arabic system.

1. ∩∩∩|||||

2. ๑๑๑๑∩∩∩||

3. ⌐⌐๑๑∩|||||

4. ⟋⟋๑∩∩∩||

5. ⌐⌐⌐๑∩∩∩∩∩|||

6. ⚥๑๑∩

7. ⌐⌐๑๑∩||||

8. ⚥⚥⟋⟋

9. ⚥⟋⌐𐤅𐤅∩

10. 𐤅𐤅∩∩∩∩|||||||||

For Exercises 11–20, write each number in the Egyptian system.

11. 7   12. 18

13. 37   14. 52

15. 168   16. 365

17. 801   18. 955

19. 1256   20. 8261

For Exercises 21–26, perform the indicated operations. Leave answers in the Egyptian system.

21. ∩∩∩∩∩||| + ∩∩∩∩∩∩|||

22. ๑๑∩∩|| + ๑∩∩∩||

23. ⟋⟋⌐⌐|| + ⟋⌐⌐⌐∩∩∩|||

24. ∩∩∩|| − ∩|||||

25. ⌐⌐๑๑๑||| − ⌐๑๑๑||||

26. ⚥⟋𐤅∩∩||| − ⟋𐤅𐤅∩∩∩||||

For Exercises 27–36, write each number in the Hindu-Arabic system.

27. ⟨𝑻𝑻

28. ⟨⟨⟨𝑻𝑻𝑻𝑻

29. ⟨⟨⟨⟨⟨𝑻

30. ⟨⟨⟨𝑻𝑻𝑻𝑻𝑻

31. ⟨❚ ⟨⟨⟨❚ ⟨❚   32. ⟨⟨❚❚ ⟨⟨⟨⟨ ⟨❚❚❚

33. ⟨⟨⟨ ⟨⟨❚ ⟨⟨❚❚❚❚   34. ⟨⟨⟨❚❚ ⟨❚ ⟨❚

35. ⟨❚❚❚ ⟨❚❚   36. ⟨⟨⟨❚ ⟨❚ ❚❚

For Exercises 37–46, write each number in the Babylonian system.

37. 32	38. 23	39. 78
40. 156	41. 292	42. 514
43. 1023	44. 1776	45. 5216
46. 8200		

For Exercises 47–56, write each number in the Hindu-Arabic system.

47. XVII	48. XCIX	49. XLIII
50. CCXXI	51. LXXXVI	52. CCXXXIII
53. CDXVIII	54. MMCMXVII	55. CDXC
56. CMVI		

For Exercises 57–66, write each number using Roman numerals.

57. 39	58. 142	59. 567
60. 893	61. 1258	62. 3720
63. 1462	64. 2170	65. 3000
66. 2222		

For Exercises 67–72, use the number 3,421,578 and find the place value of the given digit.

67. 5	68. 1	69. 2
70. 3	71. 8	72. 4

For Exercises 73–81, write each number in expanded notation.

73. 86	74. 325	75. 1812
76. 32,714	77. 6002	78. 29,300
79. 162,873	80. 200,321,416	81. 17,531,801
82. 1,326,419		

Real World Applications

Many movies use Roman numerals to indicate the date the film was made. Shown are some movies and their dates. Find the year the movie was made.

	Movie	Year
83.	*Gone with the Wind*	MCMXXXIX
84.	*Casablanca*	MCMXLII
85.	*Titanic*	MCMXCVII
86.	*West Side Story*	MCMLXI

Writing Exercises

87. Explain why the Egyptian numeration system is called an "additive" system.

88. Explain why the Babylonian numeration system is both an additive and a "positional" system.

89. Is the Roman numeral system an additive or a positional system, or is it a combination of both?

90. What are some advantages of the Hindu-Arabic numeration system as compared to the Egyptian, Babylonian, and Roman numeral systems?

Critical Thinking

91. Write a paper explaining how to multiply in the Egyptian system.

92. Write a paper showing how to add, subtract, and multiply in the Roman system.

93. Write a paper giving some applications of large numbers, such as billions, trillions, etc.

94. Make up a numeration system using your own symbols. Indicate whether it is a grouping or a positional system. Explain how to add and subtract in your numeration system.

4-2   Base Number Systems

Base Five System

Recall that in Section 4-1, the Hindu-Arabic system was called a **base** ten numeration system because each digit in a base ten number represents a power of 10. For example, in the number 382,571, the digit 2 represents thousands, or 10^3.

It is not necessary to use 10 digits to make a numeration system. A numeration system can be made with two or more digits. For example, a numeration system can be made with five digits, 0, 1, 2, 3, and 4. This system is called the base five numeration system. Each digit in a base five number represents a power of five. The place values for the digits in base five are

| etc. | six hundred twenty-five (5^4) | one hundred twenty-five (5^3) | twenty-five (5^2) | five (5^1) | one (5^0) |

Numbers written in base five use the subscript five to distinguish them from base ten numbers. The chart shows some base numbers written in base five.

Base ten number	Corresponding base five number
1	1_{five}
2	2_{five}
3	3_{five}
4	4_{five}
5	10_{five}
6	11_{five}
7	12_{five}
8	13_{five}
9	14_{five}
10	20_{five}
11	21_{five}
25	100_{five}
50	200_{five}
125	1000_{five}
625	10000_{five}

This might look somewhat complicated, but it can be clarified using the following reasoning. The numbers 1 through 4 are written the same way in both systems. The number 5 cannot be written in base five using the numeral 5 since the base five system has only the digits 0, 1, 2, 3, and 4. So it must be written as 10_{five}, meaning $1 \times 5 + 0 \times 1$, or one five and no ones. The base ten number 8 would be written as 13_{five} (in base five), meaning 1 five plus 3 ones, etc.

Converting Base Five Numbers to Base Ten Numbers

Base five numbers can be converted to base ten numbers using the place values of the base five numbers and expanded notation. For example, the number 242_{five} contains 2 twenty-fives, 4 fives, and 2 ones, and it can be written in expanded notation as $2 \times 25 + 4 \times 5 + 2 \times 1$. Hence, 242_{five} is equal to 72 in the base ten system. Example 4-12 shows how to convert numbers written in base five to numbers written in base ten.

Sidelight

Mathematics and Life on Other Planets

In order to ascertain if there is life on other planets or distant stars, astronomers have constructed giant antennas to listen for radio signals from outer space. One such antenna is located at the National Radio Observatory in Green Bank, West Virginia. The question is, though, how could astronomers decipher these messages? All agreed that if there were life on other planets, the messages received that would make most sense to intelligent life would be mathematical in nature. Mathematical facts and formulas could be exchanged and a basic vocabulary could be established between humans on earth and intelligent beings on other planets.

Example 4-12

Write each number in base ten.

(a) 42_{five}
(b) 134_{five}
(c) 4213_{five}

Solution

The place value chart for base five is used in each case.

(a) $42_{\text{five}} = 4 \times 5 + 2 \times 1 = 20 + 2 = 22$
(b) $134_{\text{five}} = 1 \times 5^2 + 3 \times 5 + 4 \times 1$
$$= 1 \times 25 + 3 \times 5 + 4 \times 1$$
$$= 25 + 15 + 4 = 44$$
(c) $4213_{\text{five}} = 4 \times 5^3 + 2 \times 5^2 + 1 \times 5 + 3 \times 1$
$$= 4 \times 125 + 2 \times 25 + 1 \times 5 + 3 \times 1$$
$$= 500 + 50 + 5 + 3 = 558$$

Try This One

4-H   Write each number in the base ten system.

(a) 302_{five}

(b) 1324_{five}

(c) 40000_{five}

For the answer, see page 157.

Converting Base Ten Numbers to Base Five Numbers

Base ten numbers can be written in the base five notation using the place values of the base five system and successive division. This method is shown in Examples 4-13, 4-14, and 4-15.

Example 4-13

Write 84 in the base five system.

Solution

Step 1   Identify the largest place value number (1, 5, 25, 125, etc.) that will divide into the base 10 number. In this case, it is 25.

Step 2   Divide 25 into 84, as shown.

$$
\begin{array}{r}
3 \\
25\overline{)84} \\
75 \\
\hline
9
\end{array}
$$

—Continued

Math Note

The answer can be checked using multiplication and addition:
$3 \times 25 + 1 \times 5 + 4 \times 1$
$= 75 + 5 + 4 = 84.$

Example 4-13 Continued—

Step 3 Divide the remainder by the next lower place value. In this case, it is 5.

$$5\overline{)9} \\ \underline{5} \\ 4$$

with quotient 1

Step 4 Continue dividing until the remainder is less than 5. In this case, it is 4, so the division process is stopped. In other words, four 1s are left. The answer, then, is 314_{five}. In 84, there are three 25s, one 5, and four 1s.

Example 4-14

Write 653 in the base five system.

Solution

Step 1 Since 625 is the largest place value that will divide into 653, it is used first.

$$625\overline{)653} \\ \underline{625} \\ 28$$

with quotient 1

Step 2 Divide by 125.

$$125\overline{)28} \\ \underline{0} \\ 28$$

with quotient 0

Even though 125 does not divide into the 28, the zero must be written to hold its place value in the base five number system.

Step 3 Divide by 25.

$$25\overline{)28} \\ \underline{25} \\ 3$$

with quotient 1

Step 4 Divide by 5.

$$5\overline{)3} \\ \underline{0} \\ 3$$

with quotient 0

Hence, the solution is 10103_{five}.

Check: $1 \times 625 + 0 \times 125 + 1 \times 25 + 0 \times 5 + 3 \times 1 = 653.$

Try This One

4-I   Write each number in the base five system.

(a) 52

(b) 486

(c) 1000

For the answer, see page 157.

Other Number Bases

As stated previously, a number system can be made up of two or more digits. For example, a base two or **binary system** (used in computers) uses only two digits, namely, 0 and 1, and the place values of the digits in the base two numeration system are

etc. | sixteen (2^4) | eight (2^3) | four (2^2) | two (2^1) | one (2^0)

The base eight or **octal system** consists of 8 digits, namely, 0, 1, 2, 3, 4, 5, 6, and 7, and the place values of the digits in the base eight system are

etc. | four thousand ninety-six (8^4) | five hundred twelve (8^3) | sixty-four (8^2) | eight (8^1) | one (8^0)

When the number base is greater than ten, new digits must be created to make the numbers. For example, base sixteen (called the **hexadecimal system**) is used in computer technology. The digits in the base sixteen are 0, 1, 2, 3, 4, 5, 6, 7, 8, 9, A, B, C, D, E, and F, where A represents 10, B represents 11, C represents 12, etc. The place values of the digits in the base sixteen are

etc. | four thousand ninety-six (16^3) | two hundred fifty-six (16^2) | sixteen (16^1) | one (16^0)

Table 4-1 shows the digits for some of the base number systems and the place values of the digits in the system. It should be pointed out that place values go on indefinitely for any base number system.

    Looking at the table, several things become apparent. *First,* the digits used in any base are less than the base number. *Second,* the place values of any base are

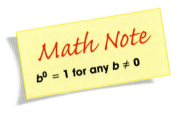

Math Note

$b^0 = 1$ for any $b \neq 0$

$$\ldots \quad \underline{b^6} \quad \underline{b^5} \quad \underline{b^4} \quad \underline{b^3} \quad \underline{b^2} \quad \underline{b^1} \quad \underline{b^0}$$

where b is the base. For example, the place values for base six are

$$\ldots \quad \underline{46656} \quad \underline{7776} \quad \underline{1296} \quad \underline{216} \quad \underline{36} \quad \underline{6} \quad \underline{1}$$

    In order to convert from numbers written in bases other than ten to base ten numbers, the expanded notation is used. This is the same procedure used in Example 4-12. Example 4-15 shows this procedure.

Table 4-1

Base Number Systems

Base two (binary system)
Digits used: 0, 1
Place values: $\underline{2^6}$ $\underline{2^5}$ $\underline{2^4}$ $\underline{2^3}$ $\underline{2^2}$ $\underline{2^1}$ $\underline{2^0}$
Numbers: 0, 1, 10 , 11, 100, 101, 110, 111, 1000, 1001, 1010, etc.

Base three
Digits used: 0, 1, 2
Place values: $\underline{3^6}$ $\underline{3^5}$ $\underline{3^4}$ $\underline{3^3}$ $\underline{3^2}$ $\underline{3^1}$ $\underline{3^0}$
Numbers: 0, 1, 2, 10, 11, 12, 20, 21, 22, 100, 101, 102, 110, etc.

Base five
Digits used: 0, 1, 2, 3, 4
Place values: $\underline{5^6}$ $\underline{5^5}$ $\underline{5^4}$ $\underline{5^3}$ $\underline{5^2}$ $\underline{5^1}$ $\underline{5^0}$
Numbers: 0, 1, 2, 3, 4, 10, 11, 12, 13, 14, 20, 21, 22, etc.

Base eight (octal system)
Digits used: 0, 1, 2, 3, 4, 5, 6, 7
Place values: $\underline{8^6}$ $\underline{8^5}$ $\underline{8^4}$ $\underline{8^3}$ $\underline{8^2}$ $\underline{8^1}$ $\underline{8^0}$
Numbers: 0, 1, 2, 3, 4, 5, 6, 7, 10, 11, 12, 13, 14, 15, 16, 17, 20, etc.

Base ten
Digits used: 0, 1, 2, 3, 4, 5, 6, 7, 8, 9
Place values: $\underline{10^6}$ $\underline{10^5}$ $\underline{10^4}$ $\underline{10^3}$ $\underline{10^2}$ $\underline{10^1}$ $\underline{10^0}$
Numbers: 0, 1, 2, 3, 4, 5, 6, 7, 8, 9, 10, 11, 12, 13, 14, 15, etc.

Base sixteen (hexadecimal system)
Digits used: 0, 1, 2, 3, 4, 5, 6, 7, 8, 9, A, B, C, D, E, F
Place values: $\underline{16^6}$ $\underline{16^5}$ $\underline{16^4}$ $\underline{16^3}$ $\underline{16^2}$ $\underline{16^1}$ $\underline{16^0}$
Numbers: 0, 1, 2, 3, 4, 5, 6, 7, 8, 9, A, B, C, D, E, F, 10, 11, etc.

Example   4-15

Write the base ten number for each.

(a) 132_{six}
(b) 10110_{two}
(c) 1532_{eight}
(d) 2102_{three}
(e) $5BD8_{sixteen}$

Solution

(a) The place values of the digits in base six are $1, 6, 36, 216, \ldots$

$$132_{six} = 1 \times 6^2 + 3 \times 6^1 + 2 \times 1$$
$$= 1 \times 36 + 3 \times 6 + 2 \times 1$$
$$= 36 + 18 + 2 = 56$$

(b) The place values of the digits in base two are $1, 2, 4, 8, 16, \ldots$

$$10110_{two} = 1 \times 2^4 + 0 \times 2^3 + 1 \times 2^2 + 1 \times 2^1 + 0 \times 1$$
$$= 1 \times 16 + 0 \times 8 + 1 \times 4 + 1 \times 2 + 0 \times 1$$
$$= 16 + 0 + 4 + 2 + 0 = 22$$

(c) The place values of the digits in base eight are $1, 8, 64, \ldots$

$$1532_{eight} = 1 \times 8^3 + 5 \times 8^2 + 3 \times 8^1 + 2 \times 1$$
$$= 1 \times 512 + 5 \times 64 + 3 \times 8 + 2 \times 1$$
$$= 512 + 320 + 24 + 2 = 858$$

(d) The place values of the digits in base three are $1, 3, 9, 27, \ldots$

$$2102_{three} = 2 \times 3^3 + 1 \times 3^2 + 0 \times 3^1 + 2 \times 1$$
$$= 2 \times 27 + 1 \times 9 + 0 \times 3 + 2 \times 1$$
$$= 54 + 9 + 0 + 2 = 65$$

(e) The place values of the digits in base sixteen are $1, 16, 256, 4096, \ldots$

$$5BD8_{sixteen} = 5 \times 16^3 + 11 \times 16^2 + 13 \times 16 + 8 \times 1$$
$$= 5 \times 4096 + 11 \times 256 + 13 \times 16 + 8 \times 1$$
$$= 20{,}480 + 2816 + 208 + 8 = 23{,}512$$

The microphone converts the sound to a voltage signal, which in turn is converted into a binary number. Each measurement is recorded as a 16-bit number.

Try This One

4-J   Write each number in base ten.

(a) 5320_{seven}

(b) 110110_{two}

(c) 32021_{four}

(d) $42AE_{sixteen}$

For the answer, see page 157.

Converting Base Ten Numbers to Other Base Numbers

Base ten numbers can be written in other bases by successive division using the place values of the given base. Examples 4-16 and 4-17 show this procedure.

Example   4-16

Write 48 in base three.

Solution

Step 1   The place values for base 3 are 1, 3, 9, 27, etc. Divide 48 by 3^3 or 27.

$$
\begin{array}{r}
1 \\
27\overline{)48} \\
27 \\
\hline
21
\end{array}
$$

Step 2   Divide the remainder by 3^2 or 9.

$$
\begin{array}{r}
2 \\
9\overline{)21} \\
18 \\
\hline
3
\end{array}
$$

Step 3   Divide the remainder by 3^1 or 3.

$$
\begin{array}{r}
1 \\
3\overline{)3} \\
3 \\
\hline
0
\end{array}
$$

Write the number in the base. It is $48 = 1210_{\text{three}}$.

Example   4-17

Write 51 in base two.

Solution

The place values for base two are 1, 2, 4, 8, 16, 32, etc. Use successive division, as shown.

$$
\begin{array}{r}
1 \\
32\overline{)51} \\
32 \\
\hline
19
\end{array}
\qquad
\begin{array}{r}
1 \\
16\overline{)19} \\
16 \\
\hline
3
\end{array}
\qquad
\begin{array}{r}
0 \\
8\overline{)3} \\
0 \\
\hline
3
\end{array}
\qquad
\begin{array}{r}
0 \\
4\overline{)3} \\
0 \\
\hline
3
\end{array}
\qquad
\begin{array}{r}
1 \\
2\overline{)3} \\
2 \\
\hline
1
\end{array}
$$

Hence, $51 = 110011_{\text{two}}$.

In Example 4-17, it was necessary to divide the remainders by all place values of base two.

Example 4-18

Write 19,443 in base sixteen.

Solution

The place values for the base sixteen are 1, 16, 256, 4096, etc. Use successive division as shown.

$$\begin{array}{r} 4 \\ 4096 \overline{)19443} \\ 16384 \\ \hline 3059 \end{array} \qquad \begin{array}{r} B \\ 256 \overline{)3059} \\ 2816 \\ \hline 243 \end{array} \qquad \begin{array}{r} F \\ 16 \overline{)243} \\ 240 \\ \hline 3 \end{array}$$

Hence, 19,433 becomes 4BF3 in base sixteen.

Try This One

4-K (a) Write 84 in base two.

(b) Write 258 in base six.

(c) Write 122 in base three.

(d) Write 874 in base sixteen.

For the answer, see page 157.

Base Numbers and Computers

Computers use three bases to perform operations. They are base two, the binary system; base eight, the octal system; and base sixteen, the hexadecimal system.

Base two is used since it contains only two characters, 0 and 1. Electric circuits can differentiate two types of pulses, on and off. Hence the early computers used one on-off vacuum tube to store one binary character. Since it is cheaper and faster to use the binary system, modern computers still use this system, and assembly language programmers must become proficient in the binary system.

The base eight system is also used by computer programmers since one bit is used for one character, and 8 bits constitute a byte. It then becomes convenient to write numbers using a series of bytes (eight characters).

The base sixteen is used for several reasons. First of all, 16 characters consist of 2 bytes. Also, 16 is 2^4, which means one hexadecimal character can replace four binary characters. This increases the speed at which the computer is able to perform numerical applications since there are fewer characters for the computer to read and fewer operations to perform. Finally, large numbers can be written in base sixteen with fewer characters than in base two or base ten, thus saving much needed space in the computer's memory.

Sidelight

Grace Murray Hopper (1906–1992)

Computer Genius

Grace Murray Hopper received a doctorate in mathematics from Yale and taught at Vassar College before joining the U.S. Navy in 1943. She was commissioned a lieutenant, junior grade, and was assigned to the Harvard Computer Laboratory in 1944. She worked with Dr. Howard Aiken on the first modern computer called the MARK I.

During her time at Harvard, she developed a computer known as "FLOWMATIC," which enabled programmers to write programs in simple languages such as COBOL and have them translated into the complicated language that the computer uses.

During her lifetime, she received honorary degrees from more than 40 colleges and universities in the United States. She retired from the U.S. Navy in 1969 at the age of 60 only to return to active duty a year later. She stayed on until the mid-1980s, when she retired for the last time as a Rear Admiral. Her retirement ceremony was held aboard the *USS Constitution*, and she was recognized as "the greatest living female authority in the computer field" at that time.

Exercise Set 4-2

Computational Exercises

For Exercises 1–20, convert each number to base ten.

1. 1011_{two}
2. 372_{twelve}
3. 53_{six}
4. $8A21_{twelve}$
5. 99_{eleven}
6. 3451_{seven}
7. 10221_{three}
8. 110011_{two}
9. 2221_{five}
10. 5320_{eight}
11. 153_{six}
12. 11001_{two}
13. 438_{nine}
14. 2561_{eight}
15. 352_{seven}
16. 11112_{five}
17. $921E_{sixteen}$
18. 2101_{three}
19. 812_{twelve}
20. 10000001_{two}

For Exercises 21–40, write each base ten number in the given base.

21. 31 in base two
22. 186 in base eight
23. 345 in base six
24. 266 in base three
25. 16 in base seven
26. 3050 in base twelve
27. 745 in base nine
28. 3217 in base four
29. 22 in base two
30. 5621 in base eleven
31. 18 in base five
32. 97 in base four
33. 2361 in base sixteen
34. 96 in base two
35. 18,432 in base five
36. 25,000 in base sixteen
37. 32 in base seven
38. 88 in base three
39. 256 in base two
40. 497 in base four

For Exercises 41–50, convert each number to base ten then change to the specified base.

41. 134_{six} to base two

42. 1011010_{two} to base eight

43. 342_{five} to base twelve

44. 4711_{nine} to base three

45. 1221_{three} to base four

46. 1521_{seven} to base three

47. 432_{eight} to base twelve

48. $3AB_{twelve}$ to base six

49. 1782_{nine} to base seven

50. 3000_{four} to base eleven

 Real World Applications

Problems similar to base number problems occur often in everyday life. Using a procedure like the one used in Examples 4-16, 4-17, and 4-18, solve each.

51. Change 87 ounces to pounds and ounces (1 lb = 16 oz).

52. Change 237 ounces to quarts, pints, and ounces (1 qt = 2 pt = 32 oz).

53. Change 1256 inches to yards, feet, and inches (1 yd = 3 ft = 36 in.).

54. Change $5.88 to quarters, dimes, nickels, and pennies using the smallest number of coins.

A process of encoding data using two symbols is called binary coding. One of the earliest codes was the Morse code. Words were encoded using dots and dashes and transmitted by sound. If we consider a dot as a zero and a dash as a one, the various letters of the alphabet can be transformed into a binary code, as follows.

A 01	J 0111	S 000
B 1000	K 101	T 1
C 1010	L 0100	U 001
D 100	M 11	V 0001
E 0	N 10	W 011
F 0010	O 111	X 1001
G 110	P 0110	Y 1011
H 0000	Q 1101	Z 1100
I 00	R 010	

For Exercises 55–60, decode the following messages.

55. 000 1 111 0110

56. 10 111 0000 111 11 0 011 111 010 101

57. 1010 0100 01 000 000 00 000 111 0001 0 010

58. 010 0 000 1 0010 111 010 0100 001 10 1010 0000

59. 00 1 00 000 010 01 00 10 00 10 110

60. 1 001 010 10 01 010 111 001 10 100

The U.S. Postal Service uses a Postnet code *on business reply forms. This code consists of a five-digit Zip code plus a four-digit extension and a check digit. This code consists of long and short vertical bars.*

Digit	Bar code	Digit	Bar code
1	ıı-ıllll	6	ıllıı
2	ıllıl	7	lııll
3	ıllll	8	lıllı
4	ıllıı	9	lıllıı
5	ıllll	0	llııı

For example, the Zip code 15131 would be encoded as follows:

ıııll ılılı ıııll ıllıı ıııll
1   5   1   3   1

Using the Postnet code, find the following five-digit Zip codes for Exercises 61–66.

61. ııılllıllıııllılılılılılllıı

62. lılıııııllllılılılılılılıllıl

63. ıllılıııllıııllııllılılılı

64. llııııllıllııllılılıllıl

65. ıllılılılılılılılılllıııılll

66. ııllılllıllılılılılllıııllılı

For Exercises 67–72, write each Zip code using the Postnet code.

67. 26135   68. 14157   69. 18423

70. 30214   71. 11672   72. 54901

Writing Exercises

73. Explain how the place values of the digits in base number systems are obtained.

74. Explain how to convert numbers in bases other than base ten to base ten numbers.

75. Explain how to convert base ten numbers into numbers in other bases.

Critical Thinking

For Exercises 76–80, use the information found in "Mathematics in Our World" found on page 156. Identify the digit shown by the bar code. State whether the digit belongs to the manufacturer's number or the product number.

76. 0111101   77. 0101111   78. 1100110

79. 1110010   80. 0111011

4-3   Operations in Base Numbers

Arithmetic operations (addition, subtraction, multiplication, and division) can be performed in other base number systems just as they are performed in base ten. For example,

$4_{\text{five}} + 3_{\text{five}} = 12_{\text{five}}$. Why? Four plus three is equal to seven in base ten, but there is no seven in base five. Seven in the base five system is written as 12_{five}. That is, 1 five and 2 ones.

An addition table can be constructed for base five, as shown. (Remember, the digits 5, 6, 7, 8, and 9 are not used in base five.) The subscript "five" has been omitted.

+	0	1	2	3	4
0	0	1	2	3	4
1	1	2	3	4	10
2	2	3	4	10	11
3	3	4	10	11	12
4	4	10	11	12	13

In order to add using the table, find one number in the left column and the other number in the top row. Then draw a horizontal and vertical line. The intersection of the lines is the sum. For example, $2_{\text{five}} + 4_{\text{five}} = 11_{\text{five}}$, as shown.

+	0	1	2	3	④
0	0	1	2	3	4
1	1	2	3	4	10
②	2	3	4	10	⑪
3	3	4	10	11	12
4	4	10	11	12	13

Addition in base five is performed the same way as it is in base ten, using the addition table and the principal of "carrying." Example 4-19 shows the procedure.

Example 4-19

Perform the indicated operation: $324_{\text{five}} + 24_{\text{five}}$.

Solution

Step 1 Add the 4 and the 4. $4_{\text{five}} + 4_{\text{five}} = 13_{\text{five}}$ (see the table). Write the 3 and carry the 1, as shown.

$$
\begin{array}{r}
1 \\
324_{\text{five}} \\
+\ 24_{\text{five}} \\
\hline
3_{\text{five}}
\end{array}
$$

Step 2 Add $1_{\text{five}} + 2_{\text{five}} + 2_{\text{five}} = 10_{\text{five}}$ (see the table). Write the 0 and carry the 1, as shown.

$$
\begin{array}{r}
1 \\
324_{\text{five}} \\
+\ 24_{\text{five}} \\
\hline
03_{\text{five}}
\end{array}
$$

Step 3 Add $1_{\text{five}} + 3_{\text{five}} = 4_{\text{five}}$ (see the table). Write the 4, as shown.

$$
\begin{array}{r}
324_{\text{five}} \\
+\ 24_{\text{five}} \\
\hline
403_{\text{five}}
\end{array}
$$

Hence, $324_{\text{five}} + 24_{\text{five}} = 403_{\text{five}}$

Math Note

The answer can be checked by converting the numbers to base ten and seeing if the answers are equal:

$$
\begin{array}{r}
324_{\text{five}} = 89_{\text{ten}} \\
+\ 24_{\text{five}} = 14_{\text{ten}} \\
\hline
403_{\text{five}} = 103_{\text{ten}}
\end{array}
$$

Example 4-20

Perform the indicated operation: $1244_{\text{five}} + 333_{\text{five}}$.

Solution

$$
\begin{array}{r}
111 \\
1244_{\text{five}} \\
+\ \ 333_{\text{five}} \\
\hline
2132_{\text{five}}
\end{array}
$$
carry row

— $4_{\text{five}} + 3_{\text{five}} = 12_{\text{five}}$ Write the 2 and carry the 1.

— $1_{\text{five}} + 4_{\text{five}} + 3_{\text{five}} = 13_{\text{five}}$ Write the 3 and carry the 1.

— $1_{\text{five}} + 2_{\text{five}} + 3_{\text{five}} = 11_{\text{five}}$ Write the 1 and carry the 1.

— $1_{\text{five}} + 1_{\text{five}} = 2_{\text{five}}$

Try This One

4-L   Perform the indicated operations.

(a) 321_{five}
    $+ 103_{\text{five}}$

(b) 4301_{five}
    $+ 2024_{\text{five}}$

For the answer, see page 157.

Addition can be performed in other bases as well as in base five. You may need to construct an addition table for the given base to help you find the answers.

Example 4-21

Perform the indicated operation: $10111_{\text{two}} + 110_{\text{two}}$.

Solution

The addition table for base two is

$$
\begin{array}{c|cc}
+ & 0 & 1 \\
\hline
0 & 0 & 1 \\
1 & 1 & 10
\end{array}
$$

Then

$$
\begin{array}{r}
11 \\
10111_{\text{two}} \\
+\ \ \ 110_{\text{two}} \\
\hline
11101_{\text{two}}
\end{array}
$$
Carry row

— $1_{\text{two}} + 0_{\text{two}} = 1_{\text{two}}$

— $1_{\text{two}} + 1_{\text{two}} = 10_{\text{two}}$ Write the 0; carry 1.

— $1_{\text{two}} + 1_{\text{two}} + 1_{\text{two}} = 11_{\text{two}}$ Write the 1 and carry 1.

— $1_{\text{two}} + 0_{\text{two}} = 1_{\text{two}}$

— bring down the 1

The addition table for the hexadecimal system is shown next. Recall that A in base $16 = 10$ in base ten, $B = 11$, $C = 12$, $D = 13$, $E = 14$, and $F = 15$.

+	0	1	2	3	4	5	6	7	8	9	A	B	C	D	E	F
0	0	1	2	3	4	5	6	7	8	9	A	B	C	D	E	F
1	1	2	3	4	5	6	7	8	9	A	B	C	D	E	F	10
2	2	3	4	5	6	7	8	9	A	B	C	D	E	F	10	11
3	3	4	5	6	7	8	9	A	B	C	D	E	F	10	11	12
4	4	5	6	7	8	9	A	B	C	D	E	F	10	11	12	13
5	5	6	7	8	9	A	B	C	D	E	F	10	11	12	13	14
6	6	7	8	9	A	B	C	D	E	F	10	11	12	13	14	15
7	7	8	9	A	B	C	D	E	F	10	11	12	13	14	15	16
8	8	9	A	B	C	D	E	F	10	11	12	13	14	15	16	17
9	9	A	B	C	D	E	F	10	11	12	13	14	15	16	17	18
A	A	B	C	D	E	F	10	11	12	13	14	15	16	17	18	19
B	B	C	D	E	F	10	11	12	13	14	15	16	17	18	19	1A
C	C	D	E	F	10	11	12	13	14	15	16	17	18	19	1A	1B
D	D	E	F	10	11	12	13	14	15	16	17	18	19	1A	1B	1C
E	E	F	10	11	12	13	14	15	16	17	18	19	1A	1B	1C	1D
F	F	10	11	12	13	14	15	16	17	18	19	1A	1B	1C	1D	1E

Example 4-22

Perform the indicated operations: $135E_{sixteen} + 21C_{sixteen}$.

Solution

$$
\begin{array}{r}
1 \qquad \text{carry column} \\
135E_{sixteen} \\
+\ 21C_{sixteen} \\
\hline
157A_{sixteen}
\end{array}
$$

$E_{sixteen} + C_{sixteen} = 1A_{sixteen}$ Write the A and carry the 1.

$1_{sixteen} + 5_{sixteen} + 1_{sixteen} = 7_{sixteen}$

$3_{sixteen} + 2_{sixteen} = 5_{sixteen}$

Bring down the 1

Try This One

4-M Perform the indicated operations.

(a) $\begin{array}{r} 1211_{three} \\ +\ 202_{three} \\ \hline \end{array}$

(b) $\begin{array}{r} 163_{eight} \\ +\ 257_{eight} \\ \hline \end{array}$

For the answer, see page 157.

Subtraction can be performed in other bases. The addition table can be used to help you find the answers for subtraction problems. For example, consider $12_{\text{five}} - 4_{\text{five}} = 3_{\text{five}}$. The answer can be found by locating 4 in the first column of the addition table then moving across the same row until 12 is located. At the number 12, move up the column until the answer is located in the top row. It is 3.

+	0	1	2	③	4
0	0	1	2	3	4
1	1	2	3	4	10
2	2	3	4	10	11
3	3	4	10	11	12
④	4	10	11	⑫	13

Remember that when the bottom digit is larger than the top digit, it is necessary to borrow from the next column.

Example 4-23

Perform the indicated operation.

$$321_{\text{five}}$$
$$-\ 123_{\text{five}}$$

Solution

Step 1   Since three is larger than one, it is necessary to borrow from the next column; hence, $11_{\text{five}} - 3_{\text{five}} = 3_{\text{five}}$.

$$
\begin{array}{rrr}
 & 1 & 11 \\
3 & \not{2} & 1_{\text{five}} \\
-\ 1 & 2 & 3_{\text{five}} \\
\hline
 & & 3_{\text{five}}
\end{array}
$$

Step 2   In the second column, $1_{\text{five}} - 2_{\text{five}}$ requires borrowing; change 3 in the third column to 2 and take $11_{\text{five}} - 2_{\text{five}}$ to get 4_{five}.

$$
\begin{array}{rrr}
2 & 11 & \\
\not{3} & \not{2} & 1_{\text{five}} \\
-\ 1 & 2 & 3_{\text{five}} \\
\hline
 & 4 & 3_{\text{five}}
\end{array}
$$

Step 3   Subtract $2_{\text{five}} - 1_{\text{five}}$ to get 1_{five}.

$$
\begin{array}{rrr}
2 & & \\
3 & 2 & 1_{\text{five}} \\
-\ 1 & 2 & 3_{\text{five}} \\
\hline
1 & 4 & 3_{\text{five}}
\end{array}
$$

Math Note

The answer can be checked by adding $143_{\text{five}} + 123_{\text{five}}$ and seeing if the answer is 321_{five}.

Hence, $321_{\text{five}} - 123_{\text{five}} = 143_{\text{five}}$.

Sidelight

Napier's Bones

Mathematicians have always been looking for easier ways to do calculations than by hand. John Napier (1550–1617) invented a method to perform multiplication called Napier's bones. The bones consist of an index bone and nine bones numbered one through nine. Each of the bones contains the multiple of the top number.

Index	0	1	2	3	4	5	6	7	8	9
1	0/0	0/1	0/2	0/3	0/4	0/5	0/6	0/7	0/8	0/9
2	0/0	0/2	0/4	0/6	0/8	1/0	1/2	1/4	1/6	1/8
3	0/0	0/3	0/6	0/9	1/2	1/5	1/8	2/1	2/4	2/7
4	0/0	0/4	0/8	1/2	1/6	2/0	2/4	2/8	3/2	3/6
5	0/0	0/5	1/0	1/5	2/0	2/5	3/0	3/5	4/0	4/5
6	0/0	0/6	1/2	1/8	2/4	3/0	3/6	4/2	4/8	5/4
7	0/0	0/7	1/4	2/1	2/8	3/5	4/2	4/9	5/6	6/3
8	0/0	0/8	1/6	2/4	3/2	4/0	4/8	5/6	6/4	7/2
9	0/0	0/9	1/8	2/7	3/6	4/5	5/4	6/3	7/2	8/1

To multiply 5 × 742, line up the bones as shown.

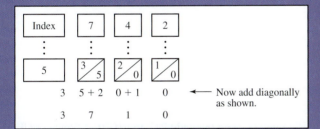

The answer is 3710.

To multiply a two-digit number, it is necessary to perform two multiplications and then add the products. Find the product 38 × 541.

First, find the product 3 × 541 as shown.

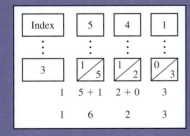

Then add a zero since you are really multiplying by 30. Hence, 30 × 541 = 16,230.

Now multiply 8 × 541.

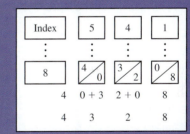

Hence, 8 × 541 = 4328. Finally, add 16,230 + 4328 = 20,558.

John Napier was born near Edinburgh, Scotland, and is credited with inventing logarithms. Logarithms were used to do multiplication and division problems by addition and subtraction and use the rules of exponents.

A set of Napier's bones were made from bone, wood, or cardboard and were carried by merchants, accountants, and navigators to do computations.

Try This One

4-N Perform the indicated operation: $7316_{\text{eight}} - 1257_{\text{eight}}$.
For the answer, see page 157.

The multiplication table for base five is shown next.

×	0	1	2	3	4
0	0	0	0	0	0
1	0	1	2	3	4
2	0	2	4	11	13
3	0	3	11	14	22
4	0	4	13	22	31

For example, $3_{\text{five}} \times 4_{\text{five}} = 22_{\text{five}}$ ($3 \times 4 = 12$ in base ten but 12 is 22_{five}).
  Multiplication is done in base five in the same way as it is done in base ten.

Example 4-24

Perform the indicated operation.

$$314_{\text{five}}$$
$$\times\ 23_{\text{five}}$$

Solution

Step 1 Multiply $3_{\text{five}} \times 4_{\text{five}} = 22_{\text{five}}$. Write the 2 and carry 2.

$$2$$
$$314_{\text{five}}$$
$$\times\ 23_{\text{five}}$$
$$2_{\text{five}}$$

Step 2 Multiply $3_{\text{five}} \times 1_{\text{five}}$ to get 3_{five} and then add 2_{five} to get 10_{five}. Write the 0 and carry the 1.

$$1$$
$$314_{\text{five}}$$
$$\times\ 23_{\text{five}}$$
$$02_{\text{five}}$$

Step 3 Multiply $3_{\text{five}} \times 3_{\text{five}} = 14_{\text{five}}$ and add 1_{five} to get 20_{five}.

$$1$$
$$314_{\text{five}}$$
$$\times\ 23_{\text{five}}$$
$$2002_{\text{five}}$$

Step 4 Multiply $2_{\text{five}} \times 4_{\text{five}} = 13_{\text{five}}$. Write the 3 and carry the 1.

$$1$$
$$314_{\text{five}}$$
$$\times\ 23_{\text{five}}$$
$$2002_{\text{five}}$$
$$30$$

—Continued

Example 4-24 Continued—

Step 5 Multiply $2_{five} \times 1_{five} = 2_{five}$ and add 1 to get 3_{five}.

$$
\begin{array}{r}
314_{five} \\
\times\ 23_{five} \\
\hline
2002_{five} \\
330
\end{array}
$$

Step 6 Multiply $2_{five} \times 3_{five}$ to get 11_{five}.

$$
\begin{array}{r}
314_{five} \\
\times\ 23_{five} \\
\hline
2002_{five} \\
11310_{five}
\end{array}
$$

Step 7 Add the partial products in base five.

$$
\begin{array}{r}
314_{five} \\
\times\ 23_{five} \\
\hline
2002_{five} \\
11330_{five} \\
\hline
13332_{five}
\end{array}
$$

Check with base 10:

$$
\begin{array}{r}
84 \\
\times 13 \\
\hline
252 \\
84 \\
\hline
1092_{ten}
\end{array}
$$

Hence, $314_{five} \times 23_{five} = 13332_{five}$.

Example 4-25

Perform the indicated operations.

$$
\begin{array}{r}
1011_{two} \\
\times\ \ 11_{two}
\end{array}
$$

Solution

The multiplication table for base two is

$$
\begin{array}{c|cc}
\times & 0 & 1 \\
\hline
0 & 0 & 0 \\
1 & 0 & 1
\end{array}
$$

Then

$$
\begin{array}{r}
1011_{two} \\
\times\ \ 11_{two} \\
\hline
1011 \\
1011 \\
\hline
100001_{two}
\end{array}
$$

Remember to add the partial products in base two.

Try This One

4-O Perform the indicated operations.

(a) 321_{four}
     $\times\ 12_{four}$

(b) 621_{eight}
     $\times\ 45_{eight}$

For the answer, see page 157.

Division is performed in the same way as long division is performed in base ten. The basic procedure for long division is

1. Divide
2. Multiply
3. Subtract
4. Bring down
5. Repeat steps 1–4

It will be helpful to look at the multiplication table for the given base in order to find the quotients.

Example 4-26

Perform the indicated operation.

$$3_{five}\overline{)2032_{five}}$$

Solution

Step 1 Using the multiplication table for base five, we need to find a product less than or equal to 20_{five} that is divisible by 3_{five}. This is done as follows.

×	0	1	2	3	4
0	0	0	0	0	0
1	0	1	2	3	4
2	0	2	4	11	13
3	0	3	11	14	22
4	0	4	13	22	31

The number is 14_{five} and $3_{five} \times 3_{five} = 14_{five}$. The first digit in the quotient is 3_{five}.

$$3_{five}\overline{)2032_{five}}^{\ \ 3}$$

Step 2 Then multiply $3_{five} \times 3_{five} = 14_{five}$ and write the quotient under 20. Subtract in base five and then bring down the next digit.

$$\begin{array}{r} 3 \\ 3_{five}\overline{)2032_{five}} \\ 14 \\ \hline 13 \end{array}$$

—Continued

Example 4-26 *Continued—*

Step 3 Next find a product smaller than or equal to 13_{five} in the table. It is 11_{five}. Since $3_{\text{five}} \times 2_{\text{five}} = 11_{\text{five}}$, write the 2 in the quotient. Then multiply $3_{\text{five}} \times 2_{\text{five}} = 11_{\text{five}}$. Write the 11 below the 13. Subtract and then bring down the 2.

$$
\begin{array}{r}
32 \\
3_{\text{five}} \overline{\smash{)}2032_{\text{five}}} \\
14 \\
\hline
13 \\
11 \\
\hline
22
\end{array}
$$

Step 4 Find a product divisible by 3_{five} that is less than or equal to 22_{five}. It is 22_{five} since $3_{\text{five}} \times 4_{\text{five}} = 22_{\text{five}}$. Write the 4 in the quotient and the 22 below the 22 in the problem. Subtract.

$$
\begin{array}{r}
324 \\
3_{\text{five}} \overline{\smash{)}2032_{\text{five}}} \\
14 \\
\hline
13 \\
11 \\
\hline
22 \\
22 \\
\hline
0
\end{array}
$$

The remainder is 0. Hence $2032_{\text{five}} \div 3_{\text{five}} = 324_{\text{five}}$.
This answer can be checked by multiplication.

$$
\begin{array}{r}
324_{\text{five}} \\
\times \quad 3_{\text{five}} \\
\hline
2032_{\text{five}}
\end{array}
$$

Example 4-27

Perform the following operation.

$$11_{\text{two}} \overline{\smash{)}100111_{\text{two}}}$$

Solution

$$
\begin{array}{r}
1101 \\
11_{\text{two}} \overline{\smash{)}100111_{\text{two}}} \\
-11 \\
\hline
11 \\
-11 \\
\hline
011 \\
-11 \\
\hline
0
\end{array}
$$

Divide 100 by 11 to get 1 and then multiply, subtract, and bring down.

Divide 11 by 11 to get 1 and then multiply, subtract, and bring down.

1 cannot be divided by 11 so bring down the last one and divide 11 by 11 to get 1.

—Continued

Example 4-27 Continued—

Hence, $100111_{two} \div 11_{two} = 1101_{two}$.

Check:

$$
\begin{array}{r}
1101_{two} \\
\times\ \ 11_{two} \\
\hline
1101 \\
1101\ \ \\
\hline
100111_{two}
\end{array}
$$

Try This One

4-P Perform the indicated operations.

(a) $10_{two} \overline{)1010_{two}}$ (b) $4_{five} \overline{)112_{five}}$

For the answer, see page 157.

Exercise Set   4-3

Computational Exercises

For Exercises 1–26, perform the indicated operations.

1. $\begin{array}{r} 11_{five} \\ +\,21_{five} \\ \hline \end{array}$

2. $\begin{array}{r} 44_{five} \\ +\,33_{five} \\ \hline \end{array}$

3. $\begin{array}{r} 3230_{four} \\ +\,1322_{four} \\ \hline \end{array}$

4. $\begin{array}{r} 8A2B_{twelve} \\ +\,191A_{twelve} \\ \hline \end{array}$

5. $\begin{array}{r} 321_{six} \\ +\,1255_{six} \\ \hline \end{array}$

6. $\begin{array}{r} 143_{five} \\ +\,432_{five} \\ \hline \end{array}$

7. $\begin{array}{r} 4344_{nine} \\ +\,2313_{nine} \\ \hline \end{array}$

8. $\begin{array}{r} 3145_{seven} \\ +\,216_{seven} \\ \hline \end{array}$

9. $\begin{array}{r} 43_{five} \\ -\,12_{five} \\ \hline \end{array}$

10. $\begin{array}{r} 143_{five} \\ -\,34_{five} \\ \hline \end{array}$

11. $\begin{array}{r} 262_{seven} \\ -\,161_{seven} \\ \hline \end{array}$

12. $\begin{array}{r} 9327_{eleven} \\ -\,7318_{eleven} \\ \hline \end{array}$

13. $\begin{array}{r} 42831_{nine} \\ -\,2781_{nine} \\ \hline \end{array}$

14. $\begin{array}{r} 6323_{seven} \\ -\,415_{seven} \\ \hline \end{array}$

15. $\begin{array}{r} 12AB_{twelve} \\ -\,93A_{twelve} \\ \hline \end{array}$

16. $\begin{array}{r} 2121_{three} \\ -\,222_{three} \\ \hline \end{array}$

17. $\begin{array}{r} 52_{six} \\ \times\,4_{six} \\ \hline \end{array}$

18. $\begin{array}{r} 241_{seven} \\ \times\,6_{seven} \\ \hline \end{array}$

19. $\begin{array}{r} 818_{nine} \\ \times\,62_{nine} \\ \hline \end{array}$

20. $\begin{array}{r} 423_{five} \\ \times\,332_{five} \\ \hline \end{array}$

21. $\begin{array}{r} AB5_{twelve} \\ \times\,42_{twelve} \\ \hline \end{array}$

22. $\begin{array}{r} 5186_{nine} \\ \times\,23_{nine} \\ \hline \end{array}$

23. $3_{nine} \overline{)1568_{nine}}$

24. $2_{three} \overline{)1202_{three}}$

25. $4_{five} \overline{)2023_{five}}$

26. $6_{seven} \overline{)1425_{seven}}$

Real World Applications

(Since the binary, octal, and hexadecimal systems are used for computers, the real-world applications will include these bases.)

For Exercises 26–42, perform the indicated operations.

27. $\begin{array}{r} 1001_{two} \\ + \ 111_{two} \\ \hline \end{array}$

28. $\begin{array}{r} 62_{eight} \\ + 145_{eight} \\ \hline \end{array}$

29. $\begin{array}{r} 3BA_{sixteen} \\ + \ 49_{sixteen} \\ \hline \end{array}$

30. $\begin{array}{r} 10111_{two} \\ + \ 1101_{two} \\ \hline \end{array}$

31. $\begin{array}{r} 1100_{two} \\ - \ 11_{two} \\ \hline \end{array}$

32. $\begin{array}{r} 732_{eight} \\ - \ 45_{eight} \\ \hline \end{array}$

33. $\begin{array}{r} 526B_{sixteen} \\ - \ 4A1_{sixteen} \\ \hline \end{array}$

34. $\begin{array}{r} 1000_{two} \\ - \ 101_{two} \\ \hline \end{array}$

35. $\begin{array}{r} 1010_{two} \\ \times \ 101_{two} \\ \hline \end{array}$

36. $\begin{array}{r} 54_{eight} \\ \times \ 2_{eight} \\ \hline \end{array}$

37. $\begin{array}{r} A25_{sixteen} \\ \times \ 4_{sixteen} \\ \hline \end{array}$

38. $\begin{array}{r} 326_{eight} \\ \times \ 21_{eight} \\ \hline \end{array}$

39. $11_{two}\overline{)1011_{two}}$

40. $6_{eight}\overline{)437_{eight}}$

41. $5_{sixteen}\overline{)37B1_{sixteen}}$

42. $10_{two}\sqrt{11111_{two}}$

Writing Exercises

43. Explain how subtraction is done on the addition table for base eight.

44. Explain how to do simple division using a multiplication table for base four.

Critical Thinking

The ASCII (American Standard Code for Information) is used by IBM. Letters are written in binary notation as follows:

A–O are prefixed by 0100 and start with A = 0001, B = 0010, C = 0011.

P–Z are prefixed by 0101 and P = 0000. Hence C = 0100 0011 and Q = 0101 0001.

For Exercises 45–48, indicate what letter of the alphabet is written in the binary code.

45. 0100 1100

46. 0101 0101

47. 0100 0111

48. 0101 1010

For Exercises 49–51, write each message in ASCII.

49. STOP

50. HURRY

51. HELLO

Summary

Section	Important Terms	Important Ideas
4-1	number numeral numeration system grouping system digit place value positional system expanded notation	**Throughout** history, people have used different numeration systems. These systems include the Egyptian, the Babylonian, and the Roman numeration systems. The system that is used in our world today is called the Hindu-Arabic numeration system. It uses the base ten and 10 symbols called digits to represent our numbers.
4-2	base binary system octal system hexadecimal system	**Numbers** can be written using different bases. For example, the base five system has only five digits. They are 0, 1, 2, 3, and 4. The place values of the numbers written in the base five system are 1s, 5s, 25s, etc.
4-3		**Operations** such as addition, subtraction, multiplication, and division can be performed in other number bases the same way they are performed in base ten.

Review Exercises

For Exercises 1–5, write each number in the Hindu-Arabic notation.

1. ⵣ◠◠∩∩∣

2. ⌒ℇℇ∩∣∣∣∣∣

3. ◁▾ ◁◁▾

4. MCXLVII

5. CDXIX

For Exercises 6–10, write each number in the system given.

6. 49 in the Egyptian system

7. 896 in the Roman system

8. 88 in the Babylonian system

9. 125 in the Egyptian system

10. 503 in the Roman system

For Exercises 11–20, write each number in base ten.

11. 1110111_{two}

12. 672_{eight}

13. $A03B_{twelve}$

14. 231_{four}

15. 14441_{five}

16. 2012_{three}

17. 6000_{seven}

18. 28645_{nine}

19. 555_{six}

20. $1A214_{eleven}$

For Exercises 21–30, write each number in the specified base.

21. 32 in base six

22. 105 in base twelve

23. 2001 in base nine

24. 81 in base three

25. 43 in base two

26. 213 in base eight

27. 19 in base four

28. 51 in base two

29. 343 in base seven

30. 899 in base twelve

For Exercises 31–48, perform the indicated operation.

31. 156_{nine}
    $+ \ \ 84_{nine}$

32. 434_{five}
    $+341_{five}$

33. 101110_{two}
    $+ \ \ \ 1101_{two}$

34. 5342_{six}
    $+1305_{six}$

35. $6A20_{twelve}$
    $+B096_{twelve}$

36. 7267_{nine}
    $- \ \ 354_{nine}$

37. 1010011_{two}
    $- \ 100111_{two}$

38. 2120_{three}
    $- \ 1212_{three}$

39. 3312_{four}
    $- \ 2321_{four}$

40. 65602_{seven}
    $- \ 46031_{seven}$

41. 371_{nine}
    $\times \ \ 51_{nine}$

42. 242_{five}
    $\times \ \ \ \ 3_{five}$

43. 1101_{two}
    $\times \ 111_{two}$

44. $6A5_{sixteen}$
    $\times \ \ \ \ 8_{sixteen}$

45. $3_{five} \overline{)1242_{five}}$

46. $7_{eight} \overline{)3426_{eight}}$

47. $10_{eight} \overline{)3426_{eight}}$

48. $5_{sixteen} \overline{)324_{sixteen}}$

Chapter Test

For Exercises 1–5, write each number in Hindu-Arabic notation.

1. ϔϔ⊚⊚⊚∩∣∣

2. ⌒⌒⌒ƒƒƒƒƒ∩∩∩∩∣∣∣∣∣

3. ⫷⫷Ⳏ ⫷Ⳏ

4. MCMLXVI

5. CDXXVI

For Exercises 6–10, write each number in the system given.

6. 93 in the Egyptian system

7. 567 in the Roman system

8. 55 in the Babylonian system

9. 521 in the Egyptian system

10. 605 in the Roman system

For Exercises 11–20, write each number in base ten.

11. 341_{five}

12. 573_{eight}

13. $A07B_{twelve}$

14. 312_{four}

15. 14411_{five}

16. 21101_{three}

17. 4000_{five}

18. 1100111_{two}

19. 463_{seven}

20. $1A436_{eleven}$

For exercises 21–30, write each number in the specified base.

21. 43 in base five

22. 183 in base twelve

23. 4673 in base nine

24. 65 in base three

25. 17 in base two

26. 316 in base eight

27. 91 in base four

28. 48 in base two

29. 434 in base seven

30. 889 in base twelve

For Exercises 31–44, perform the indicated operation.

31. 263_{nine}
    $+ \ \ 18_{nine}$

32. 341_{five}
    $+ 213_{five}$

33. 111010_{two}
    $+ \ \ \ 1101_{two}$

34. 2435_{six}
    $+ 5013_{six}$

35. $5A79_{twelve}$
    $+ B068_{twelve}$

36. 6772_{eight}
    $- \ \ 735_{eight}$

37. 11001010_{two}
    $- \ \ \ 110011_{two}$

38. 2212_{three}
    $- 1202_{three}$

39. 3213_{four}
    $- 2123_{four}$

40. 20665_{seven}
    $- 10364_{seven}$

41. 254_{six}
    $\times \ \ \ 3_{six}$

42. 413_{five}
    $\times \ \ 21_{five}$

43. $7_{eight} \overline{)1342_{eight}}$

44. $2_{three} \overline{)1012_{three}}$

Mathematics in Our World ▶ *Revisited*

Bar Codes

Bar codes consist of a series of black bars and white spaces that vary in their widths. They consist of a left margin, a start character or data message, a check digit to make sure that there is no error in it, a stop character, and a right margin. A computer scanner reads the code, which consists of letters and numbers represented by the bars and spaces and then converts them to base two numbers and letters written in binary notation. These are sent to another computer, such as a cash register, which records the information, computes the prices, and prints a receipt.

Margin   Margin

Start character   Data message and check digit   Stop character

The *Universal Product Code* (UPC) began in 1973 and consists of a 12-digit number. The first digit identifies the kind of product such as a meat product, health-related product, etc. The next five digits identify the manufacturer. The next five digits identify the product and the last digit is a check digit. The digits 0 through 9 are represented by a seven-digit number consisting of zeros and ones, as shown.

Digit	Manufacturer's Number	Product Number
0	0001101	1110010
1	0011001	1100110
2	0010011	1101100
3	0111101	1000010
4	0100011	1011100
5	0110001	1001110
6	0101111	1010000
7	0111011	1000100
8	0110111	1001000
9	0001011	1110100

Notice that each binary digit in the manufacturer's code begins with a zero and ends with a one and each digit in the product code begins with a one and ends with a zero. The reason for this is that the number can be scanned from right to left or left to right and the computer will be able to distinguish between the two numbers. Now the digit 7 is represented by 0111011 in the manufacturer's code. The manufacturer's code number is then written using bars and spaces where the digit 0 represents a space and the digit 1 represents a bar as shown.

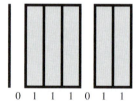

0 1 1 1 0 1 1

The check digit is used by the computer to detect or correct errors that have been made in the coding process. It can also be used to deter crime.

Projects

1. Find some information about another numeration system. Write a paper on the symbols that are used and how numbers are written in the system. Suggestions: Ionic-Greek numeration system, Chinese-Japanese numeration system, Mayan numeration system.
2. Write a paper on the evolution of the Hindu-Arabic numeration system.

3. People have always searched for easy methods to do tedious calculations such as addition, subtraction, multiplication, and division. The hand-held calculator, of course, is the easiest and most recent development. Some earlier methods are the Gelosia method (galley method), the duplation and mediation method, and finger multiplication. Select one of these methods and write a report on how the process works.

Answers to **Try This One**

4-A. (a) 456; (b) 1211; (c) 1,102,041

4-B. (a) ∩∩∩∩ |||

 (b) ᓂᓂᓂᓂᓂᓂᓂ ∩∩ |||||||

 (c) ⸠⸠⸠ᓂᓂ ∩∩∩∩∩∩∩∩ |||||

4-C. (a) ᓂᓂᓂᓂᓂᓂᓂ ∩ ||

 (b) ᓂᓂᓂᓂᓂᓂᓂ ∩∩∩∩∩ |||||||

4-D. (a) 52; (b) 1375; (c) 40,873; (d) ＜＜＜＜ ▼▼

 (e) ▼▼▼▼▼▼ ＜＜▼▼▼▼

 (f) ▼ ＜▼ ＜▼▼▼▼▼▼▼▼▼

4-E. (a) 39; (b) 1164; (c) 333

4-F. (a) LXVII; (b) CXCII; (c) CCII; (d) CMLX

4-G. (a) $5 \times 10^2 + 7 \times 10^1 + 3$

 (b) $8 \times 10^4 + 6 \times 10^3 + 4 \times 10^2 + 7 \times 10^1 + 1$

 (c) $2 \times 10^6 + 2 \times 10^5 + 1 \times 10^3 + 5 \times 10^2 + 6 \times 10^1 + 7$

4-H. (a) 77; (b) 214; (c) 2500

4-I. (a) 202_{five}; (b) 3421_{five}; (c) 13000_{five}

4-J. (a) 1876; (b) 54; (c) 905; (d) 17,070

4-K. (a) 1010100_{two}; (b) 1110_{six}; (c) 11112_{three};

 (d) $36A_{sixteen}$

4-L. (a) 424_{five}; (b) 11330_{five}

4-M. (a) 2120_{three}; (b) 442_{eight}

4-N. 6037_{eight}

4-O. (a) 11112_{four}; (b) 34765_{eight}

4-P. (a) 101_{two}; (b) 13_{five}

Chapter Five

The Real Number System

Objectives

After completing this chapter, you should be able to

1 Distinguish between a prime number and a composite number (5-1)

2 Find the factors of a number (5-1)

3 Find a prime factorization of a number (5-1)

4 Find the greatest common factor of two or more numbers (5-1)

5 Find the multiples of a number (5-1)

6 Find the least common multiple of two or more numbers (5-1)

7 Add, subtract, multiply, and divide integers (5-2)

8 Use the order of operations to simplify expressions (5-2)

9 Add, subtract, multiply, and divide rational numbers (5-3)

10 Change fractions to decimals and decimals to fractions (5-3)

11 Add, subtract, multiply, and divide irrational numbers (5-4)

12 Identify the properties of the real numbers (5-5)

13 Simplify expressions using the properties of exponents (5-6)

14 Convert large and small decimal numbers into numbers in scientific notation (5-6)

15 Convert numbers in scientific notation into numbers in decimal notation (5-6)

16 Perform multiplication and division with numbers written in scientific notation (5-6)

17 Write the terms of an arithmetic or geometric sequence (5-7)

18 Find specific terms of an arithmetic or geometric sequence (5-7)

19 Find the sum of n terms of an arithmetic or geometric sequence (5-7)

Introduction

Throughout your lifetime, you have encountered many different types of numbers. Numbers such as 0, $-5, 6.8, \frac{9}{10}, \sqrt{3}, \pi$, etc. are used often in the real world. These numbers are examples of real numbers. The real number system consists of subsets of numbers such as natural numbers, whole numbers, integers, rational numbers, and irrational numbers. This chapter explains these subsets and some of their properties and uses. ■

5-1   The Natural Numbers

Prime and Composite Numbers

The numbers that humans first used were the counting numbers. These numbers are also called the natural numbers. Formally defined,

> The set of **natural numbers** (N) consists of the numbers 1, 2, 3, ...

The natural numbers are the first ones people learn.

The basic operations for natural numbers are addition, subtraction, multiplication, and division. Every natural number can be expressed as a product of two or more natural numbers. For example,

$$8 = 2 \times 4 \qquad 12 = 3 \times 4 \qquad 17 = 1 \times 17 \qquad 36 = 2 \times 3 \times 6$$

The numbers that are multiplied to get a product are called the **factors** of the product. Since $6 \times 2 = 12$, 6 and 2 are called factors of 12. The other factors of 12 are 1, 3, 4, and 12. Hence the set of all factors of 12 consists of 1, 2, 3, 4, 6, and 12.

Notice that when a number is divided by any of its factors, the remainder is zero. For this reason, the factors of a given number are also called **divisors** of the number. Six is a factor of 12 since $12 \div 6 = 2$ remainder 0, whereas five is not a factor of 12 since $12 \div 5 = 2$ remainder 2.

In general, a number a is said to be divisible by a number b if the quotient $a \div b$ has a remainder of zero.

Mathematics in Our World

Are You Sure It's Fat Free?

Today, many health-conscious people are striving to reduce the fat content of their diets. The food industry is enthusiastically cooperating by advertising and producing food that is low in fat. For example, some labels boast that their products are "97% fat free." But is this really the case?

Nutritionists recommend that we limit our daily content of fat to 30% of the calories we consume. With this in mind, a 97% fat-free product sounds too good to be true, and it is. This chapter shows the various types of numbers and the operations and properties for these numbers that we use in today's world. Many problems can be solved using mathematics; however, it is necessary to use correct procedures to solve these problems. In the case of the food industry, incorrect mathematical procedures have been used to calculate the fat content of foods.

To find out how, see Mathematics in Our World—Revisited on page 236.

Calculator Explorations

A graphing calculator can be used to find factors of a number.

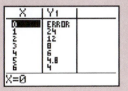

Why does an "ERROR" message appear when x = 0?

How many whole number factors of 24 are there? What are the first four whole number factors of 124?

Example 5-1

Find all the factors of 24.

Solution

In order to find the factors of a number, it is necessary to see what numbers will divide into 24 evenly. The factors of 24 are 1, 2, 3, 4, 6, 8, 12, and 24. The process is done by trial and error.

Try This One

5-A   Find all the factors of 50.
For the answer, see page 237.

In addition to the factors of a number, every natural number also has **multiples.** The multiples of a number are found by multiplying the number by 1, 2, 3, etc. For example, the multiples of 6 are 6, 12, 18, 24, 30, ... and are found by $6 \times 1, 6 \times 2, 6 \times 3$, etc.

In other words, 24 is a multiple of 6 since $6 \times 4 = 24$. The number 27 is not a multiple of 6 since there is no natural number, C, such that $6 \times C = 27$.

Some special natural numbers have only two factors, 1 and the number itself. For example, 7 has only two factors, 1 and 7. These numbers are called *prime numbers.*

A **prime number** is a natural number that has only two factors, one and itself.

The first 10 prime numbers are 2, 3, 5, 7, 11, 13, 17, 19, 23, and 29.

Numbers with more than two factors are called *composite numbers*. Numbers such as 4, 6, 8, 9, 10, etc. are composite numbers.

A **composite number** is a natural number greater than 1 that has three or more factors.

Mathematicians have been searching for a formula that will generate prime numbers. So far, no formula has been discovered that will generate all the prime numbers; however, over 2000 years ago, a Greek mathematician named Eratosthenes devised a method to generate all the prime numbers less than a given number. In order to generate all the prime numbers less than 50, a table of numbers from 1 to 50 is written. The number 1 is not a prime number so it is crossed out. The number 2 is a prime number, so it is circled. Now all multiples of 2 (4, 6, 8, etc.) are crossed out since they are composite numbers. The next number, 3, is circled since it is a prime number. Then all multiples of 3 are crossed out since they are composite numbers. Some of these numbers have already been crossed out since they are also multiples of 2. The process is continued for 5, 7, and so on. When all multiples of a number have been crossed out, the process can be terminated. In this case, when 11 is reached, all multiples of 11 (22, 33, and 44) have already been crossed out. The remaining numbers that have not been crossed out are prime numbers. See Figure 5-1.

Calculator Explorations

A graphing calculator can be used to find multiples of a number.

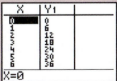

What do the values in the Y_1 column represent?

How many multiples of 6 are there?

What is the difference between factors and multiples?

Sidelight

Twin Primes

Prime numbers that differ by 2 are called twin primes. For example, 3 and 5 are twin primes, as are 17 and 19, 29 and 31, 41 and 43, and 71 and 73. One question that mathematicians have been trying to answer is, "Is there an infinite number of twin prime pairs of numbers?" It was proven by Euclid many years ago that there is an infinite number of prime numbers, but no one has, as yet, proven that there is an infinite number of twin prime pairs of numbers.

It is known by looking at a table of prime numbers that the number of twin primes decreases as the numbers get larger. For example, there are eight pairs of twin prime numbers from 1 to 100, seven prime numbers from 101 to 200, four pairs from 201 to 300. There are no pairs from 701 to 800, but there are five pairs from 801 to 900.

~~1~~	(2)	(3)	~~4~~	(5)	~~6~~	(7)	~~8~~	~~9~~	~~10~~
(11)	~~12~~	(13)	~~14~~	~~15~~	~~16~~	(17)	~~18~~	(19)	~~20~~
~~21~~	~~22~~	(23)	~~24~~	~~25~~	~~26~~	~~27~~	~~28~~	(29)	~~30~~
(31)	~~32~~	~~33~~	~~34~~	~~35~~	~~36~~	(37)	~~38~~	~~39~~	~~40~~
(41)	~~42~~	(43)	~~44~~	~~45~~	~~46~~	(47)	~~48~~	~~49~~	~~50~~

Figure 5-1

Math Note

The number 1 is neither prime nor composite since it has only one factor, itself.

Hence the set of all prime numbers less than 50 consists of 2, 3, 5, 7, 11, 13, 17, 19, 23, 29, 31, 37, 41, 43, and 47.

Prime Factorizations

Every composite number can be expressed as a product of prime numbers in only one way. For example, 24 can be written as $2 \times 2 \times 2 \times 3$ or $2^3 \times 3$. There is no other way to write 24 as a product of prime numbers. This product is called a *prime factorization* of 24. (A factorization of a number is an indicated product of factors of the number.) The *fundamental theorem of arithmetic* states this fact. (A theorem is a property that can be proved true.)

Math Note

The order of the numbers in the product is not important. For example, a prime factorization of 30 can be written in any order, such as $5 \times 3 \times 2$ or $3 \times 2 \times 5$, etc. For convenience, we usually write prime factorization in numerical order—in this case, $2 \times 3 \times 5$.

The **fundamental theorem of arithmetic** states that every composite number can be expressed as a product of prime numbers in only one way. The order of the factors is disregarded.

There are two methods that can be used to find a prime factorization of a number, the *tree method* and the *division method.*

In order to find a prime factorization of a number using the tree method, first find any factorization of the number, then continue to find factorization of the remaining composite factors until all factors are prime.

Math Note

When using the tree method, it does not matter how you start. For example, 100 could be factored as 10×10 or 4×25. The fundamental theorem of arithmetic states that you will always end with the same factors $2 \times 2 \times 5 \times 5$.

Example   5-2

Find a prime factorization of 100.

Solution

Start with any factorization of 100, say 2×50, then factor 50 as 5×10. Finally factor 10 as 2×5. This is shown using a tree.

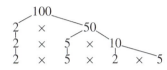

Rearrange the factors in order: $2 \times 2 \times 5 \times 5$ or $2^2 \times 5^2$.

The division method uses continued short division starting with the smallest prime factor or divisor of the number and continuing until a prime number in the quotient is obtained. Example 5-3 shows this method.

Example   5-3

Find a prime factorization of 100 using the division method.

Solution

First, divide 100 by 2 and then divide the answer by 2. Continue dividing the answer until you cannot find an answer that is not divisible by 2, then try to divide by 3, then 5, etc., as shown.

$$
\begin{array}{r}
5 \\
5\,\overline{)\,25} \\
2\,\overline{)\,50} \\
2\,\overline{)\,100}
\end{array}
$$

Math Note

To avoid mistakes, always start with the smallest prime factor and keep using it until you can't divide any further, and then move to the next prime number. Do not skip around.

Hence, $100 = 2 \times 2 \times 5 \times 5 = 2^2 \times 5^2$.

Try This One

5-B   Find a prime factorization of 360

(a) using the tree method

(b) using the division method

For the answer, see page 237.

Greatest Common Factor

Prime factoring can be used to find the greatest common factor of two or more numbers and to find the least common multiple of two or more numbers.

> The **greatest common factor (divisor) (GCF)** of two or more numbers is the largest number that is a factor or divisor of all of the numbers.

For example, the factors of 18 are 1, 2, 3, 6, 9, and 18, and the factors of 24 are 1, 2, 3, 4, 6, 8, 12, and 24. The common factors of 18 and 24 are 1, 2, 3, and 6. Hence, 6 is the GCF of 18 and 24. When the numbers are large, it is more difficult to find the GCF of two or more numbers by listing all the factors of the numbers, so the next procedure can be used.

Procedure for Finding the GCF of Two or More Numbers

Step 1 Factor the numbers into primes.
Step 2 Find the prime factors that the numbers have in common.
Step 3 Find the product of the common factors using each factor as many times as its smallest exponent.

Sidelight

Mersenne Prime Numbers

A philosopher, theologian, and mathematician, Father Marin Mersenne (1588–1648) described certain prime numbers of the form $2^p - 1$, where p is a prime number greater than 1. These numbers are called Mersenne prime numbers. For example, when $p = 2$, $2^2 - 1 = 3$. When $p = 3$, $2^3 - 1 = 7$. When $p = 5$, $2^5 - 1 = 31$, and when $p = 7$, $2^7 - 1 = 127$. Not all prime numbers generate a Mersenne prime number. For example, $2^{11} - 1 = 2047$, which is not a prime number since $23 \times 89 = 2047$.

Father Mersenne was a friend of Descartes, and both studied at Jesuit College. He corresponded with some of the greatest mathematicians of his time, and he edited the works of many of the Greek mathematicians. He also wrote on a variety of subjects.

It is an interesting fact that mathematicians have been searching for the prime numbers p that generate Mersenne prime numbers. As p gets larger, this was a very difficult task before the advent of the modern computer.

In 1903, a mathematician named Frank Adelson Cole presented a paper entitled, "On the Factorization of Large Numbers." He walked to the board and wrote $2^{67} - 1$. Then he wrote its value. Next, he proceeded to write the factors of the number showing that $2^{67} - 1$ was not a prime number. Without saying a word, he sat down. It was reported that the audience gave him a standing ovation.

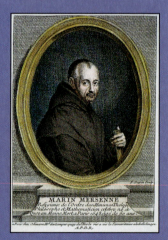

MARIN MERSENNE

Example 5-4 shows this procedure.

Example 5-4

Find the GCF of 72 and 180.

Solution

Step 1 Factor 72 and 180 into primes:
$$72 = 2 \times 2 \times 2 \times 3 \times 3 \text{ or } 2^3 \times 3^2$$
$$180 = 2 \times 2 \times 3 \times 3 \times 5 \text{ or } 2^2 \times 3^2 \times 5$$

Step 2 The factors 2 and 3 are common to 72 and 180.

Step 3 Find the product of 2^2 and 3^2. $2^2 \times 3^2 = 4 \times 9 = 36$. (The smallest exponent of the 2 is 2, and the smallest exponent of the 3 is 2.) Hence, the GCF of 72 and 180 is 36.

Math Note

When the GCF of two numbers is 1, the numbers are said to be **relatively prime.** For example, the GCF of 15 and 17 is 1; hence, 15 and 17 are relatively prime.

Try This One

5-C Find the GCF of 54 and 144.
For the answer, see page 237.

The GCF is used in reducing fractions. This use will be explained in Section 5-3.

Least Common Multiple

Prime factoring can be used to find the least common multiple (LCM) of two or more numbers.

> The **least common multiple (LCM)** of two or more numbers is the smallest number that is divisible by all the numbers.

For example, the multiples of 6 are 6, 12, 18, 24, 30, 36, . . . and the multiples of 8 are 8, 16, 24, 32, 40, ... The LCM of 6 and 8 is 24 since it is the smallest multiple that can be divided evenly by 6 and 8. There are other common multiples of 6 and 8. Some are 48, 72, and 96. These common multiples are also multiples of the LCM; however, 24 is the smallest of the common multiples of 6 and 8. The next procedure can be used to find the LCM of two or more numbers.

Procedure for Finding the LCM of Two or More Numbers

Step 1 Factor the numbers into primes.
Step 2 Write each different prime factor.
Step 3 Find the product of the different prime factors using each factor as many times as its largest exponent.

Calculator Explorations

The table feature, found on some graphing calculators, can be used to find the LCM of one or more numbers.

```
Plot1 Plot2 Plot3
\Y1■6X
\Y2■8X
\Y3=■
\Y4=
\Y5=
\Y6=
\Y7=
```

X	Y1	Y2
0	0	0
1	6	8
2	12	16
3	18	24
4	24	32
5	30	40
6	36	48

X=0

What is the smallest nonzero number that 6 and 8 share as a common multiple?

What is the LCM of 16 and 18?

One species of cicada emerges every 13 years, another every 17 years. If the species emerge together in a town this year, how long will it be until they emerge together again?

Example 5-5

Find the LCM of 24, 30, and 42.

Solution

Step 1 Find a prime factorization of each number.

$$24 = 2^3 \times 3$$
$$30 = 2 \times 3 \times 5$$
$$42 = 2 \times 3 \times 7$$

Step 2 The different prime factors are 2, 3, 5, and 7.

Step 3 Find the product of the different prime factors using each factor as many times as its largest exponent. Hence, $2^3 \times 3 \times 5 \times 7 = 840$. The LCM of 24, 30, and 42 is 840.

In Example 5-5 the factor 2 is used three times in 24, once in 30, and once in 42; hence, the largest exponent of 2 is 3. Three is used once in 24, 30, and 42; hence, the largest exponent of 3 is 1, etc.

Try This One

5-D Find the LCM of each.

(a) 40, 50

(b) 28, 35, 49

(c) 16, 24, 32

For the answer, see page 237.

Many real-world problems can be solved using the concepts of GCF and LCM as shown in Example 5-6.

Example 5-6

A person is planning to sell boxes of tea bags at a food show. She has 200 bags of regular tea and 280 bags of decaffeinated tea. She wants to package them into small boxes so that each box will contain the same number of tea bags and each box will contain only one type of tea, either regular or decaffeinated. How many boxes of each type of tea will she have, and how many tea bags will be contained in each box if she wishes to use the least number of boxes possible?

—Continued

Example   5-6   *Continued—*

Solution

In order to solve this problem, you must find the greatest common factor of 200 and 280.

$$200 = 2 \cdot 2 \cdot 2 \cdot 5 \cdot 5$$
$$280 = 2 \cdot 2 \cdot 2 \cdot 5 \cdot 7$$

(*Note:* The dot means to multiply.)

The GCF of 200 and 280 is $2^3 \cdot 5$ or 40. Hence, each box will contain 40 tea bags. Since she has 200 regular tea bags and each box will contain 40 tea bags, she will have $200 \div 40 = 5$ boxes of regular tea. Since she has 280 decaffeinated tea bags and each bag will contain 40 tea bags, she will have $280 \div 40 = 7$ boxes of decaffeinated tea bags.

Try This One

5-E   If hot dog buns are sold in packages of 8 and hot dogs are sold in packages of 10, what is the smallest number of packages of hot dog buns and packages of hot dogs you must buy so that you have the same number of hot dogs and hot dog buns?
For the answer, see page 237.

Sidelight

Fermat Numbers

A famous French mathematician named Pierre de Fermat described prime numbers of the form $2^{2n} + 1$ where n is a positive integer. For example, $2^2 + 1 = 5$, $2^4 + 1 = 17$, and $2^8 + 1 = 257$. These numbers are called Fermat numbers. It can be shown that $2^{16} + 1 = 65{,}537$ is prime; however, the numbers after that get very large very rapidly, and it was difficult in the 1600s to show whether or not there were more Fermat numbers.

Mathematicians have shown that when n is from 5 to 19, the Fermat numbers are composite. Some of the larger numbers were factored with the aid of a computer.

Fermat was a lawyer and studied mathematics in his spare time. Although he did not publish much of his findings, he contributed to many areas of mathematics through his correspondence with other mathematicians of his time.

Exercise Set 5-1

Computational Exercises

For Exercises 1–20, find all the factors of each number.

1. 16	2. 225	3. 126	4. 54
5. 32	6. 48	7. 9	8. 10
9. 96	10. 100	11. 17	12. 19
13. 64	14. 120	15. 105	16. 365
17. 98	18. 36	19. 71	20. 47

For Exercises 21–30, find five multiples of each.

21. 3	22. 7	23. 10	24. 12
25. 15	26. 20	27. 17	28. 19
29. 1	30. 25		

For Exercises 31–50, find a prime factorization of each.

31. 16	32. 18	33. 1296	34. 1960
35. 17	36. 19	37. 50	38. 64
39. 128	40. 169	41. 300	42. 500
43. 475	44. 625	45. 448	46. 77
47. 247	48. 56	49. 750	50. 825

For Exercises 51–60, find the GCF of each.

51. 3, 9	52. 10, 35	53. 7, 10	54. 6, 11
55. 12, 24, 48	56. 5, 15, 25	57. 75, 100	58. 125, 175
59. 100, 225, 350	60. 42, 56, 63		

For Exercises 61–70, find the LCM of each.

61. 5, 10	62. 12, 24	63. 50, 75	64. 60, 90
65. 70, 90	66. 195, 390	67. 4, 7, 11	68. 5, 6, 13
69. 12, 18, 36	70. 42, 48, 56		

Real World Applications

71. An amusement park has two shuttle buses. Shuttle bus A makes six stops and shuttle bus B makes eight stops. The buses take 5 minutes to go from one stop to the next. Each bus takes a different route. If they start at 10:00 A.M. from station one, at what time will they both return to station one? Station one is at the beginning and at the end of the loop and is not counted twice.

72. Four lighthouses can be seen from a boat offshore. One light blinks every 10 seconds. The second light blinks every 15 seconds. The third light blinks every 20 seconds, and the last light blinks every 30 seconds. How often will all the lights be on at the same time?

73. A teacher has 24 colored pencils and 18 pictures to be placed in groups with the same number of pencils in each group and the same number of pictures in each group. If each group must have the same type of items, how many groups of pencils and how many groups of pictures can be made? How many of each item will be in a group?

74. Two people bike on a circular trail. One person can ride around the trail in 24 minutes, and another person can ride around the trail in 36 minutes. If both start at the same place at the same time, when will they be at the starting place at the same time again?

75. Two clubs offer a "free day" (no admission charge) to recruit new members. The health club has a free day every 45 days. A nearby swimming club has a free day every 30 days. If today is a free day for both clubs, how long will it be until a person can again use both clubs for free on the same day?

76. There are 30 women and 36 men in a bowling league. The president wants to divide the members into all-male and all-female teams, each of the same size, regardless of gender. Find the number of members and the number of teams for each gender.

Writing Exercises

77. Define the set of natural numbers.
78. Explain the difference between a prime and a composite number.
79. Explain why the number 1 is neither a prime nor a composite number.
80. Describe why Eratosthenes' method identified prime numbers.
81. What does the fundamental theorem of arithmetic state about a prime factorization of a composite number?
82. Explain the difference between the greatest common factor of two numbers and the least common multiple of two numbers.
83. Can a factor of a number also be a multiple of a number? Explain your answer.
84. Why is 2 a special prime number?

Critical Thinking

85. How many prime numbers are even?
86. The prime number 17 can be made into another prime number, 71, by reversing its digits. Find all the prime number pairs less than 100 such that when the digits of the first number are reversed, the new number is also prime.
87. A German mathematician Christian Goldbach (1690–1764) made the following conjecture: Every even number greater than 2 can be expressed as the sum of two prime numbers. For example, $6 = 3 + 3$, $8 = 5 + 3$, etc. Express every even number from 4 through 20 as a sum of two prime numbers.
88. Another conjecture attributed to Goldbach is that any odd number greater than 7 can be expressed as the sum of three odd prime numbers. For example, $11 = 3 + 3 + 5$. Express every odd number from 9 through 25 as the sum of three odd prime numbers.

5-2 ## The Integers

Definition of Integers

In Section 5-1, the set of natural numbers was defined. The natural numbers consist of the numbers 1, 2, 3, 4, 5, ... A new set of numbers can be made by including zero with the natural numbers. This set is called the *whole numbers*.

> The set of **whole numbers** (W) is defined as 0, 1, 2, 3, 4, 5, ...

When the numbers $-1, -2, -3, \ldots$ are included with the set of whole numbers, a new set of numbers is formed. This set is called the *integers*. The symbol Z comes from the German word for number.

> The set of **integers** (Z) is defined as $\ldots, -3, -2, -1, 0, 1, 2, 3, \ldots$.

The integers can be depicted using a number line. See Figure 5-2.

The line is extended infinitely in both directions. The number zero is called the *origin*. The numbers to the right of zero are called the *positive integers*. The numbers to the left of zero are called the *negative integers*. *Zero is neither positive nor negative.*

Figure 5-2

Every integer has an **opposite** or additive inverse. See Figure 5-3. The opposite of 2 is -2. The opposite of -3 is 3. In general, two numbers are opposites if they are the same distance from zero on the number line but are in opposite directions from zero. The opposite of 0 is 0.

Opposites

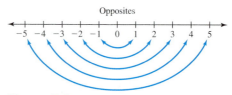

Figure 5-3

The opposite of a number x is denoted as $-(x)$. For example, the opposite of 3 is written as $-(3)$ or -3. The opposite of -3 is 3 written as $-(-3)$. Hence $-(-3) = 3$. In general, $-(-x) = x$. The expression $-x$ does not mean that the expression is negative. It depends on the value of x. For example, if x is positive, then the expression $-x$ is negative; that is, $-(+5) = -5$. However, if the value of x is negative, say -8, then the value of the expression $-x$ is positive. That is $-(-5) = +5$.

Another concept associated with the integers is the concept of absolute value. The **absolute value** of any number except zero is the number's distance from zero on a number line. The absolute value of any number except zero is always positive. Since -8 is 8 units to the left of the origin (i.e., 0), the absolute value of -8 is 8. Since $+8$ is 8 units to the right of the origin, the absolute value of $+8$ is also 8. The absolute value of zero is zero. Zero is neither positive nor negative.

Math Note

Every natural number is also a whole number. The number 5 is both a natural number and a whole number. Zero is a whole number, not a natural number.

Math Note

Every natural number is also an integer. For example, the number 5 is a natural number, a whole number, and an integer. Zero is a whole number and an integer.

Math Note

When a number (other than zero) does not have a sign in front of it, it is assumed to be positive. For example, $3 = +3, +5 = 5$, etc.

Integers are often used for applications where it makes sense to set a zero and then talk about distances to either side. For example, sea level corresponds to 0; the distance above sea level corresponds to the positive integers; and depth below sea level corresponds to negative integers.

Math Note

In mathematics, three other inequality signs are also used. They are

$\geq$ means "is greater than or equal to."

$\leq$ means "is less than or equal to."

$\neq$ means "is not equal to."

The symbol for absolute value is $|\ \ |$. For example,

$$|8| = 8 \qquad |-6| = 6 \qquad |0| = 0$$

The number line can be used to determine which of two integers is larger. When comparing two integers on the number line, the integer which is further to the right is the larger one. Hence, 10 is greater than 6 because on the number line 10 is further to the right than 6 is.

The following symbols are used to compare numbers:

$<$ means "is less than."

$>$ means "is greater than."

Hence, $10 > 6$.

Other examples that can be demonstrated graphically are

$$-5 > -7 \qquad 3 < 5 \qquad -6 < 0 \qquad +3 > -4$$

Try This One

5-F Find each.

(a) The opposite of -9

(b) The opposite of 24

(c) The opposite of 0

(d) $|-15|$

(e) $|8|$

(f) $|0|$

Use $=$, $>$, or $<$

(g) $-5 \quad -15$

(h) $-3 \quad +2$

(i) $0 \quad -6$

(j) $4 \quad -2$

For the answer, see page 237.

Addition and Subtraction of Integers

To add two positive integers such as $4 + 1$, one can use the number line. Start at 0 and count 4 units to the right to 4; then count one more unit to get 5. Hence, $4 + 1 = 5$. See Figure 5-4(a).

To add $1 + (-4)$, start at zero and count 1 unit to the right to get 1; then count 4 units to the left (since -4 is negative) to get -3. Hence, $1 + (-4) = -3$. See Figure 5-4(b).

To add $-3 + (-4)$, start at zero and count 3 units to the left to get -3. From there, count 4 units to the left to get -7. Hence, $-3 + (-4) = -7$. See Figure 5-4(c).

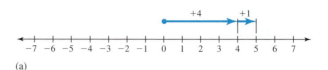

(a)

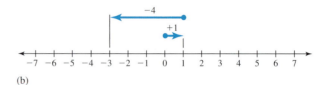

(b)

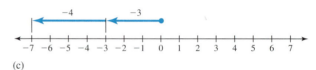

(c)

Figure 5-4

**Calculator
Explorations**

Be careful when working with signed numbers and subtraction.

```
-3-4
            -7
-3-(-4)
            1
-3--4
            1
■
```

Consider the expressions $2 - 8$ and $-8 - (-2)$.

Are they equal?

Why or why not?

Addition of integers is summarized next.

Rules for Addition of Integers

Rule 1 To add two integers with the same signs, add the absolute values of the numbers and give the answer the common sign.

Rule 2 To add two integers with different signs, subtract the number with the smaller absolute value from the number with the larger absolute value and give the answer the sign of the number with the larger absolute value.

Example 5-7

Find each.

(a) $8 + 6$

(b) $9 + 7$

(c) $-2 + (-8)$

(d) $-7 + (-4)$

(e) $-6 + 8$

(f) $-10 + 3$

(g) $12 + (-6)$

(h) $15 + (-18)$

Solution

These examples illustrate Rule 1:

(a) $8 + 6 = 14$

(b) $9 + 7 = 16$

(c) $-2 + (-8) = -10$

(d) $-7 + (-4) = -11$

These examples illustrate Rule 2:

(e) $-6 + 8 = 2$

(f) $-10 + 3 = -7$

(g) $12 + (-6) = 6$

(h) $15 + (-18) = -3$

Try This One

5-G Find each sum.

(a) $9 + 14$

(b) $(-3) + (-8)$

(c) $(-14) + 22$

(d) $19 + (-32)$

For the answer, see page 237.

In arithmetic, subtraction is thought of as "taking away." For example, $8 - 5$ is thought of as having 8 objects and taking away 5 objects with 3 objects left. Hence, $8 - 5 = 3$. However, this way of thinking gets complicated when subtracting negative numbers, and even more complicated when a negative number is subtracted from another negative number, so a new definition of subtraction is used in algebra.

> ### Math Note
>
> When performing subtraction using integers, change the sign of the number being subtracted and follow the rules of addition.

Rule for Subtraction of Integers

To subtract two integers, $a - b$, add the opposite of b to a, $a + (-b)$.

Example 5-8

Find each.

(a) $9 - 6$
(b) $-2 - (-8)$
(c) $3 - (-7)$
(d) $-2 - 6$

Solution

(a) Rewrite $9 - 6$ as $9 + (-6)$ since -6 is the opposite of 6 and use rule 2 for addition to get 3.
(b) Rewrite $-2 - (-8)$ as $-2 + (8)$ since 8 is the opposite of -8 and use rule 2 for addition to get 6.
(c) Rewrite $3 - (-7)$ as $3 + (7)$ since 7 is the opposite of -7 and use rule 1 for addition to get 10.
(d) Rewrite $-2 - 6$ as $-2 + (-6)$ since -6 is the opposite of 6 and use rule 1 for addition to get -8.

Some people have had an experience of being overdrawn on their checking accounts. For example, suppose your current checkbook balance is $65.00, and you write a check for $90.00. Your balance would be $65.00 − $90.00 = −$25.00. You would be overdrawn by $25.00!

Try This One

5-H Find each.

(a) $20 - 17$

(b) $5 - 12$

(c) $-3 - 6$

(d) $-8 - (-2)$

(e) $25 - (-6)$

(f) $-14 - 8$

For the answer, see page 237.

Multiplication and Division of Integers

Multiplication is really a shortcut for addition. For example, 3×2 really means to add 2 three times. On the number line, start at zero and count 2 three times. See Figure 5-5(a).

Now in algebra we can extend the definition of multiplication. For example, $3 \times (-2)$ means to start at 0 and add -2 three times.

$$0 + (-2) + (-2) + (-2) = -6$$

See Figure 5-5(b).

What about -3×2? In this case, one can think of *subtracting* positive 2 three times; i.e., $0 - (+2) - (+2) - (+2)$ or $0 - 2 - 2 - 2$. The rule for subtracting states: add the opposite; hence,

$$-3 \times 2 = 0 - 2 - 2 - 2 \quad \text{or} \quad 0 + (-2) + (-2) + (-2) = -6$$

See Figure 5-5(c).

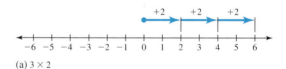

(a) 3×2

(b) $3 \times (-2)$

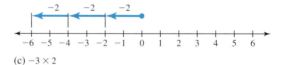

(c) -3×2

(d) $-3 \times (-2)$

Figure 5-5

Finally, what about $-3 \times (-2)$? In this case, -2 is subtracted three times, as shown.

$$0 - (-2) - (-2) - (-2) = 0 + (+2) + (+2) + (+2) = 6$$

See Figure 5-5(d).

The rules for multiplying integers are summarized next.

Rules for Multiplication of Integers

Rule 1   The product of two numbers with like signs is positive.

Rule 2   The product of two numbers with unlike signs is negative.

Example 5-9

Find each product.

(a) $(-6) \times (-4)$
(b) $3 \times (-9)$
(c) -5×16
(d) 4×32
(e) $(-6) \times (-3) \times 8 \times (-5)$ (*Hint:* Multiply left to right.)

Solution

(a) 24 using rule 1
(b) -27 using rule 2
(c) -80 using rule 2
(d) 128 using rule 1
(e) -720 using rules 1 and 2

Sidelight

Amicable Numbers

The pair of numbers 220 and 284 are called "amicable" or "friendly" numbers since the proper factors or divisors of 220 add up to 284 and the proper factors or divisors of 284 add up to 220.

$$1 + 2 + 4 + 5 + 10 + 11 + 20 + 22 + 44 + 55 + 110 = 284$$

$$1 + 2 + 4 + 71 + 142 = 220$$

This pair of numbers was the only pair of amicable numbers known to the Pythagoreans. The next pair, 17,296 and 18,416, was not discovered until 1636 by Pierre de Fermat. In 1638, Descartes discovered another pair. In 1747, Euler discovered 62 pairs of amicable numbers. Today over 600 pairs of amicable numbers are known.

Another interesting note is that in 1866, a 16-year-old Italian named Nicolo Paganini found the pair of amicable numbers 1184 and 1210 that had been overlooked by the greatest mathematicians of the time.

Try This One

5-I   Find each product.

(a) $(-16) \times 3$

(b) $(-43) \times (-56)$

(c) $12 \times (-8)$

(d) 14×13

(e) $(-2) \times (-16) \times 7 \times (-9)$

For the answer, see page 237.

Math Note

These rules can be verified by "checking" the answers. Division can be checked by multiplication. In order to check $30 \div 6 = 5$, it is necessary to multiply 5 by 6 to get 30. Hence, one can check the divisions in Example 5-10 as follows.

(a) $4 \times 8 = 32$

(b) $9 \times (-3) = -27$

(c) $-6 \times 7 = -42$

(d) $-4 \times (-5) = 20$

Rules for Division of Integers

The rules for dividing integers follow the same pattern as the rules for multiplying integers.

Rule 1   The quotient of two numbers with like signs is positive.

Rule 2   The quotient of two numbers with unlike signs is negative.

One application of negative integers is temperature. For example, if over 5 days the low temperatures in degrees Fahrenheit were $-25°$, $-21°$, $-15°$, $-23°$, $-21°$, find the average temperature by adding the integers and dividing the result by 5: $-105°/5 = -21°$.

Example   5-10

Find each quotient.

(a) $32 \div 8$
(b) $-27 \div (-3)$
(c) $-42 \div 7$
(d) $20 \div (-5)$

Solution

(a) $32 \div 8 = 4$ using rule 1
(b) $-27 \div (-3) = 9$ using rule 1
(c) $-42 \div 7 = -6$ using rule 2
(d) $20 \div (-5) = -4$ using rule 2

Suppose a merchant prices a computer at $295.00 and since it does not sell, he reduces the price by $10.00. A month later, the merchant reduces the price by another $10.00. Finally, he reduces the price by $10.00 once more. What would the final sale price be? Using integers, the reduction can be found by $(-10) + (-10) + (-10)$ or $3 \times (-10)$ or $-\$30.00$. The final price would be $\$295.00 - \$30.00 = \$265.00$.

Try This One

5-J   Find each quotient.

(a) $96 \div 6$

(b) $-48 \div (-3)$

(c) $-84 \div (-2)$

(d) $100 \div (-25)$

For the answer, see page 237.

Order of Operations

How would you interpret the following sentence? The teacher said John is tall.

Is the teacher tall or is John tall? Actually, either one could be tall depending on how the sentence is punctuated. Without punctuation, the meaning of the sentence is ambiguous.

If the teacher is tall, then you would punctuate the sentence as shown:

"The teacher," said John, "is tall."

If John is tall, then you would punctuate the sentence as shown:

The teacher said, "John is tall."

Grammar rules and punctuation symbols clarify the meaning of sentences. Mathematics also has rules and symbols that clarify the meaning of expressions. These rules are called the **order of operations.**

Steps for the Order of Operations

Step 1   Perform all calculations inside grouping symbols first. The grouping symbols used in mathematics are parentheses ( ), brackets [ ], and braces { }.

Step 2   Evaluate all exponents.

Step 3   Perform all multiplication and division in order from left to right.

Step 4   Perform all addition and subtraction in order from left to right.

> ### *Math Note*
>
> Multiplication and division are performed in the order that they appear in the problem from left to right. For example, the expression $20 \div 10 \times 2$ is evaluated by dividing 20 by 10 to get 2, then multiplying 2×2 to get 4. The same goes for addition and subtraction.

Example   **5-11**

Simplify $9 \cdot 3 - (15 \div 5)$.

Solution

$$
\begin{aligned}
9 \cdot 3 - (15 \div 5) &= 9 \cdot 3 - 3 && \text{Perform the operation in parentheses first.} \\
&= 27 - 3 && \text{Multiply before subtracting.} \\
&= 24 && \text{Subtract.}
\end{aligned}
$$

Example 5-12

Simplify $5 \cdot (8 - 10) + 2^3 \div 4$.

Solution

$$
\begin{aligned}
5 \cdot (8 - 10) + 2^3 \div 4 &= 5(-2) + 2^3 \div 4 \quad &\text{Parentheses} \\
&= 5(-2) + 8 \div 4 \quad &\text{Exponents} \\
&= -10 + 2 \quad &\text{Multiplication and division} \\
&= -8 \quad &\text{Addition}
\end{aligned}
$$

When the expression contains two or more sets of grouping symbols, start by performing the operations in the innermost set and work out.

Example 5-13

Simplify $84 \div 4 - \{3 \times [10 + (15 - 2)]\}$.

Solution

$$
\begin{aligned}
84 \div 4 - \{3 \times [10 + (15 - 2)]\} &= 84 \div 4 - \{3 \times [10 + 13]\} \\
&= 84 \div 4 - \{3 \times 23\} \\
&= 84 \div 4 - 69 \\
&= 21 - 69 \\
&= -48
\end{aligned}
$$

The order of operations is used in the real world. Suppose you want to equally divide a $12,340.00 profit among three associates and two salespersons. Would the equation $12,340.00 \div 3 + 2$ work on your calculator? The answer is no since your calculator would divide $12,340.00 by 3 and then add 2. To get the correct answer, you would need to enter the expression into your calculator as $12,340.00 \div (3 + 2)$.

Try This One

5-K   Simplify each.

(a) $8^3 - 16 \cdot 4 + 10$

(b) $216 + (4 \times 5)^2 - 13 \cdot 2$

(c) $9 \times 5 + \{[32 - (6 \times 4)] - 5\}$

For the answer, see page 237.

Exercise Set 5-2

Computational Exercises

For Exercises 1–60, perform the indicated operation(s).

1. $-6 + 5$
2. $-8 + 4$
3. $16 + (-7)$
4. $(-5) + (-7)$
5. $(-8) + (-3)$
6. $(-4) + 9$
7. $-3 + (-9)$
8. $-2 + 4$
9. $-3 + (-4) + (-6)$
10. $-5 + (-6) + (-8)$
11. $8 - (-6)$
12. $9 - 2$
13. $6 - 11$
14. $14 - 20$
15. $-3 - (-4)$
16. $-8 - (-10)$
17. $-12 - (-7)$
18. $-15 - 9$
19. $-20 - 50$
20. $-14 - 29$
21. $(5)(9)$
22. $(6)(7)$
23. $(-3)(8)$
24. $(-12)(6)$
25. $4(-9)$
26. $6(-14)$
27. $(-3)(-14)$
28. $(-7)(-14)$
29. $(-9)(0)$
30. $0(6)$
31. $64 \div 8$
32. $72 \div 9$
33. $-25 \div 5$
34. $-42 \div 7$
35. $32 \div (-8)$
36. $49 \div (-7)$
37. $-14 \div (-2)$
38. $-15 \div (-3)$
39. $-90 \div (-90)$
40. $-56 \div 4$
41. $0 \div 16$
42. $0 \div (-10)$
43. $-42 \div 6 + 7$
44. $32 \div (8 \times 2)$
45. $5^3 - 2 \cdot 7$
46. $4 \cdot 3^2 - 2 \cdot 4$
47. $9 \cdot 9 - 5 \cdot 6$
48. $32 - (-6)(4)$
49. $3^3 + 5^2 - 2^4$
50. $14^2 - 5^3 + 8^2$
51. $-3[6 + (-10) - (-2)]$
52. $-5 \cdot 4 - [-3 + 8 - (-5)]$
53. $376 - 14 \cdot 3^4$
54. $82 - 9 \cdot 6 - (-2)^2$
55. $256 - 4^3 \cdot 5 + (8 \cdot 4 - 6 \cdot 4)$
56. $6^2 + 5 \cdot 9 - (-27 + 3 \cdot 2)$
57. $-56 \div 8 - \{3 \times [-10 - (4 \times 3)]\}$
58. $(96 - 70) + [(-4 \times 9) - 32 \div 8]$
59. $32 - \{-16 + 5[25 + 9^2 + (8 - 6)]\}$
60. $2\{-5 - 6[3^2 - 7 \cdot (4 + 1)]\}$

For Exercises 61–70, insert $>$, $<$, or $=$

61. $16 \qquad 22$
62. $8 \qquad 14$
63. $-5 \qquad -10$
64. $-6 \qquad -22$
65. $0 \qquad -3$
66. $-5 \qquad 0$
67. $-9 \qquad +8$
68. $16 \qquad -32$
69. $-10 \qquad -7$
70. $-14 \qquad +3$

For Exercises 71–80, find each.

71. $|-8|$

72. $|-12|$

73. $|+10|$

74. $|+14|$

75. The opposite of -8

76. The opposite of $+27$

77. The opposite of $+10$

78. The opposite of -16

79. The opposite of 0

80. The opposite of -9

Real World Applications

81. A student's bank balance at the beginning of the month was $867. During the month, the student made deposits of $83, $562, $37, and $43. Also, the student made withdrawals of $74, $86, and $252. What was the student's bank balance at the end of the month?

82. Pike's Peak in Colorado is 14,110 feet high while Death Valley is 282 feet below sea level. Find the vertical distance from the top of Pike's Peak to the bottom of Death Valley.

83. A manager runs a hatchery for baby chicks. At the beginning of a week, there were 1286 baby chicks. The table shows the number of new chicks hatched each day of the week and the number that were sold. How many chicks were left at the end of the week?

	Mon.	Tue.	Wed.	Thur.	Fri.	Sat.
Hatched	382	494	327	778	256	641
Sold	105	850	416	237	192	965

84. A large grocery store has 354 cases of canned vegetables in the storeroom. During the past month, the store removed 87 cases, 53 cases, 42 cases, and 67 cases to put on the shelves. Also, the store received two lots of 80 cases each. How many cases are in the storeroom now?

85. The table shows the fastest growing large counties in the United States.

County	Population 1998	Population 1999
1. Sussex, Va.	10,054	12,345
2. Forsyth, Ga.	86,409	96,686
3. Douglas, Colo.	141,449	156,860
4. Loudoun, Va.	144,514	156,284
5. Henry, Ga.	104,925	113,443

Source: U.S. Census Bureau

What was the increase in population for each county?

86. The number of students per computer Internet access in Delaware is approximately 6. In North Carolina, it is approximately 25. Find the difference in the numbers.

Source: Technology in Education

87. The 30-year average snowfall in Bismarck, North Dakota, is approximately 18 inches. As of January 10, Bismarck received approximately 10 inches. How much more snow will Bismarck receive this year if this is an average year?

88. The 30-year average snowfall in Salt Lake City is about 25 inches. As of January 10, the recorded snowfall was 34 inches. How much more snow will Salt Lake City receive this year if this is an average year?

Writing Exercises

89. Explain the difference between a negative number and the opposite of a number.
90. What are two uses for the $+$ sign?
91. What are two uses for the $-$ sign?
92. Explain the difference between the "opposite" of a number and the "absolute value" of a number.
93. Explain how the set of integers differs from the set of whole numbers.

Critical Thinking

94. Why is it difficult to think of "subtraction" as "taking away" when using integers?
95. The expression $-(-x)$ is not always positive, since x can be a positive number, a negative number, or zero. What will be the value of the expression $-(-x)$ when
    (a) $x = 3$
    (b) $x = -5$
    (c) $x = -9$
    (d) $x = +10$
    (e) $x = 0$

The Rational Numbers

Definition of Rational Numbers

Although many real-world problems can be solved using integers, there are many situations where other types of numbers are necessary. For example, "The average hat size for men is $7\frac{1}{4}$ in." and "The Dow Jones industrial average closed at 10,951.24, up 154.59 from yesterday."

Numbers such as these are called *rational numbers*. The rational numbers can be found in the number line. For example, $\frac{2}{3}$ is located between 0 and 1. Likewise, -2.36 is located between -2 and -3. See Figure 5-6.

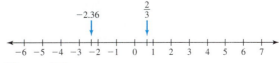

Figure 5-6

When the noninteger fractions are added to the set of integers, the new set of numbers is called the *rational numbers*.

> Any number that can be written as a fraction, $\dfrac{a}{b}$, where a and b are integers and $b \neq 0$ is called a **rational number.** The integer a is called the *numerator* of the fraction, and the integer b is called the *denominator* of the fraction.

All natural numbers are rational numbers since they can be written as fractions using 1 as a denominator. For example, $2 = \frac{2}{1}$, $8 = \frac{8}{1}$, etc. All whole numbers can be written as fractions (e.g., $0 = \frac{0}{1}$) and all integers can be written as fractions (e.g., $-5 = \frac{-5}{1}$). The number 2 then is a natural number, a whole number, an integer, and a rational number. Zero is a whole number, an integer, and a rational number. The number -5 is an integer and a rational number. The number $-\frac{5}{6}$ is a rational number. The number $-\frac{5}{6}$ is not a natural number, whole number, or an integer.

All fractions can be written as decimals and some decimals can be written as fractions. These techniques will be shown later in this section.

Fractions can be compared, reduced, changed to higher terms, added, subtracted, multiplied, and divided. The properties and rules for fractions are presented next.

There are three types of fractions. They are proper, improper, and mixed.

A **proper fraction** has a numerator whose absolute value is less than the absolute value of the denominator. For example, $\frac{3}{4}$, $\frac{1}{2}$, and $\frac{7}{8}$ are called proper fractions.

An **improper fraction** has a numerator whose absolute value is greater than or equal to the absolute value of the denominator. For example, $\frac{8}{3}$, $\frac{6}{5}$, $\frac{10}{4}$, and $\frac{6}{6}$ are called improper fractions.

Sidelight

Perfect Numbers

A *perfect number* is a number such that the sum of its proper factors is equal to the number. The proper factors of a number are all the factors of the number except the number itself. The first perfect number is 6. The factors of 6 are 1, 2, 3, and 6. The sum of the factors of 6 (excluding 6) is equal to 6: $1 + 2 + 3 = 6$. The factors of 28 are 1, 2, 4, 7, 14, and 28, and the sum of the proper factors is $1 + 2 + 4 + 7 + 14 = 28$.

Can you guess what the next perfect number is? It is 496. The sum of its proper factors is 496; i.e., $1 + 2 + 4 + 8 + 16 + 31 + 62 + 124 + 248 = 496$.

The first four perfect numbers were thought to be known by the Pythagoreans. The fifth perfect number (33,550,336) was found in 1461. Around 1600, Pietro Cataldi discovered the sixth and seventh perfect numbers. Around 1732, Leonhard Euler is said to have discovered the eighth perfect number. Since these perfect numbers are quite large, it is difficult to verify that they are indeed perfect numbers without the aid of a computer. The eighth perfect number has 19 digits!

The 9th, 10th, 11th, and 12th perfect numbers were discovered in the 1800s. The 12th perfect number contains 77 digits and is the last perfect number to be discovered without the aid of a computer. Up to 1950, only 12 perfect numbers were known.

With the advent of modern computers and as the capacity of computers has increased, larger and larger perfect numbers have been discovered.

Today over 35 perfect numbers are known, and the search continues. Note that all perfect numbers (except 28) that have been found to date end in 6. It has not been proved if other perfect numbers that do not end in 6 exist. It is not known if there are a finite number or an infinite number of perfect numbers and whether there are any odd perfect numbers.

A **mixed number** consists of a whole number and a fraction. For example, $3\frac{2}{5}$, $6\frac{1}{3}$, and $5\frac{5}{6}$ are called mixed numbers.

An improper fraction can be written as a whole number or a mixed number by dividing the numerator by the denominator and writing any remainder as a fraction. This is shown in Example 5-14.

Example 5-14

Write $\frac{5}{3}$ as a mixed number.

Solution

Divide 5 by 3 and write the remainder as a proper fraction, as shown.

$$
\begin{array}{r}
1 \quad \frac{2}{3} \\
3\,\overline{)\,5} \\
\underline{-3} \\
2
\end{array}
$$

Hence, $\frac{5}{3} = 1\frac{2}{3}$

Every nonzero fraction has three signs associated with it: the sign of the number in the numerator, the sign of the number in the denominator, and the sign in front of the fraction (called the sign of the fraction). For example, the fraction $\frac{3}{4}$ can be written as $+\frac{+3}{+4}$. Here the numbers in the numerator and denominator are positive and, as written, the sign of the fraction is positive. A rule of mathematics is that any two signs of the fraction can be changed without changing the value of the fraction. Consider the fraction $\dfrac{-10}{5}$. Here the number in the numerator is negative and the number in the denominator, as well as the sign of the fraction, are positive. By the rules of division for integers, the value of the fraction is equal to -2. Now if we change two signs, say that of the numerator and the sign of the

Fractions can be used to report stock prices.

fraction, i.e., $\dfrac{10}{-5}$, we would get $-(2)$ which is -2. (This one example is used to illustrate the rule, not to prove it.)

In general, for any nonzero fraction, $\dfrac{-a}{b} = \dfrac{a}{-b} = -\dfrac{a}{b}$. Usually when a fraction has a negative sign in either the numerator or the denominator, it is written in front of the fraction. For example, $\dfrac{-3}{4}$ is written as $-\dfrac{3}{4}$.

Reducing Fractions

Fractions can be *reduced* to **lowest terms** by dividing the numerator and denominator by the greatest common factor of both numbers. Then the fraction is said to be "simplified."

Example 5-15

Reduce $\frac{18}{24}$ to lowest terms.

Solution

Divide 18 and 24 by 6 since 6 is the GCF of 18 and 24.

$$\frac{18 \div 6}{24 \div 6} = \frac{3}{4}$$

> **Math Note**
>
> This is usually shown as $\frac{\cancel{18}^{\,3}}{\cancel{24}_{\,4}}$ and is called *canceling*. A fraction can be reduced in steps using any common factor of the numerator and denominator, as shown.
>
> $$\frac{18}{24} = \frac{9}{12} = \frac{3}{4}$$
>
> In this case, the 18 and 24 were divided by 2, then the 9 and 12 were divided by 3.

Try This One

5-L   Reduce each fraction to lowest terms.

(a) $\frac{3}{18}$

(b) $\frac{56}{64}$

(c) $\frac{-58}{87}$

(d) $\frac{60}{200}$

For the answer, see page 237.

Changing Fractions to Higher Terms

Fractions can also be changed to *higher terms* by multiplying the numerator and denominator by the same number, as shown in the next example.

Example 5-16

Change $\frac{3}{8}$ to a fraction with a denominator of 32.

Solution

Since $32 \div 8 = 4$, multiply the numerator and denominator by 4, as shown.

$$\frac{3}{8} = \frac{3 \cdot 4}{8 \cdot 4} = \frac{12}{32}$$

Try This One

5-M Change each fraction to a fraction with the denominator shown.

(a) $\dfrac{3}{8} = \dfrac{?}{24}$

(b) $\dfrac{7}{13} = \dfrac{?}{65}$

(c) $\dfrac{5}{9} = \dfrac{?}{99}$

(d) $\dfrac{4}{5} = \dfrac{?}{50}$

For the answer, see page 237.

Operations with Fractions

Fractions can be added, subtracted, multiplied, or divided using the rules found next.

Rules for Operations with Fractions

Addition and Subtraction: To add two or more fractions, find the lowest common denominator (that is, the least common multiple of the denominator numbers), change the fractions to higher terms if necessary, add or subtract the numerators, and reduce or simplify the answer, if possible.

Multiplication: To multiply two or more fractions, cancel if possible, multiply the numerators and multiply the denominators.

Division: To divide two fractions, invert the fraction after the ÷ sign and follow the rule for multiplication.

Example 5-17

Perform the indicated operations.

(a) $\frac{3}{8} + \frac{7}{8}$

(b) $\frac{1}{4} + \frac{5}{6}$

(c) $\frac{4}{9} - \frac{2}{5}$

Solution

(a) Since the denominators are the same, add the numerators and reduce or simplify.

$$\frac{3}{8} + \frac{7}{8} = \frac{3+7}{8} = \frac{10}{8} = \frac{5}{4} \text{ or } 1\frac{1}{4}$$

(b) First find the common denominator, then change to higher terms, add the numerators and reduce or simplify if possible. The lowest common denominator of 4 and 6 is 12.

$$\frac{1}{4} + \frac{5}{6} = \frac{1}{4} \cdot \frac{3}{3} + \frac{5}{6} \cdot \frac{2}{2} = \frac{3}{12} + \frac{10}{12} = \frac{3+10}{12} = \frac{13}{12} \text{ or } 1\frac{1}{12}$$

—Continued

Example 5-17 *Continued—*

(c) Find the common denominator, change to higher terms, and subtract the numerators. The lowest common denominator is 45.

$$\frac{4}{9} - \frac{2}{5} = \frac{4}{9} \cdot \frac{5}{5} - \frac{2}{5} \cdot \frac{9}{9} = \frac{20}{45} - \frac{18}{45} = \frac{2}{45}$$

Example 5-18

Perform the indicated operations.

(a) $\frac{5}{8} \times \frac{3}{5}$

(b) $\frac{3}{4} \div \left(-\frac{5}{8}\right)$

Solution

(a) First cancel, then multiply the numerators, then multiply the denominators.

$$\frac{5}{8} \times \frac{3}{5} = \frac{\cancel{5}^1}{8} \times \frac{3}{\cancel{5}^1} = \frac{1 \cdot 3}{8 \cdot 1} = \frac{3}{8}$$

(b) Invert the fraction after the $\div$ sign and follow the rule for multiplication.

$$\frac{3}{4} \div \left(-\frac{5}{8}\right) = \frac{3}{4} \cdot \left(-\frac{8}{5}\right) = \frac{3}{\cancel{4}^1} \cdot \left(-\frac{\cancel{8}^2}{5}\right) = -\frac{3 \cdot 2}{1 \cdot 5} = -\frac{6}{5} \text{ or } -1\frac{1}{5}$$

Sidelight

Mathematics and Music

Music is based on mathematics. When the string of a guitar is plucked, a note is sounded, and if that string is lengthened to make it twice as long and it is plucked again, the same note will be heard only it will be one octave below the first note. By increasing the length of the string, other notes can be sounded. For example, if a string $\frac{16}{15}$ as long as the C string is plucked, it will give the next lower note, B. If it is stretched to $\frac{6}{5}$ as long, it will sound an A note, etc. You may think that this is a modern discovery; however, the famous Greek mathematician, Pythagoras, discovered the mathematical relationship between the notes C, F, and G over 2500 years ago.

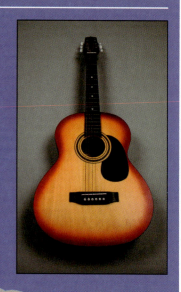

Try This One

5-N   Perform the indicated operations.

(a) $\frac{5}{12} + \frac{1}{12}$

(b) $-\frac{3}{8} + \frac{5}{6}$

(c) $\frac{9}{10} - \frac{1}{2}$

(d) $\frac{2}{15} \times \left(-\frac{5}{6}\right)$

(e) $-\frac{7}{8} \div \frac{3}{4}$

For the answer, see page 237.

Density of Rational Numbers

The set of rational numbers is said to be dense. The density property is defined in general as follows.

> A set of numbers is said to be **dense** if for any two given numbers in the set, there exists a third number in the set that lies between the two given numbers.

In order to illustrate the density property for the rational numbers, one must find a rational number between any two given rational numbers. In order to do this, add the two given rational numbers and divide the sum by 2. In this case, you will find a rational number halfway between the two given rational numbers. For example, to find a rational number halfway between $\frac{3}{4}$ and $\frac{9}{10}$, add $\frac{3}{4} + \frac{9}{10} = \frac{15}{20} + \frac{18}{20} = \frac{33}{20}$ and divide $\frac{33}{20} \div 2 = \frac{33}{40}$. Hence, $\frac{3}{4} < \frac{33}{40} < \frac{9}{10}$.

Fractions and Decimals

Fractions can be changed to decimals. You may remember that $\frac{1}{2}$ is the same as 0.5 and $\frac{3}{4}$ is the same as 0.75. Each digit in a decimal number has a **place value.** See Table 5-1.

To change a fraction to a decimal, divide the numerator by the denominator.

Gas prices are often given in thousandths of a dollar

Table 5-1

Place Value Chart for Decimals

millions	hundred-thousands	ten-thousands	thousands	hundreds	tens	ones	.	tenths	hundredths	thousandths	ten-thousandths	hundred-thousandths	millionths

Example   5-19

Write $\frac{5}{8}$ as a decimal.

Solution

Divide 5 by 8, as shown.

$$
\begin{array}{r}
0.625 \\
8\overline{\smash{)}5.000} \\
-48 \\
\hline
20 \\
-16 \\
\hline
40 \\
-40 \\
\hline
0
\end{array}
$$

Hence, $\frac{5}{8} = 0.625$

The decimal for $\frac{5}{8}$ (i.e., 0.625) is called a **terminating decimal** because a remainder of zero occurs in the division process. Not all fractions can be converted to terminating decimals, as shown in Example 5-20.

Example   5-20

Write $\frac{5}{6}$ as a decimal.

Solution

$$
\begin{array}{r}
0.8333\ldots \\
6\overline{\smash{)}5.0000} \\
-48 \\
\hline
20 \\
-18 \\
\hline
20 \\
-18 \\
\hline
20 \\
-18 \\
\hline
2
\end{array}
$$

The decimal for $\frac{5}{6}$ (i.e., 0.8333 . . .) is called a **repeating decimal.** Repeating decimals can be written by placing a line over the digits that repeat.

0.8333 . . . is written as $0.8\overline{3}$.
0.626262 . . . is written as $0.\overline{62}$.

Math Note

The decimal $0.5\overline{6}$ is not the same as $0.\overline{56}$. In the first case, only the 6 repeats (0.5666 . . .). In the second case, the 56 repeats (0.565656 . . .).

Try This One

5-O   Write each fraction as a decimal.

(a) $\frac{5}{12}$

(b) $\frac{19}{33}$

(c) $\frac{3}{8}$

(d) $\frac{2}{7}$

For the answer, see page 237.

Terminating decimals can be written as fractions by using the next procedure.

Procedure for Writing a Terminating Decimal as a Fraction

Step 1   Drop the decimal point.
Step 2   Place the resulting number in the numerator of a fraction.
Step 3   Use a denominator of 10 if there was one digit to the right of the decimal point. Use a denominator of 100 if there were two digits to the right of the decimal point. Use a denominator of 1000 if there were three digits to the right of the decimal, etc.
Step 4   Reduce the fraction if possible.

Example   5-21

Write each decimal as a fraction.

(a) 0.8
(b) 0.65
(c) 0.024

Solution

(a) $0.8 = \frac{8}{10} = \frac{4}{5}$
(b) $0.65 = \frac{65}{100} = \frac{13}{20}$
(c) $0.024 = \frac{24}{1000} = \frac{3}{125}$

Try This One

5-P   Write each decimal as a fraction.

(a) 0.4

(b) 0.48

(c) 0.325

For the answer, see page 237.

Repeating decimals such as $0.\overline{3}$ and $0.4242\ldots$ can be changed to fractions using basic algebra. Examples 5-22, 5-23, and 5-24 show this procedure.

Example 5-22

Change $0.\overline{8}$ to a fraction.

Solution

Step 1   Let $n = 0.\overline{8}$ and multiply both sides of the equation by 10 to get $10n = 8.\overline{8}$.

Step 2   Subtract the first equation from the second one as shown.

$$10n = 8.\overline{8}$$
$$-n = 0.\overline{8}$$
$$\overline{9n = 8}$$

Step 3   Solve the equation $9n = 8$ for n as shown.

$$\frac{9n}{9} = \frac{8}{9}$$
$$n = \frac{8}{9}$$

Math Note

The result can be checked by changing $\frac{8}{9}$ to a decimal.

If two digits repeat, multiply both sides of the equation by 100, then subtract. If three digits repeat, multiply both sides by 1000, etc.

Example 5-23

Change $0.\overline{63}$ to a fraction.

Solution

Step 1   Let $n = 0.\overline{63}$

then $100n = 63.\overline{63}$

Step 2 $100n = 63.\overline{63}$
$$-n = 0.\overline{63}$$
$$\overline{99n = 63}$$

Step 3 $\frac{99n}{99} = \frac{63}{99}$

$$n = \frac{63}{99} = \frac{21}{33} = \frac{7}{11}$$

Calculator Explorations

Each of the decimals shown in the display window of the calculator represents a repeating decimal. If all the decimals are repeating, why do

```
8/9
         .8888888889
63/99
         .6363636364
1/3
         .3333333333
```

you suppose the first decimal ends with a 9 rather than an 8, the second ends with a 4 rather than a 3? How do you suppose the calculator used to generate the decimals shown would display $0.\overline{64}$? $0.\overline{2}$?

Example   5-24

Change $0.8\overline{3}$ to a fraction.

Solution

Step 1   Let $n = 0.8\overline{3}$

$10n = 8.\overline{3}$

Step 2 $10n = 8.3\overline{3}$

$\underline{-n = 0.8\overline{3}}$

$9n = 7.5$

(*Note:* In order to line up the numbers, $8.\overline{3}$ is written as $8.3\overline{3}$.)

Step 3 $\dfrac{9n}{9} = \dfrac{7.5}{9}$

$n = \dfrac{7.5}{9}$

In this case, the fraction $\frac{7.5}{9}$ contains a decimal point. In order to simplify it, multiply the numerator and the denominator by 10 and reduce the fraction as shown.

$$\frac{7.5}{9} = \frac{7.5 \times 10}{9 \times 10} = \frac{75}{90} = \frac{5}{6}$$

The procedure is summarized next.

Procedure for Changing a Repeating Decimal to a Fraction

Step 1   Let n equal the decimal. Multiply both sides of the equation by 10, 100, or 1000, etc., depending on the number of different digits that repeat.

Step 2   Subtract the first equation from the second.

Step 3   Solve for n and simplify if necessary.

Try This One

5-Q   Change each decimal to a fraction.

(a) $0.\overline{4}$

(b) $0.\overline{56}$

(c) $0.6\overline{4}$

For the answer, see page 237.

Exercise Set 5-3

Computational Exercises

For Exercises 1–10, reduce each fraction to lowest terms.

1. $\frac{7}{42}$

2. $\frac{8}{24}$

3. $\frac{42}{60}$

4. $\frac{16}{20}$

5. $\frac{30}{36}$

6. $\frac{25}{75}$

7. $\frac{91}{104}$

8. $\frac{68}{119}$

9. $\frac{420}{756}$

10. $\frac{950}{2400}$

For Exercises 11–20, change each fraction to higher terms.

11. $\frac{5}{16} = \frac{?}{48}$

12. $\frac{15}{32} = \frac{?}{96}$

13. $\frac{19}{24} = \frac{?}{48}$

14. $\frac{5}{8} = \frac{?}{40}$

15. $\frac{7}{9} = \frac{?}{45}$

16. $\frac{3}{10} = \frac{?}{30}$

17. $\frac{11}{16} = \frac{?}{80}$

18. $\frac{3}{7} = \frac{?}{28}$

19. $\frac{1}{5} = \frac{?}{30}$

20. $\frac{5}{16} = \frac{?}{64}$

For Exercises 21–40, perform the indicated operations and reduce the answer to lowest terms.

21. $\frac{-5}{6} + \frac{2}{3}$

22. $\frac{3}{4} + \frac{7}{10}$

23. $\frac{-11}{12} - \frac{5}{8}$

24. $\frac{19}{24} - \frac{7}{18}$

25. $\frac{-5}{12} \times \frac{-7}{10}$

26. $\frac{5}{18} \times \frac{9}{25}$

27. $\frac{7}{9} \div \frac{2}{3}$

28. $\frac{-7}{24} \div \frac{23}{30}$

29. $\left(\frac{7}{16} \div \frac{3}{8} \right) \times \frac{3}{5}$

30. $\frac{-7}{8} \div \left(\frac{2}{3} \div \frac{15}{16} \right)$

31. $\frac{-11}{22} \times \left(\frac{1}{6} \times \frac{3}{4} \right)$

32. $\left(\frac{9}{10} - \frac{2}{3} \right) \times \frac{1}{2}$

33. $\left(\frac{5}{8} + \frac{3}{4} \right) \times \frac{2}{3}$

34. $\frac{-5}{6} \times \frac{7}{8}$

35. $\left(\frac{-3}{4} \right) \div \left(\frac{-5}{8} \right)$

36. $\left(\frac{-7}{8} \right) \div \left(\frac{-3}{4} \right)$

37. $\left(\frac{9}{14} \div \frac{3}{7} \right) \times \frac{1}{2}$

38. $\left(\frac{4}{5} + \frac{7}{8} \right) \div \frac{1}{9}$

39. $\left(\dfrac{9}{10} - \dfrac{2}{3}\right) \times \dfrac{5}{6}$

40. $\dfrac{3}{4} \div \left(\dfrac{5}{8} + \dfrac{1}{2}\right)$

For Exercises 41–52, change each fraction to a decimal.

41. $\dfrac{1}{5}$

42. $\dfrac{3}{10}$

43. $\dfrac{2}{3}$

44. $\dfrac{7}{5}$

45. $\dfrac{9}{4}$

46. $\dfrac{11}{9}$

47. $\dfrac{11}{36}$

48. $\dfrac{12}{7}$

49. $\dfrac{3}{4}$

50. $\dfrac{15}{8}$

51. $\dfrac{48}{51}$

52. $\dfrac{17}{24}$

For Exercises 53–58, change each decimal to a reduced fraction.

53. 0.875

54. 0.964

55. $0.\overline{54}$

56. $0.\overline{62}$

57. 0.375

58. $0.45\overline{3}$

Real World Applications

59. Julie and Sue are driving to a business conference. They decide to stop for gasoline, and Sue estimates that they have driven $\dfrac{3}{5}$ of the distance. The odometer shows that they have driven 285 miles. How far away is the conference?

60. An automobile race track in city A is $\dfrac{3}{8}$ of a mile long and an automobile race track in city B is $\dfrac{1}{6}$ of a mile long. If a driver drives 24 laps on each track, how many miles will the automobile travel?

61. A company uses $\dfrac{2}{7}$ of its budget for advertising. Of that, $\dfrac{1}{2}$ is spent on television advertisement. What part of its budget is spent on television advertisement?

62. According to the Census Bureau, 4 out of 25 men do not have health insurance. In a group of 250 men, about how many men would not have health insurance?

63. On a map, 1 inch represents 80 miles. If two cities are $2\dfrac{3}{8}$ inches apart, how far apart in miles are they?

64. An architect's rendering of a house plan shows that $\dfrac{1}{4}$ inch represents 1 foot. If the family room plan is 3 inches long, how long will the actual family room be?

65. A piece of wire is $\dfrac{3}{4}$ meter long. If it is cut into 10 pieces of equal length, how long will each piece be?

66. An estate was divided among five people. The first person received $\dfrac{1}{8}$ of the estate. The next two people each received $\dfrac{1}{5}$ of the estate. The fourth person received $\dfrac{1}{10}$ of the estate. What fractional part of the estate did the last person receive?

67. For a certain municipality, $\dfrac{2}{3}$ of the waste generated consisted of paper products, $\dfrac{1}{10}$ consisted of glass products, and $\dfrac{1}{5}$ consisted of plastic products. Eight thousand tons of waste were hauled. How much of it consisted of paper, glass, and plastic products?

68. A recipe calls for $2\dfrac{1}{2}$ cups of flour and $\dfrac{2}{3}$ cup of sugar. If a person wanted to cut the recipe in half, how many cups of flour and sugar are needed?

Critical Thinking

69. Two fractions $\frac{a}{b}$ and $\frac{c}{d}$ are equal if $a \cdot d = b \cdot c$. Two fractions are equal if they have the same denominators and the same numerators. Explain how the two statements are related.

70. The Egyptians wrote fractions using the symbol ⌣ to represent the inverse of the number. For example,  represents $\frac{1}{4}$ and 𝄃𝄃 represents $\frac{1}{7}$. Fractions with numbers other than one were written as the sum of two or more fractions whose numerators are one. For example, $\frac{11}{28}$ was written as $\frac{1}{4} + \frac{1}{7}$. Write each fraction as a sum of two fractions whose numerators are one.

    (a) $\frac{13}{40}$

    (b) $\frac{4}{9}$

    (c) $\frac{13}{42}$

5-4 The Irrational Numbers

Definition of an Irrational Number

In Section 5-3, it was shown that any rational number can be written either as a terminating decimal or a repeating decimal. But there are decimal numbers that neither terminate nor repeat. For example, the number 0.0100100010000100000100000001. . . neither terminates nor repeats. Although at first glance, it looks like the zeros and the ones repeat, there is no fixed sequence of numbers that repeat since between each 1, the number of 0s is increasing. Decimal numbers like these are called irrational numbers.

> The set of **irrational numbers** (I) consists of the numbers that can be written as non-terminating and nonrepeating decimals.

Two examples of irrational numbers are

$$\sqrt{2} \approx 1.41421356237\ldots$$
$$\pi \approx 3.141592653589793\ldots$$

The $\sqrt{2}$ is read as "the square root of 2." The symbol is called a *radical* sign. The $\sqrt{2}$ means to find a nonnegative number such that when it is multiplied by itself (or squared), it gives a product of 2. The square root of 2 is a number between one and two since

$$1 \times 1 = 1 \ \text{ and } \ 2 \times 2 = 4$$

You can approximate the $\sqrt{2}$ by finding a number that when squared gives an answer close to 2. For example, $(1.4)^2 = 1.96$ and $(1.5)^2 = 2.25$. Hence, the square root of 2 is a number between 1.4 and 1.5. You could guess 1.41, but $(1.41)^2 = 1.9881$. Although $(1.41)^2$ is closer to 2 than 1.4, it is still too small. Now try 1.42; $(1.42)^2 = 2.0164$. This number is too large. Hence, the $\sqrt{2}$ is between 1.41 and 1.42.

Sidelight

Pythagoras

Almost everybody remembers the Pythagorean theorem—not what it actually says, but that they studied it at one time or another in elementary school, grade school, or high school.

Pythagoras was born on the island of Samos about 570 B.C.E. He studied under another famous mathematician, Thales. Pythagoras' early life is only conjecture. He founded a society in 509 B.C.E. that was devoted to the study of mathematics. The society taught, "Numbers ruled the universe."

In order to be accepted into the society, the brethren had to undergo several degrees of initiation. For the first 5 years, the initiates to the society were not allowed to speak, wear wool, eat meat or beans, or touch a white rooster. After a brother completed the initiation, he was taught the truth about mathematics.

The Pythagoreans discovered the mathematical relationship between the length of a string of a musical instrument and its pitch. They also studied prime numbers, perfect numbers, and amicable numbers.

In addition, the Pythagoreans studied geometry. They were interested only in the study of geometric forms such as triangles, squares, and circles. They were not concerned with the measurement of lines and the numerical nature of geometry.

They did not recognize zero or the negative numbers since they could not find any physical meaning for them.

Finally, they attributed mystical powers to numbers. For example, they considered odd numbers to be masculine and even numbers to be feminine. They also believed that the future could be predicted by

numbers and a person's character could be revealed by numbers.

The members of the society were sworn to secrecy; one member, Hippasus, spoke of the discovery of irrational numbers, which the Pythagoreans refused to accept. Later when he perished in a shipwreck, the Pythagoreans believed that this was a sign that the discovery was "alagon" (unutterable)!

At the time the citizens of the community where the Pythagoreans lived became suspicious of the secret society, and when Pythagoras started to speak in public, he drew large crowds. The distrust of some then became a riot; the citizens sacked and burned the society's buildings.

Pythagoras and what was left of his followers fled to Tarentum to start a new school. His reputation, however, preceded him, and he was forced to flee again to Metapontium. At Metapontium, another riot broke out, and this time, Pythagoras was killed.

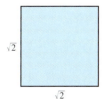

A square with an area of 2 square units would have sides $\sqrt{2}$ units long.

You can continue the process as shown.

$$1 < \sqrt{2} < 2$$
$$1.4 < \sqrt{2} < 1.5$$
$$1.41 < \sqrt{2} < 1.42$$
$$1.414 < \sqrt{2} < 1.415$$
$$1.4142 < \sqrt{2} < 1.4143$$
$$\text{etc.}$$

Notice that squaring each number on the left side of the inequality gives a number slightly smaller than 2 and squaring each number on the right side of the inequality gives a number slightly larger than 2. You could continue this process forever, but you would never get a number that when squared would give an answer that is exactly equal to 2. This fact was proved by Euclid over 3000 years ago. The square root of 2 is 1.41423562… and this has been carried out to only nine decimal places. Hence, $\sqrt{2}$ is a nonterminating and a nonrepeating decimal number or an irrational number.

The number π (pi) is the distance around a circle, the circumference, divided by the diameter of a circle. (See Chapter 10.) That is, π is the ratio of the circumference of the circle to the diameter of a circle, $C = \frac{\pi}{d}$.

It can be shown that π is an irrational number. This is beyond the scope of this textbook.

Operations with Irrational Numbers

Operations can be performed with irrational numbers, and the answers can be simplified. A radical is simplified when there is no factor of the number under the radical sign that is a perfect square if we are using square roots, a perfect cube if we are using cube roots, etc.

Simplifying Irrational Numbers

Rule 1 For any two numbers $a \geq 0$ and $b \geq 0$,

$$\sqrt{ab} = \sqrt{a} \cdot \sqrt{b}$$

Example 5-25

Simplify $\sqrt{40}$.

Solution

Find the largest perfect square that is a factor of 40, then use rule 1 to simplify the radical as shown.

$$\sqrt{40} = \sqrt{4 \cdot 10}$$
$$= \sqrt{4} \cdot \sqrt{10}$$
$$= 2\sqrt{10}$$

Example 5-26

Simplify $\sqrt{200}$.

Solution

$$\sqrt{200} = \sqrt{100 \cdot 2}$$
$$= \sqrt{100} \cdot \sqrt{2}$$
$$= 10\sqrt{2}$$

Try This One

5-R   Simplify each.

(a) $\sqrt{27}$

(b) $\sqrt{56}$

(c) $\sqrt{75}$

For the answer, see page 237.

Two square roots can be multiplied by using rule 2.

Rule 2   For any two numbers, $a \geq 0$ and $b \geq 0$,
$$\sqrt{a} \cdot \sqrt{b} = \sqrt{ab}$$

Example   5-27

Multiply $\sqrt{6} \cdot \sqrt{2}$.

Solution

Multiply the numbers under the radical signs using rule 2.
$$\sqrt{6} \cdot \sqrt{2} = \sqrt{6 \cdot 2} = \sqrt{12}$$

Then simplify $\sqrt{12}$ using rule 1.
$$\sqrt{12} = \sqrt{4 \cdot 3} = \sqrt{4} \cdot \sqrt{3} = 2\sqrt{3}$$

Example   5-28

Multiply $\sqrt{8} \cdot \sqrt{4}$.

Solution

$$\sqrt{8} \cdot \sqrt{4} = \sqrt{32}$$
$$\sqrt{32} = \sqrt{16 \cdot 2} = \sqrt{16} \cdot \sqrt{2} = 4\sqrt{2}$$

Try This One

5-S   Multiply each.

(a) $\sqrt{35} \cdot \sqrt{5}$

(b) $\sqrt{18} \cdot \sqrt{6}$

(c) $\sqrt{42} \cdot \sqrt{15} \cdot \sqrt{15}$

For the answer, see page 237.

The quotient of two square roots can be found by using rule 3.

Rule 3 For any two numbers $a \geq 0$ and $b > 0$,

$$\frac{\sqrt{a}}{\sqrt{b}} = \sqrt{\frac{a}{b}}$$

Example 5-29

Find the quotient $\dfrac{\sqrt{27}}{\sqrt{3}}$.

Solution

Use rule 3, then divide the number under the radical sign as shown.

$$\frac{\sqrt{27}}{\sqrt{3}} = \sqrt{\frac{27}{3}} = \sqrt{9} = 3$$

Example 5-30

Find the quotient $\dfrac{\sqrt{60}}{\sqrt{3}}$.

Solution

$$\frac{\sqrt{60}}{\sqrt{3}} = \sqrt{\frac{60}{3}} = \sqrt{20} = \sqrt{4 \cdot 5} = \sqrt{4} \cdot \sqrt{5} = 2\sqrt{5}$$

Try This One

5-T Find the quotients.

(a) $\dfrac{\sqrt{80}}{\sqrt{10}}$

(b) $\dfrac{\sqrt{48}}{\sqrt{3}}$

For the answer, see page 237.

> *Math Note*
>
> Caution must be used in determining whether or not two radical expressions are alike. For example $\sqrt{18}$ and $\sqrt{2}$ may at first glance both look like unlike radicals, but 18 can be simplified to $3\sqrt{2}$; hence, $\sqrt{18}$ and $\sqrt{2}$ are like radicals.

When the numbers under the radical sign (called the radicands) are the same, the expressions are said to be *like* radicals. For example, $6\sqrt{2}$ and $-\sqrt{2}$ are like radicals since radicands 2 are the same. $-8\sqrt{3}, 4\sqrt{3}$, and $\sqrt{3}$ are like radicals since they have the same radicands. $5\sqrt{6}$ and $2\sqrt{3}$ are unlike radicals since the radicands 6 and 3 are different.

Addition and subtraction of irrational numbers can be performed by using rule 4.

Rule 4 For any number $c \geq 0$, $a\sqrt{c} + b\sqrt{c} = (a + b)\sqrt{c}$ and $a\sqrt{c} - b\sqrt{c} = (a - b)\sqrt{c}$

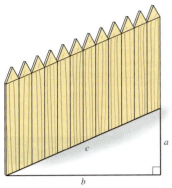

A familiar use of radicals is determining sides of right triangles with the Pythagorean theorem, $a^2 + b^2 = c^2$. If the sides of the garden a and b are 10 ft and 24 ft, respectively, how long is the fence, c?

Example 5-31

Find the sum of $6\sqrt{3} + 8\sqrt{3} + 5\sqrt{3}$.

Solution

Since all the radicals are like radicals, the sum can be found by adding the coefficients of the radicals. Hence,

$$6\sqrt{3} + 8\sqrt{3} + 5\sqrt{3} = (6 + 8 + 5)\sqrt{3} = 19\sqrt{3}$$

Example 5-32

Find the difference $3\sqrt{10} - 7\sqrt{10}$.

Solution

$$3\sqrt{10} - 7\sqrt{10} = (3 - 7)\sqrt{10} = -4\sqrt{10}$$

Example 5-33

Perform the indicated operations.

$$4\sqrt{2} - 3\sqrt{8} + 5\sqrt{32}$$

Solution

Simplify each radical first, and then add or subtract like radicals as shown.

$$-3\sqrt{8} = -3\sqrt{4 \cdot 2} = -3\sqrt{4} \cdot \sqrt{2} = -3 \cdot 2\sqrt{2} = -6\sqrt{2}$$
$$5\sqrt{32} = 5\sqrt{16 \cdot 2} = 5\sqrt{16} \cdot \sqrt{2} = 5 \cdot 4\sqrt{2} = 20\sqrt{2}$$

Hence,

$$4\sqrt{2} - 6\sqrt{2} + 20\sqrt{2} = (4 - 6 + 20)\sqrt{2} = 18\sqrt{2}$$

Math Note

The expressions $5\sqrt{2} + 6\sqrt{3}$ cannot be added since the radicands are different.

Try This One

5-U Perform the indicated operations and simplify the answers if possible.

(a) $5\sqrt{2} + 7\sqrt{2} - 10\sqrt{2}$

(b) $2\sqrt{8} + 5\sqrt{50}$

(c) $6\sqrt{28} + 4\sqrt{112} - 2\sqrt{12}$

(d) $7\sqrt{108} - 2\sqrt{75} - 4\sqrt{245}$

For the answer, see page 237.

Another method used to simplify radical expressions is called **rationalizing the denominator.**

When a radical expression contains a square root sign in the denominator of a fraction, it can be simplified by multiplying the numerator and denominator by a radical expression that will make the denominator a perfect square. This is called rationalizing the denominator. Examples 5-34, 5-35, and 5-36 show the process.

Example 5-34

Simplify $\dfrac{18}{\sqrt{3}}$.

Solution

Multiply the numerator and denominator by $\sqrt{3}$ since $\sqrt{3} \cdot \sqrt{3} = \sqrt{9} = 3$.

$$\frac{18}{\sqrt{3}} = \frac{18}{\sqrt{3}} \cdot \frac{\sqrt{3}}{\sqrt{3}} = \frac{18\sqrt{3}}{\sqrt{9}} = \frac{18\sqrt{3}}{3} = 6\sqrt{3}$$

Example 5-35

Simplify $\dfrac{6}{\sqrt{18}}$.

Solution

Multiply the numerator and denominator by $\sqrt{2}$ since $\sqrt{18} \cdot \sqrt{2} = \sqrt{36} = 6$.

$$\frac{6}{\sqrt{18}} = \frac{6}{\sqrt{18}} \cdot \frac{\sqrt{2}}{\sqrt{2}} = \frac{6\sqrt{2}}{\sqrt{36}} = \frac{6\sqrt{2}}{6} = \sqrt{2}$$

Radical expressions that contain fractions can be simplified by making the denominator of the fraction a perfect square.

Example 5-36

Simplify $\sqrt{\dfrac{5}{6}}$.

Solution

Multiply the numerator and denominator by 6 as shown.

$$\sqrt{\frac{5}{6}} = \sqrt{\frac{5}{6} \cdot \frac{6}{6}} = \sqrt{\frac{30}{36}} = \frac{\sqrt{30}}{\sqrt{36}} = \frac{\sqrt{30}}{6}$$

Math Note

Radical expressions such as $\frac{\sqrt{30}}{6}$ can also be written as $\left(\frac{1}{6}\right)\sqrt{30}$. A radical expression like $\frac{2\sqrt{3}}{5}$ can also be written as $\frac{2}{5}\sqrt{3}$.

Try This One

5-V   Simplify each.

(a) $\dfrac{3}{\sqrt{12}}$

(b) $\sqrt{\dfrac{5}{18}}$

(c) $\dfrac{\sqrt{75}}{\sqrt{8}}$

For the answer, see page 237.

Exercise Set   5-4

Computational Exercises

For Exercises 1–6, state whether each number is rational or irrational.

1. $\sqrt{49}$ 2. $\sqrt{37}$ 3. $0.232332333\ldots$

4. $\dfrac{5}{6}$ 5. π 6. 0

Simplify each.

7. $\sqrt{24}$ 8. $\sqrt{448}$ 9. $\dfrac{2}{\sqrt{72}}$

10. $3\sqrt{800}$ 11. $\sqrt{250}$ 12. $\sqrt{162}$

13. $\dfrac{1}{\sqrt{5}}$ 14. $\dfrac{3}{\sqrt{8}}$ 15. $\dfrac{3}{\sqrt{6}}$

16. $\dfrac{10}{\sqrt{20}}$ 17. $\sqrt{\dfrac{3}{28}}$ 18. $\sqrt{\dfrac{1}{9}}$

19. $\sqrt{\dfrac{2}{3}}$ 20. $\sqrt{\dfrac{7}{8}}$

Perform the indicated operations and simplify the answer.

21. $2\sqrt{3} + 5\sqrt{3} - 9\sqrt{3}$ 22. $8\sqrt{5} - 6\sqrt{5} - 7\sqrt{5}$ 23. $\sqrt{320} - \sqrt{80}$

24. $\sqrt{125} + \sqrt{20}$ 25. $6\sqrt{5} - 3\sqrt{80}$ 26. $13\sqrt{90} + 5\sqrt{40}$

27. $6\sqrt{72} - 9\sqrt{8}$ 28. $5\sqrt{10} + 2\sqrt{40}$ 29. $\sqrt{2} \cdot \sqrt{10}$

30. $\sqrt{15} \cdot \sqrt{6}$ 31. $\sqrt{18} \cdot \sqrt{15}$ 32. $\sqrt{5} \cdot \sqrt{25}$

33. $2\sqrt{6} \cdot 3\sqrt{8}$ 34. $6\sqrt{15} \cdot 2\sqrt{5}$ 35. $\dfrac{\sqrt{60}}{\sqrt{2}}$

36. $\dfrac{\sqrt{42}}{\sqrt{6}}$ 37. $\dfrac{\sqrt{64}}{\sqrt{8}}$ 38. $\dfrac{\sqrt{15}}{\sqrt{3}}$

39. $\sqrt{5}(\sqrt{75} + \sqrt{12})$ 40. $\sqrt{3}(\sqrt{48} - \sqrt{27})$

Real World Applications

Use this information for Exercises 41 and 42: The time (in seconds) it takes an object to fall h feet is given by the formula

$$t = \sqrt{\frac{2h}{32}}$$

41. How long will it take an object to fall 256 feet?

42. How long will it take an object to fall 1024 feet?

Use this information for Exercises 43 and 44: The voltage of an electric circuit can be found by the formula

$$V = \sqrt{P \cdot r}$$

where V = volts, P = power in watts, and r = resistance in ohms.

43. Find the voltage when $P = 80$ watts and $r = 5$ ohms.

44. Find the voltage when $P = 360$ watts and $r = 10$ ohms.

Use this information for Exercises 45 and 46: The period (in seconds) of a simple pendulum is given by the formula

$$t = 2\pi\sqrt{\frac{l}{32}}$$

where l is the length of the string in feet.

45. Find the period of a pendulum whose length is 128 feet.

46. Find the period of a pendulum whose length is 64 feet.

Writing Exercises

47. What is the difference between a rational number and an irrational number?

48. Explain why any fraction $\frac{a}{b}$, where a and b are whole numbers and $b \neq 0$ cannot be an irrational number.

49. A value sometimes used for π is 3.14. Explain why this value is only an approximate value.

Critical Thinking

50. Is $\sqrt{a + b} = \sqrt{a} + \sqrt{b}$? Explain your answer.

51. Are cube roots of numbers rational or irrational numbers? Explain your answer.

52. Can the sum of two irrational numbers be a rational number? Explain your answer.

53. Can the product of any two irrational numbers be a rational number? Explain your answer.

5-5 The Real Numbers

Definition of Real Numbers

The set of rational numbers and the set of irrational numbers are disjoint. That is, no number can be both rational and irrational at the same time. A number is either rational or irrational. When the set of rational numbers is combined with the set of irrational numbers, a new set of numbers is formed. This set of numbers is called the set of *real numbers.*

> The set of **real numbers** (R) consists of the union of the set of rational numbers and the set of irrational numbers.

In other words, {rational numbers} ∪ {irrational numbers} = {real numbers}.

The structure of the real number system can be shown in two ways. See Figure 5-7.

Notice that when we begin with the *natural numbers* and add the number zero, we get the set of *whole numbers.* When the set of the negatives of natural numbers is added to the set of whole numbers, a new set of numbers called the *integers* is formed.

When the terminating and repeating decimals or their fraction equivalents are added to the set of integers, the new set is called the *rational* numbers. Finally, when the set of irrational numbers is added to the set of rational numbers, the set of real numbers is obtained. See Figure 5-7(a).

Another way to illustrate the structure of the real number system is shown in Figure 5-7(b). The set of natural numbers is a subset of the set of whole numbers. The set of whole numbers is a subset of the set of integers. The set of integers is a subset of the set of rational numbers. The set of rational numbers and the set of irrational numbers are subsets of the set of real numbers.

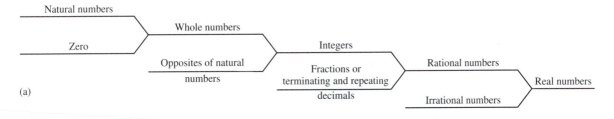

(a)

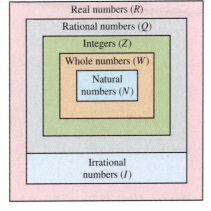

(b)

Figure 5-7

Math Note

There are other kinds of numbers that are not real numbers. For example, $\sqrt{-15}$ is called an imaginary number. Information about these kinds of numbers can be found in college algebra textbooks.

Numbers can now be classified according to type. For example, the number 2 is a natural number, a whole number, an integer, a rational number, and a real number. The number −5 is an integer, a rational number, and a real number. The number $\sqrt{3}$ is an irrational number and a real number.

Example 5-37

Classify each number according to type.

(a) 0

(b) $\sqrt{15}$

(c) $-\frac{3}{4}$

(d) $0.8\overline{6}$

(e) $\sqrt{25}$

Solution

(a) Zero is a whole number, and all whole numbers are integers, rational numbers, and real numbers.

(b) $\sqrt{15}$ is an irrational number since 15 is not a perfect square. Since all irrational numbers are real numbers, $\sqrt{15}$ is also a real number.

(c) The number $-\frac{3}{4}$ is a rational number. Since all rational numbers are real numbers, it is also a real number.

(d) The number $0.8\overline{6}$ is a repeating decimal; hence, it is a rational number and a real number.

(e) At first glance, you might think $\sqrt{25}$ is an irrational number because of the radical sign, but $\sqrt{25} = 5$, and 5 is a natural number, a whole number, an integer, a rational number, and a real number.

Math Note

No number can be classified as both rational and irrational.

Try This One

5-W   Classify each number according to type.

(a) π

(b) $-9.\overline{2}$

(c) $\sqrt{6}$

(d) -10

(e) $\sqrt{36}$

For the answer, see page 237.

Properties of Real Numbers

You might recall that when you multiply two numbers, the order in which they are multiplied does not matter. For example, $3 \times 2 = 6$ and $2 \times 3 = 6$. In other words,

Sidelight

Sophie Germain (1776–1831)

Sophie Germain was born in 1776 in Paris, France. She was the daughter of a silk merchant. Because of the French Revolution, she was confined to her home most of the time, and she read extensively. When she was 13 years old, she found a book on the history of mathematics in her father's library and began to study it. She became so fascinated with mathematics that she wanted to pursue it as a career; however, her father disapproved and forbade her to study the subject. He even punished her by taking away her light at night when she wanted to read.

At age 18, she attempted to enroll at École Polytechnique in Paris, but at that time, no women were allowed to enroll, so she studied mathematics from the notes of a male student. She also wrote a paper on mathematics and submitted it under the name of Antoine LeBlanc and impressed a professor so much that he visited her at her home.

In 1801, she began to correspond with a mathematician named Carl Friedrich Gauss on the subject of number theory. She also did work on a theorem known as Fermat's Last Theorem as well as work in physics.

Later in life, she became active in the scientific and mathematical communities of Paris.

$3 \times 2 = 2 \times 3$. However, when you subtract two numbers, order is very important. For example $5 - 3 = 2$, but $3 - 5 = -2$. Hence $5 - 3 \neq 3 - 5$.

The fact that two numbers can be multiplied in any order is called the *commutative property of multiplication.* Subtraction is *not* commutative.

There are 11 basic properties for the real numbers and these properties are explained next.

In general, the *closure property* states that whenever any two numbers of a set are selected and an operation is performed, the answer will always be found in the set. For example, the set of natural numbers is closed for addition since any two numbers can be selected and added and the answer will always be a natural number. The set of natural numbers is *not* closed under subtraction since $5 - 8 = -3$. In other words, 5 and 8 are natural numbers, but if they are subtracted in the order shown, the answer -3 is not found in the set of natural numbers.

The set of real numbers is closed under addition and multiplication since the sum or product of any two real numbers will always be a real number. Formally stated,

The closure property of addition: For any two real numbers a and b, the sum $a + b$ will be a real number.

The closure property of multiplication: For any two real numbers a and b, the product $a \cdot b$ will be a real number.

The *commutative property* for an operation states that for any two numbers, the order in which the operation is performed on the numbers does not matter. For example, addition of real numbers is commutative since $a + b = b + a$. In other words, $6 + 7 = 7 + 6$, $5 + 10 = 10 + 5$, etc.

Multiplication of real numbers is also commutative. For example, $8 \times 6 = 6 \times 8$, $15 \times 3 = 3 \times 15$, etc.

Subtraction and division are not commutative since $10 - 8 = 2$ but $8 - 10 = -2$; hence, $10 - 8 \neq 8 - 10$, and $20 \div 2 = 10$, but $2 \div 20 = 0.10$; hence $20 \div 2 \neq 2 \div 20$.

The commutative property can be stated formally.

The commutative property of addition: For any real numbers a and b, $a + b = b + a$.

The commutative property of multiplication: For any real numbers a and b, $a \times b = b \times a$.

The *associative property* for an operation states that when three numbers are selected (say, a, b, and c) and an operation is performed on all three, it does not matter which of these two pairs is handled first: a and b or b and c.

Addition of real numbers is associative. Consider the example $5 + 6 + 8$. First add the $5 + 6$ to get 11, and then add $11 + 8$ to get 19. Now try the same example, but first add the $6 + 8$ to get 14. Then add $5 + 14$ to get 19. Hence

$$(5 + 6) + 8 = 5 + (6 + 8)$$

This is true for all real numbers.

Sidelight

Mathematics and Body Fat

You can determine how much body fat you have mathematically. The process uses what is called the body mass index (BMI).

To determine your BMI, square your height (in inches), and then divide your weight in pounds by the square of your height. Finally, multiply that answer by 704.5. The result is your BMI.

For example, suppose your height is 5 foot 7 inches and your weight is 130 pounds. Your BMI is computed as follows:

$$5'7'' = 67''$$
$$67^2 = 4489$$
$$130 \div 4489 \approx 0.0289$$
$$0.0289 \times 704.5 \approx 20.36$$

Compare your BMI with the following chart.

Under 18.5	Underweight
18.5–24.9	Normal
25–29.9	Overweight
30–39.9	Obese
40 and over	Severely obese

Subtraction is *not* associative. For example, $(20 - 10) - 2 \neq 20 - (10 - 2)$. Multiplication is associative, but division is not.

> *The associative property of addition:* For any real numbers a, b, and c, $(a + b) + c = a + (b + c)$.
>
> *The associative property of multiplication:* For any real numbers a, b, and c, $(a \times b) \times c = a \times (b \times c)$.

A set is said to have an *identity* if there exists a specific number in the set such that when an operation is performed on that specific number and any number a in the set, the answer will be the number a. For example, 0 is the identity for addition since when it is added to any other number, the answer will always be that number. For example, $0 + (-8) = -8$, $0 + 5 = 5$, etc.

> *The identity property for addition:* For any real number a, there exists a real number zero called the identity for addition such that $0 + a = a$, and $a + 0 = a$. Zero is called the identity for addition.
>
> *The identity property for multiplication:* For any real number a, there exists a real number 1 called the identity for multiplication such that $1 \times a = a$ and $a \times 1 = a$. The number 1 is called the identity for multiplication.

Numbers in a set can have *inverses* for a specific operation. If an operation is performed on a number and its inverse, the answer will be the identity for that operation.

For addition, 2 and -2 are additive inverses since $2 + (-2) = 0$. Note that 0 is the identity for addition. In addition, -6 and 6 are inverses since $-6 + 6 = 0$. In algebra, the inverses for addition are called *opposites*. Every real number has an additive inverse (or opposite). The additive inverse of 0 is 0. The additive inverse for a number a is designated by $-a$.

> *Inverse property for addition:* For any real number a, there exists a real number $-a$ such that $a + (-a) = 0$ and $-a + a = 0$. $-a$ is said to be the additive inverse or opposite of a.

In order to find multiplicative inverses for real numbers, it is necessary to use the *reciprocal* of the number. For example, the reciprocal of 6 is $\frac{1}{6}$ and $6 \times \frac{1}{6} = 1$. Recall that 1 is the identity for multiplication. The multiplicative inverse of $\frac{2}{3}$ is $\frac{3}{2}$ since $\frac{2}{3} \times \frac{3}{2} = 1$. The only real number that does not have an inverse for multiplication is zero. The symbol for the multiplicative inverse of a is $\frac{1}{a}$.

> *Inverse property for multiplication:* For any real number a, except 0, there exists a real number $\frac{1}{a}$ such that $a \times \frac{1}{a} = 1$ and $\frac{1}{a} \times a = 1$.

Do not confuse inverses for addition with inverses for multiplication. To find an additive inverse for a number, change its sign. To find the multiplicative inverse for a number, use its reciprocal. The additive inverse of $-\frac{2}{3}$ is $+\frac{2}{3}$. The multiplicative inverse of $-\frac{2}{3}$ is $-\frac{3}{2}$. The sign does not change.

Math Note

Recall that the expression $-a$ does not mean that the expression is negative. It depends on the value of a. If a is positive, say 3, then $-a$ is -3 since $-(3) = -3$. However, if a is negative, say -7, then the expression $-a$ becomes $-(-7)$, which is positive since $-(-7) = 7$.

Math Note

The expression $\frac{1}{a}$ does not mean that it is a fraction. If $a = 3$, then $\frac{1}{a} = \frac{1}{3}$, but if $a = \frac{1}{5}$, then $\frac{1}{a} = 5$.

The *distributive property* of multiplication over addition states that when an indicated sum is multiplied by a number, the multiplication can be distributed over the addition. For example,

$$5(6 + 3) = 5 \cdot 6 + 5 \cdot 3$$
$$5(9) = 30 + 15$$
$$45 = 45$$

Stated formally,

> *The distributive property of multiplication over addition:* For any real numbers a, b, and c, $a \cdot (b + c) = a \cdot b + a \cdot c$.

This property will be referred to simply as the distributive property since addition cannot be distributed over multiplication, as shown.

$$5 + (2 \times 3) \neq (5 + 2) \times (5 + 3)$$
$$5 + 6 \neq 7 \times 8$$
$$11 \neq 56$$

Example   5-38

Identify the property illustrated for each.

(a) $6 + 5 = 5 + 6$
(b) $0 + 3 = 3$
(c) $4(3 + 9) = 4 \cdot 3 + 4 \cdot 9$
(d) $\frac{2}{5} \times \frac{5}{2} = 1$
(e) $(6 \cdot 7) \cdot 3 = 6 \cdot (7 \cdot 3)$

Solution

(a) Commutative property of addition
(b) Identity property for addition
(c) Distributive property
(d) Inverse property for multiplication
(e) Associative property of multiplication

The properties illustrated in Example 5-38 are straightforward. However, identifying the properties in Example 5-39 are somewhat more challenging.

Example   5-39

Identify the property illustrated for each.

(a) $(6 + 5) + 3 = (5 + 6) + 3$
(b) $1 \times 7 = 7 \times 1$
(c) $5 \cdot (6 + 2) = (6 + 2) \cdot 5$

—Continued

Example 5-39 *Continued—*

Solution

(a) This is not the associative property of addition since different numbers would have to be in parentheses. Notice that only the order of the 6 and 5 was changed. This is an example of the commutative property of addition, $6 + 5 = 5 + 6$.

(b) This is not an example of the identity property for multiplication. That property states that $1 \times 7 = 7$. This is an example of the commutative property of multiplication since $a \cdot b = b \cdot a$.

(c) This is not an example of the distributive property. The distributive property states that $5 \cdot (6 + 2) = 5 \cdot 6 + 5 \cdot 2$. Here the 5 was commuted with the $(6 + 2)$. Hence, this is an example of the commutative property of multiplication.

Try This One

5-X   Identify the property illustrated by each.

(a) $0 \cdot 6 = 6 \cdot 0$

(b) $3 \cdot 1 = 3$

(c) $4 + (2 + 3) = 4 + (3 + 2)$

(d) $5 \cdot (6 + 1) = 5 \cdot (1 + 6)$

(e) $3 \cdot \frac{1}{3} = \frac{1}{3} \cdot 3$

For the answer, see page 237.

The eleven properties are summarized in Table 5-2.

Table 5-2

Properties of the Real Number System

Name	Property	Example
(For any real numbers a, b, and c)		
Closure property of addition	$a + b$ is a real number	$8 + (-3)$ is a real number
Closure property of multiplication	$a \times b$ is a real number	-5×8 is a real number
Commutative property of addition	$a + b = b + a$	$9 + 8 = 8 + 9$
Commutative property of multiplication	$a \cdot b = b \cdot a$	$6 \cdot 8 = 8 \cdot 6$
Associative property of addition	$(a + b) + c = a + (b + c)$	$(12 + 7) + 3 = 12 + (7 + 3)$
Associative property of multiplication	$(a \cdot b) \cdot c = a \cdot (b \cdot c)$	$(5 \cdot 3) \cdot 2 = 5 \cdot (3 \cdot 2)$
Identity property for addition	$0 + a = a$	$0 + 14 = 14$
Identity property for multiplication	$1 \times a = a$	$1 \times (-3) = -3$
Inverse property for addition	$a + (-a) = 0$	$6 + (-6) = 0$
Inverse property for multiplication	$a \cdot \dfrac{1}{a} = 1,\ a \neq 0$	$4 \cdot \dfrac{1}{4} = 1$
Distributive property	$a(b + c) = a \cdot b + a \cdot c$	$6(2 + 5) = 6 \cdot 2 + 6 \cdot 5$

Exercise Set   5-5

Computational Exercises

For Exercises 1–16, classify each number by using one or more of the categories—natural, whole, integer, rational, irrational, or real.

1. -5
2. 18
3. $\frac{3}{4}$
4. $-\frac{2}{3}$
5. 6.25
6. -18.376
7. $-\sqrt{6}$
8. $-\sqrt{18}$
9. $0.03030030003\ldots$
10. $-\pi$
11. 2.8
12. 13.6
13. 33
14. -17
15. $\sqrt{9}$
16. $\sqrt{100}$

For Exercises 17–32, name the property illustrated.

17. $4 + 8$ is a real number
18. $6 \cdot 1 = 6$
19. $17 + 6 = 6 + 17$
20. $-5 \times (3 + 4) = -5(3) + (-5)(4)$
21. $4 \times 8 = 8 \times 4$
22. $6 + (-6) = 0$
23. $4 \cdot (\sqrt{5} + \sqrt{11}) = 4\sqrt{5} + 4\sqrt{11}$
24. $\sqrt{3} \times \sqrt{4}$ is a real number
25. $\frac{5}{8} \cdot \frac{8}{5} = 1$
26. $-5 + (+5) = 0$
27. $-6 \cdot (2 + 3) = -6 \cdot (3 + 2)$
28. $(16 + 3) + 5 = 16 + (3 + 5)$
29. $5 + 0 = 5$
30. $-6 + (+6) = (+6) + (-6)$
31. $\frac{3}{4} \times \frac{4}{3} = \frac{4}{3} \times \frac{3}{4}$
32. $(8 \times 4) \times 2 = (4 \times 8) \times 2$

For Exercises 33–38, determine under which operations (addition, subtraction, multiplication, division) the system is closed.

33. Natural numbers
34. Whole numbers
35. Integers
36. Rational numbers
37. Irrational numbers
38. Real numbers

Real World Applications

39. Are washing your clothes and drying your clothes commutative?
40. Are brushing your teeth, washing your face, and combing your hair associative?

Writing Exercises

41. Explain why all the numbers discussed in this chapter are real numbers.
42. Consider the set of rational numbers and the set of irrational numbers. Are they disjoint? Explain your answer.

43. Explain why every natural number is also an integer, a rational number, and a real number.

44. Explain why $-\frac{2}{3}$ is not an integer.

Critical Thinking

45. Explain why subtraction is not closed for the set of natural numbers. Is subtraction a closed operation for the set of integers? rational numbers? real numbers? Explain your answers.

46. Find the solution for the equation $x^2 + 1 = 0$. Is the answer a real number?

5-6 Exponents and Scientific Notation

Properties of Exponents

When a number is multiplied by itself n times, the multiplication can be written using **exponential notation.** For example, $5 \cdot 5 \cdot 5 \cdot 5$ can be written as 5^4. Formally defined,

> For any positive integer n,
> $$a^n = \underbrace{a \cdot a \cdot a \cdot a \cdots \cdot a}_{n \text{ factors}}$$
> where a is called the **base** and n is called the **exponent.**

The expression a^n is read as "a to the nth power." When the exponent is 2, such as 5^2, it can be read as "5 squared" or "5 to the second power." When the exponent is one, it is usually not written; that is, $a^1 = a$.

> When an exponent is negative, it is defined as follows: For any positive integer n,
> $$a^{-n} = \frac{1}{a^n}$$

For example, $5^{-3} = \frac{1}{5^3}$ or $\frac{1}{125}$. Finally, when the exponent is zero, it is defined as follows.

> For any number a, $a^0 = 1$.

That is, any nonzero number to the 0 power is equal to one. For example, $6^0 = 1$.

Example 5-40

Simplify each expression.

(a) 6^3

(b) 3^{-4}

(c) 9^0

—Continued

Example 5-40 *Continued—*

Solution

(a) $6^3 = 6 \cdot 6 \cdot 6 = 216$

(b) $3^{-4} = \dfrac{1}{3^4} = \dfrac{1}{3 \cdot 3 \cdot 3 \cdot 3} = \dfrac{1}{81}$

(c) $9^0 = 1$

Try This One

5-Y Simplify each.

(a) 3^6

(b) 2^{-6}

(c) 7^0

For the answer, see page 237.

These rules for exponents can be used to simplify expressions with exponents. For any nonzero real numbers and for any integers m and n,

1. $a^m \cdot a^n = a^{m+n}$ Product rule

2. $\dfrac{a^m}{a^n} = a^{m-n}$ Quotient rule

3. $(a^m)^n = a^{m \cdot n}$ Power rule

The product rule says that when you multiply exponential expressions with the same bases, add the exponents. For example, $3^4 \cdot 3^5 = 3 \cdot 3 \cdot 3 \cdot 3 \cdot 3 \cdot 3 \cdot 3 \cdot 3 \cdot 3 = 3^{4+5} = 3^9$.

The quotient rule says that when you divide exponential expressions with the same bases, subtract the exponents. For example, $\dfrac{6^4}{6^2} = \dfrac{6 \cdot 6 \cdot \cancel{6} \cdot \cancel{6}}{\cancel{6} \cdot \cancel{6}} = 6^{4-2} = 6^2$.

The power rule says that when you raise an exponential expression to a power, multiply the exponents. For example, $(5^2)^3 = 5^2 \cdot 5^2 \cdot 5^2 = 5^{2+2+2} = 5^6$. This is the same as $5^{2 \cdot 3}$ or 5^6.

Example 5-41

Simplify each, write the answer in exponential notation, and then evaluate the expression.

(a) $7^3 \cdot 7^2$

(b) $\frac{3^6}{3^4}$

(c) $(4^2)^5$

Solution

(a) $7^3 \cdot 7^2 = 7^5$ and $7^5 = 16,807$

(b) $\frac{3^6}{3^4} = 3^2$ and $3^2 = 9$

(c) $(4^2)^5 = 4^{10}$ and $4^{10} = 1,048,576$

Try This One

5-Z Simplify each expression and write the answer in exponential notation. Then evaluate.

(a) $6^6 \cdot 6^5$

(b) $\frac{5^6}{5^2}$

(c) $(2^4)^3$

For the answer, see page 237.

Scientific Notation

Science often uses very small and very large numbers. For example, the planet Pluto is about 3,600,000,000 miles from the sun. One red blood cell has a diameter of 0.00028 inches.

In order to write very large and very small numbers without a string of zeros, mathematicians and scientists use what is called *scientific notation*.

> A number expressed in **scientific notation** is written as a product of a number n, where n is $1 \leq n < 10$, and some power of 10.

For example, 3,600,000,000 can be written in scientific notation as

$$3.6 \times 10^9$$

and 0.00028 can be written as

$$2.8 \times 10^{-4}$$

The following table of exponents will be helpful when using scientific notation.

Powers of 10

$10^0 = 1$	$10^{-1} = 0.1$
$10^1 = 10$	$10^{-2} = 0.01$
$10^2 = 100$	$10^{-3} = 0.001$
$10^3 = 1,000$	$10^{-4} = 0.0001$
$10^4 = 10,000$	$10^{-5} = 0.00001$
$10^5 = 100,000$	$10^{-6} = 0.000001$
$10^6 = 1,000,000$	$10^{-7} = 0.0000001$
$10^7 = 10,000,000$	$10^{-8} = 0.00000001$
$10^8 = 100,000,000$	
$10^9 = 1,000,000,000$	

Writing Decimal Numbers in Scientific Notation

In order to see how scientific notation works, consider the number 30,000. It can be written as

$$30,000 = 3 \times 10,000 = 3 \times 10^4$$

The number 0.0004 can be written as

$$0.0004 = 4 \times 0.0001 = 4 \times 10^{-4}$$

To write a number in scientific notation, use the following procedure.

Procedure for Writing a Number in Scientific Notation

Step 1   Move the decimal point in the number either to the left or right until a number greater than or equal to 1 but less than 10 is obtained.

Step 2   Write the new number as a product of a power of 10 as follows:

    (a) If the point is moved to the left, the exponent of the 10 is positive and equal to the number of places that the point was moved.

    (b) If the point is moved to the right, the exponent of the 10 is negative and equal to the number of places the point was moved.

The speed of light in glass is about 2×10^8 m/s. Use $\text{time} = \dfrac{\text{distance}}{\text{speed}}$ to find the time it takes a pulse of light to travel 4000 km, or 4×10^6 m.

Example   5-42

Write each number in scientific notation.

(a) 3,572,000,000
(b) 0.000087

Solution

(a) Move the point 9 places to the left so that it falls between the 3 and 5.

$$3\,5\,7\,2\,0\,0\,0\,0\,0\,0$$
$$9\,8\,7\,6\,5\,4\,3\,2\,1$$

Hence, $3{,}572{,}000{,}000 = 3.572 \times 10^9$.

(b) Move the point to the right 5 places so that it will fall between the 8 and 7.

$$0.0\,0\,0\,0\,8\,7$$
$$1\,2\,3\,4\,5$$

Hence, $0.000087 = 8.7 \times 10^{-5}$.

Try This One

5-AA   Write each number in scientific notation.

(a) 516,000,000

(b) 0.000162

For the answer, see page 237.

Writing Scientific Notation Numbers in Decimal Notation

If a number is written in scientific notation, it can be converted to decimal notation by reversing the procedure.

Rules for Converting a Number in Scientific Notation to Decimal Notation

(a) If the exponent of the power of 10 is positive, move the decimal point to the right the same number of places as the exponent of the 10.
(b) If the exponent is negative, move the decimal point to the left the same number of places as the exponent of the 10.

Example 5-43

Write each number in decimal notation.

(a) 4.192×10^8
(b) 6.37×10^{-8}

Solution

(a) Since the exponent is positive, move the point 8 places to the right.

$$4.19200000$$
$$1\,2\,3\,4\,5\,6\,7\,8$$

Hence, $4.192 \times 10^8 = 419{,}200{,}000$.

(b) Since the exponent is negative, move the point 8 places to the left.

$$000000006.37$$
$$8\,7\,6\,5\,4\,3\,2\,1$$

Hence, $6.37 \times 10^{-8} = 0.0000000637$.

Try This One

5-BB Write each number in decimal notation.

(a) 9.61×10^{10}

(b) 2.77×10^{-6}

For the answer, see page 237.

Operations with Numbers Written in Scientific Notation

Operations such as multiplication and division can be performed on numbers written in scientific notation by using the laws of exponents. Wherever two numbers with the same bases are multiplied, the exponents are added; i.e.,

$$b^m \cdot b^n = b^{m+n}$$

For example, $2^3 \cdot 2^4 = 2 \cdot 2 \cdot 2 \cdot 2 \cdot 2 \cdot 2 \cdot 2 = 2^{3+4} = 2^7$.

Rule To multiply two numbers written in scientific notation, multiply the numbers to the left of the power of 10 and add the exponents of the 10's.

Example 5-44

Perform the indicated operation.

$$(3 \times 10^5)(2 \times 10^3)$$

Solution

$$(3 \times 10^5)(2 \times 10^3) = 6 \times 10^8$$

Multiply 3×2 and add the exponents $5 + 3$.

It may be necessary to rewrite the answer in scientific notation after multiplying when the product is greater than 10.

Example 5-45

Perform the indicated operation.

$$(5 \times 10^2)(7 \times 10^3)$$

Solution

$$(5 \times 10^2)(7 \times 10^3) = 35 \times 10^5$$

It is necessary to rewrite 35×10^5 in scientific notation as 3.5×10^6. To check your answer, check that when 35 got smaller, becoming 3.5, the exponent of the 10 got larger.

The same rule applies when the exponents are negative.

Example 5-46

Perform the indicated operations.

(a) $(3.25 \times 10^{-4})(5.1 \times 10^{-3})$
(b) $(8.6 \times 10^3)(9.7 \times 10^{-6})$

Solution

(a) $(3.25 \times 10^{-4})(5.1 \times 10^{-3}) = 16.575 \times 10^{-7} \approx 1.66 \times 10^{-6}$

Note that the number to the left of the 10 is usually rounded to one or two decimal places.

(b) $(8.6 \times 10^3)(9.7 \times 10^{-6}) = 83.42 \times 10^{-3} \approx 8.34 \times 10^{-2}$

Try This One

5-CC   Perform the indicated operations. Write each answer in scientific notation, rounding it to two decimal places.

(a) $(6 \times 10^5)(1 \times 10^4)$

(b) $(5 \times 10^3)(3.1 \times 10^6)$

(c) $(4.2 \times 10^{-4})(2 \times 10^{-2})$

(d) $(7.3 \times 10^{-9})(5.1 \times 10^6)$

(e) $(8.32 \times 10^{-3})(6.48 \times 10^5)$

For the answer, see page 237.

When dividing numbers written in scientific notation, use the following rule.

$$\frac{b^m}{b^n} = b^{m-n}$$

When dividing two numbers with the same bases, subtract the exponents.

For example, $\dfrac{2^8}{2^6} = 2^2$. $\dfrac{2 \cdot 2 \cdot \cancel{2} \cdot \cancel{2} \cdot \cancel{2} \cdot \cancel{2} \cdot \cancel{2} \cdot \cancel{2}}{\cancel{2} \cdot \cancel{2} \cdot \cancel{2} \cdot \cancel{2} \cdot \cancel{2} \cdot \cancel{2}} = 2^{8-6} = 2^2$

Rule   To divide numbers written in scientific notation, divide the two numbers to the left of the 10, subtract the exponents of the 10's and rewrite the answer in scientific notation if necessary.

Example   5-47

Perform the indicated operation.

(a) $\dfrac{9 \times 10^6}{3 \times 10^4}$

(b) $\dfrac{6 \times 10^{-3}}{2 \times 10^5}$

(c) $\dfrac{1.2 \times 10^{-5}}{4.8 \times 10^{-4}}$

Solution

(a) $\dfrac{9 \times 10^6}{3 \times 10^4} = \dfrac{9}{3} \times \dfrac{10^6}{10^4} = 3 \times 10^{6-4} = 3 \times 10^2$

(b) $\dfrac{6 \times 10^{-3}}{2 \times 10^5} = \dfrac{6}{2} \times \dfrac{10^{-3}}{10^5} = 3 \times 10^{-3-5} = 3 \times 10^{-8}$

(c) $\dfrac{1.2 \times 10^{-5}}{4.8 \times 10^{-4}} = \dfrac{1.2}{4.8} \times \dfrac{10^{-5}}{10^{-4}} = 0.25 \times 10^{-5+4} = 0.25 \times 10^{-1} = 2.5 \times 10^{-2}$

Try This One

5-DD Perform the indicated operation. Round to two decimal places.

(a) $\dfrac{3.2 \times 10^7}{8 \times 10^4}$

(b) $\dfrac{9.6 \times 10^{-5}}{1.3 \times 10^{-8}}$

(c) $\dfrac{2.2 \times 10^{-6}}{4.8 \times 10^5}$

For the answer, see page 237.

Applications of Scientific Notation

Many areas of science require the use of very large or very small numbers. When these numbers are written in scientific notation, it is possible to perform the operation on a scientific calculator. (Most calculators can only display twelve decimal places.)

Example 5-48

The Earth's orbit about the sun is approximately a circle with a radius of 1.5×10^8 km. How far does the Earth travel in one year?

Solution

The circumference of a circle is found by using the formula $C = 2\pi r$, where $\pi \approx 3.14$. Hence the distance the Earth travels in one year is

$$C = 2\pi r$$
$$= 2(3.14)(1.5 \times 10^8 \text{ km})$$
$$= 9.42 \times 10^8 \text{ km}$$

Note that one kilometer is about 0.6 of a mile. 9.42×10^8 km $\times 0.6 \approx 5.65 \times 10^8$ miles $= 565{,}000{,}000$ miles.

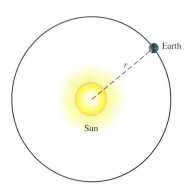

Earth

r

Sun

Try This One

5-EE The U.S. government surplus projected for 2010 is $\$5.6 \times 10^{12}$. If there are 3×10^8 people in the United States, how much money would each person receive if the government decided to give each person an equal sum of money?
For the answer, see page 237.

Exercise Set 5-6

Computational Exercises

For Exercises 1–10, evaluate each.

1. 3^5

2. 6^4

3. 8^0

4. 9^0

5. $(-5)^0$

6. $(-4)^0$

7. 3^{-5}

8. 6^{-4}

9. 2^{-6}

10. 7^{-2}

For Exercises 11–30, simplify each and write the answer in exponential notation using positive exponents. Then evaluate each.

11. $3^4 \cdot 3^2$

12. $5^3 \cdot 5^3$

13. $4^4 \cdot 4^3$

14. $2^6 \cdot 2^4$

15. $\dfrac{3^4}{3^2}$

16. $\dfrac{6^5}{6^3}$

17. $\dfrac{2^5}{2^4}$

18. $\dfrac{8^3}{8}$

19. $(5^2)^3$

20. $(4^4)^2$

21. $3^2 \cdot 3^{-4}$

22. $4^{-3} \cdot 4^5$

23. $5^{-3} \cdot 5^{-2}$

24. $6^3 \cdot 6^{-3}$

25. $\dfrac{2^5}{2^7}$

26. $\dfrac{3^4}{3^6}$

27. $\dfrac{4^4}{4^7}$

28. $\dfrac{5^2}{5^5}$

29. $\dfrac{7^2}{7^3}$

30. $\dfrac{8^2}{8^4}$

For Exercises 31–44, write each number in scientific notation.

31. 625,000,000

32. 9,910,000

33. 0.0073

34. 0.261

35. 528,000,000,000

36. 2,220,000

37. 0.00000618

38. 0.0000000077

39. 43,200

40. 56,000

41. 0.0814

42. 0.0011

43. 32,000,000,000,000

44. 43,500,000

For Exercises 45–58, write each number in decimal notation.

45. 5.9×10^4

46. 6.28×10^6

47. 3.75×10^{-5}

48. 9×10^{-10}

49. 2.4×10^3

50. 7.72×10^5

51. 3×10^{-6}

52. 4×10^{-9}

53. 1×10^3

54. 2.26×10^4

55. 8.02×10^9

56. 1×10^{-4}

57. 7×10^{12}

58. 1.33×10^2

For Exercises 59–72, perform the indicated operations. Write the answers in scientific notation. Round answer to two decimal places.

59. $(3 \times 10^4)(2 \times 10^6)$

60. $(5 \times 10^3)(8 \times 10^5)$

61. $(6.2 \times 10^{-2})(4.3 \times 10^{-6})$

62. $(1.7 \times 10^{-5})(3.8 \times 10^{-6})$

63. $(4 \times 10^4)(2.2 \times 10^{-7})$

64. $(2.2 \times 10^5)(3.6 \times 10^{-4})$

65. $(5 \times 10^{-2})(3 \times 10^{-8})$

66. $(4.3 \times 10^5)(2.2 \times 10^{-6})$

67. $\dfrac{5 \times 10^4}{2.5 \times 10^2}$

68. $\dfrac{9 \times 10^6}{3 \times 10^2}$

69. $\dfrac{4.2 \times 10^{-2}}{7 \times 10^{-3}}$

70. $\dfrac{6.4 \times 10^8}{8 \times 10^{-2}}$

71. $\dfrac{6.6 \times 10^3}{1.1 \times 10^5}$

72. $\dfrac{3 \times 10^7}{1.5 \times 10^{-5}}$

For Exercises 73–78, write each number in scientific notation and perform the indicated operations. Leave answers in scientific notation. Round the answer to two decimal places.

73. $(63,000,000)(41,000,000)$

74. $(52,000)(3,000,000)$

75. $\dfrac{600,000,000}{25,000,000}$

76. $\dfrac{32,000,000}{64,000,000}$

77. $(0.00000025)(0.000004)$

78. $\dfrac{0.0000036}{0.0009}$

Real World Applications

For Exercises 79–84, write each in decimal notation.

79. Light travels at 1.86×10^5 miles per second.

80. There are about 1×10^{14} cells in the human body.

81. The mass of a proton is 1.7×10^{-24} grams.

82. It has been estimated that there are 1×10^{20} grains of sand on the beach at Coney Island, New York.

83. The number of miles light travels in a year is called a light-year. One light-year is equal to 5.88×10^{12} miles.

84. One atom is 1×10^{-8} centimeters in length.

85. The star Gruis is 280 light-years from earth. How far is it in miles if one light-year $= 5.88 \times 10^{12}$ miles?

86. The planet Pluto is 4681 million miles from Earth. Write this number in scientific notation and in decimal notation.

87. The planet Venus is about 67,000,000 miles from the sun. How far does it travel in one revolution around the sun? Assume the orbit is circular.

88. Each red blood cell contains 250,000,000 molecules of hemoglobin. How many molecules will there be in 5×10^4 red blood cells?

89. If light travels at 1.86×10^5 miles per second, how many minutes will it take the light from the sun to reach Jupiter? Jupiter is about 480 million miles from the sun.

90. If the mass of a proton is 0.00000000000000000000000167 grams, how many protons would it take to make one ounce? One ounce is equal to 28.4 grams.

Writing Exercises

91. Explain how to write a decimal number in scientific notation.

92. What is the advantage of writing extremely small or extremely large numbers in scientific notation?

93. What is the difference between a number written in scientific notation that has a 10 with a positive exponent and a number with a 10 that has a negative exponent?

94. Why are the distances that stars are from the Earth given in light-years?

Critical Thinking

95. Extremely small numbers can be written in the metric system using micrometers, nanometers, and angstroms. Write a short paper explaining each unit and some things that these units are used to measure.

5-7   Arithmetic and Geometric Sequences

Arithmetic Sequences

The set of numbers 1, 6, 11, 16, 21, 26, . . . is an example of a *sequence*.

> A **sequence** of numbers is a list of numbers that are related to each other by a specific rule. Each number in the sequence is called a *term* of the sequence.

In the preceding example, each succeeding term can be obtained by adding five to the preceding term.

There are two common types of sequences. They are *arithmetic* sequences and *geometric* sequences. The sequence 1, 6, 11, 16, 21, 26, . . . is an example of an arithmetic sequence.

An **arithmetic sequence** is a sequence of numbers in which each succeeding term differs from the preceding term by the same amount. This amount is known as the **common difference.**

Examples of Arithmetic Sequences	Common Difference
1, 3, 5, 7, 9, 11, . . .	2
2, 7, 12, 17, 22, 27, . . .	5
100, 97, 94, 91, 88, 85, . . .	−3
$\frac{1}{2}, \frac{2}{2}, \frac{3}{2}, \frac{4}{2}, \frac{5}{2}, \frac{6}{2}, \ldots$	$\frac{1}{2}$

Example 5-49

Write the first five terms of a sequence whose first term is 9 and whose common difference is 7.

Solution

9, 16, 23, 30, 37, ...

Example 5-50

Write the first five terms of a sequence whose first term is $\frac{1}{16}$ and whose common difference is $-\frac{1}{8}$.

Solution

$\frac{1}{16}, -\frac{1}{16}, -\frac{3}{16}, -\frac{5}{16}, -\frac{7}{16}, \cdots$

The terms of a general sequence can be written as a_1, a_2, a_3, a_4, . . . , a_n, where a_1 is the first term, a_2 is the second term, etc. a_n is called the n*th term*.

The n*th term* of an arithmetic sequence can be found by using the formula

$$a_n = a_1 + (n - 1)d$$

where n = number of terms you want to find, a_1 = first term, and d = common difference.

Examples 5-51 and 5-52 show how to find the nth term of an arithmetic sequence.

Example 5-51

Find the eighth term of an arithmetic sequence that begins with 5 and has a common difference of 9.

Solution

Substitute in the formula using $a_1 = 5$, $n = 8$, and $d = 9$.

$$a_n = a_1 + (n - 1)d$$
$$a_8 = 5 + (8 - 1)(9)$$
$$= 5 + (7)(9)$$
$$= 5 + 63$$
$$= 68$$

Hence, the eighth term is 68.

Math Note

The answer can be checked by writing eight terms of the sequence as shown

5, 14, 23, 32, 41, 50, 59, 68

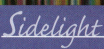

Sidelight

The Fibonacci Sequence

An Italian mathematician Leonardo of Pisa (ca. 1175–1250) discovered an amazing sequence that bears his name. The sequence is called the Fibonacci sequence because Leonardo of Pisa used that name when he began writing. Fibonacci means "Son of Bonaccio." The Fibonacci sequence is 1, 1, 2, 3, 5, 8, 13, 21, 34, 55, 89, ...

The sequence is generated by adding the two previous numbers to get the next number in the sequence: $1 + 1 = 2, 1 + 2 = 3, 2 + 3 = 5, 3 + 5 = 8$, etc.

The reason this sequence is so fascinating is that the numbers of the sequence are found in abundance in nature. For example, the number of petals of many flowers corresponds to the numbers in the sequence. Lilies and irises have 3 petals. Buttercups have 5 petals. Cosmoses have 8 petals. Marigolds have 13 petals. Daisies have 21, 34, 55, or 89 petals. The seeds on the head of a sunflower spiral out from the center in a clockwise direction and a counterclockwise direction. Some sunflowers (depending on the species) have 21 spirals in one direction and 34 in the other direction. A giant sunflower has 89 and 144 spirals.

The Fibonacci numbers appear as the number of spirals of the pinecone, the hexagonal scales of the pineapple, and in the arrangement of the leaves of plants.

The Fibonacci numbers are found in human-made objects as well. For example, one octave on the piano keyboard contains 2 black keys together, then 3 black keys together, for a total of 5 black keys. There are 8 white keys, making a total of 13 keys.

Fibonacci was considered one of the most distinguished mathematicians of the Middle Ages. He went to Arabia to study mathematics and brought back to Italy the Hindu-Arabic numeration system. His book, entitled *Liber Abbaci*, meaning *Book of the Counting*, was published in 1202 and contained arithmetic, algebra, and geometry using the Hindu-Arabic numbers.

The book also contained many problems. The problem whose solution generated the Fibonacci sequence goes something like this:

A man has a pair of rabbits in a pen surrounded by a fence. How many pairs of rabbits can be produced in

a year if every month each pair bears a new pair, and the new pair are able to reproduce on their second month after birth. It is assumed that no rabbits die and a pair is born during the first month. So during the first month, there are two pairs of rabbits. During the second month, the original pair reproduces, and the young pair reaches maturity. Hence, there are three pairs of rabbits. During the third month, the two adult pairs reproduce, and the pair born during the second month reaches maturity. The reasoning continues as shown.

Month	Adult Pairs	Young Pairs	Total Pairs
1	1	1	2
2	2	1	3
3	3	2	5
4	5	3	8
5	8	5	13
6	13	8	21

During the 12th month, there will be 377 pairs of rabbits.

Actually, Fibonacci and his contemporaries did not study or recognize the significance of his sequence. It was not until the 1800s that mathematicians began to study this sequence and realize its importance to mathematics.

Even today, mathematicians are studying the Fibonacci sequence. In fact, there is a society devoted to the study of the mathematics of Fibonacci that publishes a quarterly newsletter.

The trombone plays a different note when the musician changes the length of the instrument while setting up a vibrating column of air through the mouthpiece. For a given length of trombone L, it will resonate for notes with wavelengths $\frac{2L}{1}, \frac{2L}{2}, \frac{2L}{3}, \frac{2L}{4}, \frac{2L}{5}, \ldots$ (This is true of other brass and woodwind instruments as well.) What is the nth term of this sequence?

Example 5-52

Find the 10th term of the arithmetic sequence that begins with $\frac{1}{7}$ and whose common difference is $\frac{2}{5}$.

Solution

Substitute in the formula using $a_1 = \frac{1}{7}, n = 10$, and $d = \frac{2}{5}$.

$$a_n = a_1 + (n-1)d$$
$$a_{10} = \frac{1}{7} + (10-1)\left(\frac{2}{5}\right)$$
$$= \frac{1}{7} + (9)\left(\frac{2}{5}\right)$$
$$= \frac{131}{35}$$

Try This One

5-FF   Find the first five terms and the 10th term for each arithmetic sequence.

(a) $a_1 = 6, d = 10$

(b) $a_1 = 2, d = -5$

(c) $a_1 = \frac{1}{9}, d = \frac{1}{3}$

(d) $a_1 = -\frac{5}{6}, d = -\frac{1}{10}$

For the answer, see page 237.

Sometimes it is necessary to find the sum of the terms of an arithmetic sequence. We can use the next formula to find the sum:

$$S_n = \frac{n(a_1 + a_n)}{2}$$

where

$S_n =$ the sum of n terms of the sequence
$a_1 =$ the first term of the sequence
$a_n =$ the nth term of the sequence

Example 5-53

Find the sum of the first 10 terms of the sequence.

$$2, 4, 6, 8, 10, 12, \ldots$$

—Continued

Math Note

The answer can be checked by adding $2 + 4 + 6 + \cdots + 20 = 110.$

Example 5-53 *Continued—*

Solution

First, it is necessary to find the 10th term. Using the formula shown previously,

$$a_{10} = a_1 + (n-1)d$$

where $a_1 = 2$, $n = 10$, and $d = 2$.

$$a_{10} = 2 + (10 - 1)(2)$$
$$= 2 + 9(2)$$
$$= 20$$

Next, substitute in the formula for finding the sum.

$$S_n = \frac{n(a_1 + a_n)}{2}$$
$$= \frac{10(2 + 20)}{2}$$
$$= 110$$

The sum of the first 10 even numbers is 110.

Try This One

5-GG Find the sum of the first 12 terms of each sequence.

(a) 5, 12, 19, 26, 33, ...

(b) −1, −3, −5, −7, −9, ...

(c) $\frac{1}{5}, \frac{2}{5}, \frac{3}{5}, \frac{4}{5}, \ldots$

For the answer, see page 237.

Geometric Sequences

Another type of sequence is called a *geometric sequence*.

A **geometric sequence** is a sequence of terms in which each term after the first term is obtained by multiplying the preceding term by a nonzero number. This number is called the **common ratio.**

Geometric Sequences	Common Ratio
1, 3, 9, 27, 81, 243, ...	$r = 3$
2, 10, 50, 250, 1250, ...	$r = 5$
5, −10, 20, −40, 80, ...	$r = -2$
1, $\frac{1}{4}, \frac{1}{16}, \frac{1}{64}, \frac{1}{256}, \ldots$	$r = \frac{1}{4}$

Example 5-54

Write the first five terms for a geometric sequence whose first term is 4 and whose common ratio is -3.

Solution

The first term is 4. Then multiply by -3. $4 \cdot (-3) = -12$. $-12 \cdot (-3) = 36$, etc.

$$4, -12, 36, -108, 324, \ldots$$

Try This One

5-HH Write the first five terms of a geometric sequence whose first term is $\frac{1}{2}$ and whose common ratio is 4.
For the answer, see page 237.

The terms of a general geometric sequence can be written using exponents as

$$a_1, a_1r, a_1r^2, a_1r^3, \text{etc.}$$

where a_1 is the first term and r is the common ratio. The terms of the sequence used in Example 5-54 can also be found using this method, as shown.

$$a_1 = 4$$
$$a_2 = a_1r = 4(-3) = -12$$
$$a_3 = a_1r^2 = 4(-3)^2 = 36$$
$$a_4 = a_1r^3 = 4(-3)^3 = -108$$
$$a_5 = a_1r^4 = 4(-3)^4 = 324$$
$$\text{etc.}$$

With exponents, then, the general term of a geometric sequence is found by the next formula.

The nth term of a geometric sequence is $a_n = a_1r^{n-1}$, where a_1 is the first term and r is the common ratio.

Example 5-55

Find the fifth term of a geometric sequence whose first term is $\frac{1}{2}$ and whose common ratio is 6.

Solution

Substitute in the formula using $a_1 = \frac{1}{2}, r = 6$, and $n = 5$.

$$a_n = a_1r^{n-1}$$
$$a_5 = \tfrac{1}{2}(6)^{5-1}$$
$$= \tfrac{1}{2}(6^4)$$
$$= 648$$

Hence, the fifth term is 648.

Math Note

This can be checked by writing the first five terms of the sequence. They are $\frac{1}{2}$, 3, 18, 108, 648.

Sidelight

Sequences and the Planets

In the late 1700s, the world of scientists became very excited when two German astronomers discovered a mathematical sequence that actually predicted the average distance the then-known planets were from the sun. This distance is measured in what are called astronomical units. One astronomical unit (AU) is equal to the average distance the Earth is located from the sun (about 93 million miles).

The sequence, called the Titius-Bode law (named for its discoverers in 1777), is 0, 3, 6, 12, 24, 48, 96, 192, ... When 4 is added to each number and the sum is divided by 10, the result gives the approximate distance in AUs each planet is from the sun as shown.

Planet	Sequence	AU
Mercury	$(0 + 4) \div 10$	0.4
Venus	$(3 + 4) \div 10$	0.7
Earth	$(6 + 4) \div 10$	1.0
Mars	$(12 + 4) \div 10$	1.6
____	$(24 + 4) \div 10$	2.8
Jupiter	$(48 + 4) \div 10$	5.2
Saturn	$(96 + 4) \div 10$	10.0
____	$(192 + 4) \div 10$	19.6

Between Mars and Jupiter, no planet exists, and it looks like the sequence breaks down. However, an asteroid belt is located between Mars and Jupiter, and some astronomers thought that this was once a planet.

More amazing was the fact that in 1781 William Herchel discovered the planet Uranus, which is located at 19.2 AU from the sun!

Unfortunately, the next two planets that were discovered did not fit the sequence's pattern. Neptune's location is 30.1 AU, and Pluto's location is 39.5 AU.

Try This One

5-II   Find the ninth term of a geometric sequence whose first term is 3 and whose common ratio is (-2).
For the answer, see page 237.

The sum of the first n terms of a geometric sequence is

$$S_n = \frac{a_1(1 - r^n)}{1 - r}$$

where a_1 is the first term and r is the common ratio.

Example 5-56

Find the sum of the first six terms of a geometric sequence whose first term is 8 and whose common ratio is $-\frac{1}{2}$.

Solution

Using $a_1 = 8$, $r = -\frac{1}{2}$, and $n = 6$, substitute in the formula

$$S_n = \frac{a_1(1 - r^n)}{1 - r}$$

$$S_6 = \frac{8\left[1 - \left(-\frac{1}{2}\right)^6\right]}{\left[1 - \left(-\frac{1}{2}\right)\right]}$$

$$= 5.25 \text{ or } 5\frac{1}{4}$$

The sum is 5.25.

Try This One

5-JJ   Find the sum of the first seven terms of a geometric sequence whose first term is 6 and whose common ratio is $\frac{1}{4}$.
For the answer, see page 237.

Application of Sequences

Example 5-57

A person is offered a starting salary of $22,000 a year and a choice of $1000 raise each year or a 4% raise each year. Find the person's salary during the 10th year for both options. Under which option would the person make more money during the 10th year?

Solution

First find the person's salary during the 10th year given a $1000 a year raise. This is an example of an arithmetic progression, where $a_1 = \$22{,}000$, $n = 10$, and $d = \$1000$. Substitute in the formula and solve.

$$a_n = a_1 + (n - 1)d$$
$$a_{10} = 22{,}000 + (10 - 1)(1000)$$
$$= 22{,}000 + 9000$$
$$= \$31{,}000$$

Next find the person's salary during the 10th year given a 4% raise each year. This is an example of a geometric progression, where $a_1 = \$22{,}000$, $n = 10$, and $d = 1.04$. [The common ratio is 1.04 since the person's new salary consists of the previous

—Continued

Example 5-57 Continued—

year's salary (100%) and the 4% increase. This amount is $100\% + 4\%$ or 104% (1.04) of last year's salary.] Substitute in the formula and solve.

$$a_n = a_1 r^{n-1}$$
$$a_{10} = 22{,}000(1.04)^{10-1}$$
$$= 22{,}000(1.04)^9$$
$$= \$31{,}312.86$$

Hence, the person would make more money during the 10th year with a 4% annual raise.

Try This One

5-KK A certain automobile depreciates $\frac{1}{3}$ of its preceding value each year. If the original cost of the automobile was \$24,000, find its value after 5 years.
For the answer, see page 237.

Exercise Set 5-7

Computational Exercises

For Exercises 1–8, find each:

(a) the first term   (b) the common difference

(c) the 12th term   (d) the sum of the first 12 terms

1. 5, 13, 21, 29, 37, . . .   2. 2, 12, 22, 32, 42, . . .
3. 50, 48, 46, 44, 42, . . .   4. 12, 7, 2, -3, -8, . . .
5. $\frac{1}{8}, \frac{19}{24}, \frac{35}{24}, \frac{17}{8}, \frac{67}{24}, \ldots$ 6. $\frac{1}{2}, \frac{9}{10}, \frac{13}{10}, \frac{17}{10}, \frac{21}{10}, \ldots$
7. 0.6, 1.6, 2.6, 3.6, 4.6, . . .   8. 0.3, 0.7, 1.1, 1.5, 1.9, . . .

For Exercises 9–16, find each:

(a) the first term   (b) the common ratio

(c) the 12th term   (d) the sum of the first 12 terms

9. 4, 12, 36, 108, 324, . . .   10. 6, 12, 24, 48, 96, . . .
11. $\frac{1}{2}, \frac{1}{4}, \frac{1}{8}, \frac{1}{16}, \frac{1}{32}, \ldots$ 12. $\frac{2}{3}, \frac{2}{9}, \frac{2}{27}, \frac{2}{81}, \frac{2}{243}, \ldots$
13. -3, 15, -75, 375, -1875, . . .   14. -3, 12, -48, 192, -768, . . .
15. 1, 3, 9, 27, 81, . . .   16. 8, 2, $\frac{1}{2}, \frac{1}{8}, \frac{1}{32}, \ldots$

For Exercises 17–22, write the first five terms of the arithmetic sequence when

17. $a_1 = 1, d = 6$ 18. $a_1 = 10, d = 5$
19. $a_1 = -9, d = -3$ 20. $a_1 = -15, d = -2$
21. $a_1 = \frac{1}{4}, d = \frac{3}{8}$ 22. $a_1 = \frac{3}{7}, d = \frac{1}{7}$

For Exercises 23–28, write the first five terms of the geometric sequence when

23. $a_1 = 12, r = 2$

24. $a_1 = 8, r = 3$

25. $a_1 = -5, r = \frac{1}{4}$

26. $a_1 = -9, r = \frac{2}{3}$

27. $a_1 = \frac{1}{6}, r = -6$

28. $a_1 = \frac{3}{7}, r = -3$

For Exercises 29–32, determine whether each sequence is an arithmetic sequence or a geometric sequence.

29. $5, -15, 45, -135, 405, \ldots$

30. $42, 35, 28, 21, 14, \ldots$

31. $6, 2, -2, -6, -10, \ldots$

32. $\frac{1}{10}, \frac{3}{40}, \frac{9}{160}, \frac{27}{640}, \frac{81}{2560}, \ldots$

Real World Applications

33. Machinery at a factory originally costing $50,000 depreciates $1800 the first year, $1750 the second year, $1700 the third year, etc.

    (a) What is the amount of depreciation during the fifth year?

    (b) What is the value of the machinery at the end of the fifth year?

34. A company decided to fine its workers for parking violations on its property. The first offense carries a fine of $25, the second offense is $30, the third offense is $35, and so on. What is the fine for the eighth offense?

35. A person hired a firm to build a CB radio tower. The firm charges $100 for labor for the first 10 feet. After that, the cost of the labor for each succeeding 10 feet is $25 more than the preceding 10 feet. That is, the next 10 feet will cost $125, the next 10 feet will cost $150, etc. How much will it cost to build a 90-foot tower?

36. A ball rebounds $\frac{7}{8}$ as high as it bounced on the previous bounce and is dropped from a height of 8 feet. How high does it bounce on the fourth bounce and how far has it traveled after the fourth bounce?

37. A person deposited $500 in a savings account that pays 5% annual interest that is compounded yearly. At the end of 10 years, how much money will be in the savings account?

38. A contestant on a game show wins one chip for answering the first question correctly, two chips for answering the second question correctly, four chips for the third, eight chips for the fourth, etc. The contestant can stop anytime, but if he or she misses a question, all the accumulated chips are lost. The chips can be cashed in at the end of the game for prizes. If a contestant wishes to win a prize worth 225 chips, how many questions must the contestant answer correctly?

Writing Exercises

39. Define a sequence.

40. Explain what an arithmetic sequence is.

41. Explain what a geometric sequence is.

42. Explain the difference between a common difference and a common ratio.

Critical Thinking

43. When the famous mathematician Karl Friedrich Gauss was 10 years old, his teacher decided to give him a problem that would keep him busy and out of trouble. The problem was to find the sum of the first 100 natural numbers. By looking at the sequence and not knowing the formula for finding sums of sequences, Karl was able to find the answer in a few minutes. Explain how he may have done this.

44. Explain in words why the formula for finding the general term of an arithmetic sequence works.

45. A repeating decimal between -1 and $+1$ can be written as a sum of an infinite geometric sequence as follows

$$0.333\ldots = \frac{3}{10} + \frac{3}{100} + \frac{3}{1000} + \cdots$$

Find a_1 and r and find the sum of all of the terms using the formula $S_n = \dfrac{a_1}{1-r}$.

46. Write $0.151515\ldots$ as the sum of an infinite geometric sequence and find the sum using the formula shown in Exercise 45.

Summary

Section	Important Terms	Important Ideas
5-1	natural number factors divisors multiples prime number composite number fundamental theorem of arithmetic greatest common factor relatively prime least common multiple	**The** set of natural numbers is {1, 2, 3, 4, 5, . . .}. The set of natural numbers consists of the set of prime numbers, composite numbers, and the number 1, which is neither a prime number nor a composite number. Every composite number can be factored as a product of prime numbers (called a prime factorization) in only one way. For two or more numbers, one can find the greatest common factor (GCF) and the least common multiple (LCM).
5-2	whole number integers opposite absolute value order of operations	**The** set of whole numbers is {0, 1, 2, 3, . . .}. The set of integers is {. . . −3, −2, −1, 0, 1, 2, 3, . . .}. Each integer has an opposite. The absolute value of any integer except 0 is positive. The absolute value of 0 is 0.
5-3	rational number proper fraction improper fraction mixed number lowest terms dense place value terminating decimal repeating decimal	**Rational** numbers can be written either as fractions or decimals. The decimals are either terminating or repeating. The rules for performing operations on rational numbers are given in this section.
5-4	irrational number perfect square rationalizing the denominator	**Irrational** numbers are nonterminating nonrepeating decimals. Rules for performing operations on irrational numbers are given in this section.
5-5	real number	**A** real number is either rational or irrational. There are eleven properties for the real numbers. They are the closure properties for addition and multiplication, the commutative properties for addition and multiplication, the associative properties for addition and multiplication, the identity properties for addition and multiplication, the inverse properties for addition and multiplication, and the distributive property.
5-6	exponential notation base exponent scientific notation	**In** order to write very large or very small real numbers without a string of zeros, mathematicians and scientists use scientific notation. Using powers of 10, scientific notation simplifies operations such as multiplication and division of large or small numbers.
5-7	sequence arithmetic sequence common difference geometric sequence common ratio	**A** sequence of numbers is a list of numbers that are related to each other by a specific rule. There are two basic types of sequences. They are arithmetic sequences and geometric sequences. Many real-world problems in mathematics can be solved using sequences.

Review Exercises

For Exercises 1–6, find all the factors of each.

1. 78

2. 81

3. 45

4. 38

5. 140

6. 324

For Exercises 7–10, find five multiples of each.

7. 4

8. 32

9. 9

10. 60

For Exercises 11–16, find a prime factorization for each.

11. 96

12. 44

13. 250

14. 720

15. 600

16. 75

For Exercises 17–22, find the GCF and LCM.

17. 6, 10

18. 18, 20

19. 35, 40

20. 50, 75

21. 60, 80, 100

22. 27, 54, 72

For Exercises 23–32, perform the indicated operations.

23. $-6 + 24$

24. $18 - 32$

25. $5(-9)$

26. $32 \div (-8)$

27. $6 + (-2) - (-3)$

28. $6 \cdot 8 - (-2)^2$

29. $4 \cdot 3 \div (-3) + (-2)$

30. $100 - \{[6 + (2 \cdot 3) - 5] + 4\}$

31. $\{8 \cdot 7^3 - 55[(3 + 4) - 6]\} + 20$

32. $(-5)^3 + (-7)^2 - 3^4$

For Exercises 33–36, reduce each fraction.

33. $\frac{75}{95}$

34. $\frac{56}{64}$

35. $\frac{48}{60}$

36. $\frac{24}{30}$

For Exercises 37–48, perform the indicated operations.

37. $\frac{1}{8} + \frac{5}{6}$

38. $\frac{3}{10} - \frac{2}{5} + \frac{1}{4}$

39. $\frac{5}{9} \times \frac{3}{7}$

40. $\frac{15}{16} \div \left(-\frac{21}{40}\right)$

41. $\frac{1}{2} \div \left(\frac{2}{3} + \frac{3}{4}\right)$

42. $\frac{9}{10} \times \left(\frac{5}{6} - \frac{1}{8}\right)$

43. $\frac{2}{3}\left(\frac{3}{4} + \frac{1}{2} - \frac{1}{6}\right)$

44. $1\frac{7}{8} - \left(\frac{3}{4}\right)^2$

45. $-\frac{6}{7}\left(\frac{1}{2} + 2\frac{1}{3}\right)$

46. $\frac{9}{10} + \left(-\frac{2}{5}\right)\left(-\frac{1}{4}\right)$

47. $\frac{5}{8} - \frac{2}{3}\left(-1 + \frac{2}{5}\right)$

48. $\frac{1}{2} - \frac{3}{4} - \frac{7}{8} \cdot \frac{1}{6}$

For Exercises 49–52, change each fraction to a decimal.

49. $\frac{9}{10}$

50. $\frac{5}{16}$

51. $\frac{6}{7}$

52. $\frac{1}{9}$

For Exercises 53–56, change each decimal to a reduced fraction.

53. 0.6875

54. 0.22

55. $0.2\bar{5}$

56. $0.\overline{45}$

For Exercises 57–62, simplify each.

57. $\sqrt{48}$

58. $\sqrt{112}$

59. $\frac{7}{\sqrt{5}}$

60. $\frac{5}{\sqrt{20}}$

61. $\sqrt{\frac{3}{8}}$

62. $\sqrt{\frac{5}{12}}$

For Exercises 63–70, perform the indicated operations.

63. $\sqrt{20} + 2\sqrt{75} - 3\sqrt{5}$

64. $\sqrt{18} - 5\sqrt{2} + 4\sqrt{72}$

65. $\sqrt{27} \cdot \sqrt{63}$

66. $\sqrt{40} \cdot \sqrt{30}$

67. $\frac{\sqrt{20}}{\sqrt{5}}$

68. $\frac{\sqrt{96}}{\sqrt{16}}$

69. $\sqrt{6}(\sqrt{2} + \sqrt{5})$

70. $\sqrt{42}(\sqrt{14} - \sqrt{6})$

For Exercises 71–76, classify each number as natural, whole, integer, rational, irrational, and/or real.

71. $-\frac{5}{16}$

72. 0.86

73. $0.3\overline{7}$

74. $\sqrt{15}$

75. 0

76. 16

For Exercises 77–80, state which property of the real numbers is being illustrated.

77. $8 \cdot \frac{1}{8} = 1$

78. $3 + 5 = 5 + 3$

79. $6 + 5$ is a real number

80. $2(3 + 8) = 2 \cdot 3 + 2 \cdot 8$

For Exercises 81–90, evaluate each.

81. 4^5

82. 2^0

83. $(-3)^0$

84. 3^{-4}

85. 6^{-5}

86. $7^2 \cdot 7^4$

87. $\frac{5^6}{5^2}$

88. $(3^4)^2$

89. $2^3 \cdot 2^{-5}$

90. $6^{-2} \cdot 6^{-3}$

For Exercises 91–94, write each number in scientific notation. Round to two decimal places.

91. 3826

92. $25{,}946{,}000{,}000$

93. 0.00000327

94. 0.00048

For Exercises 95–98, write each number in decimal notation.

95. 5.8×10^{11}

96. 2.33×10^9

97. 6.27×10^{-4}

98. 8.8×10^{-6}

For Exercises 99–102, perform the indicated operations and write the answers in scientific notation.

99. $(2 \times 10^4)(4.6 \times 10^{-6})$

100. $(3.2 \times 10^{-5})(8.9 \times 10^{-7})$

101. $\dfrac{4.8 \times 10^4}{2.4 \times 10^{-6}}$

102. $\dfrac{1.8 \times 10^{-5}}{3 \times 10^2}$

For Exercises 103–106, write the first six terms of the arithmetic sequence. Find the ninth term and the sum of the first nine terms.

103. $a_1 = 8, d = 10$

104. $a_1 = 4, d = -3$

105. $a_1 = -13, d = -5$

106. $a_1 = -\frac{1}{5}, d = \frac{1}{2}$

For Exercises 107–110, write the first six terms of the geometric sequence. Find the ninth term and the sum of the first nine terms.

107. $a_1 = 7.5, r = 2$

108. $a_1 = -3, r = 3$

109. $a_1 = \frac{1}{9}, r = \frac{1}{4}$

110. $a_1 = -\frac{2}{5}, r = -\frac{1}{2}$

111. The number of people without health insurance in the United States is increasing by 1 million people per year. If there were about 24 million U.S. residents without health insurance in 1980, find the approximate number of people without health insurance in 2000.

112. The net profit of a small company is increasing by 5% each year. If the net profit for this year is $20,000, find the projected profit for the sixth year of operation and the total amount of money the company can be expected to make for 6 years.

Chapter Test

For Exercises 1–10, classify each number as natural, whole, integer, rational, irrational, and/or real.

1. -27

2. 8.6

3. $\frac{5}{9}$

4. $0.6\overline{2}$

5. $\sqrt{50}$

6. π

7. 0

8. $-\frac{13}{20}$

9. $-\sqrt{25}$

10. $\sqrt{\frac{49}{25}}$

For Exercises 11–14, find the GCF and LCM of each group of numbers.

11. 42, 56

12. 36, 45

13. 150, 175, 200

14. 80, 110, 120

For Exercises 15–20, reduce each fraction to lowest terms.

15. $\frac{15}{35}$

16. $\frac{81}{108}$

17. $\frac{112}{175}$

18. $\frac{64}{128}$

19. $\frac{49}{70}$

20. $\frac{98}{128}$

For Exercises 21–30, perform the indicated operations.

21. $-5 \cdot (-6) + 3 \cdot 2$

22. $18 - 3^2 - 4^2 + 6 \div 3$

23. $\left(\frac{5}{6} \cdot \frac{3}{4}\right) \div \frac{2}{3}$

24. $\left(\frac{1}{7} + \frac{1}{9}\right) - \frac{2}{3} \cdot \frac{3}{4}$

25. $-6 + \frac{1}{4} \div \frac{2}{3} + \sqrt{81}$

26. $[4 + (2 \times 3) - 6^2] + 18$

27. $\sqrt{27} + \sqrt{3}(2\sqrt{2} - 1)$

28. $\frac{16}{\sqrt{32}}$

29. $\frac{\sqrt{45}}{\sqrt{5}}$

30. $2\sqrt{50} - 3\sqrt{32}$

For Exercises 31–34, change each decimal into a reduced fraction.

31. 0.875

32. 0.64

33. $0.\overline{2}$

34. $0.\overline{35}$

For Exercises 35–40, state the property illustrated.

35. $0 + 15 = 15 + 0$

36. 6×7 is a real number

37. $0 + (-2) = -2$

38. $\frac{1}{5} \cdot 5 = 1$

39. $(4 \times 6) \times 10 = 4 \times (6 \times 10)$

40. $6(5 + 7) = 6 \cdot 5 + 6 \cdot 7$

For Exercises 41–45, evaluate each.

41. 8^4

42. 7^{-3}

43. 6^0

44. $4^3 \cdot 4^5$

45. $5^{-3} \cdot 5^{-2}$

46. Write 52,000,000 in scientific notation.

47. Write 0.00236 in scientific notation.

48. Write 9.77×10^3 in decimal notation.

49. Write -6×10^{-5} in decimal notation.

50. $(5.2 \times 10^8)(3 \times 10^{-5}) = $ _____ in scientific notation.

51. Divide $\dfrac{2.1 \times 10^9}{7 \times 10^5}$.

52. Write the first seven terms, the 20th term, and the sum of the first 20 terms for the arithmetic sequence where $a_1 = 1$ and $d = 2.5$.

53. Write the first seven terms, the 15th term, and the sum of the first 15 terms for the geometric sequence where $a_1 = \frac{3}{4}$ and $r = -\frac{1}{6}$.

54. A runner decides to train for a marathon by increasing the distance she runs by $\frac{1}{2}$ mile each week. If she can run 15 miles now, how long will it take her to run 26 miles?

55. A gambler decides to double his bet each time he wins. If his first bet is $20 and he wins five times in a row, how much did he bet on the fifth game? Find the total amount he won.

Projects

1. Write a paper on the history of the irrational numbers.

2. Write a paper on Fermat's last theorem.

3. Find a proof that $\sqrt{2}$ is an irrational number and explain the proof. *Hint:* Euclid proved this over 2000 years ago.

4. Explain how logarithms are used to find the square root of a number.

5. Create an algebraic procedure to find the square root of a number and explain the procedure.

Mathematics in Our World ▶ *Revisited*

Are You Sure It's Fat Free?

The reason that a 97% fat free food is too good to be true is that the food industry bases its figures on the weight of the product and not the calories the product contains. For example, suppose a 10-ounce serving of a food contains 240 calories, and the label states that it contains 9 grams of fat. The food industry then converts 10 ounces to grams by multiplying each ounce by 29 grams; hence, the total weight of the product is 290 grams, and if there are 9 grams of fat, the percentage of fat is

$$\frac{9}{290} \times 100\% = 3.1\%$$

The procedure used by the food industry is misleading. The correct way to calculate the fat content is to multiply the number of grams of fat by 9 to get the calories. (Each gram of fat is converted to 9 calories.) In this case, $9 \times 9 = 81$ calories. Next, divide the fat calories by the total calories and multiply by 100% to get the percentage of calories derived from fat. In this case, the label

stated that a 10-ounce serving contained a total of 240 calories.

$$\frac{81}{240} \times 100\% = 33.75\%$$

Hence, 33.75% of the calories come from fat, not 3% as suggested.

Answers to **Try This One**

5-A. 1; 2; 5; 10; 25; 50

5-B. (a) In ascending order:
$$2 \times 2 \times 2 \times 3 \times 3 \times 5 = 2^3 \times 3^2 \times 5^1$$

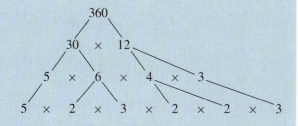

(b) $2 \times 2 \times 2 \times 3 \times 3 \times 5$ or $2^3 \times 3^2 \times 5^1$

5-C. 18

5-D. (a) 200; (b) 980; (c) 96

5-E. Five packages of hot dog buns and four packages of hot dogs since 40 is the LCM of 8 and 10.

5-F. (a) 9; (b) -24; (c) 0; (d) 15; (e) 8; (f) 0; (g) $-5 > -15$; (h) $-3 < +2$; (i) $0 > -6$; (j) $4 > -2$

5-G. (a) 23; (b) -11; (c) 8; (d) -13

5-H. (a) 3; (b) -7; (c) -9; (d) -6; (e) 31; (f) -22

5-I. (a) -48; (b) 2408; (c) -96; (d) 182; (e) -2016

5-J. (a) 16; (b) 16; (c) 42; (d) -4

5-K. (a) 458; (b) 590; (c) 48

5-L. (a) $\frac{1}{6}$; (b) $\frac{7}{8}$; (c) $-\frac{2}{3}$; (d) $\frac{3}{10}$

5-M. (a) $\frac{9}{24}$; (b) $\frac{35}{65}$; (c) $\frac{55}{99}$; (d) $\frac{40}{50}$

5-N. (a) $\frac{1}{2}$; (b) $\frac{11}{24}$; (c) $\frac{2}{5}$; (d) $-\frac{1}{9}$; (e) $-1\frac{1}{6}$

5-O. (a) $0.41\overline{6}$; (b) $0.\overline{57}$; (c) 0.375; (d) $0.\overline{285714}$

5-P. (a) $\frac{2}{5}$; (b) $\frac{12}{25}$; (c) $\frac{13}{40}$

5-Q. (a) $\frac{4}{9}$; (b) $\frac{56}{99}$; (c) $\frac{29}{45}$

5-R. (a) $3\sqrt{3}$; (b) $2\sqrt{14}$; (c) $5\sqrt{3}$

5-S. (a) $5\sqrt{7}$; (b) $6\sqrt{3}$; (c) $15\sqrt{42}$

5-T. (a) $2\sqrt{2}$; (b) 4

5-U. (a) $2\sqrt{2}$; (b) $29\sqrt{2}$; (c) $28\sqrt{7} - 4\sqrt{3}$; (d) $32\sqrt{3} - 28\sqrt{5}$

5-V. (a) $\frac{\sqrt{3}}{2}$; (b) $\frac{\sqrt{10}}{6}$; (c) $\frac{5\sqrt{6}}{4}$

5-W. (a) Irrational and real; (b) rational and real; (c) irrational and real; (d) integer, rational, and real; (e) natural, whole, integer, rational, and real

5-X. (a) Commutative property of multiplication

(b) Identity property for multiplication

(c) Commutative property of addition

(d) Commutative property of addition

(e) Commutative property of multiplication

5-Y. (a) 729; (b) $\frac{1}{64}$; (c) 1

5-Z. (a) 362,797,056; (b) 625; (c) 4096

5-AA. (a) 5.16×10^8
(b) 1.62×10^{-4}

5-BB. (a) 96,100,000,000; (b) 0.00000277

5-CC. (a) 6×10^9; (b) 1.55×10^{10}; (c) 8.4×10^{-6}; (d) 3.72×10^{-2}; (e) 5.39×10^3

5-DD. (a) 4×10^2; (b) 7.38×10^3; (c) 4.58×10^{-12}

5-EE. 1.87×10^4, or $18,700

5-FF. (a) 6; 16; 26; 36; 46; $a_{10} = 96$

(b) 2; -3; -8; -13; -18; $a_{10} = -43$

(c) $\frac{1}{9}$; $\frac{4}{9}$; $\frac{7}{9}$; $\frac{10}{9}$; $\frac{13}{9}$; $a_{10} = \frac{28}{9}$

(d) $-\frac{5}{6}$; $-\frac{14}{15}$; $-\frac{31}{30}$; $-\frac{17}{15}$; $-\frac{37}{30}$; $a_{10} = -\frac{26}{15}$

5-GG. (a) 522; (b) -144; (c) $\frac{78}{5}$

5-HH. $\frac{1}{2}$; 2; 8; 32; 128

5-II. 768

5-JJ. About 7.9995

5-KK. $3160.49

Other Mathematical Systems

Objectives

After completing this chapter, you should be able to

1 Identify the structure of a mathematical system (6-1)

2 Perform addition, subtraction, and multiplication on the 12-hour clock (6-1)

3 Determine which properties, such as closure, commutativity, etc., are true for the operations on the 12-hour clock (6-1)

4 Perform operations and determine which properties are true for modular systems (6-2)

5 Perform operations and determine which properties are true for mathematical systems without numbers (6-3)

Introduction

A **mathematical system** consists of

1. A nonempty set of elements
2. Operation(s) for the elements
3. Definitions
4. Properties of the operations

The real number system, presented in Chapter 5, is an example of a mathematical system. The elements of the real number system are the real numbers. The operations are addition, subtraction, multiplication, and division. Definitions include factor, prime number, etc. and the properties of the operations are closure, commutativity, associativity, etc. The real number system is an example of an **infinite** mathematical system since the set of real numbers is infinite.

In mathematics, there are also **finite** mathematical systems. Finite mathematical systems have a finite number of elements. This chapter explains some finite mathematical systems including **clock arithmetic** systems, modular arithmetic systems, and mathematical systems without numbers. ■

6-1 Clock Arithmetic

The mathematics used in the 12-hour clock is an example of a finite mathematical system since the elements of the system are the numbers 1 through 12. The basic operations are addition and multiplication. The symbol + will be used to denote addition on the clock. For example, if it is 2 o'clock now, 3 hours from now it will be 5 o'clock. This is denoted as $2 + 3 = 5$. If it is 9 o'clock now, 5 hours from now it will be 2 o'clock, so $9 + 5 = 2$.

Addition on the Twelve-Hour Clock

The answers to addition problems on the 12-hour clock can be found by counting clockwise around the face of a clock. If it is 7 o'clock now, in 8 hours, it will be 3 o'clock, as shown in Figure 6-1.

Mathematics in Our World

Game or Mathematical System?

When we were children, we learned how to play a game called "tic-tac-toe," also called "Xs and Os." The game is played by two players using a grid consisting of nine spaces. One player uses an X and the other uses an O. Each player takes turns placing his or her letter in a space. To win the game, one player must get either three Xs or three Os in a vertical row, a horizontal row, or a diagonal row. Some games are shown next.

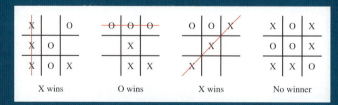

Actually, this game and other games such as football, baseball, Monopoly, chess, and Scrabble can be thought of as mathematical systems. After studying this chapter, you will see why in Mathematics in Our World—Revisited.

Figure 6-1

Some other examples of addition on the 12-hour clock are:

$$7 + 6 = 1 \qquad 6 + 11 = 5$$
$$8 + 7 = 3 \qquad 4 + 3 = 7$$

Example 6-1

Using the 12-hour clock, find these sums.

(a) $9 + 12$
(b) $6 + 5$
(c) $8 + 8$
(d) $3 + 11$
(e) $5 + 7$

Solution

(a)   9. Start at 9 and count 12 hours clockwise, ending at 9.
(b) 11. Start at 6 and count 5 hours clockwise, ending at 11.
(c)   4. Start at 8 and count 8 hours clockwise, ending at 4.
(d)   2. Start at 3 and count 11 hours clockwise, ending at 2.
(e) 12. Start at 5 and count 7 hours, ending at 12.

An addition table for the 12-hour clock can be constructed using all possible combinations of the numbers 1 through 12, as shown here.

+	1	2	3	4	5	6	7	8	9	10	11	12
1	2	3	4	5	6	7	8	9	10	11	12	1
2	3	4	5	6	7	8	9	10	11	12	1	2
3	4	5	6	7	8	9	10	11	12	1	2	3
4	5	6	7	8	9	10	11	12	1	2	3	4
5	6	7	8	9	10	11	12	1	2	3	4	5
6	7	8	9	10	11	12	1	2	3	4	5	6
7	8	9	10	11	12	1	2	3	4	5	6	7
8	9	10	11	12	1	2	3	4	5	6	7	8
9	10	11	12	1	2	3	4	5	6	7	8	9
10	11	12	1	2	3	4	5	6	7	8	9	10
11	12	1	2	3	4	5	6	7	8	9	10	11
12	1	2	3	4	5	6	7	8	9	10	11	12

Multiplication on the Twelve-Hour Clock

Multiplication can also be done on the clock. In order to multiply 6×3, start at 12 o'clock and count around the clock in 3-hour sections six times. That is, $3 + 3 + 3 + 3 + 3 + 3$. You will end up on 6 since

$$3 + 3 = 6 \text{ o'clock}$$
$$6 + 3 = 9 \text{ o'clock}$$
$$9 + 3 = 12 \text{ o'clock}$$
$$12 + 3 = 3 \text{ o'clock}$$
$$3 + 3 = 6 \text{ o'clock}$$

Hence, $6 \times 3 = 6$ on the 12-hour clock.

There is an easier way to compute products on the clock. First, multiply 3×6 to get 18; then divide 18 by 12 and use the remainder, 6, as your answer, as shown.

$$12\overline{)18} \quad \begin{array}{r} 1R6 \\ \underline{12} \\ 6 \end{array}$$

Hence, $6 \times 3 = 6$.

Note that 1R6 can be read as "1 revolution plus 6."

Example 6-2

Perform these multiplications on the 12-hour clock.

(a) 5×8
(b) 12×9
(c) 6×4
(d) 11×5
(e) 10×9

Solution

(a) $5 \times 8 = 40$ and $40 \div 12 = 3$, remainder 4. Hence, $5 \times 8 = 4$.
(b) $12 \times 9 = 108$ and $108 \div 12 = 9$, remainder 0. Since 0 corresponds to 12 on the 12-hour clock, the answer is 12. Hence, $12 \times 9 = 12$.
(c) $6 \times 4 = 24$ and $24 \div 12 = 2$, remainder 0. Hence, $6 \times 4 = 12$.
(d) $11 \times 5 = 55$ and $55 \div 12 = 4$, remainder 7. Hence, $11 \times 5 = 7$.
(e) $10 \times 9 = 90$ and $90 \div 12 = 7$, remainder 6. Hence, $10 \times 9 = 6$.

Calculator Explorations

A calculator can be used to show how the whole number 47 can be converted to its 12-hour clock equivalent.

```
47/12
        3.916666667
Ans-3
         .9166666667
Ans*12
              11
■
```

What is the equivalent number to 47 on the 12-hour clock? What is the equivalent number to 23 on the 12-hour clock?

A table can be constructed for multiplication in the same manner as the addition table was constructed. The multiplication table is shown here.

×	1	2	3	4	5	6	7	8	9	10	11	12
1	1	2	3	4	5	6	7	8	9	10	11	12
2	2	4	6	8	10	12	2	4	6	8	10	12
3	3	6	9	12	3	6	9	12	3	6	9	12
4	4	8	12	4	8	12	4	8	12	4	8	12
5	5	10	3	8	1	6	11	4	9	2	7	12
6	6	12	6	12	6	12	6	12	6	12	6	12
7	7	2	9	4	11	6	1	8	3	10	5	12
8	8	4	12	8	4	12	8	4	12	8	4	12
9	9	6	3	12	9	6	3	12	9	6	3	12
10	10	8	6	4	2	12	10	8	6	4	2	12
11	11	10	9	8	7	6	5	4	3	2	1	12
12	12	12	12	12	12	12	12	12	12	12	12	12

Try This One

6-A   Perform the following operations on the 12-hour clock.

(a) $4 + 6$

(b) $10 + 11$

(c) 7×5

(d) 8×9

(e) $7(6 + 9)$

For the answer, see page 266.

Figure 6-2

Subtraction on the Twelve-Hour Clock

Subtraction can be performed on the 12-hour clock by counting counterclockwise (backward). For example, $8 - 10$ means that if it is 8 o'clock now, what time was it 10 hours ago? Figure 6-2 shows that if you start at 8 and count counterclockwise, you will end at 10 o'clock; hence, $8 - 10 = 10$.

Example   6-3

Perform these subtraction operations on the 12-hour clock.

(a) $2 - 10$
(b) $12 - 7$
(c) $5 - 9$
(d) $6 - 12$
(e) $4 - 11$

Solution

(a) Starting at 2 on the clock and counting backward 10 numbers, you will get 4.
(b) Starting at 12 and counting seven numbers backward, you will get 5.
(c) Starting at 5 and counting nine numbers backward, you will get 8.
(d) Starting at 6 and counting 12 numbers backward, you will get 6.
(e) Starting at 4 and counting 11 numbers backward, you will get 5.

Try This One

6-B   Perform the following subtractions on the 12-hour clock.

(a) $7 - 2$

(b) $3 - 10$

(c) $9 - 12$

For the answer, see page 266.

Sidelight

Evariste Galois

Evariste Galois (1811–1832) was a brilliant young mathematician who lived a short life with incredible bad luck. When he was 17 years old, he submitted a manuscript on the solvability of algebraic equations to the French Academy of Sciences. Augustin-Louis Cauchy, a famous mathematician of his time, was appointed as a referee to read it. However, Cauchy apparently lost the manuscript. A year later, Galois submitted a revised version to the academy. A new referee was appointed to read it but died before he could read it, and the manuscript was lost a second time.

A year later, Galois submitted the manuscript again for a third time. After a 6-month delay, the new referee, Simon-Dennis-Poisson, rejected it, saying it was too vague and recommended that Galois rewrite it in more detail.

Galois at that time was considered a dangerous political radical, and he was provoked into a duel. (Some feel that the challenger was hired by local police to eliminate him.) Galois realized that he would probably die the next morning, so he spent the night trying to revise his manuscript as well as some other papers he had written, but he did not have enough time to finish everything.

The next day, he was shot and killed. Eleven years after his death, Galois's writings were found by Joseph Liouville, who studied them, realized their importance, and published them. At last, Galois was given credit for his work.

> ### Math Note
>
> In order to verify properties using specific examples, all possible combinations of all numbers on the clock must be shown to be true. Since this would require a lot of time and effort, only a few selected examples are used, and one can reason inductively that the property is true.

Properties of the Twelve-Hour Clock

The 11 properties for addition and multiplication of real numbers were explained in Chapter 5. Some of these properties were the commutative property of addition, the closure property for multiplication, etc. The system of the 12-hour clock also has some of the same basic properties for addition and multiplication as the real numbers. These properties are explained next.

The closure property for addition and multiplication can be verified by looking at the two operation tables shown previously. The answers for every combination of addition problems and for every combination of multiplication problems are numbers on the clock.

The commutative properties for addition and multiplication can be verified by noting that for addition or multiplication, the order of operating on the numbers does not matter. For example, $6 + 7$ gives the same answer on the clock as $7 + 6$. The same is true for multiplication.

The associative properties for addition and multiplication are also true on the 12-hour clock system. In order to verify these properties, you would need to show that $(a + b) + c = a + (b + c)$ and $(a \cdot b) \cdot c = a \cdot (b \cdot c)$ is true for all the numbers on the 12-hour clock. Since this would be very time-consuming, only an example of the associative property is shown.

$$(6 + 9) + 10 \stackrel{?}{=} 6 + (9 + 10)$$

$$3 + 10 \stackrel{?}{=} 6 + 7$$

$$1 = 1$$

The identity element for addition is 12 since adding 12 hours to any number on the clock brings you back to the same number:

$$1 + 12 = 1$$
$$2 + 12 = 2$$
$$3 + 12 = 3$$

etc.

The identity property for multiplication is 1 since

$$1 \times 1 = 1$$
$$2 \times 1 = 2$$
$$3 \times 1 = 3$$

etc.

The inverse property for addition on the 12-hour clock is also true. Recall that if a number and its inverse are added, one gets the identity for addition. Since the identity for addition is 12 on the 12-hour clock, the inverse for 1 is 11, the inverse for 2 is 10, the inverse for 3 is 9, etc. In other words,

$$1 + 11 = 12$$
$$2 + 10 = 12$$
$$3 + 9 = 12$$

etc.

The inverse property for multiplication is not true for all numbers on the 12-hour clock. Some numbers do have multiplicative inverses. For example, the multiplicative inverse for 5 is 5 since $5 \times 5 = 1$ (recall that if you multiply a number by its inverse, you get the identity). The multiplicative inverse of 11 is 11 since $11 \times 11 = 1$. But there is no multiplicative inverse for 4 since there is no number such that $4 \times \underline{} = 1$.

The distributive property for multiplication over addition is also valid for the operations performed on the 12-hour clock. Again, all cases of $a \cdot (b + c) = a \cdot b + a \cdot c$ would need to be checked to verify this property. Only one case is shown.

$$5 \times (6 + 8) \overset{?}{=} 5 \times 6 + 5 \times 8$$

$$5 \times 14 \overset{?}{=} 30 + 40$$

$$5 \times 2 \overset{?}{=} 70$$

$$10 = 10$$

Recall that a mathematical system consists of a set of elements, operations for elements, definitions, and properties of each operation. The operations used on the systems presented in this chapter are called **binary operations** since they are performed on two elements of the set. Addition, subtraction, multiplication, and division are binary operations.

Some mathematical systems are called *groups*. Besides mathematics, the theory of groups is used in chemistry, physics, and other areas such as the secret codes that were used in the enigma machine. In particle theory, an example of a binary operation is when particles are combined to form new particles.

A mathematical system is called a **group** if it has these properties:

1. The set of elements is *closed* for the binary operation.
2. There exists an *identity* element for the set.
3. Any three elements in the set are associative for the binary operation.
4. Every element has an inverse.

Sidelight

Niels Henrik Abel

Niels Henrik Abel (1802–1829) was a famous Norwegian mathematician who made many contributions to the theory of equations. At age 19, he tried to find a general solution to a quintic equation using radicals, but he was unable to do so. Later, he proved that it could be done in general. He tried to publish his findings, but due to a lack of funds, he had to condense his work to a six-page pamphlet. Because the condensation was difficult to follow, contemporary mathematicians dismissed his work.

Later, he was given a chance to publish his work in a series of articles in a journal, and after his death, he was given credit for his findings. The term abelian is derived from his name.

Notice that the definition of a group does not include the commutative property. When the elements of the set satisfy the commutative property, the group is said to be a *commutative* or **abelian group.**

The 12-hour clock system is an abelian group under the binary operation of addition. However, it is not a group under multiplication since there is not an inverse element for each given element.

The 12-hour clock is an example of a finite mathematical system. In Section 6-2, other clock systems, called modular systems, will be shown.

Exercise Set 6-1

Computational Exercises

For Exercises 1–12, find the equivalent number on the 12-hour clock.

1. 27	2. 92	3. 155
4. 334	5. 18	6. 42
7. 259	8. 3230	9. −5
10. −10	11. −3	12. −20

For Exercises 13–24, perform the additions on the 12-hour clock.

13. $5 + 9$	14. $10 + 8$	15. $11 + 11$
16. $9 + 7$	17. $12 + 3$	18. $4 + 8$
19. $10 + 20$	20. $9 + 6$	21. $(6 + 5) + 12$
22. $8 + (10 + 9)$	23. $3 + (11 + 8)$	24. $(5 + 7) + 2$

For Exercises 25–36, perform the subtractions on the 12-hour clock.

25. $8 - 6$ 26. $12 - 10$ 27. $9 - 11$

28. $10 - 12$ 29. $0 - 6$ 30. $6 - 10$

31. $3 - 12$ 32. $0 - 8$ 33. $4 - 5$

34. $12 - 8$ 35. $3 - 11$ 36. $2 - 7$

For Exercises 37–48, perform the multiplications on the 12-hour clock.

37. 3×2 38. 10×10 39. 8×6

40. 9×7 41. 2×5 42. 12×6

43. 3×7 44. 4×5 45. $5 \times (6 \times 9)$

46. $3 \times (2 \times 9)$ 47. $(6 \times 4) \times 7$ 48. $(8 \times 3) \times 5$

For Exercises 49–58, find the additive inverse for each number.

49. 12   50. 3   51. 5

52. 8   53. 2   54. 9

55. 7   56. 4   57. -5

58. -6

For Exercises 59–64, find the multiplicative inverse if it exists for each number.

59. 4   60. 7   61. 12

62. 9   63. 1   64. 10

For Exercises 65–70, give an example of each using the 12-hour clock.

65. Associative property of addition

66. Commutative property of multiplication

67. Identity property of addition

68. Inverse property of multiplication

69. Distributive property

70. Commutative property of addition

For Exercises 71–80, find the value of y using the 12-hour clock.

71. $5 + y = 3$ 72. $9 + y = 2$ 73. $y - 5 = 8$

74. $y + 6 = 2$ 75. $4 \times (2 + y) = 4$ 76. $8 \times 2 = y$

77. $6 \times 9 = y$ 78. $9 \times 4 = y$ 79. $3 \times (4 + 10) = y$

80. $5 \times (6 - 11) = y$

Real World Applications

81. A computer simulation takes 3 hours to run. If it is now 11 o'clock A.M., what time will it be when the simulation has run 10 times?

Time in the military is based on a 24-hour clock. From midnight to noon is designated as 0000 to 1200. From noon until midnight, time is designated as 1200 to 2359 where the first two digits indicate the hour and the last two digits represent the minutes. For example, 1824 means 6:24 P.M. For Exercises 82–88, translate military time into standard time.

82. 0948   83. 0311   84. 0500

85. 1542   86. 1938   87. 2218

88. 2000

For Exercises 89–96, change the standard times into military times.

89. 6:56 A.M.   90. 3:52 A.M.   91. 4:00 A.M.

92. 11:56 A.M.   93. 5:27 P.M.   94. 8:06 P.M.

95. 11:42 P.M.   96. 9:36 P.M.

Find the standard time for each.

97. 0627 + 3 hours and 42 minutes

98. 2342 + 5 hours and 6 minutes

99. 1540 − 1 hour and 4 minutes

100. 1242 − 2 hours and 20 minutes

Writing Exercises

101. Define a mathematical system.

102. What is the difference between a finite mathematical system and an infinite mathematical system?

103. Explain how to find an inverse for addition for a number on the 12-hour clock.

104. Explain how you can tell if the commutative property for an operation is valid for a mathematical system.

Critical Thinking

105. Write a short paragraph and explain how the operation of division might be performed on the clock. (*Hint:* 8 ÷ 4 = ___ can be rewritten as 4 × ___ = 8)

106. Explain how division can be performed using the multiplication table.

107. What are the answers to 4 × 4, 4 × 7, and 4 × 10? Explain why these answers are all the same.

6-2   Modular Systems

In Section 6-1, the mathematics of the 12-hour clock was explained. This section will explain **modular systems**. Modular systems with a specific number of elements are analogous to the 12-hour clock system. For example, a modular 5 system, denoted as mod 5 would have five elements: 0, 1, 2, 3, and 4 and use the clock shown in Figure 6-3(a), whereas a mod 3 system would have the elements 0, 1, and 2 and use the clock shown in Figure 6-3(b). A mod 8 system contains the elements 0, 1, 2, 3, 4, 5, 6, and 7 and uses the

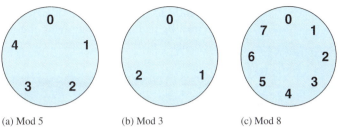

(a) Mod 5   (b) Mod 3   (c) Mod 8

Figure 6-3

clock shown in Figure 6-3(c). In general, a mod m system consists of n numbers starting with zero and concluding with $m - 1$. The number n is called the *modulus*.

Operations for the Modular Systems

Operations in modular systems can be performed using their corresponding clocks, as was shown for the 12-hour clock system. For example, $4 + 2 = 1$ in mod 5 since starting at 4 on the 5-hour clock and counting clockwise two numbers gives you an answer of 1. Performing operations this way can be very time-consuming since a different clock would be needed for every modular system; however, some rules can be formulated and used with any modular system.

Rule 1: Any number a can be changed to a number b in a specific modular system m by dividing a by m and taking the remainder, b. This is written as $a \equiv b \pmod{m}$, and we say a is congruent to b in the modulus m.

Calculator Explorations

A calculator can be used to perform modular system operations. Here, 19 is converted to its congruent value in the mod 5 system.

```
19/5
            3.8
Ans-3
             .8
Ans*5
              4
```

What number is congruent to 19 in the mod 5 system? What number is congruent to 9 in the mod 5 system? 18?

Example 6-4

Find the number that is congruent to 19 in the mod 5 system.

Solution

Divide 19 by 5 and take the remainder, as shown.

$$\begin{array}{r} 3\text{R}4 \\ 5)\overline{19} \\ \underline{15} \\ 4 \end{array}$$

Hence, $19 = 4 \pmod 5$. This answer can be verified by starting at 0 on the mod 5 clock and counting around 19 numbers.

Example 6-5

Find the number that is congruent to 25 in the mod 3 system.

Solution

Divide 25 by 3 and take the remainder as shown.

$$\begin{array}{r} 8\text{R}1 \\ 3)\overline{25} \\ \underline{24} \\ 1 \end{array}$$

Hence, $25 = 1 \pmod 3$.

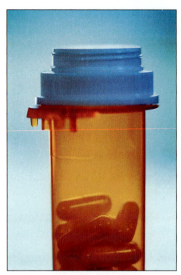

Suppose that you need to take antibiotics three times a day for 10 days. Your first pill was a morning dose. You can't remember if you took your second pill today or not. To find out, count the remaining pills and convert to mod 3: if the answer is 1, you took your pill. (The one remaining pill is your evening dose.) If the answer is 2, you missed a dose. If the answer is 0, you missed two doses.

Rule 2:   The operations of addition and multiplication in modular systems can be performed by adding or multiplying the numbers as usual, then converting the answers to equivalent numbers in the specified system using rule 1.

Example   6-6

Evaluate 4×4 in the mod 5 system.

Solution

$4 \times 4 = 16$ and $16 \div 5 = 3$ remainder 1; hence, $4 \times 4 = 1$ (mod 5).

Example   6-7

Evaluate $6 + 5$ in the mod 7 system.

Solution

$6 + 5 = 11$ and $11 \div 7 = 1$ remainder 4; hence, $6 + 5 = 4$ (mod 7).

Example   6-8

Evaluate each.

(a) 5×6 in mod 9
(b) $9 + 7$ in mod 11
(c) 3×7 in mod 8

Solution

(a) $5 \times 6 = 30$ and $30 \div 9 = 3$ remainder 3; hence, $5 \times 6 = 3$ (mod 9).
(b) $9 + 7 = 16$ and $16 \div 11 = 1$ remainder 5; hence, $9 + 7 = 5$ (mod 11).
(c) $3 \times 7 = 21$ and $21 \div 8 = 2$ remainder 5; hence, $3 \times 7 = 5$ (mod 8).

Try This One

6-C   Evaluate each in the mod 6 system:

(a) $2 + 4$

(b) $5 + 3$

(c) 4×5

(d) 2×4

(e) $3 \times (4 + 4)$

For the answer, see page 266.

Subtraction is performed by counting backward (counterclockwise) in the same manner as shown for the 12-hour clock. For example, $2 - 7$ in the mod 9 system would be -5, which would be equivalent to 4 (mod 9).

Properties of the Modular Systems

Using these two rules (or using a clock), addition and multiplication tables for specific modular systems can be constructed. For example, the addition and multiplication tables for mod 3 are shown here.

+	0	1	2
0	0	1	2
1	1	2	0
2	2	0	1

×	0	1	2
0	0	0	0
1	0	1	2
2	0	2	1

From the tables, the various properties such as closure, commutativity, associativity, etc., can be checked to see if they are true for the operations in the specific modular systems.

For example, addition in the mod 3 system is closed, commutative, and associative. The identity for addition is 0, and each number has an inverse for addition.

Math Note

Closure can also be determined by looking at the addition table. If every element in the body of the table is also in the margins, then addition is closed. That is, there are no new elements appearing in the body that are not in the margins.

Example 6-9

Show that the closure property of addition is true for the mod 3 system.

Solution

To determine if a system is closed, one must perform all possible additions using two numbers and verifying that the solution is indeed a number in the modular system.

$$0 + 0 = 0 \qquad 1 + 0 = 1 \qquad 2 + 0 = 2$$
$$0 + 1 = 1 \qquad 1 + 1 = 2 \qquad 2 + 1 = 0$$
$$0 + 2 = 2 \qquad 1 + 2 = 0 \qquad 2 + 2 = 1$$

Since all the solutions are elements of the mod 3 system, we can say the system is closed under addition.

Example 6-10

Show that addition in the mod 3 system is commutative.

Solution

To show that addition is commutative, you need to perform all possible operations of the form $a + b = b + a$ for any two numbers in the system. For example, $0 + 1 = 1 + 0$, $2 + 1 = 1 + 2$, etc.

Example 6-11

What is the identity for addition in the mod 3 system?

Solution

Zero is the identity for addition in the mod 3 system, since when 0 is added to any number a, $0 + a = a$ for $a = 0, 1, 2$.

Example 6-12

What are the inverses for addition of 0, 1, and 2 in the mod 3 system?

Solution

Recall that the inverse property states that an element + inverse = identity for the operation, and 0 is the identity for addition. Hence, the inverse of 0 is 0 since $0 + 0 = 0$. The inverse of 1 is 2 since $1 + 2 = 0$, and the inverse of 2 is 1.

A lamp with a four-way switch (off, low, medium, and high) can be represented by addition in the mod 4 system (0, 1, 2, 3). Turning the knob is like doing addition—two turns moves you up two notches, and so on. What is the inverse of 1 in the system? (That is, if the lamp is on low, how many turns does it take to turn it off?)

Properties for multiplication in the mod 3 system and for other modular systems can be verified in the same manner as Examples 6-9 through 6-12 have shown.

Try This One

6-D   Construct an addition table for a mod 2 system and verify these properties.

(a) Closure

(b) Commutativity

(c) Identity

(d) Inverse

For the answer, see page 266.

Application of Modular Systems

Money orders, checks, Federal Express bills, and other items that require legal transactions contain tracking or identification numbers. These numbers contain "checking digits." The checking digit provides for the security of the documents and keeps people from altering the documents. For example, if a person decides to forge a driver's license, he needs to create an identification number. When the number is entered into a computer, the computer checks it to see if it is a valid driver's license number. This is done by using modular arithmetic. Here's how.

A security system based on modular systems is used for this check.

Calculator Explorations

When applying modular systems to large numbers using a calculator, rounding must be considered.

```
320476566/7
       45782366.57
Ans-45782366
          .571429
Ans*7
        4.000003
```

What number is congruent to 320476566 in the mod 7 system? 5547721?

Suppose a driver's license contained the identification number 3204765664. In order to verify that this is a valid driver's license number, the computer would divide the number 320476566 (the last digit is omitted) by a certain number and compare the remainder with the checking digit. In this case, the checking digit is the last digit of the original number, which is 4. The divisor is only known by the state that issues the driver's license.

For this example, assume that the divisor is 7. Hence $320476566 \div 7 = 45782366$ with a remainder of 4. When the remainder matches the check digit, the identification number on the driver's license is valid. In other words, $320476566 \equiv 4 \pmod{7}$.

If the checking digit does not match the remainder or the congruent number in the mod 7 system, the driver's license is a forgery. Since only the state knows which digit is the checking digit and which modulus is being used, it becomes very difficult to select a valid license number by chance alone.

Modular systems are only a few types of systems that are found in mathematics. In Section 6-3, you will see mathematical systems that do not use numbers.

Exercise Set 6-2

Computational Exercises

For Exercises 1–30, perform the following operations in the specified mod system.

1. $4 + 3 = $ (mod 5)
2. $8 + 6 = $ (mod 9)
3. $3 + 3 = $ (mod 4)
4. $5 + 6 = $ (mod 7)
5. $5 \times 8 = $ (mod 9)
6. $3 \times 7 = $ (mod 8)
7. $3 \times 3 = $ (mod 4)
8. $4 \times 6 = $ (mod 7)
9. $3 - 8 = $ (mod 9)
10. $5 - 7 = $ (mod 10)
11. $2 - 3 = $ (mod 4)
12. $1 - 9 = $ (mod 11)
13. $(3 + 5) + 2 = $ (mod 7)
14. $(4 + 4) + 4 = $ (mod 6)
15. $2 + (3 + 5) = $ (mod 8)
16. $2 + (3 + 4) = $ (mod 5)
17. $4 \times (2 \times 3) = $ (mod 6)
18. $(2 \times 2) \times 2 = $ (mod 3)
19. $7 \times (3 \times 5) = $ (mod 9)
20. $(2 \times 6) + 4 = $ (mod 7)
21. $6 \times (2 - 5) = $ (mod 8)
22. $5 \times (8 + 3) = $ (mod 9)
23. $7 \times (3 - 5) = $ (mod 10)
24. $4 \times (1 - 7) = $ (mod 8)
25. $2 - (3 - 5) = $ (mod 6)
26. $3 - (1 - 4) = $ (mod 5)
27. $(4 - 7) - 3 = $ (mod 9)
28. $(2 - 10) - 1 = $ (mod 11)
29. $8 - (2 - 5) = $ (mod 12)
30. $(1 - 1) - 1 = $ (mod 2)

For Exercises 31–40, find the values of each number in the given mod system.

31. $32 = $ (mod 6)
32. $51 = $ (mod 4)
33. $135 = $ (mod 7)
34. $48 = $ (mod 5)
35. $16 = $ (mod 9)
36. $92 = $ (mod 10)
37. $326 = $ (mod 3)
38. $451 = $ (mod 5)
39. $987 = $ (mod 8)
40. $1656 = $ (mod 11)

For Exercises 41–50, find the value for y in each equation.

41. $3 + y = 1 \pmod 6$

42. $y + 3 = 2 \pmod 4$

43. $1 - y = 6 \pmod 8$

44. $3 - y = 5 \pmod 9$

45. $7 \times y = 6 \pmod 8$

46. $9 + y = 8 \pmod{10}$

47. $y + 4 = 1 \pmod 5$

48. $y - (-1) = 7 \pmod 8$

49. $4 \times y = 6 \pmod 7$

50. $1 \times y = 6 \pmod 7$

Real World Applications

The days of the week can be thought of as a modular system using 0 = Sunday, 1 = Monday, 2 = Tuesday, etc. Using this system, find the answer to each and give it as the day of the week. (See the Critical Thinking Exercises.)

51. Sunday + 30 days

52. Monday + 5 days

53. Tuesday + 45 days

54. Friday + 120 days

55. Saturday + 360 days

56. Wednesday + 20 days

For Exercises 57–60, use the last digit in the identification number as the checking digit and modulus 9 to see if the number is valid.

57. 76241382

58. 5374193

59. 134804354

60. 215805671

Writing Exercises

61. What is the identity for addition in a modular system?

62. Explain why rule 1 given in this section works.

63. Explain why rule 2 given in this section works.

64. Are all modular systems closed under addition and multiplication? Explain your answer.

Critical Thinking

65. Complete the addition and multiplication tables for mod 7.

+	0	1	2	3	4	5	6
0							
1							
2							
3							
4							
5							
6							

×	0	1	2	3	4	5	6
0							
1							
2							
3							
4							
5							
6							

66. Find the additive inverse for each element in mod 7.
67. What properties (closure, commutative, etc.) are true for addition in mod 7?
68. What properties are true for multiplication in mod 7?

6-3 Mathematical Systems without Numbers

Mathematical systems do not need numbers or operations such as addition or multiplication. It is possible to create a mathematical system using any set of symbols and made-up operations. For example, consider the elements w, x, y, and z, and the operation $*$. The operation $*$ is defined by

$*$	w	x	y	z
w	x	y	z	w
x	y	z	w	x
y	z	w	x	y
z	w	x	y	z

Operations for Systems without Numbers

Operations are performed using this table in the same manner as they are performed in the modular systems. For example, to find $x * y$ using the table shown, find x on the vertical axis and y on the horizontal axis, and then draw a vertical line across from and a horizontal line down from y. The point of intersection is the answer, w.

$*$	w	x	y	z
w				
x			w	
y				
z				

Hence, $x * y = w$.

Example 6-13

Perform the following operations using the system just described.

(a) $w * y$
(b) $z * x$
(c) $y * y$
(d) $z * (w * x)$
(e) $(w * w) * y$

Solution

(a) $w * y = z$
(b) $z * x = x$
(c) $y * y = x$
(d) $w * x = y$ and $z * y = y$; hence, $z * (w * x) = y$
(e) $w * w = x$ and $x * y = w$; hence, $(w * w) * y = w$

Try This One

6-E Using the system just shown, perform these operations.

(a) $y * z$

(b) $z * w$

(c) $x * (y * z)$

For the answer, see page 266.

Properties for Systems without Numbers

The system shown in Figure 6-4(a) has these properties:

1. Closure
2. Commutative
3. Associative
4. Identity property
5. Inverse property

Some of these properties are readily discernable by looking at the table. Recall that a system is closed under an operation if all the elements in the body of the table appear in the margins of the table. See Figure 6-4(a). The system in Figure 6-4(a) is closed for the operation while the system shown in Figure 6-4(b) is not closed for the operation.

The system shown in Figure 6-4(b) is not closed since it contains an element in the body of the table that is not found in the margins.

A system can be checked for commutativity by looking at the body of the table. If the elements are symmetrical with respect to the main diagonal, then the commutative property for the operation is true. See Figure 6-4(c).

*	*a*	*b*	*c*
a	*a*	*b*	*c*
b	*b*	*c*	*a*
c	*c*	*a*	*b*

(a) Closed

*	*a*	*b*	*c*
a	*a*	*b*	*c*
b	*b*	*c*	*a*
c	*c*	*a*	*d*

(b) Not closed

*	*a*	*b*	*c*
a	*a*	*b*	*c*
b	*b*	*c*	*a*
c	*c*	*a*	*b*

(c) Commutative

*	*a*	*b*	*c*
a	*a*	*c*	*b*
b	*b*	*c*	*b*
c	*c*	*a*	*b*

(d) Not commutative

*	*a*	*b*	*c*
a	*a*	*b*	*c*
b	*b*	*c*	*a*
c	*c*	*a*	*b*

(e) Identity (*a*)

*	*a*	*b*	*c*
a	*c*	*b*	*a*
b	*b*	*c*	*a*
c	*a*	*c*	*b*

(f) No identity

Figure 6-4

The system shown in Figure 6-4(d) is not commutative. For example, we see that $c * a \neq a * c$.

Finally, a system has an identity element if there is a row and a column of elements in the table that are identical to the row and column outside. See Figure 6-4(e).

The system shown in Figure 6-4(f) does not have an identity element that can be used for all the elements in the system.

Sidelight

Magic Squares

Have you ever heard of a magic square? These squares have fascinated and intrigued people ever since the first one was discovered more than 4000 years ago.

The sum of the numbers in each row is equal to 15:

$$4 + 9 + 2 = 15$$
$$3 + 5 + 7 = 15$$
$$8 + 1 + 6 = 15$$

The sum of the numbers in each column is also 15:

$$4 + 3 + 8 = 15$$
$$9 + 5 + 1 = 15$$
$$2 + 7 + 6 = 15$$

And the sum of the two diagonals is also 15:

$$4 + 5 + 6 = 15$$
$$2 + 5 + 8 = 15$$

A magic square!

4	9	2
3	5	7
8	1	6

The first magic square appears in the Chinese classic *I-King* and is called *lo-shu*. It is said that the Emperor Yu saw the square engraved on the back of a divine tortoise on the bank of the Yellow River in 2200 B.C.E. The square was the same as the one shown in the above figure except that the numerals were indicated by black and white dots, the even numbers in black and the odd numbers in white.

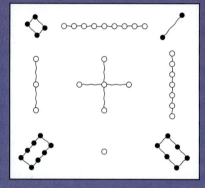

Since some sort of magic was attributed to these squares, they were frequently used to decorate the abodes of gypsies and fortune tellers. They were very popular in India, and they were eventually brought to Europe by the Arabs.

Albrecht Dürer, a 16th-century artist from Germany, used the magic square in his famous woodcut entitled "Melancholia." This square is indeed magic, for not only are the sums of the rows, columns, and diagonals all equal to 34, but also the sum of the four corners $(16 + 13 + 4 + 1)$ is 34. Furthermore, the sum of the four center cells $(10 + 11 + 6 + 7)$ is also 34, and the sum of the slanting squares $(2 + 8 + 9 + 15 \text{ and } 3 + 5 + 12 + 14)$ is 34. Finally, the year in which Dürer made the woodcut appears in the bottom center squares (1514).

16	3	2	13
5	10	11	8
9	6	7	12
4	15	14	1

A great mathematician, Leonhard Euler (1707–1783), constructed the magic square shown next. The sum of the rows and columns is 260. Stopping

—Continued

Continued—

halfway on each row or column gives a sum of half of 260 or 130. Finally, a knight from a chess game can start on square 1 and proceed in L-shaped moves and come to rest on all squares in numerical order.

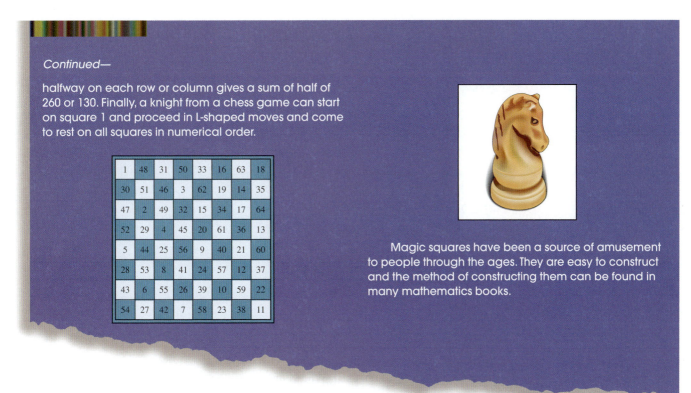

1	48	31	50	33	16	63	18
30	51	46	3	62	19	14	35
47	2	49	32	15	34	17	64
52	29	4	45	20	61	36	13
5	44	25	56	9	40	21	60
28	53	8	41	24	57	12	37
43	6	55	26	39	10	59	22
54	27	42	7	58	23	38	11

Magic squares have been a source of amusement to people through the ages. They are easy to construct and the method of constructing them can be found in many mathematics books.

The other properties need to be checked using the elements of the system and the operation.

Example 6-14

Find the inverse of the elements in the system defined as

*	w	x	y	z
w	x	y	z	w
x	y	z	w	x
y	z	w	x	y
z	w	x	y	z

Solution

Since the identity is z, the inverse of each element can be found by solving these equations.

$$w * \underline{\quad} = z$$
$$x * \underline{\quad} = z$$
$$y * \underline{\quad} = z$$
$$z * \underline{\quad} = z$$

$w * y = z$, so y is the inverse of w.
$x * x = z$, so x is the inverse of x.
$y * w = z$, so w is the inverse of y.
$z * z = z$, so z is the inverse of z.

The game scissors, paper, rock forms a system, as shown here. 0 means a tie, and the operation is "game," called **g** here. What properties hold for this system?

Try This One

6-F   Answer these questions using the system shown here.

△	a	b	c
a	c	a	b
b	a	b	c
c	b	c	a

(a) Is the system closed for △?

(b) Is the system commutative?

(c) What is the identity element?

(d) What is the inverse of *a*?

For the answer, see page 266.

Mixing paint using the three primary colors, red (R), blue (B), and yellow (Y), can be thought of as a mathematical system with a binary operation. When red and blue are mixed, the color purple (P) is obtained. When blue and yellow are mixed, green (G) is obtained. When red and yellow are mixed, orange (O) is obtained. Mixing paint can be set up in table form, as shown.

Let be the operation of mixing two colors of paint.

Ⓜ	R	B	Y
R	R	P	O
B	P	B	G
Y	O	G	Y

What properties does the system have?

As shown in this section, mathematical systems need not be a set of numbers with operations such as addition or multiplication. They can use any set of elements and any clearly-defined operations.

Exercise Set   6-3

Computational Exercises

For Exercises 1–15, use the elements C, D, E, and F, and the operation ? as defined by

?	C	D	E	F
C	D	F	C	E
D	F	E	D	C
E	C	D	E	F
F	E	C	F	D

1. $C ? E$

2. $F ? D$

3. $E ? E$

4. $F ? F$

5. $C ? F$

6. $(D ? E) ? D$

7. $E ? (C ? C)$

8. $F ? (D ? C)$

9. $(E ? D) ? E$

10. $C ? (D ? E)$

11. Is the system closed under the operation?

12. Is there an identity for the operation?

13. Is the operation commutative?

14. What are the inverses for each element?

15. Find the value for x when $E ? x = D$.

For Exercises 16–30, use the elements and the operation ∗ as defined by

∗	△	□	○
△	△	□	○
□	□	☆	△
○	○	△	□

16. $△ ∗ ○$

17. $□ ∗ □$

18. $□ ∗ ○$

19. $○ ∗ ○$

20. $△ ∗ △$

21. $△ ∗ (□ ∗ ○)$

22. $□ ∗ (○ ∗ △)$

23. $(○ ∗ ○) ∗ □$

24. $(□ ∗ ○) ∗ △$

25. $(△ ∗ △) ∗ □$

26. Is the system closed under ∗?

27. Is ∗ commutative?

28. Is there an identity for the system?

29. Is ∗ associative?

30. What is the inverse of $○$?

31. Given a universal set U, a specified set A, and the null set $\varnothing$, construct a table for the operation of the union of sets. For example, $U \cup A = U$, $A \cup \varnothing = A$, etc.

∪	U	A	$\varnothing$
U	U		
A			A
$\varnothing$			

For Exercises 32–34, use the elements and operation shown in Exercise 31.

32. Is the operation commutative?

33. Is the system closed under set union?

34. Is there an identity for the operation?

35. For the universal set U, a specified set A, and the null set, construct a table for the operation of intersection of two sets.

∩	U A $\varnothing$
U	
A	
$\varnothing$	

For Exercises 36–38, use the elements and operations shown for ∩.

36. Is the operation commutative?

37. Is the system closed under intersection?

38. Is there an identity for the operation?

Writing Exercises

39. What four things are necessary to create a mathematical system?

40. Using the operation table, describe how you can tell if a system is closed for the operation.

41. Using the operation table, describe how you can tell if a system is commutative for the operation.

42. Using the operation table, describe how you can find the inverse of an element if it exists.

Critical Thinking

A truth table similar to one shown in Chapter 3 for the conjunction is shown in Figure 6-5(a). This can be converted to a mathematical system using T and F as the elements and ∧ as the operation. This is shown in Figure 6-5(b).

p	q	$p \wedge q$
T	T	T
T	F	F
F	T	F
F	F	F

∧	T	F
T	T	F
F	F	F

(a)   (b)

Figure 6-5

43. Construct a table for a mathematical system for $p \vee q$ using $\vee$ as the operation.

44. Construct a table for a mathematical system for $p \rightarrow q$. What properties are valid for this system?

45. Construct a table for a mathematical system for $p \leftrightarrow q$. What properties are valid for this system?

Summary

Section	Important Terms	Important Ideas
6-1	mathematical system infinite system finite system clock arithmetic binary operation group abelian group	**A** mathematical system consists of a nonempty set of elements, operations on the elements, definitions, and properties of the operations. A finite mathematical system has a specific number of elements, whereas an infinite mathematical system has an unlimited number of elements. A finite mathematical system using the 12-hour clock is explained in this section. When an operation is performed using two elements of a system, it is called a binary operation. A mathematical system is called a group if it is closed under the operation, has an identity element, satisfies the associative property for the operation, and every element has an inverse. If the operation is commutative, then the group is called an abelian group.
6-2	modular system	**A** modular system uses principals similar to those shown for the 12-hour clock. It can have as few as two elements.
6-3		**It** is possible to create mathematical systems using elements other than numbers. Some of these systems are shown in this section.

Review Exercises

For Exercises 1–20, find the equivalent number for the given mod system.

1. $67 = \quad (\bmod\ 5)$

2. $41 = \quad (\bmod\ 3)$

3. $532 = \quad (\bmod\ 8)$

4. $861 = \quad (\bmod\ 6)$

5. $22 = \quad (\bmod\ 4)$

6. $10 = \quad (\bmod\ 2)$

7. $37 = \quad (\bmod\ 10)$

8. $999 = \quad (\bmod\ 7)$

9. $56 = \quad (\bmod\ 9)$

10. $80 = \quad (\bmod\ 5)$

11. $173 = \quad (\bmod\ 9)$

12. $45 = \quad (\bmod\ 7)$

13. $250 = \quad (\bmod\ 10)$

14. $64 = \quad (\bmod\ 3)$

15. $18 = \quad (\bmod\ 3)$

16. $1235 = \quad (\bmod\ 6)$

17. $4721 = \quad (\bmod\ 8)$

18. $856 = \quad (\bmod\ 11)$

19. $1000 = \quad (\bmod\ 12)$

20. $25 = \quad (\bmod\ 4)$

For Exercises 21–40, perform the indicated operation for the given mod system.

21. $5 + 9 = \quad (\bmod\ 11)$

22. $2 - 10 = \quad (\bmod\ 12)$

23. $6 \times 6 = \quad (\bmod\ 7)$

24. $7 + 8 = \quad (\bmod\ 9)$

25. $3 - 7 = \quad (\bmod\ 8)$

26. $4 \times 5 = \quad (\bmod\ 6)$

27. $3 + 2 = \quad (\bmod\ 4)$

28. $5 - 12 = \quad$ (mod 13)

29. $6 \times 7 = \quad$ (mod 10)

30. $10 \times 10 = \quad$ (mod 12)

31. $3 - 4 = \quad$ (mod 5)

32. $5 \times 5 = \quad$ (mod 6)

33. $5 \times (3 + 7) = \quad$ (mod 8)

34. $2 \times (2 + 9) = \quad$ (mod 12)

35. $3 - (3 - 5) = \quad$ (mod 6)

36. $(10 - 6) - 9 = \quad$ (mod 11)

37. $5 \times (7 - 9) = \quad$ (mod 12)

38. $8 + 8 + 8 = \quad$ (mod 10)

39. $4 \times 3 \times 5 = \quad$ (mod 9)

40. $3 \times (4 + 5) = \quad$ (mod 7)

For Exercises 41–50, find the value of y in each equation.

41. $6 + y = 2$ (mod 8)

42. $y + 7 = 1$ (mod 10)

43. $y + 7 = 1$ (mod 9)

44. $3 - y = 6$ (mod 8)

45. $y - 2 = 5$ (mod 6)

46. $3 \times y = 6$ (mod 8)

47. $y + 2 = 1$ (mod 12)

48. $5 - y = 6$ (mod 9)

49. $3 \times 5 = y$ (mod 7)

50. $5 \times (2 + y) = 1$ (mod 12)

For Exercises 51–70, use this system:

$\cdot$	i	-1	$-i$	1
i	-1	$-i$	1	i
-1	$-i$	1	i	-1
$-i$	1	i	-1	$-i$
1	i	-1	$-i$	1

51. $i \cdot i$

52. $-1 \cdot i$

53. $i \cdot 1$

54. $i \cdot (i \cdot i)$

55. $(-i \cdot i) \cdot (-1)$

56. $(1 \cdot 1) \cdot (-i)$

57. i^2

58. i^3

59. i^{10}

60. Find the value of y when $i \cdot y = 1$.

61. Find the value of y when $y \cdot (-1) = i$.

62. Find the value of y when $i \cdot (y \cdot i) = 1$.

63. Is $(-i \cdot 1) \cdot i = -i \cdot (1 \cdot i)$?

64. Is $i^3 = i^7$?

65. Is the system closed under $\cdot$?

66. Is the system commutative?

67. What is the identity for the system?

68. What is the inverse of i?

69. What is the inverse of -1?

70. What is the inverse of $-i$?

Chapter Test

For Exercises 1–6, find the equivalent number for the given mod system.

1. $43 = \quad$ (mod 6)

2. $518 = \quad$ (mod 3)

3. $15 = \quad$ (mod 2)

4. $56 = \quad$ (mod 12)

5. $-6 = \quad$ (mod 4)

6. $-15 = \quad$ (mod 5)

For Exercises 6–12, perform the indicated operation for the given mod system.

7. $8 + 6 = \quad$ (mod 10)

8. $3 + 7 = \quad$ (mod 9)

9. $4 - 6 = \quad$ (mod 7)

10. $8 - 10 = \quad$ (mod 12)

11. $5 \times 9 = \quad (\bmod\ 11)$

12. $4 \times 7 = \quad (\bmod\ 10)$

For Exercises 13–18, find the value of y in each equation.

13. $2 + y = 4 \ (\bmod\ 6)$

14. $y + 8 = 6 \ (\bmod\ 10)$

15. $y - 3 = 7 \ (\bmod\ 8)$

16. $y - 2 = 4 \ (\bmod\ 5)$

17. $3 \times y = 0 \ (\bmod\ 12)$

18. $3 \times y = 0 \ (\bmod\ 5)$

For Exercises 19–26, use the system shown.

*	x	y	z
x	x	y	z
y	y	x	z
z	z	z	s

19. $x * z$

20. $(y * x) * z$

21. $z * z$

22. $x * (z * y)$

23. $(z * x) * x$

24. What is the inverse of y?

25. Is the operation $*$ commutative?

26. Is the system closed?

Projects

A mathematical system can be created by rotating a five-pointed star about its center. See Figure 6-6(a). The operation of rotating the star is designated by the symbol Ⓡ. The operation Ⓡ is defined as follows. The first element is the starting position, and the second element is the number of degrees the star is rotated. For example, B Ⓡ C means to start at B and rotate the star $144°$. B will then be in the position of D; hence, B Ⓡ $C = D$. See Figure 6-6(b). C Ⓡ E means to start at C and rotate the star $288°$; hence, C Ⓡ $E = B$.

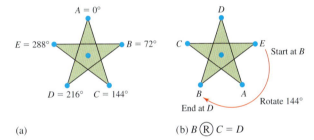

(a)   (b) B Ⓡ $C = D$

Figure 6-6

Mathematics in Our World ▶ *Revisited*

Game or Mathematical System?

As stated previously, games can be thought of as mathematical systems. Recall that a mathematical system consists of a set of elements, operations on elements, definitions and properties for the operations and elements. In the case of "tic-tac-toe," the elements are Xs and Os. The operation is placing an X or an O in a space on the grid. A *win* by either player is defined as getting three Xs or three Os in a row, column, or diagonal. A *draw* is defined as neither X nor O winning. There are many properties that can be stated for this game. A few are given here.

1. There are nine opening moves. (Actually, three if symmetry is considered.)

2. If X begins and there is no winner, X has five moves.
3. If X begins and there is no winner, O has four moves.
4. There are eight ways to win.
5. There are 126 possible distinct games.

There are two properties that might surprise you.

6. Out of the 126 possible distinct games, X wins 120 when X begins.
7. If X begins, O can (using the right strategy) end every game in a draw.

Other games, such as tennis, chess, and some card games can be considered as mathematical systems.

Adapted from *An Introduction to the Elements of Mathematics* by John N. Fujii, published by John Wiley and Sons, 1961.

Answer these questions:

1. Construct a table for the operation.

2. Perform each operation.

   (a) $B \circledR B$ (d) $E \circledR C$

   (b) $C \circledR A$ (e) $A \circledR B$

   (c) $D \circledR E$ (f) $C \circledR B$

3. Is the system commutative?

4. Is there an identity? If so, what is it?

5. What is the inverse of B?

6. What is the inverse of C?

7. Is the system closed?

Answers to **Try This One**

6-A. (a) 10; (b) 9; (c) 11; (d) 12; (e) 9

6-B. (a) 5; (b) 5; (c) 9

6-C. (a) $6 = 0 \pmod 6$; (b) $8 = 2 \pmod 6$;
(c) $20 = 2 \pmod 6$; (d) $8 = 2 \pmod 6$;
(e) $24 = 0 \pmod 6$

6-D.

+	0	1
0	0	1
1	1	0

(a) $0 + 0 = 0 =$ member of table;

$0 + 1 = 1 =$ member of table

$1 + 0 = 1 =$ member of table

$1 + 1 = 0 =$ member of table

(b) $0 + 0 = 0 = 0 + 0$;

$0 + 1 = 1 = 1 + 0$;

$1 + 1 = 0 = 1 + 1.$

(c) The identity is 0 since

$0 + 0 = 0$

$1 + 0 = 1$

(d) The inverse of 0 is 0 since

$0 + 0 = 0$

The inverse of 1 is 1 since

$1 + 1 = 0$

6-E. (a) y; (b) w; (c) w

6-F. (a) Yes; (b) Yes; (c) b; (d) c

Chapter

Seven

Topics in Algebra

Outline

Objectives

After completing this chapter you should be able to

1 Simplify algebraic expressions by combining like terms and using the distributive property (7-1)

2 Evaluate algebraic expressions and formulas (7-1)

3 Solve linear equations in one variable (7-2)

4 Identify equations that have no solution or infinitely many solutions (7-2)

5 Translate verbal expressions into mathematical symbols (7-3)

6 Solve real-world problems using linear equations (7-3)

7 Solve linear inequalities and graph the solutions on the number line (7-4)

8 Solve real-world problems using linear inequalities (7-4)

9 Write ratios as fractions (7-5)

10 Simplify ratios (7-5)

11 Solve proportions (7-5)

12 Solve real-world problems using proportions and variation (7-5)

13 Solve quadratic equations using factoring or the quadratic formula (7-6)

14 Solve real-world problems using quadratic equations (7-6)

Introduction

This chapter explains some of the basic concepts of algebra. Algebra can be used to solve many mathematical problems found in physics, chemistry, engineering, science, and business as well as other areas including everyday life.

The chapter begins by explaining the concept of an algebraic expression and how to evaluate an expression and a formula. The next sections show how to solve an equation and give real-life applications of equations. Ratios and proportions are used extensively in mathematics and are explained in Section 7-5. In the last section, linear inequalities are explained. ■

7-1   Fundamental Concepts of Algebra

Basic Definitions

Algebra is a branch of mathematics which generalizes the concepts of arithmetic. It was developed by the Babylonians around 1800 B.C.E. The word "algebra" is derived from the Arabic word "al-jabr" meaning "the science of reduction and cancellation." Although many real-world problems can be solved using arithmetic, algebra can facilitate in finding their solutions.

In algebra, we use *variables*.

> A **variable,** usually a letter, can represent different numerical values.

For example, in the formula $d = rt$, d, r, and t are variables, and d represents the distance an automobile can go in miles, r is the rate in miles per hour or the speed, and t is the time in hours the automobile travels.

> An **algebraic expression** consists of any meaningful combination of variables, numbers, operation symbols, and grouping symbols.

Some examples of algebraic expressions are

$$3x + 2, \quad 8x^2, \quad 7, \quad \frac{9}{5}C + 32$$

> **Math Note**
>
> An algebraic expression does *not* contain an equal sign. Formulas and equations have equal signs.

Mathematics in Our World

Medicine and Children

When physicians prescribe medicine for infants and children, they must take into account the age and weight of the child. For example, an adult may receive 600 mg of a certain medication; however, this amount of medication might be too much for a 6-year-old child. In order to calculate the proper amount of medication for a 6-year-old child, a mathematical proportion is used. Writing proportions, evaluating formulas, and solving equations will be explained in this chapter. Once these skills are learned, you will be able to solve many real-world problems.

To see how to compute the proper dosage for a child, see page 330.

An algebraic expression is made up of one or more *terms*. Terms are connected by addition or subtraction signs. For example, the expression $3x + 2$ has two terms, namely, $3x$ and 2. The expression $8x^2 + 6x - 3$ has three terms, namely, $8x^2$, $6x$, and -3. The expression 7 has one term.

A term has a *numerical coefficient* (number) and may have one or more variables (letters). For the term $-17x^2y$, -17 is the numerical coefficient and x and y are the variables.

The Distributive Property

Recall from Chapter 5 that one of the properties of real numbers was the distributive property.

> The **distributive property** *of multiplication over addition* states that, for any real numbers a, b, and c, $a(b + c) = ab + ac$.

The distributive property states that when an expression in parentheses is multiplied by a term outside the parentheses, the multiplication can be *distributed* over the addition. For example, $5(6 + 8) = 5 \cdot 6 + 5 \cdot 8$, as shown.

$$5(6 + 8) = 5 \cdot 6 + 5 \cdot 8$$
$$5(14) = 30 + 40$$
$$70 = 70$$

When the expression contains variables such as $5(2x + 7)$, this can be written as $5 \cdot 2x + 5 \cdot 7$ or $10x + 35$.

Math Note

In the expression $8x^2 + 6 - 3$, the negative sign in front of the 3 is usually considered as part of the term. This expression has three terms, $8x^2$, $6x$, and -3. Recall from Chapter 5 that subtraction can be considered as adding the opposite; hence, $8x^2 + 6x - 3$ can be written as $8x^2 + 6x + (-3)$.

Math Note

When the numerical coefficient of a term is 1, it is usually not written. For example, $1xy = xy$. As another example, the numerical coefficient of $-xy$ is -1; i.e., $-1xy = -xy$.

Example 7-1

Use the distributive property to multiply each.

(a) $5(3x + 7)$
(b) $-3(6A - 7B + 10)$

Solution

(a) $5(3x + 7) = 5 \cdot 3x + 5 \cdot 7 = 15x + 35$
(b) $-3(6A - 7B + 10) = -3 \cdot 6A - (-3)(7B) + (-3)(10) = -18A + 21B - 30$.

Try This One

7-A Use the distributive property to multiply each.

(a) $7(4x - 20)$

(b) $5(3x - 7y + 18)$

For the answer, see page 331.

Table 7-1

Like Terms and Unlike Terms

Like Terms		Unlike Terms	
$6x$	$-10x$	$6x$	$-10x^2$
$8x^3$	$6x^3$	$8x^3$	$6x^2$
$2x^2y$	$-5x^2y$	$2x^2y$	$-5xy^2$
5	12	x	5

Combining Like Terms

Like terms have the same variables with the same exponents for the variables. For example, $3x$ and $-5x$ are like terms, while $3x$ and $-5y$ are unlike terms. Table 7-1 shows some examples of like terms and unlike terms.

Like terms can be added or subtracted by using the reverse of the distributive property. For example, $3x + 5x = (3+5)x = 8x$ and $-2x^2y + 3x^2y + 9x^2y = (-2+3+9)x^2y = 10x^2y$. Unlike terms cannot be added or subtracted.

In other words, to add or subtract like terms (i.e., combine like terms), add or subtract the numerical coefficients of the like terms. Unlike terms cannot be combined by addition or subtraction.

Example 7-2

Combine like terms for each, if possible.

(a) $9x - 20x$
(b) $3x^2 + 8x^2 - 2x^2$
(c) $6x + 8x^2$

Solution

(a) $9x - 20x = (9 - 20)x = -11x$ (Found by subtracting $9 - 20$)
(b) $3x^2 + 8x^2 - 2x^2 = (3 + 8 - 2)x^2 = 9x^2$ (Found by adding and subtracting $3 + 8 - 2$)
(c) $6x + 8x^2$ (These terms cannot be combined since they are not like terms.)

An expression can have many terms. When *simplifying an expression* with many terms, it is necessary to search out the like terms and then combine them. For example, in the expression $-6x + 3y - 12 + 8y - 2x + 10$, $-6x$ and $-2x$ are like terms, $3y$ and $8y$ are like terms, and -12 and 10 are like terms. Hence, when the expression is simplified, one gets $-8x + 11y - 2$.

Example 7-3

Simplify each.

(a) $9x - 7y + 18 - 27 + 6y - 10x$
(b) $3x^3 + 4x^2 - 6x + 10 - 7x^2 + 4x^3 + 2x - 6$

—Continued

Example 7-3 *Continued—*

Solution

(a) Combine like terms:

$$9x - 10x = -x$$
$$-7y + 6y = -y$$
$$18 - 27 = -9$$

The answer is $-x - y - 9$.

(b) Combine like terms:

$$3x^3 + 4x^3 = 7x^3$$
$$4x^2 - 7x^2 = -3x^2$$
$$-6x + 2x = -4x$$
$$10 - 6 = 4$$

The answer is $7x^3 - 3x^2 - 4x + 4$.

Try This One

7-B Simplify each.

(a) $2x - 6 + 3y - 7x + 8y - 12$

(b) $9y^3 + 7y - 2y^2 + 6 - 8 + 8y^3 - 7y + 12y^2$

(c) $3a - 2b + 4c - 7a + 3b + 2c$

For the answer, see page 331.

Simplifying Algebraic Expressions

Algebraic expressions can be simplified by first eliminating parentheses using the distributive property, and then combining like terms. Example 7-4 shows this procedure.

Example 7-4

Simplify the expression $8(3x - 2) - 6x + 5$.

Solution

Remove parentheses: $8(3x - 2) - 6x + 5 = 24x - 16 - 6x + 5$. Combine like terms: $24x - 16 - 6x + 5 = 18x - 11$.

Try This One

7-C Simplify each expression.

(a) $5(7x + 10) - 4x + 8$

(b) $-2(4a + 6b) + 5c - 12a$

For the answer, see page 331.

Sidelight

History of Algebra

Algebra had its beginnings when people attempted to solve mathematical riddles. One of the earliest books which contained algebraic problems (i.e., riddles) was a collection of 85 problems copied by an Egyptian priest, Ahmes, around 1650 B.C.E. This manuscript later became known as the *Rhind Papyrus.*

The next important development in algebraic thinking came around 250 C.E. when a famous mathematician named Diophantus wrote a book called *Arithmetica,* which contained about 130 algebraic problems and a number of algebraic principles called theorems. *Arithmetica* contained problems that were solved using first-degree and second-degree equations in one unknown. Diophantus studied mathematics in Alexandria in northern Egypt where Euclid and Hypatia had also lived.

Diophantus is known as the "Father of Algebra" because he was the first mathematician to use symbols to represent mathematical concepts. Prior to Diophantus, all mathematical concepts were written out in words.

A Persian mathematician, Al-khwārizmī, wrote a book on algebra, and when it was translated into Latin, the word "al-jabr," which later became "algebra," was used in the title of the translation. The word means the science of reduction and cancellation.

One of the first mathematicians to realize the existence of negative numbers was Leonardo de Pisa, called Fibonacci (1170–1250 C.E.), who used them to solve financial problems. The use of the plus sign (+) and minus sign (−) first appeared in print in an arithmetic textbook written by Johann Widman in 1489. Prior to this, plus and minus signs were used by

merchants to mark bales or barrels that were weighed at warehouses and compared to a standard weight. If the barrels were heavier than the standard weight, they were marked with a plus sign. If they weighed less than the standard weight, they were marked with a minus sign.

In 1494, a Franciscan friar, Luca Pacioli, published a book called *Summa de Arithmetica,* which contained all the known algebra to date. A French mathematician, François Vieta (1540–1603), was the first person to use letters to represent quantities. He used vowels for variables or unknowns and consonants for constants or knowns.

René Descartes in 1637 used a variation of Vieta's symbolization. He used the first letters of the alphabet to represent constants or knowns and the letters at the end of the alphabet to represent variables or unknowns. When an equation had one unknown, Descartes used the letter "x" to represent it. Descartes also used the letters "x" and "y" to represent the coordinates of a point.

Many other mathematicians made contributions to algebra as it is known today.

Evaluating Algebraic Expressions and Formulas

An algebraic expression is **evaluated** when a specific value is substituted for each variable in the expression. Then, following the order of operations (see Section 5-2), the expression is simplified. Example 7-5 shows this process.

Example 7-5

Evaluate $9x - 3$ when $x = 5$.

Solution

Substitute 5 for the value of x, then simplify.

$$9x - 3 = 9(5) - 3$$
$$= 45 - 3$$
$$= 42$$

Hence, the value of $9x - 3$ when $x = 5$ is 42.

Expressions can contain more than one variable. These expressions are evaluated in the same way.

Example 7-6

Evaluate $5x^2 - 7y + 2$ when $x = -3$ and $y = 6$.

Solution

Substitute -3 for x and 6 for y in the expression, and then simplify.

$$5x^2 - 7y + 2 = 5(-3)^2 - 7(6) + 2$$
$$= 5(9) - 7(6) + 2$$
$$= 45 - 42 + 2$$
$$= 5$$

Hence, the value of $5x^2 - 7y + 2$ when $x = -3$ and $y = 6$ is 5.

Try This One

7-D Evaluate each expression.

(a) $9x - 17$ when $x = 3$

(b) $2x^2 - 3x + 5$ when $x = -10$

(c) $6x + 8y - 15$ when $x = -5$ and $y = 7$

For the answer, see page 331.

Algebraic expressions are useful in the real world. Consider these examples.

Suppose a person receives a salary of $600 a month and a 10% commission of all sales he or she makes; this means to multiply the total sales, x, by 0.10 and add 600. The

expression $0.10x + 600$ then can be used to compute the person's income when the sales amount is given for x. Example 7-7 shows how to use this expression.

In another situation, you can find the total cost of an item when there is a 7% sales tax by multiplying the price, x, of an item by 1.07. The expression is $1.07x$.

Finally, if there is a 40% decrease in the price of an item, you can find the reduced cost by multiplying the price, x, by $(1 - 0.4)$ or 0.6. The expression is $(1 - 0.4)x$ or $0.6x$.

Example 7-7

The monthly income a person earns is given by the expression $0.10x + 600$, where x is the amount of her net sales (i.e., 10% of sales plus a $600 salary). If the person's net sales for July are $8240, find the income for the month.

Solution

Substitute $8240 for x in the expression $0.10x + \$600$ and simplify.

$$0.10x + 600 = 0.10(8240) + 600$$
$$= 824 + 600$$
$$= 1424$$

Hence, the salesperson's income for July is $1424.

Calculator Explorations

After entering the single variable expression in Y=, you can, by using the Ask feature in the Table Setup menu and entering *x* values, see the corresponding *y* values (incomes).

```
Plot1 Plot2 Plot3
\Y1■0.10X+600
\Y2=
\Y3=
\Y4=
\Y5=
\Y6=
\Y7=
```

```
TABLE SETUP
TblStart=-3
∆Tbl=1
Indpnt: Auto Ask
Depend: Auto Ask
```

```
  X    │ Y1  │
 8240  │1424 │
       │     │
X=
```

What is a person's income if the net sales for the month are $15,250?

Try This One

7-E For a certain occupation, it was found that the relationship between the number of hours, x, a person works per week and the number of accidents, n, the person has per year is given by the expression $n = 0.5x - 17$. Find the number of yearly accidents for a person who works 40 hours per week.
For the answer, see page 331.

Formulas are used extensively in mathematics as well as in other areas such as chemistry, physics, etc. For example, the formula $A = \frac{1}{2}bh$ is the formula for finding the area of a triangle given the base and the height. The formula $I = prt$ is the formula for finding the simple interest on a sum of money known as the principal (p) when the interest rate per year (r) and the time (t) in years are given. The same procedure that is used to evaluate an algebraic expression is used to evaluate a formula. Example 7-8 shows the procedure.

Example 7-8

The distance in miles an automobile travels is given by the formula $D = rt$, where r is the rate in miles per hour and t is the time in hours. How far will an automobile travel in 6 hours at a rate of 55 miles per hour?

—*Continued*

Math Note

Formulas can be evaluated using units in addition to numbers. For example, in the previous problem, if the units were included, the formula would look like this:

$$D = \frac{55 \text{ miles}}{\text{hour}} \cdot \frac{6 \text{ hours}}{1}$$

$$= 330 \text{ miles}$$

The hours cancel out.

Example 7-8 *Continued—*

Solution

In the formula $D = rt$, substitute 55 for r and 6 for t and evaluate.

$$D = rt$$

$$D = 55(6)$$

$$D = 330$$

Hence, the automobile will travel 330 miles in 6 hours.

Try This One

7-F In Canada and in other countries where the metric system is used, the temperature is given in Celsius instead of Fahrenheit. If the temperature in Montreal is 36°C, find the corresponding Fahrenheit temperature. The formula is $F = \frac{9}{5}C + 32°$. *For the answer, see page 331.*

Exercise Set 7-1

Computational Exercises

For Exercises 1–20, simplify each expression.

1. $5x + 12x - 6x$
2. $3x^2 + 8x^2 - 15x^2$
3. $4y - 10y - 12y$
4. $8A - 15A + 2A$
5. $3p + 2q - 7 + 6p - 3q - 10$
6. $5x - 8y + 9 + 4y - 27 + 2x$
7. $8x^2 + 6x - 10 + 15 - 7x + 3x^2$
8. $-9x^2 - 2x - 7 + 3 - 5x + 21x$
9. $5(6x - 7)$
10. $9(3x + 8)$
11. $-4(12x - 10)$
12. $-8(4m + 7)$
13. $3(2x + 6) - 5x + 9$
14. $4(6x - 3) - 7 - 10x$
15. $-7(3x + 8) - 5x + 6$
16. $-10(4x + 11) - 15x + 19$
17. $3x + 7 + 5(x - 6)$
18. $9b + 12 + 8(2b + 3)$
19. $4x - 17 - 5(x - 6)$
20. $14x + 9 - 6(3x - 2)$

For Exercises 21–40, evaluate each.

21. $5x - 7$ when $x = 18$
22. $-3x + 8$ when $x = 5$
23. $-2c + 10$ when $c = -3$
24. $4x + 9$ when $w = -12$
25. $3x^2 + 2x - 6$ when $x = 5$
26. $8x^2 - 7x + 4$ when $x = 16$
27. $9r^2 - 5r - 10$ when $r = -7$
28. $14x^2 - 6x + 30$ when $x = -7$
29. $5x + 18y + 10$ when $x = 8$ and $y = 3$
30. $9a + 18b - 5$ when $a = 7$ and $b = 2$

31. $3x^2 - 2y^2 + 6x$ when $x = -8$ and $y = 2$

32. $5x^2 - 7x + 2y^2$ when $x = -1$ and $y = 5$

33. $13y^2 - 6x^2 + 7y - 6x + 1$ when $x = -5$ and $y = 9$

34. $5x^2 - 4x + 3 - 2y$ when $x = 7$ and $y = -3$

35. $9x^2 + 7y^2 + 6x + 2y + 5$ when $x = 1$ and $y = 5$

36. $10y^3 + 10y^2 + 7x - 6$ when $x = -3$ and $y = 10$

37. $x + \dfrac{3y}{2}$ when $x = 8$ and $y = 6$

38. $3x - \dfrac{7y}{6}$ when $x = 7$ and $y = 3$

39. $8x^2 - \dfrac{5}{2y}$ when $x = 4$ and $y = 6$

40. $6x^2 - \dfrac{10}{3y}$ when $x = -5$ and $y = 15$

For Exercises 41–50, evaluate each formula.

41. $A = 2\pi rh$ when $\pi = 3.14$, $r = 6$ in., and $h = 10$ in.

42. $P = 2l + 2w$ when $l = 10$ feet and $w = 5$ feet

43. $A = P(1 + RT)$ when $P = \$5000$, $R = 0.07$ per year, and $T = 3$ years

44. $S = \frac{1}{2}gt^2$ when $g = 32$ ft/sec^2 and $t = 20$ seconds

45. $V = \frac{4}{3}\pi r^3$ when $r = 4$ mm and $\pi = 3.14$

46. $S = 4\pi r^2$ when $r = 7$ and $\pi = 3.14$

47. $FV = P(1 + R)^N$ when $P = \$20,000$, $R = 0.06$, and $N = 8$

48. $v = V + gt$ when $V = 50$, $g = 32$, and $t = 8$

49. $SA = 2\pi rh + 2\pi r^2$ when $\pi = 3.14$, $r = 6$, and $h = 12$

50. $T = 2\pi \sqrt{\dfrac{L}{g}}$ when $\pi = 3.14$, $L = 10$, and $g = 32$

Real World Applications

51. The simple interest (I) on a certain amount of money (p) that is invested at a specified interest rate (r) for a specific period of time (t) can be found by the formula $I = prt$. Find the interest on a principal of $500 invested at 5% yearly for 4 years.

52. For a particular occupation, a person's hourly income can be estimated by using this expression: $11.2 + 1.88x + 0.547y$, where x is the number of years of experience on the job and y is the number of years of higher education completed. Find the income of a person who has completed 4 years of college and has worked for the company for 5 years.

53. A manufacturer found that the number of defective items can be estimated by the expression $2.2x - 1.08y + 9.6$, where x is the number of hours an employee worked on a shift and y is the total number of items produced. Find the number of defective items produced for a person who has worked 9 hours and produced 24 items.

54. A real estate agent found that the value of a farm in thousands of dollars can be estimated by $7.56x - 0.266y + 44.9$, where x is the number of acres on the farm and y is the number of rooms in the farmhouse. Predict the value of a farm that has 371 acres and a farmhouse with six rooms.

55. Find the Celsius temperature that corresponds to a Fahrenheit temperature of $50°$. $C = \frac{5}{9}(F - 32)$.

56. Find the electric current, I, delivered by battery cells connected in a series given by the formula $I = \dfrac{nE}{R} + nr$ when $n = 4$, $E = 2$ volts, $R = 12$ ohms, and $r = 0.2$ ohms.

57. The kinetic energy (KE) in ergs of an object is given by the formula $KE = \dfrac{mv^2}{2}$ where m is the mass of the object and v is the velocity. Find the kinetic energy when $m = 30$ g and $v = 200$ cm/sec.

58. The heat energy from electricity is given by the formula $E = 0.238I^2Rt$. Find E when $I = 25$ amps, $R = 12$ ohms, and $t = 175$ sec.

59. The resistance of an electrical conductor is given by the formula $R = \dfrac{kl}{d^2}$. Find R when $k = 8$, $l = 100$ feet, and $d = 30$ mil.

60. The volume of a sphere is given by the formula $V = \frac{4}{3}\pi r^3$. Find V when $\pi = 3.14$ and $r = 4$ inches.

61. The future value (FV) of a compound interest investment (P) at a specific interest rate (R) for a specific number of periods, N, is found by the formula $FV = P(1 + R)^N$. Find the future value of $9000 invested at 8% compounded annually for 6 years.

Writing Exercises

62. What is a variable?

63. Describe how variables are used in mathematics.

64. What is an algebraic expression?

65. What are like terms?

66. Explain why like terms can be added or subtracted, but unlike terms cannot be added or subtracted.

67. What is the distributive property?

68. What does it mean to evaluate an algebraic expression?

69. Explain why formulas are important in mathematics.

Critical Thinking

70. The amount of medication a person receives is sometimes based on what is called *body surface area* (BSA) in square meters (m^2). The formula for BSA is BSA = (weight in kg) × (height in cm/3600). If an order of medication is 50 mg/m^2, how much medication should be given to a person who weighs 88 kg and has a height of 150 cm?

7-2 Solving Linear Equations

Basic Definitions

The basic concepts for expressions were explained in Section 7-1. This section explains how to solve an equation.

> An **equation** is a statement of equality of two algebraic expressions.

The basic difference between an expression and an equation is that an equation has an equal sign. Table 7-2 shows some expressions and some equations.

There are two types of equations, open and closed.

> An **open equation** contains at least one variable.

The equation $x + 6 = 10$ is an **open equation** since it contains the variable x. The equation $8 + 5 = 13$ is a **closed equation** since it does not have any variables. Closed equations can be either true or false. The equation $8 + 5 = 13$ is a closed *true* equation, whereas the equation $8 + 5 = 14$ is a closed *false* equation.

A *solution* for an open equation is a number that when substituted for the variable in the equation makes the equation a closed true equation. For example, a solution for the equation $x + 6 = 10$ is $x = 4$ since when 4 is substituted for the value of x in the equation, you get a closed true equation, $4 + 6 = 10$. This process is called *checking* an equation.

Equations can have no solutions, one solution, or more than one solution. For this reason, solutions to equations are sometimes written in set notation. In this case, the **solution set** to $x + 6 = 10$ is {4}.

Solving the Four Basic Types of Equations

The equations in this section are called first-degree equations or linear equations because the exponents of the variables are equal to one. There are four basic types of first-degree equations, and there are four properties of equality that can be used to solve them.

The equation $x - 5 = 9$ is an example of the first type of equation. It can be solved by using the *addition property of equality*.

Math Note

A solution to an equation is not valid because you have solved it by some process. The solution is valid because it checks.

Table **7-2**

Expressions and Equations

Expressions	Equations
$5x + 7$	$5x + 7 = 15$
$3x + 4y + 7$	$3x + 4y + 7 = 10$
$8 + 5$	$8 + 5 = 13$
$4(x + 6)$	$4(x + 6) = 20$

The **addition property of equality** states that the same real number or algebraic expression can be added to both sides of an equation without changing the solution set for the equation; i.e., if $a = b$, then $a + c = b + c$.

The type 1 equation $x - 5 = 9$ can be solved by adding 5 to both sides, as shown.

$$x - 5 = 9$$
$$x - 5 + 5 = 9 + 5$$
$$x + 0 = 14$$
$$x = 14$$

The solution set is {14}.

Check:

$$x - 5 = 9$$
$$14 - 5 \overset{?}{=} 9$$
$$9 = 9$$

An example of the second type of equation is $x + 10 = 17$. This type of equation can be solved by using the *subtraction property of equality*.

The **subtraction property of equality** states that the same real number or algebraic expression can be subtracted from both sides of an equation without changing the solution set for the equation; i.e., if $a = b$, then $a - c = b - c$.

The equation $x + 10 = 17$ then can be solved by subtracting 10 from both sides of the equation, as shown.

$$x + 10 = 17$$
$$x + 10 - 10 = 17 - 10$$
$$x + 0 = 7$$
$$x = 7$$

The solution set is {7}.

Check:

$$x + 10 = 17$$
$$7 + 10 \overset{?}{=} 17$$
$$17 = 17$$

The equation $\dfrac{x}{6} = 3$ is an example of the third type of equation. It can be solved by using the *multiplication property of equality*.

The **multiplication property of equality** states that the same nonzero real number can be multiplied to both sides of the equation without changing the solution set for the equation; i.e., if $a = b$ and $c \neq 0$, then $ac = bc$.

The equation $\dfrac{x}{6} = 3$ can be solved by multiplying both sides of the equation by 6, as shown.

$$\frac{x}{6} = 3$$

$$\frac{6}{1} \cdot \frac{x}{6} = 3 \cdot \frac{6}{1}$$

$$x = 18$$

The solution set is {18}.

Check:

$$\frac{x}{6} = 3$$

$$\frac{18}{6} \overset{?}{=} 3$$

$$3 = 3$$

An example of the fourth type of equation is $5x = 30$. This equation can be solved by the *division property of equality*.

> The **division property of equality** states that both sides of an equation can be divided by the same nonzero real number without changing the solution set of the equation; i.e., if $a = b$ and $c \neq 0$, then $\dfrac{a}{c} = \dfrac{b}{c}$.

The equation $5x = 30$ can be solved by dividing both sides of the equation by 5, as shown.

$$5x = 30$$

$$\frac{5x}{5} = \frac{30}{5}$$

$$x = 6$$

The solution set is {6}.

Check:

$$5x = 30$$

$$5(6) \overset{?}{=} 30$$

$$30 = 30$$

Example 7-9

Solve each equation.

(a) $x - 18 = 25$
(b) $3x = 42$
(c) $x + 16 = 20$
(d) $\dfrac{x}{4} = 24$

—Continued

Example 7-9 *Continued—*

Solution

(a) $\qquad x - 18 = 25$

$\qquad x - 18 + 18 = 25 + 18 \quad$ Add 18 to both sides.

$\qquad\qquad x = 43$, or the solution set is $\{43\}$

(b) $\quad 3x = 42$

$\qquad \dfrac{3x}{3} = \dfrac{42}{3} \quad$ Divide both sides by 3.

$\qquad x = 14$, or the solution set is $\{14\}$

(c) $\qquad x + 16 = 20$

$\qquad x + 16 - 16 = 20 - 16 \quad$ Subtract 16 from both sides.

$\qquad\qquad x = 4$, or the solution set is $\{4\}$

(d) $\qquad \dfrac{x}{4} = 24$

$\qquad \dfrac{4}{1} \cdot \dfrac{x}{4} = 24 \cdot \dfrac{4}{1} \quad$ Multiply both sides by 4.

$\qquad\qquad x = 96$, or the solution set is $\{96\}$

Calculator Explorations

A graphing calculator can assist you in viewing the graphical representation of the solution of an algebraic equation.

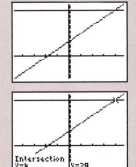

Which of the numbers shown as the intersection is the solution?

Try This One

7-G Solve each equation.

(a) $5x = 45$

(b) $x + 17 = 42$

(c) $x - 3 = 20$

(d) $\dfrac{x}{6} = 7$

For the answer, see page 331.

Solving Equations in General

More complex equations can be solved by the next procedure.

Procedure for Solving Equations

Step 1 Remove parentheses.

Step 2 Combine like terms on each side of the equation.

Step 3 Get the variables on one side of the equation and the numbers on the other side of the equation by using the addition and/or subtraction properties of equality.

Step 4 Combine like terms.

Step 5 Use the multiplication or division property of equality to solve for the variable.

Examples 7-10, 7-11, and 7-12 show this procedure.

Example 7-10

Solve the equation $5x + 9 = 29$.

Solution

$$5x + 9 = 29$$
$$5x + 9 - 9 = 29 - 9 \quad \text{Subtract 9 from both sides of the equation.}$$
$$5x = 20 \quad \text{Combine like terms.}$$
$$\frac{5x}{5} = \frac{20}{5} \quad \text{Divide both sides by 5.}$$
$$x = 4, \text{ or the solution set is } \{4\}$$

Check:

$$5x + 9 = 29$$
$$5(4) + 9 \stackrel{?}{=} 29$$
$$29 = 29$$

Example 7-11

Solve $6x - 10 = 4x + 8$.

Solution

$$6x - 10 = 4x + 8$$
$$6x - 4x - 10 = 4x - 4x + 8 \quad \text{Subtract } 4x \text{ from both sides of the equation.}$$
$$2x - 10 = 8 \quad \text{Combine like terms.}$$
$$2x - 10 + 10 = 8 + 10 \quad \text{Add 10 to both sides of the equation.}$$
$$2x = 18 \quad \text{Combine like terms.}$$
$$\frac{2x}{2} = \frac{18}{2} \quad \text{Divide both sides by 2.}$$
$$x = 9, \text{ or the solution set is } \{9\}$$

Check:

$$6x - 10 = 4x + 8$$
$$6(9) - 10 \stackrel{?}{=} 4(9) + 8$$
$$54 - 10 \stackrel{?}{=} 36 + 8$$
$$44 = 44$$

Calculator Explorations

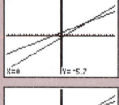

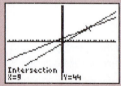

Here is the solution for Example 7-11 using the graphing calculator.

Calculator Explorations

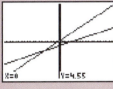

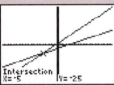

Here is the solution for Example 7-12 using the graphing calculator.

Example 7-12

Solve $3(2x + 5) - 10 = 3x - 10$.

Solution

$$3(2x + 5) - 10 = 3x - 10$$
$$6x + 15 - 10 = 3x - 10 \qquad \text{Remove parentheses.}$$
$$6x + 5 = 3x - 10 \qquad \text{Combine like terms.}$$
$$6x - 3x + 5 = 3x - 3x - 10 \qquad \text{Subtract } 3x \text{ from both sides of the equation.}$$
$$3x + 5 = -10 \qquad \text{Combine like terms.}$$
$$3x + 5 - 5 = -10 - 5 \qquad \text{Subtract 5 from both sides of the equation.}$$
$$3x = -15 \qquad \text{Combine like terms.}$$
$$\frac{3x}{3} = \frac{-15}{3} \qquad \text{Divide both sides by 3.}$$
$$x = -5, \text{ or the solution set is } \{-5\}$$

Check:

$$3(2x + 5) - 10 = 3x - 10$$
$$3(2 \cdot (-5) + 5) - 10 \overset{?}{=} 3(-5) - 10$$
$$3(-10 + 5) - 10 \overset{?}{=} -15 - 10$$
$$3(-5) - 10 \overset{?}{=} -25$$
$$-15 - 10 = -25$$
$$-25 = -25$$

Try This One

7-H   Solve each equation.

(a) $8x - 27 = 3x + 33$

(b) $2(x - 7) + 5 = 3x - 10$

(c) $-5(2x - 8) + 6 = 4x - 32$

For the answer, see page 331.

Solving Equations Containing Fractions

Whenever an equation contains fractions, it is necessary to "clear" the fractions from the equation. In order to clear the equation of fractions, multiply each term in the equation by the lowest common denominator of all the fractions in the equation. Example 7-13 shows this procedure.

Example   7-13

Solve the equation $\dfrac{2x}{3} + \dfrac{x}{5} = \dfrac{26}{3}$.

Solution

The lowest common denominator of the fractions with denominators of 3 and 5 is 15. Hence, the first step is to multiply each term in the equation by 15.

$$\frac{15}{1} \cdot \frac{2x}{3} + \frac{15}{1} \cdot \frac{x}{5} = \frac{15}{1} \cdot \frac{26}{3}$$

$$5 \cdot 2x + 3x = 5 \cdot 26$$

$$10x + 3x = 130$$

$$13x = 130$$

$$\frac{13x}{13} = \frac{130}{13}$$

$$x = 10, \text{ or the solution set is } \{10\}$$

Try This One

7-I   Solve the equation $\dfrac{3x}{5} - \dfrac{1}{4} = \dfrac{3x}{10}$.

For the answer, see page 331.

Solving an Equation or Formula for a Specific Variable

An equation with two variables can be solved for one of the variables in terms of the other variable using the same methods as shown previously. For example, the equation $5x + 6y = 18$ can be solved for y in terms of x by treating y as the variable, as shown.

$$5x + 6y = 18$$

$$5x - 5x + 6y = 18 - 5x \qquad \text{Subtract } 5x \text{ from both sides.}$$

$$6y = 18 - 5x$$

$$\frac{6y}{6} = \frac{18 - 5x}{6} \qquad \text{Divide both sides by 6.}$$

Hence,

$$y = \frac{18 - 5x}{6}$$

This can be simplified further, as shown.

$$y = \frac{18}{6} - \frac{5x}{6} \qquad \text{Divide each term in the numerator by 6.}$$

$$y = 3 - \frac{5x}{6}$$

Formulas may contain two or more variables. Often it is necessary to rewrite the formula in terms of another variable that is used in the right side of the formula. For example, the formula $C = 2\pi r$ is the formula for the circumference of a circle in terms of the radius.

The formula $I = \dfrac{V}{R}$ is used to find the electrical current of a circuit, where I represents the current, V represents the voltage, and R represents the resistance.

One could rewrite the formula for the radius of a circle in terms of the circumference by dividing both sides by 2π, as shown.

$$C = 2\pi r$$

$$\frac{C}{2\pi} = \frac{2\pi r}{2\pi}$$

$$\frac{C}{2\pi} = r$$

or

$$r = \frac{C}{2\pi}$$

Here, given the circumference of a circle, one could find the radius by dividing the circumference C by 2π.

Another example would be to solve $I = \dfrac{V}{R}$ for R. In order to do this, one would multiply both sides by R, as shown.

$$I = \frac{V}{R}$$

$$I \cdot R = \frac{V}{R} \cdot R$$

$$IR = V$$

$$\frac{IR}{I} = \frac{V}{I}$$

$$R = \frac{V}{I}$$

Example   7-14

The formula $F = \frac{9}{5}C + 32$ gives the Fahrenheit temperature for a temperature in Celsius. Solve for C in terms of F.

Solution

$$F = \frac{9}{5}C + 32$$

$$5F = 5 \cdot \frac{9}{5}C + 5 \cdot 32 \qquad \text{Multiply both sides by 5.}$$

$$5F = 9C + 160$$

$$5F - 160 = 9C + 160 - 160 \qquad \text{Subtract 160 from both sides.}$$

$$5F - 160 = 9C$$

$$\frac{5}{9}F - \frac{160}{9} = \frac{9C}{9} \qquad\qquad \text{Divide both sides by 9.}$$

$$\frac{5}{9}F - \frac{160}{9} = C$$

Hence,

$$C = \frac{5}{9}F - \frac{160}{9}$$

Calculator Explorations

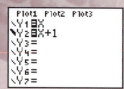

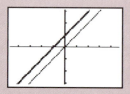

Will these lines (the graphs of the left side and right side of the equation) intersect? Is there a solution to this equation?

Infinite solutions can be illustrated graphically using the animator feature on the graphing calculator.

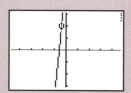

What is the solution set when the graphs of the left side and right side of the equation are the same?

Try This One

7-J

(a) Solve the equation $x + 2y = 18$ for y.

(b) Solve the formula $A = p(1 + rt)$ for r.

For the answer, see page 331.

Some equations have no solution. For example, the equation $x = x + 1$ has no solution since the equation is saying: find a number that, when 1 is added to it, you will get the same number. There is no real number that will satisfy this equation since when 1 is added to any real number, the new number will always be larger than the original number. If you try to solve the equation, you will get a false statement, as shown.

$$x = x + 1$$
$$x - x = x - x + 1$$
$$0 = 1$$

Hence, the solution to $x = x + 1$ is the empty set designated by $\{\}$ or $\varnothing$.

Other equations have an infinite number of solutions. That is, any real number will satisfy the equation. For example, the equation $2(5x + 8) - 10 = 10x + 6$ has an infinite number of solutions. As you attempt to solve the equation, a true statement occurs, as shown.

$$2(5x + 8) - 10 = 10x + 6$$
$$10x + 16 - 10 = 10x + 6$$
$$10x + 6 = 10x + 6$$
$$10x - 10x + 6 = 10x - 10x + 6$$
$$6 = 6$$

In this case, any real number substituted for x in the original equation always gives you a true closed equation. Hence, the solution set is the set of all real numbers or $\{x \mid x \text{ is a real number}\}$.

In summary, if the variable is eliminated when solving an equation and the resulting equation is false, then the solution set is the empty set. When the variable is eliminated and the resulting equation is true, the solution set is the set of all real numbers.

Example 7-15

Indicate whether the equation has a solution set of all real numbers or the empty set.

(a) $3(x - 6) + 2x = 5x - 18$

(b) $6x - 4 + 2x = 8x - 10$

—Continued

Calculator Explorations

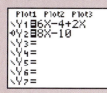

Here is the solution for Example 7-15a using the graphing calculator.

Here is the solution for Example 7-15b using the graphing calculator.

Example 7-15 *Continued—*

Solution

(a) $3(x - 6) + 2x = 5x - 18$

$$3x - 18 + 2x = 5x - 18$$

$$5x - 18 = 5x - 18$$

$$-18 = -18$$

Since the resulting equation is true, the solution set is $\{x \mid x \text{ is a real number}\}$.

(b) $6x - 4 + 2x = 8x - 10$

$$8x - 4 = 8x - 10$$

$$8x - 8x - 4 = 8x - 8x - 10$$

$$-4 = -10$$

Since the resulting equation is false, the equation has no solution. Hence, the solution set is $\varnothing$.

Try This One

7-K For each equation, determine whether the solution set is all real numbers or the empty set.

(a) $13x - 6 = 2(5x + 4) + 3x$

(b) $5(x + 6) - 5x = 30$

For the answer, see page 331.

Exercise Set 7-2

Computational Exercises

Solve each equation.

1. $x + 6 = 32$
2. $7 + x = 43$
3. $x - 5 = 54$
4. $36 = x - 9$
5. $-5 = x - 2$
6. $12 = x + 18$
7. $9x = 27$
8. $6x = 42$
9. $-3x = 36$
10. $-42 = -7x$
11. $6x + 12 = 48$
12. $10x - 30 = -5$
13. $-5x + 25 = -55$
14. $-3x + 18 = 42$
15. $2x + 10 = 4x - 30$
16. $5x - 6 = 2x - 24$
17. $x = 6x - 55$
18. $9x - 18 = 7x + 4$
19. $-2x = 15 - 5x$
20. $5x + 8 = 10x - 32$
21. $-6x + 15 = 4x - 25$

22. $9 - 2x = 7 - x$

23. $2(x + 8) = 40$

24. $2(x - 6) = 2$

25. $3(x + 2) = 26$

26. $7(x - 3) = 42$

27. $6 + 3(x - 5) = 2(x - 3)$

28. $-2(4x - 7) = 3x - 8$

29. $6 + 7(x - 3) = 2x + 10$

30. $5(9 - x) = 4(x + 6)$

31. $12(x - 2) - 10(x + 7) = 14$

32. $-2x + 3 + 4(x - 6) = 18$

33. $6(-x + 5) = 2(x + 8) - 10$

34. $-\frac{3}{7}x = 21$

35. $\frac{5}{6}x = 30$

36. $-\frac{1}{4}x = 2$

37. $\frac{5}{8}x = 40$

38. $-\frac{1}{2}x = 25$

39. $\frac{3}{4}x + 2 = 21$

40. $\frac{1}{8}x - 10 = -16$

41. $\dfrac{5x}{6} + \dfrac{x}{3} = 30$

42. $\dfrac{3x}{4} + \dfrac{7x}{2} = 18$

43. $\dfrac{7x}{3} + \dfrac{4x}{2} = 28$

44. $\dfrac{4x}{6} + \dfrac{x}{5} = \dfrac{2}{3} - \dfrac{x}{5}$

45. $\dfrac{7x}{3} + 5 = \dfrac{4x}{8} + 10$

46. $\dfrac{3x}{2} + \dfrac{1}{2} = \dfrac{4x}{5} + \dfrac{3}{5}$

47. $\dfrac{x}{6} + \dfrac{3x}{2} = \dfrac{4}{5} - \dfrac{2x}{15}$

48. $\dfrac{5}{12} - \dfrac{1}{3}x = \dfrac{2}{3}x - \dfrac{7}{6}$

49. $\dfrac{9x}{5} - \dfrac{2x}{7} = \dfrac{3}{5} - \dfrac{6}{7}$

50. $\dfrac{6x}{8} + \dfrac{4x}{2} - 5 = \dfrac{3x}{4}$

For Exercises 51–56, solve each equation for the specified variable.

51. $3x + 8 = 2y + 4$ for y

52. $5y = 3x + 2$ for x

53. $2 + 5x - 7y = 18$ for x

54. $5y - 3x + 2 = 10$ for y

55. $7x + 2y = 9$ for y

56. $3y + 6 = 2x + 8$ for x

Determine whether each equation has a solution set $\{x \mid x \text{ is a real number}\}$ or $\varnothing$.

57. $8x - 5 + 2x = 10x - 10 + 5$

58. $3x + 7 - x = 2x + 21$

59. $5(x - 3) + 2 = 5x - 8$

60. $4(x + 2) + 6 = 2x + 2x + 14$

Real World Applications

61. The electrical resistance for a conductor can be found by the formula $R = \frac{kL}{d^2}$. Solve the formula for L.

62. The illumination of a light can be found by the formula $I = \frac{C}{D^2}$. Solve the formula for C.

63. The volume of a cylinder can be found by the formula $V = \pi r^2 h$. Solve the formula for h.

64. The formula for the perimeter of a rectangle is $P = 2l + 2w$. Solve the formula for w.

65. The formula for the volume of a rectangular solid is $V = lwh$. Solve the formula for h.

66. The formula for the area of a trapezoid is $A = \frac{1}{2}h(a + b)$. Solve the formula for h.

67. The formula for converting mass to energy is $E = mc^2$. Solve the formula for m.

68. The formula for the distance traveled during an acceleration period is $d = \frac{1}{2}at^2$. Solve the formula for a.

69. The formula for the area of a triangle is $A = \frac{1}{2}bh$. Solve the formula for h.

70. The formula for the average a of two numbers b and c is $a = \frac{b+c}{2}$. Solve for b.

71. The centripetal force of an object can be found by using the formula $F = \frac{mv^2}{r}$. Solve the formula for r.

Writing Exercises

72. What is the difference between an equation and an expression?

73. What is the difference between an open equation and a closed equation?

74. List the four properties that can be used to solve basic first-degree equations.

75. What is meant by a solution set for an equation?

76. List the steps for solving an equation in general.

77. Explain what is meant by "clearing" fractions.

78. When does an equation have a solution set that is empty?

Critical Thinking

79. In this section, four properties of equality are explained. Explain why only two properties, namely, the addition property of equality and the multiplication property of equality, are necessary to solve equations.

80. Explain why both sides of an equation cannot be multiplied by zero.

7-3 # Applications of Linear Equations

Translating Verbal Statements into Symbols

Many types of real-life problems can be solved using equations. For example, a car rental company might charge a customer $25 a day plus $0.10 per mile for each mile driven over 100 miles. How many miles can a person drive if the person wants to rent the car for 2 days and spend only $100?

In order to solve problems using equations, the next procedure is used.

Procedure for Solving Word Problems Using Equations

Step 1 Read the problem carefully.

Step 2 Let x represent an unknown quantity.

Step 3 Write the equation based on the information given in the problem.

Step 4 Solve the equation for x.

Step 5 Check the solution.

Table 7-3

Common Phrases that Represent Operations

Phrases that represent addition

Six more than a number	$6 + x$
A number increased by 8	$x + 8$
Five added to a number	$5 + x$
The sum of a number and 17	$x + 17$

Phrases that represent subtraction

18 decreased by a number	$18 - x$
6.5 less than a number	$x - 6.5$
3 subtracted from a number	$x - 3$
The difference between a number and 5	$x - 5$

Phrases that represent multiplication

8 times a number	$8x$
Twice a number	$2x$
A number multiplied by 4	$4x$
The product of a number and 19	$19x$
$\frac{2}{3}$ of a number	$\frac{2}{3}x$

Phrases that represent division

A number divided by 5	$x \div 5$
35 divided by a number	$35 \div x$
The quotient of a number and 6	$x \div 6$

The step that students find most difficult is step 3. In order to write the equation for a word problem, one must translate the verbal statement into a symbolic statement. For example, when the verbal statement is "six more than three times a number" it is written symbolically as "$6 + 3x$" or "$3x + 6$."

In order to help you translate verbal statements into symbolic statements, Table 7-3 will be helpful. You should become familiar with these statements.

Example 7-16

Translate each verbal statement into symbols.

(a) 14 times a number
(b) 3 less than 4 times a number
(c) 6 times the sum of a number and 18
(d) a number divided by 7
(e) 6 is subtracted from 10 times a number

Solution

(a) $14x$
(b) $4x - 3$
(c) $6(x + 18)$
(d) $\frac{x}{7}$
(e) $10x - 6$

Try This One

7-L   Translate each verbal statement into a symbolic statement.

(a) 5 plus the product of a number and 7

(b) 51 minus 3 times a number

(c) $\frac{5}{8}$ of a number

For the answer, see page 331.

Solving Word Problems

After reading the problem and letting x represent the unknown, write the equation. This involves translating the verbal information in the problem into symbols as previously shown.

Example   7-17

If eight times a number plus three is 27, find the number.

Solution

Let $x =$ the number. Then 8 times a number plus three is written as $8x + 3$. Finally, the equation is written, as shown.

$$8x \quad\quad + \quad 3 \;=\; 27$$

eight times a number   plus   3   is   27

Solve for x

$$8x + 3 = 27$$
$$8x + 3 - 3 = 27 - 3$$
$$8x = 24$$
$$\frac{8x}{8} = \frac{24}{8}$$
$$x = 3$$

The number is 3.

Check: Eight times a number plus three should equal 27; 8 times 3 is 24 and $24 + 3 = 27$. Hence, the solution is correct.

Sometimes when there are two unknowns in a word problem, one unknown can be represented in terms of the other. For example, if I know that one number is 5 more than another number, then the first number can be represented by x and the second number can be represented by $x + 5$. Example 7-18 shows this concept.

Example   7-18

Mary is 8 inches taller than her younger brother. The sum of their heights is 92 inches. Find the height of each person.

—Continued

Example 7-18 *Continued—*

Solution

Let x = the height of Mary's brother. Then $x + 8$ = Mary's height.

$$\underset{\text{brother's height}}{x} \quad \underset{\text{sum}}{+} \quad \underset{\text{Mary's height}}{(x+8)} \quad = \quad \underset{\text{total}}{92}$$

$$x + x + 8 = 92$$
$$2x + 8 = 92$$
$$2x = 84$$
$$x = 42$$
$$x + 8 = 50$$

Hence, Mary's brother is 42 inches tall and Mary's height is 50 inches.

Check: $42 + 50 = 92$.

Example 7-19

In a child's bank, there are only dimes and quarters. There are three times as many dimes as there are quarters. If the total amount of money contained in the bank is $2.20, how many dimes and how many quarters are there?

Solution

Let x = the number of quarters and $3x$ = the number of dimes. Since each quarter is worth $0.25, the value of the quarters is $0.25x$; and since each dime is worth $0.10, the value of the dimes is $0.10(3x)$.

The equation, then, is

$$\underset{\text{value of quarters}}{0.25x} \quad + \quad \underset{\text{value of dimes}}{0.10(3x)} \quad = \quad \underset{\text{total value}}{2.20}$$

$$0.55x = 2.20$$
$$\frac{0.55x}{0.55} = \frac{2.20}{0.55}$$
$$x = 4$$
$$3x = 12$$

Hence, there are 4 quarters and 12 dimes.

Check:
$$4 \text{ quarters} = 4 \times \$0.25 = \$1.00$$
$$+ 12 \quad \text{dimes} = 12 \times \$0.10 = \underline{\$1.20}$$
$$\text{Total} = \$2.20$$

Try This One

7-M In 1999 there were 65 female officials in Congress and there were 47 more female members of the House of Representatives than female senators. Find the number of females in each house of Congress.

For the answer, see page 331.

Exercise Set 7-3

Computational Exercises

For Exercises 1–20, write each phrase in symbols.

1. 3 less than a number
2. A number decreased by 17
3. A number increased by 9
4. 6 increased by a number
5. 11 decreased by a number
6. 8 more than a number
7. 9 less than a number
8. 6 subtracted from a number
9. 7 minus a number
10. Seven times a number
11. A number multiplied by 8
12. One-half a number
13. Five more than three times a number
14. The quotient of three times a number and 6
15. 3 more than five times a number
16. Four times a number
17. Double a number
18. Four less than six times a number
19. The quotient of a number and 14
20. A number divided by 8

For Exercises 21–29, solve each.

21. Six times a certain number plus the number is equal to 56. Find the number.
22. The sum of a number and the number plus 2 is equal to 20. Find the number.
23. Twice a number is 32 less than four times the number. Find the number.
24. The larger of two numbers is 10 more than the smaller number. The sum of the number is 42. Find the numbers.
25. The difference of two numbers is 6. The sum of the numbers is 28. Find the numbers.
26. Five times a number is equal to the number increased by 12. Find the number.
27. Twice a number is 24 less than four times the number. Find the number.
28. The difference between one-half a number and the number is 8. Find the number.
29. Twelve more than a number is divided by 2. The result is 20. Find the number.

Real World Applications

30. A mathematics class containing 57 students was divided into two sections. One section has three more students than the other. How many students were in each section?
31. For a certain year, the combined revenues for PepsiCo and Coca-Cola were $47 billion. If the revenue for PepsiCo was $11 billion more than Coca-Cola, how much was the revenue of each company?
32. The cost, including sales tax, of a Ford Focus LX is $13,884.94. If the sales tax is 6%, find the cost of the automobile before the tax was added.

33. Pete is three times as old as Bill. The sum of their ages is 48. How old is each?

34. An electric bill for September is $2.32 less than the electric bill for October. If the total bill for the 2 months was $119.56, find the bill for each month.

35. If a person invested half of her money at 8% and half at 6% and received $210 interest, find the total amount of money invested.

36. A basketball team played 32 games and won 4 more games than it lost. Find the number of games the team won.

37. The population of White Oak, Pennsylvania, decreased by 6% between the years of 1990 and 1998. If the total population for the two years was 16,984, find the population in 1990 and 1998.

38. If a television set is marked $\frac{1}{3}$ off and sells for $180, what was the original price?

39. A carpenter wanted to cut a 6-foot board into three pieces such that each piece is 6 inches longer than the preceding one. Find the length of each piece.

40. The Halloween Association reported that for a specific year, Americans spent $0.68 billion more on candy than costumes for Halloween. If the total spent by Americans for both items was $3.18 billion, how much did Americans spend on each item?

41. A landowner planted $\frac{1}{2}$ of his land with corn and $\frac{1}{4}$ of his land with beans. If he had 40 acres left, how large is his farm?

42. A ball bounces half as far as it did on the preceding bounce and traveled a total distance of 12 feet on two bounces. How far did it go on the first bounce?

43. A person sold her house for $82,000 and made a profit of 20%. How much did she pay for her home?

44. A father left $\frac{1}{2}$ of his estate to his son, $\frac{1}{3}$ of his estate to his granddaughter, and the remaining $6000 to charity. What was his total estate?

Writing Exercises

45. Explain the process used to solve word problems.

46. What specific words can be used to represent addition?

47. What specific words can be used to represent subtraction?

48. What specific words can be used to represent multiplication?

49. What specific words can be used to represent division?

Critical Thinking

50. The temperature and the wind combine to cause body surfaces to lose heat. Meteorologists call this effect "the wind chill factor." For example, if the actual temperature outside is 10°F and the wind speed is 20 miles per hour, it will feel like it is −20°F outside. Thus, −20°F is called the wind chill. When it is 25°F outside and the wind speed is 40 miles per hour, it will feel like it is −35°F outside. From the information given, derive a linear equation for determining the wind chill using the actual temperature and the wind speed.

7-4 Solving Linear Inequalities

In the previous sections, we studied linear equations. In this section, you will learn how to solve **linear inequalities.** Recall that an equation contains an *equal* sign, whereas an inequality contains an *inequality* sign. The inequality signs are shown in Table 7-4.

Graphing Inequalities on the Number Line

An inequality of the form $x \geq 5$ means that x can be any real number from 5 on up. The number 5 is included in the solution set. The graph for the solution set of $x \geq 5$ looks like this:

To show that the number 5 is included in the solution set, a *solid* or *closed dot* (•) is used. On the other hand, the solution set for the inequality $x > 5$ includes all real numbers greater than 5; however, the actual number 5 is not in the solution set. The graph for the solution set of $x > 5$ looks like this:

To show that the number 5 is not included in the solution set, an *open circle* (○) is used. The same reasoning is used for inequalities containing $\leq$ and $<$.

Sometimes it is necessary to graph an inequality of the form

$$-2 \leq x < 4$$

In this case, -2 is less than or equal to x *and* x is less than 4. What this means is that x can be any number between -2 and 4 and -2 is included in the solution set but not 4. The graph for the solution set would look like this:

A summary of the inequalities is shown in Table 7-5. The solution set for inequalities can be written in set notation.

Solving Inequalities

Inequalities such as $2x + 10 \leq 30$ can be solved like equations; however, there is one exception.

> To solve a linear inequality, proceed as if you were solving a linear equation except that when multiplying or dividing by a *negative* number, you must *reverse* the inequality sign.

Table 7-4

Inequality Signs

Symbol	Name
$\geq$	Greater than or equal to
$>$	Greater than
$\leq$	Less than or equal to
$<$	Less than

Table 7-5

Graphs of Solution Sets for Linear Inequalities

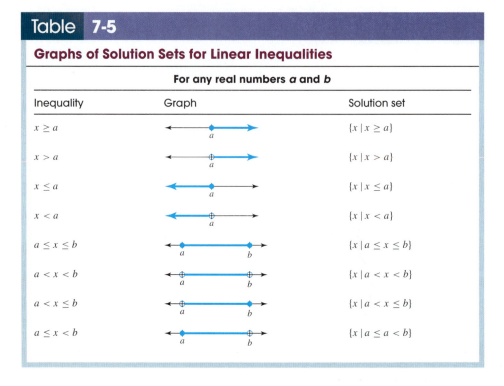

	For any real numbers *a* and *b*	
Inequality	Graph	Solution set
$x \geq a$		$\{x \mid x \geq a\}$
$x > a$		$\{x \mid x > a\}$
$x \leq a$		$\{x \mid x \leq a\}$
$x < a$		$\{x \mid x < a\}$
$a \leq x \leq b$		$\{x \mid a \leq x \leq b\}$
$a < x < b$		$\{x \mid a < x < b\}$
$a < x \leq b$		$\{x \mid a < x \leq b\}$
$a \leq x < b$		$\{x \mid a \leq a < b\}$

Math Note

Reversing an inequality means that $\geq$ becomes $\leq$, $>$ becomes $<$, $\leq$ becomes $\geq$, and $<$ becomes $>$.

In other words, the same number or expression can be added or subtracted from both sides of the inequality without changing the solution of the inequality. Both sides of an inequality can be multiplied or divided by a *positive* number without changing the solution of the inequality. Finally, if you are multiplying or dividing an inequality by a negative number, you must reverse the inequality sign.

Examples 7-20, 7-21, and 7-22 show how to solve linear inequalities.

Example 7-20

Solve and graph the solution set for $5x - 9 \geq 21$.

Solution

$$5x - 9 \geq 21$$
$$5x - 9 + 9 \geq 21 + 9 \quad \text{Add 9 to both sides.}$$
$$5x \geq 30$$
$$\frac{5x}{5} \geq \frac{30}{5} \quad \text{Divide both sides by 5.}$$
$$x \geq 6$$

The solution set is $\{x \mid x \geq 6\}$. The graph of the solution set is

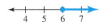

Math Note

An inequality cannot be checked for the exact answer as an equation can; however, an approximate check can be made by selecting some number in the solution set for x, substituting it in the inequality and seeing if a true closed inequality results. In Example 7-20, select $x = 8$. (You may choose any value for x as long as it is six or larger.)

$$5x - 9 \geq 21$$
$$5(8) - 9 \stackrel{?}{\geq} 21$$
$$40 - 9 \stackrel{?}{\geq} 21$$
$$31 \geq 21 \quad \text{True}$$

Example 7-21

Solve and graph the solution set for $16 - 3x > 40$.

Solution

$$16 - 3x > 40$$
$$16 - 16 - 3x > 40 - 16 \quad \text{Subtract 16 from both sides.}$$
$$-3x > 24$$
$$\frac{-3x}{-3} < \frac{24}{-3} \qquad \text{Divide both sides by } -3 \text{ and reverse the inequality sign.}$$
$$x < -8$$

Hence, the solution set is $\{x \mid x < -8\}$ and the graph of the solution set is

It is important to check your answer when you reverse the inequality sign. For example, $x \leq 8$ will be true for $x = 8$ even if it should read $x \geq 8$; checking $x = 6$ would be better.

Example 7-22

Solve and graph the solution set for $4(x + 3) < 2x - 26$.

Solution

$$4(x + 3) < 2x - 26$$
$$4x + 12 < 2x - 26 \qquad \text{Remove parentheses.}$$
$$4x - 2x + 12 < 2x - 2x - 26 \quad \text{Subtract } 2x \text{ from both sides.}$$
$$2x + 12 < -26$$
$$2x + 12 - 12 < -26 - 12 \qquad \text{Subtract 12 from both sides.}$$
$$2x < -38$$
$$\frac{2x}{2} < \frac{-38}{2} \qquad \text{Divide both sides by 2.}$$
$$x < -19$$

Hence, the solution set is $\{x \mid x < -19\}$ and the graph is

Try This One

7-N   Solve and graph the solution set for each inequality.

(a) $9x + 29 \geq 1$

(b) $10x - 30 < 7x + 21$

(c) $6(2x - 7) \leq 3x + 8$

(d) $14 - 7x < 56$

For the answer, see page 332.

Applications of Inequalities

There are many real-life applications of inequalities. These applications involve problems in which it is necessary to determine such things as "at most" or "at least." For example, suppose you wanted to go on a vacation, and you know the motel room would cost you $65 a night. Furthermore, you had decided that additional expenses would cost you a total of $100. If you wanted to spend *at most* $490 on the entire trip, how many nights could you plan on staying?

The procedure for solving an inequality word problem is the same as the one used to solve word problems for equations given on page 289. Some students have difficulty translating phrases such as "at most," "not more than," etc. into symbols. If so, Table 7-6 may help.

Example   7-23

Solve the vacation problem at the beginning of this subsection.

Solution

Let x represent the number of nights on vacation and $65x$ represent the cost of the motel. The inequality can be written as

$$65x + 100 \leq 490$$

Solve the inequality.

$$65x + \$100 \leq 490$$
$$65x + 100 - 100 \leq 490 - 100$$
$$65x \leq 390$$
$$\frac{65x}{65} \leq \frac{390}{65}$$
$$x \leq 6$$

Hence, you could spend at most 6 nights on vacation.

Check:

$$65x + 100 \leq 490$$
$$65(6) + 100 \overset{?}{\leq} 490$$
$$390 + 100 \overset{?}{\leq} 490$$
$$490 \leq 490$$

Table 7-6

Common Phrases Used in Inequality Word Problems

>	<
Greater than	Less than
Above	Below
Higher than	Lower than
Longer than	Shorter than
Larger than	Smaller than
Increased	Decreased

$\geq$	$\leq$
Greater than or equal to	Less than or equal to
At least	Is at most
Not less than	Not more than

Example 7-24

Mike is going to buy lunch for his coworkers. He decides to buy hamburgers and fries. Hamburgers cost $1.00 and fries cost $0.80, and he also needs 5% of the total for sales tax. What is the maximum number of items that can be purchased if he wants to buy the same number of hamburgers as fries and he has $10.00 to spend?

Solution

Let x represent the number of hamburgers and the number of fries purchased. Then ($1.00)$x$ represents cost of the hamburgers and ($0.80)$x$ represents cost of the fries.

$$0.05(1.00x + 0.80x) = \text{tax}$$
$$1.00x + 0.80x + 0.05(1.00x + 0.80x) \leq 10.00$$
$$1.00x + 0.80x + 0.05x + 0.04x \leq 10.00$$
$$1.89x \leq 10.00$$
$$\frac{1.89x}{1.89} \leq \frac{10.00}{1.89}$$
$$x \leq 5.29$$

Hence, he can purchase at most five hamburgers and five fries. We rounded 5.29 down to 5 since he must purchase a whole number of items.

Check:

$$\begin{aligned}
5 \text{ hamburgers cost } 5 \times \$1.00 = \quad &\$5.00 \\
5 \text{ fries cost } 5 \times \$0.80 = \quad &+4.00 \\
\text{Total cost of food} \quad &9.00 \\
\text{Tax: } 0.05 \times 9.00 = 0.45 \quad &+0.45 \\
\text{Total bill} \quad &9.45
\end{aligned}$$

Try This One

7-O Jay wishes to purchase a computer and monitor. He wants to spend at most $1000. He finds that there is a 10% rebate on all computers purchased at a certain store. If the monitor costs $100 (no rebate), what is the most he can pay for the computer? (Ignore the sales tax.)
For the answer, see page 332.

Exercise Set 7-4

Computational Exercises

For Exercises 1–10, show the solutions using a graph.

1. $x \geq 3$	2. $x < -2$	3. $x < -4$
4. $x \geq 0$	5. $x \leq -9$	6. $x \leq 1$
7. $-3 < x < 7$	8. $4 \leq x \leq 10$	9. $2 < x \leq 5$
10. $-3 \leq x < 0$		

For Exercises 11–40, solve each inequality and graph the solution set on the number line.

11. $x + 6 < 11$	12. $x - 2 \leq 15$	13. $x - 7 \leq 23$
14. $x + 9 > 20$	15. $3x \geq 18$	16. $5x < 30$
17. $7 - x > 42$	18. $9 - x \leq 20$	19. $\frac{2}{3}x < 18$
20. $\frac{3}{4}x \geq 36$	21. $-10x < 30$	22. $-25x \geq 100$
23. $2x + 8 \geq 32$	24. $5x - 6 < 39$	25. $-3x + 12 \leq 36$
26. $5 - 2x > 25$	27. $6(x - 12) \leq 54$	28. $-3(2x + 7) < -16$
29. $-5(3 - x) \leq 27$	30. $9(4x - 1) > 71$	31. $16 - 5x \geq 22$
32. $5 - 3x < 25$	33. $3(x + 1) - 10 < 2x + 7$	34. $4(x - 8) - 2x < -22$
35. $9 - 5(x + 6) \geq 32$	36. $18 - 6(x + 2) < 41$	37. $6(2x + 3) \geq 5(2x - 15)$
38. $-x \geq -15$	39. $6x - 7 \geq 5(x - 2) - 17$	40. $3x + 6 < -8x + 7$

Real World Applications

41. Mary wishes to purchase a used car. She wishes to spend at most $8000. The sales tax rate in her state is 7%. Title and license plate fee is $120. What is the maximum amount she can spend for an automobile?

42. Bill has three test grades of 95, 84, and 85 so far. If the final examination, still to come, counts for two test scores, what is the lowest he can score on the final exam and still get an A for the course? He needs at least 450 points for an A.

43. In order to get a C for her sociology course, Betsy needs at least a 70% average. On exam 1 she scored 78% and on exam 2 she scored 68%. What is the lowest score she can get on the last exam?

44. A husband and wife wish to sell their house and make at least a 10% profit. The real estate agent's commission is 7% and closing costs are $1000. If they paid $150,000 for their home, what is the minimum price they should ask for their house?

The list shown here represents the average number of tornados per year for the given states. Let x represent the average number of tornados per year. For Exercises 45–50, write the name or names of the states that are described by the solution to the inequality.

State	Average number of tornados per year
Texas	168
Florida	79
Kansas	75
Colorado	58
West Virginia	30
Missouri	26
Pennsylvania	22
California	14
Maryland	11
Arizona	5
Delaware	2
Maine	1

Source: USA TODAY

45. $x \geq 58$

46. $x < 22$

47. $26 \leq x < 58$

48. $2 < x \leq 22$

49. $11 \leq x \leq 30$

50. $30 < x < 168$

Writing Exercises

51. Explain the difference between using an open dot and a closed dot when graphing a solution set to a linear inequality.

52. List some differences between solving a linear inequality and a linear equation.

53. Explain an approximate method that you can use to check the solution to a linear inequality.

54. Explain why the method used to check a linear inequality using a specific number does not guarantee that the solution is correct.

Critical Thinking

55. To earn a B in a course, a student needs a grade average between 80% and 89%. If the student has previous grades of 77%, 82%, and 86%, what grade must the student earn on the final exam to obtain a B? The final exam counts double.

56. To earn a C for a course, a student needs to obtain a grade average between 70% and 79%. If the student's grades are 68%, 72%, 83%, and 58%, what is the lowest possible score the student can make on the last exam?

7-5   Ratio, Proportion, and Variation

Many real-life problems can be solved by using the concepts of ratio and proportion.

Ratios

A **ratio** is a comparison of two quantities using division.

For example, the U.S. Census Bureau reported that 45% of joggers were female and 55% were male. Hence, a comparison for the genders of the joggers can be made. The ratio of female joggers to male joggers is 45 to 55.

For two nonzero numbers, a and b, the ratio of a to b is written as $a{:}b$ (read a to b) or $\dfrac{a}{b}$.

Ratios can be written using a colon or a fraction as shown in the definition. In mathematics, the fraction is used more often.

> ### Math Note
>
> To set up a correct ratio, whatever number comes first in the ratio statement must be placed in the numerator of the fraction and whatever number comes second in the ratio statement must be placed in the denominator of the fraction.

Example   7-25

The *USA TODAY* Snapshot shows the number (in millions) of people who participate in various recreational activities. Find each.

(a) The ratio of recreational swimmers to recreational bikers

(b) The ratio of people who fish to people who bowl

USA TODAY Snapshots®

Participation sports attract millions

Swimming might not get the coverage of baseball, basketball or football, but in terms of participation, it is America's most popular sport based on the U.S. population 6 years of age or older[1]:

What sports Americans participate in:

	(in millions)
Recreational swimming	95.1
Recreational walking	84.1
Recreational biking	56.2
Bowling	52.6
Freshwater fishing	44.5

1–participated at least once in 1999

Source: Sporting Goods Manufacturers

By Ellen J. Horrow and Quin Tian, USA TODAY

Source: USA TODAY. Copyright 2000. USA TODAY. Reprinted with permission.

Solution

(a) $\dfrac{\text{number of swimmers}}{\text{number of bikers}} = \dfrac{95.1}{56.2}$

(b) $\dfrac{\text{number of people who fish}}{\text{number of people who bowl}} = \dfrac{44.5}{52.6}$

Since ratios can be expressed as fractions, they can be simplified by reducing the fraction. For example, the ratio of 10 to 15 is written 10:15 or $\frac{10}{15}$ and the fraction $\frac{10}{15}$ can be reduced to $\frac{2}{3}$. Hence, the ratio 10:15 is the same as 2:3.

Sometimes ratios can be written as fractions where the numerator and the denominator have the same units of measure. Example 7-26 shows how to do this.

Example 7-26

Find the ratio of 18 inches to 2 feet.

Solution

First write the ratio as $\dfrac{18 \text{ inches}}{2 \text{ feet}}$. Then change 2 feet to inches and substitute in the fraction as shown.

$$1 \text{ foot} = 12 \text{ inches; hence, } 2 \text{ feet} = 2 \cdot 12 = 24 \text{ inches}$$

$$\frac{18 \text{ inches}}{24 \text{ inches}}$$

Finally reduce the fraction.

$$\frac{18}{24} = \frac{3}{4} \quad \text{(The inches will cancel out.)}$$

Hence, the ratio of 18 inches to 2 feet can be written as $\frac{3}{4}$.

Proportions

When two ratios are equal, they can be written as a *proportion*.

A **proportion** is a statement of equality of two ratios.

For example, the ratio of 4:7 and 8:14 can be written as a proportion as shown.

$$\frac{4}{7} = \frac{8}{14}$$

Two fractions, $\dfrac{a}{b}$ and $\dfrac{c}{d}$, are equal if $ad = bc$. The product of the numerator of one fraction and the denominator of the other fraction is called a cross product. For example, $\frac{3}{4} = \frac{6}{8}$ since $3 \cdot 8 = 4 \cdot 6$ or $24 = 24$.

Two ratios form a proportion if the cross products of their numerators and denominators are equal. For example, the two ratios $\frac{5}{6}$ and $\frac{15}{18}$ can be written as a proportion since

$$\frac{5}{6} \,\diagup\!\!\!\!\diagdown\, \frac{15}{18}$$

$$5 \cdot 18 = 6 \cdot 15$$

$$90 = 90$$

Thus, we can write 5:6 = 15:18.

Solving Proportions

When three of the four numbers of a proportion are known, the fourth number can be found by setting up a proportion using x as the unknown number, cross multiplying, and solving the resulting equation for x. This procedure is shown in Examples 7-27 and 7-28.

The height of the person and the statue are in proportion. If we know the ratio of the heights and the height of the person, we can find the height of the statue.

Example 7-27

Solve the proportion for x.

$$\frac{12}{48} = \frac{3}{x}$$

Solution

$$\frac{12}{48} \bowtie \frac{3}{x} \qquad \text{Cross multiply.}$$

$$12x = 3 \cdot 48$$

$$12x = 144$$

$$\frac{12x}{12} = \frac{144}{12} \qquad \text{Divide by 12 to solve for } x.$$

$$x = 12$$

Example 7-28

Solve the proportion.

$$\frac{x - 5}{10} = \frac{x + 2}{20}$$

Solution

$$\frac{x - 5}{10} \bowtie \frac{x + 2}{20} \qquad \text{Cross multiply.}$$

$$20(x - 5) = 10(x + 2) \qquad \text{Remove parentheses.}$$

$$20x - 100 = 10x + 20 \qquad \text{Solve for } x.$$

$$20x - 10x - 100 = 10x - 10x + 20$$

$$10x - 100 = 20$$

$$10x - 100 + 100 = 20 + 100$$

$$10x = 120$$

$$\frac{10x}{10} = \frac{120}{10}$$

$$x = 12$$

—Continued

Example 7-28 *Continued—*

Proportions can be checked by substituting the value for x in the proportion and simplifying. For example, in Exercise 7-28, substitute $x = 12$ and simplify both sides as shown.

$$\frac{x-5}{10} = \frac{x+2}{20}$$

$$\frac{12-5}{10} \overset{?}{=} \frac{12+2}{20}$$

$$\frac{7}{10} \overset{?}{=} \frac{14}{20}$$

$$\frac{7}{10} = \frac{7}{10}$$

Try This One

7-P Solve each problem:

(a) $\dfrac{x}{7} = \dfrac{22}{25}$

(b) $\dfrac{5}{12} = \dfrac{x}{48}$

(c) $\dfrac{x+6}{15} = \dfrac{x-2}{5}$

For the answer, see page 332.

Application of Proportions

Many word problems can be solved using proportions. The procedure is given next.

Procedure for Solving Word Problems Using Proportions

Step 1 Read the problem and find the ratio statement.

Step 2 Write the ratio using a fraction.

Step 3 Set up the proportion using x for the unknown number.

Step 4 Solve the proportion for x.

Example 7-29

A person used 12 gallons of gasoline to drive 228 miles. How many gallons of gasoline will be needed to drive an additional 380 miles?

Solution

Step 1 Identify the ratio statement. The ratio statement is "12 gallons to drive 228 miles."

Step 2 Write the ratio as a fraction. The ratio then is $\dfrac{12 \text{ gallons}}{228 \text{ miles}}$.

Step 3 Set up the proportion. Let x represent the number of gallons that are needed to drive 380 miles; then the proportion is

$$\frac{12 \text{ gallons}}{228 \text{ miles}} = \frac{x \text{ gallons}}{380 \text{ miles}}$$

Step 4 Solve the proportion: $\dfrac{12}{228} = \dfrac{x}{380}$.

$$\frac{12}{228} \diagup\hspace{-0.9em}\diagdown \frac{x}{380}$$

$$228x = 12 \cdot 380$$

$$228x = 4560$$

$$\frac{228x}{228} = \frac{4560}{228}$$

$$x = 20$$

Hence, 20 gallons will be needed.

> **Math Note**
>
> When setting up a proportion, be sure to put like quantities in the numerators and like quantities in the denominators. In Example 7-29, gallons were placed in the numerators and miles in the denominators.

An interesting example of the use of proportions is in estimating wildlife populations of a specific region. Example 7-30 explains how biologists use this procedure.

Example 7-30

A biologist wishes to estimate how many fish live in a small lake. She catches a sample of 35 fish, tags them, and then releases them into the lake. A week later, she catches 80 fish and finds 16 of them are tagged. Approximately how many fish live in the lake?

Solution

Let x represent the approximate number of the total fish in the lake. The ratio of tagged fish to the total number of fish caught is $\frac{16}{80}$ and originally 35 fish were

—Continued

Example 7-30 Continued—

tagged, so the proportion becomes

$$\frac{16 \text{ tagged}}{80 \text{ caught}} = \frac{35 \text{ tagged}}{x}$$

$$\frac{16}{80} = \frac{35}{x}$$

$$16x = 80 \cdot 35$$

$$16x = 2800$$

$$\frac{16x}{16} = \frac{2800}{16}$$

$$x = 175$$

Hence, there are approximately 175 fish in the lake.

Try This One

7-Q A student can type 18 pages in 10 minutes. At the same rate, how long will it take the student to type 72 pages?
For the answer, see page 332.

Variation

One variable can vary *directly* with another variable. For example, if a person earns $100 a day, his or her income varies directly with the number of days he or she works. That is, the more the person works, the more he or she gets paid. This is an example of what is called a **direct variation.** An *equation of variation* can be written using a *constant, k,* which is the result of a ratio. In this case, $k = \dfrac{\$100}{1 \text{ day}}$.

y is said to vary *directly* with *x* if there is some nonzero constant *k* such that $y = kx$.

Example 7-31

A person earns $100 per day. Find the person's earnings if the person works

(a) 6 days
(b) 15 days

Solution

Let y = the amount of earnings
x = the number of days the person works
$k = \$100$ per day, then
$y = 100x$ is the variation equation.

—Continued

Example 7-31 *Continued—*

(a) $y = 100x$. For $x = 6$ days,

$$y = 100(6)$$
$$y = \$600$$

(b) $y = 100x$. For $x = 15$ days,

$$y = 100(15)$$
$$y = \$1500$$

Hence, if the person works 6 days, the person's earnings will be $600. If the person works 15 days, the person's earnings will be $1500.

Sometimes it is necessary to find the value of k when given certain information and then solve the problem. The next example shows this.

Example 7-32

The weight of a certain type of cable varies directly with its length. If 20 feet of cable weighs 4 pounds, find k and determine the weight of 75 feet of cable.

Solution

Step 1   Write the equation of variation.

$$y = kx \quad \text{where} \quad y = \text{the weight}$$
$$x = \text{length of cable in feet}$$
$$k = \text{the constant}$$

$$4 \text{ lbs} = k \cdot 20 \text{ ft}$$

Step 2   Solve for k.

$$4 \text{ lbs} = k \cdot 20 \text{ ft}$$

$$\frac{4 \text{ lbs}}{20 \text{ ft}} = \frac{k \cdot 20 \text{ ft}}{20 \text{ ft}}$$

$$k = 0.2 \text{ lbs/ft}$$

Hence, the equation of variation can be written as $y = 0.2x$.

Step 3   Solve the problem for the new values of x and y using $k = 0.2$.

$$y = 0.2x$$
$$y = 0.2 \cdot 75$$
$$y = 15 \text{ pounds}$$

Hence, 75 feet of cable will weigh 15 pounds.

Try This One

7-R The weight (in pounds) of a hollow statue varies directly with the square of its height (in feet); i.e., $y = kx^2$, where $y =$ the weight and $x =$ the height. If a statue that is 4 feet tall weighs 2 pounds, find the weight of a statue that is 6 feet tall.
For the answer, see page 332.

A variable can also vary *inversely* with another variable. For example, the time it takes to drive a certain distance to a vacation home varies inversely with the rate of speed of the automobile. That is, if a person drives on average 55 miles per hour as opposed to 40 miles per hour, the person will get there in less time. This is called **inverse variation.**

> y is said to vary *inversely* with x if there is some nonzero constant k such that $y = \dfrac{k}{x}$.

Example 7-33

A person has a vacation home that is 378 miles from their residence. Find the time it takes to drive the distance if she drives on average

(a) 55 miles per hour
(b) 40 miles per hour

Solution

Let $y =$ the time it takes to drive the distance
$\quad x =$ the average speed of the automobile
$\quad k = 378$ miles
$\quad y = \dfrac{k}{x}$ is the variation equation

Then

(a) $y = \dfrac{k}{x}$

$\quad y = \dfrac{378}{55}$ for an average speed of 55 miles per hour

$\quad y \approx 6.87$ hours

(b) $y = \dfrac{k}{x}$

$\quad y = \dfrac{378}{40}$

$\quad y = 9.45$ hours

Hence, if a person averages 55 miles per hour, it will take about 6.87 hours, and if a person averages 40 miles per hour, it will take about 9.45 hours to reach the destination.

Example 7-34

If the temperature of a gas is held constant, the pressure the gas exerts on a container varies inversely with its volume. If a gas has a volume of 38 cubic inches and exerts a pressure of 8 pounds per square inch, find the volume when the pressure is 64 pounds per square inch.

Solution

Step 1 Write the equation of variation.

$$y = \frac{k}{x} \quad \text{where} \quad \begin{aligned} y &= \text{the volume of the gas in cubic inches} \\ x &= \text{the pressure of the gas in pounds per square inch} \\ k &= \text{the constant} \end{aligned}$$

Step 2 Solve for k.

$$y = \frac{k}{x}$$

$$38 = \frac{k}{8}$$

$$38 \cdot 8 = k$$

$$304 = k$$

Step 3 Solve the problem for the new values of x and y by using $k = 304$.

$$y = \frac{304}{x}$$

$$y = \frac{304}{64}$$

$$y = 4.75 \text{ cubic inches}$$

Hence, when the pressure of the gas is 64 pounds per square inch, its volume will be 4.75 cubic inches.

Try This One

7-S The strength of a beam varies inversely with the cube of its length (i.e., $y = \frac{k}{x^3}$, where y = strength in pounds and x = length in feet). If a 12-foot beam can support 1800 pounds, how many pounds can a 15-foot beam support?
For the answer, see page 332.

Exercise Set 7-5

Computational Exercises

For Exercises 1–10, write each ratio statement as a fraction and reduce to lowest terms if possible.

1. 18 to 28
2. 5 to 12
3. 14:32
4. 40:75
5. 12 cents to 15 cents
6. 18 inches to 42 inches
7. 3 weeks to 8 weeks
8. 2 pounds to 12 ounces
9. 5 feet to 30 inches
10. 12 years to 18 years

For Exercises 11–20, solve each proportion.

11. $\dfrac{3}{x} = \dfrac{14}{45}$

12. $\dfrac{x}{2} = \dfrac{18}{6}$

13. $\dfrac{5}{6} = \dfrac{x}{42}$

14. $\dfrac{9}{8} = \dfrac{45}{x}$

15. $\dfrac{x-6}{12} = \dfrac{1}{3}$

16. $\dfrac{x+3}{5} = \dfrac{35}{25}$

17. $\dfrac{2}{x-3} = \dfrac{5}{x+8}$

18. $\dfrac{4}{x-3} = \dfrac{16}{x-2}$

19. $\dfrac{x-3}{4} = \dfrac{x+6}{20}$

20. $\dfrac{x}{10} = \dfrac{x-2}{20}$

Real World Applications

21. The Information Resources Institute reports that one out of every five people who buys ice cream buys vanilla ice cream. If a store sells 75 ice cream cones in one day, about how many will be vanilla?

22. The U.S. Department of Agriculture reported that 57 out of every 100 milk drinkers drink skim milk. If a storeowner orders 25 gallons of milk, how many should be skim?

23. Under normal conditions, 1.5 feet of snow will melt into 2 inches of water. After a recent snowstorm, there were 3.5 feet of snow. How many inches of water will there be when the snow melts?

24. The Travel Industry Association of America reports that 4 out of every 35 people who travel do so by air. If there are 180 faculty members who will travel this year, about how many of them will fly?

25. A gallon of paint will cover 640 square feet of wall space. If a person has to paint a room whose walls measure 2560 square feet, how many gallons of paint will the person need?

26. The American Dietetic Association reported that 31 out of every 100 people wish to maintain or lose weight. If there are 384 people at a convention, approximately how many will want to lose or maintain their weight?

27. The U.S. Census Bureau reported that 9 out of every 20 joggers are female. On a trail, there were 220 joggers on July 4. Approximately how many were female?

28. If a person drives 4000 miles in 8 months, how many miles will that person drive every 2 years?

29. A quality control inspector found that out of every 50 calculators manufactured, 2 were defective. In a lot of 1000 calculators, about how many will be defective?

30. The American Dietetic Association states that 11 out of every 25 people do not eat breakfast. If there are 175 students in a large lecture hall, about how many of them did not eat breakfast?

31. At South Campus, the student-faculty ratio is about 16 to 1. If 128 students enroll in Statistics 101, how many sections should be offered?

32. If a 10-foot pole casts a shadow of 4 feet, how tall is a tree whose shadow is 7 feet?

33. A small college has 1200 students and 80 professors. The college is planning to increase enrollment to 1500 students next year. How many new professors should be hired, assuming they want to maintain the same ratio?

34. The taxes on a house assessed at $64,000 are $1600 a year. If the assessment is raised to $80,000 and the tax rate did not change, how much would the taxes be now?

35. If five small cans of paint can cover 20 square feet of wood siding, how many cans of paint should be purchased to cover $13\frac{1}{3}$ square feet of siding?

36. If you need a minimum of 27 correct out of 30 to get an A on a test, how many correct answers would you need to get an A on an 80-point test assuming that the same ratio is used for scoring purposes?

37. The amount of simple interest on a specific amount of money varies directly with the time the money is kept in a savings account when the interest rate is constant. Find the amount of interest on a $5000 savings account, if the interest rate is 6%, and the money has been invested for 4 years.

38. The number of tickets purchased for a prize varies directly with the amount of the prize. For a prize of $1000, 250 tickets are purchased. Find the approximate number of tickets that will be purchased on a prize worth $5000.

39. The diameter of a circle varies inversely with the circumference. The constant is $\pi = 3.14$. Find the diameter of a circle whose circumference is 32 inches.

40. The strength of a beam varies inversely with the square of its length. If a 10-foot beam can support 500 pounds, how many pounds can a 12-foot beam support?

Writing Exercises

41. What is the definition of a ratio?

42. Give an example of a real-life ratio taken from a newspaper or magazine.

43. What is a proportion?

44. Explain how to solve a proportion when three of the four members are known.

Critical Thinking

Stores are required by law to display unit prices. A unit price is the ratio of the total price to the number of units. For example, if a 40-pound bag of sand costs $5.00, then the unit price per pound would be

$$unit\ price = \frac{price}{number\ of\ pounds}$$

$$= \frac{\$5.00}{40\ pounds}$$

$$= \$0.125$$

or 0.12\frac{1}{2}$ per pound.

For Exercises 45–49, find the unit price and then decide which is a better buy. These prices were obtained from actual foods.

45. Flour: 10 pounds for $3.39 or 25 pounds for $7.49

46. Candy: 20 ounces for $1.50 or 24 ounces for $1.75

47. Potato sticks: 7 ounces for $1.99 or 1.5 ounces for $0.50

48. Cookies: 7 ounces for $0.99 or 14 ounces for $1.50

49. Coffee: 11.5 ounces for $2.75 or 34.5 ounces for $7.49

7-6 Solving Quadratic Equations

The Standard Form of a Quadratic Equation

The equations that you solved in Section 7-2 are called first-degree or linear equations. Recall that an equation is of the first degree if the exponents of the variables are always 1. In this section, you will learn how to solve *second-degree* or **quadratic equations.** An equation is called a quadratic equation or a second-degree equation when the largest exponent of the variable is two. (Assume the equation has only one variable.)

Quadratic equations can be written in *standard form*.

> The **standard form** of a quadratic equation is $ax^2 + bx + c = 0$, where a, b, and c are real numbers and $a \neq 0$.

The standard form means to place all the terms on the left side of the equation and place the x^2 term first followed by the x term and then by the constant term. The equation $2x^2 + 3x = 6$ can be written in standard form as $2x^2 + 3x - 6 = 0$. In this equation $a = 2$, $b = 3$, and $c = -6$.

The quadratic equation $5x - 3x^2 = 10$ is written in standard form as $-3x^2 + 5x - 10 = 0$. In this case, $a = -3$, $b = 5$, and $c = -10$. When a is negative, both sides of the equation can be multiplied by -1 to make the coefficient of the x^2 term positive. In this case,

$$-1(-3x^2 + 5x - 10) = -1(0)$$

$$3x^2 - 5x + 10 = 0$$

Then $a = 3$, $b = -5$, and $c = 10$. The reasoning is that it is usually easier to work with a quadratic equation when the coefficient of the x^2 term is positive.

Example   7-35

Write each equation in standard form and identify a, b, and c.

(a) $7 + 9x^2 = 3x$
(b) $4x - 15 = 3x^2$
(c) $5x^2 = 25$

Solution

(a) $9x^2 - 3x + 7 = 0$; $a = 9$, $b = -3$, $c = 7$
(b) $-3x^2 + 4x - 15 = 0$; $a = -3$, $b = 4$, $c = -15$

Note that if both sides of the equation are multiplied by -1, the resulting equation is $3x^2 - 4x + 15 = 0$, and $a = 3$, $b = -4$, and $c = 15$.

(c) $5x^2 - 25 = 0$; $a = 5$, $b = 0$, $c = -25$

Try This One

7-T   Write each quadratic equation in standard form and identify a, b, and c.

(a) $6 + 8x - x^2 = 0$

(b) $6x - 6x^2 = 0$

(c) $4x + 5x^2 = 0$

For the answer, see page 332.

Multiplying Binomials

The solution to a quadratic equation can be found by using factoring or by using the quadratic formula. Before explaining the factoring method, it is necessary to explain the FOIL method for multiplying two binomials.

A binomial expression (called a **binomial**) has two terms. Examples of binomials are

$$x - 5 \qquad 2x + 3 \qquad -6x + 4$$

The product of two binomials can be found by using the **FOIL method. F** represents the product of the *first* terms of the binomial. **O** represents the product of the *outer* terms. **I** represents the product of the inner terms. **L** represents the product of the *last* terms of the binomials.

The product of two binomials using the FOIL method is

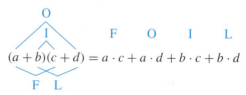

$$\overset{O}{\overset{I}{(a + b)(c + d)}} = \overset{F}{a \cdot c} + \overset{O}{a \cdot d} + \overset{I}{b \cdot c} + \overset{L}{b \cdot d}$$

Examples 7-36 and 7-37 show how to multiply binomials using the FOIL method.

Example 7-36

Multiply $(x - 8)(x + 3)$.

Solution

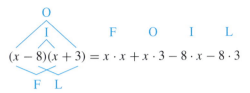

$$\underset{F\ L}{\underbrace{(x - 8)(x + 3)}} = \overset{F}{x \cdot x} + \overset{O}{x \cdot 3} - \overset{I}{8 \cdot x} - \overset{L}{8 \cdot 3}$$

Multiply the *first* terms: $x \cdot x = x^2$
Multiply the *outer* terms: $x \cdot 3 = 3x$ Like terms
Multiply the *inner* terms: $-8 \cdot x = -8x$ Like terms
Multiply the *last* terms: $(-8) \cdot (+3) = -24$

    Notice that $3x$ and $-8x$ are like terms and when combined equal $-5x$. Hence, the product is $x^2 - 5x - 24$.

Example 7-37

Multiply $(2x - 5)(3x - 8)$.

Solution

$$(2x - 5)(3x - 8) = \overset{F}{2x \cdot 3x} - \overset{O}{2x \cdot 8} - \overset{I}{5 \cdot 3x} - \overset{L}{5(-8)}$$

$$= 6x^2 - 16x - 15x + 40$$

$$= 6x^2 - 31x + 40$$

Hence, the product is $6x^2 - 31x + 40$.

Try This One

7-U   Find the product of each.

(a) $(x + 7)(x + 9)$

(b) $(4x - 9)(2x + 5)$

(c) $(3x - 8)(5x - 2)$

For the answer, see page 332.

Factoring Trinomials When *a* = 1

The terms on the left side of a complete (i.e., all nonzero coefficients) quadratic equation written in standard form make up an expression called a **trinomial** (i.e., three terms). The expression $x^2 + 9x + 20$ is called a trinomial. It is the product of two binomials $(x + 4)(x + 5)$. When an expression is written as a product of its factors, the process is called *factoring*. For example, the trinomial $x^2 - x - 12$ can be factored as $(x - 4)(x + 3)$. To show this is correct, use the FOIL method to multiply $(x - 4)(x + 3)$.

Now, how does one factor trinomials? Using

$$x^2 + 9x + 20 = (x + 4)(x + 5)$$

as an example, notice that the first term of the trinomial, x^2, is obtained from multiplying the first terms of the two binomial factors, namely, x and x. Also, the last term of the trinomial, 20, is obtained from multiplying the last two numbers of the two binomial factors, namely, 4 and 5. Finally, the middle term of the trinomial, $9x$, is the result of combining the products of the outer and inner factors of the binomials, namely, $4 \cdot x$ and $5 \cdot x$. Trinomials can be factored by reversing the FOIL method, so to speak. Example 7-38 shows how this is done.

Sidelight

Mathematics and Photography

When a photographer takes a picture, he or she wants to have as much of the picture (foreground and background) "in focus" as possible. In order to accomplish this, the settings of the camera must be adjusted accordingly.

The hyperfocal distance of the camera's lens determines what objects will be in focus in a photograph. When the hyperfocal distance is known, then everything from infinity to half of the hyperfocal distance will be in focus in the photograph. For example, if the camera's lens is set so that the hyperfocal distance is 20 feet, then every object from 10 feet to infinity will be in focus. Objects closer than 10 feet will be blurred.

The hyperfocal distance is determined mathematically and depends on the focal length of the lens and the f-stop. This is why a photographer changes lenses in order to get different types of pictures. Some types of lenses that are used are telephoto lenses, close-up lenses, and wide-angle lenses.

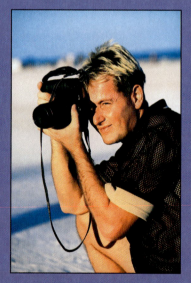

An autofocus camera emits an electronic beam that determines the distance the object to be photographed is from the camera and then adjusts the camera's settings automatically.

Calculations to determine the hyperfocal distance of a lens can be found in photographic manuals.

Example 7-38

Factor $x^2 + 10x + 16$.

Solution

Step 1   Write all the possible two-factor products of the first and last term.

$$x^2 \quad + 10x \, + 16$$

$$\text{Factors} \quad x \cdot x \qquad 4 \cdot 4$$
$$8 \cdot 2$$
$$16 \cdot 1$$

Step 2   Look at the factors of the last term and see which pair, when added, gives the coefficient of the middle term, in this case 10. It is $8 \cdot 2$.

Step 3   Set up the product as $x^2 + 10x + 16 = (x + 8)(x + 2)$

Hence, the solution is $(x + 8)(x + 2)$.

   The solution can be checked by using the FOIL method to multiply $(x + 8)(x + 2)$.

$$(x + 8)(x + 2) = x^2 + 10x + 16$$

Example 7-39

Factor $x^2 + 11x + 24$.

Solution

Step 1

$$x^2 \quad + 11x \, + 24$$

$$\text{Factors} \quad x \cdot x \qquad 24 \cdot 1$$
$$12 \cdot 2$$
$$8 \cdot 3$$
$$6 \cdot 4$$

Step 2   Since the sum of 8 and 3 is 11, these factors are the correct ones to use.

Step 3 $x^2 + 11x + 24 = (x + 8)(x + 3)$

Check:

$$(x + 8)(x + 3) = x^2 + 3x + 8x + 24 = x^2 + 11x + 24$$

   In Examples 7-38 and 7-39, the sign of the third term or constant term was positive, and the sign of the second term or the x was positive; hence, the signs of the factors of the third term must both be positive since the product of two positive numbers is positive and the sum of two positive numbers is positive.

   A second case occurs when the sign of the constant term is positive and the sign of the x term is negative; for example, $x^2 - 7x + 12$. In this case, the two factors of 12 must be negative since the product of two negative numbers is positive and the sum of two negative numbers is negative.

Example 7-40

Factor $x^2 - 7x + 12$.

Solution

Step 1
$$x^2 \quad - 7x \quad + 12$$
$$x \cdot x \qquad (-12)(-1)$$
$$(-6)(-2)$$
$$(-3)(-4)$$

Step 2 Since $(-4) + (-3) = -7$, these are the correct numbers.

Step 3 $x^2 - 7x + 12 = (x - 4)(x - 3)$

Check:
$$(x - 4)(x - 3) = x^2 - 3x - 4x + 12$$
$$= x^2 - 7x + 12$$

The third possibility is when the sign of the constant term is negative. For example, in the expressions $x^2 - 2x - 8$ and $x^2 + 2x - 8$, the signs of the constant terms are negative. When this occurs, one of the factors of the constant term will be positive and the other factor will be negative. The reason is that the product of a positive number and a negative number is a negative number.

Example 7-41

Factor $x^2 - 2x - 8$.

Solution

Step 1
$$x^2 \quad - 2x \quad - 8$$
$$x \cdot x \qquad (-8)(+1)$$
$$(+8)(-1)$$
$$(-4)(+2)$$
$$(+4)(-2)$$

Step 2 The correct numbers are -4 and $+2$ since $(-4) + (+2) = -2$ (the coefficient of the middle term).

Step 3 $x^2 - 2x - 8 = (x - 4)(x + 2)$

Check:
$$(x - 4)(x + 2) = x^2 + 2x - 4x - 8$$
$$= x^2 - 2x - 8$$

The rule of signs for factoring a trinomial when $a = 1$ is summarized next.

Rule of Signs for Factoring Trinomials

If the sign of the third term (i.e., the constant term) of the trinomial is *positive*, then the signs of its factors are both positive if the sign of the second term (i.e., the x term) is positive, or both negative if the sign of the second term is negative.

If the sign of the third term of the trinomial is negative, then the sign of one of its factors will be positive and the sign of the other factor will be negative.

Math Note

When none of the sums of the factors gives the coefficient of the middle or *x* term, the trinomial cannot be factored using integers.

Try This One

7-V　Factor each trinomial.

(a)　$x^2 + 13x + 36$

(b)　$x^2 - 12x + 20$

(c)　$x^2 - 3x - 40$

(d)　$x^2 + x - 30$

For the answer, see page 332.

Factoring a Trinomial When $a \neq 1$

In the previous subsection, the coefficient of the x^2 term was 1; however, the coefficient could be any number other than 0. For example, in the trinomial $3x^2 + 7x + 2$, $a = 3$. These trinomials are somewhat more difficult to factor since the factors of the x^2 terms must be considered along with the factors of the constant term. Notice that the sum of 7 in the trinomial $3x^2 + 7x + 2$ results from the sum of the products of the combinations of the factors of 3 and 2 as shown using the FOIL method.

$$3x^2 + 7x + 2 = (3x + 1)(x + 2)$$

$$1x$$
$$6x$$

This can be done by listing all the possibilities when $3x$ and x are the factors of $3x^2$ and 2 and 1 are the factors of 2.

Possible factors	Sum of the products of the outer and inner terms
$(3x + 2)(x + 1)$	$3x + 2x = 5x$
$(3x + 1)(x + 2)$	$6x + 1x = 7x$

Math Note

You can also use the rule of signs shown above when factoring these trinomials, as long as *a* is positive.

Since the second pair of binomials results in $7x$, this pair is the correct answer. The solution can be checked using the FOIL method.

There were only two possible answers since the 3 and 2 can only be factored in one way. However, Examples 7-42 and 7-43 show that sometimes many combinations must be checked before finding the right one.

Example 7-42

Factor $14x^2 - 33x + 10$.

Solution

The factors of 14 are 14 and 1 and 7 and 2. The factors of 10 are 10 and 1 and 2 and 5. The signs between the terms of the binomials must both be negative.
  The possible factors are

Possible factors	Sum of the products of the inner and outer terms
$(14x - 10)(x - 1)$	$14x(-1) + (-10)x = -14x + (-10x) = -24x$
$(14x - 1)(x - 10)$	$14x(-10) + (-1)x = -140x - 1x = -141x$
$(14x - 5)(x - 2)$	$14x(-2) + (-5)x = -28x + (-5x) = -33x^*$
$(14x - 2)(x - 5)$	
$(7x - 10)(2x - 1)$	
$(7x - 1)(2x - 10)$	
$(7x - 5)(2x - 2)$	
$(7x - 2)(2x - 5)$	

*When the right combination is found, it is not necessary to continue. The solution is $(14x - 5)(x - 2)$.

Example 7-43

Factor $2x^2 - 9x - 5$.

Solution

The factors of 2 are 2 and 1. The factors of 5 are 1 and 5. Note that -5 can either be $5(-1)$ or $(-5)(1)$.
  The possible factors are

Possible factors	Sum of the products of the factors of the outer and inner terms
$(2x + 5)(x - 1)$	$-2x + 5x = +3x$
$(2x - 5)(x + 1)$	$2x + (-5x) = -3x$
$(2x - 1)(x + 5)$	$10x + (-1x) = +9x$
$(2x + 1)(x - 5)$	$-10x + 1x = -9x$

The right combination and solution is $(2x + 1)(x - 5)$. Note that once the right combination is found, there is no need to continue.

Try This One

7-W Factor each trinomial.

(a) $6x^2 + 25x + 21$

(b) $3x^2 - 11x + 6$

(c) $2x^2 - x - 6$

For the answer, see page 332.

Solving Quadratic Equations by Factoring

One method of solving quadratic equations is by factoring. The procedure is given next.

Procedure for Solving Quadratic Equations by Factoring

Step 1 Write the quadratic equation in standard form.

Step 2 Factor the left side.

Step 3 Set both factors equal to zero.

Step 4 Solve each equation for x.

Examples 7-44 and 7-45 show how to use this procedure.

Example 7-44

Solve $x^2 - 13x = -36$.

Solution

Step 1	$x^2 - 13x + 36 = 0$	Write in standard form.
Step 2	$(x - 9)(x - 4) = 0$	Factor the left side.
Step 3	$x - 9 = 0$ or $x - 4 = 0$	Set both factors equal to zero.

Step 4

$$\begin{array}{c|c} x - 9 = 0 & x - 4 = 0 \quad \text{Solve each equation for } x. \\ x - 9 + 9 = 0 + 9 & x - 4 + 4 = 0 + 4 \\ x = 9 & x = 4 \end{array}$$

The solution set is $\{9, 4\}$.

The solutions can be checked by substituting the values in the original equation, as shown.

$$\begin{array}{ll} \text{For } x = 9, & \text{For } x = 4, \\ x^2 - 13x = -36 & x^2 - 13x = -36 \\ 9^2 - 13(9) \overset{?}{=} -36 & 4^2 - 13(4) \overset{?}{=} -36 \\ 81 - 117 \overset{?}{=} -36 & 16 - 52 \overset{?}{=} -36 \\ -36 = -36 & -36 = -36 \end{array}$$

Example 7-45

Solve $6x^2 - 6 = -5x$.

Solution

Step 1 $6x^2 + 5x - 6 = 0$ Write in standard form.

Step 2 $(3x - 2)(2x + 3) = 0$ Factor.

—Continued

Example 7-45   *Continued—*

Step 3 $3x - 2 = 0$ $2x + 3 = 0$ Set each factor equal to 0.

Step 4

$$3x - 2 = 0 \qquad\qquad 2x + 3 = 0$$
$$3x - 2 + 2 = 0 + 2 \qquad\qquad 2x + 3 - 3 = 0 - 3$$
$$3x = 2 \qquad\qquad 2x = -3$$
$$\frac{3x}{3} = \frac{2}{3} \qquad\qquad \frac{2x}{2} = \frac{-3}{2}$$
$$x = \frac{2}{3} \qquad\qquad x = \frac{-3}{2}$$

Hence, the solution set is $\left\{ \dfrac{2}{3}, \dfrac{-3}{2} \right\}$.

Try This One

7-X   Solve each.

(a) $6x^2 - 4x = 2$

(b) $x^2 + 1 = -2x$

For the answer, see page 332.

Solving Quadratic Equations Using the Quadratic Formula

When a quadratic equation cannot be solved by factoring or when factoring involves checking a large number of possible solutions, the *quadratic* formula can be used.

The formula

$$x = \frac{-b \pm \sqrt{b^2 - 4ac}}{2a}$$

is called the **quadratic formula** and can be used to solve any quadratic equation written in standard form, $ax^2 + bx + c = 0, a \neq 0$.

Example 7-46

Solve $2x^2 - x - 8 = 0$.

Solution
Identify a, b, and c.

$$a = 2, b = -1, \text{ and } c = -8$$

—Continued

Calculator Explorations

```
(1+√(65))/4
       2.265564437
(1-√(65))/4
      -1.765564437
■
```

Why do we insert parentheses around the numerator?

Example 7-46 Continued—

Substitute in the quadratic formula.

$$x = \frac{-b \pm \sqrt{b^2 - 4ac}}{2a}$$

$$= \frac{-(-1) \pm \sqrt{(-1)^2 - 4(2)(-8)}}{2(2)}$$

$$= \frac{1 \pm \sqrt{1 + 64}}{4}$$

$$x = \frac{1 + \sqrt{65}}{4} \quad \text{or} \quad x = \frac{1 - \sqrt{65}}{4}$$

$$\approx \frac{1 + 8.06}{4} \qquad\qquad \approx \frac{1 - 8.06}{4}$$

$$\approx 2.265 \qquad\qquad\quad \approx -1.77$$

$$\{2.265, -1.77\}$$

Solutions to quadratic equations can be left in the radical form; however, sometimes this can be simplified as shown

$$\frac{9 \pm \sqrt{27}}{6} = \frac{9 \pm \sqrt{9 \cdot 3}}{6} = \frac{9 \pm 3\sqrt{3}}{6} = \frac{\overset{1}{\cancel{3}}(3 \pm \sqrt{3})}{\underset{2}{\cancel{6}}} = \frac{3 \pm \sqrt{3}}{2}$$

First $\sqrt{27}$ was simplified to $3\sqrt{3}$ and a 3 was factored out of the numerator. Finally, a factor of 3 was cancelled out of the numerator and denominator.

Math Note

When the number under the radical sign is negative, such as −6, there is no real solution to the quadratic equation.

Try This One

7-Y Solve each quadratic equation using the quadratic formula.

(a) $3x^2 - 3x = 1$

(b) $x = x^2 - 13$

For the answer, see page 332.

Applications of Quadratic Equations

Real-life problems can be solved using quadratic equations using the same procedure shown in Section 7-3.

Example 7-47

A woodworker wishes to build a rectangular table whose length is 2 feet longer than its width and the table's area is to be 15 square feet. Find the dimensions of the table.

—Continued

Calculator Explorations

A graphing calculator can assist you in viewing a graphical representation and determining the solution(s) of the algebraic equation.

ZOOM 4 ZOOM 0 ZOOM 3 ENTER to view both intersections, the solutions.

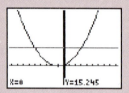

Calculate the intersections.

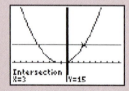

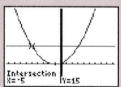

Will the intersection values be the same when graphing the equation $x^2 + 2x - 15 = 0$?

Example 7-47 *Continued—*

Solution

Let x = width of the table and let $x + 2$ = length of the table.

Since the area of a rectangle = lw, the equation is $x(x + 2) = 15$. The equation is solved as follows:

$$x^2 + 2x = 15 \qquad \text{Use the distributive property.}$$
$$x^2 + 2x - 15 = 0 \qquad \text{Write in standard form.}$$
$$(x + 5)(x - 3) = 0 \qquad \text{Solve for } x \text{ by factoring.}$$
$$x + 5 = 0 \quad \bigg| \quad x - 3 = 0$$
$$x = -5 \quad \bigg| \quad x = 3$$

The value -5 must be disregarded since the width cannot be negative; hence, the width is $x = 3$ feet and the length is $x + 2 = 3 + 2 = 5$ feet. When the problem has a context, you can use what you know about the range of possible answers to eliminate some.

Check:

$$5 \text{ ft} \times 3 \text{ ft} = 15 \text{ sq ft}$$

Try This One

7-Z   The formula for the distance that an object falls freely to the ground is $d = rt + 16t^2$, where:

d is the distance it falls (in feet);

r is the rate at which the object starts to fall; and

t is the number of seconds the object falls.

How long will it take an object that is dropped from the top of the Sony Building to hit the ground? The Sony Building, 1 Madison Square Plaza, New York City, is 576 feet tall. (*Hint:* Since the object is dropped, $r = 0$.)
For the answer, see page 332.

Exercise Set 7-6

Computational Exercises

For Exercises 1–10, use the FOIL method to multiply the two binomials.

1. $(x + 7)(x + 9)$
2. $(x - 8)(x - 12)$
3. $(x - 7)(x - 10)$
4. $(x + 4)(x + 2)$

5. $(x - 15)(x + 8)$ 6. $(x + 10)(x - 3)$

7. $(2x - 7)(7x - 9)$ 8. $(4x - 1)(4x - 1)$

9. $(5x + 7)(3x - 8)$ 10. $(2x - 5)(3x + 8)$

For Exercises 11–30, solve each quadratic equation by factoring.

11. $x^2 + 5x + 6 = 0$ 12. $x^2 + 9x + 20 = 0$

13. $x^2 + x - 12 = 0$ 14. $x^2 - 3x - 10 = 0$

15. $x^2 - 14x = 51$ 16. $x^2 - x = 20$

17. $x^2 + 24x = 81$ 18. $x^2 - 12x = 64$

19. $x^2 + 15 = 8x$ 20. $x^2 + 20 - 12x = 0$

21. $2x^2 - x - 21 = 0$ 22. $5x^2 + 27x - 18 = 0$

23. $6x^2 - x - 12 = 0$ 24. $4x^2 + 13x - 12 = 0$

25. $6x^2 - 12 = x$ 26. $10x^2 + 21x = 10$

27. $7x - 6 = -5x^2$ 28. $5x^2 - 18 = 27x$

29. $12 - x = 6x^2$ 30. $6x^2 + 6 = 13x$

For Exercises 31–40, solve each quadratic equation by using the quadratic formula.

31. $3x^2 + x - 1 = 0$ 32. $4x^2 - 7x = 2$

33. $2x^2 - 5x = 12$ 34. $x^2 + 5x - 12 = 0$

35. $3x^2 + 5x + 1 = 0$ 36. $5x^2 + 2x = 3$

37. $x^2 - 8x - 9 = 0$ 38. $6x^2 + x = 35$

39. $x^2 + 5x = 3$ 40. $6x - 1 = 4x^2$

Real World Applications

41. How long will it take an object to hit the ground if it is dropped from a height of 1296 feet? (Use $d = rt + 16t^2$.) Assume the object initially is held still.

42. The product of two consecutive even integers is 288. Find the numbers. (*Hint:* Consecutive even integers can be written as x and $x + 2$.)

43. The product of two consecutive integers is 156. Find the numbers. (*Hint:* Consecutive integers can be written as x and $x + 1$.)

44. If the height of a triangle is 6 inches longer than its base and the area of the triangle is 8 square inches, find the lengths of the base and height. Use the formula $A = \frac{1}{2}bh$.

45. A person planted two square plots, one for tomatoes and one for cabbage, carrots, and lettuce. If the side of the tomato plot is 6 feet longer than the length of the side of the other plot and the sum of the areas of both plots is 116 square feet, find the dimensions of each square. Use the formula $A = s^2$.

46. Mary is 5 years older than Bill. The product of their present ages is three times what the product of their age was 5 years ago. Find their present ages.

Critical Thinking

47. It can be shown using algebra that $2 = 1$. Don't believe it? Look at the proof and find the error.

Let $x = 1$, then multiply both sides by x:

$x^2 = x$, then subtract one from each side:

$x^2 - 1 = x - 1$, then factor the left side:

$(x + 1)(x - 1) = x - 1$; divide both sides by $x - 1$:

$$(x + 1)\frac{(x - 1)}{(x - 1)} = \frac{(x - 1)}{(x - 1)}$$

$x + 1 = 1$; substitute 1 for x:

$$1 + 1 = 1$$
$$2 = 1$$

Summary

Section	Important Terms	Important Ideas
7-1	variable algebraic expression distributive property like terms evaluate formula	**Algebra** uses expressions and equations. Expressions can be simplified by using the distributive property and combining like terms. Expressions can be evaluated by substituting the values for the variables and using the order of operations to simplify the expression. Many real-world problems can be solved by using specific formulas that apply to the given situation.
7-2	equation open equation closed equation solution set addition property of equality subtraction property of equality multiplication property of equality division property of equality	**An** equation is a statement of equality of two algebraic expressions. An open equation contains at least one variable. There are four basic types of equations that can be solved by using one of the four properties of equality. More complex equations can be solved by using the procedure shown in this section.
7-3		**Many** real-world problems can be solved by writing an appropriate equation for the problem and then solving the equation.
7-4	linear inequality	**Inequalities** are similar to equations but use an inequality sign instead of an equal sign. Inequalities can be solved by using the same principles as equations with one exception: If you multiply or divide both sides of the inequality by a negative number, the inequality sign must be reversed.
7-5	ratio proportion direct variation inverse variation	**Two** quantities can be compared by using a ratio. Two equal ratios constitute a proportion. Proportions can also be used to solve many real-world problems.
7-6	quadratic equation standard form binomial FOIL method trinomial quadratic formula	**An** equation is called a quadratic equation when the largest exponent of the variable is 2. Some quadratic equations can be solved by factoring. When the trinomial cannot be factored, the quadratic formula can be used. Quadratic equations can also be used to solve many real-world problems.

Review Exercises

For Exercises 1–6, simplify each algebraic expression.

1. $6x + 3y - 10 + 2y - 8x + 3$

2. $4x - 9 - 2x + 7y - 3y + 16$

3. $5(x - 6) + 2(x - 3)$

4. $-9(2x + 4) - 3(x - 2)$

5. $2x + 7(x - 3) + 4x$

6. $6x + 7 - 3(2x - 8) + 3x$

For Exercises 7–12, evaluate each algebraic expression or formula.

7. $2x^2 + 5x - 3$ when $x = 6$

8. $3x - 5 + x^2$ when $x = -5$

9. $6(x - 8) - 10$ when $x = -2$

10. $2(x - 4) + 3x$ when $x = 9$

11. $d = rt$ when $r = 8$ and $t = 15$

12. $A = P(1 + rt)$ when $P = 3000$, $r = 0.08$, and $t = 5$

For Exercises 13–22, solve each equation for x.

13. $4x + 8 = -32$

14. $5x + 6 = 36$

15. $8x - 3 = 6x + 37$

16. $2x - 10 = 7x + 55$

17. $5(x + 9) = -20$

18. $3(x - 6) = 33$

19. $6(x + 8) - 4x = 3x - 19$

20. $9(x - 7) - 5x = 2x + 10$

21. $5x + 3 = 16x + 47$

22. $9(2x - 4) = 15x - 27$

For Exercises 23–26, write each statement in symbols.

23. eight times a number decreased by 4

24. the product of 6 and two times a number

25. 3 added to four times a number

26. a number increased by 5

27. A person takes 8 hours to drive to and back from a conference. She averages 40 miles per hour out and 50 miles an hour back. Find the time it took her to get there and the time it took her to return. (*Hint:* Use distance = rate × time.)

28. A merchant mixed nuts for $1.20 a pound and beer nuts for $1.80 a pound. If the merchant sold two more pounds of beer nuts than mixed nuts and the total bill was $21.60, how many pounds of each kind were sold?

29. Tickets for a school play sold for $8.00, $10.00, and $12.00. Twice as many $8.00 tickets were sold as $10.00 tickets, and 10 more of the $12.00 tickets than the $10.00 tickets were sold. If the total revenue was $3122, how many of each denomination were sold?

30. On a train excursion trip, the adult fare was $12.00 and the child's fare was $6.00. The number of passengers was 400, and the total revenue for the trip was $4020. How many adults were there on the train?

For Exercises 31–36, solve each inequality.

31. $7x + 10 > 80$

32. $3x + 6 \leq 2x - 14$

33. $4 - 5x < -31$

34. $4(x - 6) > 3(x - 15)$

35. $6x - 3 \geq 5(2x + 18)$

36. $2x + 7 \leq 6(2x + 9) - 20$

37. Membership for a health club is $75 to join and $32.50 per month. How many months can a person sign up for if the person wishes to spend at most $1000?

38. The chairperson of a school club wants to purchase flowers for the auditorium stage for a show. The club has authorized the chairperson to spend at most $200. How many potted plants can be purchased if each plant costs $12.95 plus 6% tax?

For Exercises 39–42, write each ratio as a fraction.

39. 82 miles to 15 gallons of gasoline

40. 16 ounces cost $2.37

41. 4 months to 2 years

42. 18 minutes to 2 hours

For Exercises 43–46, solve each proportion for x.

43. $\dfrac{2}{x} = \dfrac{14}{63}$

44. $\dfrac{16}{5} = \dfrac{x}{2.5}$

45. $\dfrac{8}{24} = \dfrac{24}{x}$

46. $\dfrac{5}{11} = \dfrac{6.2}{x}$

47. If a person burns 300 calories when exercising for 12 minutes, how many calories will the person burn for exercising for 30 minutes?

48. The U.S. Center for Disease Control reported that 4 out of 10 people with incomes between $15,000 and $24,999 exercise regularly. About how many people exercise regularly in a group of 85 people who are in the preceding income bracket?

49. In his will, a man's estate was divided according to a ratio of three parts for his wife and two parts for his son. If his estate amounted to $18,000, how much did each receive?

50. A professor states that if a student misses a unit exam (worth 30 points), he will use the score on the final exam (100 points), proportionally reduced, for the score on the unit exam. If a student scored 85 on the final exam, what would be the student's score on the unit test?

51. The cost of building a deck varies with the area of the deck. If a 6-foot by 9-foot deck costs $2160, find the cost of building a 9-foot by 12-foot deck.

52. The amount of paint needed to paint a spherical object varies directly with the square of the diameter. If 3 pints of paint are needed to paint a spherical object with a diameter of 36 inches, how much paint must be purchased to paint a spherical object with a diameter of 60 inches?

53. The amount of amperage in amps of electricity passing through a wire varies inversely with the resistance in ohms of the wire when the potential remains the same. If the resistance is 20 ohms when the amperage is 10 amps, find the amperage when the resistance is 45 ohms.

54. The cost of producing an item varies inversely with the square root of the number of items produced (i.e., $y = \dfrac{k}{\sqrt{x}}$, where y = the cost of the item and x = the number of items produced). Find the cost of producing 1600 items if the cost of producing 900 items is $600.

For Exercises 55–60, solve each equation by factoring.

55. $x^2 - 6 = x$

56. $x^2 + 11x - 26 = 0$

57. $x^2 - 4x - 21 = 0$

58. $2x^2 + 5x = 3$

59. $2 = x + 3x^2$

60. $2x^2 + 9 = 9x$

For Exercises 61–66, solve each equation using the quadratic formula.

61. $x^2 - 5x = 7$

62. $5x^2 - 7x - 4 = 0$

63. $8x^2 + 14x + 4 = 0$

64. $9x^2 - 12x = 7$

65. $4x^2 - 5 = -14x$

66. $5x^2 + 5 = 12x$

67. If the product of two consecutive numbers is 132, find the numbers.

68. How long will it take an object to fall a distance of 1024 feet? Use $d = rt + 16t^2$.

Chapter Test

For Exercises 1 and 2, simplify each.

1. $3x - 7y + 2x - 3y + 5$

2. $5(x - 6) + 2x - 10$

For Exercises 3 and 4, evaluate each.

3. $3x^2 - 2x + 6$ when $x = -5$

4. $E = 0.2381I^2Rt$ when $I = 30$, $R = 5$, and $t = 80$

For Exercises 5 and 6, solve each.

5. $3x - 5(2x + 10) = -59$

6. $2(x - 6) = 7 + 4(x + 16)$

7. Solve: $F = \dfrac{mv^2}{r}$ for r

8. Solve for y: $3x + 2y = 10$

For Exercises 9 and 10, solve each.

9. $4 - 3x \geq x + 10$

10. $2(x - 3) < 5x + 12$

For Exercises 11 and 12, solve each proportion.

11. $\dfrac{x}{9} = \dfrac{16}{36}$

12. $\dfrac{3}{7} = \dfrac{15}{x}$

For Exercises 13 and 14, find the product.

13. $(x - 8)(2x + 3)$

14. $(3x - 5)(4x - 7)$

For Exercises 15–17, solve by factoring.

15. $x^2 - 14x - 51 = 0$

16. $x^2 + 12x - 40 = -27$

17. $6x^2 + x = 12$

For Exercises 18 and 19, solve each using the quadratic formula.

18. $3x^2 - x - 1 = 0$

19. $5x^2 + 2x = 3$

20. A person has invested part of $5000 in stocks paying a 4% dividend and the rest in stocks paying a 6% dividend. If the total of the dividends was $270, how much did the person invest in each stock?

21. A person invested part of $3000 at 6% interest and the rest at 8% interest. If the interest was $190, how much money was invested at each rate?

22. If a person purchases two uniforms at $57, how many uniforms can be purchased for $228?

23. If you can bike 2 miles in $12\frac{1}{2}$ minutes, how many hours will it take to bike 210 miles without counting rest stops?

24. The number of vibrations per second of a metal string varies directly with the square root of the tension when all other factors remain unchanged. Find the number of vibrations per second of a string under a tension of 64 pounds when a string under a tension of 25 pounds makes 125 vibrations per second.

25. The number of days it takes to do a certain job varies inversely with the number of people working on the

Mathematics in Our World ▶ Revisited

Medicine and Children

There are two ways to compute the proper dosage of medication for children. The first method is based on the child's age and uses the following proportions:

$$\frac{\text{child's age in months}}{\text{adult's age in months}} = \frac{\text{proper dosage}}{\text{adult's dosage}}$$

An adult is considered to be 150 months of age, and a 6-year-old child is considered to be 72 months old. The adult dosage of medication is 600 mg. Substitute in the proportion and solve for x:

$$\frac{72 \text{ months}}{150 \text{ months}} = \frac{x \text{ mg}}{600 \text{ mg}}$$

$$\left(\frac{72}{150} \diagdown \frac{x}{600} \quad \text{cross multiply} \right)$$

$$150x = 72 \cdot 600$$

$$\frac{150x}{150} = \frac{43{,}200}{150}$$

$$x = 288 \text{ mg}$$

Based on this formula, a 6-year-old child should receive 288 mg.

Another method used to compute the proper dosage of a medicine for a child is to use the child's weight. The proportion is

$$\frac{\text{child's weight}}{\text{adult's weight}} = \frac{\text{child's dosage}}{\text{adult dosage}}$$

An adult is considered to be 150 pounds. If the child weighs 56 pounds, then the proportion is

$$\frac{56 \text{ pounds}}{150 \text{ pounds}} = \frac{x \text{ mg}}{600 \text{ mg}}$$

$$\left(\frac{56}{150} \diagdown \frac{x}{600} \quad \text{cross multiply} \right)$$

$$150x = 56 \cdot 600$$

$$\frac{150x}{150} = \frac{33{,}600}{150}$$

$$x = 224$$

Solving for x, one gets 224 mg of the medication.

job. If it takes 12 people to complete a job in 8 hours, how many hours will it take 8 people to complete the job?

26. The product of two consecutive even numbers is 624. Find the numbers. (*Hint:* Consecutive even numbers can be represented by x and $x + 2$.)

27. A board game is played on a rectangular board whose area is 96 square inches. If the length is 4 inches longer than the width, find its dimensions.

Projects

1. An interesting method for solving quadratic equations came from India. The steps are

   (a) Move the constant term to the right side of the equation.

   (b) Multiply each term in the equation by four times the coefficient of the x^2 term.

   (c) Square the coefficient of the original x term and add it to both sides of the equation.

   (d) Take the square root of both sides.

   (e) Set the left side of the equation equal to the positive square root of the number on the right side and solve for x.

   (f) Set the left side of the equation equal to the negative square root of the number on the right side of the equation and solve for x.

Example: Solve $x^2 + 3x - 10 = 0$.

$$x^2 + 3x = 10$$
$$4x^2 + 12x = 40$$
$$4x^2 + 12x + 9 = 40 + 9$$
$$4x^2 + 12x + 9 = 49$$
$$2x + 3 = \pm 7$$

$2x + 3 = 7$	$2x + 3 = -7$
$2x = 4$	$2x = -10$
$x = 2$	$x = -5$

Try these.

(a) $x^2 - 2x - 13 = 0$

(b) $4x^2 - 4x + 3 = 0$

(c) $x^2 + 12x - 64 = 0$

(d) $2x^2 - 3x - 5 = 0$

2. Mathematicians have been searching for a formula that yields prime numbers. One such formula was $x^2 - x + 41$. Select some numbers for x, substitute them in the formula, and see if prime numbers occur. Try to find a number for x that when substituted in the formula yields a composite number.

Answers to **Try This One**

7-A. (a) $28x - 140$
     (b) $15x - 35y + 90$

7-B. (a) $-5x + 11y - 18$
     (b) $17y^3 + 10y^2 - 2$
     (c) $-4a + b + 6c$

7-C. (a) $31x + 58$
     (b) $-20a - 12b + 5c$

7-D. (a) 10
     (b) 235
     (c) 11

7-E. 3

7-F. 96.8

7-G. (a) $\{9\}$
     (b) $\{25\}$
     (c) $\{23\}$
     (d) $\{42\}$

7-H. (a) $\{12\}$
     (b) $\{1\}$
     (c) $\left\{ \dfrac{39}{7} \right\}$

7-I. $\left\{ \dfrac{5}{6} \right\}$

7-J. (a) $y = \dfrac{18 - x}{2}$ or $9 - \dfrac{x}{2}$
     (b) $r = \dfrac{A}{pt} - \dfrac{1}{t}$ or $\dfrac{1}{t}\left(\dfrac{A}{p} - 1 \right)$

7-K. (a) $\varnothing$
     (b) $\{x \mid x \text{ is a real number}\}$

7-L. (a) $5 + 7x$
     (b) $51 - 3x$
     (c) $\frac{5}{8}x$

7-M. 9 in Senate; 56 in House of Representatives

7-N. (a) $\left\{x \mid x \geq -3\frac{1}{9}\right\}$

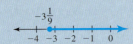

(b) $\{x \mid x < 17\}$

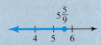

(c) $\left\{x \mid x \leq 5\frac{5}{9}\right\}$

(d) $\{x \mid x > -6\}$

7-O. \$1000

7-P. (a) $\dfrac{154}{25}$ or $6\dfrac{4}{25}$

(b) 20

(c) 6

7-Q. 40 minutes

7-R. 4.5 lbs

7-S. 921.6 lbs

7-T. (a) $-x^2 + 8x + 6 = 0; a = -1; b = 8; c = 6$ or
$x^2 - 8x - 6 = 0; a = 1; b = -8; c = -6$

(b) $-6x^2 + 6x = 0; a = -6; b = 6; c = 0$ or
$6x^2 - 6x = 0; a = 6; b = -6; c = 0$

(c) $5x^2 + 4x = 0; a = 5; b = 4; c = 0$

7-U. (a) $x^2 + 16x + 63$

(b) $8x^2 + 2x - 45$

(c) $15x^2 - 46x + 16$

7-V. (a) $(x + 9)(x + 4)$

(b) $(x - 10)(x - 2)$

(c) $(x - 8)(x + 5)$

(d) $(x + 6)(x - 5)$

7-W. (a) $(6x + 7)(x + 3)$

(b) $(3x - 2)(x - 3)$

(c) $(2x + 3)(x - 2)$

7-X. (a) $\left\{-\dfrac{1}{3}, 1\right\}$

(b) $\{-1\}$

7-Y. (a) $\dfrac{3 \pm \sqrt{21}}{6} \approx \{1.26, -0.26\}$

(b) $\dfrac{1 \pm \sqrt{53}}{2} \approx \{4.14, -3.14\}$

7-Z. 6 seconds

Chapter Eight

Additional Topics in Algebra

Outline

Objectives

After completing this chapter, you should be able to

1 Graph points on the Cartesian plane (8-1)

2 Graph lines on the Cartesian plane (8-1)

3 Find the slope of a line given two points (8-1)

4 Write an equation of a line in slope-intercept form and identify the slope and intercepts (8-1)

5 Solve a linear system of equations in two variables by three methods: graphing, substitution, and addition/subtraction (elimination) (8-2)

6 Determine whether a system of linear equations is consistent, inconsistent, or dependent (8-2)

7 Solve real-world problems involving a system of linear equations (8-2)

8 Solve a system of linear inequalities in two variables (8-3)

9 Solve real-world problems using linear programming (8-4)

10 Determine the domain and range of a relation (8-5)

11 Determine whether or not a relation is a function (8-5)

12 Determine the vertex, axis, and intercepts of a parabola (8-5)

13 Graph a parabola (8-5)

14 Graph an exponential function (8-5)

15 Solve real-world problems using quadratic or exponential functions (8-5)

Introduction

The last chapter presented the basic concepts of algebra such as evaluating algebraic expressions, solving linear equations, and quadratic equations.

This chapter continues the development of algebra beginning with the rectangular coordinate system including the graphs of lines, linear inequalities, parabolas, and exponential functions. Graphic methods can be used to solve a system of linear equations and linear inequalities. The chapter concludes with an explanation of relations and functions. ■

8-1 The Rectangular Coordinate System and the Line

The **rectangular coordinate system** is also called the **Cartesian plane.** On this plane, two perpendicular axes are drawn. The horizontal axis is a number line and is called the *x* **axis.** The vertical axis is also a number line and is called the *y* **axis.** The point of intersection of the axes is called the **origin.** The plane is divided into four regions called **quadrants.** They are designated by I, II, III, and IV. See Figure 8-1.

A point on the plane can be named by a capital letter and by an *ordered* pair of numbers (x, y). The numbers x and y are called the **coordinates** of the point. The coordinates of the origin are $(0, 0)$. A point P whose x coordinate is 2 and whose y coordinate is 5 is written as $P(2, 5)$. It is plotted by starting at the origin and moving two units right and five units up, as shown in Figure 8-2.

Mathematics in Our World

Stopping Distance of an Automobile

An important safety consideration while driving is the stopping distance of an automobile. During what was called an "Energy Crisis," the U.S. government passed a law that all states have a maximum speed limit of 55 miles per hour. This was done to cut down on gasoline usage. An interesting side effect was that there were fewer accidents. (The law was later repealed.) One reason for the reduction in accidents is that the slower an automobile goes, the quicker the driver can stop it when a danger is perceived.

There are two components to the stopping distance of an automobile. They are reaction time and braking time. The reaction time is the time it takes a driver to put his or her foot on the brake from the time the driver perceives the danger. Braking time is the time it takes the automobile to stop once the brake is applied. The stopping distance of an automobile then consists of the distance the automobile has traveled during the reaction time period plus the distance the automobile has traveled during the braking time. The stopping distance of an automobile can be computed mathematically by using two mathematical functions for the distance the automobile will travel during these two times. The problem then is to find the stopping distance for an automobile traveling at 70 miles per hour, and next to find the stopping distance for an automobile traveling at 55 miles per hour, and then to compare the results.

The solutions using functions, which will be explained in this chapter, can be found in Mathematics in Our World Revisited on page 398.

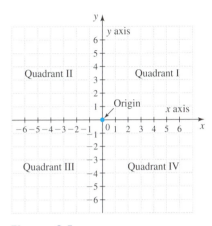

Figure 8-1

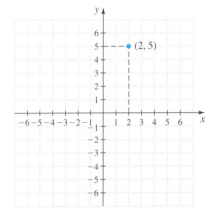

Figure 8-2

If the value of the x coordinate is negative, start at the origin and move left. If the value of the y coordinate is negative, move down.

Sidelight

Descartes and Analytic Geometry

René Descartes was born near Tours, France, in 1596. Because of delicate health, he developed a habit of resting in bed until late morning. He used these hours to meditate on subjects such as mathematics, philosophy, and science. In 1616, he graduated from the University of Portieres with a degree in law. He was very dissatisfied with the education he received, saying that it was antiquated.

At the age of 20, Descartes decided to revise the philosophy of thinking and developed a new branch of mathematics, which is now called "analytic geometry." Here he combined arithmetic, algebra, and geometry into one subject. Numbers became points on a graph and equations became lines and curves. Analytic geometry became the foundation of higher mathematics.

In 1617 Descartes became a soldier, and while meditating on a cold day, he received a revelation that all knowledge should be devised using mathematical reasoning. It wasn't until 18 years later that Descartes revealed his knowledge to the public. In 1637, he published a book entitled *Discoveries on*

René Descartes (1596–1650)

the Method of Rightly Conducting the Reason. Today this book is regarded as a major work of philosophy. It is ironic that Descartes's ideas on analytic geometry consisted of a 106-page footnote in his philosophy book. Over the next three centuries, this footnote became of greater significance than the philosophy in his book. As a tribute to Descartes, mathematicians call the rectangular coordinate plane the Cartesian plane.

Archaeological digs use a rectangular coordinate system to track where objects are found.

Example 8-1 shows how to plot points.

Example   8-1

Plot the points $A(5, -3)$, $B(0, 4)$, $C(-3, -2)$, $D(-2, 0)$, and $E(2, 6)$.

Solution

To plot each point, start at the origin and move left or right according to the x value, and then up or down according to the y value. See Figure 8-3.

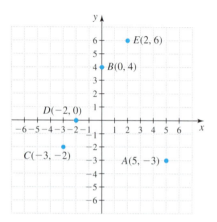

Figure 8-3

The interface between the mouse and the computer uses a coordinate system to match the cursor's motion to the mouse's motion.

Given a point on the plane, its coordinates can be found by drawing a vertical line back to the x axis and a horizontal line back to the y axis. For example, the coordinates of point C shown in Figure 8-4 are $(-3, 4)$.

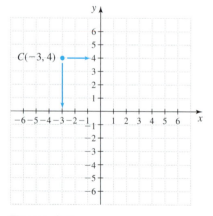

Figure 8-4

Example 8-2

Find the coordinates of each point shown in Figure 8-5.

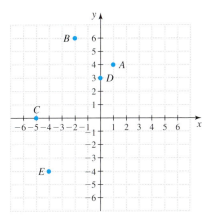

Figure 8-5

Solution

$A(1, 4)$
$B(-2, 6)$
$C(-5, 0)$
$D(0, 3)$
$E(-4, -4)$

Try This One

8-A (a) Plot the points whose coordinates are $A(2, 6)$, $B(-1, 5)$, $C(0, 4)$, $D(-3, 0)$, and $E(-4, -2)$.

(b) Find the coordinates of the points shown.

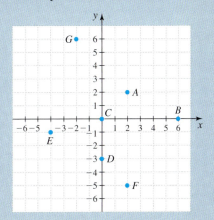

For the answer, see page 399.

The Line

An equation of the form $ax + by = c$, when a, b, and c are constants, is called a linear equation in two variables. These equations are called linear because all the coordinate pairs of numbers that satisfy the equation fall in a straight line. Before graphing a linear equation in two variables, the coordinates of at least two points that lie on the line must be found. The ordered pair of coordinates of a point that is on a line is called a *solution* to the equation. For example, a solution for the equation $3x + y = 6$ is $(1, 3)$ since when 1 is substituted for x in the equation and 3 for y, a true closed equation is obtained.

$$3x + y = 6$$

$$3(1) + 3 \overset{?}{=} 6$$

$$3 + 3 \overset{?}{=} 6$$

$$6 = 6$$

There is an unlimited number of solutions for a linear equation in two variables since a line contains an infinite number of points.

In order to find the graph of a linear equation in two variables, at least two solutions (i.e., the coordinates of two points) must be found. It is even better to use three points in case a mistake is made when finding the coordinates of one of the points. One way to do this is by inspection. For example, another solution for the equation $3x + y = 6$ is $(3, -3)$, since $3(3) + (-3) = 6$.

Another method for finding a solution is to select any number for one of the variables, substitute it into the equation, and solve the equation for the other variable. For example, if one selects $x = 2$, then

$$3x + y = 6$$

$$3(2) + y = 6$$

$$6 + y = 6$$

$$y = 0$$

Hence, the ordered pair $(2, 0)$ is a solution to the equation.

To draw the graph of $3x + y = 6$, plot the three points $(1, 3)$, $(3, -3)$, and $(2, 0)$ on the Cartesian plane and draw a straight line through these points. This is shown in Figure 8-6.

Math Note

When selecting a number for one of the variables and finding the corresponding value for the other variable, one may get a fraction or decimal for the second variable.

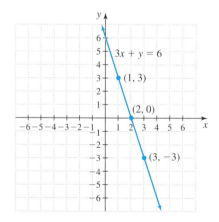

Figure 8-6

Calculator Explorations

The Table feature can help you graph $x + 2y = 5$.

When choosing solutions to graph, which appear to be simplest to plot on the graph?

Example 8-3

Graph $x + 2y = 5$.

Solution

Select at least three points to plot. In this case, we choose $x = 5$, $x = 1$, and $x = -1$. (*Note:* Any values for x can be selected.) Substitute each value in the equation $x + 2y = 5$ and solve each equation for y as shown previously. The three points are $(5, 0)$, $(1, 2)$, and $(-1, 3)$. Plot the points on the plane and draw a straight line through them. See Figure 8-7.

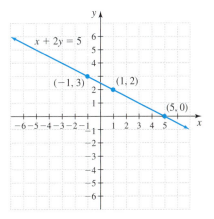

Figure 8-7

Try This One

8-B Graph $2x - y = 10$.
For the answer, see page 399.

Intercepts

The point where a line crosses the x axis is called the **x intercept.** The point where the line crosses the y axis is called the **y intercept.** The next rule can be used to find the intercepts.

Finding Intercepts

To find the x intercept, substitute 0 for y and solve the equation for x.
To find the y intercept, substitute 0 for x and solve the equation for y.

Example   8-4

Find the intercepts for $2x - 3y = 6$.

Solution

To find the x intercept, let $y = 0$ and solve for x.

$$2x - 3y = 6$$
$$2x - 3(0) = 6$$
$$2x = 6$$
$$x = 3$$

Hence, the x intercept has the coordinates $(3, 0)$.
   To find the y intercept, let $x = 0$ and solve for y.

$$2x - 3y = 6$$
$$2(0) - 3y = 6$$
$$-3y = 6$$
$$y = -2$$

Hence, the y intercept has the coordinates $(0, -2)$.

An equation can be graphed by finding the intercepts, plotting the points, and drawing a straight line through the intercepts. The graph for $2x - 3y = 6$ is shown in Figure 8-8.

Try This One

8-C   For each equation, find the intercepts and draw the graph for the equation using the intercepts.

(a) $x + 5y = 10$
(b) $4x - 3y = 12$

For the answer, see page 399.

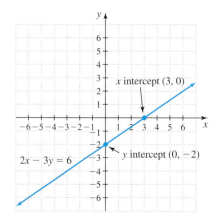

Figure 8-8

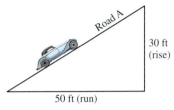

Slope = 0.6

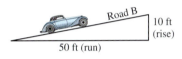

Slope = 0.2

Figure 8-9

The slope of this road is 0, because the rise is 0.

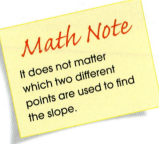

Math Note

It does not matter which two different points are used to find the slope.

Slope

The slope of a line on a plane is analogous to the slope of a road. It is the "steepness" of the road or line. Consider the two roads shown in Figure 8-9.

The "slope" of a road can be defined as the "rise" (vertical height) divided by the "run" (horizontal distance) or as the change in y with respect to the change in x. In road A, we have

$$\frac{30 \text{ ft}}{50 \text{ ft}} = 0.6$$

That is, for every 50 feet horizontally the road rises a height of 30 feet. Road B has a slope of

$$\frac{10 \text{ ft}}{50 \text{ ft}} = 0.2$$

Since the slope of road A is larger than the slope of road B, we say that road A is steeper than road B.

On the Cartesian plane, the slope is defined as follows:

The **slope** of a line (designated by m) is

$$m = \frac{y_2 - y_1}{x_2 - x_1}$$

where (x_1, y_1) and (x_2, y_2) are two points on the line.

In other words, the slope of a line can be determined by subtracting the y coordinates (the vertical height) of two points and dividing the difference by the difference obtained by subtracting the x coordinates (the horizontal distance) of the same two points. See Figure 8-10.

Example 8-5 shows the procedure for finding the slope of a line given two points.

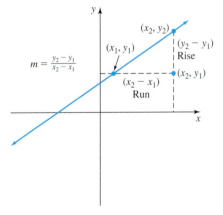

Figure 8-10

Math Note

If the line goes "uphill" from left to right, the slope will be positive. If a line goes "downhill" from left to right, the slope will be negative. The slope of a vertical line is *undefined*. The slope of a horizontal line is 0.

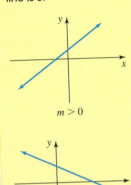

$m > 0$

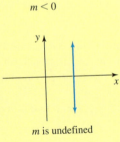

$m < 0$

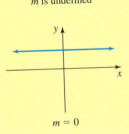

m is undefined

$m = 0$

Example   8-5

Find the slope of a line passing through the points (2, 3) and (5, 8).

Solution

Designate the two points as follows

$$(2,\ 3) \quad \text{and} \quad (5,\ 8)$$
$$\downarrow\ \downarrow \qquad\qquad \downarrow\ \downarrow$$
$$(x_1, y_1) \qquad\quad (x_2, y_2)$$

Substitute in the formula

$$m = \frac{y_2 - y_1}{x_2 - x_1} = \frac{8 - 3}{5 - 2} = \frac{5}{3}$$

Hence the slope of the line is $\frac{5}{3}$. That means the line is rising five feet vertically for every three feet horizontally.

It does not matter which point is selected for x_1 and y_1; however, the order of subtraction must be the same in the numerator and denominator (i.e., $y_2 - y_1$ and $x_2 - x_1$).

If you have the equation of a line, you can get the slope by finding any two different points on the line and substituting in the formula for slope. Example 8-6 shows this procedure.

Example   8-6

Find the slope for the line $5x - 3y = 15$.

Solution

Find the coordinates of any two points on the line. In this case, we can select (3, 0) and (0, −5). Then subtract in the formula.

$$m = \frac{y_2 - y_1}{x_2 - x_1} = \frac{-5 - 0}{0 - 3} = \frac{-5}{-3} = \frac{5}{3}$$

The slope of the line $5x - 3y = 15$ is $\frac{5}{3}$.

When you are given the equation of a line, its slope can be found by solving the equation for y. The coefficient of x will be the slope of the line.

The **slope-intercept form** for an equation in two variables is $y = mx + b$, where m is the slope and $(0, b)$ is the point where the line crosses the y axis.

Example 8-7

Find the slope of the line $5x - 3y = 15$.

Solution

Solve the equation for y.

$$5x - 3y = 15$$
$$-3y = -5x + 15$$
$$\frac{-3y}{-3} = \frac{-5x}{-3} + \frac{15}{-3}$$
$$y = \frac{5}{3}x - 5$$

Hence, the slope of the line $5x - 3y = 15$ is $\frac{5}{3}$, which agrees with the result of the preceding example. Also note that the coordinates of the y intercept are $(0, -5)$.

The graph of a line in the slope-intercept form can be drawn by using the y intercept as a point, then plotting the "run," which is the denominator of the slope in fraction form, and then the rise, which is the numerator, as shown in Example 8-8.

Example 8-8

Plot the graph for the line $y = \frac{5}{3}x - 6$.

Solution

Use $(0, -6)$ as the first point since these are the coordinates of the y intercept. Then from this point, move vertically up 5 units for the rise, and move 3 units to the right for the run. This will be the second point $(3, -1)$. Then draw a line through the two points, as shown in Figure 8-11. Notice that $(3, -1)$ does satisfy the original equation.

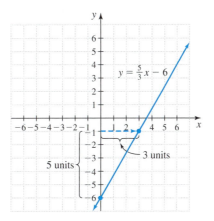

Figure 8-11

When a slope of a line is negative such as $-\frac{2}{3}$, start at the y intercept and move 2 units up and 3 units to the left to get the second point or move 2 units down and 3 units to the right to get the second point.

Consider this example of an application. Suppose you decide to bike home at 6 miles per hour from a distance of 15 miles. The equation that describes the situation is $y = -6x + 15$, where y is the distance in miles and x is the time in hours. In this case, then, the slope of the line would be the rate at which you biked, 6 miles per hour. It is negative since your distance from home is decreasing. Remember that the slope is the rate of change. The y intercept is your starting point, 15 miles from home. The x intercept is the time you arrive at home, which is $y = 0$. See Figure 8-12.

Try This One

8-D Find the slope of and graph the line $2x - 5y = 10$.

(a) Using two points.

(b) By writing the line in slope-intercept form.

For the answer, see page 399.

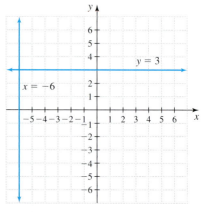

Horizontal and Vertical Lines

When an equation has only one variable, such as $x = -6$ or $y = 3$, the graph will be a line parallel to the y axis or the x axis, respectively. For example, the graph of $y = 3$ is a horizontal line passing through the point $(0, 3)$ on the y axis and is parallel to the x axis, as shown in Figure 8-13. For every point on the horizontal line, $y = 3$. The graph of $x = -6$ is a vertical line passing through $(-6, 0)$ on the x axis and parallel to the y axis. For every point on the vertical line, $x = -6$. See Figure 8-13.

> **Math Note**
>
> The graph of $x = 0$ is the y axis and the graph of $y = 0$ is the x axis.

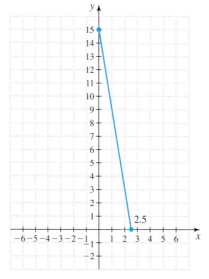

Figure 8-12

Figure 8-13

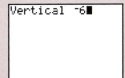
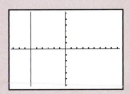
Example   8-9

Graph each line.

(a) $x = 5$
(b) $y = -3$

Solution

(a) The graph of line $x = 5$ is a vertical line passing through $(5, 0)$ on the x axis. See Figure 8-14.
(b) The graph of $y = -3$ is a horizontal line passing through $(0, -3)$ on the y axis. See Figure 8-14.

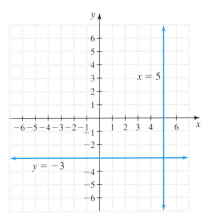

Figure 8-14

Applications of Linear Equations

A variety of real-life problems can be solved using linear equations in two variables. These equations are usually written in the slope-intercept form, $y = mx + b$. Example 8-10 shows this.

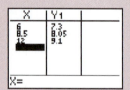
Example 8-10

The fare for a taxi is $5.50 plus $0.30 a mile. How much would it cost to take a taxi ride for 6 miles? 8.5 miles? 12 miles?

Solution

Let y = the value for the fare. Let x = the number of miles traveled. Since it costs $0.30 per mile, the cost of x miles is $0.30x$. This, together with the initial cost of $5.50, gives us the equation $y = 0.3x + 5.50$. (*Note:* This is an equation of a line.) Substitute each value for x in the equation.

For $x = 6$ miles,

$$y = 0.3(6) + 5.50$$
$$y = \$7.30$$

For $x = 8.5$ miles,

$$y = 0.3(8.5) + 5.50$$
$$y = \$8.05$$

For $x = 12$ miles,

$$y = 0.3(12) + 5.50$$
$$y = \$9.10$$

Try This One

8-E The cost of a medium cheese pizza is $6.75 and each additional topping costs $0.35. Find the cost of a medium cheese pizza with

(a) Three toppings.

(b) Five toppings.

For the answer, see page 399.

Exercise Set 8-1

Computational Exercises

For Exercises 1–14, plot each point on the Cartesian plane.

1. $(-2, -5)$
2. $(-3, -8)$
3. $(-6, 4)$
4. $(3, -7)$
5. $(6, 0)$
6. $(-4, 0)$
7. $(0, 3)$
8. $(0, -2)$
9. $(5.6, -3.2)$
10. $(-4.8, 7.3)$
11. $(-6, -10)$
12. $(-4, -9)$
13. $(0, 0)$
14. $(5, 5)$

For Exercises 15–24, draw the graph for each equation by finding at least two points on the line.

15. $5x + y = 20$
16. $x + 4y = 24$
17. $3x - y = 15$
18. $2x - y = 10$
19. $4x + 7y = 28$
20. $3x - 8y = 24$
21. $2x + 7y = -12$
22. $5x - 3y = -18$
23. $6x - 4y = 28$
24. $2x - 7y = 28$

For Exercises 25–32, find the slope of the line passing through the two points.

25. $(-3, -2), (6, 7)$
26. $(4, 0), (3, -5)$
27. $(2, 10), (4, 9)$
28. $(6, 3.5), (4.2, 6)$
29. $(-4, -5), (-9, -2)$
30. $(3.8, -1.2), (2.2, 3.1)$
31. $(2, -6.1), (3.4, -2.8)$
32. $(4, 0), (0, 7)$

For Exercises 33–40, find the coordinates for the x intercept and the y intercept for each line.

33. $3x + 4y = 24$
34. $-2x + 7y = -28$
35. $-5x - 6y = 30$
36. $x + 6y = 10$
37. $2x - y = 18$
38. $9x + 4y = -36$
39. $5x - 2y = 15$
40. $9x - 7y = 18$

For Exercises 41–48, write the equation in the slope-intercept form, then find the slope and the y intercept. Finally, draw the graph of the line.

41. $7x + 5y = 35$
42. $-2x + 7y = 14$
43. $x - 4y = 16$
44. $4x - 8y = 15$
45. $8x - 3y = 24$
46. $3x - 7y = 14$
47. $2x - y = 19$
48. $3x - 9y = 20$

For Exercises 49–52, draw the graph for each.

49. $x = -3$ 50. $y = 2$

51. $y = 6$ 52. $x = 7$

Real World Applications

53. A newspaper advertisement costs $6.50 per week to run the ad plus a setup charge of $50.00. Find the cost of running the ad for

    (a) 3 weeks.

    (b) 5 weeks.

    (c) 10 weeks.

54. A painter's labor charges are $50 plus $40 per room to paint the interior of a house. Find the cost of painting

    (a) a five-room house.

    (b) a seven-room house.

    (c) a nine-room house.

55. The cost of renting an automobile is $40 per day plus $1.10 per mile. Find the cost of renting the automobile for one day if it is driven

    (a) 63 miles.

    (b) 42 miles.

    (c) 127 miles.

56. The number in billions of pieces of mail delivered in the United States is approximately determined by the equation $y = 10x + 190$, where x is equal to the number of years from now. Find the number of pieces of mail that will be delivered 5 years from now.

57. The number in thousands of civilian staff in the military can be approximated by the equality $y = 10.5x$, where x is the number of years from now. Find the number of civilians in the military during the year that is 3 years from now.

58. The percentage of alcohol-related traffic deaths can be approximated by $y = 1.2x$, where x is the number of years from now. What will be the percentage of alcohol-related traffic deaths during the year that is 2 years from now?

59. The number in millions of Americans over age 65 can be approximated by the equation $y = 0.5x + 35.3$, where x is the number of years from now. Find the number of people over 65 living 5 years from now.

60. The percentage of the population of the United States with less than 12 years of school can be approximated by the equation $y = (18 - 1.1x)$, where x represents a specific year from now. Find the percentage of the population with less than 12 years of school in the year that is 4 years from now.

Writing Exercises

61. Explain why the quadrant where a point is located can be determined by looking at the signs of the coordinates.

62. Explain what the slope of a line means.

63. Explain how to find the intercepts for a line.

64. Explain how to determine if a line is vertical by looking at its equation.

65. Explain how to determine if a line is horizontal by looking at its equation.

66. Explain how to find the slope of a line without finding two points on the line. Assume the line is neither vertical nor horizontal.

Critical Thinking

67. Show why the slope of a vertical line is said to be undefined. (*Hint:* Select two points on a vertical line and calculate the slope.)

68. Why is the slope of a horizontal line zero?

69. Find the formula for the distance between two points on a line. (*Hint:* Use the Pythagorean theorem. See Chapter 10.)

70. Find the formulas for the coordinates for a midpoint when given two points on a line. (A midpoint is halfway between the two given points.)

8-2 Systems of Linear Equations

Section 8-1 explained that the graph of an equation of the form $ax + by = c$ was a straight line. When two linear equations of the form $ax + by = c$ are paired, the pair of equations is called a *system of two linear equations* in two variables. Formally defined,

> A **system of two linear equations** in two variables can be represented as
> $$a_1x + b_1y = c_1$$
> $$a_2x + b_2y = c_2$$

An example of a linear system is

$$x + 3y = 8$$
$$2x - y = 9$$

When solving a system of linear equations, it is necessary to find the set of points, if any, that are solutions to both equations. There are several ways to solve a linear system of two equations in two variables. Three ways will be shown here. They are solving systems graphically, solving systems by substitution, and solving systems by addition/subtraction.

Solving a Linear System Graphically

The first method that can be used to solve a linear system of two equations in two variables is the graphic method.

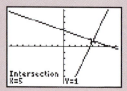

Procedure for Solving a System of Equations Graphically

Step 1 Draw the graphs of the equations on the same Cartesian plane.

Step 2 Find the point or points of intersection of the two lines if they exist.

Example 8-11

Solve this system graphically:

$$x + 3y = 8$$
$$2x - y = 9$$

Solution

Step 1 Draw the graphs for both equations on the Cartesian plane.

$x + 3y = 8$

x	y
8	0
2	2
−1	3

$2x - y = 9$

x	y
1	−7
4	−1
5	1

Step 2 Find the point of intersection of the two lines. In this case, it is (5, 1). Hence, the solution is {(5, 1)}. See Figure 8-15.

Math Note

The solution can be checked by substituting the solution in both equations and seeing if two closed true equations result:

$x + 3y = 8$ $2x - y = 9$

$5 + 3(1) \stackrel{?}{=} 8$ $2(5) - 1 \stackrel{?}{=} 9$

$5 + 3 \stackrel{?}{=} 8$ $10 - 1 \stackrel{?}{=} 9$

$8 = 8$ $9 = 9$

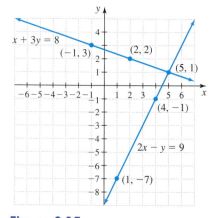

Figure 8-15

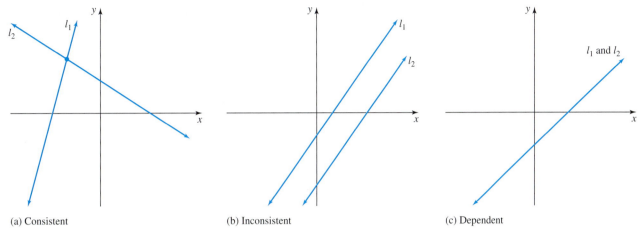

(a) Consistent   (b) Inconsistent   (c) Dependent

Figure 8-16

There are three possibilities to consider when you are finding the solution to a system of two linear equations in two variables.

1. *The lines intersect at a single point.* In this case, there is only one solution, and that is the point of intersection of the lines. See Figure 8-16(a). In this case, the system is said to be **consistent** and **independent.**
2. *The lines are parallel.* In this case, there would be no solution, or the solution would be the empty set, $\varnothing$, since parallel lines do not intersect. See Figure 8-16(b). In this case, the system is said to be **inconsistent** and *independent.*
3. The lines *coincide.* In this case, the graph of both equations is the same line; hence, any point on the line will satisfy both equations. There are infinitely many solutions. In this case, the system is said to be *consistent* and **dependent.** The solution set is written as $\{(x, y) \mid ax + by = c\}$, where $ax + by = c$ is either one of the two equations. See Figure 8-16(c).

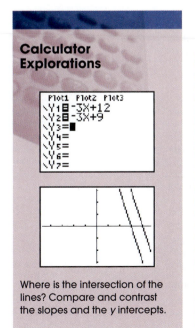

Calculator Explorations

Where is the intersection of the lines? Compare and contrast the slopes and the y intercepts.

Example 8-12

Solve this system graphically:

$$3x + y = 12$$
$$6x + 2y = 18$$

Solution

Step 1 Graph both equations.

$3x + y = 12$			$6x + 2y = 18$	
x	y		x	y
4	0		3	0
3	3		0	9
2	6		2	3

—Continued

8-19

Example 8-12 *Continued—*

Step 2 Find the point or points of intersection. In this case, the lines are parallel and there is no point of intersection. Hence, the solution set is $\varnothing$, and the system is inconsistent. See Figure 8-17.

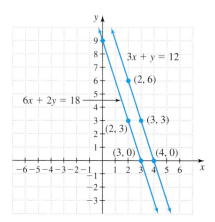

Figure 8-17

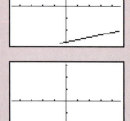

Where is the intersection of the lines? Compare and contrast the slopes and the y intercepts.

Example 8-13

Solve this system graphically:

$$x - 5y = 15$$
$$2x - 10y = 30$$

Solution

Step 1 Graph both equations.

	$x - 5y = 15$		$2x - 10y = 30$	
	x	y	x	y
	-5	-4	-5	-4
	5	-2	0	-3
	0	-3	5	-2

Step 2 Find the point or points of intersection. In this case, the lines coincide; hence, the solution set is any point on either line. This is written as

—Continued

Example 8-13 Continued—

$\{(x, y) \mid x - 5y = 15\}$. The system is said to be dependent. See Figure 8-18.

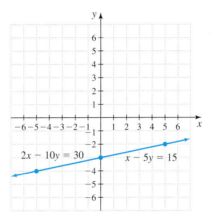

Figure 8-18

As an application, suppose Mary leaves home at 9:00 and starts hiking around the lake at 3 miles per hour. At 10:00, Dave follows her, jogging at 5 miles per hour. When and how far from home do they meet?

Two equations can be used to model the two situations. Use 10:00 when Dave leaves as the start time. Let x = time in hours and y = distance in miles. For Mary, after 1 hour, when Dave leaves, she has traveled (3 mph)(1 hr) = 3 miles. So at the starting time of 10:00, her starting point on the y axis is at 3. Her rate or speed is 3. Here the equation is $y = 3x + 3$. Dave starts from home, so his y axis is 0. His rate is 5. Hence, $y = 5x$. When graphing the two equations, they intersect at (1.5, 7.5); hence, they meet at 11:30 and at 7.5 miles from home. See Figure 8-19.

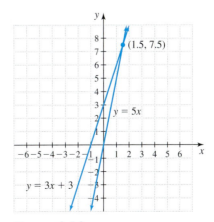

Figure 8-19

Try This One

8-F Solve each system graphically.

(a) $x - 4y = 8$
     $2x - y = -5$

(b) $3x + 4y = 12$
     $6x + 8y = 18$

(c) $2x - y = 9$
     $6x - 3y = 27$

For the answer, see page 399.

Solving a Linear System by Substitution

Since solving systems graphically can be imprecise, algebraic methods are used. One algebraic method is known as **substitution.**

Math Note

You can use either equation and either variable, so look at the equations and determine which equation to use and which variable is easier to solve for. To check your answer, substitute the values for x and y in the original equations and see if they are true.

Procedure for Solving a System of Equations by Substitution

Step 1 Select one equation and solve it for one variable (either *x* or *y*) in terms of the other variable (see Math Note).

Step 2 Substitute the expression containing the other variable that you found in step 1 into the *other* equation.

Step 3 Solve the equation for the unknown (it now has only one variable).

Step 4 Select one of the original equations, substitute the value found in step 3 for the variable, and solve it for the value of the other variable.

Example 8-14 shows this procedure.

Example   8-14

Solve the following system by substitution.

$$x + 3y = 8$$
$$2x - y = 9$$

Solution

Step 1 Select the first equation and solve it for *x* in terms of *y*.

$$x + 3y = 8$$
$$x = 8 - 3y$$

—Continued

Example 8-14 *Continued—*

Step 2 Substitute the expression for x (i.e., $8 - 3y$) in the second equation.

$$2x - y = 9$$
$$2(8 - 3y) - y = 9$$

Step 3 Solve the equation for y.

$$2(8 - 3y) - y = 9$$
$$16 - 6y - y = 9$$
$$-7y = 9 - 16$$
$$-7y = -7$$
$$\frac{-7y}{-7} = \frac{-7}{-7}$$
$$y = 1$$

Step 4 Substitute $y = 1$ in either equation and solve for x.

$$x + 3y = 8$$
$$x + 3(1) = 8$$
$$x + 3 = 8$$
$$x = 5$$

Hence, the solution is $\{(5, 1)\}$. Note that this is the same solution obtained by graphing, as shown in Example 8-11.

Example 8-15

Solve the following system by substitution.

$$3x - 4y = 10$$
$$2x + 3y = 1$$

Solution

Step 1 Select one equation and solve for one variable.

$$3x - 4y = 10$$
$$3x = 10 + 4y$$
$$x = \frac{10}{3} + \frac{4y}{3}$$

Step 2 Substitute the expression for x in the other equation.

$$2x + 3y = 1$$
$$2\left(\frac{10}{3} + \frac{4y}{3}\right) + 3y = 1$$

—Continued

Example **8-15** *Continued—*

Step 3 Solve for y.

$$\frac{20}{3} + \frac{8y}{3} + 3y = 1$$

$$\frac{3}{1} \cdot \frac{20}{3} + \frac{3}{1} \cdot \frac{8y}{3} + 3 \cdot 3y = 3 \cdot 1 \qquad \text{Multiply each term by the LCM of the denominators.}$$

$$20 + 8y + 9y = 3$$

$$17y = -17$$

$$\frac{17y}{17} = -\frac{17}{17}$$

$$y = -1$$

Step 4 Substitute $y = -1$ in one equation and solve for x.

$$3x - 4y = 10$$

$$3x - 4(-1) = 10$$

$$3x + 4 = 10$$

$$3x = 6$$

$$\frac{3x}{3} = \frac{6}{3}$$

$$x = 2$$

Hence, the solution set is $\{(2, -1)\}$.

Check:

$$3x - 4y = 10 \qquad\qquad 2x + 3y = 1$$

$$3(2) - 4(-1) \stackrel{?}{=} 10 \qquad 2(2) + 3(-1) \stackrel{?}{=} 1$$

$$6 + 4 \stackrel{?}{=} 10 \qquad\qquad 4 - 3 \stackrel{?}{=} 1$$

$$10 = 10 \qquad\qquad\qquad 1 = 1$$

Try This One

8-G Solve this system by substitution:

$$5x + y = 8$$

$$x + 2y = 7$$

For the answer, see page 399.

Solving a System by Addition/Subtraction (Elimination)

Another algebraic method that is used to solve a system of linear equations in two variables is called the **addition/subtraction method** or the *elimination* method. The steps are shown next.

Procedure for Solving a System of Equations Using the Addition/Subtraction (Elimination) Method

Step 1 If necessary, write both equations in the form $ax + by = c$.

Step 2 Multiply one or both equations by numbers so that the absolute values of either the coefficients of the x terms or the y terms are alike.

Step 3 Eliminate one of the variables by adding the equations if the signs of the coefficients of the variable are different. Subtract the equations if the signs of the coefficients of the variable are the same.

Step 4 Solve the resultant equation for the remaining variable.

Step 5 Select one equation from the original two equations, substitute the value of the variable found in step 4, and solve for the other variable.

Example 8-16

Solve the following system by the addition/subtraction method:

$$2x - 3y = -6$$
$$x = 7 - y$$

Solution

Step 1 Write both equations in the form $ax + by = c$.

$$2x - 3y = -6$$
$$x + y = 7$$

Step 2 Multiply the second equation by 2 in order to make the coefficients of the x terms equal.

$$2x - 3y = -6$$
$$2(x + y) = 2 \cdot 7$$

which gives

$$2x - 3y = -6$$
$$2x + 2y = 14$$

Step 3 Subtract the second equation from the first equation to eliminate the x variable.

$$2x - 3y = -6$$
$$\underline{-2x - 2y = -14}$$
$$-5y = -20$$

Step 4 Solve the equation for y.

$$-5y = -20$$
$$\frac{-5y}{-5} = \frac{-20}{-5}$$
$$y = 4$$

—Continued

Example 8-16 *Continued—*

Step 5 Select one equation and substitute 4 for y and solve for x.

$$2x - 3y = -6$$
$$2x - 3(4) = -6$$
$$2x - 12 = -6$$
$$2x = 6$$
$$\frac{2x}{2} = \frac{6}{2}$$
$$x = 3$$

Hence, the solution set is $\{(3, 4)\}$.

Check:

$$2x - 3y = -6 \qquad x = 7 - y$$
$$2(3) - 3(4) \overset{?}{=} -6 \qquad 3 \overset{?}{=} 7 - 4$$
$$6 - 12 \overset{?}{=} -6 \qquad 3 = 3$$
$$-6 = -6$$

Sometimes it is necessary to multiply each equation by a different number in order to eliminate one of the variables.

Example 8-17

Solve this system by using the addition/subtraction method:

$$5x - 2y \qquad = 10$$
$$3x + 5y + 56 = 0$$

Solution

Step 1 Write the equations in the form $ax + by = c$.

$$5x - 2y = 10$$
$$3x + 5y = -56$$

Step 2 Multiply the first equation by 5 and the second equation by 2 to make the coefficients of the y terms equal in absolute value.

$$5 \cdot (5x - 2y) = 5 \cdot 10$$
$$2 \cdot (3x + 5y) = 2 \cdot (-56)$$

which yields

$$25x - 10y = 50$$
$$6x + 10y = -112$$

—Continued

Example 8-17 *Continued—*

Step 3 Add the equations.

$$25x - 10y = 50$$
$$6x + 10y = -112$$
$$\overline{31x \qquad\quad = -62}$$

Step 4 Solve the resulting equation for x.

$$\frac{31x}{31} = \frac{-62}{31}$$
$$x = -2$$

Step 5 Substitute $x = -2$ in one of the equations and solve for y.

$$5x - 2y = 10$$
$$5(-2) - 2y = 10$$
$$-10 - 2y = 10$$
$$-2y = 20$$
$$\frac{-2y}{-2} = \frac{20}{-2}$$
$$y = -10$$

Hence, the solution set is $\{(-2, -10)\}$.

Check:

$$5x - 2y = 10 \qquad\qquad 3x + 5y + 56 = 0$$
$$5(-2) - 2(-10) \overset{?}{=} 10 \qquad 3(-2) + 5(-10) + 56 \overset{?}{=} 0$$
$$-10 + 20 \overset{?}{=} 10 \qquad\qquad -6 - 50 + 56 \overset{?}{=} 0$$
$$10 = 10 \qquad\qquad\qquad\qquad 0 = 0$$

Try This One

8-H Solve this system by the addition/subtraction method:

$$3x - 4y = 4$$
$$4x - 3y = 3$$

For the answer, see page 399.

Recall that a linear system can also be inconsistent (i.e., the lines are parallel) or dependent (i.e., the lines coincide). When you try to solve an inconsistent or dependent system by substitution or addition/subtraction, both variables will be eliminated. *If the resulting equation is false, the system is inconsistent, and if the resulting equation is true, the system is dependent.*

Example 8-18

Solve this system:

$$3x + y = 12$$
$$6x + 2y = 15$$

Solution

Solving by substitution, one gets

$$3x + y = 12$$
$$y = 12 - 3x$$
$$6x + 2(12 - 3x) = 15$$
$$6x + 24 - 6x = 15$$
$$24 = 15$$

Since the resulting equation, $24 = 15$, is false, the system is inconsistent. The lines are parallel, and the solution set is $\varnothing$.

Sidelight

Gabrielle-Emilie du Chatelet (1706–1749)

In France during the 1700s, women were expected to be reticent. When Emilie du Chatelet was born, her father thought that because of her looks, she would never marry, so he decided that she should instead be educated. However, at age 19, Emilie turned out to be a beautiful young lady, and after she received her formal education, she married an older French aristocrat. During the marriage, she had three children.

Her love of mathematics began in school and continued throughout her lifetime. When her husband was away on business, which he often was, she studied mathematics, and in 1733, she befriended a Frenchman named Voltaire. Together, they studied the works of Sir Isaac Newton on physics. She also wrote a textbook on physics entitled *Institutions de Physique*.

Emilie received some recognition for her work from Frederick the Great, who was the king of Prussia, and at the time that was an accomplishment since people believed that women were unable to understand science and mathematics.

One time she was refused admittance to a gathering of mathematicians because she was a woman. She later returned dressed as a man and participated in the discussion.

During the final years of her life, scientists and mathematicians came to visit her and study with her. She died in childbirth at the age of 43.

Example 8-19

Solve this system:

$$5x + y = 9$$
$$10x + 2y = 18$$

Solution

Solve the system by substitution.

$$5x + y = 9$$
$$y = 9 - 5x$$
$$10x + 2y = 18$$
$$10x + 2(9 - 5x) = 18$$
$$10x + 18 - 10x = 18$$
$$18 = 18$$

Since the variables are eliminated and the equation $18 = 18$ is true, the system is dependent. The solution set is $\{(x, y) \mid 5x + y = 9\}$.

Try This One

8-I Determine whether each system is inconsistent or dependent.

(a) $4x = 2y + 9$
     $6y = 12x - 27$

(b) $3x + 8y = 15$
     $9x + 24y = 32$

For the answer, see page 399.

Applications of Linear Systems

Most real-world problems involve many variables and many equations. These problems can be solved by using systems of equations. In this book, though, we will be limiting the discussion to solving systems of two equations using two variables.

Example 8-20

A person wishes to invest $12,000 in two funds with the goal of generating $1100 interest. One fund (a high-risk fund) pays 10% interest, and the other fund (a safer investment) pays 8%. How much should be invested in each account to earn $1100 interest?

—Continued

Example 8-20 *Continued—*

Solution

Let $x =$ the amount of money invested at 8%. Let $y =$ the amount of money invested at 10%. The first equation is $x + y = \$12,000$ since this is the total amount of money invested. The second equation is $0.08x + 0.10y = \$1100$, since this is the amount of interest earned from 8% of the investment amount x and 10% of the investment amount y.

The system is

$$x + y = 12,000$$
$$0.08x + 0.10y = 1100$$

Solving the system by substitution:

$$x + y = 12,000$$
$$y = 12,000 - x$$
$$0.08 + 0.10y = 1100$$
$$0.08x + 0.10(12,000 - x) = 1100$$
$$0.08x + 1200 - 0.1x = 1100$$
$$-0.02x + 1200 = 1100$$
$$-0.02x + 1200 - 1200 = 1100 - 1200$$
$$-0.02x = -100$$
$$\frac{-0.02x}{-0.02} = \frac{-100}{-0.02}$$
$$x = \$5000$$

Hence, $5000 should be invested at 8%.

$$x + y = 12,000$$
$$5000 + y = 12000$$
$$5000 - 5000 + y = 12000 - 5000$$
$$y = \$7000$$

Hence, $7000 should be invested at 10%.

Try This One

8-J An office used 200 stamps. Some were 35¢ stamps and some were 20¢. If the total cost was $62.95, how many of each did the office use?
For the answer, see page 399.

Exercise Set 8-2

Computational Exercises

For Exercises 1–8, solve each system by graphing.

1. $2x - y = 14$
   $x = y + 7$

2. $x - 3y = 6$
   $4x + 3y = 9$

3. $2x - y = -4$
   $x + y = -2$

4. $x + y = 2$
   $3x - 2y = -9$

5. $4x + 3y = 2$
   $3x + 5y = -4$

6. $x + 2y = 0$
   $2x - y = 0$

7. $x - 2y = 10$
   $x + y = 4$

8. $3x - 5y = -2$
   $10y - 6x = 4$

For Exercises 9–16, solve each system by substitution.

9. $x + y = 0$
   $x - 2y + 9 = 0$

10. $x = 2 - y$
    $3x - 2y = 1$

11. $x - 3y = 7$
    $4x + 3y = 13$

12. $x + 2y = 11$
    $2x - y = 7$

13. $4x + 3y = 24$
    $3x + 5y = 22$

14. $3x = 5y + 16$
    $5y - 3x = -16$

15. $2x - 3y = 1$
    $x = y + 2$

16. $2x + 3y = -1$
    $x = y - 13$

For Exercises 17–24, solve each system by addition/subtraction.

17. $3x = 2y + 10$
    $9y = 3x - 7$

18. $5x - y = 10$
    $x - 2y = 18$

19. $3x + 4y = 2$
    $y = x - 3$

20. $x = 3y - 6$
    $x - 3y = 12$

21. $5x - 2y = 11$
    $15x - 6y = 33$

22. $x = 3y + 2$
    $x + 3y = 14$

23. $x = 3y - 2$
    $5x - 4y = 12$

24. $2x - 3y = -2$
    $5x = 2y - 9$

For Exercises 25–32, solve each system by any method; state whether the system is consistent, inconsistent, or dependent; and give the solution set.

25. $4x + y = 2$
    $7x + 3y = 1$

26. $8x - 2y = -2$
    $3x - 5y = 4$

27. $x - 6y = 19$
    $2x + 7y = 0$

28. $4x - y = 11$
    $7x + 3y = 14$

29. $3x - 5y = 5$
    $5x - 7y = 1$

30. $x = 3y + 2$
    $x + 3y = 14$

31. $4x - y = 11$
    $7x + 3y = 8$

32. $4x - y = 3$
    $8x - 2y = 6$

Real World Applications

33. The sum of the ages of two sisters is 32. The difference between four times the older sister's age and two times the younger sister's age is 38. Find the age of each sister.

34. A grocer mixes coffee that sells for $3.20 a pound with coffee that sells for $5.40 a pound. If she wishes to have 20 pounds of coffee to sell at $4.52 a pound, how many pounds of each would she have to use?

35. Adult tickets for a tourist train ride cost $12.00 and child tickets cost $8.00. If there were 40 passengers on the train and the net revenue was $420.00, how many adults and how many children rode the train?

36. If a man is three times as old as his daughter and the sum of their ages is 52, find each person's age.

37. The difference between the ages of two students is 4 years. If the sum of their ages is 42 years, find each student's age.

38. A candy store owner mixes spearmint candy selling for $0.99 a pound with cinnamon candy selling for $0.89 a pound. If there are 10 pounds of the mixture selling for $0.94 per pound, how many pounds of each type were mixed together?

39. At a fast food restaurant, three chicken sandwiches and two large orders of French fries cost $8.87, and five chicken sandwiches and four large orders of French fries cost $15.55. How much does each item cost?

40. A person wishes to invest $2400.00, part at 9% and part at 6%. If the total interest desired at the end of the year is $189.00, how much should be invested at each rate?

41. At a school concert, student tickets cost $5.00 and general admission tickets for non-students were $8.00. If the total revenue was $3034.00 and 500 tickets were sold, how many students attended the concert? How many general admission tickets were sold?

42. At a flea market, a person purchased some VCR tapes and CDs. The CDs sold for $5.00 each, and the VCR tapes sold for $3.00 each. If the total cost of the items was $78.00 and the total number of products sold was 18, find how many of each item the person purchased.

Writing Exercises

43. Explain what is meant when a system of linear equations is said to be consistent.

44. Explain what is meant when a system of equations is inconsistent.

45. Explain what is meant when a system of equations is dependent.

46. Explain how to solve a system of equations graphically.

47. Explain how to solve a system of equations by substitution.

48. Explain how to solve a system of equations by using the addition/subtraction (elimination) method.

49. When solving a system by substitution or by addition/subtraction (elimination), explain how to determine if the system is inconsistent.

50. When solving a system by substitution or by addition/subtraction (elimination), explain how to determine if the system is dependent.

Critical Thinking

51. Write a system of equations which has $(3, -5)$ as a solution. (There are many possible answers.)

52. A system of three equations with three unknowns can be solved by selecting a pair of equations and eliminating one variable and then selecting another pair and eliminating the same variable. The resulting two equations will have two variables. This system can be solved by methods shown in this chapter. Solve this system:

$$3x + 2y - 2z = 1$$
$$x + 3y + z = 10$$
$$2x - 4y + z = -3$$

8-3 Systems of Linear Inequalities

Recall that equation $ax + by = c$ is called a linear equation in two variables and its graph is a straight line. When the equal sign is replaced by $\geq, >, \leq$, or $<$, the equation becomes a **linear inequality.** Examples of linear inequalities are $2x + y > 6$, $3x - 8y \leq 20$, and $x - y \geq 3$.

In order to draw the graph of a linear inequality, first replace the inequality sign with an equal sign, and then draw the graph of the line. Draw a solid line if the inequality sign is $\geq$ or $\leq$, and a dashed line if the inequality sign is $<$ or $>$. The reason is that when the inequality sign is $\leq$ or $\geq$, all the points that are on the line are included in the solution set of the linear inequality. A dashed line means that the points on the line are not included in the solution set of the equation. After the line is drawn, it separates the plane into two *half planes.*

> A **half plane** is the set of points on the Cartesian plane that are on one side of a line.

Math Note

A good test point to use is the origin since its coordinates are $(0, 0)$. However, if the line passes through the origin, another point must be selected as the test point.

The solution set for a linear inequality includes the set of points on the line and the points on one side of the line (i.e., the half plane) if the inequality sign is $\geq$ or $\leq$, or the points on one side of the line (i.e., the half plane) if the inequality sign is $>$ or $<$. In order to determine which half plane is included in the solution set, a test point not on the line is selected and substituted in the linear inequality. If the resulting inequality is true, then the half plane containing the test point is the one that is included in the solution set. If the resulting inequality is false, then the other half plane is the one included in the solution set.

Examples 8-21 and 8-22 show how to find the graphic solution for a linear inequality.

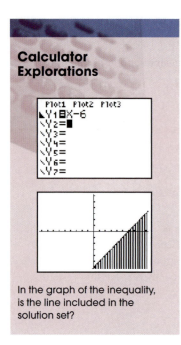

In the graph of the inequality, is the line included in the solution set?

Example 8-21

Graph $x - y \geq 6$.

Solution

Step 1 Graph the line $x - y = 6$ using two points on the line. In this case, the intercepts can be used. They are $(6, 0)$ and $(0, -6)$. Draw a solid line through the points because they are contained in the solution set since the inequality sign is $\geq$. See Figure 8-20.

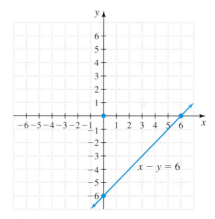

Figure 8-20

Step 2 Select a test point not on the line, substitute in the equation, and see if a true or false statement results.

Select $(0, 0)$; then

$$x - y \geq 6$$

$$0 - 0 \overset{?}{\geq} 6$$

$$0 \geq 6 \qquad \text{False}$$

The test point is not in the correct half plane since the resulting inequality is false.

Step 3 Shade the half plane that does not contain $(0, 0)$. See Figure 8-21.

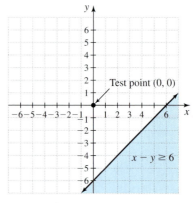

Figure 8-21

Sidelight _____

Maria Gaetana Agnesi (1718–1799)

Agnesi was born into a wealthy family of mathematicians and scientists. Her father taught at the University of Bologna, Italy, and he made sure that she was well educated. By the time she was 13, she spoke seven languages.

To help her younger brother with his mathematics, Agnesi wrote an informal textbook on algebra and geometry entitled *Analytical Instructions.* The book was written so clearly and concisely that it was published and used as an academic textbook in England, France, and Italy.

In addition to her contributions to mathematics, she took sick and downtrodden people into her home and cared for them.

When she died, she was buried in a common grave with 15 outcasts, and no fancy marker was placed on her grave.

Try This One

8-K   Graph $2x - y < 8$.
For the answer, see page 399.

Systems of Linear Inequalities

Systems of linear equations were solved graphically by graphing each line and finding the point of the intersection of the lines. **Systems of linear inequalities** can be solved graphically by graphing each inequality and finding the intersection of the half planes.

Example   8-22

Solve the system graphically.

$$x + 3y \geq 6$$
$$2x - y < 10$$

Solution

Graph each inequality. For $x + 3y \geq 6$, use the points $(6, 0)$ and $(0, 2)$. Draw a solid line through the two points since the inequality sign is $\geq$. Select $(0, 0)$ as a test point.

$$x + 3y \geq 6$$
$$0 + 3(0) \overset{?}{\geq} 6$$
$$0 \geq 6 \quad \text{False}$$

—Continued

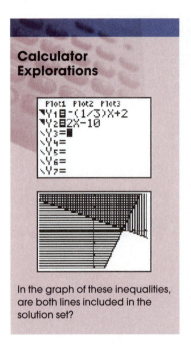
Example 8-22 *Continued—*

Since this is a false statement, shade the area above the line. See Figure 8-22(a).

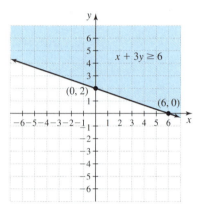

$x + 3y \geq 6$

$(0, 2)$

$(6, 0)$

Figure 8-22a

For $2x - y < 10$, use the points $(5, 0)$ and $(4, -2)$, and draw a dashed line through the points since the inequality sign is $<$. Select $(0, 0)$ as a test point.

$$2x - y < 10$$

$$2(0) - 0 \overset{?}{<} 10$$

$$0 < 10$$

Since this is a true statement, shade the area that contains the test point. See Figure 8-22(b).

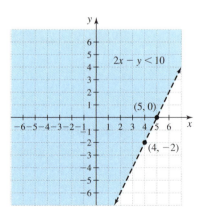

$2x - y < 10$

$(5, 0)$

$(4, -2)$

Figure 8-22b

—Continued

Example 8-22 *Continued—*

The solution is the intersection of the two half planes and part of the line for the equation $x + 3y = 6$. See Figure 8-22(c).

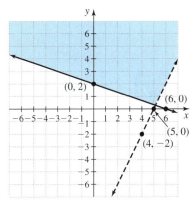

Figure 8-22c

Example 8-23

Solve the system graphically.

$$x > -3$$
$$y \leq 2$$

Solution

To graph $x > -3$, use $x = -3$ and draw a dashed vertical line passing through -3 on the x axis. Then use $(0, 0)$ as a test point, and shade the half plane to the right of the line. See Figure 8-23.

To graph $y \leq 2$, use $y = 2$ and draw a solid horizontal line passing through 2 on the y axis. Then use $(0, 0)$ as a test point, and shade the area below the line. See Figure 8-23.

The solution is the intersection of the two half planes and part of the line $y = 2$. See Figure 8-23.

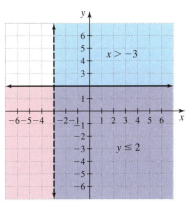

Figure 8-23

Try This One

8-L   Find the solution set for the system.

$$4x - 3y \geq 12$$
$$5x - 2y < 10$$

For the answer, see page 400.

Exercise Set   8-3

Computational Exercises

For Exercises 1–10, graph each linear inequality.

1. $2x - y \geq 5$
2. $x + 4y < -3$
3. $3x - 4y \leq 12$
4. $12x - 3y > 10$
5. $x \geq 5$
6. $y < -6$
7. $y \leq 0$
8. $x > -3$
9. $3x + y < -6$
10. $-2x + y \geq 0$

For Exercises 11–20, find the solution set for the system of linear inequalities.

11. $2x - y \leq 6$
    $x + y > 3$
12. $x - 3y < 10$
    $3x - 2y \geq 5$
13. $x \geq -2$
    $y < -3$
14. $x < -3$
    $y \geq 0$
15. $5x - 3y \geq 15$
    $3x - 8y \geq 12$
16. $x - y < 6$
    $2x - y \geq 3$
17. $x + y \geq 7$
    $x - y < -5$
18. $3x - y \leq -6$
    $2x + y > 6$
19. $4x - 2y < 6$
    $8x - 4y > -3$
20. $x + 7y < -5$
    $7x - y \leq 10$

Writing Exercises

21. Explain the difference between using a solid line and a dashed line when graphing a linear inequality.
22. Explain how the test point is used to determine which half plane is used for the solution set.

Critical Thinking

23. When is the point of intersection of the lines included in a solution of a system of linear inequalities?
24. Can the solution set for a system of linear inequalities be the empty set? Explain your answer.
25. Solve this system:

$$x + y > 6$$
$$x + y < -3$$

8-4 Linear Programming

Section 8-3 showed how to graph a system of two linear inequalities. One practical application of linear systems is called **linear programming.** Linear programming was developed to handle military logistical problems such as transporting equipment and personnel during wartime. Today the techniques of linear programming are used by business and industry to make decisions and to find cost-effective solutions to many problems. Basically, a linear programming problem makes restrictions on variables called **constraints** and represents them as a system of linear inequalities.

The system is graphed on the Cartesian plane, and in most cases, a **polygonal region** is formed. The coordinates of the vertices of the polygon are substituted into what is called the **objective function.** This function relates the variables of the problem.

The objective function must be a linear function, and the goal is to either maximize or minimize the objective function. For example, suppose the objective function $P = 3x + 8y$ represents the profit P from manufacturing x footballs and y basketballs. Now we want to select values for x and y that would maximize the profit. These values could be subject to certain *constraints*. A constraint in this case might be that a maximum of 50 footballs can be manufactured in an 8-hour day.

Finally, a decision is made based on the evaluation of the objective function. Using advanced mathematics beyond the scope of this book, it has been mathematically proven that any maximum or minimum of the objective function will occur at a vertex of a certain polygonal region. Example 8-24 shows the procedure for solving the problem using linear programming.

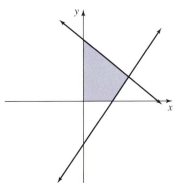

A typical polygonal region in a linear programming problem. The vertices are where the lines intersect.

Example 8-24

The owner of a small business manufactures children's playground sets and children's playhouses. The process involves two steps. First, the lumber must be cut, and second, the wood must be drilled and the product assembled. For a playground set, it takes a worker 2 hours to cut the lumber and another worker 1 hour to drill and assemble the set. For the playhouse, it takes a worker 1 hour to cut the lumber and another worker 1.5 hours to drill and assemble the playhouse. Workers work at most 8 hours per day. The owner makes $15.00 profit on every playground set that is sold and $10.00 profit on every playhouse that is sold. Using linear programming techniques, how many of each product should be manufactured in one day in order to maximize profit?

Solution

Step 1 Write the objective function.

Since the owner makes a profit of $15.00 on each playground set sold and a profit of $10.00 on each playhouse sold, the objective function can be written as

$$P = 15x + 10y$$

where $x =$ the number of playground sets sold, $y =$ the number of playhouses sold, and $P =$ the profit made.

—Continued

Example 8-24 *Continued—*

Step 2 Write the constraints.

Listing the information in a table makes it somewhat easier to find the constraints.

Item	Cut (hours)	Assemble (hours)	Profit
Playground set	2	1	$15.00
Playhouse	1	1.5	$10.00
Time limit	8	8	

Since it takes 2 hours to cut the lumber for the playground set and 1 hour to cut the lumber for the playhouse, the first constraint is

$$2x + y \leq 8$$

Since it takes 1 hour to assemble the playground set and 1.5 hours to assemble the playhouse, the second constraint is

$$x + 1.5y \leq 8$$

In addition, x and y cannot be less than zero. Hence, there are two additional constraints:

$$x \geq 0$$
$$y \geq 0$$

Step 3 Graph the linear system.

$$2x + y \leq 8 \qquad x \geq 0$$
$$x + 1.5y \leq 8 \qquad y \geq 0$$

See Figure 8-24.

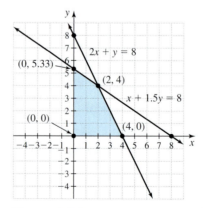

Figure 8-24

—Continued

Example 8-24 *Continued—*

Step 4 Find the vertices of the polygonal region. The vertices of the polygon are (0, 0), (0, 5.33), (2, 4), (4, 0).

The vertex (0, 0) is the origin. The vertex (2, 4) is found by determining the coordinates of the intersection of the two lines $2x + y = 8$ and $x + 1.5y = 8$. The vertex (0, 5.33) is found by substituting 0 for x and solving for y in the equation $x + 1.5y = 8$. Finally, the vertex (4, 0) is found by substituting 0 for y and solving for x in the equation $2x + y = 8$.

Step 5 Substitute the vertices in the objective function and find the maximum value. (Recall that the maximum or minimum values of the objective function will occur at a vertex of the polygonal region.)

$$P = 15x + 10y$$

$$(0, 0) \quad P = 15(0) + 10(0) = 0 + 0 = 0$$
$$(0, 5.33) \quad P = 15(0) + 10(5.33) = 0 + 53.3 = 53.3$$
$$(2, 4) \quad P = 15(2) + 10(4) = 30 + 40 = 70$$
$$(4, 0) \quad P = 15(4) + 10(0) = 60 + 0 = 60$$

Hence, the vertex (2, 4) produces the maximum value 70. The solution is to make two playground sets and four playhouses for a maximum profit of $70.00 per day.

The procedure for solving problems using linear programming is summarized next.

Procedure for Using Linear Programming

Step 1 Write the objective function.

Step 2 Write the constraints.

Step 3 Graph the constraints.

Step 4 Find the vertices of the polygonal region.

Step 5 Substitute the coordinates of the vertices into the objective function and find the maximum or minimum value.

(*Note:* The solutions will not always be integers.)

Example 8-25

An automobile dealer has room for no more than 100 automobiles on his lot. The dealer sells two models, convertibles and sedans, and he sells at least three times as many sedans as convertibles. If he makes a profit of $1000 on a convertible and a profit of $1500 on a sedan, how many of each automobile should he have in his lot in order to maximize his profit?

—Continued

Example 8-25   *Continued—*

Solution

Step 1   Write the objective function.

Since the dealer makes a profit of $1000 on each convertible sold and $1500 on each sedan sold, the objective function is

$$P = 1000x + 1500y$$

where x = the number of convertibles sold, y = the number of sedans sold, and P = profit made.

Step 2   Write the constraints.

$$x + y \leq 100$$
$$3x \leq y$$

In addition, x and y cannot be negative; hence,

$$x \geq 0$$
$$y \geq 0$$

Step 3   Graph the system.

$$x + y \leq 100 \quad x \geq 0$$
$$3x \leq y \quad\quad y \geq 0$$

See Figure 8-25.

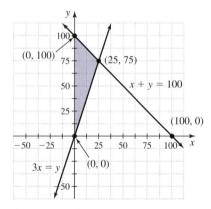

Figure 8-25

Step 4   Find the vertices of the polygonal region. They are (0, 0), (0, 100), and (25, 75). The vertex (0, 0) is the origin. The vertex (25, 75) is found by finding the coordinates of the point of the intersection of the two lines $x + y = 100$ and $3x = y$. The vertex (0, 100) is found by substituting $x = 0$ and solving for y in the equation $x + y = 100$.

Step 5   Substitute in the objective function and find the maximum value.

$$P = 1000x + 1500y$$

(0, 0) $P = 1000(0) + 1500(0) = 0 + 0 = 0$

(0, 100) $P = 1000(0) + 1500(100) = 0 + 150{,}000 = 150{,}000$

(25, 75) $P = 1000(25) + 1500(75) = 25{,}000 + 112{,}500 = 137{,}500$

Hence, the dealer should stock 0 convertibles and 100 sedans for a maximum profit.

Try This One

8-M A craftsperson makes leather purses and cloth purses. She can make at most 20 purses per month. She wishes to make at least 5 leather purses and no more than 15 cloth purses per month. Her profit on a leather purse is $5.00, and her profit on a cloth purse is $10.00. How many of each type of purse should she make to maximize her profit?

For the answer, see page 400.

Exercise Set 8-4

Computational Exercises

For Exercises 1–10, write each constraint or objective function.

1. A company makes calculators. It manufactures at most 50 calculators in one day. Write the constraint.

2. The profit on a scientific calculator is $10.00, and the profit on a graphing calculator is $12.50. Write the objective function for the profit as a function of the number of scientific calculators sold and the number of graphing calculators sold.

3. A company wishes to manufacture twice as many pens as pencils. Write the constraint.

4. The manager of a store wishes to purchase at least 20 more desktop computers than laptop computers. Write the constraint.

5. The profit on a desktop computer is $85.00, and the profit on a laptop computer is $130.00. Write the profit function as a function of the number of desktop computers and the number of laptop computers sold.

6. A food store owner buys turkeys and hams. The owner wishes to purchase at most a total of 50 products. Write a constraint.

7. The profit on each turkey sold is $4.00, and the profit on each ham sold is $3.00. Write the profit function as a function of the number of turkeys sold and the number of hams sold.

8. The food store owner wishes to purchase at most 10 more turkeys than hams. Write the constraint.

9. It takes a person 2 hours to assemble a VCR and 3 hours to assemble a CD player. Each employee can work at most 6 hours a day. Write a constraint for the problem.

10. The profit on a CD player is $32.00, and the profit on a VCR is $45.00. Write a profit function as a function of the number of CD players sold and the number of VCRs sold.

For Exercises 11–16, evaluate the profit function for each vertex and determine which vertex gives the maximum profit.

11. $P = \$30x + \$40y$; $(0, 0)$, $(5, 6)$, $(10, 0)$, $(0, 8)$

12. $P = \$85x + \$132y$; $(0, 0)$, $(9, 0)$, $(0, 3)$, $(7, 2)$

13. $P = \$15x + \$5y$; $(0, 0)$, $(4, 8)$, $(0, 6)$, $(7, 0)$

14. $P = \$5x + \$20y$; $(0, 0)$, $(2, 7)$, $(0, 10)$, $(9, 0)$

15. $P = \$120x + \$340y$; $(0, 0)$, $(20, 50)$, $(0, 62)$, $(43, 0)$

16. $P = \$92x + \$45y$; $(0, 0)$, $(20, 13)$, $(0, 18)$, $(25, 0)$

Real World Applications

17. A store owner sells televisions and VCRs, but at most, he can stock 30 of such appliances, in any combination. The profit on the sale of a television set is $40.00, and the profit on a VCR is $60.00. The owner wishes to carry at least twice as many television sets as VCRs. How many of each should the owner stock to maximize the profit?

18. A small bakery makes cakes and pies every business day. They can make a total of 30 items of which at least 5 must be cakes and at least 10 must be pies for their restaurant customers. The profit on each cake is $1.50, and the profit on each pie is $2.00. How many of each should be made in order to maximize the profit?

19. For Easter, a person raises rabbits and chicks. He can raise at most 50 animals. It costs him $4.00 to raise a rabbit and $3.00 to raise a chick. He has saved $360.00 for the costs of raising the animals. When he sells the animals, he makes a profit of $5.00 on each rabbit and $3.00 on each chick. How many of each should he raise in order to maximize his profit?

20. A toy maker makes wooden toy trains and wooden toy trucks. It takes 5 hours to cut and assemble a toy train and 2 hours to paint it by hand. It takes 3 hours to cut and assemble a toy truck and 1 hour to paint it by hand. When he sells a train, he makes a profit of $12.00, and when he sells a truck, he makes a profit of $6.00. How many of each should he make within a 40-hour week, if he does not want any unfinished toys at the end of the week? (Assume he wants a maximum profit.)

21. On a small 50-acre farm, a person grows beans and corn. It costs the person $40.00 to grow an acre of beans and $30.00 to grow an acre of corn. The owner has set aside $3200.00 to cover the cost of growing the items. The profit for selling the beans grown on 1 acre is $65.00, and the profit for selling the corn grown on 1 acre is $40.00. How many acres of each vegetable should be grown in order to maximize the profit?

22. A school theater has 120 seats. The theater manager wishes to reserve some seats for students and some seats for nonstudents (i.e., the general public). Students pay $4.00 per ticket, and nonstudents pay $8.00 per ticket. The school has a rule that at least twice as many seats for students have to be reserved as seats for nonstudents. In order to maximize the profit, how many student seats should be reserved?

Writing Exercises

23. Explain the purpose of linear programming.

24. What is a constraint?

25. What is an objective function?

26. What does a polygonal region show?

27. Explain how the vertices of the polygonal region are used to find the maximum or minimum profit.

Critical Thinking

28. Consider a scenario where you would have to use linear programming to solve a problem in your field of study. Write the constraints and a profit function. Then solve the problem as shown in this section.

8-5 Functions

Relations

Many applications of mathematics use the concepts of *relations* and *functions*.

> A **relation** is a set of ordered pairs of elements.

The set $\{(0, 1), (1, 2), (2, 3)\}$ is an example of a relation. The elements used are numbers. The set $\{(a, b), (a, c), (a, d), (a, e)\}$ is also a relation, and the elements are letters.

The **domain** of a relation is the set of first elements of each of the ordered pairs. The **range** of a relation is the set of second elements of each of the ordered pairs. In the relation $\{(0, 1), (1, 2), (2, 3)\}$, the domain is $\{0, 1, 2\}$ and the range is $\{1, 2, 3\}$. In the relation $\{(a, b), (a, c), (a, d), (a, e)\}$, the domain is $\{a\}$, and the range is $\{b, c, d, e\}$.

In addition to expressing relations as a set of ordered pairs of elements, relations can also be expressed by using equations or by using graphs. For example, the relation $\{(0, 1), (1, 2), (2, 3)\}$ can be expressed as $y = x + 1$, where $x \in \{0, 1, 2\}$. In other words, when 0, 1, and 2 are substituted in the equation for x, the corresponding y values are 1, 2, and 3. This relation can also be shown by a graph. See Figure 8-26.

Functions

A special type of relation is called a *function*.

> A relation is a **function** if for each element in the domain there is a unique element in the range.

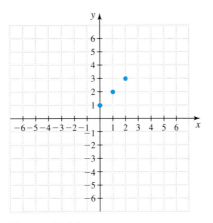

Figure 8-26

The relation $\{(0, 1), (1, 2), (2, 3)\}$ is a function since for each element in the domain, there is a unique element in the range. On the other hand, the relation $\{(a, b), (a, c), (a, d), (a, e)\}$ is *not* a function since for the first element, a, there are several second elements, namely, b, c, d, and e.

Example   8-26

Which of these relations are functions?

(a) $\{(2, 10), (5, 15), (7, 20)\}$
(b) $\{(a, 1), (b, 2), (c, 3), (a, 2)\}$
(c) $\{(3, 7)\}$
(d) $\{(4, 1), (4, 2), (4, 3)\}$
(e) $\{(1, a), (2, a), (3, a)\}$

Solution

(a) Function
(b) Not a function since "a" is paired with 2 elements, namely, 1 and 2.
(c) Function
(d) Not a function since 4 is paired with 1, 2, and 3.
(e) Function

Try This One

8-N   Which of these relations are functions?

(a) $\{(a, 5), (a, 6), (b, 1), (b, 2)\}$

(b) $\{(a, 10), (b, 10), (c, 10)\}$

(c) $\{(1, 1), (2, 2), (3, 3), (4, 4), (5, 5)\}$

For the answer, see page 400.

Since functions are relations, the set of first elements of the ordered pairs is called the *domain* of the function, and the set of second elements of the set of ordered pairs is called the *range* of the function.

Example   8-27

State the domain and range of each function.

(a) $\{(12, 24), (14, 26), (16, 28), (18, 30)\}$
(b) $\{(a, b), (c, d), (e, f)\}$
(c) $\{(a, 3), (b, 3), (c, 3), (d, 3)\}$

Solution

(a) Domain $= \{12, 14, 16, 18\}$; range $= \{24, 26, 28, 30\}$
(b) Domain $= \{a, c, e\}$; range $= \{b, d, f\}$
(c) Domain $= \{a, b, c, d\}$; range $= \{3\}$

If your grade on a test is a function of how long you spend studying, then the domain is the possible study times, and the range is the possible exam grades.

The Vertical Line Test

Recall that some relations are functions. The **vertical line test** can be used to determine whether the graph of a relation is a function.

The Vertical Line Test for Functions

If no vertical line can intersect the graph of a relation at more than one point, then the relation is a function.

A relation is a function if for each value x in the domain, there is only one value y in the range. By looking at the graph, if a vertical line intersects the relation at two or more places, then there are two or more values in the range for one specific value x. Hence, the relation cannot be a function. For example, the relation shown in Figure 8-27(a) is a function since a vertical line can be drawn through each point of the relation and does not intersect any other point of the relation at the same time. The relation shown in Figure 8-27(b) is not a function since it is possible to draw a vertical line through at least one point of the relation that intersects another point of the relation at the same time.

Like relations, some functions can be expressed using set-builder notation (see Chapter 2) and equations. For example $G = \{(x, y) \mid y = 3x\}$. In this case, the function consists of all ordered pairs of numbers (x, y) that satisfy the equation $y = 3x$. Some ordered pairs that satisfy the condition are $(0, 0)$, $(1, 3)$, $(2, 6)$, $(-1, -3)$, and $(-2, -6)$. *Unless otherwise specified, the domain is the set of real numbers that when substituted for x gives real number values for y.* In this case, the domain would be all real numbers, since any real number value for x would give a real number value for y.

The domain of the function $H = \left\{(x, y) \mid y = \dfrac{1}{x}\right\}$ would be all real numbers except $x = 0$, since substituting 0 for x in the equation yields an undefined value for y.

When equations are used to express functions, instead of using set-builder notation, a special notation called *functional notation* is used. Some equations can be expressed as functions by solving for y in terms of x and then replacing y by $f(x)$, which is read "f of x." To express $2x + y = 1$ as a function, solve for y, as shown.

$$2x + y = 1$$
$$2x - 2x + y = -2x + 1$$
$$y = -2x + 1$$

> **Math Note**
>
> The function notation $f(x)$ does *not* mean f times x; it is used to designate a function, and it means that y is a function of x. That is, y is defined in terms of x.

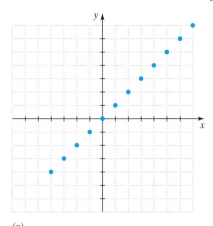

(a)

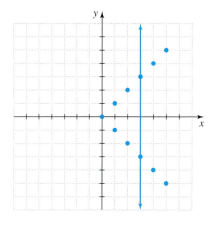

Figure 8-27

Then replace y by $f(x)$.

$$f(x) = -2x + 1$$

Hence, instead of using set-builder notation, $F = \{(x, y) \mid y = -2x + 1\}$, the notation $f(x) = -2x + 1$ is used.

A function can be evaluated for a specific value of x by substituting the value for x in the function and simplifying using the order of operations. Example 8-28 shows this.

Example 8-28

Evaluate the function $f(x) = -2x + 1$ for

(a) $x = 2$.
(b) $x = -5$.
(c) $x = 0$.

Solution

(a) Substitute 2 for x and simplify as shown

$$f(x) = -2x + 1$$
$$f(2) = -2(2) + 1$$
$$= -4 + 1$$
$$= -3$$

Hence, $f(2) = -3$. The notation $f(2)$ means the value of the function when $x = 2$.

(b) Substitute -5 for x and simplify

$$f(x) = -2x + 1$$
$$f(-5) = -2(-5) + 1$$
$$= 10 + 1$$
$$= 11$$

Hence, $f(-5) = 11$.

(c) Substitute 0 for x and simplify

$$f(x) = -2x + 1$$
$$f(0) = -2(0) + 1$$
$$= 0 + 1$$
$$= 1$$

Hence, $f(0) = 1$.

Graphing Linear Functions

The graph of a **linear function** can be drawn by using the same procedures as drawing a line.

Procedure for Graphing a Linear Function

Step 1 Select at least 3 values for *x*. (Only two values are necessary. The third is used as a check.)

Step 2 Substitute them in the function and find the corresponding values for *f*(*x*).

Step 3 Plot the points (*x*, *f*(*x*)) on the Cartesian plane using the *y* axis as the *f*(*x*) axis.

Step 4 Draw a line through the points.

Example 8-29

Draw the graph for the function $f(x) = 5x - 3$.

Solution

Step 1 Select $x = 2$, $x = 0$, and $x = -1$.

Step 2 Find $f(x)$ for $x = 2$, $x = 0$, and $x = -1$.

$$f(2) = 5(2) - 3 \qquad f(0) = 5(0) - 3 \qquad f(-1) = 5(-1) - 3$$
$$f(2) = 10 - 3 \qquad f(0) = 0 - 3 \qquad f(-1) = -5 - 3$$
$$f(2) = 7 \qquad f(0) = -3 \qquad f(-1) = -8$$

Step 3 Plot the points $(2, 7)$, $(0, -3)$, and $(-1, -8)$ on the graph.

Step 4 Draw a line through the points. See Figure 8-28.

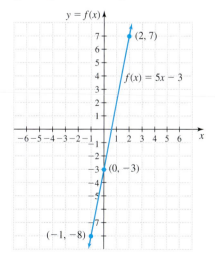

Figure 8-28

Try This One

8-O Draw a graph for $f(x) = -3x + 4$.
For the answer, see page 400.

Quadratic Functions

In Chapter 7, quadratic equations were explained. A quadratic equation has the form $ax^2 + bx + c = 0$. A quadratic function is defined as follows.

> An equation of the form $f(x) = ax^2 + bx + c$, where a, b, and c are real numbers and $a \neq 0$, is called a **quadratic function.** This equation can also be written as $y = ax^2 + bx + c$. The graph of a quadratic function is called a **parabola.**

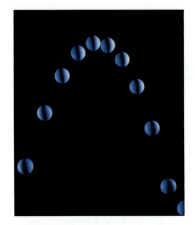

Parabolas often occur in problems about the motion of a falling object.

A parabola opens upward if $a > 0$ (i.e., a is positive); it opens downward if $a < 0$ (i.e., a is negative). See Figure 8-29. The lowest point on a parabola that opens upward is called the **vertex.** If a parabola opens downward, then the vertex is the highest point on a parabola. The point or points, if they exist where a parabola crosses the x axis, are called the x *intercepts*. The point where a parabola crosses the y axis is called the y *intercept*. A vertical line drawn through the vertex of a parabola is called the **axis of symmetry.** The axis of symmetry is not actually a part of a parabola; however, we say that the graph of a parabola is symmetric about the axis, which means one side is a mirror image of the other side. See Figure 8-29.

Graphing the Parabola

The graph of a parabola can be drawn by plotting enough points to determine its shape and then drawing a smooth curve through the points. However, it is easier to use the vertex, intercepts, and axis and a few selected points.

The vertex of a parabola can be found by substituting the values of a and b into the formula $x = \dfrac{-b}{2a}$, then substituting this value in the equation and finding the corresponding value for y. Recall that a is the coefficient of the x^2 term, and b is the coefficient of the x term. For example, the vertex of the parabola $y = x^2 - 6x + 5$, where $a = 1$ and $b = -6$, can be found as follows:

$$x = \frac{-b}{2a}$$
$$= \frac{-(-6)}{2(1)}$$
$$= 3$$

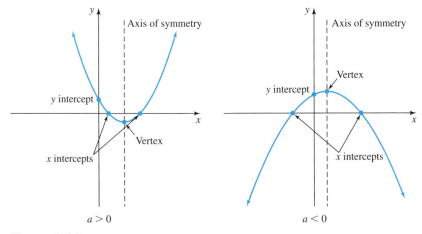

Figure 8-29

The y coordinate of the vertex can be found by substituting the value 3 in the equation $y = x^2 - 6x + 5$.

$$y = (3)^2 - 6(3) + 5$$
$$= 3^2 - 6 \cdot 3 + 5$$
$$= 9 - 18 + 5$$
$$= -4$$

Hence, the vertex is $(3, -4)$.

To find the y intercept of the parabola, substitute 0 for x in the equation and find the value for y.

$$y = x^2 - 6x + 5$$
$$= 0^2 - 6(0) + 5$$
$$= 5$$

Hence, the y intercept is $(0, 5)$.

[*Note:* This is the same as using the coordinates $(0, c)$ for the y intercept.]

To find the x intercept, set $y = 0$ and solve the quadratic equation for x. There will either be two, one, or no x intercepts, depending on the nature of the parabola.

For the parabola $y = x^2 - 6x + 5$, the intercepts can be found by solving the equation $0 = x^2 - 6x + 5$. Factoring as shown can solve this quadratic equation.

$$0 = x^2 - 6x + 5$$
$$0 = (x - 5)(x - 1)$$

$0 = x - 5$	$0 = x - 1$
$5 = x$ or	$1 = x$ or
$x = 5$	$x = 1$

Hence, the x intercepts are $(5, 0)$ and $(1, 0)$.

Next, a few select points can be found, in order to help determine the shape of the parabola, by selecting values for x, substituting in the equation, and solving for y. Select $x = 2$.

$$y = x^2 - 6x + 5$$
$$= (2)^2 - 6(2) + 5$$
$$= 4 - 12 + 5$$
$$= -3$$

Select $x = 4$.

$$y = (4)^2 - 6(4) + 5$$
$$= 16 - 24 + 5$$
$$= -3$$

Now plot the points $(3, -4)$, $(0, 5)$, $(5, 0)$, $(1, 0)$, $(2, -3)$, and $(4, -3)$. Also, because $a = 1$, the parabola opens upward (i.e., $a > 0$). Finally, draw a smooth curve through the points. See Figure 8-30.

The procedure for graphing a quadratic function is summarized next.

Math Note

Since the axis of symmetry is a vertical line and passes through the vertex, its equation is $x = 3$.

Math Note

If the equation cannot be solved by factoring, use the quadratic formula. If $b^2 - 4ac$ is negative, the parabola has no x intercepts since the square root of a negative number is not a real number.

Calculator Explorations

What can be said about the points on either side of the vertex in this parabola?

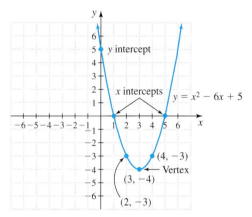

Figure 8-30

Procedure for Graphing the Quadratic Function $y = ax^2 + bx + c$

Step 1 Identify a, b, and c, then find the vertex using $x = \frac{-b}{2a}$ to get the x coordinate. Then substitute this value in the equation $y = ax^2 + bx + c$ to get the y coordinate.

Step 2 Find the y intercept by substituting $x = 0$ in the equation and solving for y or use $(0, c)$.

Step 3 Find the x intercepts by substituting 0 for y and solving the equation for x, either by factoring or by using the quadratic formula.

Step 4 Find several other points in order to determine the shape.

Step 5 Determine whether the parabola opens upward ($a > 0$) or downward ($a < 0$). Plot the points and draw a smooth curve through the points.

Example 8-30

Find the vertex and the x intercepts for $y = x^2 + 16x + 8$.

Solution

$a = 1$, $b = 16$, and $c = 8$. The x coordinate of the vertex is $x = \dfrac{-b}{2a}$.

$$x = \frac{-16}{2 \cdot 1} = \frac{-16}{2} = -8$$

The y coordinate is

$$y = (-8)^2 + 16(-8) + 8$$
$$= 64 - 128 + 8$$
$$= -56$$

The coordinates of the vertex are $(-8, -56)$. To find the x intercepts, substitute $y = 0$ in the equation and solve for x.

$$y = x^2 + 16x + 8$$
$$0 = x^2 + 16x + 8$$

—Continued

Example 8-30 Continued—

Since the right side of the equation cannot be factored, x intercepts are found by using the quadratic formula. (See Chapter 7.)

$$x = \frac{-b \pm \sqrt{b^2 - 4ac}}{2a}$$

$$x = \frac{-16 \pm \sqrt{(16)^2 - 4(1)(8)}}{2}$$

$$x = \frac{-16 \pm \sqrt{224}}{2}$$

$$x \approx \frac{-16 + 14.97}{2} \quad \text{or} \quad x \approx \frac{-16 - 14.97}{2}$$

$$x \approx -0.515 \quad \text{or} \quad x \approx -15.485$$

Try This One

8-P   Graph $y = x^2 - 3x - 10$.
For the answer, see page 400.

Sidelight

Mathematics and Sea Life

One of the most unusual seashells is the one created by the chambered nautilus, a sea creature that lives in the South Pacific Ocean. When the animal outgrows its shell, it adds a new larger chamber to the old shell, creating a near-perfect mathematical spiral. It lives in the outer chamber. Lines drawn from the center to the ends of the chambers form equal angles.

One of the earliest mathematicians to study spirals was Archimedes (287–212 B.C.E.), who wrote a book on spirals.

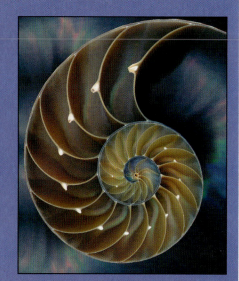

Applications of Quadratic Functions

In order to solve real-life problems using quadratic functions, the first thing to do is to write the equation for the function in the form $f(x) = ax^2 + bx + c$. If the problem asks for a maximum or minimum value, use the vertex $\left(\dfrac{-b}{2a}, f\left(\dfrac{-b}{2a}\right)\right)$ to find this value. Other problems have solutions that involve finding the x intercepts of the function.

Example 8-31

A husband and wife wish to fence in a rectangular area in their backyard for their children to play. The family has 60 feet of fencing, and only three sides will be fenced in since the house will provide the fourth side. Find the quadratic function that describes the situation and then find the dimensions so that the maximum area will be enclosed.

Solution

Let x = the length of each of the two equal sides. Then $60 - 2x$ will be the length of the third side since the total amount of fencing is 60 feet and $2x$ feet will be needed for the other two sides. See Figure 8-31(a). The area of the plot (length × width) is given by the function $f(x) = x(60 - 2x)$.

To find the maximum area, find the coordinates of the vertex since it is the highest point on the graph. See Figure 8-31(b).

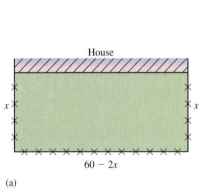

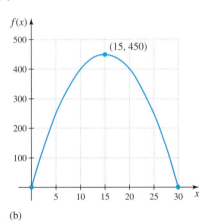

(a)   (b)

Figure 8-31

$$f(x) = x(60 - 2x)$$
$$= 60x - 2x^2$$
$$= -2x^2 + 60x$$

The x coordinate of the vertex is

$$x = -\frac{b}{2a}$$

$$= -\frac{60}{2 \cdot (-2)} = \frac{60}{4}$$

$$= 15$$

—Continued

Example   8-31   *Continued—*

Hence, 15 is the length of each of the two equal sides. The other side can be found by finding the value of

$$60 - 2x = 60 - 2(15)$$
$$= 60 - 30$$
$$= 30$$

Hence the dimensions of the area would be 15 feet, 30 feet, and 15 feet.

To find the area, substitute 15 in the equation and evaluate.

$$f(x) = -2x^2 + 60x$$
$$f(15) = -2(15)^2 + 60(15)$$
$$= -450 + 900$$
$$= 450$$

The maximum area fenced in is 450 square feet.

Calculator Explorations

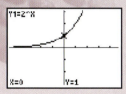

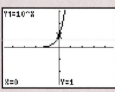

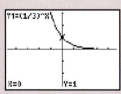

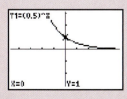

What is the *y* intercept for each graph?

Try This One

8-Q   A wholesaler sells calculators to a retail store. The profit on the calculators is given by the function $f(x) = 36x - 0.09x^2$, where x is the number of calculators purchased. What is the size of the order that would produce maximum profit for the wholesaler?

For the answer, see page 400.

Exponential Functions

Exponential functions can be used to describe many real-life situations, such as investments, growth and decay problems, depreciation, and population changes.

> An **exponential function** has the form $f(x) = a^x$, where a and x are real numbers such that $a > 0$ but $a \neq 1$.

Examples of exponential functions are

$$f(x) = 2^x \qquad\qquad f(x) = 10^x$$
$$f(x) = \left(\tfrac{1}{3}\right)^x \qquad f(x) = (0.5)^x$$

The graph of an exponential function has two forms.

1. When $a > 1$, the function increases as x increases. See Figure 8-32(a).
2. When $0 < a < 1$, the function decreases as x increases. See Figure 8-32(b).

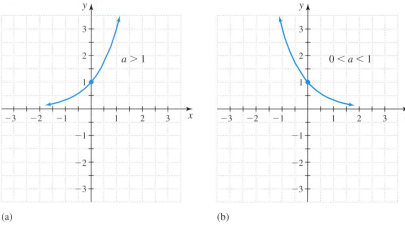

(a)   (b)

Figure 8-32

Notice in both cases the curve passes through $(0, 1)$. That is, the point $(0, 1)$ is the y intercept. Also, the curve approaches the x axis but never touches it. When this occurs, the x axis is said to be a *horizontal* **asymptote.**

    To draw the graph of an exponential function, select several points for x, find $f(x)$, then plot the graph using a shape similar to those shown in Figure 8-32.

Example   8-32

Draw the graph for each.

(a) $f(x) = 5^x$
(b) $f(x) = 3^{x+1}$

Solution

(a) Select several numbers for x and find $f(x)$.

    Let $x = -2$, then $f(-2) = 5^{-2} = \frac{1}{5^2} = \frac{1}{25} = 0.04$.

    Let $x = -1$, then $f(-1) = 5^{-1} = \frac{1}{5} = 0.2$.

    Let $x = 0$, then $f(0) = 5^0 = 1$.

    Let $x = 2$, then $f(2) = 5^2 = 25$.

    Plot the points, and then connect them with a smooth curve. See Figure 8-33(a).

(b) Select several numbers for x and find $f(x)$.

    Let $x = -3$, then $f(x) = 3^{-3+1} = 3^{-2} = \frac{1}{9}$.

    Let $x = -2$, then $f(x) = 3^{-2+1} = 3^{-1} = \frac{1}{3}$.

    Let $x = -1$, then $f(x) = 3^{-1+1} = 3^0 = 1$.

    Let $x = 0$, then $f(x) = 3^{0+1} = 3^1 = 3$.

    Let $x = 1$, then $f(x) = 3^{1+2} = 3^2 = 9$.

—Continued

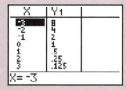

What do the two graphs have in common? How are they different?

Example 8-32 *Continued—*

Plot the points and draw the graph as shown in Figure 8-33(b).

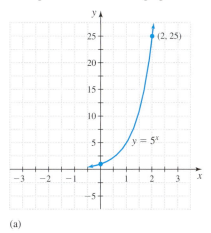

(a)

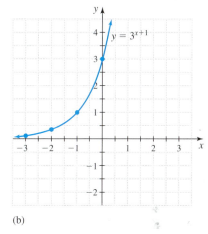

(b)

Figure 8-33

Example 8-33

Draw the graph for $f(x) = \left(\frac{1}{2}\right)^x$.

Solution

Select several points for x and find $f(x)$.

For $x = -2$, $f(-2) = \left(\frac{1}{2}\right)^{-2} = 4$.

For $x = -1$, $f(-1) = \left(\frac{1}{2}\right)^{-1} = 2$.

For $x = 0$, $f(0) = \left(\frac{1}{2}\right)^{0} = 1$.

For $x = 1$, $f(1) = \left(\frac{1}{2}\right)^{1} = 0.5$.

For $x = 2$, $f(2) = \left(\frac{1}{2}\right)^{2} = 0.25$.

Plot the points and draw the graph as shown in Figure 8-34.

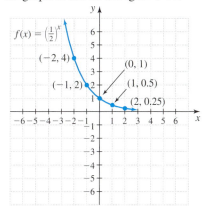

Figure 8-34

Try This One

8-R Draw the graph for each function.

(a) $f(x) = 2^x$

(b) $f(x) = \left(\frac{1}{3}\right)^x$

For the answer, see page 400.

Applications of Exponential Functions

As stated previously, exponential functions can be used to describe many real-life situations, including investments, growth and decay problems, depreciation, and population changes. Examples 8-34, 8-35, and 8-36 show some applications of exponential functions.

Example 8-34

The population growth of a certain geographic region follows a pattern $f(x) = A_0 (1.2)^t$, where A_0 is the population when $t = 0$, and t is the time in decades (i.e., 10 years). If the current population is 2 million, find the population in 30 years.

Solution

Let $A_0 = 2,000,000$

$\qquad t = 3$ (i.e., 30 years = 3 decades)

then

$$f(3) = 2,000,000(1.2)^3$$
$$= 3,456,000$$

In 30 years, the population will be 3,456,000.

It is often said that people would save more if they understood how compound interest worked. Use the formula in Example 8-35 to solve the following: At age 25, you invest your savings of $10,000 at 8% interest compounded quarterly. How much do you have when you are 65?

Example 8-35

Interest on a savings account can be compounded annually, semiannually, quarterly, or daily. The formula for compound interest is given by

$$A = P\left(1 + \frac{r}{n}\right)^{nt}$$

where A is the amount of money, which includes the principal plus the earned interest, P is the principal (amount initially invested), r is the yearly interest rate, n is the number of times the interest is compounded a year, and t is the time in years that the principal has been invested. If $3000 is invested at 8% per year compounded quarterly, find the value of the investment (amount) at the end of 5 years.

Solution

$$P = \$3000$$
$$r = 8\% = 0.08$$
$$t = 5 \text{ years}$$

—Continued

Example 8-35 *Continued—*

$n = 4$, since the interest is compounded quarterly or four times a year.

$$A = P\left(1 + \frac{r}{n}\right)^{nt}$$

$$= 3000\left(1 + \frac{0.08}{4}\right)^{4 \cdot 5}$$

$$= \$4457.84$$

At the end of 5 years, the $3000 will grow to $4457.84.

Example 8-36

An isotope of antimony decays (i.e., disintegrates) according to the function $F(x) = A_0 2^{-0.0167x}$, where A_0 is the initial mass of the element and x is the time in days. If there is one milligram of antimony now, find the number of milligrams that will be left after 60 days.

Solution

Substitute $A_0 = 1$ and $x = 60$ into the function.

$$F(x) = A_0 2^{-0.0167(x)}$$

$$= 1 \cdot 2^{-0.0167(60)}$$

$$\approx 0.499 \text{ mg}$$

Radioactive decay is used to date objects.

Try This One

8-S Find the value of a $5000 investment compounded semiannually (i.e., twice a year) for 2 years at 6% interest.

For the answer, see page 400.

Exercise Set 8-5

Computational Exercises

For Exercises 1–8, find the domain and range for each relation and state whether or not the relation is a function.

1. $\{(5, 8), (6, 9), (7, 10), (8, 11)\}$
2. $\{(0, 1), (0, 2), (0, 3), (0, 4), (0, 5)\}$
3. $\{(6, 11), (7, 11), (8, 11), (9, 11)\}$

4. $\{(-2, -3)\}$

5. $\{(0, 0)\}$

6. $\{(-2, 2), (-3, 4), (-4, 5), (-5, 6)\}$

7. $\{(-10, 20), (-10, 40), (-40, 60), (-60, 60)\}$

8. $\{(5, 8), (5, 10), (5, 12)\}$

For Exercises 9–16, evaluate each function for the specific value.

9. $f(x) = 3x + 8$ for $x = 3$

10. $f(x) = -2x + 5$ for $x = -5$

11. $f(x) = 4x - 8$ for $x = -10$

12. $f(x) = 6x^2 - 2x + 5$ for $x = 2$

13. $f(x) = 8x^2 + 3x$ for $x = 0$

14. $f(x) = -3x^2 + 5$ for $x = 1.5$

15. $f(x) = x^2 + 4x + 7$ for $x = -3.6$

16. $f(x) = -x^2 + 6x - 3$ for $x = 0$

For Exercises 17–20, graph each function.

17. $f(x) = 7x - 8$

18. $f(x) = 6x$

19. $f(x) = -3x + 2$

20. $f(x) = -5x + 1$

For Exercises 21–28, graph each parabola.

21. $y = x^2$

22. $y = x^2 - 6x$

23. $y = x^2 + 6x + 9$

24. $y = 4x^2 - 4x + 1$

25. $y = -x^2 + 12x - 36$

26. $y = -2x^2 + 3x + 4$

27. $y = -10x^2 + 20x$

28. $y = -3x^2 + 5x + 2$

For Exercises 29–32, graph each exponential function.

29. $y = 5^x$

30. $y = 3^{x+1}$

31. $y = \left(\frac{1}{2}\right)^{x-2}$

32. $y = \left(\frac{1}{4}\right)^{x+1}$

For Exercises 33–38, use the vertical line test to determine whether or not each relation is a function.

33.

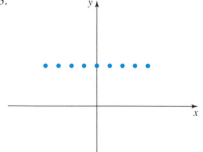

34.

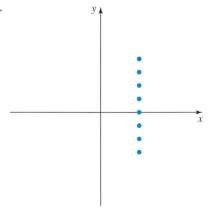

35.
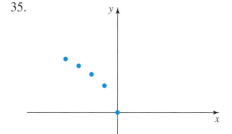

36.
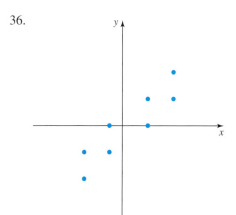

37.
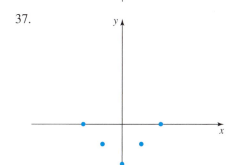

38.
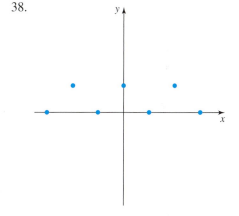

Real World Applications

39. If a ball is thrown vertically up from a height of 6 feet with an initial velocity of 60 feet per second, then the height of the ball t seconds after it is thrown is given by $f(t) = -16t^2 + 60t + 6$. Find the maximum height that the ball will attain and find the number of seconds that will elapse from the moment the ball is thrown to the moment it hits the ground.

40. A vendor sells boxes of computer paper. The amount of revenue made per week is given by the function $f(x) = 12x - 0.1x^2$, where x is the number of boxes sold. How many boxes should be sold if the vendor is to make a maximum profit?

41. Find the length of the sides of a gutter consisting of three sides that can be made from a piece of aluminum that is 16 inches wide in order for it to carry the maximum capacity of water. (The shape of the gutter is ⊔.)

42. A stone is dropped from a bridge, and 3 seconds later a splash is heard. How high is the bridge? Use $f(x) = -16x^2$, where $f(x)$ is the height of the bridge.

43. The population growth of a certain geographic region is defined by the function $f(t) = A_0(1.4)^t$, where A_0 is the present population and t is the time in decades. If the present population is 4,000,000, find the population in 5 years.

44. Using the formula $A = P\left(1 + \dfrac{r}{n}\right)^{nt}$ find the worth of a $500 bond that collects interest at a rate of 6% calculated quarterly and is held for 3 years.

45. An isotope decays according to the function $F(x) = A_0 2^{-0.23x}$, where A_0 is the present mass of the element and x is the time in days. If there are 5 milligrams of the isotope, find the number of milligrams after 20 days.

46. Interest can be compounded daily. The formula for interest which is compounded daily is $I = P\left(1 + \dfrac{R}{365}\right)^N - P$, where P is the principal, R is the rate, and N is the number of days. Using this formula, find the interest on $1200 at 7% for 30 days.

Writing Exercises

47. Explain the difference between a relation and a function.
48. What is meant by the domain of a function?
49. What is meant by the range of a function?
50. What is a linear function?
51. What is a parabola?
52. What is the vertex of a parabola?
53. How can you tell if a parabola opens upward?
54. What is a quadratic function?
55. Explain two applications of exponential functions.

Critical Thinking

For Exercises 56 and 57, graph each relation by plotting points and explain the nature of the graph.

56. $x = y^2 - 6y - 7$
57. $x = -y^2 + 2y + 4$

Summary

Section	Important Terms	Important Ideas
8-1	rectangular coordinate system Cartesian plane x axis y axis origin quadrants coordinates x intercept y intercept slope slope-intercept form	**The** rectangular coordinate system (also called the Cartesian plane) consists of two number lines, one vertical axis called the y axis and one horizontal axis called the x axis. The axes divide the plane into four quadrants. A point is located on the plane by its coordinates that consist of an ordered pair of numbers (x, y). A line on the plane can be represented by an equation of the form $ax + by = c$. The slope of a line is defined as the rise divided by the run. The slope of a horizontal line is zero. The slope of a vertical line is undefined. The point where a line crosses the x axis is called the x intercept. The point where a line crosses the y axis is called the y intercept.
8-2	system of linear equations consistent system independent system inconsistent system dependent system substitution method addition/subtraction method	**A** system of linear equations consists of two or more linear equations in two variables. The solution to a system of linear equations is a value of x and a value of y that, when substituted in both equations, makes them closed true equations. Systems can be solved graphically by finding the point of intersection of the lines, or algebraically using the substitution method or the addition/subtraction method. If a system has only one solution, it is said to be consistent, and the solution is the coordinates of the point of intersection of the lines. If the lines are parallel, the system is said to be inconsistent, and the solution is the empty set. If the lines coincide, the system is said to be dependent and every point on the line is a solution.
8-3	linear inequality half plane system of linear inequalities	**A** system of linear inequalities consists of two or more linear inequalities in two variables. The solution to a system of linear inequalities is found by graphing the linear inequalities and finding the intersection of the half planes.
8-4	linear programming constraint polygonal region objective function	**Linear** programming is one real application that uses a system of linear inequalities. Solving a problem using linear programming consists of representing the constraints as linear inequalities and the relationship of the variables using an objective function. When the inequalities are graphed, a polygonal region is formed and the coordinates of the vertices are substituted in the objective function. A decision is then made on the maximum or minimum values of the objective function.
8-5	relations domain range function vertical line test linear function quadratic function parabola vertex axis of symmetry exponential function asymptote	**A** relation is a set of ordered pairs of elements. A relation is a function if for each element in the domain, there is a unique element in the range. The vertical line test can be used to determine if a relation is a function. Various functions such as linear, quadratic, and exponential functions are explained here. The graph of a quadratic function is called a parabola. The parabola has a vertex and an axis of symmetry.

Review Exercises

For Exercises 1–6, draw the graph for each line.

1. $4x - y = 8$

2. $y = 2x + 6$

3. $x = -5$

4. $y = 8$

5. $y = -3x + 10$

6. $x - 6y = 11$

For Exercises 7–12, find the slope of the line containing the two given points.

7. $A(-8, 6)$, $B(4, 3)$

8. $A(3, 8)$, $B(-2, 6)$

9. $A(-5, -6)$, $B(-3, 8)$

10. $A(2, 6)$, $B(2, 10)$

11. $A(5, 9)$, $B(-3, 9)$

12. $A(-3, -8)$, $B(6, -2)$

For Exercises 13–16, write the equation in slope-intercept form and find the slope, x intercept, and y intercept, and graph the line.

13. $3x + y = 12$

14. $-2x + 8y = 15$

15. $4x - 7y = 28$

16. $x - 3y = 9$

17. Mary has $16.72 more than Betty. Together they have $32.00. How much does each have?

18. A person invested a certain sum of money at 5% and twice as much at 8%. If the combined interest is $89.00, how much money does the person have in each investment?

For Exercises 19–22, solve each system graphically.

19. $x - 2y = 6$
    $2x - y = 18$

20. $x + y = 9$
    $x - y = 1$

21. $2x + y = 12$
    $5x - 2y = 21$

22. $4x - y = 10$
    $8x - 2y = 20$

For Exercises 23–26, solve each system by substitution.

23. $x + 6y = 4$
    $x - y = 11$

24. $3x - y = 10$
    $x - 2y = 5$

25. $x - y = 3$
    $x - 3y = -3$

26. $3x + 2y = 10$
    $2x - 3y = 11$

For Exercises 27–30, solve each system by addition/-subtraction (elimination).

27. $3x - y = 12$
    $x + 2y = 4$

28. $5x - 6y = 15$
    $10x - 12y = 21$

29. $4x - 2y = 11$
    $5x - y = 10$

30. $6x - y = 15$
    $2x + 5y = 21$

31. Three pounds of coffee and 4 pounds of tea cost $14.50. Five pounds of coffee and 2 pounds of tea cost $16.00. What is the cost per pound of the coffee and the tea?

32. A merchant has invested money in the partial ownership of two stores. Last year, store 1 earned 10% of the money invested in it while store 2 lost 8% of the amount invested in it. The net total gain was $1300.00. This year, the respective quantities were +6%, +5%, and $2250.00. Find the merchant's investment for each store.

For Exercises 33–36, graph each.

33. $x - 19y \geq 5$

34. $2x + 7y < 14$

35. $-3x + 10y \leq -15$

36. $x \geq -5$

For Exercises 37–40, find the solution set to each system by graphing.

37. $5x - y > 10$
    $2x + 3y \leq 12$

38. $6x + 2y \leq 15$
    $3x - y > 6$

39. $12x + 3y \geq 24$
    $-x + y \leq 5$

40. $3x + y \geq -4$
    $x - y < 0$

41. A manufacturer produces children's scooters in two models. Model A takes 6 hours to manufacture and 2 hours to paint. Model B takes 5 hours to manufacture and 1 hour to paint. The assembly plant works 160 hours and the painting department works 100 hours. If the profit for model A is $20.00 and for model B is $10.00, how many models of each should be produced to maximize the profit?

42. An appliance store owner stocks two types of refrigerators and has warehouse space for at most 36 units. The owner keeps at least twice as many of model I as model II. If the profit on model I is $24.00 and the profit on model II is $36.00, how many of each model should the owner carry to maximize profit?

For Exercises 43–46, find the domain and range for each relation and state whether or not the relation is a function.

43. $\{(2, 5), (5, -7), (6, -10)\}$

44. $\{(-1, 5), (2, 6), (-1, 3)\}$

45. $\{(4, 10), (5, 10), (6, 10)\}$

46. $\{(2, -3)\}$

For Exercises 47–50, evaluate each function for the specific value.

47. $f(x) = 5x - 12$ for $x = 5$

48. $f(x) = -2x + 10$ for $x = -3$

49. $f(x) = x^2 + 7x + 10$ for $x = -10$

50. $f(x) = 2x^2 - 3x$ for $x = 8$

For Exercises 51–54, graph each parabola.

51. $y = x^2 + 10x + 25$ 52. $y = 3x^2 - 4x - 4$

53. $y = -x^2 + 25$ 54. $y = -6x^2 + 12x$

For Exercises 55 and 56, graph each.

55. $y = -3^x$ 56. $y = 6^x$

57. A ball is thrown up from a height of 4 feet with a velocity of 80 feet per second. Find the maximum height that the ball will attain.

58. An isotope decays according to the function $F(x) = A_0 2^{-0.5x}$. Find the number of units left after 10 days if there are 200 units to begin with ($A_0 = 200$ and $x = 10$).

Chapter Test

For Exercises 1–6, draw the graph for each.

1. $3x - 5y = -15$ 2. $y = 8x - 24$

3. $x = 0$ 4. $y = 0$

5. $2x + 3y = -8$ 6. $-4x - 5y = 20$

For Exercises 7 and 8, find the slope of the line containing the two points.

7. $A(-5, -10), B(2, 7)$ 8. $A(3, -6), B(4, 1)$

For Exercises 9 and 10, write each equation in the slope-intercept form and find the slope, x intercept, and y intercept and graph the line.

9. $x + 5y = 20$ 10. $2x - 11y = 22$

For Exercises 11–16, solve by any method you choose.

11. $x - 7y = 10$
    $2x + y = 5$

12. $4x + y = 15$
    $x - y = 8$

13. $3x + 4y = 19$
    $9x + 12y = 57$

14. $5x - y = 18$
    $2y = 10x + 36$

15. $x + 6y \geq 10$
    $6x - y \leq 23$

16. $x + y > 10$
    $x - y < 0$

For Exercises 17 and 18, find the domain and range and state whether or not each relation is a function.

17. $\{(-4, 6), (-10, 18), (12, 5)\}$

18. $\{(3, 7), (4, 8), (5, 9), (3, 6)\}$

For Exercises 19 and 20, evaluate each function for the specific value.

19. $f(x) = -3x + 10$ for $x = -15$

20. $f(x) = 2x^2 + 6x - 5$ for $x = 3$

For Exercises 21–24, graph each.

21. $y = x^2 - 9x + 14$ 22. $y = 2x^2 + x - 3$

23. $y = 2^{1.5x}$ 24. $y = -3^{-0.8x}$

25. A college algebra class of 57 students was divided into two sections. There were 5 more students in one section than in the other section. How many students were in each section?

26. A newsstand sells a certain number of *Daily News* newspapers for $0.35 and *Tribune* newspapers for $0.50. The gross amount made on both papers was $38.70. If the stand would sell twice as many *Tribunes* and half as many *Daily News* papers, the gross amount made would be $44.85. Find the number of each paper sold.

27. The wages per hour for three cooks and eight food servers is $66.50. The cooks receive $2.00 an hour more than the food servers. How much does each earn per hour?

28. A craftsperson makes Christmas wreaths and decorated snowmen. She takes 2 hours to assemble a wreath and 1 hour to decorate it. She takes 1 hour to assemble a snowman and 1 hour to decorate it. If she can work at most 24 hours a week and make a profit of $5.00 on a wreath and $3.00 on a snowman, how many of each should she make in a week in order to maximize her profit?

29. A contractor needs to make a rectangular drain consisting of three sides from a piece of aluminum that is 24 inches wide. Find the length of each side if it needs to carry the maximum capacity of water. The drain is ⊔ shaped.

30. Find the compound interest on a $4000 investment held for 5 years, if the rate is 8.5% compounded semiannually. Use $A = P\left(1 + \dfrac{r}{n}\right)^{nt}$.

Mathematics in Our World ▶ *Revisited*

Stopping Distance of an Automobile

The stopping distance in feet of an automobile consists of the sum of the distance an automobile travels during the reaction time and during the braking time. The distance an automobile travels during the reaction time can be found by using the formula

$$d_1 = 1.47rs$$

where r = reaction time in seconds
s = speed of the automobile in miles per hour

The distance an automobile travels during braking time can be found by using the formula

$$d_2 = \frac{(1.47S)^2}{2R}$$

where S = speed of the automobile in miles per hour
R = retardation of the automobile

The retardation of an automobile is defined as the number of feet per second that it takes the automobile to slow down once the brakes have been applied. The retardation factor depends on several factors. Two of the most important factors are the efficiency of the brakes and the condition of the road surface. (The computation of R is beyond the scope of this course and its value will be given.)

An average reaction time is 0.75 of a second, so the stopping distance for reaction time at 55 miles per hour is

$$d_1 = 1.47rs$$
$$= 1.47(0.75)(55)$$
$$\approx 60.6 \text{ feet}$$

Using 22 for R, the stopping distance for braking time at 55 miles per hour is

$$d_2 = \frac{(1.47S)^2}{2R}$$
$$= \frac{(1.47(55))^2}{2(22)}$$
$$\approx 148.6 \text{ feet}$$

Hence, the total stopping distance for an automobile moving at 55 miles per hour is about 60.6 feet + 148.6 feet = 209.2 feet.

For an automobile traveling at 70 miles per hour, the stopping distance for reaction time is

$$d_1 = 1.47rs$$
$$= 1.47(0.75)(70)$$
$$\approx 77.2 \text{ feet}$$

The stopping distance for braking time is

$$d_2 = \frac{(1.47S)^2}{2R}$$
$$= \frac{(1.47(70))^2}{2(22)}$$
$$\approx 240.6 \text{ feet}$$

Hence, the total stopping distance for an automobile moving at 70 miles per hour is about 77.2 feet + 240.6 feet = 317.8 feet.

In conclusion, the stopping distance of an automobile traveling at 70 miles per hour is over 100 feet more than the stopping distance for an automobile traveling at 55 miles per hour.

Projects

1. Using various resources such as almanacs, scientific or medical journals, the Internet, etc., select a topic such as the growth of world population, the occurrence of AIDS etc. and find a function that approximates the variable. Write a paper describing the function, draw an appropriate graph, and predict what will happen 5 years from now if the present growth rate continues.

2. Select an area of interest to you and explain how linear programming can be used to solve problems. Give a specific example.

3. Find a textbook and investigate how *matrices* can be used to solve a system of linear equations in two unknowns. Select an exercise in this chapter and solve it using matrices.

Answers to **Try This One**

8-A. (a)

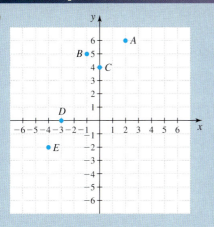

(b) $A = (2, 2)$; $B = (6, 0)$; $C = (0, 0)$;
$D = (0, -3)$; $E = (-4, -1)$; $F = (2, -5)$;
$G = (-2, 6)$

8-B.

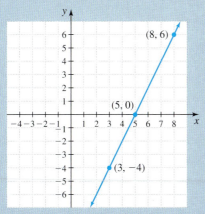

8-C. (a) and (b)

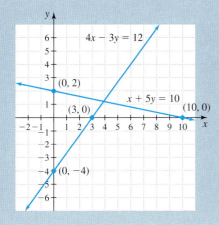

8-D. (a)

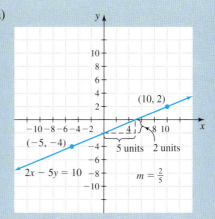

(b)

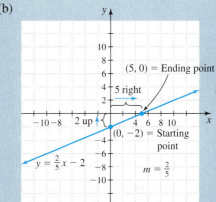

8-E. (a) \$7.80; (b) \$8.50

8-F. (a) $\{(-4, -3)\}$; (b) no solution: lines are parallel;
(c) $\{(x, y) \mid 2x - y = 9\}$; both equations yield the
same line.

8-G. $\{(1, 3)\}$

8-H. $\{(0, -1)\}$

8-I. (a) dependent; (b) inconsistent

8-J. 47 of the 20¢ stamps; 153 of the 35¢ stamps.

8-K.

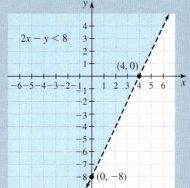

8-L. Solution set = dotted region, plus part of line
$4x - 3y = 12$

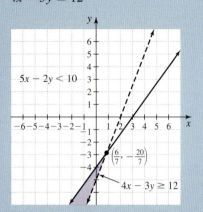

8-M. 5 leather ones and 15 cloth ones

8-N. Relations b and c are functions

8-O.

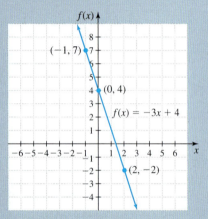

8-P.

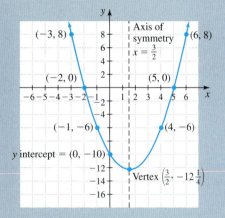

8-Q. 200

8-R. (a)

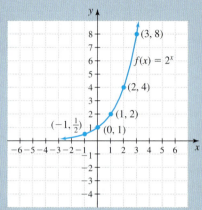

(b)

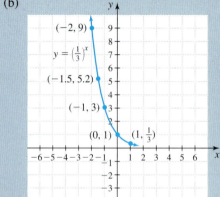

8-S. $5627.54

Chapter Nine

Consumer Mathematics

Objectives

After completing this chapter, you should be able to

1 Convert percents to fractions and decimals (9-1)

2 Convert fractions and decimals to percents (9-1)

3 Solve the three types of percent problems (9-1)

4 Solve word problems that use percents (9-1)

5 Find the simple interest on a loan or savings (9-2)

6 Find the principal or rate or time given the simple interest and the other two variables (9-2)

7 Find the compound interest and maturity value for a savings account (9-2)

8 Find the effective rate when interest is compounded for a specific stated rate (9-2)

9 Find the future value of an annuity (9-2)

10 Compute the finance charge and new balance for a credit card statement using the unpaid balance method (9-3)

11 Compute the finance charge and new balance for a credit card statement using the average daily balance method (9-3)

12 Find the annual percentage rate for an installment loan using the constant ratio formula (9-3)

13 Find the refund (unearned interest) using the Rule of 78's when an installment loan is paid off early (9-3)

14 Find the monthly payment and total interest for a mortgage (9-4)

15 Compute an amortization schedule for the payments on a mortgage (9-4)

16 Find the property tax on a home (9-4)

17 Find the markup on cost and selling price for an item sold at a retail store (9-5)

18 Find the markup on the selling price on an item sold at a retail store (9-5)

19 Find the markup rate on an item sold at a retail store (9-5)

20 Find the selling price when an item is marked down (9-5)

Introduction

Introductory rate on small business loans

T his chapter is about something that is very important to almost everyone—money! As we go through life, we usually become involved in complex financial transactions such as buying on credit (i.e., using our credit cards) or obtaining a loan to purchase an automobile or to pay for higher education, or to purchase a home. Learning the "ins and outs" of money management can help us make better decisions. Most of the financial world operates on the concept of "percent." Interest rates for loans are given in percents. Sales taxes are based on a percent of the cost of the item, and commissions are found by taking a percent of the sales. For this reason, this chapter opens with a presentation on the basic concepts of percent. The remaining sections show some of the many applications of percent, such as borrowing and saving money, computing financial charges on credit cards, purchasing a home, and marking up an item to be sold or marking down an item for a special sale.

After completing this chapter, you will be more knowledgeable about the intricacies of money. ∎

9-1 Percent

Percents are used extensively in the world of business.

> **Percent** means hundredths or part of a hundred; i.e., $1\% = \frac{1}{100}$.

For example, the International Mass Retail Association reports that 20% of adults plan to buy Valentine's Day cards. Of those who plan to buy cards, 26% of the men and 36% of the women plan to buy romantic cards.

In this situation, 20% means 20 out of every 100 adults plan to purchase Valentine's Day cards.

Percent Conversions

In order to perform mathematical operations such as multiplication and division with percents, you must convert the percents to decimals or fractions. Likewise, when answers to

Mathematics in Our World

How Much Can You Afford to Pay for a Home?

A home may be one of the most expensive items that you will ever purchase. Many homebuyers want to know how much they can afford to pay for a home.

The answer depends on several factors. The most important one, of course, is your gross (before deductions) monthly income. Another factor that you need to consider is the amount of money that you can put down on a house (the down payment). Finally, you must consider the interest rate and the term of the mortgage, that is, how long it will take you to pay off the loan.

Suppose your yearly income is $36,000, and you have saved $10,000 for a down payment. How much can you afford to pay for a house if you qualify for a 7% mortgage for 25 years? See Mathematics in Our World—Revisited for the answer.

problems require percents, these answers are computed in decimals or fractions and then they must be converted to percents.

Converting Percents to Decimals

In order to change a percent to a decimal, drop the % sign and move the decimal point two places to the left.

Example 9-1

Change each percent to a decimal.

(a) 84%
(b) 5%
(c) 37.5%
(d) 172%

Solution

(Drop the % sign and move the decimal point two places to the left.)

(a) $84\% = 84.\% = 0.84$
(b) $5\% = 5.\% = 0.05$
(c) $37.5\% = 0.375$
(d) $172\% = 172.\% = 1.72$

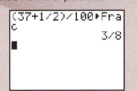

Math Note

Moving the decimal point two places to the left is the same as dividing by 100.

Converting Percents to Fractions

A percent can be converted to a fraction by dropping the percent sign and using the percent number as the numerator of a fraction whose denominator is 100.

Be sure to reduce fractions to lowest terms when possible.

Calculator Explorations

A graphing calculator will assist you in viewing the fractional representation of mixed numeral percents.

```
(37+1/2)/100▶Fra
c
             3/8
■
```

Is this the same answer you got working the exercise by hand?

Example 9-2

Change each percent to a fraction.

(a) 42%
(b) 6%
(c) $37\frac{1}{2}\%$
(d) 15.8%

Solution

(a) $42\% = \frac{42}{100} = \frac{21}{50}$
(b) $6\% = \frac{6}{100} = \frac{3}{50}$

—Continued

Example   9-2   *Continued—*

(c) $37\frac{1}{2}\% = \dfrac{37\frac{1}{2}}{100} = 37\frac{1}{2} \div 100 = \dfrac{3}{8}$

or

$\dfrac{37\frac{1}{2}}{100} \times \dfrac{10}{10} = \dfrac{375}{1000} = \dfrac{3}{8}$

(d) $15.8\% = \dfrac{15.8}{100} = \dfrac{15.8}{100} \cdot \dfrac{10}{10} = \dfrac{158}{1000} = \dfrac{79}{500}$

Converting a Decimal to a Percent

To change a decimal to a percent, move the decimal point two places to the right and add a percent sign.

Example   9-3

Change each decimal to a percent.

(a) 0.74
(b) 0.05
(c) 0.327
(d) 5.463

Solution

(a) $0.74 = 74\%$
(b) $0.05 = 5\%$
(c) $0.327 = 32.7\%$
(d) $5.463 = 546.3\%$

> *Math Note*
> Moving the decimal point two places to the right is the same as multiplying by 100.

Changing a Fraction to a Percent

To change a fraction to a percent, first change the fraction to a decimal, and then change the decimal to a percent.

Example   9-4

Convert each fraction to a percent.

(a) $\frac{7}{8}$
(b) $\frac{3}{4}$
(c) $\frac{5}{6}$
(d) $1\frac{1}{2}$

> *Math Note*
> Recall that to change a fraction to a decimal, we divide the numerator by the denominator.

—Continued

Example 9-4   *Continued—*

Solution

(a) $\frac{7}{8} = 7 \div 8 = 0.875 = 87.5\%$

(b) $\frac{3}{4} = 3 \div 4 = 0.75 = 75\%$

(c) $\frac{5}{6} = 0.83\overline{3} = 83.\overline{3}\%$

(d) $1\frac{1}{2} = 1.5 = 150\%$

Try This One

9-A   Change each percent to a decimal.

(a) 62.5%

(b) 3%

(c) 250%

Change each percent to a fraction.

(d) 90%

(e) 16.5%

(f) 130%

Change each decimal or fraction to a percent.

(g) 0.97

(h) 0.378

(i) $\frac{5}{16}$

(j) $1\frac{3}{4}$

For the answer, see page 455.

Three Types of Percent Problems

There are three types of percent problems. Each uses the words *base, rate,* and *part*. The **base** is the whole, total, or 100%. For example, if the total number of students in a class is 40, then the base is 40 students.

The **rate** is expressed as a percent. For example, 5% of the students in the class are absent.

The **part** is a portion of the base. For example, if 5% of the students are absent, then the part of students that are absent is found by multiplying the rate by the base: $5\% \times 40 = 0.05 \times 40 = 2$. The part of the class that is absent then is 2 students.

In percent problems, the part can be larger than the base or whole when the percent is greater than 100. For example, if the value of a stock is $100 a share, and it increases $200, the base is $100 and the part is $200. This is a 200% increase.

The three types of percent problems can be solved by this formula:

Part = Rate × Base $P = R \times B$

I. Finding a Part

To find a part, change the percent to a decimal or fraction and multiply. Use $P = R \times B$.

Example 9-5

Find 32% of 66.

Solution

Change the percent to a decimal and then multiply.

$$32\% = 0.32$$
$$P = R \times B$$
$$0.32 \times 66 = 21.12$$

Hence, 32% of 66 is 21.12.

II. Finding a Percent

To find what percent one number is of another number, substitute in the formula $P = R \times B$ and solve for R. Be sure to change the decimal into a percent.

There are 5 cats, and 2 of them are black. To find the percent of cats who are black, use 5 as the base and 2 as the part: the rate is $\frac{2}{5}$, which can also be expressed as 0.4 or 40%.

Example 9-6

45 is what percent of 60?

Solution

$$P = 45 \quad \text{and} \quad B = 60$$
$$P = R \times B$$
$$45 = R \times 60$$
$$\frac{45}{60} = \frac{R \times 60}{60}$$
$$\frac{45}{60} = R$$
$$0.75 = R$$

Hence, $R = 0.75$ or 75%.

III. Finding a Base

To find a base when a percent of it is known, substitute in the formula $P = R \times B$ and solve for B. Be sure to change the percent to a decimal or fraction before dividing.

Example 9-7

70% of what number is 63?

Solution

$$P = 63, R = 70\% = 0.7$$
$$P = R \times B$$
$$63 = 0.7 \times B$$
$$\frac{63}{0.7} = \frac{0.7 \times B}{0.7}$$
$$90 = B$$

Try This One

9-B   Solve each percent problem.

(a) 42 is what percent of 50?

(b) Find 18% of 54.

(c) 16 is 40% of what number?

For the answer, see page 455.

Applications of Percents

Many aspects of consumer mathematics deal with finding parts of a whole. For example, you may want to leave a 15% tip or you may want to figure out a 33% markup or you may want to calculate an 8% commission on sales. Some applications are shown in Examples 9-8 through 9-10.

Example 9-8

The sales tax in Allegheny County in Pennsylvania is 7%. What is the tax on a calculator that costs $89.95? What is the total amount paid?

Solution

Find 7% of $89.95.

$$0.07 \times \$89.95 = \$6.30 \text{ rounded}$$

Hence the sales tax is $6.30. The total amount paid is

$$\$89.95 + \$6.30 = \$96.25$$

> *Math Note*
>
> In Example 9-8, you can find the total by multiplying the cost by 1.07 (i.e., 107%).
> 1.07($89.95) = $96.25 rounded.

Percent increases can be greater than 100%, as shown in Example 9-9.

Example 9-9

Based on information from *Amusement Business* magazine, attendance at amusement parks in the United States increased from 97 million people in 1986 to 300 million in 1997. Find the percent of increase in attendance.

Solution

Find the increase.

$$300 - 97 = 203 \text{ million people}$$

Find the percent increase.

$$\text{Percent increase} = \frac{203}{97} \approx 2.09 = 209\%$$

The reasoning is as follows:

$$B = 97 \text{ and } P = 203$$
$$P = R \times B$$
$$203 = R \times 97$$
$$\frac{203}{97} = \frac{R \times 97}{97}$$
$$\frac{203}{97} = R$$
$$2.09 \approx R$$

Hence, attendance increased by 209%.

Math Note

Whenever a problem asks you to find the percent of increase or decrease, always use the amount of the increase or decrease for the part (P) and the *old* or *original* number for the base (B) in the formulas.

Try This One

9-C The U.S. Census Bureau reported that the number of people over age 65 living in the United States increased from 3.1 million in 1900 to 35.3 million in 2000. Find the percent of increase.
For the answer, see page 455.

Example 9-10

A real estate agent receives a 7% commission on all home sales. How expensive was the home if she received a commission of $5775.00?

—Continued

Sidelight

Money, Banks, and Credit Cards

Long before money existed, people bartered for goods and services. For example, if you needed the roof of your hut fixed, you would pay the repair person two chickens. The first known coins were made over 2500 years ago in Western Turkey. They consisted of a mixture of gold and silver and were stamped to guarantee uniformity. These coins were first accepted by the merchants of the area. Also around that time, coins were made in India and China.

Paper money was first made in China about 1400 years ago; however, when Marco Polo brought the idea to Europe, it was rejected by the people. It wasn't until the 1600s that banks in Europe began to issue paper money to their depositors and borrowers.

In the United States, early settlers used tobacco, beaver skins, and foreign coins as currency. A popular coin was the Spanish dollar, called "pieces of eight." For purchases of less than one dollar, the coin was cut into eight pieces. Each piece was called a "bit" and was worth $12\frac{1}{2}$ cents. Hence 25 cents became known as "two bits," 50 cents as "four bits," etc.

During the Revolutionary War, the Continental Congress authorized the printing of paper money to pay war debts. The government printed more money than it could back up with gold and silver, and the dollar became virtually worthless. The phrase "not worth a Continental dollar" is still used today.

The U.S. Mint opened on April 2, 1792, in Washington, D.C., to mint coins. Gold was used for the $10.00, $5.00, and $2.50 coins. Silver was used for the $1.00, $0.50, $0.25, $0.10, and the $0.05 coins, and copper was used for the $0.01 and $0.001/2 coins.

Most paper money was issued by state banks until 1863 when Congress established national banks to issue currency notes. In 1913, the Federal Reserve System was established to issue notes, which became our standard currency. These people are pictured on our bills.

$1.00	Washington
$2.00	Jefferson
$5.00	Lincoln

$10.00	Hamilton
$20.00	Jackson
$50.00	Grant
$100.00	Franklin
$500.00	McKinley
$1000.00	Cleveland
$10,000.00	Chase

No bills over $100.00 have been used since 1969.

Banking began in ancient Babylonia about 2000 B.C.E. when people kept their money in temples. They thought that if the temples were robbed, the gods would punish the robbers. In medieval times, money was kept in vaults in castles and protected by the armies of the nobles. The first bank was established in 1148 in Genoa, Italy. It was called the Bank of San Giorgio. The first bank in the United States was established in 1781 in Philadelphia and was called the Bank of North America.

Credit cards were first issued by large hotels in the early 1900s. These cards were considered to be prestigious and were issued only to customers who spent a lot of money at the hotel. Department stores and gasoline companies began to issue credit cards around 1915. During World War II, the United States forbade the use of credit cards. Banks began to issue credit cards in the 1950s. Finally, in the late 1960s, banks agreed to sponsor credit cards such as Master Card, Visa, etc.

Example 9-10 *Continued—*

Solution

In this case, the problem can be written as $5775.00 is 7% of what number? This is an example of the third type of percent problem.

$$R = 7\% = 0.07 \quad \text{and} \quad P = 5775$$
$$P = R \times B$$
$$5775 = 0.07 \times B$$
$$\frac{5775}{0.07} = \frac{0.07 \times B}{0.07}$$
$$82{,}500 = B$$

The home was purchased for $82,500.00.

Try This One

9-D A sales clerk receives a 9% commission on all sales. Find the total sales the clerk made if his or her commission was $486.00.
For the answer, see page 455.

Exercise Set 9-1

Computational Exercises

For Exercises 1–12, express each as a percent.

1. 0.63	2. 0.87	3. 0.025
4. 0.0872	5. 1.56	6. 3.875
7. $\frac{1}{5}$	8. $\frac{5}{8}$	9. $\frac{2}{3}$
10. $\frac{1}{6}$	11. $1\frac{1}{4}$	12. $2\frac{3}{8}$

For Exercises 13–20, express each as a decimal.

13. 18%	14. 23%	15. 6%
16. 2%	17. 62.5%	18. 75.6%
19. 320%	20. 275%	

For Exercises 21–30, express each as a fraction or mixed number.

21. 24%	22. 36%	23. 9%
24. 4%	25. 236%	26. 520%
27. $\frac{1}{2}\%$	28. $12\frac{1}{2}\%$	29. $16\frac{2}{3}\%$
30. $4\frac{1}{6}\%$		

Real World Applications

31. Find the sales tax and total cost of a diamond pendant that costs $299.99. The tax rate is 5%.

32. Find the sales tax and total cost of a CD boom box that costs $59.95. The tax rate is 7%.

33. Find the sales tax and total cost of a 13-inch TV/VCR that costs $149.99. The tax rate is 6%.

34. Find the sales tax and total cost of a video game that costs $19.99. The tax rate is 4.5%.

35. A 14K diamond ring was reduced from $999.99 to $399.99. Find the percent of the reduction in the price.

36. An electric shaver was reduced from $109.99 to $99.99. Find the percent of the reduction in price.

37. A 36-inch television set is on sale for $249.99. It was reduced $80.00 from the original price. Find the percent of the reduction in price.

38. The sale price of a vacuum cleaner was $179.99, and the advertisement said "Save $20.00." Find the percent of the reduction in price.

39. A luggage set was selling for $159.99, and the ad states that it has now been reduced 40%. Find the sale price.

40. If a sales clerk receives a 7% commission on all sales, find the commission the clerk receives on the sale of a computer system costing $1799.99.

41. If the cost of a gas grill is $199.99 and it is on sale for 25% off, find the sale price.

42. If the commission for selling a 52-inch high-definition television set is 12%, find the commission on a television set that costs $2499.99.

43. Milo receives a commission of 6% on all sales. If his commission on a sale was $75.36, find the cost of the item he sold.

44. The sales tax in Pennsylvania is 6%. If the tax on an item is $96.00, find the cost of the item.

45. For a certain year, 19% of all books sold were self-help books. If a bookstore sold 12,872 books, about how many were self-help books?

46. A person saved $200.00 on a diamond bracelet. If this was a 60% savings from the original price, find the original cost of the bracelet.

Writing Exercises

47. What is the meaning of percent?

48. How are percents changed to decimals? Fractions?

49. How are fractions changed to percents?

50. How are decimals changed to percents?

51. Explain the three basic types of percent problems and how each one can be solved.

Critical Thinking

52. A store has a sale with 25% off every item. When you enter the store, you receive a coupon that states that you receive an additional 10% off. Is this equal to a 35% discount? Explain your answer.

53. You purchase a stock at $100.00 per share. It drops 30% the next day; however, a week later, it increases in value by 30%. If you sell it, will you break even? Explain your answer.

9-2 Interest

Interest is a fee paid for the use of money. For example, if you obtain a loan from a bank in order to purchase an automobile, you must not only pay back the amount of money that you borrowed, but you must also pay an additional amount, called the interest, for the use of the bank's money. On the other hand, if you deposit money in a savings account, the bank will pay you interest for saving money since they will be using your money to provide loans, mortgages, etc. to people who are borrowing money. The stated rate of interest is generally given as a yearly value.

There are two kinds of interest: *simple interest* and *compound interest*. Simple interest is computed on the amount of the loan or savings whereas compound interest is computed on the amount of the loan or savings and the previous interest accrued.

Simple Interest

In order to compute simple interest, three figures (amounts) are needed. They are the *principal,* the *rate,* and the *time.*

> **Interest** is the fee charged or paid for the use of money.
>
> The **principal** is the amount of money borrowed or placed in a savings account.
>
> **Rate** is the percent of the principal that is paid for the use of the money. (Rates are usually given for a year.)
>
> **Time** or **term** is the duration that the money is borrowed or invested for or has been invested. When the time is given in days or months, it must be converted to years by dividing by 365 or 12, respectively.
>
> **Maturity value** is the amount of the loan or investment or savings (principal) plus the interest.

The basic formulas for computing interest use the principal, rate, and the time as follows:

> ### Formulas for Computing Simple Interest and Maturity Value
>
> Interest = Principal × Rate × Time
>
> or
>
> $I = PRT$
>
> Maturity Value = Principal + Interest
>
> or
>
> $MV = P + I$ or $MV = P(1 + RT)$

Example 9-11

Find the interest on a loan of $3600.00 for 3 years at a rate of 8% per year.

Solution

Change the rate to a decimal and substitute in the formula $I = PRT$:

$$8\% = 0.08$$
$$I = PRT$$
$$= (\$3600.00)(0.08)(3)$$
$$= \$864.00$$

The interest on the loan is $864.00.

Example 9-12

Find the maturity value for the loan in Example 9-11.

Solution

Substitute in the formula $MV = P + I$

$$MV = P + I$$
$$= \$3600.00 + \$864.00$$
$$= \$4464.00$$

The total amount of money to be paid back is $4464.00.

Alternate Solution

Substitute in the formula $MV = P(1 + RT)$

$$MV = P(1 + RT)$$
$$= \$3600(1 + 0.08 \cdot 3)$$
$$= \$4464.00$$

When the time of a loan or investment is given in months, it must be changed to years by dividing the number of months by 12.

Example 9-13

United Ceramics Inc. needed to borrow $2000.00 at 4% for 3 months. Find the interest.

Solution

Change 3 months to years by dividing by 12 since the interest rate is for one year, and change the rate to a decimal. Substitute in the formula $I = PRT$

$$I = (\$2000.00)(0.04)\left(\tfrac{3}{12}\right)$$
$$= \$20.00$$

The interest is $20.00.

Sometimes a simple interest loan is paid off in monthly installments. To find the monthly payment, divide the maturity value of the loan by the number of months in the term of the loan.

Example 9-14

Admiral Chauffeur Services borrowed $600.00 at 9% for $1\frac{1}{2}$ years to repair a limousine. Find the interest, maturity value, and the monthly payment.

Solution

Step 1 Find the interest.

$$I = PRT$$
$$= (\$600.00)(0.09)\left(1\frac{1}{2}\right)$$
$$= 81$$

The interest is $81.00.

Step 2 Find the maturity value of the loan.

$$MV = P + I$$
$$= \$600.00 + \$81.00$$
$$= \$681.00$$

Step 3 Divide the maturity value of the loan by the number of months. Since $1\frac{1}{2}$ years = 18 months, divide $681.00 by 18 to get $37.83. The monthly payment is $37.83.

Try This One

9-E The Lookout Restaurant took out a loan for $5000.00. The interest rate was 6.5%, and the term of the loan was 3 years. Find the interest, maturity value, and monthly payment.
For the answer, see page 455.

Finding the Principal, Rate, and Time

In addition to finding the interest and maturity value for a loan or investment, the principal, the rate, and the time period can also be found by substituting in the formula $I = PRT$ and solving for the unknown.

Examples 9-15, 9-16, and 9-17 show how to find the principal, rate, and time.

Example 9-15

Phillips Beauty Spa is replacing one of its workstations. The interest on a loan secured by the spa was $93.50. The money was borrowed at 5.5% for 2 years. Find the principal.

—Continued

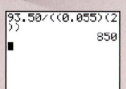
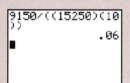
Math Note

Be sure to change the decimal to a percent since rates are given in percents.

Example 9-15 *Continued—*

Solution

$$I = \$93.50, \; R = 5.5\% = 0.055, \text{ and } T = 2$$

$$I = PRT$$

$$93.50 = P(0.055)(2)$$

$$\frac{93.50}{(0.055)(2)} = \frac{P(0.055)(2)}{(0.055)(2)}$$

$$P = 850$$

Hence, the amount of the loan is $850.00.

The same formulas can be used for investments as well. Example 9-16 shows this.

Example 9-16

R & S Furnace Company invested $15,250.00 for 10 years and received $9150.00 in interest. What was the rate that the investment paid?

Solution

$$P = 15,250, \; T = 10, \text{ and } I = 9150$$

$$I = PRT$$

$$9150 = (15,250)(R)(10)$$

$$\frac{9150}{(15,250)(10)} = \frac{(15,250)(R)(10)}{(15,250)(10)}$$

$$0.06 = R$$

$$R = 0.06 \text{ or } 6\%$$

Hence, the interest paid on the investment is 6%.

Example 9-17

Pryor Furnace Company borrowed $4500.00 at $8\frac{3}{4}\%$ to upgrade its equipment. The company had to pay $2756.25 interest. Find the term of the loan.

Solution

$$P = 4500, \; R = 8\frac{3}{4}\% = 0.0875, \text{ and } I = 2756.25$$

$$I = PRT$$

$$2756.25 = (4500)(0.0875)T$$

$$\frac{2756.25}{(4500)(0.0875)} = \frac{(4500)(0.0875)T}{(4500)(0.0875)}$$

$$7 = T$$

Hence, the term of the loan is 7 years.

Try This One

9-F Find the missing value for each.

Principal	Rate	Time	Interest
(a) $8000	____	2.5 years	$1000
(b) ____	6%	3 years	$76.50
(c) $750	4%	____	$150

For the answer, see page 455.

Compound Interest

When interest is computed on the principal alone as previously shown, it is called simple interest. When interest is computed on the principal and any previous interest, it is called **compound interest.** Suppose $5000.00 was placed in a savings account for 3 years at 8% interest. The simple interest on the amount would be found by using the formula

$$I = PRT$$
$$= \$5000.00 \times 0.08 \times 3$$
$$= \$1200.00$$

The simple interest is $1200.00 or $400.00 per year. Compound interest on that amount can be computed by finding the interest for the first year and then adding it on to the principal, then finding the interest for the second year using the new principal, and finally finding the interest for the third year using the principal plus the interest for the first 2 years as the principal. The total interest is found by adding the interest for each of the 3 years. Example 9-18 shows this.

Example 9-18

Find the compound interest on $5000.00 invested for 3 years at 8%.

Solution

Step 1 Find the interest for the first year.

$$I = PRT$$
$$= \$5000.00 \times 0.08 \times 1$$
$$= \$400.00$$

The interest for the first year is $400.00.

Step 2 Find the interest for the second year using $5000.00 + $400.00 = $5400.00 for the principal.

$$I = PRT$$
$$= \$5400.00 \times 0.08 \times 1$$
$$= \$432.00$$

The interest for the second year is $432.00.

—Continued

Example 9-18 Continued—

Step 3 Find the interest for the third year using $5400.00 + $432.00 = $5832.00 for the principal.

$$I = PRT$$
$$= \$5832.00 \times 0.08 \times 1$$
$$= \$466.56$$

Hence, the interest for the third year is $466.56. The total compound interest for 3 years is $400.00 + $432.00 + $466.56 = $1298.56.

A simple-interest savings account would earn $1200.00 interest while a compound-interest savings account would earn $1298.56, almost $100.00 ($98.56) more over a 3-year period.

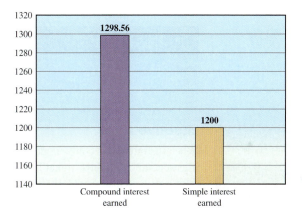

Interest can be compounded yearly (once a year), semiannually (twice a year), quarterly (four times a year), or daily (every day). It would be very tedious to calculate compound interest for each period as shown in Example 9-18 on a long-term savings account. For example, if the interest is compounded quarterly for 4 years, 16 calculations would be required. Instead, a formula can be used.

Math Note

When interest is compounded yearly, $n = 1$; semiannually, $n = 2$; quarterly, $n = 4$; and daily, $n = 365$.

Formula for Computing Compound Interest

$$MV = P\left(1 + \frac{r}{n}\right)^{nt}$$

where MV is the maturity value (Principal + Interest)

r = the yearly interest rate

n = number of periods the interest is compounded per year

t = term of the loan in years.

Calculator Explorations

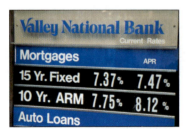

A graphing calculator will assist you in viewing the solution when evaluating formulas.

```
7000(1+0.03/4)^(
4*5)
        8128.288996
■
```

What is the answer rounded to the nearest cent?

Example 9-19

Find the interest on $7000.00 compounded quarterly at 3% for 5 years.

Solution

$$P = \$7000.00, \ r = 3\%, \ n = 4, \ t = 5$$

$$\text{MV} = P\left(1 + \frac{r}{n}\right)^{nt}$$

$$= \$7000.00\left(1 + \frac{0.03}{4}\right)^{4 \cdot 5}$$

$$= \$8128.29$$

To find the interest, subtract the principal from the maturity value.

$$I = \$8128.29 - \$7000.00$$

$$= \$1128.29$$

Hence the interest is $1128.29.

Try This One

9-G Find the interest on $600.00 compounded semiannually at 4.5% for 6 years.
For the answer, see page 455.

Effective Rate

When the interest is compounded more often than once a year, the interest earned on a savings account is actually higher than the stated rate. For example, consider a savings account with a principal of $5000.00 and an interest rate of 4% compounded semiannually. The interest is compounded twice a year at 2%. Using the formula shown in Example 9-19, the actual interest for one year is $202.00. Even though the stated rate is 4%, the actual rate is found by dividing $202.00 by $5000.00, and it is 4.04%. This rate is called the *effective rate* or *annual yield*.

> The **effective rate** (also known as the **annual yield**) is the simple interest rate which would yield the same maturity value over one year as the compound interest rate.

The next formula can be used to calculate the effective interest rate.

Formula for Effective Interest Rate

$$E = \left(1 + \frac{r}{n}\right)^{n} - 1$$

where

E = effective rate

n = number of periods per year the interest is calculated

r = interest rate per year (i.e., stated rate)

Example 9-20 shows this formula.

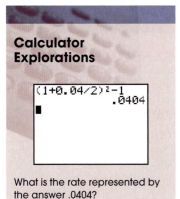
Example 9-20

Find the effective interest rate when the stated rate is 4% and the interest is compounded semiannually.

Solution

Let $R = 0.04$ and $n = 2$ and then substitute in the formula.

$$E = \left(1 + \frac{r}{n}\right)^n - 1$$

$$= \left(1 + \frac{0.04}{2}\right)^2 - 1$$

$$= 0.0404 = 4.04\%$$

Hence, the effective rate is 4.04%. (This is the same rate that was calculated previously.)

Try This One

9-H Find the effective rate for a stated interest rate of 8% compounded quarterly.
For the answer, see page 455.

Annuities

An **annuity** is a savings investment whereby an individual or business makes the same payment each period (i.e., annually, semiannually, or quarterly) into a compound-interest account where the interest does not change during the term of the investment. For example, an individual may pay $500 annually for 3 years into an account that yields 6% interest compounded annually. The total amount accumulated (payments plus interest) is called the **future value** of the annuity. Annuities are set up by individuals to pay for college expenses, vacations, or retirement. Annuities are set up by businesses to pay future expenses such as purchasing new equipment, expanding their business, etc. The payments are made at the end of the period.

Example 9-21 shows how an annuity works.

Example 9-21

Find the future value of an annuity where a $500 payment is made annually for 3 years at 6%.

Solution

The interest rate is 6% and the payment is $500 each year for 3 years.

  I. End of the first year $500 (payment)
 II. End of the second year

—Continued

Example **9-21**   *Continued—*

The $500 collected 6% interest and a $500 payment is made; hence, the value of the annuity at the end of the second year is

$$\$500(0.06) = \$\ 30 \qquad \text{Interest}$$
$$\$500 \qquad \text{Principal paid at end of first year}$$
$$\underline{+\$500} \qquad \text{Payment at the end of the second year}$$
$$\$1030$$

III. End of the third year

During the third year, the $1030 earns 6% interest and a payment of $500 is made at the end of the third year. Hence the annuity is worth

$$\$1030(0.06) = \$\ \ 61.80 \qquad \text{Interest}$$
$$\$1030.00 \qquad \text{Principal at end of second year}$$
$$\underline{+\$500.00} \qquad \text{Payment at end of third year}$$
$$\$1591.80$$

Hence, the future value of the annuity at the end of the three years is $1591.80.

Finding the future value of an annuity by the method shown in Example 9-21 could become quite lengthy if the annuity is taken out for a long term. For example, if payments were made quarterly for 10 years on an annuity, 40(4 × 10) calculations would be required to compute the future value. In order to avoid this situation, the next formula can be used for calculating the future value of an annuity.

Formula for Finding the Future Value of an Annuity

$$FV = P\left(\frac{(1 + R)^N - 1}{R}\right)$$

where

FV is the future value of the annuity
P is the payment
R is the interest rate per period (year, quarter, etc.)
N is the number of payments (periods in a year times the number of years)

Before showing how the formula is used, two things have to be explained.

First, N is the number of periods over which payments are made. When payments are made annually, one payment is made per year. When payment is made semiannually, two payments are made per year; when payments are made quarterly, four payments are made per year.

To find N, then, it is necessary to multiply the number of times per year payments are made by the term of the annuity or the number of years payments are made. For example, if payments are made quarterly for 10 years, then $N = 4 \times 10 = 40$. If payments are made semiannually for six years, then $N = 2 \times 6 = 12$.

Sidelight

Doubling Your Money

People will often ask, "If I put my money in a savings account, how long will it take me to double it?" The answer, of course, is it depends on the interest rate. Bankers use a simple rule to find the answer in what is called the *rule of 72*.

All that is necessary to find the answer is to divide 72 by the current interest rate. For example, if the interest rate is 3%, then it will take you 72 ÷ 3 or 24 years. If the interest rate is 10%, it will take you 72 ÷ 10 or approximately 7.2 years.

There are several things to consider, though. First, the answer is only an approximation. Second, it is assumed that the interest is compounded yearly. When interest is compounded semiannually, quarterly, or daily, the answer found by the rule is less precise. Other simple rules can be used to find how to triple your money, quadruple your money, etc.

As you know, the higher the interest rate, the faster your money will grow. The growth rate of your money also depends on how often the interest rate is compounded. Usually, interest is compounded yearly (once a year), semiannually (two times a year), or quarterly (four times a year). There are two other ways to compound interest. They are daily (365 times a year) and continuously (all the time). You might think that interest compounded continuously will make you a lot of money, but in reality, continuous interest pays only slightly more money than daily interest. The comparison is shown in the chart.

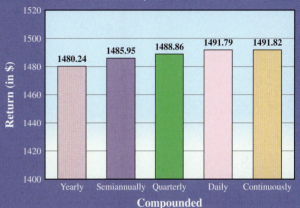

Investment of $1000 at 4% for 10 Years.

Investment of $1000 at 4% for 10 years.

Compounded	Return
Yearly	$1480.24
Semiannually	$1485.95
Quarterly	$1488.86
Daily	$1491.79
Continuously	$1491.82

As you can see, you will earn only approximately $0.03 more from continuous interest than you would earn with daily compounding.

Next, R is the interest rate per period; so if the interest is compounded yearly, divide the yearly rate by 1 to get R. If the interest is compounded semiannually, divide the yearly rate by 2 to get R. If the interest is compounded quarterly, divide the interest rate by 4 to get R.

Now the future value of the annuity in the previous example can be determined by the formula where $R = \frac{6\%}{1} = 0.06$.

$$N = 3 \times 1 = 3 \quad \text{and} \quad P = \$500$$

Hence,

$$\text{FV} = P\left(\frac{(1+R)^N - 1}{R}\right)$$
$$= 500\left(\frac{(1+0.06)^3 - 1}{0.06}\right)$$
$$= \$1591.80$$

As you can see, this is the same value obtained in Example 9-21.

Example 9-22

Find the future value of an annuity when the payment is $800 semiannually, the interest rate is 5% compounded semiannually, and the term is 4 years.

Solution

$$P = \$800$$

$$R = \frac{5\%}{2} = 2.5\% = 0.025$$

$$N = 4 \text{ years} \times 2 \text{ (semiannually)} = 8 \text{ payments}$$

Then

$$\text{FV} = P\left(\frac{(1 + R)^N - 1}{R}\right)$$

$$= \$800\left(\frac{(1 + 0.025)^8 - 1}{0.025}\right)$$

$$= \$6988.89$$

Hence, the future value of the annuity at the end of 4 years is $6988.89.

Try This One

9-I   Find the future value of an annuity when the payment is $275 quarterly, the interest rate is 6.5% compounded quarterly, and the term is 3 years.
For the answer, see page 455.

Exercise Set 9-2

Computational Exercises

For Exercises 1–20, find the missing value.

	Principal	Rate	Time	Simple Interest
1.	$12,000	6%	2 years	_____
2.	$25,000	8.5%	6 months	_____
3.	$1800	10%	_____	$360
4.	$600	4%	_____	$72
5.	$4300	_____	6 years	$1290
6.	$200	_____	3 years	$45
7.	_____	9%	4 years	$354.60

8. _____	15%	7 years	$65,625
9. $500	_____	2.5 years	$40
10. $1250	5%	_____	$375
11. $900	$9\frac{1}{2}\%$	18 months	_____
12. $420	_____	30 months	$31
13. $660	12%	_____	$514.80
14. $1975	7.2%	$3\frac{1}{2}$ years	_____
15. $14,285	_____	6 years	$8571
16. $650	15%	_____	$877.50
17. $325	_____	8 years	$156
18. $15,000	11%	_____	$742.50
19. $700	$6\frac{3}{4}\%$	_____	$141.75
20. $135	7%	6.5 years	_____

For Exercises 21–28, find the compound interest and maturity value for each.

	Principal	Rate	Compounded	Time
21.	$825	4%	Annually	10 years
22.	$3250	2%	Annually	5 years
23.	$75	3%	Semiannually	6 years
24.	$1550	5%	Semiannually	7 years
25.	$625	8%	Quarterly	12 years
26.	$2575	4%	Quarterly	2 years
27.	$1995	5%	Semiannually	6 years
28.	$460	6%	Quarterly	7 years

Real World Applications

29. To purchase a refrigerated showcase, Georgetown Florists borrowed $8000.00 for 6 years. The simple interest is $4046.40. Find the rate.

30. Wayward Singing Telegrams borrowed $15,000.00 for 12 years to pay for a new vehicle. The simple interest is $18,000.00. Find the rate.

31. To take advantage of a going-out-of-business sale, Pleasant Valley Novelty had to borrow some money. They paid $150.00 on a 6-month loan at 12%. Find the principal.

32. To purchase two new industrial ovens, the Oak Tree Bakery paid $1350.00 simple interest on a 9% loan for 3 years. Find the principal.

33. To train employees to use new equipment, Williams Muffler Repair had to borrow $4500.00 at $9\frac{1}{2}\%$. The company paid $1282.50 in simple interest. Find the term of the loan.

34. Berger Car Rental borrowed $8650.00 at 6.8% interest to cover the increasing cost of auto insurance. Find the term of the loan if the simple interest is $441.15.

35. To pay for new supplies, Jiffy Photo Company borrowed $9325.00 at 8% and paid $3170.50 in simple interest. Find the term of the loan.

36. Mary Beck earned $216.00 simple interest on a savings account at 8% over 2 years. Find the principal.

37. John White has a savings of $4300.00, which earned $9\frac{3}{4}$% simple interest for 5 years. Find the interest.

38. Ed Bland had a savings of $816.00 invested at $4\frac{1}{2}$% for 3 years. Find the simple interest.

39. A couple decides to set aside $5000.00 in a savings account for a second honeymoon trip. It is compounded quarterly for 10 years at 9%. Find the amount of money they will have in 10 years.

40. In order to pay for college, the parents of a child invest $20,000 in a bond that pays 8% interest compounded semiannually. How much money will there be in 18 years?

41. A 25-year-old person plans to retire at age 50. She decided to invest an inheritance of $60,000.00 at 7% interest compounded semiannually. How much will she have at age 50?

42. To pay for new machinery in 5 years, a company owner invests $10,000.00 at $7\frac{1}{2}$% compounded quarterly. How much money will be available in 5 years?

43. Find the effective rate when the stated rate is 6% and interest is compounded quarterly.

44. Find the effective rate when the stated rate is 10% and the interest is compounded semiannually.

45. Find the effective rate when the stated rate is 6.5% and the interest is compounded quarterly.

46. Find the effective rate when the stated rate is 9.55% and the interest is compounded semiannually.

47. A husband and wife wish to save money for their daughter's college education in 4 years. They decide to purchase an annuity with a semiannual payment earning 7.5% compounded semiannually. Find the future value of the annuity in 4 years if the semiannual payment is $2250.

48. A business owner decided to purchase an annuity to pay for new copy machines in 3 years. The payment is $600 quarterly at 8% interest compounded quarterly. Find the future value of the annuity in 3 years.

49. Find the future value of an annuity if you invest $200 quarterly for 20 years at 5% interest compounded quarterly.

50. A person decides to save money for a vacation in 2 years. The person purchases an annuity with semiannual payments of $200 at 9% interest compounded semiannually. Find the amount of money the person can spend for the vacation.

51. The owner of Rusty Recording Company wishes to build a second recording studio in 5 years. He purchases an annuity which pays 10.5% interest compounded annually. If the payment is $4000 a year, find the future value of the annuity in 5 years.

52. Mary Lee, owner of Mary Lee's Attic, wishes to purchase an annuity that pays 4% interest compounded quarterly for 4 years. If the quarterly payment is $160, find the future value of the annuity.

53. In order to plan for their retirement, a married couple decides to purchase an annuity that pays 8% interest compounded semiannually. If the semiannual payment is $2000, how much will they have saved in 10 years?

54. The Red River Machine Company will need to replace two drill presses in 3 years. If the owner purchases an annuity at 6% interest compounded annually with an annual payment of $800, find the value of the annuity in 3 years.

Writing Exercises

55. What is interest?

56. What is the difference between simple interest and compound interest?

57. What is meant by the term of a loan?

58. What is meant by the maturity value of a loan?

59. What is meant by the effective rate of a savings?

60. When will the effective rate be the same as the stated rate?

Critical Thinking

61. Susan would like to purchase a new car. Which loan would have the higher interest amount: a personal loan of $10,000.00 at 9% for 6 years or an auto loan of $10,000.00 at 8% for 60 months? Why?

62. Sea Drift Motel is converting its rooms into privately owned condominiums. The interest on a $1,000,000.00, 20-year construction loan is $98,000.00. What is the rate of interest? Does the rate seem unreasonable?

63. The Laurel Township Fire Department must decide whether to purchase a new tanker truck or repair the one they now use. For a new truck loan, the interest rate on $25,000.00 is 18% for a 10-year period; to repair the existing truck, the department must borrow $18,000.00 at $12\frac{1}{2}$% for 8 years. Which loan is less expensive?

64. A local miniature golf course owner must recarpet his fairways and replace the golf clubs his customers use. A bank will lend the owner the necessary $7800.00 at 9.5% interest over 48 months. A savings and loan company will lend the owner $7800.00 at 8.5% interest for 54 months. Which loan will be less expensive for the golf course owner to assume?

9-3 Installment Buying

Oftentimes people purchase items with a credit card. They prefer to use a credit card to avoid carrying large sums of cash. When purchasing items over the phone or online, businesses often prefer the customer to use a credit card. Credit cards can also be used to purchase items when a person is short of cash. A credit card purchase can be thought of as an **installment loan.** This is actually a way to borrow money.

In order for the credit card company or lending institution to make money, they charge the credit card holder or the person who borrowed the money interest. The Truth in Lending Act of 1969 requires the lending institution to disclose the cost of borrowing the money. This cost is called the **finance charge** and includes the amount of interest and any additional fees for borrowing the money. There are two methods that credit card companies use to compute the finance charge: the *unpaid balance method* and the *average daily balance method.*

Math Note

The billing cycle for credit cards and loans can begin and end on any day of the month; however, for the examples given in this section, we will assume the cycle ends on the last day of the month.

Unpaid Balance Method

With the **unpaid balance method,** interest is charged only on the balance from the previous month. Example 9-23 shows how to find the interest on the unpaid balance.

Example 9-23

For the month of April, a person had an unpaid balance of $356.75 at the beginning of the month and made purchases of $436.50. A payment of $200.00 was made during the month. The interest on the unpaid balance is 1.8% per month. Find the finance charge and the balance on May 1.

Solution

Step 1 Find the finance charge on the unpaid balance using 1.8%.

$$I = PRT$$
$$= \$356.75 \times 0.018 \times 1$$
$$= \$6.42 \text{ (rounded)}$$

Hence, the finance charge is $6.42.
(*Note:* Since the interest rate is given per month, the time is always equal to 1 month.)

Step 2 To the unpaid balance, add the finance charge and the purchases for the month; then subtract the payment to get the new balance.

$$\text{New balance} = \$356.75 + \$6.42 + \$436.50 - \$200$$
$$= \$599.67$$

Hence, the new balance as of May 1 is $599.67.

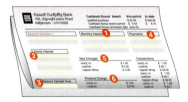

Try This One

9-J For the month of January, a person has an unpaid balance of $846.50 from December. Purchases for the month were $532.86 and a payment of $350.00 was made during the month. If the interest on the unpaid balance is 2% per month, find the finance charge and the balance on February 1.
For the answer, see page 455.

Average Daily Balance Method

Because the finance charge is computed only on the unpaid balance, some people pay off the balance on their credit cards each month; hence, they have essentially free use of the credit card institution's money. In order to make money, then, the credit card institution may charge a yearly fee such as $25.00, $50.00, etc. for the use of its card.

Another way to make money for these institutions is to use the **average daily balance method** for computing the finance charge. Here the balance for each day of the month is used to compute the average daily balance for the month. Then the interest rate is computed using this balance. Example 9-24 shows how to compute interest using the average daily balance method.

Example 9-24

Betty's credit card statement showed the following transactions during the month of August.

August 1	Previous balance	$165.50
August 7	Purchases	59.95
August 12	Purchases	23.75
August 18	Payment	75.00
August 24	Purchases	107.43

(a) Find the average daily balance.
(b) Find the finance charge for the month. The interest rate is 1.5% per month on the average daily balance.
(c) Find the new balance on September 1.

Solution

Step 1 Find the balance for each transaction.

August 1	$165.50
August 7	$165.50 + $59.95 = $225.45
August 12	$225.45 + $23.75 = $249.20
August 18	$249.20 − $75.00 = $174.20
August 24	$174.20 + $107.43 = $281.63

Step 2 Find the number of days for each balance.

Date	Balance	Days	Calculations
August 1	$165.50	6	$(7 - 1 = 6)$
August 7	$225.45	5	$(12 - 7 = 5)$
August 12	$249.20	6	$(18 - 12 = 6)$
August 18	$174.20	6	$(24 - 18 = 6)$
August 24	$281.63	8	$(31 - 24 + 1 = 8)$

> **Math Note**
>
> One must be added since the transaction period starts on August 24 and ends on the last day of the month which must be included. The total number of days must equal the number of days in the given month.

Step 3 Multiply each balance by the number of days, and add these products.

Date	Balance	Days	Calculations
August 1	$165.50	6	$165.50(6) = $993.00
August 7	$225.45	5	$225.45(5) = $1127.25
August 12	$249.20	6	$249.20(6) = $1495.20
August 18	$174.20	6	$174.20(6) = $1045.20
August 24	$281.63	8	$281.63(8) = $2253.04
		31	$6913.69

Step 4 Divide the total by the number of days in the month to get the average daily balance.

$$\text{Average daily balance} = \frac{\$6913.69}{31} = \$223.02$$

Hence, the average daily balance is $223.02.

Step 5 Find the finance charge. Multiply the average daily balance by the rate:

$$\text{Finance charge} = \$223.02 \times 0.015 = \$3.35.$$

—Continued

Example 9-24 *Continued—*

Step 6 Find the new balance. Add the finance charge to the balance as of the last transaction.

$$\text{New balance:} \quad \$281.63 + \$3.35 = \$284.98$$

Hence, the average daily balance is $223.02. The finance charge is $3.35, and the new balance is $284.98.

The procedure for finding the average daily balance is summarized next.

Procedure for the Average Daily Balance Method

Step 1 Find the balance for each transaction.

Step 2 Find the number of days for each balance.

Step 3 Multiply the balances by the number of days and find the sum.

Step 4 Divide the sum by the number of days in the month.

Step 5 Find the finance charge (multiply the average daily balance by the monthly rate).

Step 6 Find the new balance (add the finance charge to the balance as of the last transaction).

Some people misuse credit cards. For example, they may charge more than they can afford to pay off, and with the interest charged by the credit card company, the debt becomes even more of a financial burden.

When the credit is based on the unpaid balance method, the cardholder can, if he or she is able, pay the full amount of the bill at the end of the month, thus avoiding any interest. However, if the credit card company computes interest using the average daily balance method, you will pay interest no matter when you pay off your bill.

Try This One

9-K A credit card statement for the month of November showed the following transactions.

November 1	Previous balance	$937.25
November 4	Purchases	$531.62
November 13	Payment	$400.00
November 20	Purchases	$89.95
November 28	Payment	$100.00

(a) Find the average daily balance.

(b) Find the finance charge. The interest rate is 1.9% per month on the average daily balance.

(c) Find the new balance on December 1.

For the answer, see page 455.

Annual Percentage Rate

When a loan is paid back in installments such as monthly payments, the borrower does not have use of the full amount of the principal for the term of the loan. Because of this fact, the interest that is paid on the loan is higher than the stated rate. The Truth in Lending Law of 1969 requires the lending institution to disclose the true interest rate for the loan. This rate is known as the **annual percentage rate** or APR.

Since the computations involved in computing the true APR are somewhat complicated, they are beyond the scope of this book.

An approximate APR can be found by using the *constant ratio formula.*

Math Note

The constant ratio formula gives only an approximate APR. In order to get a more accurate estimate of the APR, the U.S. government provides tables for lending institutions to use.

The Constant Ratio Formula for APR

$$\text{APR} \approx \frac{2NI}{P(T+1)}$$

where

> $N =$ number of payments per year (usually 12 since most loans are paid back in monthly payments)
>
> $I =$ finance charge (i.e., total interest plus any additional charges)
>
> $P =$ principal
>
> $T =$ total number of payments

Example 9-25

In order to purchase a television set, Mary Lou borrowed $300.00. The loan was to be paid back in 18 monthly installments of $18.00. Approximate the APR using the constant ratio formula.

Rent-to-own

No credit check!
No down payment!

Total Price: $1200 or just $59.99/month

Call Now! 555-1678

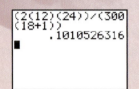
Example 9-25 Continued—

Solution

Step 1 Find the total interest. Since 18 payments of $18.00 each are made, the total amount of money paid back is $18 \times \$18.00 = \$324.00$. Since the loan is for $300.00, the interest is $\$324.00 - \$300.00 = \$24.00$.

Step 2 Substitute in the formula using $N = 12$, $I = \$24.00$, $P = \$300.00$, and $T = 18$.

$$\text{APR} \approx \frac{2NI}{P(T+I)}$$

$$\approx \frac{2(12)(24)}{300(18+1)} = \frac{576}{5700} = 0.101$$

$$\approx 10.1\%$$

Hence, the annual percentage rate is approximately 10.1%.

Try This One

9-L A loan for $800.00 is to be paid back in 24 monthly payments of $38.67. Find the APR using the constant ratio formula.
For the answer, see page 455.

Rule of 78s

When a loan is paid off early, the borrower can save money on the interest. There are several ways to compute the savings. One method is called the **rule of 78s.** This rule is used mainly on short-term loans such as automobile loans.

Formula for the Rule of 78s

$$u = \frac{fk(k+1)}{n(n+1)}$$

where

$u =$ unearned interest (i.e., amount saved)

$f =$ finance charge

$k =$ number of remaining monthly payments

$n =$ original number of payments

The rule of 78s is used to compute the unearned interest because the lending institution collects more interest in the beginning months of the loan. For example, on a 12-month loan, the lending institution collects

$\frac{12}{78}$ of the total interest for the first month

$\frac{11}{78}$ of the total interest for the second month

$\frac{10}{78}$ of the total interest for the third month

.
.
.

$\frac{1}{78}$ of the total interest for the last month.

If you add $\frac{12}{78} + \frac{11}{78} + \frac{10}{78} + \cdots + \frac{2}{78} + \frac{1}{78}$, you will get $\frac{78}{78}$; hence, this is the reason that the rule is called the rule of 78s.

Example 9-26

A $5000-automobile loan is to be paid off in 36 monthly installments of $172.00. The borrower decides to pay off the loan after 24 payments have been made. Find the amount of interest saved by paying the loan off early. Use the rule of 78s.

Solution

Find the finance charge (i.e., total interest).

$$\$172.00 \times 36 = \$6192.00$$
$$\$6192.00 - \$5000.00 = \$1192.00$$

Substitute in the formula using $f = \$1192.00$, $n = 36$, and $k = 36 - 24 = 12$.

$$u = \frac{fk(k+1)}{n(n+1)}$$
$$= \frac{1192(12)(12+1)}{36(36+1)}$$
$$= \$139.60$$

Hence $139.60 is saved by paying off the loan.

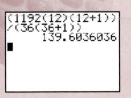
Try This One

9-M A $16,500 automobile loan is to be paid off in 24 monthly installments of $797.50. The borrower decides to pay off the loan after 20 payments have been made. Find the amount of interest saved. Use the rule of 78s.
For the answer, see page 455.

Exercise Set 9-3

Real World Applications

1. For the month of January, a person had an unpaid balance on a credit card statement of $832.50 at the beginning of the month and made purchases of $675.00. A payment of $400.00 was made during the month. If the interest rate was 2% per month on the unpaid balance, find the finance charge and the new balance on February 1.

2. For the month of July, the unpaid balance on Sue's credit card statement was $1131.63 at the beginning of the billing cycle. She made purchases of $512.58. She also made a payment of $750.00 during the month. If the interest rate was 1.75% per month on the unpaid balance, determine the finance charge and the new balance on the first day of the August billing cycle.

3. Sam is redecorating his apartment. On the first day of his credit card billing cycle, his balance was $2364.79. He has recently made purchases totaling $1964.32. He was able to make a payment of $1000.00 during this billing cycle. If his interest rate is 1.67% per month on the unpaid balance, what is the finance charge and what will Sam's new balance be on the first day of the next billing cycle?

4. Janine has recently accepted a position with an upscale clothing store. On the first day of her March credit card billing cycle, her unpaid balance was $678.34. She has made clothing purchases totaling $3479.03. She was able to make one payment of $525.00 during the billing cycle. If the interest rate is 2.25% per month on the unpaid balance, find the finance charge and the new balance on the first day of the April billing cycle.

5. Joe's credit card statement on the first day of the May billing cycle shows a balance of $986.53. During this billing cycle, he charged $186.50 to his account and made a payment of $775.00. At 1.35% interest per month on the unpaid balance, what is the finance charge? Also, find the balance on the first day of the next billing cycle.

6. Frank's credit card statement shows a balance of $638.19 on the first day of the billing cycle. If he makes a payment of $475.00 and charges $317.98 during this billing period, what will his finance charge be (the interest rate is 1.50% of the unpaid balance per month)? What will his beginning balance be at the start of the next billing cycle?

7. Mary's credit card statement showed these transactions during September:

September 1	Previous balance	$627.75
September 10	Purchase	$87.95
September 15	Payment	$200.00
September 27	Purchases	$146.22

(a) Find the average daily balance.

(b) Find the finance charge for the month. The interest rate is 1.2% per month on the average daily balance.

(c) Find the new balance on October 1.

8. Sam's credit card statement showed these transactions during March:

March 1	Previous balance	$2162.56
March 3	Payment	$800.00
March 10	Purchases	$329.27
March 21	Payment	$500.00
March 29	Purchases	$197.26

(a) Find the average daily balance.

(b) Find the finance charge for the month. The interest rate is 2% per month on the average daily balance.

(c) Find the new balance on April 1.

9. Mike's credit card statement showed these transactions during the month of June:

June 1	Previous balance	$157.95
June 5	Purchases	$287.62
June 20	Payment	$100.00

(a) Find the average daily balance.

(b) Find the finance charge for the month. The interest rate is 1.4% per month on the average daily balance.

(c) Find the new balance on July 1.

10. Charmaine's credit card statement showed these transactions during the month of December:

December 1	Previous balance	$1325.65
December 15	Purchases	$287.62
December 16	Purchases	$439.16
December 22	Payment	$700.00

(a) Find the average daily balance.

(b) Find the finance charge for the month. The interest rate is 2% per month on the average daily balance.

(c) Find the new balance on January 1.

11. Ruth's credit card statement showed these transactions for the month of July:

July 1	Previous balance	$65.00
July 2	Purchases	$720.25
July 8	Payment	$500.00
July 17	Payment	$100.00
July 28	Purchases	$343.97

(a) Find the average daily balance.

(b) Find the finance charge for the month. The interest rate is 1.1% per month.

(c) Find the new balance on August 1.

12. Tamera's credit card statement showed these transactions for the month of September:

September 1	Previous balance	$50.00
September 13	Purchases	$260.88
September 17	Payment	$100.00
September 19	Purchases	$324.15

(a) Find the average daily balance.

(b) Find the finance charge for the month. The interest rate is 1.9% per month on the average daily balance.

(c) Find the new balance on October 1.

For Exercises 13–18, use the constant ratio formula to find the annual percentage rate.

13. The Hole-In-One golf course borrowed $4000.00 to remodel the clubhouse. It is to be paid back in 48 monthly installments of $110.00. Find the annual percentage rate.

14. Country Rabbit Crafts store borrowed $300.00 to buy extra supplies for December. It is to be paid back in 18 monthly installments of $18.17. Find the annual percentage rate.

15. Penn's Glass Company borrowed $850.00 to fix their delivery truck. It is to be paid back in 72 monthly installments of $19.80. Find the annual percentage rate.

16. Murphy's Heating and Air Conditioning borrowed $1325.00 to buy parts for furnaces. It is to be paid back in 24 monthly installments of $68.46. Find the annual percentage rate.

17. Max's Tux Shop borrowed $3256.00 to repaint and remodel their store. It is to be paid back in 36 monthly payments of $116.00. Find the annual percentage rate.

18. Alexander's Transmission Shop borrowed $10,000.00 for 10 years. The monthly payment is $129.17. Find the annual percentage rate.

For Exercises 19–24, use the rule of 78s.

19. A $4200.00 loan is to be paid off in 36 monthly payments of $141.17. The borrower decides to pay off the loan after 20 payments have been made. Find the amount of interest saved.

20. Fred borrowed $150.00 for 1 year. His payments are $13.75 per month. If he decides to pay off the loan after 6 months, find the amount of interest that he will save.

21. Greentree Limousine Service borrowed $200.00 to repair a limousine. The loan was to be paid off in 18 monthly installments of $13.28. After a good season, they decide to pay off the loan early. If they pay off the loan after 10 payments, how much interest do they save?

22. Household Lighting Company borrowed $600.00 to purchase items from another store that was going out of business. The loan required 24 monthly payments of $29.50. After 18 payments were made, the company decided to pay off the loan. How much interest was saved?

23. Lydia needed to have her roof repaired. She borrowed $950.00 for 10 months. The monthly payments were $99.75 each. After seven payments, she decided to pay off the balance of the loan. How much interest did she save?

24. The owners of Scottdale Village Inn decided to remodel the dining room at a cost of $3250.00. They borrowed the money for 1 year and repaid it in monthly payments of $292.50. After eight payments were made, the owners decided to pay off the loan. Find the interest saved.

Writing Exercises

25. Explain what is meant by an installment loan.
26. What is a finance charge?
27. Explain the difference between the unpaid balance method and the average daily balance method for computing interest.
28. What is the annual percentage interest rate? Why is it different from the stated interest rate?
29. Explain the rule of 78s.
30. Why do lending institutions use the rule of 78s to compute the unearned interest when a loan is paid off early?

Critical Thinking

31. Find the principal on a loan at 8% for 4 years when the monthly payments are $100.00 per month.
32. A person borrowed $3000.00 at 6% for 5 years and paid it back in monthly installments. Find the annual percentage rate.

33. A person borrowed $800.00 for 1 year at 12% interest. Payments were made monthly. After eight payments were made, the person decided to pay it off. Find the interest that was saved if it was computed equally over 12 months. Then find the interest saved using the rule of 78s. Explain which is a better deal for the borrower.

 **9-4**

Home Ownership

Most of you have purchased or will purchase at least one home in your lifetime. Usually this will be the most costly purchase for you. If you are lucky, you may be able to pay cash for a home; however, most of you will have to borrow money from a bank or savings and loan association. A loan to purchase a home is called a **mortgage.** There are several types of mortgages. A **fixed-rate mortgage** means that the rate of interest remains the same for the entire term of the loan. The payments (usually monthly) stay the same. An **adjustable-rate mortgage** means that the rate of interest may fluctuate (i.e., increase and decrease) during the period of the loan. Some lending institutions will allow you to make **graduated payments.** This means that even though the interest does not change for the period of the loan, you can make smaller payments in the first few years and larger payments at the end of the loan period.

Finding the Monthly Payments and Total Interest

In order to find the monthly payments for a fixed-rate mortgage, Table 9-1 can be used. This table gives the monthly payment for each $1000.00 of a mortgage.

Table 9-1

Monthly Payment per $1000 of Mortgage (Includes Principal and Interest)

Rate (%)	Number of Years					
	15	20	25	30	35	40
6.5	$8.71	$7.46	$6.75	$6.32	$6.04	$5.85
7	8.99	7.75	7.70	6.65	6.39	6.21
7.5	9.28	8.06	7.39	6.99	6.74	6.58
8	9.56	8.36	7.72	7.34	7.10	6.95
8.5	9.85	8.68	8.05	7.69	7.47	7.33
9	10.15	9.00	8.40	8.05	7.84	7.72
9.5	10.45	9.33	8.74	8.41	8.22	8.11
10	10.75	9.66	9.09	8.70	8.60	8.50
10.5	11.06	9.98	9.44	9.15	8.98	8.89
11	11.37	10.32	9.80	9.52	9.37	9.28
11.5	11.69	10.66	10.16	9.90	9.76	9.68
12	12.01	11.01	10.53	10.29	10.16	10.08
12.5	12.33	11.36	10.90	10.67	10.55	10.49
13	12.66	11.72	11.28	11.06	10.95	10.90
13.5	12.99	12.07	11.66	11.45	11.35	11.30
14	13.32	12.44	12.04	11.85	11.76	11.71

Procedure for Finding the Monthly Payment for a Fixed-Rate Mortgage

Step 1 Find the down payment.

Step 2 Subtract the down payment from the cost of the home to find the principal of the mortgage.

Step 3 Divide the principal by 1000.

Step 4 Find the number in the table that corresponds to the interest rate and the term of the mortgage.

Step 5 Multiply that number by the number obtained in step 3 to get the monthly payment.

Example 9-27 shows this procedure.

Calculator Explorations

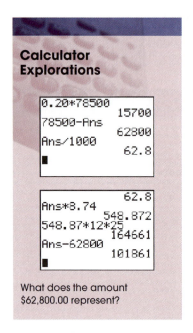

What does the amount $62,800.00 represent?

Example 9-27

Find the monthly payment for a $78,500.00 home with a 20% down payment. The terms of the mortgage are 9.5% for 25 years.

Solution

Step 1 Find the down payment.
$$20\% \times \$78,500.00 = 0.20 \times \$78,500.00 = \$15,700.00$$

Step 2 Subtract the down payment from the cost of the home.
$$\$78,500.00 - \$15,700.00 = \$62,800.00$$

Step 3 Divide by 1000.
$$\frac{62,800}{1000} = 62.8$$

Step 4 Find the value in Table 9-1 for a 25-year mortgage at 9.5%. It is 8.74.

Step 5 Multiply 62.8 by 8.74 to get the monthly payment.
$$62.8 \times 8.74 = \$548.87$$

Hence, the monthly payment is $548.87.

To find the total interest that will be paid on the loan, it is necessary to multiply the monthly payment by 12 times the number of years of the loan and then subtract the principal.

Example 9-28

Find the total amount of interest paid on the loan in Example 9-27.

Solution

Multiply the monthly payment by 12×25.

$$\$548.87 \times 12 \times 25 = \$164,661.00$$

Subtract $\$164,661.00 - \$62,800.00 = \$101,861.00$.

Hence, the total amount of interest is $\$101,861.00$.

When purchasing a home, you should consider your option of choosing the term of your mortgage. Although a shorter term will require higher monthly payments, it offers two advantages. First, the loan will be paid off sooner, and second, you will save a considerable sum of money on the interest. For example, let's look at a $150,000 mortgage at 7% interest. If you take a 30-year mortgage, your monthly payments will be $997.50, and the total interest you will pay is $209,100. Whereas, if you borrow the same amount at the same interest rate for 20 years, your monthly payment will be $1162.50, and the total interest you will pay is $129,000. So if you can afford the extra $165 a month for 20 years, you can save $80,000 in interest!

Try This One

9-N A home, priced at $92,000.00, is purchased with a 10% down payment and a loan. The mortgage rate is 8% and the term of the mortgage is 15 years.

(a) Find the down payment.

(b) Find the amount of the mortgage.

(c) Find the amount of the monthly payment.

For the answer, see page 455.

In addition to borrowing money for a home, the buyer is usually required to pay points. A point is 1% of the principal of the mortgage. There may also be an attorney fee, a title search fee, and other various costs.

Computing an Amortization Schedule

After securing a mortgage, the lending institution will prepare an **amortization schedule.** This schedule shows what part of the monthly payment is paid on the principal and what part of the monthly payment is paid in interest.

In order to prepare an amortization schedule, the next procedure should be used.

Math Note

Be sure to subtract any down payment from the cost of the home before beginning the amortization table.

Procedure for Computing an Amortization Schedule

Step 1 Find the interest for the first month. Use $I = PRT$, where $T = \frac{1}{12}$. Enter this value in the column labeled Interest.

Step 2 Subtract the interest from the monthly payment to get the amount paid on the principal. Enter this amount in the column labeled Payment on Principal.

Step 3 Subtract the amount of the payment on principal found in step 2 from the principal to get the balance of the loan. Enter this in the column labeled Balance of Loan.

Step 4 Repeat the steps using the amount of the balance found in step 3 for the new principal.

Calculator Explorations

```
62800.30*0.095*1
/12
        497.1690417
548.87-497.17
            51.7
62800-Ans
        62748.3
■
```

What does the value $62,800.00 represent?

What does the value 0.095 represent?

What does the value $548.87 represent?

Example 9-29

Compute the amortization schedule for the first 2 months for the mortgage used in Example 9-27.

Solution

The value of the mortgage is $62,800.00, the interest rate is 9.5%, and the monthly payment is $548.87.

Step 1 Find the interest for month 1.

$$I = PRT$$
$$= \$62{,}800.00 \cdot 0.095 \times \tfrac{1}{12}$$
$$= \$479.17$$

Enter this value in the column labeled Interest.

Step 2 Subtract the interest from the monthly payment.

$$\$548.87 - \$497.17 = \$51.70$$

Enter this value in the column labeled Payment on Principal.

Step 3 Subtract $51.70 from $62,800.00.

$$\$62{,}800.00 - \$51.70 = \$62{,}748.30$$

Enter this value in the column labeled Balance of the Loan.

Step 4 Repeat steps 1 through 3 using $62,748.30.

$$I = \$62{,}748.30 \times 0.095 \times \tfrac{1}{12}$$
$$= \$496.76$$

Step 5 $548.87 - $496.76 = $52.11

Step 6 $62,748.30 - $52.11 = $62,696.19

—Continued

Example 9-29 Continued—

The amortization schedule would be

Payment Number	Interest	Payment on Principal	Balance of Loan
1	$497.17	$51.70	$62,748.30
2	$496.76	$52.11	$62,696.19

Try This One

9-O Make an amortization schedule for the first 3 months for the home purchased in Try This One 9-N.
For the answer, see page 455.

In many areas citizens must vote to approve certain changes to property taxes. Voters might choose to increase taxes to fund new projects, or roll back taxes and cut spending if they feel they are too high.

Property Tax

Owners of real estate have to pay real estate or **property taxes** to the municipality, school district, and county where they reside. There are several methods used to compute property taxes. One method uses a *millage rate*. A **mill** is $\frac{1}{1000}$ of $1.00 or $0.001, and a millage rate is the number of mills per each dollar the house is worth. In other words, if the millage rate is 45 mills, then the home owner would pay 45 mills for each $1.00 of the value of the house. It is easier to think of the millage rate in terms of $1000 rather than $1.00; that is, a millage rate of 45 mills is the same as saying the owner would pay $45 in tax for each $1000 of the value of the house since the ratio is $\frac{0.045}{1.00} = \frac{45}{1000}$.

A home has two values, a *market value* and an *assessed value*. The **market value** of a home is the value of the home if it were to be sold today. The **assessed value** of the home is a percentage of the market value. For example, in one county, the assessed value of a home is 25% of its market value. In order to find the assessed value of a home in that county, multiply the market value of the home by 0.25 or 25%. In order to find the property tax, divide the assessed value of the home by 1000 and multiply the answer by the millage rate.

Example 9-30

Find the property tax on a home with a market value of $105,000.00 that is assessed at 25% of market value with a millage rate of 20 mills.

Solution

Find the assessed value.

$$\$105,000.00 \times 0.25 = \$26,250.00$$

Find the property tax.

$$\$26,250.00 \div 1000 \times 20 = \$525.00$$

Hence the property tax is $525.00.

Try This One

9-P   Find the property tax on a home that has a market value of $159,000.00 that is assessed at 30% of market value and the tax rate is 45 mills.
For the answer, see page 455.

Exercise Set 9-4

Real World Applications

1. A house sells for $145,000.00 and a 15% down payment is made. A mortgage was secured at 7% for 25 years.

   (a) Find the down payment.
   (b) Find the amount of the mortgage.
   (c) Find the monthly payment.
   (d) Find the total interest paid.

2. A house sells for $82,500.00 and a 5% down payment is made. A mortgage is secured at 10% for 15 years.

   (a) Find the down payment.
   (b) Find the amount of the mortgage.
   (c) Find the monthly payment.
   (d) Find the total interest paid.

3. A building sells for $200,000.00 and a 40% down payment is made. A 30-year mortgage at 11% was obtained.

   (a) Find the down payment.
   (b) Find the amount of the mortgage.
   (c) Find the monthly payment.
   (d) Find the total interest paid.

4. An ice cream store sells for $125,000.00 and a 12% down payment is made. A 25-year mortgage at 8.5% was obtained.

   (a) Find the down payment.
   (b) Find the amount of the mortgage.
   (c) Find the monthly payment.
   (d) Find the total interest paid.

5. An auto parts store sells for $52,500.00 and a 5% down payment is made. A 40-year mortgage at 12.5% was obtained.

   (a) Find the down payment.
   (b) Find the amount of the mortgage.
   (c) Find the monthly payment.
   (d) Find the total interest paid.

6. A beauty shop sells for $75,000.00 and a 22% down payment is made. A 20-year mortgage at 6.5% was obtained.

   (a) Find the down payment.
   (b) Find the amount of the mortgage.
   (c) Find the monthly payment.
   (d) Find the total interest paid.

7. An appliance store sold for $200,000.00. The buyer made a 30% down payment and secured a 20-year mortgage on the balance at 11.5%.

   (a) Find the down payment.
   (b) Find the amount of the mortgage.
   (c) Find the monthly payment.
   (d) Find the total interest paid.

8. A grocery store sells for $550,000.00 and a 25% down payment is made. A 40-year mortgage at 10% was obtained.

   (a) Find the down payment.
   (b) Find the amount of the mortgage.
   (c) Find the monthly payment.
   (d) Find the total interest paid.

9. Compute an amortization schedule for the first three months for the house purchased in Exercise 1 of this section.

10. Compute an amortization schedule for the first three months for the house purchased in Exercise 2 of this section.

11. Compute an amortization schedule for the first three months for the appliance store purchased in Exercise 7 of this section.

12. Compute an amortization schedule for the first three months for the grocery store purchased in Exercise 8 of this section.

13. Find the property tax paid on a house whose market value is $160,000.00 that is assessed at 25% of market value. The tax rate is 75 mills.

14. Find the property tax paid on a building whose market value is $127,000.00 that is assessed at 30% of market value. The tax rate is 85 mills.

15. Find the property tax paid on a building whose market value is $375,000.00 that is assessed at 50% of market value. The tax rate is 36.5 mills.

16. Find the property tax on a house whose market value is $49,000.00 that is assessed at 75% of market value. The tax rate is 15 mills.

17. Find the property tax on a house whose market value is $110,000.00 that is assessed at 40% of market value. The tax rate is 15 mills.

18. Find the property tax on a house whose market value is $52,800.00 that is assessed at 20% of market value. The tax rate is 62.5 mills.

19. Find the property tax on a house whose market value is $275,000.00 that is assessed at 80% of market value. The tax rate is 105 mills.

20. Find the property tax on a pharmacy whose market value is $250,000.00 that is assessed at 50% of market value. The tax rate is 110 mills.

Writing Exercises

21. Explain the difference between a fixed-rate mortgage and an adjustable-rate mortgage.

22. What is a down payment?

23. Explain how to find the total interest paid on a mortgage.

24. What is a property tax?

25. What is the difference between the market value of a home and the assessed value of a home?

26. What is a mill?

Critical Thinking

27. You decide to purchase an $80,000.00 home. If you make a 25% down payment, you can get a 20-year mortgage at 9%, but if you can make a 10% down payment, you can get a 25-year mortgage at 7%. Which is the better option for you?

28. Which mortgage would cost you less, a 30-year mortgage at 8.5% or a 15-year mortgage at 12%?

9-5   Markup and Markdown

In order to make a profit, a retail store must sell merchandise for more than it pays for it. In addition, the owner must also pay for his or her overhead. This includes among other things the salary of the employees, and the cost of utilities, taxes, and insurance.

The difference between the cost of an item and the selling price of an item is called the *markup*. In addition to retail stores, wholesalers, distributors, and manufacturers use markup to make a profit.

The basic definitions are

> The **markup** (M) for an item is the difference between the cost and the selling price of an item.
> The **cost** (C) of an item is the price that the merchant pays for the item.
> The **selling price** (S) of an item is the price for which the merchant sells the item.

The basic formulas are

> Markup = Selling price − Cost   or $M = S - C$
> Selling price = Cost + Markup   or $S = C + M$
> Cost = Selling price − Markup   or $C = S - M$

There are two kinds of markup. The markup for an item can be based on the cost or on the selling price of the item.

Markup on Cost

When finding the markup based on the cost of an item, a percent of the cost is found and then it is added on to the cost to get the selling price. In this case, the cost is the base or 100% and the markup is a percent of the cost.

The formula shown in Section 9-1 can be used here. In this case, the cost is the base (B). The rate (R) is the percent. The percentage (P) is the part of the base or the amount of the markup. Hence $P = R \times B$, $R = \frac{P}{B}$, and $B = \frac{P}{R}$.

> **Math Note**
>
> Be sure to change the rate or percent to a fraction or decimal before multiplying or dividing.

Example   9-31

A gift store sells salt and pepper shakers for a 40% markup on cost. If the salt and pepper shakers cost the merchant $2.00, find the selling price.

Solution

Multiply $2.00 by 40% to get the markup ($P = R \times B$).

$$0.40 \times \$2.00 = \$0.80$$

Add the markup to the cost to get the selling price ($S = C + M$).

Hence, the item will sell for $2.80.

Sometimes it is necessary to find the rate or percent of the markup. Example 9-32 shows this.

Example 9-32

If an item costs $200.00 and sells for $250.00, find the markup rate.

Solution

First, find the markup using $M = S - C$.

$$\$250.00 - \$200.00 = \$50.00$$

Next find the rate using $R = \frac{P}{B}$.

$$R = \frac{50}{200} = 0.25 = 25\%$$

Hence, there is a 25% markup based on cost.

Example 9-33

The markup on an item is $32.00. If it is marked up 80% on cost, find the cost and the selling price.

Solution

Divide the markup by 80% to get the cost. $B = \frac{P}{R}$.

$$\frac{32}{80\%} = \frac{32}{0.8} = \$40.00$$

Add the cost and the markup to get the selling price.

$$S = C + M$$
$$= \$40.00 + \$32.00$$
$$= \$72.00$$

Try This One

9-Q

(a) Find the markup and selling price for an item that costs $12.00 and is marked up 60% on cost.

(b) If an item costs $48.00 and the markup is $15.00, find the markup rate based on the cost.

(c) The markup on an item is $72.00 and the markup rate is 90% on cost. Find the cost and selling price.

For the answer, see page 455.

Sidelight

Mathematics and Economics

Economists use mathematics in many ways. One of these ways is to predict the effects of inflation. Inflation is defined as an increase in the quantity of money or credit that produces a rise in prices.

One way to approximate the effect of inflation on the cost of an item 5 years from now is to change the inflation rate to a decimal, add 1, raise this number to the fifth power, and multiply by the cost of the item today. For example, an item that costs $100.00 today will cost $133.82 five years from now at an annual inflation rate of 6%. (i.e., $100(1.06)^5 = \$133.82$.)

The assumption here is that the inflation rate will remain at 6% every year. In real life, the rate changes each year; however, one must assume that the average rate will be 6% per year to arrive at the answer.

Markup on Selling Price

When finding the markup based on the selling price, the selling price becomes the base (B) or 100%. The markup amount then is the percentage (P), and is based on the selling price. The rate (R) is the percent. The same formulas are used, but the selling price becomes the base.

Example 9-34

An item sells for $32.00 and is marked up 40% on the selling price. Find the markup and cost of the item.

Solution

The markup is $P = B \times R$.

$$\$32.00 \times 40\% = 32 \times 0.40 = \$12.80$$

The cost is

$$C = S - M$$
$$= \$32.00 - \$12.80$$
$$= \$19.20$$

Example 9-35

The markup on the selling price of an item is $48.00. If the item costs $64.00, find the markup rate.

Solution

Since the markup is based on the selling price, the selling price must be found first.

$$S = C + M$$
$$= \$64.00 + \$48.00$$
$$= \$112.00$$

—Continued

Example 9-35 *Continued—*

Next find the rate.

$$R = \frac{B}{P} = \frac{\$48.00}{\$112.00} \approx 0.43 = 43\%$$

The markup rate on the selling price is 43%.

Example 9-36

Find the selling price of an item that is marked up $8.00 when the markup rate based on the selling price is 16%.

Solution

$$B = \frac{P}{R} = \frac{\$8.00}{16\%} = \frac{8}{0.16} = 50$$

The selling price is $50.00.

Math Note

The problem in Example 9-36 can be checked by finding 16% of $50.00. Thus $0.16 \times 50 = 8$. Hence the answer $50.00 is correct.

Try This One

9-R   Find the missing values for each.

	Cost	Selling Price	Markup on	Markup Rate	Markup Amount
(a)	$300.00	_____	cost	75%	_____
(b)	_____	$75.00	selling price	15%	_____
(c)	$90.00	$130.00	cost	_____	_____
(d)	$25.00	$40.00	selling price	_____	_____
(e)	_____	_____	cost	60%	$36.00
(f)	_____	$150.00	selling price	20%	_____

For the answer, see page 456.

Conversions

Given the markup rate on cost, one can find the markup rate on the selling price, or given the markup on the selling price, one can find the markup rate on cost. The appropriate formulas follow.

Markup rate on selling price: $\dfrac{\text{markup rate on cost}}{1 + \text{markup rate on cost}}$

Markup rate on cost: $\dfrac{\text{markup rate on selling price}}{1 - \text{markup rate on selling price}}$.

Be sure to convert the percents to decimals before substituting in the formulas.

Example   9-37

Find the markup rate on the selling price if the markup rate on the cost is 100%.

Solution

Change 100% to 1.00 and substitute in the formula.

$$\text{Markup rate on selling price} = \frac{\text{markup rate on cost}}{1 + \text{markup rate on cost}}$$

$$= \frac{1.00}{1 + 1.00} = \frac{1}{2} = 0.5 = 50\%$$

Hence, a 100% markup on the cost is equivalent to a 50% markup on the selling price.

Example   9-38

Find the markup on cost if the markup on the selling price is 30%.

Solution

Change 30% to 0.30 and substitute in the formula.

$$\text{Markup rate on cost} = \frac{\text{markup rate on selling price}}{1 - \text{markup rate on selling price}}$$

$$= \frac{0.30}{1 - 0.30} = \frac{0.30}{0.70} \approx 0.429 = 42.9\%$$

Hence, a 30% markup on the selling price is equivalent to a 42.9% markup on the cost.

Try This One

9-S

(a) Find the markup rate on cost equivalent to a 25% markup on the selling price.

(b) Find the markup rate on the selling price equivalent of a markup rate of 48% on cost.

For the answer, see page 456.

Markdown

In business, merchants sometimes have to mark down an item in order to sell it. There are several reasons items are marked down. When the seasons change, a merchant will want to get rid of summer clothes in order to sell fall clothing. A manufacturer is bringing out a new model and will not manufacture the old models any longer. The store may be moving or going out of business. *When computing **markdown**, always use the selling price as the base!*

Example   9-39

An item sells for $30.00 and is marked down 20%. Find the reduced price for a sale.

Solution

Find the amount of the markdown.

$$P = B \times R$$
$$= \$30.00 \times 20\% = \$30.00 \times 0.20 = \$6.00$$

The markdown is $6.00.

Find the reduced price

$$\text{Reduced price} = \$30.00 - \$6.00 = \$24.00$$

Hence, the reduced price is $24.00.

Sometimes items are marked up and marked down in a sequence. In this case, perform each operation as it occurs in the problem.

Example   9-40

Bill's Sport Shop paid $600.00 for a treadmill. It was marked up 40% on cost on November 8. It was marked down 15% for a Thanksgiving sale. On December 1, it was marked up 20% on the reduced price and on December 18 marked down 10% for a holiday sale. Find the final selling price.

Solution

Find the first markup.

$$P = B \times R$$
$$= \$600.00 \times 40\% = 600 \times 0.40 = \$240.00$$
$$S = C + M$$
$$= \$600.00 + \$240.00$$
$$= \$840.00$$

Find the markdown using $840.00 as the base.

$$P = B \times R$$
$$= \$840.00 \times 15\% = 840 \times 0.15$$
$$= \$126.00, \text{ then } \$840.00 - \$126.00 = \$714.00$$

Find the markup using $714.00 as the base.

$$P = B \times R$$
$$= \$714.00 \times 20\% = 714 \times 0.20$$
$$= \$142.80, \text{ then } \$714.00 + \$142.80 = \$856.80$$

—Continued

Example **9-40**   *Continued—*

Find the last markdown.

$$P = B \times R$$
$$= \$856.80 \times 10\% = 856.80 \times 0.10 = \$85.68$$

Then $856.80 − $85.68 = $771.12. Hence, the final selling price of the treadmill is $771.12.

Try This One

9-T   Murry's Hardware Store purchased a cordless drill for $70.00. It was marked up 90% on cost. Two months later, it was marked down 30% for a spring fix-up sale. After the sale, it was marked up 40%. Find the final selling price.
For the answer, see page 456.

Exercise Set   9-5

Computational Exercises

For Exercises 1–16, find the missing numbers.

	Cost	Selling Price	Markup on	Markup Rate	Markup Amount
1.	$2250.00	_____	cost	55%	_____
2.	$475.00	_____	cost	125%	_____
3.	_____	$80.00	selling price	10%	_____
4.	_____	$1250.00	selling price	30%	_____
5.	$60.00	$85.00	cost	_____	_____
6.	$800.00	$1250.00	cost	_____	_____
7.	$130.00	$200.00	selling price	_____	_____
8.	$1650.00	$2000.00	selling price	_____	_____
9.	_____	$84.00	selling price	_____	$40.00
10.	_____	$196.00	selling price	_____	$100.00
11.	_____	_____	cost	45%	$90.00
12.	_____	_____	cost	60%	$32.00
13.	$180.00	_____	cost	200%	_____
14.	_____	$3000.00	selling price	75%	_____
15.	_____	_____	cost	150%	$50.00
16.	_____	_____	selling price	10%	$90.00

Real World Applications

17. A keyboard was purchased at a cost of $30.00 and is marked up at 20% of cost. Find the amount of the markup and selling price.

18. A 600-watt microwave oven costs a merchant $425.00. At the store, the price was marked up 80% on cost. Find the amount of the markup and the selling price.

19. Two 14-channel radios were each marked up $20. This is 30% of the cost. Find the cost and the selling price of each radio.

20. A Hi-Fi VCR was marked up 80% on cost. If the amount of the markup is $20.00, find the cost and the selling price.

21. A CD boom box costs $50.00 and sells for $70.00. Find the markup amount and the percent markup on the cost.

22. A cordless phone costs $60.00 and sells for $80.00. Find the amount of the markup and the percent markup on the cost.

23. Gold earrings sell for $520.00 a pair. This is a 50% markup on the selling price. Find the amount of the markup and the cost.

24. A sterling silver bracelet sells for $56.00 and is marked up 60% on the selling price. Find the amount of the markup and the cost.

25. A combination necklace and bracelet is marked up $150.00, and this is equivalent to an 80% markup on the selling price. Find the selling price and the cost.

26. A wristwatch is marked up $40.00, and this is equivalent to a 20% markup on the selling price. Find the amount of the selling price and the cost.

27. Bedroom slippers cost $8.00 and sell for $15.00. Find the percent markup on the selling price.

28. A decorative holiday candle costs $10.00 and sells for $30.00. Find the percent markup on the selling price.

29. A picture frame sells for $56.99 and is reduced by $20.00. Find the percent of the markdown.

30. A men's sport shirt originally sells for $42.00 and is marked down to $21.00. Find the percent of the markdown.

31. A woman's bathrobe sells for $60.00 and is marked down to $36.00. Find the percent of the markdown.

32. A pair of shoes sells for $70.00 and is marked down to $45.00. Find the percent of the markdown.

33. A home pizza oven sells for $98.00 and is marked down 40%. Find the amount of the markdown and the reduced price.

34. A high speed blender is priced at $42.00 and is marked down 30%. Find the amount of the markdown and the reduced price.

35. A set of glassware sells for $15.00 and is marked down 60%. Find the amount of the markdown and the reduced price.

36. A toaster oven sells for $30.00 and is marked down 25%. Find the amount of the markdown and the reduced price.

37. On October 1, the Bright Light Candle Store marked up a $40 candle by 20%. On November 3, the same item was marked down 10%. It was marked up 30% on sales

on December 1. Finally, it was marked down 25% for Christmas sales. Find the final selling price.

38. At 1-month intervals, and for various reasons, the price of a chest of drawers changed from $178 as follows: up by 15%, down by 20%, down by 10%, up by 28%. Find the final price.

39. Find the markup rate on the selling price that is equivalent to a 55% markup cost.

40. Find the markup rate on the selling price that is equivalent to a 75% markup on cost.

41. Find the markup rate on cost that is equivalent to a 65% markup on the selling price.

42. Find the markup rate on cost that is equivalent to a 28% markup on the selling price.

Writing Exercises

43. Explain the difference between the markup on cost and the markup on selling price.

44. Explain why a store might mark down an item that would sell for less than the owner paid for it.

Critical Thinking

45. Why can't an item be marked up 110% on the selling price?

46. If an item was marked up 100% of the selling price, how much would it cost?

47. If an item is marked down 20%, what is the percent markup that would bring the item back up to the original selling price?

Summary

Section	Important Terms	Important Ideas
9-1	percent base rate part	**Percent** means hundredths. That is 45% means 45 hundredths. In order to use percents when doing calculations, the percents must be changed to fractions or decimals. There are three types of basic percent problems. They are (1) finding a percent of a number; (2) finding what percent one number is of another; and (3) finding a number when a percent of it is known.
9-2	interest simple interest principal rate time (term) maturity value compound interest effective rate annual yield annuity future value	**Percents** are used quite frequently in the business world. One application of percent is in the banking world. When you borrow money, you pay a fee for its use. This fee is called interest. Likewise, when you put money into any type of savings account, the money earns interest. There are two types of interest, depending on how it is computed. Simple interest is computed on the principal alone while compound interest is computed on the principal plus any previously accumulated interest. Compound interest investments earn more money than simple interest investments if the rates are the same. Since the rate of return is higher than the stated rate when interest is compounded more than once a year, the true interest rate is called the effective rate or annual yield. An annuity is a savings investment whereby an individual or business makes the same payment each period into a compound interest account where the rate remains the same for the term of the annuity.
9-3	installment loan finance charge unpaid balance method average daily balance method annual percentage rate rule of 78s	**Credit card** companies also charge interest. There are two ways the companies compute interest. One method is computing interest on the unpaid balance. Hence you are charged interest only on last month's balance. The other method is called the average daily balance. Here the interest is computed on the average balance on all of the days of the month. This includes any purchases and payments made during the month. When a loan is paid back in monthly payments, it is called an installment loan. Because you pay back some of the principal each month, you do not have the full use of the money for the term of the loan. This means that the actual interest is higher than the stated interest rate. This actual interest rate is called the annual percentage rate and can be computed approximately by the constant ratio formula. When an installment loan is paid off early, the amount of interest saved can be determined by the rule of 78s.
9-4	mortgage fixed-rate mortgage adjustable-rate mortgage graduated payments amortization schedule property tax mill market value assessed value	**Purchasing** a home is an important factor in one's life, and since most people cannot afford to pay the cash for a home, they will be required to get a loan. A loan for a home is called a mortgage, and the schedule that is used when the loan is paid back is called an amortization schedule. Most home owners are required to pay a property tax. The home has a market value, which is an estimate of the value of the home if it were sold today, and also an assessed value, which is usually a percentage of the market value. The property tax is based on the assessed value and is given in mills. A mill is 0.001 of a dollar.
9-5	markup cost selling price markdown	**Markup** is used by stores to make a profit and is the difference between the cost and selling price of an item. Markup rate can be used on the cost of the item or the selling price of an item. An item is marked down as an incentive to sell it.

Review Exercises

For Exercises 1–10, find the missing value.

	Fraction	Decimal	Percent
1.	$\frac{7}{8}$	_____	_____
2.	_____	0.54	_____
3.	_____	_____	80%
4.	$\frac{5}{12}$	_____	_____
5.	_____	_____	185%
6.	_____	0.06	_____
7.	$5\frac{3}{4}$	_____	_____
8.	_____	1.55	_____
9.	_____	_____	45.5%
10.	$\frac{3}{8}$	_____	_____

11. Find 72% of 96.

12. 18 is what percent of 60?

13. 25% of what number is 275?

14. If the sales tax is 5% on a calculator, find the tax and the total cost if the calculator is $19.95.

15. If the sales tax on a coffee table is $3.60, find the cost of the table if the tax rate is 6%.

16. Marcia received a commission of $2275.00 for selling a small home. If she receives a 7% commission, find the price of the home.

For Exercises 17–24, find the missing value.

	Principal	Rate	Time	Simple Interest
17.	$4300.00	9%	6 years	_____
18.	$16,000.00	_____	3 years	$1920.00
19.	$875.00	12%	_____	$262.50
20.	$50.00	6%	18 months	_____
21.	$230.00	_____	6.5 years	$104.65
22.	_____	3%	5 years	$63.75
23.	_____	14%	2 years	$385.00
24.	$785.00	_____	12 years	$1130.40

For Exercises 25–28, find the compound interest and maturity value.

	Principal	Rate	Compounded	Time
25.	$1775.00	5%	annually	6 years
26.	$200.00	4%	semiannually	10 years
27.	$45.00	8%	quarterly	3 years
28.	$21,000.00	6%	quarterly	7 years

29. Ace Auto Parts borrowed $6000.00 at 6% for 5 years to enlarge its display area. Find the simple interest and maturity value of the loan.

30. Sam's Sound Shack borrowed $13,450.00 at 8% for 15 years to remodel its existing store. Find the simple interest and maturity value of the loan.

31. Julie earned $60.48 in simple interest on a savings account balance of $4320.00 over a 12-month period. Find the rate of interest.

32. John has an opportunity to purchase a new boat. He must borrow $5300.00 at 11% simple interest for 36 months. Find the monthly payment.

33. Find the effective rate when the stated rate is 12% and the interest is computed quarterly.

34. The Evergreen Landscaping Company will need to purchase a new backhoe in 7 years. The owner purchases an annuity that pays 8.3% interest compounded semiannually. If the semiannual payment is $4000.00, find the future value of the annuity in 7 years.

35. Mike and Marie wish to take an African vacation in 3 years. In order to save money for the trip, they purchase an annuity that pays 3% interest compounded quarterly. Find the future value of the annuity in 3 years if their quarterly payment is $650.00.

36. For the month of February, Pete had an unpaid balance of $563.25 at the beginning of the month. He had purchases of $563.25 and made a payment of $350.00 during the month. Find the finance charge if the interest rate is 1.75% per month on the unpaid balance and find the new balance on March 1.

37. Sid's Used Cars had these transactions on its credit card statement:

April 1	Unpaid balance	$5628.00
April 10	Purchases	$2134.60
April 22	Payment	$900.00
April 28	Purchases	$437.80

Find the finance charge if the interest rate is 1.8% on the average daily balance and find the new balance for May 1.

38. Mark must borrow $20,000.00 to purchase an automobile. The money is to be paid back in 36 monthly installments of $660.00. Find the annual percentage rate using the constant ratio formula.

39. A loan for $1500.00 is to be paid back in 30 monthly installments of $61.25. The borrower decides to pay off the balance after 24 payments have been made. Find the amount of interest saved. Use the rule of 78s.

40. A home is purchased for $145,000.00 with a 20% down payment. The mortgage rate is 8.5% and the term of the mortgage is 25 years.

   (a) Find the amount of the down payment.
   (b) Find the amount of the mortgage.
   (c) Find the monthly payment.
   (d) Compute an amortization schedule for the first 2 months.

41. Find the property tax on a $135,000.00 home if it is assessed at 30% of the market value and the tax rate is 70 mills.

For Exercises 42–46, find the missing values.

Cost	Selling Price	Markup on	Markup Rate	Markup Amount
42. $425.00	_____	cost	95%	_____
43. _____	$25.00	selling price	18%	_____
44. $1950.00	$2950.00	cost	_____	_____
45. $675.00	$900.00	selling price	_____	_____
46. _____	_____	cost	80%	$480.00

47. Find the markup rate on cost equivalent to a 42% markup on the selling price.

48. A student desk that sells for $98.00 is marked down 30%. Find the reduced price.

Chapter Test

1. Change $\frac{5}{16}$ to a percent.

2. Write 0.63 as a percent.

3. Write 28% as a fraction in lowest terms.

4. Change 16.7% to a decimal.

5. 32 is what percent of 40?

6. Find 87.5% of 48.

7. 45% of what number is 135?

8. Find the sales tax on a toaster oven that sells for $29.95. The tax rate is 8%.

9. If a person receives a 15% commission on all merchandise sold, find the amount sold if the person's commission is $385.20.

10. On the first day of math class, 28 students were present. The next day, 7 more students enrolled in the class because the other section was canceled. Find the percent increase in enrollment.

11. Find the simple interest on $1350.00 at 12% for 3 years.

12. Find the rate for a principal of $200.00 invested for 15 years if the simple interest earned is $150.00.

13. Ron's Detailing Service borrowed $435.00 at 3.75% for 6 months to purchase new equipment. Find the simple interest and maturity value of the loan and the monthly payment.

14. The Express Delivery borrowed $1535.00 at 4.5% for 3 months to purchase safety equipment for its employees. Find the simple interest and maturity value of the loan and the monthly payment.

15. Benson Electric borrowed $1800.00 at 12% for 1 year from a local bank. Find the interest and maturity value of the loan and the monthly payment.

16. Find the interest and maturity value for a principal of $500.00 invested at 6.5% compounded semiannually for 4 years.

17. Find the interest and maturity value on a principal of $9750.00 invested at 10% compounded quarterly for 6 years.

18. In order to purchase a motorcycle, John borrowed $12,000.00 at 9.5% for 4 years. Find his monthly payment.

19. Find the effective rate when the stated interest rate is 8% and the interest is compounded semiannually.

20. In order to open a new branch of her business in 3 years, the owner of Quick Fit Fitness Center purchases an annuity that pays 4.5% interest compounded semiannually. If her semiannual payment is $3,000, find the future value of the annuity in 3 years.

21. For the month of November, Harry had an unpaid balance of $1250.00 on his credit card. During the month, he made purchases of $560.00 and a payment of $800.00. Find the finance charge if the interest rate is 1.6% per month on the unpaid balance and find the new balance on December 1.

22. Rhonda's credit card statement for the month of May shows these transactions.

May 1	Unpaid balance	$474.00
May 11	Payment	$300.00
May 20	Purchases	$86.50
May 25	Purchases	$120.00

Find the finance charge if the interest rate is 2% on the average daily balance and find the new balance on June 1.

23. Tamara borrowed $800.00 for tuition. She is to pay it back in 12 monthly installments of $70.70. Find the annual percentage rate.

24. A loan for $2200.00 is to be paid off in 24 monthly installments of $111.85. The borrower decides to pay off the loan after 20 payments have been made. Find the amount of interest saved.

25. A home is purchased for $80,000.00 with a 5% down payment. The mortgage rate is 9% and the term is 30 years.

   (a) Find the amount of the down payment.
   (b) Find the amount of the mortgage.
   (c) Find the monthly payment.
   (d) Compute an amortization schedule for the first 2 months.

26. A home has a market value of $160,000.00 and is assessed at 40% of the market value. Find the assessed value of the house and the property tax if the rate is 70 mills.

27. If a telephone costs $15.00 and is sold for $25.00, find the markup rate on the cost and the markup rate on the selling price.

28. Find the selling price on an item that costs $65.00 when there is a 13% markup on the cost.

29. Find the markup rate on the cost that is equivalent to a 20% markup on the selling price.

30. An ice crusher that sells for $50.00 is marked down 35%. Find the reduced price.

Mathematics in Our World ▶*Revisited*

How Much Can You Afford to Pay for a Home?

Experts suggest that a person can afford to pay 28% of his or her gross monthly income for a home mortgage. Given this assumption and using Table 9-1 on page 435, you can figure out how much you can afford to pay for a home, as shown.

First find the monthly income.

$$\$36,000.00 \div 12 = \$3000.00$$

Next find 28% of the monthly income.

$$0.28 \times \$3000.00 = \$840.00$$

Hence you can afford a monthly mortgage payment of $840.00.

Now to see what you can afford to borrow, look up the number corresponding to 7% and 25 years in Table 9-1 on page 435. It is 7.70. Set up an equation and solve for x.

$$7.70x = \$840$$
$$\frac{7.70x}{7.70} = \frac{840}{7.70}$$
$$x = \$109.09$$

Multiply x by 1000 since the monthly payments in Table 9-1 are per $1000.00 of the mortgage.

$$\$109.09 \times 1000 = \$109,090$$

Hence, you can afford a mortgage of $109,090. Finally, add your down payment of $10,000.00.

$$\$109,090.00 + \$10,000.00 = \$119,090.00$$

You can purchase a home costing about $119,090.00. (The process used here is the reverse of the process used in Example 9-27 in this chapter.)

Projects

1. You have $1000.00 to invest. Investigate the advantages and disadvantages of each type of investment.

   (a) Checking account
   (b) Money market account
   (c) Passbook savings account
   (d) Certificate of deposit

   Write a short paper indicating which type of account you have chosen and why you chose that account.

2. Banks use two ways to compute the terms of a loan. They are ordinary time and exact time. Find a resource that explains the difference between the two methods and write a short paper on each method.

3. Money that will be needed in the future has a value today. It is called the *present value*. Find a resource that explains the idea of the present value of money and write a short paper explaining the concept.

4. Another way to save money is called an *annuity*. There are two types of annuities, ordinary annuities and annuities due. Find a resource that explains annuities and write a short paper on the two types of annuities.

5. There are many fees involved when buying or selling a home. Some of these include an appraisal fee, survey fee, etc. Consult a real estate agency to see what is necessary to purchase a home in your area and write a short paper on the necessary closing costs.

Answers to **Try This One**

9-A. (a) 0.625
     (b) 0.03
     (c) 2.50
     (d) $\frac{9}{10}$
     (e) $\frac{33}{200}$
     (f) $\frac{13}{10}$
     (g) 97%
     (h) 37.8%
     (i) 31.25%
     (j) 175%

9-B. (a) 84%
     (b) 9.72
     (c) 40

9-C. 1039%

9-D. $5400

9-E. $975; $5975; $165.97

9-F. (a) 5%
     (b) $425
     (c) 5 years

9-G. $183.63

9-H. 8.24%

9-I. $3611.51

9-J. Finance charge = $16.93; balance on Feb. 1 = $1046.29

9-K. (a) $1198.69
     (b) $22.78
     (c) $1081.60

9-L. 15.4%

9-M. $88

9-N. (a) $9200
     (b) $82,800
     (c) $791.57

9-O.

Payment Number	Interest	Payment on Principal	Balance of Loan
1	$552	$239.57	$82,560.43
2	$550.40	$241.17	$82,319.26
3	$548.80	$242.77	$82,076.49

9-P. $2146.50

9-Q. (a) $M = \$7.20$; $S = \$19.20$
     (b) 31.25%
     (c) $C = \$80$; $S = \$152$

9-R. (a) $S = \$525$; $M = \$225$
(b) $C = \$63.75$; $M = \$11.25$
(c) $R \approx 0.44 = 44\%$; $M = \$40$
(d) $R = 37.5\%$; $M = \$15$
(e) $C = \$60$; $S = \$96$
(f) $C = \$120$; $M = \$30$

9-S. (a) 33.3%
(b) 32.4%

9-T. $130.34

Chapter

Ten

Geometry

Outline

Objectives

After completing this chapter, you should be able to

1 Identify the basic geometric figures such as the line, half line, ray, etc. (10-1)

2 Name an angle in four different ways (10-1)

3 Classify angles as acute, right, obtuse, and straight (10-1)

4 Find the measures of complements and supplements of angles (10-1)

5 Find the measures of angles formed by parallel lines and transversals (10-1)

6 Classify triangles according to the measures of their sides or angles (10-2)

7 Given the measures of two angles of a triangle, find the measure of the third angle (10-2)

8 Solve problems using the Pythagorean theorem (10-2)

9 Solve problems involving similar triangles (10-2)

10 Name polygons according to the number of sides (10-3)

11 Identify the different types of quadrilaterals (10-3)

12 Find the perimeter of polygons (10-3)

13 Find the area of polygons (10-4)

14 Find the circumference and the area of a circle (10-4)

15 Find the surface area and volume of a solid figure (10-5)

16 Define the three trigonometric ratios: sine, cosine, and tangent (10-6)

17 Solve right triangle problems using the three trigonometric ratios (10-6)

18 Define and solve problems using angle of elevation and angle of depression (10-6)

19 Determine whether or not a given network is traversable (10-7)

Introduction

This chapter explains the basic concepts of geometry, the mathematics used to describe and analyze shapes and space. **Geometry** includes the study of points, lines, angles, plane figures (triangles, squares, etc.), and solids (cubes, cylinders, etc.). The word "geometry" is derived from two Greek words meaning "earth measure." The Egyptians used geometry in land measurement and in architecture.

The study of geometry includes much more information than is presented in this chapter. An early Greek mathematician, Euclid, collected all that was known about geometry and wrote a treatise called *Elements* in 300 B.C.E. This treatise consisted of 13 books; the first six contained information on plane geometry, and the others contained information on solid geometry and arithmetic.

We begin by studying the basic geometric figures such as points, lines, angles, and planes. This is followed by an explanation of the triangle and other polygons. The next sections include an explanation of the concepts of perimeter, area, and volume. This is followed by a section on the basic concepts of right triangle trigonometry. The chapter concludes with a presentation of network theory. ■

10-1 Points, Lines, Planes, and Angles

Points, Lines, and Planes

The basic geometric figures consist of the point, the line, and the plane. These figures are theoretical and cannot be specifically defined. A **point** is represented by a dot, but theoretically it has no dimensions. A **line** is a set of connected points that has an infinite length but no width. Lines are assumed to be straight unless otherwise defined and are determined by two specific points. A **plane** is a flat surface that is infinite in length and width but has no thickness.

A point is named by a capital letter and is symbolized by a dot. A line is named by two points that determine the line and is symbolized by $\overleftrightarrow{AB}$. Lines can also be named by lowercase letters. See Figure 10-1.

Points and lines can be used to make other geometric figures. See Figure 10-2. A *line segment* consists of two points called endpoints and the part of a line between the two endpoints.

Basic Geometric Figures

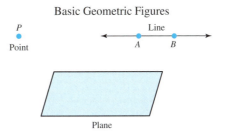

Figure 10-1

The symbol for a line segment is $\overline{AB}$. When the endpoints are not included in the segment, it is called an **open segment** and it is represented by the symbol $\overset{\circ\!-\!\circ}{AB}$. When the segments contain one endpoint and not the other, it is called a **half open segment** and is represented by $\overset{\bullet\!-\!\circ}{AB}$ or $\overset{\circ\!-\!\bullet}{AB}$.

The half open line segment symbolized by $\overset{\bullet\!-\!\circ}{AB}$ means the segment contains point A but not point B. The half open segment symbolized by $\overset{\circ\!-\!\bullet}{AB}$ means that the segment contains the point B but not the point A.

A point on a line separates the line into two half lines. The symbol for a **half line** is $\overset{\circ\!-\!\!\rightarrow}{AB}$. The half line does not contain the endpoint. When the figure consists of a half line and its endpoint, it is called a **ray.** The symbol for a ray is $\overset{\bullet\!-\!\!\rightarrow}{AB}$. See Figure 10-2.

> Two rays with a common endpoint form an **angle.** The rays are called the *sides* of the angle and the endpoint is called the **vertex.**

A ray of light starts at a point and continues out.

An angle can be named in several different ways. The symbol for an angle is $\angle$. The angle shown in Figure 10-3 can be named as $\angle ABC$, $\angle CBA$, $\angle B$, or $\angle 1$.

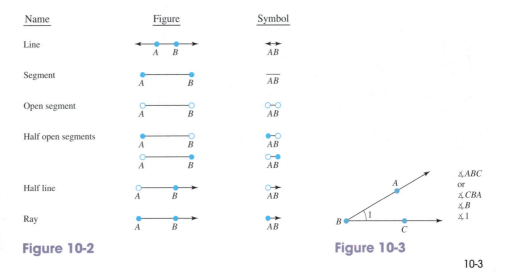

Name	Figure	Symbol
Line		$\overset{\leftrightarrow}{AB}$
Segment		$\overline{AB}$
Open segment		$\overset{\circ\!-\!\circ}{AB}$
Half open segments		$\overset{\bullet\!-\!\circ}{AB}$
		$\overset{\circ\!-\!\bullet}{AB}$
Half line		$\overset{\circ\!-\!\!\rightarrow}{AB}$
Ray		$\overset{\bullet\!-\!\!\rightarrow}{AB}$

Figure 10-2

Figure 10-3

Example 10-1

Name the angle shown here in four different ways.

Solution

$\angle RST$, $\angle TSR$, $\angle S$, and $\angle 3$.

An angle is measured in **degrees,** symbolized by $°$, and the instrument that is used to measure an angle is called a **protractor.** Figure 10-4 shows how to measure an angle using a protractor. The center of the base of the protractor is placed at the vertex of the angle, and the bottom of the protractor is placed on one side of the angle. The opening is measured where the other side falls on the scale. The angle shown in Figure 10-4 has a measure

Sidelight

Euclid ca. 323–ca. 285 B.C.E.

Very little is known about Euclid's life and personality, except that he was a professor of mathematics at the University of Alexandria. This university was built by Ptolemy to serve as the center of knowledge of the then-known world. It had lecture rooms, laboratories, gardens, and museums. The library contained over 600,000 papyrus scrolls. Scholars from various fields were invited to come and live there as long as they wanted. They received free room and board, were granted salaries, and were free from paying taxes. In return, they were only expected to give regular lectures.

Euclid wrote at least 10 works, but the most famous of these is the one entitled *Elements,* which consists of 13 books on geometry, number theory, and algebra, which pertains to geometry. What Euclid did was to collect and arrange systematically the works of early writers, and then he added his own contributions. He also used definitions and axioms to prove 465 theorems, or as he called them, propositions.

Only one book, the *Bible,* has been more widely translated, edited, and used. *Elements* remained basically the same as Euclid wrote it until the 1800s,

when mathematicians found defects in the logical structure of the books and set about correcting them. All modern editions of *Elements* are based on a revision prepared by Theon of Alexandria in about 400 C.E. Even Abraham Lincoln boasted that at the age of 40, he studied the first six books of *Elements* in order to train his mind to think logically.

A story told about Euclid is that the Egyptian ruler Ptolemy asked Euclid to show him a shortcut to geometry. Euclid replied, "There is no royal road to geometry."

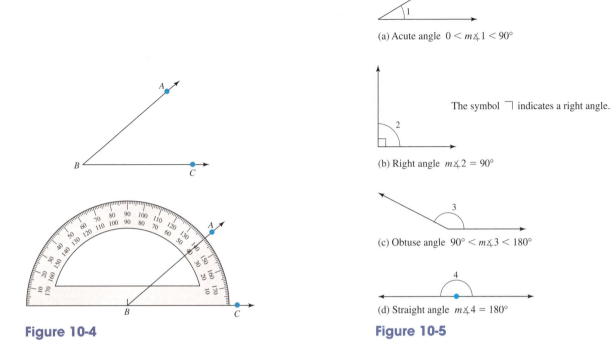

(a) Acute angle $0 < m\angle 1 < 90°$

The symbol ⌐ indicates a right angle.

(b) Right angle $m\angle 2 = 90°$

(c) Obtuse angle $90° < m\angle 3 < 180°$

(d) Straight angle $m\angle 4 = 180°$

Figure 10-4

Figure 10-5

of 40°. The symbol for the measure of an angle is $m\angle$; hence, $m\angle ABC = 40°$. Angles can be classified by their measurement.

> An **acute angle** has a measure between 0° and 90°.
>
> A **right angle** has a measure of 90°.
>
> An **obtuse angle** has a measure between 90° and 180°.
>
> A **straight angle** has a measure of 180°.
>
> See Figure 10-5.

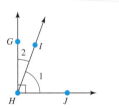

(a) Adjacent angles

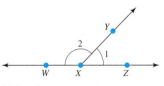

(b) Complementary angles
$m\angle 1 + m\angle 2 = 90°$

(c) Supplementary angles
$m\angle 1 + m\angle 2 = 180°$

Figure 10-6

Pairs of Angles

Pairs of angles have various names depending on how they are related.

> Two angles are called *adjacent angles* if they have a common vertex and a common side. Figure 10-6(a) shows a pair of adjacent angles, $\angle ABC$ and $\angle CBD$. The common vertex is B and the common side is $\overline{BC}$.
>
> Two angles are said to be **complementary** if the sum of their measures is 90°. Figure 10-6(b) shows two complementary angles. The sum of the measures of $\angle GHI$ and $\angle IHJ$ is 90°.
>
> Two angles are said to be **supplementary** if the sum of the measures of each is equal to 180°. Figure 10-6(c) shows two supplementary angles. The sum of the measures of $\angle WXY$ and $\angle YXZ$ is 180°.

Example   10-2

If $\angle FEG$ and $\angle GED$ are complementary and $m\angle FEG$ is 28°, find the $m\angle GED$.

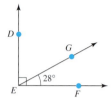

Solution

Since the two angles are complementary, the sum of their measures is 90°; therefore,

$$m\angle GED = 90° - m\angle FEG$$

$$= 90° - 28°$$

$$= 62°$$

Hence, the measure of $\angle GED = 62°$.

Example   10-3

If $\angle RQS$ and $\angle SQP$ are supplementary and $m\angle RQS = 135°$, find the measure of $\angle SQP$.

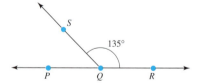

Solution

Since $\angle RQS$ and $\angle SQP$ are supplementary, the sum of their measures is 180°; therefore,

$$m\angle SQP = 180° - m\angle RQS$$

$$= 180° - 135°$$

$$= 45°$$

Hence, the measure of $\angle SQP$ is 45°.

Try This One

10-A Find the measures of each angle.

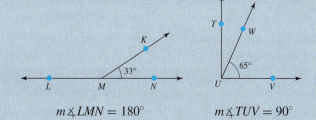

$$m\angle LMN = 180°$$ $$m\angle TUV = 90°$$

(a) $\angle KML$

(b) $\angle TUW$

For the answer, see page 522.

Example 10-4

If two adjacent angles are supplementary and one angle is three times as large as the other, find the measures of each.

Solution

Since the sum of the measures of the angles is 180° and one angle is three times as large as the other, an equation can be written for the relationships.

$$m\angle x + 3m\angle x = 180°$$

$$4m\angle x = 180°$$

$$\frac{4m\angle x}{4} = \frac{180°}{4}$$

$$m\angle x = 45°$$

The measure of one angle is 45°. The measure of the other angle is three times as large; hence, the measure of the other angle is

$$3m\angle x = 3 \cdot 45$$

$$= 135°$$

Hence, the measure of one angle is 45° and the measure of the other angle is 135°.

When two lines intersect, four angles are formed. See Figure 10-7.

The opposite angles formed by two intersecting lines are called **vertical angles.** The measures of vertical angles are equal.

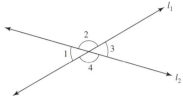

$\angle 1$ and $\angle 3$ are vertical angles; $m\angle 1 = m\angle 3$
$\angle 2$ and $\angle 4$ are vertical angles; $m\angle 2 = m\angle 4$

Figure 10-7

Figure 10-7 shows that when line l_1 intersects with line l_2, four angles are formed. $\angle 1$ and $\angle 3$ are called vertical angles, and $\angle 2$ and $\angle 4$ are called vertical angles. The measure of $\angle 1$ is equal to the measure of $\angle 3$. Likewise, the measure of $\angle 2$ is equal to the measure of $\angle 4$.

If the measure of one of the four angles is known, then the measure of the other three angles can be found. This is shown in Example 10-5.

Example   10-5

Find $m\angle 2$, $m\angle 3$, and $m\angle 4$ when $m\angle 1 = 40°$.

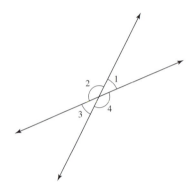

Solution

Since $\angle 1$ and $\angle 3$ are vertical angles and $m\angle 1 = 40°$, $m\angle 3 = 40°$. Since $\angle 1$ and $\angle 2$ form a straight angle (180°).

$$m\angle 2 = 180° - m\angle 1$$
$$= 180° - 40°$$
$$= 140°$$

Finally, since $\angle 2$ and $\angle 4$ are vertical angles, $m\angle 4 = 140°$.

Try This One

10-B   Find $m\angle 2$, $m\angle 3$, and $m\angle 4$ when $m\angle 1 = 75°$.

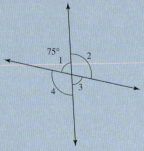

For the answer, see page 522.

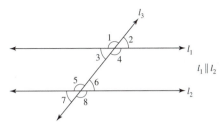

Pairs of alternate interior angles
$\angle 3$ and $\angle 6$ $m\angle 3 = m\angle 6$
$\angle 4$ and $\angle 5$ $m\angle 4 = m\angle 5$

Pairs of alternate exterior angles
$\angle 1$ and $\angle 8$ $m\angle 1 = m\angle 8$
$\angle 2$ and $\angle 7$ $m\angle 2 = m\angle 7$

Pairs of corresponding angles
$\angle 1$ and $\angle 5$ $m\angle 1 = m\angle 5$
$\angle 2$ and $\angle 6$ $m\angle 2 = m\angle 6$
$\angle 3$ and $\angle 7$ $m\angle 3 = m\angle 7$
$\angle 4$ and $\angle 8$ $m\angle 4 = m\angle 8$

Figure 10-8

Two straight lines in the same plane are **parallel** if they do not intersect. The symbol for parallel lines is ∥. (Remember that lines are infinite in length.)

When two parallel lines are cut by a third line called a **transversal,** eight angles are formed. There are three different kinds of pairs of angles. See Figure 10-8.

> **Alternate interior angles** are the angles formed between two parallel lines on the opposite sides of the transversal that intersects the two lines. Alternate interior angles have equal measures.
>
> **Alternate exterior angles** are the opposite exterior angles formed by the transversal that intersects two parallel lines. Alternate exterior angles have equal measures.
>
> **Corresponding angles** consist of one exterior and one interior angle on the same side of the transversal that intersects two parallel lines. Corresponding angles have equal measures.

When the measure of one of the eight angles is known, the measures of the other seven angles can be found.

Example   10-6

Find the measures of all the angles shown when the measure of $\angle 2$ is $50°$.

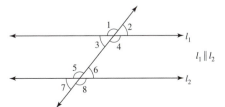

—Continued

Parallels and transversals can be found in many city plans.

Example 10-6 *Continued—*

Solution

If $m\angle 2 = 50°$, then $m\angle 3 = 50°$ since they are vertical angles. $\angle 3$ and $\angle 6$ are alternate interior angles and $m\angle 6 = 50°$. Finally, $m\angle 7 = 50°$ since $\angle 3$ and $\angle 7$ are corresponding angles.

Since $\angle 1$ and $\angle 2$ are supplementary, $m\angle 1 = 180° - m\angle 2 = 180° - 50° = 130°$. $m\angle 4 = 130°$ since $\angle 1$ and $\angle 4$ are vertical angles. $m\angle 5 = 130°$ since $\angle 4$ and $\angle 5$ are alternate interior angles.

Finally, $m\angle 8 = 130°$ since $\angle 5$ and $\angle 8$ are vertical angles.

Try This One

10-C Find the measures of all the angles shown when the measure of $\angle 1$ is 165°.

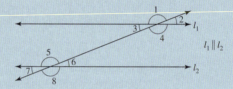

$l_1 \parallel l_2$

For the answer, see page 522.

Exercise Set 10-1

Computational Exercises

For Exercises 1–8, identify and name each figure.

1. $\underset{P}{\circ}\!\!-\!\!-\!\!\underset{Q}{\circ}$

2. $\underset{A}{\bullet}\!\!-\!\!-\!\!\underset{B}{\bullet}\!\!\rightarrow$

3. $\leftarrow\!\!\underset{R}{\bullet}\;\underset{S}{\bullet}\!\!\rightarrow$

4. $\leftarrow\!\!-\!\!-\!\!-\!\!\rightarrow\, l$

5. $\underset{C}{\bullet}\!\!-\!\!-\!\!\underset{D}{\circ}$

6. $\bullet\, P$

7. $\underset{T}{\bullet}\!\!-\!\!-\!\!\underset{U}{\bullet}$

8. $\underset{E}{\circ}\!\!-\!\!-\!\!\underset{F}{\bullet}\!\!\rightarrow$

For Exercises 9–10, name each angle in four different ways.

9.

10.

For Exercises 11–14, classify each angle as acute, right, obtuse, or straight.

11.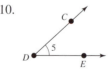

12.

13.    14.

For Exercises 15–22, identify each pair of angles as alternate interior, alternate exterior, corresponding, or vertical.

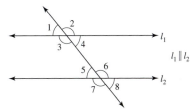

$l_1 \parallel l_2$

15. ∡1 and ∡4   16. ∡3 and ∡6   17. ∡2 and ∡6
18. ∡5 and ∡8   19. ∡1 and ∡5   20. ∡2 and ∡7
21. ∡1 and ∡8   22. ∡4 and ∡8

For Exercises 23–28, find the measure of the complement of each angle.

23. 8°   24. 24°   25. 32°
26. 56°   27. 78°   28. 84°

For Exercises 29–34, find the measure of the supplement for each angle.

29. 156°   30. 90°   31. 62°
32. 143°   33. 120°   34. 38°

For Exercises 35–38, find the measures of ∡1, ∡2, and ∡3.

35. 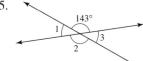   36.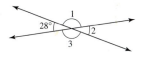

37. 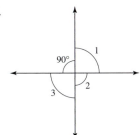   38.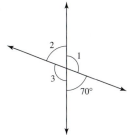

For Exercises 39 and 40, find the measure of ∡1 through ∡7.

39. 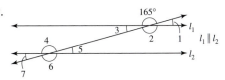   40.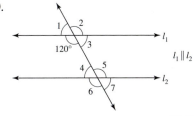

Real World Applications

For Exercises 41–44, identify the measure of the angle made by the hands of a clock at these times:

41. 3 o'clock   42. 6 o'clock

43. 2 o'clock   44. 4 o'clock

Writing Exercises

45. Explain the concepts of a point, a line, and a plane.

46. Explain how to name an angle four different ways.

47. Explain the difference between a ray and a half line.

48. Explain how to measure angles.

49. Explain the difference between complementary and supplementary angles.

50. Name and define three different types of pairs of angles that are found when two parallel lines are cut by a transversal.

Critical Thinking

51. Find the measure of $\angle 3$ and $\angle 4$ if $m\angle 1 = m\angle 2$ when line l_1 is parallel to line l_2.

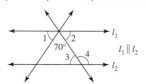

52. Find the $m\angle 1$ if line l_1 is parallel to line l_2.

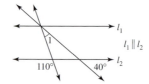

10-2 Triangles

The basic geometric figures such as the point, line, ray, angle, etc. were explained in Section 10-1. A geometric figure is said to be **closed** when one can start at one point, trace the entire figure and end at the starting point without lifting the pencil from the paper. See Figure 10-9. This section explains the geometric figure called a *triangle*.

A **triangle** is a closed geometric figure consisting of three sides and three angles.

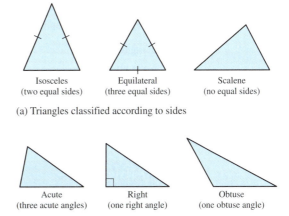

(a) Triangles classified according to sides

Isosceles (two equal sides)   Equilateral (three equal sides)   Scalene (no equal sides)

Acute (three acute angles)   Right (one right angle)   Obtuse (one obtuse angle)

(b) Triangles classified according to angles

Figure 10-10

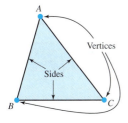

Figure 10-9

The **sides** of the triangle are line segments, and the **vertices** consist of the points of the intersection of the sides.

Types of Triangles

Triangles are often named by the vertices. The symbol for triangle is $\triangle$. The triangle shown in Figure 10-9 is named $\triangle ABC$.

Triangles can be classified according to the measures of their sides or the measures of their angles. Classification of triangles according to sides is as follows:

> An **isosceles triangle** has two sides of equal length.
> An **equilateral triangle** has three sides of equal length.
> A **scalene triangle** has no two sides of equal length.
> See Figure 10-10(a).

Classification of triangles according to angles is as follows:

> An **acute triangle** has three acute angles.
> An **obtuse triangle** has one obtuse angle.
> A **right triangle** has one right angle.
> See Figure 10-10(b).

Math Note

Because an isosceles triangle has two equal sides, it also has two equal angles. The equilateral triangle has three equal angles. The dash marks on the sides of the triangles indicate the sides have the same length.

Example 10-7

Identify the type of triangle.

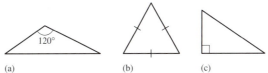

(a)   (b)   (c)

—Continued

10-13

Example   **10-7**   *Continued—*

Solution

(a) Obtuse triangle since one angle is greater than 90°.
(b) Equilateral triangle since all sides are equal, or acute triangle since all angles are acute.
(c) Right triangle since one angle is 90°.

Try This One

10-D   Identify the type of triangle.

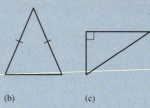

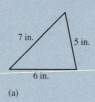

(a)   (b)   (c)

For the answer, see page 522.

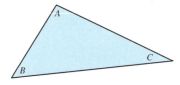

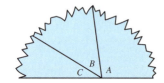

Figure 10-11

Another important property of triangles deals with the measures of the angles.

The sum of the measures of the angles of a triangle is 180°.

This property can be proved mathematically; however, a demonstration will show that it is true. Draw a triangle on a sheet of paper, and then tear off the angles. Finally place these together as shown in Figure 10-11. They will form a straight line, which is an 180° angle.

Example   **10-8**

Find the measure of angle C in the triangle.

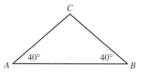

Solution

Since the sum of the measures of the angles of a triangle is 180°,

$$m\angle A + m\angle B + m\angle C = 180°$$
$$40° + 40° + m\angle C = 180°$$
$$80° + m\angle C = 180°$$
$$m\angle C = 180° - 80°$$
$$= 100°$$

Hence, the measure of angle C is 100°.

Several ancient societies understood at least some cases of the Pythagorean theorem. Egyptian builders used a rope divided into sections 3, 4, and 5 units long to check that angles were right angles—if the rope wouldn't stretch into a taut triangle with sides of length three and four against the supposedly right edges, they weren't at a right angle.

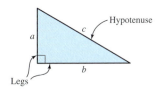

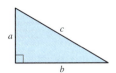

Pythagorean theorem $c^2 = a^2 + b^2$

Figure 10-12

Try This One

10-E Find the measure of angle B.

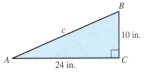

For the answer, see page 522.

The Pythagorean Theorem

A theorem about the relationship of the sides of a right triangle is attributed to the Greek mathematician and philosopher *Pythagoras* (ca. 585–ca. 500 B.C.E.), and it bears his name. The side opposite the right angle in a right triangle is called the *hypotenuse*. The other two sides are called the *legs* of the triangle.

> The **Pythagorean theorem** states that for any right triangle, the sum of the squares of the length of the legs of a right triangle is equal to the square of the length of the hypotenuse (the side opposite the right angle).

Symbolically, the relationship for the triangle shown in Figure 10-12 is $a^2 + b^2 = c^2$.

Example 10-9

For the right triangle shown, find the length of the hypotenuse (i.e., side c).

Solution

$$c^2 = a^2 + b^2$$
$$= 24^2 + 10^2$$
$$= 576 + 100$$
$$= 676$$
$$c = \sqrt{676}$$
$$= 26$$

Hence, the hypotenuse is 26 inches.

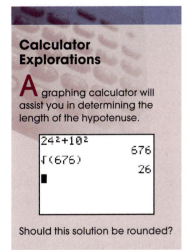

Calculator Explorations

A graphing calculator will assist you in determining the length of the hypotenuse.

```
24²+10²
           676
√(676)
            26
■
```

Should this solution be rounded?

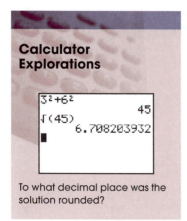

Example   10-10

To build the frame for a roof of a shed, the carpenter must cut a 2 × 4 to fit on the diagonal. If the length of the horizontal beam is 12 feet and the height is 3 feet, find the length of the diagonal beam.

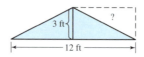

Solution

The frame forms two right triangles with legs of 3 feet and 6 feet. The length of the diagonal beam can be found using the Pythagorean theorem.

$$c^2 = a^2 + b^2$$
$$= 3^2 + 6^2$$
$$= 9 + 36$$
$$= 45$$
$$c = \sqrt{45}$$
$$\approx 6.7 \text{ feet}$$

Hence, the length of the beam should be about 6.7 feet.

Try This One

10-F   The frame for a large sign is 10 feet long and 8 feet high. Find the length of a diagonal beam that will be used for bracing.

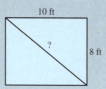

For the answer, see page 522.

Similar Triangles

When two triangles have the same shape but not necessarily the same size, they are called **similar triangles.** Consider the triangles in Figure 10-13. Since the measure of angle A and the measure of angle A' are equal, they are called *corresponding angles*. Likewise, angle B and angle B' are corresponding angles since they have the same measure. Finally, angle C and angle C' are corresponding angles. When two triangles are similar, then their corresponding angles will have the same measure, and conversely, two triangles that have equal corresponding angles are similar.

The sides that are opposite the corresponding angles are called *corresponding sides*. When two triangles are similar, the ratios of the corresponding sides are equal. For the two

Similar triangles

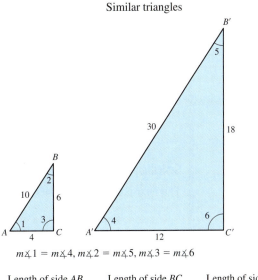

$$m \angle 1 = m \angle 4, m \angle 2 = m \angle 5, m \angle 3 = m \angle 6$$

$$\frac{\text{Length of side } AB}{\text{Length of side } A'B'} = \frac{\text{Length of side } BC}{\text{Length of side } B'C'} = \frac{\text{Length of side } AC}{\text{Length of side } A'C'}$$

Figure 10-13

triangles shown in Figure 10-13,

$$\frac{\text{length of side } AB}{\text{length of side } A'B'} = \frac{10}{30} = \frac{1}{3}$$

$$\frac{\text{length of side } AC}{\text{length of side } A'C'} = \frac{4}{12} = \frac{1}{3}$$

$$\frac{\text{length of side } BC}{\text{length of side } B'C'} = \frac{6}{18} = \frac{1}{3}$$

In addition, the lengths of corresponding sides of similar triangles are proportional. For example, in the two triangles shown in Figure 10-13,

$$\frac{\text{length of side } AB}{\text{length of side } A'B'} = \frac{\text{length of side } AC}{\text{length of side } A'C'}$$

or

$$\frac{10}{30} = \frac{4}{12}$$

$$\frac{\text{length of side } BC}{\text{length of side } B'C'} = \frac{\text{length of side } AC}{\text{length of side } A'C'}$$

or

$$\frac{6}{18} = \frac{4}{12}$$

etc.

If triangle ABC is similar to triangle $A'B'C'$, then

$$\frac{\text{length of side } AB}{\text{length of side } A'B'} = \frac{\text{length of side } AC}{\text{length of side } A'C'} = \frac{\text{length of side } BC}{\text{length of side } B'C'}$$

Example 10-11

Given that $\triangle ABC$ is similar to $\triangle A'B'C'$, find the length of $\overline{B'C'}$.

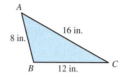

Solution

The next proportion can be written.

$$\frac{\text{length of side } AB}{\text{length of side } A'B'} = \frac{\text{length of side } BC}{x}$$

where x is the length of side $B'C'$.

$$\frac{8}{6} = \frac{12}{x}$$

Solving for x, one gets

$$8x = 72$$

$$\frac{8x}{8} = \frac{72}{8}$$

$$x = 9 \text{ in.}$$

Hence, length of side $B'C' = 9$ in.

Similar triangles can be used to measure things indirectly when direct measurement would be difficult. The next real-life example shows how to use indirect measurement.

Example 10-12

If a tree casts a shadow 12 feet long and at the same time a person who is 5 feet 10 inches tall casts a shadow of 5 feet, find the height of the tree.

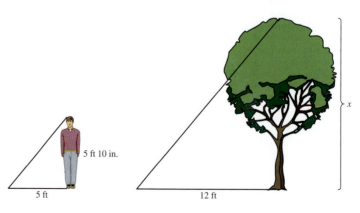

—Continued

Example **10-12** *Continued—*

Solution

Since the sun is in the same position in both instances, similar triangles are formed. A proportion can be set up as follows.

$$\frac{\text{height of person}}{\text{length of person's shadow}} = \frac{\text{height of tree}}{\text{length of tree's shadow}}$$

$$\frac{5\frac{10}{12}}{5} = \frac{x}{12}$$

$$5x = 70$$

$$x = 14 \text{ feet}$$

Hence, the height of the tree is 14 feet.

Try This One

10-G Find the length of a pole if it casts a 20-foot shadow at the same time that a man 6 feet tall casts a 15-foot shadow.
For the answer, see page 522.

Exercise Set **10-2**

Computational Exercises

For Exercises 1–6, classify each triangle.

1.

2.

3.

4.
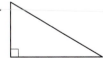

5.
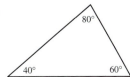

6.
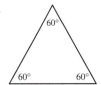

For Exercises 7–10, find the measure of angle C.

7.
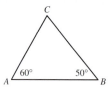

8.
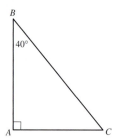

9.
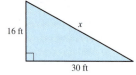

10.
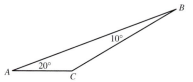

For Exercises 11–16, use the Pythagorean theorem to find the measure of side x.

11.
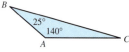

12.
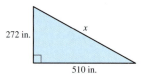

13.
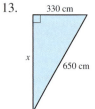

14.

15.
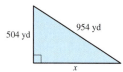

16.
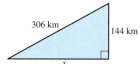

For Exercises 17–22, use the proportional property of similar triangles to find the measure of x. *(The two triangles in each exercise are similar.)*

17.
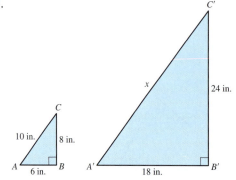

18.

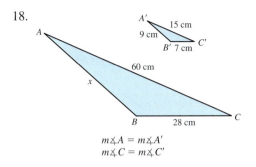

$$m\angle A = m\angle A'$$
$$m\angle C = m\angle C'$$

19.

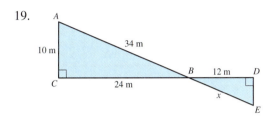

20.

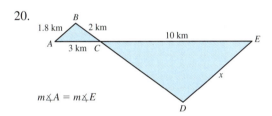

$$m\angle A = m\angle E$$

21.

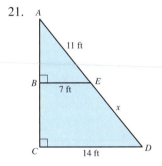

22.

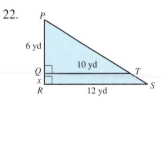

Real World Applications

23. In little league baseball, the distance from home plate to first base is 60 feet. Find the distance from home plate to second base.

24. The screen of a television set or computer monitor is measured by the length of the diagonal. If a screen is 17 inches long and 12 inches high, what is the measure of the diagonal?

25. Find the width of the lake shown.

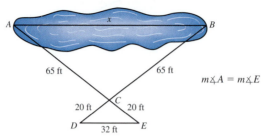

$m \angle A = m \angle E$

26. Find the width of the river shown.

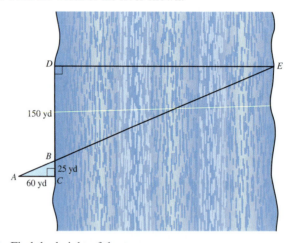

27. Find the height of the tree.

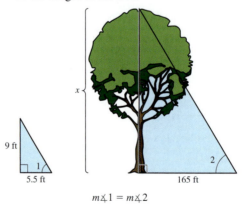

$m \angle 1 = m \angle 2$

28. Find the height of the tower.

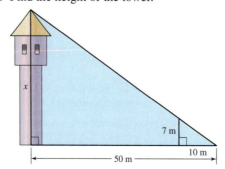

29. For a play, a diagonal beam is to be placed across an 8-foot by 10-foot room on the stage. Find the length of the beam.

30. How high up on a wall is the top of a 20-foot ladder if its bottom is 6 feet from the base of the wall?

31. A plane flies 175 miles north then 120 miles due east. How far diagonally is the plane from its starting point?

32. Two trees are 20 feet tall and 38 feet tall and 30 feet apart. How far is it from the top of one tree to the other?

Writing Exercises

33. Explain how triangles are classified according to the measures of their sides.

34. Explain how triangles are classified according to their angles.

35. Explain how it can be shown that the sum of the measures of the three angles of a triangle is 180°.

36. State the Pythagorean theorem.

37. What are similar triangles?

38. What is meant by indirect measurement?

Critical Thinking

39. Draw a right triangle; then draw a semicircle on each side. Find the areas of the three semicircles. Describe how the areas are related.

40. In the right triangle shown, it can be shown that the lengths of the segments have these properties:

$$\frac{\text{length of side } AD}{\text{length of side } CD} = \frac{\text{length of side } CD}{\text{length of side } DB}$$

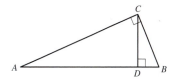

Show the proportion is true by using similar triangles.

10-3 Polygons and Perimeter

In Section 10-2, the concepts of a triangle, one type of polygon, were explained. This section will explain other types of polygons.

Polygons

Closed geometric figures whose sides are line segments are classified according to the number of sides. These figures are called *polygons*. Table 10-1 shows the number of sides

Table 10-1

Basic Polygons

Name	Number of sides	
Triangle	3	
Quadrilateral	4	
Pentagon	5	
Hexagon	6	
Heptagon	7	
Octagon	8	
Nonagon	9	
Decagon	10	
Dodecagon	12	
Icosagon	20	

Many shapes found in nature are polygons.

and shapes of these polygons: triangle, **quadrilateral, pentagon, hexagon, heptagon, octagon, nonagon, decagon, dodecagon,** and **icosagon.**

It was shown in Section 10-2 that the sum of the measures of the angles of a triangle is equal to 180°. The sum of the measures of the angles of any geometric figure can be found by using the next formula.

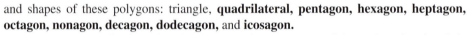

The sum of the measures of the angles of a polygon with n sides is $(n - 2)180°$.

Example 10-13

Find the sum of the measures of the angles of a heptagon.

Solution

Since a heptagon has seven sides, the sum of the measures of the angles of the heptagon is

$$(n - 2)180° = (7 - 2)180°$$
$$= 5 \cdot 180°$$
$$= 900°$$

Hence, the sum of the measures of the angles of a heptagon is 900°.

Sidelight

History of Geometry

Knowledge of rudimentary geometric facts began before recorded history. Egyptian and Babylonian geometry consisted of isolated concepts such as how to find areas of circles and triangles. The circumference of a circle was found by using $\pi = 3$. The Babylonians divided the circle into 360 parts. Geometric problems dating to 1800 B.C.E. were found on papyrus tables. The Egyptians used practical geometry to build the pyramids.

One of the first Greek mathematicians to study geometry was Thales of Miletus who lived in the 6th century B.C.E. Thales is credited with discovering that the diameter of a circle bisects the circle, the base angles of an isosceles triangle are equal, and the vertical angles formed by intersecting lines are equal.

The next Greek mathematician to make significant contributions to geometry was Pythagoras, who was born about 585 B.C.E. Although he is credited with proving the theorem that bears his name, it is believed that perhaps one of his followers did the work. Also, there is evidence that people knew of the theorem before the time of Pythagoras. However, Pythagoras and his followers made many contributions to geometry.

The most famous of the Greek geometers was Euclid, who wrote the book called *Elements* around 300 B.C.E. Euclid used deductive proofs in his book, and his book remained unchanged until the 19th century when some of Euclid's work was refined.

In the 1800s, three mathematicians, Carl Gauss, Nikolai Lobachevski, and János Bolyai discovered independently that other logically correct geometries existed besides Euclid's geometry. These geometries are known as non-Euclidean geometries.

Try This One

10-H   Find the sum of the measures of the angles of an icosagon.
For the answer, see page 522.

Quadrilaterals

There are several different types of quadrilaterals (closed geometric figures with four sides) that have special names.

A **trapezoid** is a quadrilateral that has only two parallel sides. See Figure 10-14(a).

A **parallelogram** is a quadrilateral in which opposite sides are parallel and equal in measure. See Figure 10-14(b).

A **rectangle** is a parallelogram with four right angles. See Figure 10-14(c).

A **rhombus** is a parallelogram in which all sides are equal in length. See Figure 10-14(d).

A **square** is a rhombus with four right angles. See Figure 10-14(e).

Types of Quadrilaterals

(a) Trapezoid

(b) Parallelogram

(c) Rectangle

(d) Rhombus

(e) Square

Figure 10-14

The next diagram explains the relationships of the quadrilaterals.

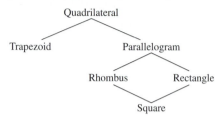

Looking at the relationships, you can see that a square is also a rectangle and a rhombus. A rhombus and a rectangle are also parallelograms.

A *regular* polygon is a polygon in which all sides have the same length and all the angles are equal in measure. The square is an example of a regular quadrilateral.

Example 10-14

Find the measure of each angle of a regular hexagon.

Solution

First find the sum of the measures of the angles for the hexagon. The formula is $(n - 2) \cdot 180°$. Since the hexagon has six sides, the sum of the measures of the angles is $(6 - 2) \cdot 180° = 720°$. Next, divide the sum by 6 since the hexagon has six angles: $720 \div 6 = 120°$.

Hence, each angle of a regular hexagon has a measure of $120°$.

Try This One

10-I Find the measure of each angle of a regular pentagon.
For the answer, see page 522.

Perimeter

The **perimeter** of a polygon is the sum of the lengths of its sides. The perimeter of a triangle with sides of lengths a, b, and c is $P = a + b + c$. The perimeter of a square whose side has a measure of s is $P = 4 \cdot s$. The perimeter of a rectangle with a length l and a width w is $P = 2l + 2w$. See Figure 10-15.

Perimeter

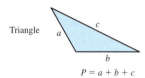

Triangle

$P = a + b + c$

Square

$P = 4s$

Rectangle

$P = 2l + 2w$

Figure 10-15

Example 10-15

How much fence would be needed to enclose a rectangular garden whose length is 30 feet and whose width is 12 feet?

Solution

$$P = 2l + 2w$$
$$= 2 \cdot 30 + 2 \cdot 12$$
$$= 60 + 24 = 84$$

Hence, 84 feet of fence is needed.

Try This One

10-J Find the perimeter of each geometric figure.

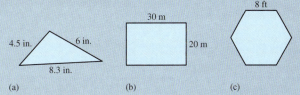

(a)   (b)   (c)

For the answer, see page 522.

Exercise Set 10-3

Computational Exercises

For Exercises 1–6, identify each polygon and find the sum of the measures of the angles.

1.

2.

3.

4.

5.

6.

For Exercises 7–10, identify each quadrilateral.

7.

8.

9.

10.

For Exercises 11–20, find the perimeter.

11.
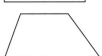
22 yd
16 yd

12.

15 in.
7 in.

13.
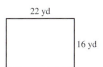
3 ft
7 ft
8 ft
10 ft

14.
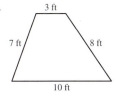
9 cm

15.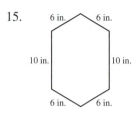

16.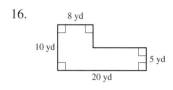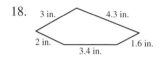

17.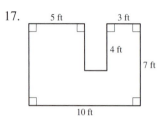

18.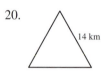

19.

20.

Real World Applications

21. At least how far does a major league player run when he hits a home run? It is 90 feet between the bases.

22. How many feet of fence are needed to fence in a rectangular plot 24 feet by 18 feet if an opening of 3 feet is left for a gate?

23. How many feet of hedges will be needed to enclose a triangular display at an amusement park if the sides measure 62 feet, 85 feet, and 94 feet?

24. Log rail fencing comes in 8-foot sections. Each section costs $24.00. Find the cost to fence in a rectangular yard 42 feet by 56 feet.

25. How much molding in length will be needed to frame an 11×14-inch picture if there is to be a 2-inch mat around the picture?

26. A carpenter needs to put baseboard around the room shown. How many feet are needed? Each door is 30 inches wide.

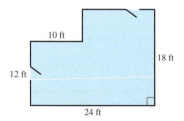

27. How many times must a person walk around a football field in order to walk a mile? The dimensions of a football field are 360 feet by 160 feet. One mile = 5280 feet.

28. How many times must a person walk around a soccer field in order to walk a mile? The dimensions of a soccer field are 345 feet by 223 feet.

Writing Exercises

29. Explain the similarities and the differences between a rectangle and a rhombus.
30. Explain the similarities and differences between a trapezoid and a parallelogram.
31. Explain why a square is both a rectangle and a rhombus.
32. Name three types of parallelograms.
33. Explain how to find the sum of the measures of the angles of a polygon.
34. Define a regular polygon.

Critical Thinking

35. For the triangle shown, $\angle BCD$ is called an exterior angle. If you know the measures of $\angle A$ and $\angle B$, explain how the measure of $\angle BCD$ can be found.

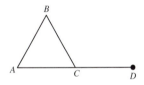

36. What is the measure of each exterior angle of a regular pentagon?

10-4 Areas of Polygons and the Circle

Areas of Polygons

The **area** of a geometric figure is a measure of the region bounded by its sides. Area is measured in square units such as square feet, square meters, etc., abbreviated ft², m², etc. For example, if one wishes to tile a kitchen that measures 8 feet by 10 feet, the person would need to purchase a certain number of tiles in order to cover the floor. Tiles are usually square and measure 12 inches on a side. Each tile represents a square foot. Figure 10-16

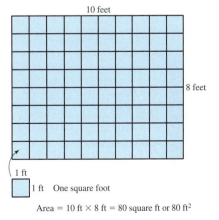

10 feet

8 feet

1 ft

1 ft   One square foot

Area = 10 ft × 8 ft = 80 square ft or 80 ft²

Figure 10-16

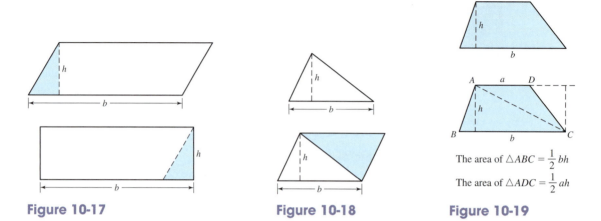

Figure 10-17   **Figure 10-18**   **Figure 10-19**

The area of $\triangle ABC = \frac{1}{2}bh$

The area of $\triangle ADC = \frac{1}{2}ah$

shows a kitchen that measures 8 feet by 10 feet and is divided into square feet. It would take 80 tiles to cover the floor; therefore, the area is 80 square feet.

The area of a rectangle then can be found by multiplying the length times the width.

$$A = l \cdot w$$

A square is a rectangle with equal sides and its area can be found by multiplying its length by its width. Since the length is equal to the width, the area of a square is

$$A = s \cdot s = s^2$$

The area of a parallelogram is equal to the base times the height. This can be shown by cutting off the triangular part of the parallelogram on one side and placing it on the other side. The resulting figure is a rectangle whose area is the length times the width, which, in this case, is the same as the base times the height for the parallelogram.

$$A = bh$$

See Figure 10-17.

The area of a triangle with a base b and a height h can be found by making an identical triangle and placing it next to the original triangle, as shown in Figure 10-18. The resulting figure is a parallelogram whose area is equal to the base times the height. Since the parallelogram is twice as large as the original triangle, the area of the triangle then is equal to $\frac{1}{2}$ of the base times the height.

$$A = \tfrac{1}{2}bh$$

The area of a trapezoid can be found by cutting the trapezoid as shown in Figure 10-19 and forming two triangles.

The area of triangle $ABC = \frac{1}{2}bh$, and the area of triangle ADC is $\frac{1}{2}ah$. The area of the trapezoid then is

$$A = \tfrac{1}{2}ah + \tfrac{1}{2}bh$$

$$= \tfrac{1}{2}h(a + b)$$

The formulas for the areas of the geometric figures are summarized in Table 10-2.

How would you find the area of the front of this house, to estimate the amount of paint needed to paint it?

Table 10-2

Formulas for Areas of Selected Geometric Figures

Rectangle

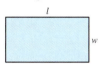

$A = lw$

Square

$A = s^2$

Parallelogram

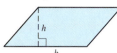

$A = bh$

Triangle

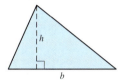

$A = \frac{1}{2}bh$

Trapezoid

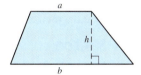

$A = \frac{1}{2}h(a + b)$

Example 10-16

Find the area of the triangle shown.

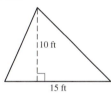

Solution

The area of a triangle is

$$A = \tfrac{1}{2}bh$$
$$= \tfrac{1}{2}(15)(10) = 75 \text{ square feet or } 75 \text{ ft}^2$$

The area of the triangle then is 75 square feet.

Try This One

10-K   Find the area of the trapezoid shown.

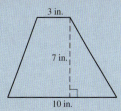

For the answer, see page 522.

Example 10-17

A person wishes to carpet an **L**-shaped living room, as shown.

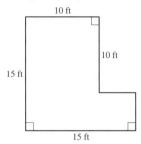

Find the total cost of the carpet if it is priced at $25.00 per square yard.

Solution

Divide the figure into two figures as shown and find the sum of the two areas.

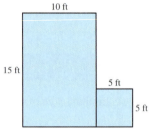

$$A = lw \qquad\qquad A = s^2$$
$$= 15 \cdot 10 \qquad\qquad = 5^2$$
$$= 150 \text{ square feet} \qquad = 25 \text{ square feet}$$

The total area is 150 square feet + 25 square feet = 175 square feet. Next, change square feet into square yards (1 square yard = 9 square feet); $175 \div 9 = 19.44$ square yards. Find the cost of the carpet.

$$\$25.00 \times 19.44 = \$486.00$$

Hence, the carpet will cost $486.00.

Try This One

10-L   A person wishes to place sod around his house, as shown.

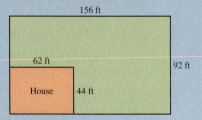

If sod costs $3.98 per square yard, find the total cost.
For the answer, see page 522.

Sidelight

Mathematics and Bees

When bees build honeycombs to store honey, they make the cells in the shape of a regular hexagon. This shape provides the most amount of storage space using the least amount of wax. In addition, it provides a very strong structure that keeps predators out.

Pappus, a Greek mathematician who lived in the 4th century, said God gave humans the best understanding of mathematics, but He allotted bees a share of mathematical wisdom.

The Circle

Another common geometric figure is the circle.

> A **circle** is the set of all points in a plane equidistant from a fixed point called the *center*.

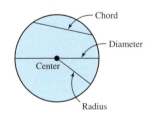

Math Note

The center of a circle is inside the circle but is not actually part of the circle.

A line segment from the center of a circle to any point on a circle is called a **radius.** Every radius of a given circle has the same length and is denoted by r. A line segment from one point on a circle to another point on the circle and passing through the center of the circle is the **diameter** (d). Every diameter of a given circle has the same length and is denoted by d. The diameter is equal to 2 times the radius (i.e., $d = 2r$), and the radius is equal to the diameter divided by 2 $\left(\text{i.e., } r = \dfrac{d}{2} \right)$. A line segment connecting two points on a circle is called a **chord.** All diameters are also chords. The distance around the outside of a circle (i.e., perimeter) is called the **circumference** (C) of the circle. See Figure 10-20. In order to find the circumference and the area of a circle, an important number called π **(pi)** is needed.

The ratio of the circumference of a circle to its diameter is the same for all circles and is a number called π.

$$\frac{C}{d} = 3.1415926535897933238\ldots$$

The number π is an irrational number, which means that it is an infinite nonrepeating decimal. In order to use the formulas for the circumference and the area of a circle, π is usually approximated by 3.14 or $\frac{22}{7}$.

Figure 10-20

(Figure labels: Chord, Diameter, Center, Radius)

Sidelight

Mathematics and the Orbits of Comets

In the past, humans have always been fascinated and mystified by the appearance of comets. However, once astronomers used mathematics to unlock their secrets, they were able to predict comets' appearances.

One such comet that appears regularly is known as Halley's Comet. The appearance of this comet remained unexplained until the 16th century when an astronomer named Edmund Halley began studying it in 1704. He noticed that earlier records showed that the comet appeared in the same regions of the sky in 1456, 1531, and 1607. He reasoned that the comet became visible about every 75 or 76 years. He correctly predicted that the comet would appear in 1758.

The orbit of Halley's Comet is elliptical in nature, and by using the mathematics of the ellipse, scientists can determine its path, speed, and appearance dates. Since this discovery was made, many other comets in our solar system have been discovered, and scientists can now determine their paths and when they will appear in the sky.

The formula for the circumference of a circle is

$$C = 2\pi r \qquad \text{or} \qquad C = \pi d$$

Example   10-18

Find the circumference of a circular track when the radius is 7 m. Use $\pi = 3.14$.

Solution

$$
\begin{aligned}
C &= 2\pi r \\
&= 2 \cdot (3.14) \cdot 7 \\
&= 43.96 \text{ m}
\end{aligned}
$$

Hence, the circumference of a circular track with $r = 7$ m is 43.96 m.

Try This One

10-M   Find the circumference of a circle whose diameter is 13 inches. Use 3.14 for π.
For the answer, see page 522.

The formula for the area of a circle is $A = \pi r^2$.

Example 10-19

Find the area of a circle whose diameter is 10 feet. Use 3.14 for π.

Solution

First find the radius.

$$r = \frac{d}{2}$$

$$= \frac{10}{2}$$

$$= 5 \text{ feet}$$

Next find the area.

$$A = \pi r^2$$

$$= 3.14 \cdot 5^2$$

$$= 78.5 \text{ square feet or } 78.5 \text{ ft}^2$$

The area of a circle whose diameter is 10 feet is 78.5 square feet.

Try This One

10-N   Find the area of a circle whose radius is 35 cm. Use $\pi = \frac{22}{7}$.
For the answer, see page 522.

Exercise Set   10-4

Computational Exercises

For Exercises 1–12, find the area of each figure.

1.
17 in.

2.
22 ft

12 ft

3.
15 yd

30 yd

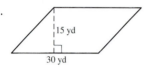

4.
105 cm

150 cm

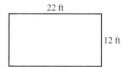

5.
20 m

20 m

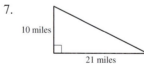

6.
10 ft

14 ft

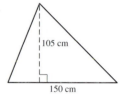

7.
10 miles

21 miles

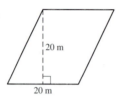

8.
9 in.

9 in.

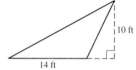

9.
20 in.

17 in.

35 in.

10.
13 km

19 km

24 km

11.
6 in.

10 in.

10 in.

3 in.

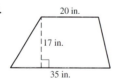

12.
20 yd

6 yd

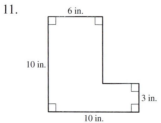

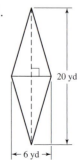

For Exercises 13–18, find the circumference and the area of each circle. Use π = 3.14.

13.

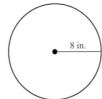

14.

15.

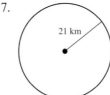

16.

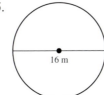

17.

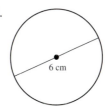

18.

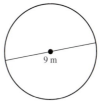

Real World Applications

19. How many square yards of vinyl linoleum are needed to cover a square kitchen that measures 10 feet on each side? (1 square yard = 9 square feet.)

20. Find the cost of coating a rectangular driveway that measures 11 feet by 21 feet, at $2.50 per square foot.

21. How many tickets that measure 2 inches by 5 inches can be cut from a poster board that measures 24 by 25 inches?

22. Find the amount and the cost of artificial turf needed to cover a football field that measures 360 feet by 160 feet. The cost of the turf is $20.00 per square foot.

23. A stage floor shaped like a trapezoid with bases of 60 feet and 75 feet and a height of 40 feet is to be covered with plywood for a play. What would be the cost for the plywood if it sells for $0.60 a square foot?

24. A children's play area in a nursery school that is shaped like a trapezoid with bases of 8 feet and 9.6 feet and a height of 10 feet is to be covered with carpet. Find the cost if the carpet sells for $18.00 a square yard. (*Hint:* One square yard is equal to 9 square feet.)

25. A triangular-shaped shelf in a student's dormitory room that is placed in a corner has a base of 5 feet and a height of 3 feet. Find the area of the shelf.

26. How many square yards of fabric are needed to make a triangular-shaped display flag whose base is 6 feet and whose height is 6 feet?

27. Find the cost of the outdoor carpeting needed to cover the floor of a circular gazebo whose diameter is 12 feet? Carpet costs $15.00 a square yard. Use π = 3.14.

28. How much more pizza do you get in a large pizza that has a diameter of 15 inches than you do in a small pizza that has a diameter of 10 inches? Use π = 3.14.

29. A college radio station can broadcast over an area with a radius of 60 miles. How much area is covered? Use $\pi = 3.14$.

30. Find the distance around a track which has the dimensions shown below. Use $\pi = 3.14$.

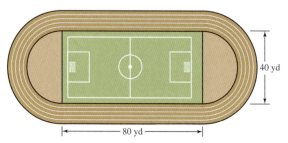

31. A cloth flag whose dimensions are shown is being sewn to hang from an ice cream shop. Find the area of the material that is used to make the flag. Use $\pi = 3.14$.

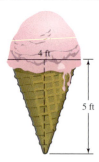

32. Find the area of the walkway (24 inches wide) that will be installed around a circular jacuzzi with a diameter of 6 feet. Use $\pi = 3.14$.

Writing Exercises

33. In your own words, explain the difference between the perimeter and the area of a polygon.

34. Explain why the area of a triangle is exactly $\frac{1}{2}$ the area of a parallelogram.

35. Explain how the number π is derived.

36. Explain why π is not exactly equal to $\frac{22}{7}$ or 3.14.

Critical Thinking

37. The area of a triangle can be found if the measures of the three sides are known. This formula was discovered about 100 B.C.E. by a Greek mathematician known as Heron. The formula is

$$A = \sqrt{s(s-a)(s-b)(s-c)}$$

where $s = \frac{1}{2}(a + b + c)$ and a, b, and c are the measures of the lengths of the sides of the triangle. Using the formula, find the area of the triangle if the sides are 5 in., 12 in., and 13 in.

38. The triangle in Problem 37 is also a right triangle. Find the area using the formula $A = \frac{1}{2}bh$ and see if you get the same answer.

10-5 Surface Area and Volume

The geometric figures discussed in Section 10-4 were two-dimensional or plane figures. This section describes three-dimensional or solid figures. A three-dimensional figure has a surface area and a volume. For example, if the hide of a baseball were taken off the ball and laid out on a flat surface, the area of the hide would be called the **surface area.** The amount of water or antifreeze that your car radiator holds is an example of **volume.** Area is measured in square units, i.e., square inches, square feet, square meters, etc. Volume, on the other hand, is measured in cubic units such as cubic inches, cubic feet, etc. abbreviated in.^{3}, ft^3, etc.

Figure 10-21 shows a cube whose side measures 2 inches. If the cube is taken apart and flattened out, it consists of six square sides with each side measuring 2 inches. Hence, the surface area of a cube is $6s^2$ or $6 \cdot 2^2 = 24$ square inches.

Again consider the cube that measures 2 inches on each side. If it is taken apart again but this time using solid cubes that have 1-inch sides, it consists of eight smaller cubes. Hence the volume of the original cube is 8 cubic inches. See Figure 10-22.

Table 10-3 shows some common solid geometric figures and the formulas for their surface areas and volumes.

The relation between surface area and volume is important when determining how something will absorb or lose heat. Ice absorbs heat through its surface. When you break up a large piece of ice, the volume stays the same, but the surface area is greatly increased. So small ice cubes melt quickly, and large blocks of ice last much longer.

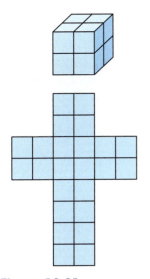

Figure 10-21

Table 10-3

Surface Area and Volume Formulas for Solid Figures

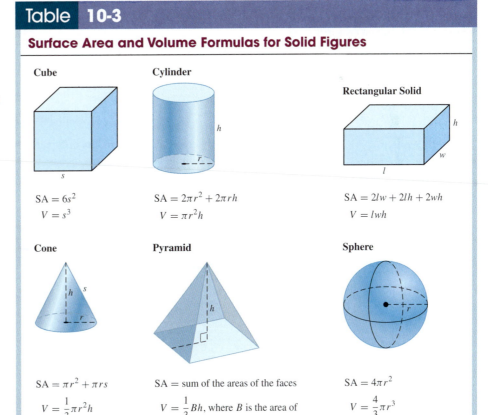

Cube

$$SA = 6s^2$$
$$V = s^3$$

Cylinder

$$SA = 2\pi r^2 + 2\pi rh$$
$$V = \pi r^2 h$$

Rectangular Solid

$$SA = 2lw + 2lh + 2wh$$
$$V = lwh$$

Cone

$$SA = \pi r^2 + \pi rs$$
$$V = \frac{1}{3}\pi r^2 h$$

Pyramid

$$SA = \text{sum of the areas of the faces}$$
$$V = \frac{1}{3}Bh, \text{ where } B \text{ is the area of the base}$$

Sphere

$$SA = 4\pi r^2$$
$$V = \frac{4}{3}\pi r^3$$

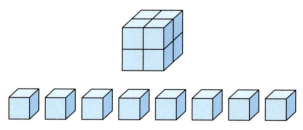

Figure 10-22

Calculator Explorations

A graphing calculator will assist you in viewing the solution when evaluating formulas.

```
4(3.14)(5²)
              314
4/3(3.14)(5^3)
         523.3333333
■
```

How are the formulas different?
How are the formulas alike?

Example 10-20

Find the surface area and the volume of a sphere whose radius is 5 inches.

Solution

The surface area of a sphere is $SA = 4\pi r^2$.

$$SA = 4\pi r^2$$
$$= 4(3.14)(5^2)$$
$$= 314 \text{ square inches or } 314 \text{ in.}^2$$

The volume of a sphere is $V = \frac{4}{3}\pi r^3$.

$$V = \frac{4}{3}\pi r^3$$
$$= \frac{4}{3}(3.14)(5^3)$$
$$\approx 523.33 \text{ cubic inches or } 523.33 \text{ in.}^3$$

Hence, for a sphere which has a radius of 5 inches, the surface area is 314 square inches and the volume is 523.33 cubic inches.

Try This One

10-O Find the surface area and the volume of the cone shown.

8 yd

5 yd

For the answer, see page 522.

There are many real-life applications involving the solid geometric figure.

Non-Euclidean Geometry

The geometry presented in this chapter is called Euclidean geometry, and it is based on an assumption (postulate) in *Elements* written about 300 B.C.E. This postulate is called *Euclid's Fifth* or *Parallel Postulate*. It states that given a point not on a given line, there is one and only one line that is parallel to the given line.

Up until the late 1600s and early 1700s, mathematicians thought that they could prove this postulate and make it a theorem. An Italian mathematician named Girolamo Saccheri (1667–1733) attempted to prove this postulate by using what is called a proof by contradiction. That is, he assumed that if the postulate were false, it would lead to a contradiction. There are two ways to contradict this postulate. First, it can be assumed that there is no line that can be drawn through the point that is parallel to the given line, or, second, that there are at least two lines (possibly an infinite number of lines) that can be drawn through the point that are parallel to the given line. Saccheri was unable to find a contradiction using either of the two assumptions.

About 100 years later, two mathematicians, Nikolai Lobachevsky (1792–1856) and János Bolyai (1802–1860), working independently, attempted to find a contradiction to *Euclid's Fifth Postulate*. They decided to assume that an infinite number of lines could be drawn through the point that are parallel to the given line. They were able to prove many unusual theorems without finding any contradictions to Euclid's geometry. Bolyai published his work in 1832 and Lobachevsky published his work in 1829.

At first the works were not well received by the mathematical community; however, mathematicians later accepted the "new" type of geometry and called

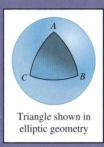

Triangle shown in elliptic geometry

Triangle shown in hyperbolic geometry

it *hyperbolic geometry*. Hyperbolic geometry can be represented on a shape called a pseudosphere. Many unusual theorems exist in hyperbolic geometry. For example, the shortest distance between two points is a curved line and the sum of the measures of the angles of a triangle is *less* than 180°.

After the discovery of hyperbolic geometry, another mathematician, Bernhard Riemann (1826–1866) decided to assume that no lines could be drawn through the given point that are parallel to the given line. Riemann then developed another alternative geometry, which is called *elliptic geometry*. This type of geometry can be represented on a circle or more generally, an ellipse. Here there are no parallel lines and the sum of the measures of the angles of a triangle is *greater* than 180°. Albert Einstein used the principles of elliptical geometry to explain his theory of the universe.

The most universal thing about the discoveries of the non-Euclidian geometries is that Euclid in 300 B.C.E. somehow recognized that his *Fifth Postulate* was unable to be proved using his geometry!

Example   10-21

How many gallons of water will a cylindrical tank hold if it has a radius of 0.75 feet and a height of 4 feet? One cubic foot of water is about 7.46 gallons.

—Continued

Example 10-21 *Continued—*

Solution

First find the volume of the tank.

$$V = \pi r^2 h$$

$$= 3.14(0.75)^2 4$$

$$\approx 7.07 \text{ cubic feet or } 7.07 \text{ ft}^3$$

Since one cubic foot of water is about 7.46 gallons, multiply 7.46 gallons by 7.07 cubic feet to get the capacity of the tank.

$$7.46 \times 7.07 \approx 52.7 \text{ gallons}$$

The tank will hold 52.7 gallons of water.

Try This One

10-P Find the cost of "ready mix" concrete for constructing a patio that is 20 feet long by 12 feet wide by 8 inches deep. The cost of the ready mix concrete is $70.00 per cubic yard.

For the answer, see page 522.

Exercise Set 10-5

Computational Exercises

For Exercises 1–12, find the surface area and volume for each figure.

1.

5 in.

2.
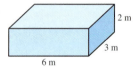

2 m
3 m
6 m

3.
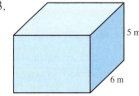

5 m
6 m
7 m

4.

2 cm

5.

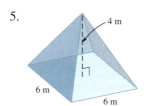

4 m

6 m

6 m

6.

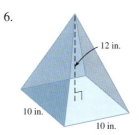

12 in.

10 in.

10 in.

7.

32 cm

28 cm

8.

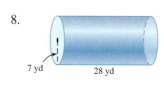

7 yd

28 yd

9.

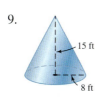

15 ft

8 ft

10.

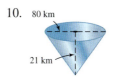

80 km

21 km

11.

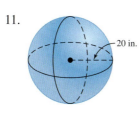

20 in.

12.

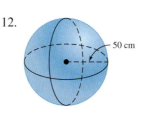

50 cm

For Exercises 13 and 14, find the volume of the shaded area.

13.

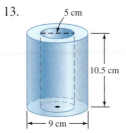

5 cm

10.5 cm

9 cm

14.

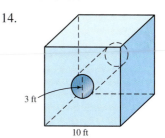

3 ft

10 ft

Real World Applications

15. How many cubic feet of dirt must be removed to build an in-ground swimming pool whose dimensions are 18 feet in length, 12 feet in width, and 3 feet in depth?

16. How many square feet of plywood are needed to build a stage prop that is shaped like a cube with each side measuring 6 feet?

17. A pyramid found in Central America has a square base measuring 932 feet on each side and a height of 657 feet. Find the volume of the pyramid.

18. A child's tent is shaped like a pyramid with a square base measuring 6 feet on a side and a height of 4 feet. Find the volume of the tent.

19. A cylindrical-shaped gasoline tank has a 22-inch diameter and is 36 inches long. Find its volume.

20. Find the surface area of a can of green beans with a diameter of 8 cm and a height of 8.5 cm.

21. Find the surface area of a cone-shaped funnel whose base diameter is 7 inches and whose height is 11 inches.

22. Find the volume of a cone-shaped Christmas tree that is 6 feet tall and has a base diameter of 4 feet.

23. Find the volume of the planet Mars if its diameter is 4200 miles.

24. Find the surface area of Jupiter if its diameter is 87,000 miles.

Writing Exercises

25. What is meant by the surface area of a solid figure?
26. Explain the meaning of the volume of a solid.

Critical Thinking

27. Twelve rubber balls with 3-inch diameters are placed in a box with dimensions 12 inches × 9 inches × 3 inches. Find the volume of the space that is left over.

28. A rain gutter (cross section shown) is 24 feet long. How much water will it hold when it is full? One cubic foot of water is equal to 7.46 gallons.

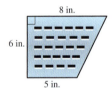

8 in.

6 in.

5 in.

10-6 Right Triangle Trigonometry

Trigonometry means "triangle measurement" and uses the various relationships between the sides and angles of the triangle to solve problems. Trigonometry is used in many areas of mathematics as well as navigation, architecture, aviation, etc.

The basic principles of trigonometry use the right triangle. Recall that a right triangle has a 90° angle. The right triangle is labeled in a specific way. The angles are labeled using capital letters at their vertices, and the right angle is labeled angle C. The other two angles are labeled A and B. The sides are labeled with lowercase letters and the letters correspond to the opposite angles. For example, the side opposite the right angle C is labeled side c. The side opposite angle A is labeled a, and the side opposite angle B is labeled b. The lengths of the sides are designated by a, b, and c also. Finally, recall that the longest side of the triangle, which is opposite the right angle, is called the hypotenuse. See Figure 10-23.

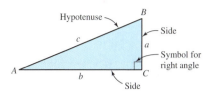

Figure 10-23

Trigonometry is used extensively in surveying.

The Trigonometric Ratios

There are three basic trigonometric ratios. They are called the **sine** (abbreviated sin), the **cosine** (abbreviated cos), and the **tangent** (abbreviated tan). The trigonometric ratios are defined as follows:

$$\sin A = \frac{\text{length of side opposite angle } A}{\text{length of hypotenuse}} = \frac{a}{c}$$

$$\cos A = \frac{\text{length of side adjacent to angle } A}{\text{length of hypotenuse}} = \frac{b}{c}$$

$$\tan A = \frac{\text{length of side opposite angle } A}{\text{length of side adjacent to angle } A} = \frac{a}{b}$$

Since these ratios use the *side opposite* the specific angle, the *side adjacent* to the specific angle, and the *hypotenuse,* the three trigonometric ratios remain constant for an angle regardless of the lengths of the sides of the triangle. This is true because the lengths of the sides of similar triangles are proportional. For example, the three triangles shown in Figure 10-24 are similar because their corresponding angles are equal. Notice that the sine

Sidelight

Mathematics and Billiards

The path of a billiard (i.e., pool) ball on the billiard table can be determined mathematically. The angle at which the ball hits the side of the table has the same measure as the angle at which it rebounds. Thus, an expert player can control the path of the ball. Although these expert players don't actually perform the mathematics in their heads, they know from much practice how to hit the Q-ball so that it travels where they want it to go.

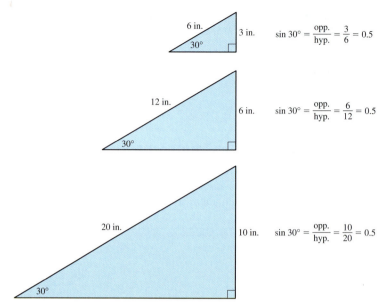

Figure 10-24

of 30° will always be 0.5 because the ratio of the side opposite 30° to the hypotenuse will always be the same (i.e., 0.5) even though the lengths of the sides of the triangles are different. In this case, $\sin 30° = \frac{3}{6} = \frac{6}{12} = \frac{10}{20} = 0.5$.

The table in Appendix B shows the values to four decimal places for the three trigonometric functions—sine, cosine, and tangent—for the angles from 0° to 90°. For example, the tangent of 42° (written tan 42°) is 0.9004. The cosine of 53° (written cos 53°) is 0.6018.

In order to solve problems involving right triangles, the next procedure can be used.

Three steps can be used to solve right triangle trigonometric problems:

Step 1 Draw and label the angles of the right triangle and the measures of the sides.
Step 2 Select the appropriate formula and substitute the values (you may have to use the table in Appendix B).
Step 3 Solve the equation for the unknown.

Example 10-22

In the right triangle ABC, find the length of side a when $m \angle A = 30°$ and $b = 200$ feet.

Solution

Step 1 Draw and label the figure.

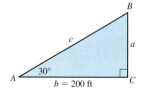

—Continued

Example   10-22   *Continued—*

Step 2   Select the appropriate formula and substitute. Since we are given side b and are asked to find side a, we can use the tangent of $30°$.

$$\tan A = \frac{a}{b}$$

$$\tan 30° = \frac{a}{200}$$

Step 3   Solve for a.
From the table, $\tan 30° = 0.5774$.

$$0.5774 = \frac{a}{200}$$

$$(0.5774)(200) = \frac{a}{\cancel{200}}(\cancel{200})$$

$$115.48 = a$$

Hence, the length of side a is 115.48 feet.

Example   10-23

In the right triangle ABC, find the measure of side c when $m\angle B = 72°$ and the length of side a is 24 feet.

Solution

Step 1   Draw and label the figure.

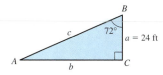

Step 2   Use the cosine function.

$$\cos B = \frac{a}{c}$$

$$\cos 72° = \frac{24}{c}$$

$$0.3090 = \frac{24}{c}$$

$$0.3090c = 24$$

$$\frac{0.3090c}{0.3090} = \frac{24}{0.3090}$$

$$c = 77.67 \text{ feet}$$

Try This One

10-Q   In the right triangle ABC, find the measure of side b when $m\angle A = 53°$ and the hypotenuse (side c) is 18 cm.
For the answer, see page 522.

There are many real world problems that can be solved by using trigonometry.

Example   10-24

A 12-foot ladder is resting against a building at an angle of $65°$. How far is the bottom of the ladder from the base of the wall of the building?

Solution

Step 1   Draw and label the figure.

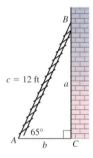

Step 2   Select the appropriate formula and substitute the values for the variables. Since we need to find the length of side b and we are given the measures of angle A and side c, the appropriate formula is the cosine function.

$$\cos A = \frac{b}{c}$$

$$\cos 65° = \frac{b}{12}$$

Step 3   Solve for b. The cosine of $65°$ is 0.4226.

$$0.4226 = \frac{b}{12}$$

$$\frac{12}{1}(0.4226) = \frac{b}{12} \cdot \frac{12}{1}$$

$$5.07 \text{ feet} = b$$

Hence, the ladder is approximately 5.07 feet from the base of the wall.

Angle of Elevation and Angle of Depression

Many applications of right angle trigonometry use what is called the *angle of elevation* or the *angle of depression*.

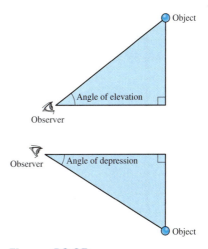

Angle of elevation

Observer

Observer Angle of depression

Object

Figure 10-25

The **angle of elevation** of an object is the measure of the angle from a horizontal line at the point of an observer upward along the line of sight to the object. The **angle of depression** is the measure of an angle from a horizontal line at the point of an observer downward along the line of sight to the object. See Figure 10-25.

Example 10-25

In order to determine the height of a building an observer who is 6 feet tall measures the angle of elevation from his head to the top of a pole to be 32°. The distance the person is from the building is 200 feet. Find the height of the building.

Solution

Step 1   Draw and label the figure.

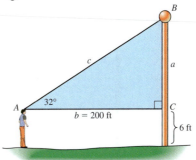

Step 2   Select the appropriate formula and substitute the values for the variables. Since we are given the measure of angle A and the measure of side b, we can use the tangent function.

$$\tan A = \frac{a}{b}$$

$$\tan 32° = \frac{a}{200}$$

—*Continued*

Example 10-25 *Continued—*

The tangent of 32° is 0.6249.

$$0.6249 = \frac{a}{200 \text{ ft}}$$

$$200(0.6249) = a$$

$$124.98 = a$$

Now, to find the height of the building, you must add 6 feet to 124.98 since the line of sight was 6 feet above the ground. Hence, the building is $124.98 + 6 = 130.98$ feet tall.

Try This One

10-R   A person standing on top of a 150-foot cliff sights a boat at an angle of depression of 24°. How far is the boat from the base of the cliff? Assume the observer is 5.6 feet tall.
For the answer, see page 522.

Given the measures of two sides of a right triangle, the measure of the third side and the measures of the angles can be determined as shown in Example 10-26.

Example 10-26

In the right triangle *ABC*, side *c* (the hypotenuse) measures 25 inches and side *b* measures 24 inches. Find the measure of side *a* and the measures of angle *A* and angle *B*.

Solution

Draw the figure and label each part. The measure of side *a* can be found by using the Pythagorean theorem.

$$c^2 = a^2 + b^2$$

$$25^2 = a^2 + 24^2$$

$$625 = a^2 + 576$$

$$625 - 576 = a^2$$

$$49 = a^2$$

$$\sqrt{49} = a$$

$$7 \text{ inches} = a$$

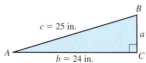

—Continued

Example 10-26 *Continued—*

The measures of the angles can be found by using one of the trigonometric functions.

$$\sin A = \frac{a}{c}$$
$$= \frac{7}{25}$$
$$= 0.28$$

Now look in the table under the sine column for the value closest to 0.28. It is 0.2756. Find the measure of the angle whose sine is 0.2756. It is 16°. Hence, the measure of angle A is about 16°.

Since the sum of the angles of a triangle is 180°, the measure of angle B can be found by subtracting $180° - 16° - 90° = 74°$. (*Note:* Angle C is a right angle whose measure is 90°.)

Try This One

10-S Find the measures of the angles and the measure of side c for the right triangle ABC, when the measure of side a is 42 inches and the measure of side b is 18 inches. *For the answer, see page 522.*

Exercise Set 10-6

Computational Exercises

For Exercises 1–20, use the right triangle ABC (where C is the right angle) and the appropriate trigonometric function to find the measure of each:

1. Side a if $m\angle B = 72°$ and side $c = 300$ cm
2. Side b if $m\angle A = 41°$ and side $c = 200$ yd
3. Side c if $m\angle A = 76°$ and side $b = 18.6$ in.
4. Side a if $m\angle A = 30°$ and side $c = 40$ km
5. Side b if $m\angle B = 8°$ and side $c = 10$ mm
6. Side c if $m\angle B = 62°$ and side $a = 313$ miles
7. Side a if $m\angle A = 60°$ and side $b = 71$ in.
8. Side b if $m\angle B = 85°$ and side $a = 110$ yd
9. Side c if $m\angle A = 28°$ and side $a = 872$ ft
10. Side a if $m\angle A = 22°$ and side $b = 27$ ft
11. Side b if $m\angle B = 53°$ and side $c = 97$ mi
12. Side c if $m\angle A = 15°$ and side $b = 1250$ ft
13. $m\angle A$ if side $b = 183$ ft and side $c = 275$ ft
14. $m\angle B$ if side $b = 104$ yd and side $c = 132$ yd
15. $m\angle A$ if side $b = 18$ mi and side $c = 36$ mi

16. $m\angle B$ if side $a = 529$ mi and side $c = 1000$ mi

17. $m\angle A$ if side $a = 306$ in. and side $b = 560$ in.

18. $m\angle B$ if side $a = 1428$ ft and side $c = 1800$ ft

19. $m\angle B$ if side $a = 413$ ft and side $b = 410$ ft

20. $m\angle A$ if side $a = 532$ yd and side $c = 780$ yd

Real World Applications

21. An airplane is flying at an altitude of 6780 feet and sights the angle of depression to a control tower at the next airport to be 16°. Find the horizontal distance the plane is from the control tower. (Disregard the height of the tower.)

22. If the angle of elevation from ground level to the top of a building is 38° and the observer is 500 feet from the building, find the height of the building.

23. How high is a weather balloon if the angle of elevation from the ground is 82° and the observation point is 700 meters from the point where the balloon is vertically over the ground?

24. Find the angle of inclination of a 6-foot treadmill if the front is 4 inches above the ground.

25. Find the length of the lake shown.

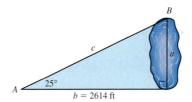

26. Find the height of the flagpole shown.

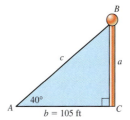

27. If a tree 20 feet tall casts a shadow of 8 feet, find the angle of elevation of the sun at that time of day.

28. A 25-foot ladder leaning against a wall makes an angle of 63° with the level ground. How far up the wall is the top of the ladder?

29. Find the horizontal distance a ship is from the base of a lighthouse if the angle of depression to the ship from the top of the lighthouse is 21°. The lighthouse is 138 feet tall. (Disregard the dimensions of the ship.)

30. Find the length of a shadow cast by a 6-foot person if the angle of elevation of the sun is 44°.

31. The light from a searchlight is reflected on a cloud at an angle of elevation of 58°. If the horizontal distance from the light to the base of the cloud is 2456 feet, find the height of the cloud.

32. What is the angle of elevation of a railroad track that rises 349 feet over a horizontal distance of 1000 feet?

Writing Exercises

33. What does the word "trigonometry" mean?

34. Name and define the three trigonometric ratios that exist using a right triangle.

35. Explain why the tangent of 30° always has the same value even though the sides of the right triangle with a 30° angle may differ in length?

36. What is meant by an angle of elevation?

37. What is meant by an angle of depression?

38. If you are given the measures of two sides of a right triangle, explain how to find the measure of the third side and the measures of the angles. (Answers may vary.)

Critical Thinking

39. From the top of a building 300 feet high, the angle of elevation to a plane is 33° and the angle of depression to an automobile directly below the plane is 24°. Find the height of the plane and the distance the automobile is from the base of the building.

40. The angle of elevation to the top of a tree sighted from ground level is 18°. If the observer moves 80 feet closer, the angle of elevation from ground level to the top of the tree is 38°. Find the height of the tree.

10-7 Networks

A branch of mathematics called *network theory* grew out of a real-life problem. The town of Königsberg, East Prussia (now Kaliningrad, Russia), was built on the Pregel River and consisted of two islands connected to each other and to the mainland by seven bridges, as shown in Figure 10-26(a).

On Sundays, the local residents liked to stroll across the bridges. They tried to cross all seven bridges only once before returning to their homes. No matter where they started, they

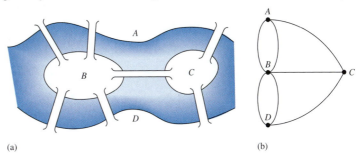

(a)   (b)

Figure 10-26

found that it was impossible to accomplish the task without crossing at least one bridge twice. Perplexed, some residents brought the problem to the attention of a Swiss mathematician named Leonard Euler. Euler was in residence in 1735 in the court of Russian Empress Catherine the Great in St. Petersburg. Euler drew a diagram of the bridge system similar to the one shown in Figure 10-26(b).

After studying the problem, Euler proved that it was impossible to traverse all the bridges without crossing at least one bridge twice. (The solution will be shown later.)

Each of the three graphs shown in Figure 10-27 is called a **network.** A network consists of a set of line segments connected by points. The points are called *vertices,* and the line segments are called *paths* or *arcs.*

Vertices can be *even* or *odd.* An *even vertex* has an even number of paths emanating from it. See Figure 10-28(a).

An *odd vertex* has an odd number of paths emanating from it. See Figure 10-28(b).

> A network is **traversable** if it is possible to pass through or trace each path exactly once without lifting your pencil. A vertex can be crossed more than once.

For example, the network shown in Figure 10-29(a) can be traversed by starting at point A and continuing through the vertices C, D, E, B, D, and A, and ending at vertex B. See Figure 10-29(b).

Now try to traverse the network shown in Figure 10-30. After a few tries, you will find that it is impossible to traverse the network without passing over at least one path twice.

After studying networks, Euler showed that they have these properties:

1. The number of odd vertices in any network must always be even.
2. A network in which all vertices are even is traversable.
3. If a network has exactly two odd vertices, it is traversable providing that one starts at one of the odd vertices and ends at the other odd vertex.
4. If a network has more than two odd vertices, it is impossible to traverse.

Using the four properties, you can determine whether or not a network is traversable. For example, the network shown in Figure 10-31 has five even vertices, and it is traversable since property 2 states that if all the vertices of a network are even, it can be traversed. You can start at any vertex.

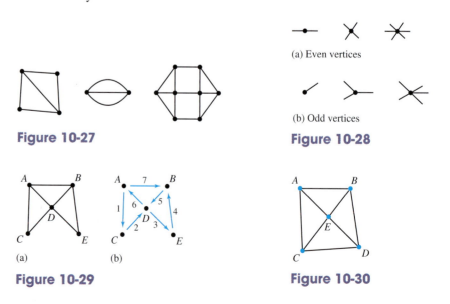

(a) Even vertices

(b) Odd vertices

Figure 10-27

Figure 10-28

Figure 10-29

Figure 10-30

Sidelight

The Mobius Strip

An unusual mathematical shape can be created from a flat strip of paper. After cutting a strip, give it a half twist and tape the ends together to make a closed ring as shown.

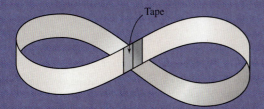

The shape you have made is called a Mobius strip, named for the German mathematician Augustus Ferdinand Mobius (1790–1868) who created it. What is unusual about the Mobius strip is that it has only one side. You can draw a line through the center of the strip, starting anywhere and ending where you started without crossing over an edge. You can color the entire strip with one color by starting anywhere and ending where you started without crossing over an edge.

The Mobius strip has more unusual properties. Take a scissors and cut through the center of the strip following the line you drew as shown.

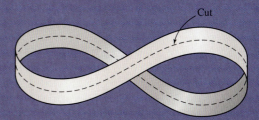

You will not get two strips as you might suspect, but you will have one long continuous two-sided strip.

Finally, make a new strip and cut it one-third of the way from the edge. You will find that you can make two complete trips around the strip using one continuous cut. The result of this cut will be two distinct strips: one will be a two-sided strip, and the other will be another one-sided Mobius strip.

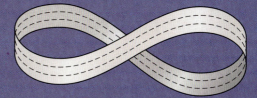

The Mobius strip is a shape that belongs to a recently-developed branch of mathematics called *topology*. Topology is a kind of geometry in which solid shapes are bent or stretched into different shapes. The properties of the shape, which remain unchanged in the new shape, are then studied.

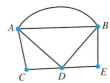

Figure 10-31

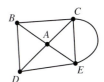

Figure 10-32

The network shown in Figure 10-32 is also traversable since it has exactly two odd vertices, *B* and *D*. Property 3 states that any network with exactly two odd vertices can be traversed by starting at one odd vertex and ending at the other odd vertex.

Example   10-27

Determine which vertices in the network shown are even and which are odd.

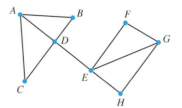

Solution

A and G are odd, and B, C, D, E, F, and H are even.

Example   10-28

Is the network shown in Example 10-27 traversable? If it is, show a possible route.

Solution

Since there are exactly two odd vertices, the network is traversable. One possible path starting at vertex A and ending at vertex G is shown.

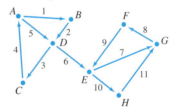

Try This One

10-T   Which networks shown are traversable?

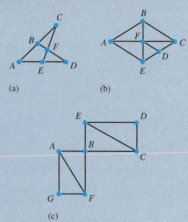

(a)   (b)

(c)

For the answer, see page 522.

The Königsberg bridge problem shown in Figure 10-26 can now be solved. Since the network has four odd vertices A, B, C, and D, property 4 states it cannot be traversed; hence, it is not possible to traverse each bridge only once.

Network theory is used in chemistry and physics to show how atoms in a molecule are linked together. It is also used in determining paths for airline flights connecting airports, in determining highway construction, and a variety of other places where paths link vertices.

Exercise Set 10-7

Computational Exercises

For Exercises 1–10, identify the number of odd vertices and the number of even vertices for each network. Based on your answer, state whether or not the network is traversable.

1.

2.

3.

4.

5.

6.

7.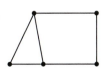

8.

9.

10.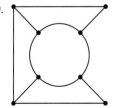

Real World Applications

11. Draw a figure like the one of the Königsberg problem (Figure 10-26) and add or eliminate a bridge to make the network traversable.

Writing Exercises

12. What two elements are necessary for a network?

13. What is meant by a *traversable* network? (Do not make any reference to odd and even vertices.)

14. Explain the difference between an odd vertex and an even vertex.

15. Draw a network that consists of exactly 6 odd vertices and exactly 1 even vertex.

16. What are the conditions (in terms of the number of even and odd vertices) for a network to be traversable?

Critical Thinking

17. Design a park with islands and bridges and be sure the network is traversable.

Summary

Section	Important Terms	Important Ideas
10-1	geometry point line plane open segment half open segment half line ray angle vertex degree protractor acute angle right angle obtuse angle straight angle complementary angles supplementary angles vertical angles parallel lines alternate interior angles alternate exterior angles corresponding angles	**The** principals of geometry have long been useful in helping people understand the physical world. Geometric principles were used to survey land and build structures such as the great pyramids of Egypt. The basic geometric figures are the point, the line, and the plane. From these figures, other figures such as segments, rays, and half lines can be made. When two rays have a common endpoint, they form an angle. Angles can be measured in degrees using a protractor. When lines intersect, several different types of angles are formed. They are vertical, corresponding, alternate interior, and alternate exterior. Two adjacent angles are called complementary angles if the sum of their measures is equal to 90°. Two adjacent angles are called supplementary if the sum of their measures is equal to 180°.
10-2	triangle sides vertices isosceles triangle equilateral triangle scalene triangle acute triangle obtuse triangle right triangle Pythagorean theorem similar triangles	**A** closed geometric figure with three sides is called a triangle. A triangle can be classified according to the lengths of its sides or according to the measures of its angles. For a right triangle, the Pythagorean theorem states that $c^2 = a^2 + b^2$, where c is the length of the hypotenuse and a and b are the lengths of the sides. Two triangles with the same shape are called similar triangles.
10-3	quadrilateral pentagon hexagon heptagon octagon nonagon decagon dodecagon icosagon trapezoid	**A** polygon is classified according to the number of sides. A quadrilateral has four sides; a pentagon has five sides, etc. The sum of the measures of the angles of a polygon with n sides is $(n - 2)180°$. Special quadrilaterals such as the trapezoid, parallelogram, rhombus, and square are used in this chapter. The distance around the outside of a polygon is called the perimeter.

—Continued

Summary   *Continued—*

Section	Important Terms	Important Ideas
	parallelogram rectangle rhombus square perimeter	
10-4	area circle radius diameter chord circumference π (pi)	**The** measure of the portion of the plane enclosed by a geometric figure is called the area of the geometric figure. A circle is a closed geometric figure in which all the points are the same distance from a fixed point called the center. A segment connecting the center with any point on the circle is called a radius. A segment connecting two points on the circle and passing through the center of the circle is called a diameter.
10-5	surface area volume cube cylinder rectangular solid cone pyramid sphere	**Geometry** is also the study of solid figures. Some of the familiar solid figures are the cube, the rectangular solid, the pyramid, the cone, the cylinder, and the sphere. The surface area of a solid figure is the area of the faces or surfaces of the figure. The volume of a solid geometric figure is the amount of space that is enclosed by the surfaces of the figure.
10-6	trigonometry sine cosine tangent angle of elevation angle of depression	**Trigonometry** is the study of the relationship between the angles and the sides of a triangle. The trigonometric ratios of sine, cosine, and tangent are used with a right triangle. The concepts of right triangle trigonometry can be used to solve many problems in navigation measurement, engineering, physics, and many other areas.
10-7	network traversable	**Network** theory grew out of a real-life problem studied by Leonard Euler in 1735. It deals with the study of traversable paths in a system that consists of vertices and paths. Today, engineers are able to map traffic, chemists are able to model complex molecules, and business managers are able to make organizational charts using network theory.

Review Exercises

Note: Use $\pi = 3.14$ where needed.
For Exercises 1–10, identify each figure.

1.
    A B

2. R S

3.
  C D

4.
   5

5.

6.

7.

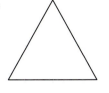

8.

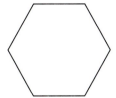

9.

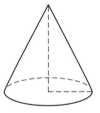

10.

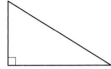

For Exercises 11–16, identify each type of angle or pairs of angles.

11.

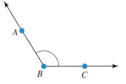

12.

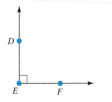

13.

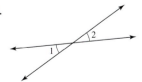

∡1 and ∡2

14.

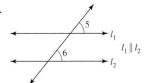

$l_1 \parallel l_2$

∡5 and ∡6

15.

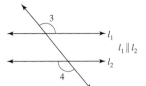

$l_1 \parallel l_2$

∡3 and ∡4

16.

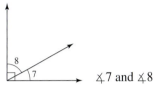

∡7 and ∡8

17. Find the complement of each angle

(a) 27°

(b) 88°

18. Find the supplement of each angle

(a) 172°

(b) 13°

19. Find the measure of the third angle of a triangle if the measures of the other two angles are 95° and 42°.

20. Find the measure of the hypotenuse of a right triangle if the lengths of the two legs are 8 inches and 15 inches.

21. Find the height of a water tower if its shadow is 24 feet when a 6-foot fence pole casts a shadow of 2 feet.

22. Find the measures of ∡1, ∡2, and ∡3.

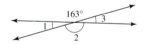

23. Find the measures of ∡1 through ∡7.

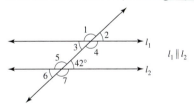

$l_1 \parallel l_2$

24. Find the sum of the angles of a decagon.

25. Find the perimeter of the figure shown.

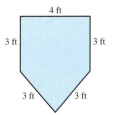

For Exercises 26–29, find the area of each figure.

26.

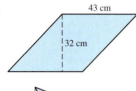

43 cm

32 cm

27.

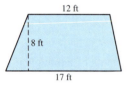

7.6 m

16 m

28.

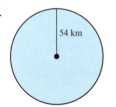

54 km

29.

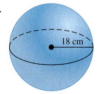

12 ft

8 ft

17 ft

30. Find the circumference and area of a circle whose diameter is 16.5 yards.

For Exercises 31–34, find the volume.

31.

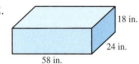

18 cm

32.

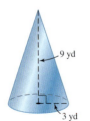

18 in.

24 in.

58 in.

33.

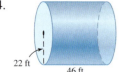

9 yd

3 yd

34.

22 ft

46 ft

For Exercises 35 and 36, find the surface area.

35.

$r = 5.2$ in.

36.

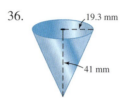

19.3 mm

41 mm

37. If the diameter of a bicycle wheel is 26 inches, how many revolutions will the wheel make if the rider rides 1 mile (1 mile = 5280 feet)?

38. Find the area of the kite shown.

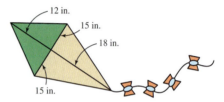

12 in.

15 in.

18 in.

15 in.

39. Find the volume of a ball if the diameter is 4.2 cm.

40. Find the surface area for the walls of the building shown. (Include openings but do not include the area of the roof.)

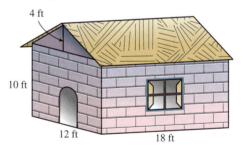

4 ft

10 ft

12 ft

18 ft

41. If a tree 32 feet tall casts a shadow of 40 feet, find the angle of elevation of the sun.

42. A pole is leaning against a wall and is 15 feet from the base of the wall. If the angle of elevation from the ground to the base of the pole is $63°$, find the length of the pole.

For Exercises 43–47, state which networks are traversable.

43.

44.

45.

46.

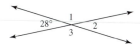

47.

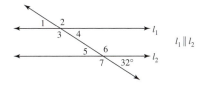

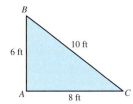

Chapter Test

Note: Use $\pi = 3.14$ where needed.

1. Find the complement of an angle whose measure is 73°.

2. Find the supplement of an angle whose measure is 149°.

3. Find the measures of ∡1, ∡2, and ∡3.

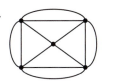

4. Find the measures of ∡1 through ∡7.

5. If two angles of a triangle have a measure of 85° and 47°, respectively, find the measure of the third angle.

6. If the hypotenuse of a right triangle is 29 inches and the length of one leg is 21 inches, find the length of the other leg.

7. Find the measure of side x. The two triangles are similar.

8. A 6-foot pole casts a shadow of 2.7 feet. Find the height of a tree if its shadow is 13.9 feet.

9. Find the sum of the angles of an octagon.

10. Find the perimeter of the figure.

For Exercises 11–15, find the area.

11.

12.

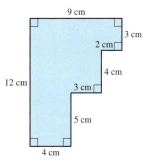

13.

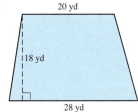

14.

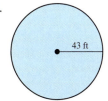

15.

18.6 miles

16. Find the circumference of a circle whose radius is 13 cm.

For Exercises 17–22, find the volume of each.

17.

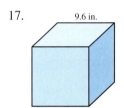

9.6 in.

18.

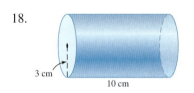

3 cm
10 cm

19.

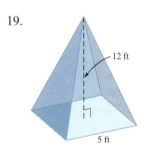

12 ft

5 ft

20.

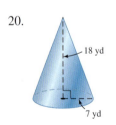

18 yd

7 yd

21.

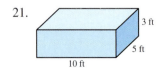

3 ft
5 ft
10 ft

22.

20 m

For Exercises 23–25, find the surface area of each figure.

23.

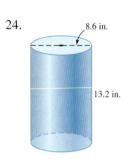

20 ft   24 ft

24.

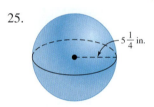

8.6 in.

13.2 in.

25.

$5\frac{1}{4}$ in.

26. A city lot is shaped like a trapezoid with the front and back lines parallel and 157 feet apart. The front is 32 feet and the back is 48 feet. Find the area of the lot.

27. How high up on a wall does a 25-foot ladder reach if it is 5 feet from the base of the wall?

28. Find the cost of fringe to sew on the border of a quilt that measures 70 inches by 100 inches. The fringe costs $1.35 per yard.

29. If lead weighs 0.41 pounds per cubic inch, find the weight of a lead sinker that is the shape of a cone with a diameter of 1 inch and a height of 2 inches.

30. A person digs a hole that is the shape of a cylinder that has a diameter of 3 feet and a depth of 5 feet. What is the weight of the dirt that has been removed if 1 cubic foot of dirt weighs 98 pounds?

31. From the top of a 200-foot lighthouse, the angle of depression to a boat is 33°. How far is the boat from the lighthouse?

For Exercises 32–35, state whether or not the networks shown are traversable.

32.

33.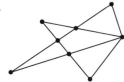

34.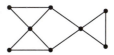

35.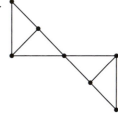

Mathematics in Our World ▶ Revisited

The Force of the Wind

The force of the wind against a surface can be determined by the formula

$$F = 0.004\,As^2$$

where F is the force of the wind in pounds per square foot, A is the area of the surface in square feet, and s is the speed of the wind in miles per hour.

The area of the sign is equal to length × width or $8 \times 10 = 80$ square feet.

The force against the sign for a wind speed of 20 miles per hour is

$$F = 0.004 \cdot 80 \cdot 20^2 = 128 \text{ pounds}$$

For a wind speed of 40 miles per hour, the force is

$$F = 0.004 \cdot 80 \cdot 40^2 = 512 \text{ pounds}$$

For a wind speed of 60 miles per hour, the force is

$$F = 0.004 \cdot 80 \cdot 60^2 = 1152 \text{ pounds}$$

Hence, the sign could withstand a wind speed of 40 miles per hour.

Projects

1. The number pi (π) has an interesting history. For example, in the *Bible* π was defined as 3. Write a brief paper on the history of π.

2. Mathematicians have attempted to compute the value of π in many different ways. Research several methods of computing the value of π and write a paper on two or three methods.

3. There are many idiosyncrasies based on the value of π. For example David Blatner in his book *The Joy of π* relates that the sequence 123456789 first appears at the 523,551,502nd digit. Find several other idiosyncrasies and write a short report on them.

4. The numbers 3, 4, and 5 are called Pythagorean triples since $3^2 + 4^2 = 5^2$. The numbers 5, 12, and 13 are also Pythagorean triples since $5^2 + 12^2 = 13^2$. Can you find any other Pythagorean triples? Actually, there is a set of formulas that will generate an infinite number of Pythagorean triples. Research the topic of Pythagorean triples and write a brief report on the subject.

5. There have been many "proofs" of the Pythagorean theorem. One was formulated by President James A. Garfield. Write a paper on several of the proofs for the Pythagorean theorem.

6. A mathematician named Fermat stated that the equation $x^n + y^n = z^n$ has solutions that are integers only when $n = 2$. He did not prove this theorem. Mathematicians have been looking for a proof since he died. The theorem was only recently proved. Write a brief report on Fermat's theorem. (*Note:* It is called Fermat's last theorem.)

7. The geometric concepts presented in this chapter are based on what is called Euclidean geometry. There are other types of geometries that differ from Euclidean geometry. Two other types are hyperbolic geometry and elliptic geometry. Write a brief report on each type and explain how these geometries differ from Euclidean geometry.

Answers to Try This One

10-A. (a) 147°
      (b) 25°

10-B. $m\angle 2 = 105°$; $m\angle 3 = 75°$; $m\angle 4 = 105°$

10-C. 165° for each of angles 1, 5, 4, and 8; 15° for each of angles 2, 3, 6, and 7

10-D. (a) Scalene; also acute
      (b) Isosceles; also acute
      (c) Right

10-E. 60°

10-F. 12.8 ft

10-G. 8 ft

10-H. 3240°

10-I. 108°

10-J. (a) 18.8 in.
      (b) 100 m
      (c) 48 ft

10-K. 45.5 square inches

10-L. $5140.39

10-M. 40.82 in.

10-N. 3850 cm²

10-O. SA = 226.61 yd²; V = 209.33 yd³

10-P. $414.82 or $415.10 (depending on rounding)

10-Q. 10.83 cm

10-R. 349.51 ft

10-S. 45.69 for c; 67° for A; 23° for B; 90° for C

10-T. (a) Traversable
      (b) Not traversable
      (c) Not traversable

Chapter Eleven

Probability and Counting Techniques

Outline

Objectives

After completing this chapter, you should be able to

1 Determine sample spaces and find the probability of an event using classical probability or empirical probability (11-1)

2 Using a tree diagram or table, find the sample space for a sequence of events, then find the probability of various events (11-2)

3 Given the probability of an event, find the odds of the event (11-3)

4 Given the odds for an event, find the probability of the event (11-3)

5 Find the expected value of an event (11-3)

6 Find the probability of two or more events using the addition rule (11-4)

7 Find the probability of two or more events using the multiplication rule (11-5)

8 Find the conditional probability of an event (11-5)

9 Use the counting rule to determine the total number of outcomes for a sequence of events (11-6)

10 Use the permutation rule to determine the number of different ways r objects can be selected from n objects when the order is important (11-6)

11 Use the combination rule to determine the number of ways r objects can be selected from n objects when order is not important (11-7)

12 Use the counting rule, the permutation rule, and the combination rule to find the number of outcomes in a sample space, then determine the probabilities of various events of the sample space (11-8)

Introduction

Most people are familiar with probability from observing or playing games of chance, such as card games, slot machines, or lotteries. In addition to being used in games of chance, probability theory is used in the fields of insurance, investments, and weather forecasting, and in many other areas of everyday life.

From the time a person awakes until he or she goes to bed, that person makes decisions regarding the possible events that are governed at least in part by chance. For example, should I carry an umbrella to work today? Will my car battery last until spring? Should I accept that new job?

The basic concepts of probability are explained in this chapter. These concepts include *probability experiments, sample spaces, the addition and multiplication rules,* and *rules for counting the number of ways events can occur.* ■

How often do you think "What are the chances of that?" In this chapter you'll learn how to find out.

11-1 Basic Concepts of Probability

The theory of probability grew out of the study of various games of chance using coins, dice, and cards. Since these devices lend themselves well to the application of concepts of probability, they will be used in this chapter for some of the examples. This section begins with an explanation of some basic concepts of probability.

Sample Spaces

Processes such as flipping a coin, rolling a die, or drawing a card from a deck are called *probability experiments.*

Mathematics in Our World

Would You Bet Your Life?

Humans engage in all sorts of gambles, not just betting money at a casino or in a lottery, but they also bet their lives by engaging in unhealthy activities such as smoking, drinking, using drugs, and exceeding the speed limit when driving. Many people don't seem to care about the risks involved in these activities since they do not understand the concepts of probability. On the other hand, people may fear activities that involve little risk to health or life because these activities have been sensationalized by the press and media.

In his book *Probabilities in Everyday Life* (Ivy Books, 1986), author John D. McGervey states:

When people have been asked to estimate the frequency of death from various causes, the most overestimated categories are those involving pregnancy, tornadoes, floods, fire, and homicide. The most underestimated categories include deaths from diseases such as diabetes, strokes, tuberculosis, asthma, and stomach cancer (although cancer in general is overestimated).

The question is: Would you feel safer if you flew across the United States on a commercial airline or if you drove? How much greater is the risk of one way to travel over the other? See Mathematics in Our World—Revisited for the answer.

A **probability experiment** is a process that leads to well-defined results called outcomes. An **outcome** is the result of a single trial of a probability experiment.

A trial means flipping a coin once, rolling one die once, drawing a single card from a deck, etc. When a coin is tossed, there are two possible outcomes: head or tail. (*Note:* We exclude the possibility of a coin landing on its edge.) In the roll of a single die, there are six possible outcomes: 1, 2, 3, 4, 5, or 6. In a probability experiment, we can predict what outcomes are possible, but we cannot predict with certainty which one will occur. We say that the outcomes occur at random. In any experiment, the set of all possible outcomes is called the *sample space.*

A **sample space** is the set of all possible outcomes of a probability experiment.

Some sample spaces for various probability experiments are shown here.

Experiment	Sample Space
Toss one coin	{head, tail}
Roll a die	{1, 2, 3, 4, 5, 6}
Answer a true–false question	{true, false}
Toss two coins	{head/head, tail/tail, head/tail, tail/head}

It is important to realize that when two coins are tossed, there are *four* possible outcomes. Consider tossing a quarter and a dime at the same time. Both coins could fall heads up. Both coins could fall tails up. The quarter could fall heads up and the dime could fall tails up and, finally, the quarter could fall tails up and the dime could fall heads up. The situation is the same even if the coins are indistinguishable.

When finding probabilities, it is sometimes necessary to consider several outcomes of a probability experiment. For example, when a die is rolled, we may want to consider obtaining an odd number, i.e., 1, 3, or 5. Getting an odd number when rolling a die is an example of an event.

An **event** is any subset of the sample space of a probability experiment.

Experiment: draw a card. Sample space: 52 cards. Event: drawing an ace.

Classical Probability

Sample spaces are used in classical probability to determine the numerical probability that an event will happen. One does not actually have to perform the experiment to determine that probability. Classical probability is so named because it was the first type of probability studied by mathematicians in the 17th and 18th centuries.

Classical probability assumes that all outcomes in the sample space are equally likely to occur. For example, when a single die is rolled, each outcome is assumed to have the same probability of occurring. Since there are six outcomes, each outcome has a probability of $\frac{1}{6}$. When a card is selected from an ordinary deck of 52 cards, one assumes that the deck has been shuffled, and each card has the same probability of being selected. In this case, it is $\frac{1}{52}$.

A die roll has six outcomes. If E = roll a 2, then $P(E) = \frac{1}{6}$. If $P(E)$ = roll an even number, then $P(E) = \frac{3}{6} = \frac{1}{2}$.

Formula for Classical Probability

The probability of any event E is

$$\frac{\text{number of outcomes in } E}{\text{total number of outcomes in the sample space } S}$$

This probability is denoted by

$$P(E) = \frac{n(E)}{n(S)}$$

This probability, called **classical probability,** is based on a sample space S.

Probabilities can be expressed as fractions, decimals, or—where appropriate—percents. If one asks, "What is the probability of getting a head when a coin is tossed?" An appropriate response could be any of these three:

    "One half."

    "Point five."

    "Fifty percent."

These answers are all equivalent. In most cases, the answers to examples and exercises given in this chapter are expressed as fractions or decimals, but percents are used where appropriate.

Example   **11-1**

A die is rolled. Find the probability of getting

(a) A 2
(b) A number less than 5
(c) An odd number

Solution

In this case, since the sample space is 1, 2, 3, 4, 5, and 6, there are six outcomes; i.e., $n(S) = 6$.

(a) There is one possible outcome that gives a 2, so $P(2) = \frac{1}{6}$.
(b) There are four possible outcomes for the event of getting a number less than 5—namely, 1, 2, 3, or 4; hence, $n(E) = 4$ and

$$P(\text{a number less than 5}) = \frac{n(E)}{n(S)} = \frac{4}{6} = \frac{2}{3}$$

—Continued

Example 11-1 *Continued—*

(c) There are three possible outcomes for the event of getting an odd number, namely, 1, 3, or 5; hence, $n(E) = 3$

$$P(\text{odd number}) = \frac{n(E)}{n(S)} = \frac{3}{6} = \frac{1}{2}$$

Example 11-2

Two coins are tossed. Find the probability of getting

(a) Two heads
(b) At least one head
(c) At most one head

Solution

The sample space is {HH, HT, TH, TT}; therefore, $n(S) = 4$.

(a) There is only one way to get two heads; i.e., HH. Hence,

$$P(\text{two heads}) = \frac{n(E)}{n(S)} = \frac{1}{4}$$

(b) "At least one head" means one or more heads; i.e., one head or two heads. There are three ways to get at least one head, HT, TH, and HH. Hence, $n(E) = 3$; then

$$P(\text{at least one head}) = \frac{n(E)}{n(S)} = \frac{3}{4}$$

(c) "At most one head" means no heads or one head; i.e., TT, TH, HT. Hence, $n(E) = 3$; then

$$P(\text{at most one head}) = \frac{n(E)}{n(S)} = \frac{3}{4}$$

Math Note

A good problem-solving strategy to use is to make a list of all possible outcomes in the sample space before computing the probabilities of events.

Try This One

11-A Each number from one to nine is written on a card and placed in a box. If a card is selected at random, find the probability that the number on the card is

(a) A 7

(b) An odd number

(c) A number less than four

(d) A number greater than seven

For the answer, see page 593.

A 10% chance of rain is not the same as no chance of rain. You would expect it to rain on roughly 1 in 10 days when a 10% chance of rain was forecast.

The probability of an event will always be a number between and including zero and one. This is denoted mathematically as $0 \le P(E) \le 1$. Probabilities cannot be negative or greater than one.

When an event cannot occur (i.e., the event contains no members in the sample space), the probability is zero. For example, the probability of selecting a former president of the United States who is female is zero since to date all presidents of the United States have been men.

When an event is certain to occur, the probability is one. For example, if one removes all the red cards from a deck and only 26 black cards remain, the probability of selecting a black card is $\frac{26}{26}$ or 1.

The sum of the probabilities of all the outcomes in the sample space will always be one. For example, when a die is rolled, each of the six outcomes has a probability of $\frac{1}{6}$, and the sum of the probabilities of the six outcomes will be one.

When the probability of an event is close to zero, the event is very unlikely to occur. When the probability of an event is close to $\frac{1}{2}$ or 0.5, the event has approximately a 50% chance of occurring. When the probability of an event is near one, the event is almost certain to occur. See Figure 11-1.

In addition to finding the probability that an event will occur, it is sometimes necessary to find the probability that the event will *not* occur. For example, if a die is rolled, the probability that a 4 will not occur, symbolized by $P(\overline{4})$, is $\frac{5}{6}$ since there are five ways that a 4 will not occur; i.e., 1, 2, 3, 5, 6. The solution can also be found by finding the probability of getting a 4 and subtracting it from one; the sum of the probabilities of the outcomes in the sample space is

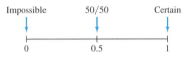

Impossible   50/50   Certain

0   0.5   1

Figure 11-1

$$P(\overline{4}) = 1 - P(4)$$
$$= 1 - \tfrac{1}{6}$$
$$= \tfrac{5}{6}$$

The rule can be generalized as follows

For any event, E, $P(\overline{E}) = 1 - P(E)$.

Probability and Sets

The theory of probability is related to the theory of sets discussed in Chapter 2. For a given probability experiment, the sample space can be considered the universal set, and an event E can be considered as a subset of the universal set.

For example, when rolling a die, the sample space is $\{1, 2, 3, 4, 5, 6\}$; hence, the universal set, $U = \{1, 2, 3, 4, 5, 6\}$. Let the event E be getting an odd number; i.e., 1, 3, or 5. In sets, $E = \{1, 3, 5\}$. A Venn diagram can now be drawn illustrating this example. See Figure 11-2.

Notice that set E contains 1, 3, and 5 while the numbers 2, 4, and 6 are in the universal set but not in E. Hence, $\overline{E} = \{2, 4, 6\}$. Now recall from set theory that $E \cup \overline{E} = U$. As stated previously, the sum of the probabilities of the outcomes in the sample space is one, so $P(U) = 1$. $\overline{E}$ represents the elements in U but not in E and is written in probability theory as $\overline{E}$. Therefore, $P(E) + P(\overline{E}) = 1$. Subtracting $P(E)$ from both sides, one gets $P(\overline{E}) = 1 - P(E)$. Additional relationships between probability theory and set theory will be shown in other sections of this chapter.

U

1   4
3
5   6
E
2

$E \cup \overline{E} = U$
$P(E) + P(\overline{E}) = U$

Figure 11-2

Empirical Probability

Probabilities can also be computed for situations not related to gambling games. For example, suppose a researcher surveyed 25 students in a classroom and asked each one

Sidelight

History of Probability

There is evidence that the Greek scientist Aristotle (384–322 B.C.E.) investigated simple probability concepts. Paintings in tombs excavated in Egypt show that the Egyptians played games of chance based on probability. One game called "Hounds and Jackals" (similar to the present-day game of "Snakes and Ladders") was known to be played in 1800 B.C.E.

Ancient Greeks and Romans made crude dice from animal bones, various stones, minerals, and ivory. When these recently found artifacts were tested mathematically, they were found to be quite accurate.

Not much else was done, however, until the 16th century, when Girolamo Cardan (1501–1576) wrote the first book on probability entitled *The Book on Chance and Games.* Cardan was an astrologer, philosopher, physician, mathematician, and gambler. In his book, he included techniques on how to cheat and how to catch others at cheating.

He also listed correctly the sample space for the outcomes when two dice are rolled. Cardan is thought to be the first mathematician to formulate a definition of classical probability.

During the mid-1600s, a professional gambler named Chevalier de Mere made a considerable amount of money on a gambling game. He would bet unsuspecting patrons that in four rolls of a die, he could get at least one 6. He was so successful that word got out, and people refused to play the game. He decided to invent a new game to keep on making a living. He reasoned (incorrectly) that if he rolled two dice 24 times, he could get at least one double 6. However, he began to lose systematically.

Unable to figure out why, he contacted the renowned mathematician Blaise Pascal to find out why he was losing.

Girolamo Cardan (1501–1576)

Pascal became interested in studying probability theory formally and began a correspondence with a French government official and fellow mathematician, Pierre de Fermat (1601–1665). Together, the two were able to solve de Mere's dilemma and formulated the beginnings of probability theory.

In 1657, a Dutch mathematician named Christiaan Huygens wrote a treatise on the Pascal-Fermat correspondence and introduced the idea of mathematical expectation.

A professor and mathematician, Augustine Louis Cauchy (1789–1857), wrote a book on probability while teaching at the Military School of Paris. One of his students was Napolean Bonaparte.

In 1895, the Fey Manufacturing Company, located in San Francisco, invented the first three-reel automatic slot machine.

Today, probability theory is used in the insurance industry, the gambling industry, and various other fields such as war games and risk-taking ventures.

whether he or she liked the flavor of a new soft drink. Suppose 17 said, "Yes," and 8 said, "No." Now if a person is selected at random from the classroom, the probability that the person liked the soft drink is $\frac{17}{25}$ since 17 out of 25 people liked the drink. This type of probability is called **empirical probability** and is based on *observed frequencies.* In this case, the observed frequency of those who said, "Yes" is 17, and the observed frequency of those who said "No" is 8. The sum of the observed frequencies is $17 + 8 = 25$.

Formula for Empirical Probability

$$P(E) = \frac{\text{observed frequency of the specific event } (f)}{\text{sum of frequencies } (n)}$$

$$= \frac{f}{n}$$

In this coin toss, the empirical probability of heads was $\frac{6}{10}$, or $\frac{3}{5}$. With more tosses, you would expect $P(\text{heads})$ to approach $\frac{1}{2}$.

The information in the previous problem can be written in the form of a *frequency distribution* which consists of classes and frequencies for the classes, as shown

Response (class)	Frequency
Yes	17
No	8
Total	25

This technique may be helpful when working out problems.

Type O negative blood is the least likely to react badly with a patient's blood. If it is not possible to test the donor blood before a transfusion, then O negative is used.

Example 11-3

In a sample of 50 people, 21 had type O blood, 22 had type A blood, 5 had type B blood, and 2 had type AB blood. Set up a frequency distribution and find these probabilities for a person selected at random from the sample.

(a) The person has type O blood.
(b) The person has type A or type B blood.
(c) The person has neither type A nor type O blood.
(d) The person does not have type AB blood.

Source: Based on American Red Cross figures

—Continued

Example **11-3** *Continued—*

Solution

Type (class)	Frequency
A	22
B	5
AB	2
O	21
Total	50

(a) $P(O) = \frac{f}{n} = \frac{21}{50}$

(b) $P(A \text{ or } B) = \frac{22}{50} + \frac{5}{50} = \frac{27}{50}$

(Add the frequencies of the two classes, A and B.)

(c) $P(\text{neither A nor O}) = \frac{5}{50} + \frac{2}{50} = \frac{7}{50}$

(Neither A nor O means that a person has either type B or type AB blood.)

(d) $P(\text{not AB}) = 1 - P(AB) = 1 - \frac{2}{50} = \frac{48}{50} = \frac{24}{25}$

(Find the probability of not AB by subtracting the probability of type AB from 1.)

Math Note

The solution to part c could also be done as
P(neither A nor O) =
1 − P(A or O):

$1 - \left(\frac{22}{50} + \frac{21}{50}\right)$
$= 1 - \frac{43}{50} = \frac{7}{50}.$

Try This One

11-B A bag of Hershey's Assorted Miniatures contains 18 Hershey Milk Chocolate bars, 9 Mr. Goodbars, 9 Krackel bars, and 8 Hershey's Special Dark Chocolate bars. If a bar is selected at random from the bag, find the probability that it is

(a) A Mr. Goodbar

(b) A Krackel or a Special Dark Chocolate bar

(c) Not a Milk Chocolate bar

For the answer, see page 593.

It is important to understand the relationship of classical probability and empirical probability in certain situations. Classical probability, the probability of rolling a 3 when a die is thrown, is found by looking at the sample space and is $\frac{1}{6}$. To find the probability of getting a 3 when a die is thrown using empirical probability, one would actually toss a die a specific number of times and count the number of times a 3 was obtained; then divide that number by the number of times the die was rolled. For example, suppose a die was rolled 60 times, and a 3 occurred 12 times. Then the empirical probability of getting a 3 would be $\frac{12}{60} = \frac{1}{5}$. Most of the time, the probability obtained from empirical methods will differ from that obtained using classical probability. The question is, then, "How many times should I roll the die when using empirical probability?" There is no specific answer except to say that the more times the die is tossed, the closer the results obtained from empirical probability will be to those of classical probability. Classical probability is also called theoretical probability.

In summary, then, classical probability uses sample spaces and assumes the outcomes are equally likely. Empirical probability uses observed frequencies and the total of the number of frequencies.

Exercise Set 11-1

Real World Applications

1. If a die is rolled one time, find the probability of
   (a) Getting a 4
   (b) Getting an even number
   (c) Getting a number greater than 4
   (d) Getting a number less than 7
   (e) Getting a number greater than 0
   (f) Getting a number greater than 3 or an odd number
   (g) Getting a number greater than 3 and an odd number

2. A couple has two children. Find the probability that
   (a) Both children are girls
   (b) At least one child is a girl
   (c) Both children are of the same gender

3. A spinner for a children's game consists of the numbers 1 through 7, with equally sized regions for each of these numbers. If the player spins once, find the probability that the number is
   (a) A 6   (b) An even number
   (c) A number greater than 4   (d) A number less than 8
   (e) A number greater than 7

4. A box contains five red, two white, and three green marbles. If a marble is selected at random, find these probabilities.
   (a) That it is red   (b) That it is green
   (c) That it is red or white   (d) That it is not green
   (e) That it is not red

5. In an office there are five women and four men. If one person is selected at random, find the probability that the person is a woman.

6. If there are 50 tickets sold for a raffle and one person buys seven tickets at random, what is the probability of that person winning the prize?

7. In an office there are seven women and nine men. If one person is promoted at random, find the probability that the person is a man.

8. A survey found that 53% of Americans think U.S. military forces should be used to "protect the interest of U.S. corporations" in other countries. If an American is selected at random, find the probability that he or she will disagree or have no opinion on the issue.

9. The U.S. Bureau of Justice Statistics reported that in 1995, 12,249 men and women escaped from state prisons. Of these, 12,166 were captured. If a 1995 escapee is selected at random, find the probability that the escapee was captured.

10. Thirty-nine of 50 states are currently under court order to alleviate overcrowding and poor conditions in one or more of their prisons. If a state is selected at random, find the probability that it is currently under such a court order.
    Source: Harper's Index

11. In a survey, 16% of American children said they used flattery to get their parents to buy them things. If a child is selected at random, find the probability that the child said he or she does not use parental flattery.
Source: Harper's Index

12. Among 100 students at a small school, 50 are mathematics majors, and 20 are history majors. If a student is selected at random, find the probability that he or she is neither a math major nor a history major.

13. For a drawing the numbers 1 through 15 are placed in a hat and mixed. A number is selected at random. Find the probability that the number is

(a) An odd number
(b) A number that is divisible by 3
(c) A number that is a multiple of 5
(d) A number that is greater than 10
(e) A number that is less than 4

14. In a box there are nine black checkers and seven red checkers. If a checker is selected at random, find the probability that it is

(a) Black
(b) Red

15. A box contains a one-dollar bill, a five-dollar bill, a ten-dollar bill, a twenty-dollar bill, and a fifty-dollar bill. If a person selects a bill at random, find the probability of that person receiving

(a) A twenty-dollar bill

(b) A bill larger than five dollars

(c) A bill that contains the digit 5 in its numerical value

(d) A bill that does not contain a 0 in its numerical value

16. The distribution of ages of CEOs is as follows:

Age	Frequency
21–30	1
31–40	8
41–50	27
51–60	29
61–70	24
71 or older	11

Source: USA Today

If a CEO is selected at random, find the probability that his or her age is
(a) Between 31 and 40
(b) Under 41
(c) Over 30 and under 51
(d) Under 31 or over 60

Writing Exercises

17. What is a probability experiment?
18. Define *sample space*.
19. What is the difference between an outcome and an event?
20. What are equally likely outcomes?

21. What is the range of the values of a probability event?

22. When an event is certain to occur, what is its probability?

23. If an event cannot happen, what value is assigned to its probability?

24. What is the sum of the probabilities of all the outcomes in a sample space?

25. If the probability that it will rain tomorrow in your town is 0.45, explain why the probability that it will not rain tomorrow in your town is 0.55.

26. Explain the difference between classical probability and empirical probability.

27. A probability experiment is conducted. Identify any of these that cannot be considered a probability of an event.

    (a) $\frac{1}{3}$ (b) $-\frac{1}{3}$ (c) 0.80   (d) -0.59 (e) 0

    (f) 1.45   (g) 1   (h) 33%   (i) 112%

Critical Thinking

28. A person flipped a coin 100 times and obtained 73 heads. Can the person conclude that the coin was unbalanced?

29. A medical doctor stated that with a certain treatment, a patient has a 50% chance of recovering without surgery. That is, "Either he will get well or he won't get well." Comment on this statement.

30. The wheel spinner shown below is spun twice. Find the sample space, and then determine the probability of the following events.

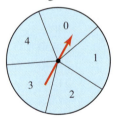

    (a) An odd number on the first spin and an even number on the second spin. (*Note*: 0 is considered even.)

    (b) A sum greater than 4

    (c) Even numbers on both spins

    (d) A sum that is odd

    (e) The same number on both spins

31. If three dice are rolled, find the probability of getting any triple; that is, 111, 222, etc.

11-2 Tree Diagrams, Tables, and Sample Spaces

In the previous examples, the sample spaces were found by observation and reasoning; however, two devices can be used to find all possible outcomes when two or more experiments are performed together or in a sequence. One device is a *tree diagram,* and the other is a *table* of outcomes.

Tree Diagrams

A **tree diagram** consists of branches corresponding to the outcomes of two or more probability experiments that are done in sequence.

When constructing a tree diagram, use branches emanating from a single point to show the outcomes for the first experiment, and then show the outcomes for the second experiment using branches emanating from each branch that was used for the first experiment, etc. Example 11-4 shows how to use a tree diagram to determine the outcomes for probability experiments.

Which event does this family represent?

Example 11-4

Use a tree diagram to find the sample space for the genders of three children in a family.

Solution

There are two possibilities for the first child, boy or girl, two for the second, boy or girl, and two for the third, boy or girl. Hence, the tree diagram can be drawn as shown in Figure 11-3.

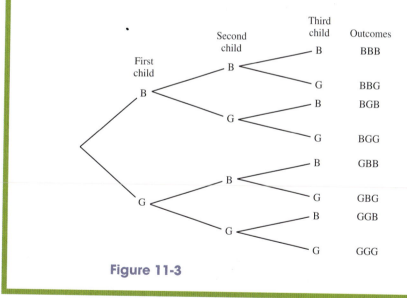

Figure 11-3

After a tree diagram is drawn, the outcomes can be found by tracing through all of the branches. In this case, the sample space would be {BBB, BBG, BGB, BGG, GBB, GBG, GGB, GGG}.

Once a tree diagram is drawn and the sample space is found, you can compute the probabilities for various events.

Example 11-5

If a family has three children, find the probability that all three children are the same gender; that is, all boys or all girls.

Solution

The sample space shown in Example 11-4 has eight outcomes, and there are two possible ways to have three children of the same gender, BBB or GGG. Hence, the probability of the three children being of the same gender is $\frac{2}{8}$ or $\frac{1}{4}$.

Example 11-6

A coin is tossed, and then a die is rolled. Use a tree diagram to find the sample space.

Solution

The coin can land two ways, head up or tail up, and the die can be tossed so that there are six outcomes: 1, 2, 3, 4, 5, or 6. The tree diagram can be drawn as shown in Figure 11-4.

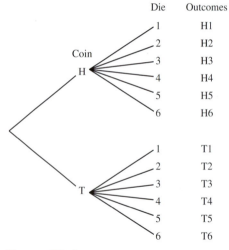

Figure 11-4

Hence, the sample space is {H1, H2, H3, H4, H5, H6, T1, T2, T3, T4, T5, T6}.

Example 11-7

When a coin is tossed and a die is rolled, find the probability of getting a head on the coin and an even number on the die.

Solution

The total number of outcomes for the experiment is 12. The number of ways to get a head on the coin and an even number on the die is 3: H2, H4, or H6. Hence, the probability of getting a head and an even number when a coin is tossed and a die is rolled is $\frac{3}{12}$ or $\frac{1}{4}$.

Try This One

11-C In order to collect information for a student survey, a researcher classifies students according to eye color (blue, brown, green), gender (male, female), and class rank (i.e., freshman, sophomore). A folder for each classification is then made up (e.g., freshman/female/green eyes). Find the sample space for the folders using a tree diagram. If a folder is selected at random, find the probability that

(a) It includes students with blue eyes.

(b) It includes students who are female.

(c) It includes students who are male freshmen.

For the answer, see page 594.

When constructing tree diagrams, all branches do not have to be the same length. For example, suppose two players, Alice and Bill, play chess, and the first one to win two games wins the tournament. The tree diagram would be like the one shown in Figure 11-5. A means that Alice wins, and B means that Bill wins the game.

Notice that if Alice wins the first two games, the tournament is over. Hence the first branch is shorter than the second one. However, if Alice wins the first game and Bill wins

Sidelight
_____

Mathematics and the Draft Lottery

Today there are many state lotteries that offer patrons a chance to win money. At one time, there was a national lottery not to win money but to see who would be drafted for military service. During the Vietnam War, men were drafted based on their age and their birthday. The oldest men were drafted first. Men born in the last months of the year could avoid the draft since they were younger in months and days than those born earlier in the year.

In 1970, the United States decided to change the system and draft men based on a lottery. The days of the year were numbered one through 366 (leap year), placed in a barrel, and mixed up. Then numbers corresponding to birth dates were selected one at a time and men were drafted on the basis of their draft number.

It was assumed that the drawing was random. However, mathematicians found that the drawing was not random since those born later in the year had a higher chance of being drafted than those born earlier in the year. It was concluded that the capsules containing the numbers were not thoroughly mixed and that the numbers placed in the barrel last were on the top of the pile. They were selected first.

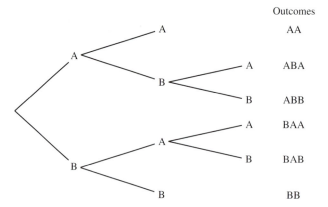

Figure 11-5

	A	2	3	4	5	6	7	8	9	10	J	Q	K
♥	A♥	2♥	3♥	4♥	5♥	6♥	7♥	8♥	9♥	10♥	J♥	Q♥	K♥
♦	A♦	2♦	3♦	4♦	5♦	6♦	7♦	8♦	9♦	10♦	J♦	Q♦	K♦
♠	A♠	2♠	3♠	4♠	5♠	6♠	7♠	8♠	9♠	10♠	J♠	Q♠	K♠
♣	A♣	2♣	3♣	4♣	5♣	6♣	7♣	8♣	9♣	10♣	J♣	Q♣	K♣

Figure 11-6

the second game, they need to play a third game in order to decide who wins the tournament. Similar reasoning can be applied to the rest of the branches.

Tables

Another way of determining a sample space is by making a **table.** Consider the sample space of selecting a card from a standard deck of 52 cards. (The cards are assumed to be shuffled to make sure that the selection occurs at random.) There are four suits—hearts, diamonds, spades, and clubs, and 13 cards of each suit consisting of the denominations ace (A), 2, 3, 4, 5, 6, 7, 8, 9, 10, and 3 picture or face cards—jack (J), queen (Q), and king (K). The sample space is shown in Figure 11-6.

Example 11-8

A card is drawn from an ordinary deck. Use the sample space shown in Figure 11-6 to find the probabilities of getting

(a) A jack
(b) The 6 of clubs
(c) A 3 or a diamond

—Continued

Example 11-8 *Continued—*

Solution

(a) Refer to the sample space in Figure 11-6. There are four jacks and 52 possible outcomes. Hence,

$$P(\text{jack}) = \frac{4}{52} = \frac{1}{13}$$

(b) Since there is only one 6 of clubs, the probability of getting a 6 of clubs is

$$P(\text{6 of clubs}) = \frac{1}{52}$$

(c) There are four 3s and 13 diamonds, but the 3 of diamonds is counted twice in this listing. Hence, there are 16 possibilities of drawing a 3 or a diamond, so

$$P(\text{3 or diamond}) = \frac{16}{52} = \frac{4}{13}$$

Try This One

11-D A single card is drawn at random from a well-shuffled deck. Using the sample space shown in Figure 11-6, find the probability that the card is

(a) An ace

(b) A queen

(c) A club

(d) A 4 or a diamond

(e) A 6 and a spade (i.e., the 6 of spades)

For the answer, see page 594.

When two dice are rolled, there are 36 outcomes in the sample space. Assume for the purpose of illustration that one die is red and the other is blue. Now how many outcomes are there? The sample space for the red die is $\{1, 2, 3, 4, 5, 6\}$, and the sample space for the blue die is also $\{1, 2, 3, 4, 5, 6\}$. Now if the outcomes for each die are combined, the sample space will consist of 36 outcomes, as shown in Figure 11-7.

		Blue die				
	1	**2**	**3**	**4**	**5**	**6**
1	(1, 1)	(1, 2)	(1, 3)	(1, 4)	(1, 5)	(1, 6)
2	(2, 1)	(2, 2)	(2, 3)	(2, 4)	(2, 5)	(2, 6)
3	(3, 1)	(3, 2)	(3, 3)	(3, 4)	(3, 5)	(3, 6)
4	(4, 1)	(4, 2)	(4, 3)	(4, 4)	(4, 5)	(4, 6)
5	(5, 1)	(5, 2)	(5, 3)	(5, 4)	(5, 5)	(5, 6)
6	(6, 1)	(6, 2)	(6, 3)	(6, 4)	(6, 5)	(6, 6)

(Red die)

Figure 11-7

The outcome $(3, 6)$ means the 3 was obtained on the red die and a 6 was obtained on the blue die. The sum of the spots on the faces then would be $3 + 6 = 9$.

Example 11-9

When two dice are rolled, find the probability of getting

(a) A sum of 8
(b) Doubles
(c) A sum less than 5

Solution

Using the sample space shown in Figure 11-7, there are 36 possible outcomes.

(a) There are five ways to get a sum of 8. They are $(2, 6)$, $(3, 5)$, $(4, 4)$, $(5, 3)$, and $(6, 2)$. Hence $n(E) = 5$ and $n(S) = 36$; then,

$$P(\text{sum of 8}) = \frac{n(E)}{n(S)} = \frac{5}{36}$$

(b) There are six ways to get doubles. They are $(1, 1)$, $(2, 2)$, $(3, 3)$, $(4, 4)$, $(5, 5)$, and $(6, 6)$. Hence, $n(E) = 6$ and $n(S) = 36$; then,

$$P(\text{doubles}) = \frac{n(E)}{n(S)} = \frac{1}{6}$$

(c) A sum less than 5 means a sum of 4 or 3 or 2. The number of ways this can occur is 6, as shown.

Sum of 4: $(1, 3)$, $(2, 2)$, $(3, 1)$
Sum of 3: $(1, 2)$, $(2, 1)$
Sum of 2: $(1, 1)$
Hence, $n(E) = 6$ and $n(S) = 36$; then,

$$P(\text{sum less than 5}) = \frac{6}{36} = \frac{1}{6}$$

Try This One

11-E Two dice are rolled. Use the sample space shown in Figure 11-7 to find the probability of

(a) Getting a sum of 9

(b) Getting a sum that is an even number

(c) Getting a sum greater than 6

For the answer, see page 594.

Tree diagrams and tables are useful devices for finding sample spaces for more complex probability problems.

Exercise Set   11-2

Real World Applications

1. Using the sample space for the genders of three children in the family shown in Example 11-4, find the probability that
   (a) The family will have exactly two girls.
   (b) The family will have three boys.
   (c) The family will have at least one girl.
   (*Note:* At least one girl means one, two, or three girls.)

2. Using the sample space for tossing a coin and rolling a die shown in Example 11-6, find these probabilities.
   (a) A head on the coin and an odd number on the die
   (b) A head on the coin and a prime number on the die (*Note:* 2, 3, and 5 are prime numbers.)
   (c) A tail on the coin, and a number less than 5 on the die

3. Using the sample space for tossing a coin and rolling a die shown in Example 11-6, find these probabilities:
   (a) A tail on the coin and a number greater than 1 on the die
   (b) A head on the coin and an even number on the die
   (c) A tail on the coin and a number divisible by 3 on the die

4. Draw a tree diagram for all possible answer keys for a true–false quiz consisting of three questions. Find the probability that
   (a) All answers will be true or all answers will be false.
   (b) The answers will alternate (i.e., TFT or FTF).
   (c) Exactly two answers will be true.

5. A box contains a one-dollar bill, a five-dollar bill, and a ten-dollar bill. A bill is selected and its value is noted, then it is replaced in the box. A second bill is then selected. Draw a tree diagram to determine the sample space and find the probability that
   (a) Both bills have the same value.
   (b) The second bill is larger than the first bill selected.
   (c) The value of each of the two bills is even.
   (d) The value of exactly one of the bills is odd.
   (e) The sum of the values of both bills is less than $10.

6. Draw a tree diagram to determine the sample space when four coins are tossed and find the probability that
   (a) Exactly three coins land heads up.
   (b) All coins land tails up.
   (c) Two or more coins land heads up.
   (d) No more than two coins land tails up.
   (e) At least one coin lands tails up.

7. Mark and Bill play a chess tournament consisting of three games. They are equal in ability. Draw a tree diagram to determine the sample space and find the probability that

   (a) Either Mark or Bill win all three games.

   (b) Either Mark or Bill win two out of three games.

   (c) Mark wins only two games in a row.

   (d) Bill wins the first game, loses the second game, and wins the third game.

8. A box contains a red pen, a blue pen, a green pen, and a black pen. A person selects a pen, notes its color, and replaces it. Then the person selects a second pen. Draw a tree diagram to determine the sample space and find the probability that

   (a) Both pens are the same color.

   (b) Pens differ in color.

   (c) At least one pen is green.

   (d) Both pens are red.

   (e) One pen is green and the other pen is blue.

9. Five balls are numbered one through five and placed in a box. Two balls are selected without replacement. Draw a tree diagram to determine the sample space and find the probability that

   (a) The sum of the numbers is odd.

   (b) The number on the second ball is larger than the number on the first ball.

   (c) The sum of the numbers on both balls is greater than 4.

10. A person is purchasing a new automobile and can select one option from each category.

Model	Type	Color
Panther	Hard top	White
Wildcat	Soft top	Blue
Tiger		Red

    Draw a tree diagram and find the sample space for all possible choices. Find the probability that the automobile if chosen at random is

    (a) A soft top   (b) A white Panther

    (c) A red hard top Wildcat   (d) Blue

    (e) Either a red or blue Tiger

11. Kimberly decides to have a computer custom made. She can select one option from each category.

MB	Monitor	Color
64	15 in.	Tan
128	17 in.	Ivory
256		

    Draw a tree diagram to determine the sample space and find the probability that her computer if chosen at random will have

    (a) A 17-in. monitor

    (b) A 128 MB and a 15-in. monitor

    (c) A 15-in. Ivory monitor

12. A box contains five envelopes with one coin inside each envelope. The coins are a penny, a nickel, a dime, a quarter, and a half dollar. Two envelopes are selected in succession without replacement. Draw a tree diagram to determine the sample space and find the probability that

   (a) The amount of the first coin is less than the amount of the second coin.

   (b) Neither coin is a quarter.

   (c) One coin is a penny and the other coin is a nickel or a dime.

   (d) The sum of the amounts of both coins is even.

   (e) The sum of the amounts of both coins is less than $0.40.

13. Using the sample space for drawing a single card from an ordinary deck of 52 cards shown in Figure 11-6, find the probability of getting

   (a) A 10   (b) A club

   (c) An ace of hearts   (d) A 3 or a 5

   (e) A 6 or a spade   (f) A queen or a club

   (g) A diamond or a club   (h) A red king

   (i) A black card or an 8   (j) A red 10

14. Using the sample space for drawing a single card from an ordinary deck of 52 cards shown in Figure 11-6, find the probability of getting

   (a) A 6 of clubs   (b) A black card

   (c) A queen   (d) A black 10

   (e) A red card or a 3   (f) A club and a 6

   (g) A 2 or an ace   (h) A club or a diamond or a spade

   (i) A diamond face card   (j) A red ace

15. Using the sample space for rolling two dice shown in Figure 11-7, find the probability of getting a

   (a) Sum of 5

   (b) Sum of 7 or 11

   (c) Sum greater than 9

   (d) Sum less than or equal to 5

   (e) Three on one die or on both dice

   (f) Sum that is odd

   (g) A prime number on one or both dice (*Note:* 2, 3, and 5 are prime numbers.)

   (h) A sum greater than 1

16. Using the sample space for rolling two dice shown in Figure 11-7, find the probability of getting a

   (a) Sum of 8

   (b) Sum that is prime (2, 3, 5, 7, and 11 are prime)

   (c) Five on one or both dice

   (d) Sum greater than or equal to 7

   (e) Sum that is less than 3

   (f) Sum greater than 12

   (g) Six on one die and 3 on the other die

Writing Exercises

17. Explain how to draw a tree diagram.

18. Give an example of a tree diagram where all the branches of a tree diagram do not have the same lengths.

Critical Thinking

19. Consider the sample space when three dice are rolled. How many different outcomes would there be?

20. When three dice are rolled, how many ways can a sum of 6 be obtained?

21. When three dice are rolled, find the probability of getting a sum of 6.

11-3 Odds and Expectation

Casinos set the odds slightly in their favor, so the house makes a profit.

Many times probabilities are expressed in terms of **odds.** Odds are used by casinos, racetracks, and other gambling establishments to determine the payoffs when bets are made or lottery tickets are purchased. For example, if a person rolled a die and won every time he or she rolled a 6, then the person would win on average once every six times. In order to make the game fair, odds of five to one would be given. That means if a person bets $1.00 and wins, the person would win $5.00. On average, the person would win $5.00 once in six rolls and lose $1.00 on the other five rolls—hence the term *fair game.*

In most gambling games, the odds given are not fair. For example, if the odds of winning are really 20 to 1, the house might offer odds of 15 to 1 in order to make a profit.

Odds can be expressed as a fraction or ratio such as $\frac{5}{1}$ or 5:1 or 5 to 1. In any probability experiment, such as a gambling game, there are two ways to express the odds. They are "odds in favor of the event" (i.e., odds of winning) and the "odds against the event" (i.e., odds of losing).

The formulas for odds are

$$\text{odds in favor} = \frac{P(E)}{1 - P(E)}$$

$$\text{odds against} = \frac{P(\overline{E})}{1 - P(\overline{E})}$$

where $P(E)$ is the probability that event E occurs and $P(\overline{E})$ is the probability that the event E does not occur.

Example   11-10

A card is drawn from a deck.

(a) Find the odds in favor of getting an ace.
(b) Find the odds against getting an ace.

Solution

(a) There are four aces in a deck of 52 cards $P(A) = \frac{4}{52}$; hence, the odds in favor of getting an ace are

$$\text{odds in favor} = \frac{P(A)}{1 - P(A)}$$

$$= \frac{\frac{4}{52}}{1 - \frac{4}{52}} = \frac{\frac{4}{52}}{\frac{48}{52}}$$

$$= \frac{4}{52} \div \frac{48}{52}$$

$$= \frac{4^{1}}{52} \cdot \frac{52}{48_{12}}$$

$$= \frac{1}{12}$$

Hence, the odds in favor are 1:12.

(b) Since there are 48 ways not to get an ace, the odds against an ace are

$$\text{odds against} = \frac{\frac{48}{52}}{1 - \frac{48}{52}}$$

$$= \frac{48}{52} \div \frac{4}{52}$$

$$= \frac{48^{12}}{52_{1}} \times \frac{52^{1}}{4^{1}}$$

$$= \frac{12}{1}$$

Hence, the odds against an ace are 12:1.

Math Note

The odds in favor of an event is the ratio of the number of winning outcomes to the number of losing outcomes. The odds against an event is the ratio of the number of losing outcomes to the number of winning outcomes.

Try This One

11-F   Two dice are rolled. Find

(a) The odds in favor of obtaining a sum of 9.

(b) The odds against getting a sum of 12.

For the answer, see page 594.

In Example 11-10, the odds for an event were found when the probabilities were known. When the odds of an event are given, the probability of an event can be found.

> If the odds in favor of an event E are $a{:}b$, then the probability that the event will occur is
> $$P(E) = \frac{a}{a + b}$$
> If the odds against an event E are $c{:}d$, then the probability that E will not occur is
> $$P(\overline{E}) = \frac{c}{c + d}$$

Math Note

When the odds are 1:1, the game is said to be fair. That is, both parties have an equal chance of winning or losing.

Sports games are often described with odds, as in "New Jersey is the 3 to 1 favorite to win over New England."

Example 11-11

If the odds that an event will occur are 3:5, find the probability that the event will occur.

Solution

In this case, $a = 3$ and $b = 5$. Hence,

$$P(E) = \frac{3}{3 + 5} = \frac{3}{8}$$

Math Note

If the odds in favor of an event are $a{:}b$, the odds against the event are $b{:}a$.

Try This One

11-G When two dice are rolled, the odds in favor of getting a sum of 9 are 1:8. Find the probability of not getting a sum of 9 when two dice are rolled.
For the answer, see page 594.

Expectation

Another concept that is related to probability and odds is the concept of **expectation** or **expected value.** Expected value is used in various games of chance, in insurance, and in other areas, such as decision theory when the outcomes are numerical.

Expected value is used to determine what happens in a probability situation over the long run. For example, if a person flips a balanced coin many times, we can expect that "heads" will occur approximately one half of the time.

> The expected value for the outcomes of a probability experiment is
> $$E = X_1 \cdot P(X_1) + X_2 \cdot P(X_2) + \cdots + X_n \cdot P(X_n)$$
> where the X's correspond to the outcomes and the $P(X)$'s are the corresponding probabilities of the outcomes.

Math Note

In Example 11-12, a die cannot show 3.5 spots. It can only show 1, 2, 3, 4, 5, or 6 spots. However, in this case, the expected value would be the long run average—that is, if one adds up the number of spots, then divides by the number of times the die is rolled, the average would be close to 3.5.

Example 11-12

When a die is rolled, find the expected value of the outcome.

Solution

Since each number, 1 through 6, has a probability of $\frac{1}{6}$, the expected value is

$$E = 1 \cdot \tfrac{1}{6} + 2 \cdot \tfrac{1}{6} + 3 \cdot \tfrac{1}{6} + 4 \cdot \tfrac{1}{6} + 5 \cdot \tfrac{1}{6} + 6 \cdot \tfrac{1}{6} = \tfrac{21}{6} = 3.5$$

In gambling games, the expected value is found by multiplying the amount won and lost by the corresponding probabilities and then finding the sum.

Example 11-13

One thousand tickets are sold at $1 each for a color television valued at $350. What is the expected value of the gain if a person purchases one ticket?

Solution

The problem can be set up as follows:

	Win	Lose
Gain, X	$349	−$1
Probability, $P(X)$	$\frac{1}{1000}$	$\frac{999}{1000}$

Two things should be noted. First, for a win, the net gain is $349, since the person does not get the cost of the ticket ($1) back. Second, for a loss, the gain is represented by a negative number, in this case, −$1. The solution, then, is

$$E(X) = \$349 \cdot \tfrac{1}{1000} + (-\$1) \cdot \tfrac{999}{1000} = -\$0.65$$

Expected value problems of this type can also be solved by finding the overall gain and subtracting the cost of the tickets, as shown:

$$E(X) = \$350 \cdot \tfrac{1}{1000} - \$1 = -\$0.65$$

Here, the overall gain ($350) must be used.

Note that the expectation is −$0.65. This does not mean that a person loses $0.65, since the person can only win a television set valued at $350 or lose $1 on the ticket. What this expectation means is that the average of the losses is $0.65 for each of the 1000 ticket holders. Here is another way of looking at this situation: If a person purchased one ticket each week over a long period of time, the average loss would be $0.65 per ticket, since theoretically, on average, that person would win the television set once for each 1000 tickets purchased.

Sidelight

Mathematics and Slot Machines

Today most slot machines are run electronically, much the same as video games are. However, early slot machines were mechanical in nature. The first slot machines were invented by the Fey Manufacturing Company of San Francisco in 1895. They consisted of three large wheels, which were spun when the handle on the side of the machine was pulled. In order to control the number of wins and the payoffs, each wheel contains 20 symbols. However, the number of the same symbols is not the same on each wheel. For example, there may be two oranges on wheel 1, six oranges on wheel 2, and no oranges on wheel 3. When a person gets two orange symbols, he may think that he almost won; i.e., 2 out of 3. However, since there is no orange symbol on the third wheel, the probability of getting three oranges is zero! The higher the probability of getting three oranges, the lower the payoff. Using probability theory, the owner of the machines can determine his long-run profit.

A stock you bought two years ago with high hopes is now selling for less than you paid, and things look grim for the company. Do you sell, or hold on and hope it will come back to the original price before you sell?

A model economists use for such situations is a game no one wants to play: Suppose you have a choice: lose $100, or take a $\frac{50}{50}$ chance between losing nothing, and losing $300. Which do you choose? Find the expected value for each strategy over 10 trials.

Example   11-14

One thousand tickets are sold at $1 each for four prizes of $100, $50, $25, and $10. What is the expected value if a person purchases two tickets?

Solution

Find the expected value if the person purchases one ticket.

Gain, x	$99	$49	$24	$9	−$1
Probability, $P(x)$	$\frac{1}{1000}$	$\frac{1}{1000}$	$\frac{1}{1000}$	$\frac{1}{1000}$	$\frac{996}{1000}$

$$E(x) = \$99 \cdot \tfrac{1}{1000} + \$49 \cdot \tfrac{1}{1000} + \$24 \cdot \tfrac{1}{1000} + \$9 \cdot \tfrac{1}{1000} - \$1 \cdot \tfrac{996}{1000} = -\$0.815$$

Now multiply by 2 since two tickets were purchased.

$$-\$0.815(2) = -\$1.63$$

An alternative solution is

$$E(x) = \$100 \cdot \tfrac{1}{1000} + \$50 \cdot \tfrac{1}{1000} + \$25 \cdot \tfrac{1}{1000} + \$10 \cdot \tfrac{1}{1000} - \$1.00 = -\$0.815$$

Multiply by 2.

$$-\$0.815(2) = -\$1.63$$

In gambling games, if the expected value of the gain is 0, the game is said to be fair. If the expected value of the gain of a game is positive, then the game is in favor of the player. That is, the player has a better-than-even chance of winning. If the expected value of the gain is negative, then the game is said to be in favor of the house. That is, in the long run, the players will lose money.

Try This One

11-H A ski resort loses $70,000 per season when it does not snow very much and makes $250,000 profit when it does snow a lot. The probability of it snowing at least 75 inches (i.e., a good season) is 40%. Find the expectation for the profit.
For the answer, see page 594.

Exercise Set 11-3

Real World Applications

1. When two dice are tossed, find the odds
   (a) In favor of getting a sum of 10
   (b) In favor of getting a sum of 12
   (c) Against getting a sum of 7
   (d) Against getting a sum of 3
   (e) In favor of getting doubles

2. When a single die is tossed, find the odds
   (a) In favor of getting a 3
   (b) In favor of getting a 6
   (c) Against getting an odd number
   (d) Against getting an even number
   (e) In favor of getting a prime number

3. When a single card is drawn from a shuffled deck of cards, find the odds
   (a) In favor of getting a queen
   (b) In favor of getting a face card
   (c) Against getting a club
   (d) In favor of getting an ace
   (e) In favor of getting a black card

4. When three coins are tossed, find the odds
   (a) In favor of getting exactly three heads
   (b) In favor of getting exactly three tails
   (c) Against getting exactly two heads
   (d) Against getting exactly one tail
   (e) In favor of getting at least one tail

5. Find the probability that the event E will occur given these odds
   (a) 7:4 in favor of E
   (b) 2:5 against E
   (c) 3:1 in favor of E
   (d) 1:4 against E

6. Find the probability that the event E will occur given these odds
   (a) 3:4 in favor of E
   (b) 1:7 against E
   (c) 5:4 in favor of E
   (d) 6:5 in favor of E

7. If the odds against a horse winning a race are 9:5, find the probability that the horse will win the race.

8. A person rolls two dice and wins if he or she throws doubles. What are the odds in favor of the event? What are the odds against the event?

9. A cash prize of $5000 is to be awarded as a fundraiser. If 2500 tickets are sold at $5 each, find the expected value of the gain.

10. A box contains ten $1 bills, five $2 bills, three $5 bills, one $10 bill, and one $100 bill. Find the expectation if one bill is selected.

11. If a person rolls doubles when he tosses two dice, he wins $5. For the game to be fair, how much should the person pay to play the game?

12. If a player rolls two dice and gets a sum of 2 or 12, the person wins $20. If the person gets a 7, the person wins $5. The cost to play the game is $3. Find the expectation of the game.

13. A lottery offers one $1000 prize, one $500 prize, and five $100 prizes. One thousand tickets are sold at $3 each. Find the expectation if a person buys one ticket.

14. In Exercise 13, find the expectation if a person buys two tickets.

15. For a daily lottery, a person selects any three-digit number from 000 to 999. If a person plays for $1, the person can win $500. Find the expectation. In the same daily lottery, if a person boxes a number, the person can win $80. Find the expectation if the number 123 is played for $1 and boxed. (When a number is "boxed," it can win when the digits occur in any order.)

16. If a 60-year-old buys a $1000 life insurance policy at a cost of $60 and has a probability of 0.972 of living to age 61, find the expectation of the policy until the person reaches 61.

Writing Exercises

17. Explain the difference between the odds in favor of an event and the odds against an event.

18. Explain the numerical relationship between the odds in favor of an event and the odds against an event.

19. Explain the meaning of odds in a gambling game.

20. Explain how to find the probability of an event occurring when given the odds in favor of an event.

21. Explain what is meant by the expected value of an event.

Critical Thinking

22. Consider the following problem: A con man has three coins. One coin has been specially made and has a head on each side. A second coin has been specially made and on each side it has a tail. Finally, a third coin has a head and a tail on it. All coins are of the same denomination. The con man places the coins in his pocket, selects

one at random, and shows you one side. It is heads. He is willing to bet you even money that it is the two-headed coin. His reasoning is that it can't be the two-tailed coin since a head is showing; therefore, there is a 50–50 chance of it being the two-headed coin. Should you take the bet?

23. Chevalier de Mere, a famous gambler, won money when he bet unsuspecting patrons that in four rolls of a die, he could get at least one 6, but he lost money when he bet that in 24 rolls of two dice, he could get a double 6. Using the probability rules, find the probability of each event and explain why he won the majority of the time on the first game but lost the majority of the time when playing the second game.

24. A roulette wheel has 38 numbers: 1 through 36, 0, and 00. A ball is rolled, and it falls into one of the 38 slots, giving a winning number. If a player bets $1 on a number and wins, the player gets $36 plus his $1 back. Otherwise, he loses the $1 he bet. Find the expected value of the game.

11-4 The Addition Rules for Probability

Many problems involve finding the probability of two or more events. For example, at a large political gathering, one might wish to know, for a person selected at random, the probability that the person is a female or is a Republican. In this case, there are three possibilities to consider:

1. The person is a female.
2. The person is a Republican.
3. The person is both a female and a Republican.

Consider another example. At the same gathering there are Republicans, Democrats, and Independents. If a person is selected at random, what is the probability that the person is a Democrat or an Independent? In this case, there are only two possibilities:

1. The person is a Democrat.
2. The person is an Independent.

The difference between the two examples is that in the first case, the person selected can be a female and a Republican at the same time. In the second case, the person selected cannot be both a Democrat and an Independent at the same time. In the second case, the two events are said to be mutually exclusive; in the first case, they are not mutually exclusive.

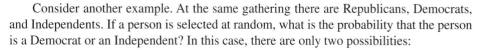

Two events are **mutually exclusive** if they cannot occur at the same time (i.e., they have no outcomes in common).

In another situation, the events of getting a 4 and getting a 6 when a single card is drawn from a deck are mutually exclusive events, since a single card cannot be both a 4 and a 6. On the other hand, the events of getting a 4 and getting a heart on a single draw are not mutually exclusive, since one can select the 4 of hearts when drawing a single card from an ordinary deck.

The probability of two or more events occurring can be determined by using the **addition rules.** The first addition rule is used when the events are mutually exclusive.

If you select a T-shirt at random in the morning, the probability that it is blue or red is $\frac{1}{5} + \frac{1}{5}$.

Addition Rule 1

When two events are mutually exclusive, the probability that *A* or *B* will occur is

$$P(A \text{ or } B) = P(A) + P(B)$$

Math Note

Sometimes it is better not to reduce probability fractions at first, since you may need to find a common denominator in order to add, as was done in Example 11-15. It is usually preferred that the final answer, if expressed as a fraction, be reduced to lowest terms.

Example 11-15

A restaurant has three pieces of apple pie, five pieces of cherry pie, and four pieces of pumpkin pie in its dessert case. If a customer selects at random one kind of pie for dessert, find the probability that it will be either cherry or pumpkin.

Solution

The events are mutually exclusive. Since there is a total of 12 pieces of pie, five of which are cherry and four of which are pumpkin,

$$P(\text{cherry or pumpkin}) = P(\text{cherry}) + P(\text{pumpkin})$$
$$= \tfrac{5}{12} + \tfrac{4}{12} = \tfrac{9}{12} = \tfrac{3}{4}$$

Example 11-16

A card is drawn from a deck. Find the probability of getting an ace or a queen.

Solution

The events are mutually exclusive. There are four aces and four queens; therefore,

$$P(\text{ace or queen}) = P(\text{ace}) + P(\text{queen})$$
$$= \tfrac{4}{52} + \tfrac{4}{52} = \tfrac{8}{52} = \tfrac{2}{13}$$

Try This One

11-I At a political rally, there are 20 Republicans, 13 Democrats, and 6 Independents. If a person is selected, find the probability that he or she is either a Democrat or an Independent.
For the answer, see page 594.

In summary, when two events, A and B, are mutually exclusive, the probability that either event A or event B will occur is found by adding the sum of the individual probabilities of event A and event B.

The addition rule for mutually exclusive events can be extended to three or more events as shown in Example 11-17.

Example 11-17

A card is drawn from a deck. Find the probability that it is either a club or a diamond or a heart.

Solution

In the deck of 52 cards there are 13 clubs, 13 diamonds, and 13 hearts. Hence,

$$P(\text{club or diamond or heart}) = P(\text{club}) + P(\text{diamond}) + P(\text{heart})$$
$$= \tfrac{13}{52} + \tfrac{13}{52} + \tfrac{13}{52} = \tfrac{39}{52} = \tfrac{3}{4}$$

In Example 11-17, $P(A \text{ or } B \text{ or } C) = P(A) + P(B) + P(C)$.

When two events are not mutually exclusive, one must subtract the probability of the outcomes that are common to both events since they have been counted twice. That is, one can use addition rule 2.

Math Note

This rule can also be used when the events are mutually exclusive, because $P(A \text{ and } B)$ will always equal 0 when the events are mutually exclusive. However, it is important to make a distinction between the two situations when thinking about probabilities.

Addition Rule 2

If A and B are not mutually exclusive, then
$$P(A \text{ or } B) = P(A) + P(B) - P(A \text{ and } B)$$

Example 11-18

A single card is drawn from a deck of cards. Find the probability that it is a king or a club.

Solution

In this case, there are four kings and 13 clubs. However, the king of clubs has been counted twice since the two events are not mutually exclusive. When finding the probability of getting a king or a club, the probability of getting the king of clubs (i.e., a king and a club) must be subtracted, as shown.

$$P(\text{king or club}) = P(\text{king}) + P(\text{club}) - P(\text{king and club})$$
$$= \tfrac{4}{52} + \tfrac{13}{52} - \tfrac{1}{52} = \tfrac{16}{52} = \tfrac{4}{13}$$

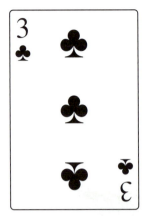

Example   11-19

Two dice are rolled. Find the probability of getting doubles or a sum of 6.

Solution

Using the sample space shown in Section 11-2, there are six ways to get doubles—i.e., $(1, 1), (2, 2), (3, 3), (4, 4), (5, 5), (6, 6)$; hence, $P(\text{doubles}) = \frac{6}{36}$. There are five ways to get a sum of 6—i.e., $(1, 5), (2, 4), (3, 3), (4, 2), (5, 1)$; hence, $P(\text{sum of 6}) = \frac{5}{36}$. Notice that there is one way of getting doubles and a sum of 6; hence, $P(\text{doubles and a sum of 6}) = \frac{1}{36}$. Finally,

$$P(\text{doubles or a sum of 6}) = P(\text{doubles}) + P(\text{sum of 6}) - (\text{doubles and sum of 6})$$
$$= \frac{6}{36} + \frac{5}{36} - \frac{1}{36} = \frac{10}{36} = \frac{5}{18}$$

Try This One

11-J   A card is drawn from an ordinary deck. Find the probability that it is a heart or a face card.
For the answer, see page 594.

In many cases, the information in probability problems can be arranged in table form in order to make it easier to compute the probabilities for various events. Example 11-20 shows this technique.

Example   11-20

In a hospital there are eight nurses and five physicians. Seven nurses and three physicians are females. If a staff person is selected, find the probability that the subject is a nurse or a male.

—Continued

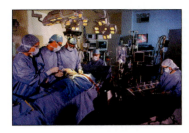

Example 11-20 Continued—

Solution

The sample space can be written in table form.

Staff	Females	Males	Total
Nurses	7	1	8
Physicians	3	2	5
Total	10	3	13

Looking at the table, one sees that there are 8 nurses and 3 males, and there is one person that is both a male and a nurse. The probability is

$$P(\text{nurse or male}) = P(\text{nurse}) + P(\text{male}) - P(\text{male and a nurse})$$

$$= \tfrac{8}{13} + \tfrac{3}{13} - \tfrac{1}{13} = \tfrac{10}{13}$$

Sidelight

Blaise Pascal (1623–1662)

Child Genius

Pascal was born in the French province of Auvergne in 1623. His father, Etienne Pascal, was a mathematician. Blaise Pascal was a child prodigy when it came to mathematics. At the age of 13, he deduced and proved many of the theorems of Euclid's geometry without ever seeing a geometry book. At the age of 14, he participated in weekly meetings of the mathematicians of the French Academy. At the age of 16, he invented a mechanical adding machine that would carry using gears.

In addition to mathematics, he made contributions to mechanics and physics. In 1650, because of frail health, Pascal gave up the study of mathematics and physics and devoted himself to the study of religion and metaphysics. However, in 1654, he returned to the study of mathematics, and at that time he began his correspondence with Fermat concerning probability. When he narrowly escaped death by getting out of the way of runaway horses, he took that as a sign to return to his religious contemplations.

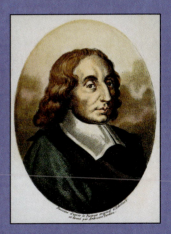

In 1658, while suffering from a toothache, he returned to mathematics and developed the geometry of the cycloid in eight days. Because of poor health, Pascal spent most of his life in physical pain and mental torment, and some feel that he did not develop to his full potential for mathematics. He died at the age of 39 in 1662.

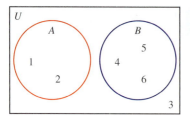

(a) Mutually exclusive events
$A \cap B = \phi$
$P(A \cup B) = P(A) + P(B)$

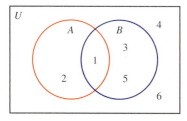

(b) Events not mutually exclusive
$A \cap B \neq \phi$
$P(A \cup B) = P(A) + P(B) - P(A \cap B)$

Figure 11-8

Try This One

11-K In a class of students, there are 15 freshmen and 10 sophomores. Six of the freshmen are males and four of the sophomores are males. If a student is selected at random, find the probability that the student is a sophomore or a male.
For the answer, see page 594.

Probability and Sets

The addition rules can be represented by Venn diagrams. First, consider two mutually exclusive events. For example, we will use the probability experiment of rolling a die. Let A be the event of getting a number less than 3 and B be the event of getting a number greater than 3. In set theory, $U = \{1, 2, 3, 4, 5, 6\}$, $A = \{1, 2\}$, and $B = \{4, 5, 6\}$. The Venn diagram for mutually exclusive events is shown in Figure 11-8(a). Notice $A \cup B = \{1, 2\} \cup \{4, 5, 6\} = \{1, 2, 4, 5, 6\}$; now $P(A \cup B) = P(A) + P(B)$ or in probability language, $P(A \text{ or } B) = P(A) + P(B)$. Since $P(A) = \frac{2}{6}$, and $P(B) = \frac{3}{6}$, then $P(A \text{ or } B) = \frac{5}{6}$.

Next, consider two events that are not mutually exclusive. Let A be the event of getting a number less than 3 and B be the event of getting an odd number. The sample space for A is $\{1, 2\}$. The sample space for B is $\{1, 3, 5\}$. The Venn diagram for this example is shown in Figure 11-8(b).

Notice that $A \cap B = \{1\}$.

Hence, $P(A \cup B) = P(A) + P(B) - P(A \cap B)$.

$$P(A \cup B) = \frac{2}{6} + \frac{3}{6} - \frac{1}{6} = \frac{4}{6} = \frac{2}{3}.$$

In probability language, this is equivalent to $P(A \text{ or } B) = P(A) + P(B) - P(A \text{ and } B)$.

Exercise Set   11-4

Real World Applications

1. A furniture store decides to select a month for its annual sale. Find the probability that it will be April or May. Assume that all months have an equal opportunity of being selected.

2. In a large department store, there are two managers, four department heads, 16 clerks, and four stockers. If a person is selected at random, find the probability that the person is either a clerk or a manager.

3. At a convention there are seven mathematics instructors, five computer science instructors, three statistics instructors, and four science instructors. If an instructor is selected at random, find the probability of getting a science instructor or a math instructor.

4. An automobile dealer has 10 Fords, 7 Buicks, and 5 Plymouths on his used car lot. If a salesperson selects a used car at random to use as a demo, find the probability that it is a Ford or a Buick.

5. A store has eight red pens, nine blue pens, and three black pens. If a student selects a pen at random, find the probability that it is

   (a) A red or a blue pen   (b) A red or a black pen

   (c) A blue or a black pen

6. In the mathematics department, there are two full professors, three associate professors, five assistant professors, three instructors, and six adjunct faculty. If a member of the department is selected at random, find the probability that the person is

   (a) Either an instructor or an adjunct professor

   (b) Either a full professor or an assistant professor

   (c) Either a full professor, instructor, or adjunct faculty

7. On a small college campus, there are five English professors, four mathematics professors, two science professors, three psychology professors, and three history professors. If a professor is selected at random, find the probability that the professor is

   (a) An English or psychology professor

   (b) A mathematics or science professor

   (c) A history, science, or mathematics professor

   (d) An English, mathematics, or history professor

8. A single card is drawn from a deck. Find the probability of selecting

   (a) A 4 or a diamond   (b) A club or a diamond

   (c) A jack or a black card

9. In a statistics class there are 18 juniors and 10 seniors; 6 of the seniors are females, and 12 of the juniors are males. If a student is selected at random, find the probability of selecting

   (a) A junior or a female   (b) A senior or a female

   (c) A junior or a senior

10. A large department store purchases 400 boomboxes for a sale. 250 of them have CD players and 150 of them have cassette players. 140 of the boomboxes with CD players are black, and the rest are white. 80 of the boomboxes with cassette players are black, and the rest are white. If a boombox is selected at random, find the probability that it will

    (a) Be black or have a CD player

    (b) Be white or have a cassette player

11. A women's clothing store owner buys from three companies: A, B, and C. The most recent purchases are shown here.

Product	Company A	Company B	Company C
Dresses	24	18	12
Blouses	13	36	15

    If one item is selected at random, find these probabilities:

    (a) It was purchased from Company A or is a dress.

    (b) It was purchased from Company B or Company C.

    (c) It is a blouse or was purchased from Company A.

12. In a recent study, the following data were obtained in response to the question, "Do you favor the school combining the elementary and middle school students in one building?"

	Yes	No	No opinion
Males	72	81	5
Females	103	68	7

If a person is selected at random, find these probabilities:

(a) The person has no opinion.

(b) The person is a male or is against the issue.

(c) The person is a female or favors the issue.

13. A grocery store employs cashiers, stock clerks, and deli personnel. The distribution of employees according to marital status is shown next.

Marital status	Cashiers	Stock clerks	Deli personnel
Married	8	12	3
Not married	5	15	2

If an employee is selected at random, find these probabilities:

(a) The employee is a stock clerk or married.

(b) The employee is not married.

(c) The employee is a cashier or is not married.

14. In a certain geographic region, newspapers are classified as being published daily morning, daily evening, and weekly. Some have a comics section and some do not. The distribution is shown next.

Have comics section	Morning	Evening	Weekly
Yes	2	3	1
No	3	4	2

If a newspaper is selected at random, find these probabilities:

(a) The newspaper is a weekly publication.

(b) The newspaper is a daily morning paper and has comics.

(c) The newspaper is published weekly or does not have comics.

15. Three cable channels (6, 8, and 10) air quiz shows, comedies, and dramas. The numbers of shows aired are shown here.

Type of show	Channel 6	Channel 8	Channel 10
Quiz show	5	2	1
Comedy	3	2	8
Drama	4	4	2

If a show is selected at random, find these probabilities:

(a) The show is a quiz show or it is shown on Channel 8.

(b) The show is a drama or a comedy.

(c) The show is shown on Channel 10 or it is a drama.

16. A local postal carrier distributed first-class letters, advertisements, or magazines. For a certain day, she distributed the following number of each type of item.

Delivered to	First-class letters	Ads	Magazines
Home	325	406	203
Business	732	1021	97

If an item of mail is selected at random, find these probabilities:

(a) The item went to a home.

(b) The item was an ad or it went to a business.

(c) The item was a first-class letter or it went to a home.

17. If one card is drawn from an ordinary deck of cards, find the probability of getting

   (a) A king or a queen or a jack   (b) A club or a heart or a spade

   (c) A king or a queen or a diamond   (d) An ace or a diamond or a heart

   (e) A 9 or a 10 or a spade or a club

18. Two dice are rolled. Find the probability of getting

   (a) A sum of 6 or 7 or 8

   (b) Doubles or a sum of 4 or 6

   (c) A sum greater than 9 or less than 4 or equal to 7

19. An urn contains six red balls, two green balls, one blue ball, and one white ball. If a ball is drawn, find the probability of getting a red or a white ball.

20. Three dice are rolled. Find the probability of getting

   (a) Triples   (b) A sum of 5

Writing Exercises

21. Define *mutually exclusive events.*

22. Determine whether these events are mutually exclusive.

   (a) Roll a die: Get an even number, or get a number less than 3.

   (b) Roll a die: Get a prime number (2, 3, 5), or get an odd number.

   (c) Roll a die: Get a number greater than 3, or get a number less than 3.

   (d) Select a student in your class: The student has blond hair, or the student has blue eyes.

   (e) Select a student in your college: The student is a sophomore, or the student is a business major.

   (f) Select any course: It is a calculus course, or it is an English course.

   (g) Select a registered voter: The voter is a Republican, or the voter is a Democrat.

Critical Thinking

23. The probability that a customer selects a pizza with mushrooms or pepperoni is 0.55, and the probability that the customer selects mushrooms only is 0.32. If the probability that he or she selects pepperoni only is 0.17, find the probability of the customer selecting both items.

24. In building new homes, a contractor finds that the probability of a homeowner selecting a two-car garage is 0.70 and of selecting a one-car garage is 0.20. Find the probability that the buyer will select no garage. The builder does not build houses with garages for three or more cars.

25. In Exercise 24, find the probability that the buyer will not want a two-car garage.

11-5 The Multiplication Rules and Conditional Probability

Section 11-4 showed how the addition rules are used to compute probabilities. This section introduces two more rules for finding probabilities called the multiplication rules.

The **multiplication rules** can be used to find the probability of two or more events that occur in sequence. For example, if a coin is tossed and then a die is rolled, one can find the probability of getting a head on the coin *and* a 4 on the die. These two events are said to be *independent* since the outcome of the first event (tossing a coin) does not affect the probability outcome of the second event (rolling a die).

> Two events, A and B, are **independent** if the fact that A occurs does not affect the probability of B occurring.

Here are other examples of independent events:

Rolling a die and getting a 6, and then rolling a second die and getting a 3.

Drawing a card from a deck and getting a queen, replacing it, and drawing a second card and getting a queen.

On the other hand, when the occurrence of the first event changes the probability of the occurrence of the second event, the two events are said to be *dependent*. For example, suppose a card is drawn from a deck and *not* replaced, and then a second card is drawn. The probability of the second card is changed since the sample space contains only 51 cards when the first card is not replaced.

> When the outcome or occurrence of the first event affects the outcome or occurrence of the second event in such a way that the probability of the second event is changed, the events are said to be **dependent.**

Here are some examples of dependent events:

Drawing a card from a deck, not replacing it, and then drawing a second card.

Selecting a ball from an urn, not replacing it, and then drawing a second ball.

Parking in a no-parking zone and getting a parking ticket.

In order to find the probability of two independent events occurring, one must find the probability of each event occurring separately and then multiply the answers. For example, if a coin is tossed twice, the probability of getting two heads is $\frac{1}{2} \cdot \frac{1}{2} = \frac{1}{4}$ since the probability of getting a head on each coin is $\frac{1}{2}$. This result can be verified by looking at the sample space, {HH, HT, TH, TT}. Then $P(\text{HH}) = \frac{1}{4}$.

Each day a professor selects a student at random to work out a problem from the homework. If the student is selected regardless of who already went, the events are independent. If the student is selected from the pool of students who haven't gone yet, the events are dependent.

Multiplication Rule 1

When two events are independent, the probability of both occurring is

$$P(A \text{ and } B) = P(A) \cdot P(B)$$

Example 11-21

A coin is flipped and a die is rolled. Find the probability of getting a head on the coin and a 4 on the die.

Solution

$$P(\text{head and } 4) = P(\text{head}) \cdot P(4) = \tfrac{1}{2} \cdot \tfrac{1}{6} = \tfrac{1}{12}$$

Note that the sample space for the coin is {H, T}, and for the die it is {1, 2, 3, 4, 5, 6}.

The problem in Example 11-21 can also be solved by using the sample space:

{H1 H2 H3 H4 H5 H6 T1 T2 T3 T4 T5 T6}

The solution is $\tfrac{1}{12}$, since there is only one way to get the head-4 outcome. (See Section 11-2 for the tree diagram of the experiment.)

Example 11-22

An urn contains three red balls, two blue balls, and five white balls. A ball is selected and its color is noted. Then it is replaced. A second ball is selected and its color is noted. Find the probability of each of these.

(a) Selecting two blue balls.
(b) Selecting a blue ball and then a white ball.
(c) Selecting a red ball and then a blue ball.

Solution

Remember selection is done with replacement, which makes the events independent.

(a) The probability of selecting a blue ball on each trial is $\tfrac{2}{10}$; hence,

$$P(\text{blue and blue}) = P(\text{blue}) \cdot P(\text{blue}) = \tfrac{2}{10} \cdot \tfrac{2}{10} = \tfrac{4}{100} = \tfrac{1}{25}$$

(b) The probability of selecting a blue ball is $\tfrac{2}{10}$ and the probability of selecting a white ball is $\tfrac{5}{10}$; hence,

$$P(\text{blue and white}) = P(\text{blue}) \cdot P(\text{white}) = \tfrac{2}{10} \cdot \tfrac{5}{10} = \tfrac{10}{100} = \tfrac{1}{10}$$

(c) The probability of selecting a red ball is $\tfrac{3}{10}$ and the probability of selecting a blue ball is $\tfrac{2}{10}$; hence,

$$P(\text{red and blue}) = P(\text{red}) \cdot P(\text{blue}) = \tfrac{3}{10} \cdot \tfrac{2}{10} = \tfrac{6}{100} = \tfrac{3}{50}$$

Try This One

11-L A card is drawn from a deck and replaced; then a second card is drawn. Find the probability of getting a queen and then an ace.
For the answer, see page 594.

Math Note

Multiplication rule 1 can be extended to three or more independent events by using the formula

$P(A$ and B and C and ... $K)$
$= P(A) \cdot P(B) \cdot P(C) \cdot \cdots \cdot P(K)$

Example   11-23

Three cards are drawn from a deck. After each card is drawn, its denomination is noted and it is replaced before the next card is selected. Find the probability of getting

(a) Three kings
(b) Three clubs

Solution

(a) Since there are four kings, the probability of selecting a king on each draw is $\frac{4}{52}$ or $\frac{1}{13}$. Hence, the probability of selecting a king with replacement is

$$\frac{1}{13} \cdot \frac{1}{13} \cdot \frac{1}{13} = \frac{1}{2197}$$

(b) Since there are 13 clubs, the probability of selecting a club is $\frac{13}{52}$ or $\frac{1}{4}$. Hence, the probability of selecting three clubs in a row is

$$\frac{1}{4} \cdot \frac{1}{4} \cdot \frac{1}{4} = \frac{1}{64}$$

Try This One

11-M   A die is rolled four times. Find the probability of getting

(a) Four sixes

(b) Four odd numbers

For the answer, see page 594.

When a few subjects are selected from a large number of subjects in a sample space, and the subjects are not replaced, the probability of the event occurring changes so slightly that, for the most part, it is considered to remain the same. Example 11-24 illustrates this concept.

Example   11-24

A Harris poll found that 46% of Americans say they suffer great stress at least once a week. If three people are selected at random, find the probability that all three will say that they suffer great stress at least once a week.
Source: 100% American by Daniel Evan Weiss (Poseidon Press, 1988).

Solution

Let S denote stress. Then

$$P(S \text{ and } S \text{ and } S) = P(S) \cdot P(S) \cdot P(S)$$
$$= (0.46)(0.46)(0.46) \approx 0.097$$

Try This One

11-N The probability that a specific medical test will show positive is 0.32. If four people are tested, find the probability that all four will show positive.
For the answer, see page 594.

In order to find probabilities when events are dependent, we use the multiplication rule with a modification. For example, when two cards are drawn at random from a deck without replacement, the probability of getting an ace on the first draw is $\frac{4}{52}$, and the probability of getting an ace on the second draw is $\frac{3}{51}$ because the first card was not replaced before drawing the second card. Using the multiplication rule, the probability of both events occurring is

$$\frac{4}{52} \cdot \frac{3}{51} = \frac{12}{2652} = \frac{1}{221}$$

Sidelight

Probability and Risk Taking

An area where people fail to understand probability is risk taking. Actually, people fear situations or events that have a relatively small probability of happening rather than those events that have a greater likelihood of occurring. For example, a recent *USA Weekend* magazine poll (August 22–24, 1997) showed that 9 out of 10 Americans answered "No" when asked the question, "Is the world a safer place than when you were growing up?" However, in his book entitled *How Risk Affects Your Everyday Life* (Merritt Publishing, Santa Monica, California, 1996), author James Walsh states: "Despite widespread concern about the number of crimes committed in the United States, FBI and Justice Department statistics show that the national crime rate has remained fairly level for 20 years. It even dropped slightly in the early 1990's."

He further states, "Today most media coverage of risk to health and well-being focuses on shock and outrage." Shock and outrage make good stories and can scare us about the wrong dangers. For example, the author states that if a person is 20% overweight, the loss of life expectancy is 900 days (about 3 years), but loss of life expectancy from exposure to radiation emitted by nuclear power plants is 0.02 days. As you can see, being overweight is much more of a threat than being exposed to radioactive emission.

Many people gamble daily with their lives—for example, using tobacco, drinking and driving, riding motorcycles, etc. When people are asked to estimate the probabilities or frequencies of death from various causes, they tend to overestimate causes such as accidents, fires, and floods and underestimate the probabilities of death from diseases (other than cancer), strokes, etc. For example, people think that their chances of dying of a heart attack are 1 in 20 when in fact it is almost 1 in 3; the chances of dying by pesticide poisoning are 1 in 200,000 (*True Odds* by James Walsh). The reason people think this way is that the news media sensationalize deaths resulting from catastrophic events and rarely mention deaths from disease.

When dealing with life-threatening catastrophes such as hurricanes, floods, automobile accidents, or smoking, it is important to get the facts. That is, get the actual numbers from accredited statistical agencies or reliable statistical studies, and then compute the probabilities and make decisions based on your knowledge of probability and statistics.

In summary, then, when you make a decision or plan a course of action based on probability, make sure that you understand the true probability of the event occurring. Also, find out how the information was obtained (i.e., from a reliable source). Weigh the cost of the action and decide if it is worth it. Finally, look for other alternatives or courses of action with less risk involved.

This example illustrates the fact that getting an ace on the second draw has taken into consideration that an ace was drawn the first time.

The probability that event B occurs after event A has already occurred is written as $P(B \mid A)$. This notation does not mean that B is divided by A; rather, it means the probability that event B occurs given that event A has occurred. In this case, the second card was an ace given that the first card is an ace, and the probability is equal to $\frac{3}{51}$ since the first card was not replaced.

Multiplication Rule 2

When two events are dependent, the probability of both occurring is

$$P(A \text{ and } B) = P(A) \cdot P(B \mid A)$$

Example   11-25

In a shipment of 25 microwave ovens, three are defective. If two ovens are randomly selected and tested, find the probability that both are defective. The first oven is *not* replaced after it has been tested.

Solution

Since there are three defective ovens and a total of 25 ovens, the probability of selecting a defective oven on the first draw is $\frac{3}{25}$. After the first one is selected and tested, there are 24 remaining ovens of which two are defective; hence, the probability of selecting a defective oven on the second draw is $\frac{2}{24}$. Finally, we want to find the probability that both are defective, meaning the first one is defective and the second one is defective given that the first one is defective. Hence,

$$P(D_1 \text{ and } D_2) = P(D_1) \cdot P(D_2 \mid D_1)$$
$$= \frac{3}{25} \cdot \frac{2}{24} = \frac{6}{600} = \frac{1}{100}$$

Try This One

11-O   A box contains three red balls, two blue balls, and one white ball. A ball is selected at random and its color is noted. The ball is *not* replaced and a second ball is selected and its color is noted. Find the probability that

(a) Both balls are blue

(b) A red ball is selected first, and a blue ball is selected second

For the answer, see page 594.

Multiplication rule 2 can be extended to three or more events as shown in Example 11-26.

Example   11-26

Three cards are drawn from an ordinary deck and not replaced. Find the probability of

(a) Getting three jacks
(b) Getting an ace, a king, and a queen in order
(c) Getting a club, a spade, and a heart in order
(d) Getting three clubs

Solution

(a) $P(\text{three jacks}) = \dfrac{4}{52} \cdot \dfrac{3}{51} \cdot \dfrac{2}{50} = \dfrac{24}{132{,}600} = \dfrac{1}{5525}$

(b) $P(\text{ace and king and queen}) = \dfrac{4}{52} \cdot \dfrac{4}{51} \cdot \dfrac{4}{50} = \dfrac{64}{132{,}600} = \dfrac{8}{16{,}575}$

(c) $P(\text{club and spade and heart}) = \dfrac{13}{52} \cdot \dfrac{13}{51} \cdot \dfrac{13}{50} = \dfrac{2197}{132{,}600} = \dfrac{169}{10{,}200}$

(d) $P(\text{three clubs}) = \dfrac{13}{52} \cdot \dfrac{12}{51} \cdot \dfrac{11}{50} = \dfrac{1716}{132{,}600} = \dfrac{11}{850}$

Conditional Probability

If the probability of an event B occurring is affected by an event A occurring, we say that a *condition* has been imposed on event B, and the probability of B then is written $P(B \mid A)$ as shown previously. This is read the **conditional probability** of event B occurring given that event A has occurred.

The conditional probability of an event can be found by dividing both sides of the equation for multiplication rule 2 by $P(A)$, as shown:

$$P(A \text{ and } B) = P(A) \cdot P(B \mid A)$$

$$\frac{P(A \text{ and } B)}{P(A)} = \frac{P(A) \cdot P(B \mid A)}{P(A)}$$

$$= P(B \mid A)$$

Formula for Conditional Probability

The probability that the second event B occurs given that the first event A has occurred can be found by dividing the probability that both events occurred by the probability that the first event has occurred. The formula is

$$P(B \mid A) = \frac{P(A \text{ and } B)}{P(A)}$$

Example 11-27 illustrates the use of this rule.

Military strategies (and other types of strategies) use conditional probability. If A occurs, how likely is B? How likely is B if A doesn't occur?

Example 11-27

A die is rolled. Find the probability of getting a 3 if it is known that an odd number occurred.

Solution

Let event A be an odd number (i.e., 1, 3, or 5); then $P(A) = \frac{3}{6} = \frac{1}{2}$. Let event B be a 3; then $P(A \text{ and } B) = \frac{1}{6}$ since events A and B have one element in common. Then

$$P(B \mid A) = \frac{P(A \text{ and } B)}{P(A)} = \frac{\frac{1}{6}}{\frac{3}{6}} = \frac{1}{6} \div \frac{3}{6} = \frac{1}{6} \cdot \frac{6}{3} = \frac{1}{3}$$

Example 11-28

Two dice are rolled. If it is known that the sum of the numbers that occurred is even, find the probability that one number was a 4.

Solution

Referring to the sample space in Section 11-2, there are 18 ways to get a sum that is even; hence, $P(\text{even sum}) = \frac{18}{36} = \frac{1}{2}$. In addition, there are 5 ways to have one die turn up a 4 and the sum is even [i.e., (2, 4), (4, 4), (6, 4), (4, 2), and (4, 6)]. Hence, $P(\text{even sum and a 4 occurs on one die}) = \frac{5}{36}$. Then

$$P(\text{one number is a 4} \mid \text{sum is even}) = \frac{\frac{5}{36}}{\frac{18}{36}} = \frac{5}{36} \div \frac{18}{36} = \frac{5}{36} \cdot \frac{36}{18} = \frac{5}{18}.$$

Notice that when the condition A having occurred is imposed on event B, the sample space is reduced to the outcomes in A; therefore, another method for computing $P(B \mid A)$ is to count the number of outcomes that the A and B have in common and divide that number by the number of outcomes in A.

In Example 11-28, there were five outcomes in the event of getting a 4 and getting an even sum, and there were 18 ways to get a sum that is even. Hence, $P(\text{getting a 4 on one die} \mid \text{an even sum}) = \frac{5}{18}$.

Try This One

11-P   A card is drawn from a deck. Find the probability that it is a queen given that the card is a red card.

For the answer, see page 594.

Probability and Sets

Conditional probability can be explained using set notation. Consider the information given in Example 11-28. Let A = the set of sums that are even and B = set of outcomes that contain a 4. Then $A \cap B$ = set of outcomes that are even and contain a 4. The Venn diagram for this situation is shown in Figure 11-9.

$$U$$

A

(1, 1)	(5, 1)
(1, 3)	(5, 3)
(1, 5)	(5, 5)
(2, 2)	(6, 2)
(2, 6)	
(3, 1)	(6, 6)
(3, 3)	
(3, 5)	

(4, 2)
(4, 4)
(4, 6)
(2, 4)
(6, 4)

B

(4, 1)
(4, 3)
(4, 5)
(1, 4)
(3, 4)
(5, 4)

(1, 2)	(1, 6)	(3, 2)		(3, 6)
(2, 1)	(2, 3)	(2, 5)	(5, 2)	(5, 6)
(6, 1)	(6, 3)	(6, 5)		

$$P(B \mid A) = \frac{P(A \cap B)}{P(A)}$$

Figure 11-9

Notice that $P(B \mid A) = \dfrac{P(A \cap B)}{P(A)}$, which can be written in probability notation as

$P(B \mid A) = \dfrac{P(A \text{ and } B)}{P(A)}$. Figure 11-9 shows $\dfrac{P(A \cap B)}{P(A)} = \dfrac{\frac{5}{36}}{\frac{18}{36}} = \dfrac{5}{18}$. In other words,

$\dfrac{\text{number of pairs in } A \text{ and } B}{\text{number of pairs in } A} = P(B \mid A)$.

Exercise Set 11-5

Real World Applications

1. If 18% of all Americans are underweight, find the probability that if three Americans are selected at random, all will be underweight.

   Source: 100% American by Daniel Evan Weiss (New York: Poseidon Press, 1988)

2. A national study of patients who were overweight found that 56% also have elevated blood pressure. If two overweight patients are selected, find the probability that both have elevated blood pressure.

3. The Gallup Poll reported that 52% of Americans used a seat belt the last time they got into a car. If four people are selected at random, find the probability that they all used a seat belt the last time they got into a car.

   Source: 100% American by Daniel Evan Weiss (New York: Poseidon Press, 1988)

4. An automobile saleswoman finds that the probability of making a sale is 0.23. If she talks to four customers today, find the probability that she will sell four cars.

5. If 25% of U.S. federal prison inmates are not U.S. citizens, find the probability that two randomly selected federal prison inmates will not be U.S. citizens.

   Source: Harper's Index 290, no 1740 (May 1955), p. 11.

6. If two people are selected at random, what is the probability that they were both born in December?

7. If two people are selected at random, find the probability that they were born in the same month.

8. If three people are selected, find the probability that all three were born in March.

9. If half of Americans believe that the federal government should take "primary responsibility" for eliminating poverty, find the probability that three randomly selected Americans will agree that it is the federal government's responsibility to eliminate poverty.

10. What is the probability that a husband, wife, and daughter have the same birthday?

11. A flashlight has six batteries, two of which are defective. If two are selected at random without replacement, find the probability that both are defective.

12. In Exercise 11, find the probability that the first battery tests good and the second one is defective.

13. The U.S. Department of Justice reported that 6% of all American murders are committed without a weapon. If three murder cases are selected at random, find the probability that a weapon was not used in any one of them.

   Source: 100% American by Daniel Evan Weiss (New York: Poseidon Press, 1988)

14. In a department store there are 120 customers, 90 of whom will buy at least one item. If five customers are selected at random, one by one, find the probability that all will buy at least one item.

15. Three cards are drawn from a deck *without* replacement. Find these probabilities.

   (a) All are jacks   (b) All are clubs

   (c) All are red cards

16. In a scientific study there are eight guinea pigs, five of which are pregnant. If three are selected at random without replacement, find the probability that all are pregnant.

17. In Exercise 16, find the probability that none are pregnant.

18. In a class consisting of 15 men and 12 women, two different homework papers were selected at random. Find the probability that both papers belonged to women.

19. A coin is tossed and then a die is rolled. Find the probability of getting a 3 on the die given that the coin landed heads up.

20. A card is selected from an ordinary deck. Find the probability that the card was a king given that it was a black card.

21. Two dice are rolled. Find the probability that the sum was a 7 if it is known that one of the numbers was a 6.

22. A coin is tossed and then a die is rolled. Find the probability of getting a tails on the coin given that the number on the die was odd.

23. A card is selected from an ordinary deck. Find the probability of getting a diamond given that the card was a face card.

24. Two dice are rolled. Find the probability that the sum obtained was greater than 8 given that one number on the die was a 6.

Use this information for Exercises 25–28:
Three red cards are numbered 1, 2, and 3. Three black cards are numbered 4, 5, and 6. The cards are placed in a box and one card is selected at random.

25. Find the probability that a red card was selected given that the number on the card was an odd number.

26. Find the probability that a number less than 5 was selected given that the card was a black card.

27. Find the probability that a number less than 5 was selected given that the card was red.

28. Find the probability that a black card was selected given that the number on the card was an even number.

Use the following information for Exercises 29–32:
A survey shows the average number of minutes that people talk on their cell phones each month.

	Less than 200	**200–399**	**400–599**	**600 or more**
Men	56	18	10	16
Women	61	18	13	8

If a person is selected at random, find these probabilities:

29. The person talked less than 200 minutes if it was known that the person was a woman.

30. The person talked 600 or more minutes if it was known that the person was a man.

31. The person was a woman if it is known that the person talked between 200 and 399 minutes.

32. The person was a man if it is known that the person talked between 200 and 599 minutes.

Writing Exercises

33. What is the difference between independent and dependent events? Give an example of each.

34. State which events are independent and which are dependent.
    (a) Tossing a coin and drawing a card from a deck
    (b) Drawing a ball from an urn, not replacing it, and then drawing a second ball
    (c) Getting a raise in salary and purchasing a new car
    (d) Driving on ice and having an accident
    (e) Having a large shoe size and having a high I.Q.
    (f) A father being left-handed and a daughter being left-handed
    (g) Smoking excessively and having lung cancer
    (h) Eating an excessive amount of ice cream and smoking an excessive amount of cigarettes

Critical Thinking

35. When three dice are rolled, find the probability of getting a sum of 9.

36. When five coins are tossed, find the probability of getting at least one head.

37. Find the probability that three people selected at random will have the same birthday. (Ignore the year.)

11-6

The Fundamental Counting Rule and Permutations

Many problems in probability and statistics require knowing the total number of ways a sequence of events can occur. For example, if a person designs a new license plate for a state, that person must make sure that the design consisting of letters and numbers provides enough different license plates to ensure that every registered automobile has a different plate number.

In order to determine the total number of outcomes for a sequence of events, three rules can be used. They are the **fundamental counting rule,** *the permutation rule, and the combination rule*. The first two rules are explained in this section and the combination rule is explained in Section 11-7.

> ## Math Note
>
> The occurrence of the first event in no way affects the occurrence of the second event, which in turn, does not affect the occurrence of the third event, etc.

Fundamental Counting Rule

In a sequence of n events in which the first event can occur in k_1 ways and the second event can occur in k_2 ways and the third event can occur in k_3 ways and so on, the total number of ways the sequence can occur is

$$k_1 \cdot k_2 \cdot k_3 \cdot \cdots \cdot k_n$$

Example   11-29

There are four blood types, A, B, AB, and O. Blood can also be Rh$^+$ and Rh$^-$. Finally, a blood donor can be male or female. How many different ways can a donor have his or her blood labeled?

Solution

Since there are four possibilities for blood type, two possibilities for the Rh factor, and two possibilities for the gender of the donor, there are

$$4 \cdot 2 \cdot 2 = 16$$

different classification categories.

Try This One

11-Q   A paint manufacturer wishes to manufacture several different paints. The categories include

Color   Red, blue, white, black, green, brown, yellow

Type   Latex, oil

Texture   Flat, semigloss, high gloss

Use   Outdoor, indoor

How many different kinds of paint can be made if a person can select one color, one type, one texture, and one use?
For the answer, see page 594.

When determining the number of different ways a sequence of events can occur, you must know whether or not repetitions are permitted. The next two examples show the difference between the two situations.

Example 11-30

The letters A, B, C, D, and E are to be used in a four-letter ID card. How many different cards are possible if repetitions are permitted?

Solution

Since there are four spaces to fill and five choices for each space, the solution is

$$5 \cdot 5 \cdot 5 \cdot 5 = 5^4 = 625$$

Example 11-31

Using the letters A, B, C, D, and E, how many different ID cards are possible if repetitions are not permitted?

Solution

If repetitions are not permitted, the first letter can be chosen in five ways. But the second letter can be chosen in only four ways, since there are only four letters left, etc. Thus, the solution is

$$5 \cdot 4 \cdot 3 \cdot 2 = 120$$

How many three-digit codes are possible if repetition is not permitted?

A similar situation occurs when one is drawing balls from an urn or cards from a deck. If the ball or card is replaced before the next one is selected, then repetitions are permitted, since the same one can be selected again. However, if the selected ball or card is not replaced, then repetitions are not permitted since the same ball or card cannot be selected the second time.

Try This One

11-R   An urn contains four balls whose colors are red, blue, black, and white. A ball is selected, its color is noted, and it is replaced. Then a second ball is selected, and its color is noted.

(a) How many different color schemes are possible?

(b) If the first ball is not replaced, how many different outcomes are there?

For the answer, see page 594.

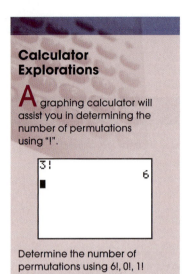
Factorial Notation

Permutations and combinations (explained later) use the **factorial notation.** The symbol for a factorial is the exclamation mark (!). In general, $n!$ means to multiply the whole numbers from n down to 1. For example,

$$1! = 1 = 1$$
$$2! = 2 \cdot 1 = 2$$
$$3! = 3 \cdot 2 \cdot 1 = 6$$
$$4! = 4 \cdot 3 \cdot 2 \cdot 1 = 24$$
$$5! = 5 \cdot 4 \cdot 3 \cdot 2 \cdot 1 = 120$$

The formal definition of factorial notation is given next.

For any natural number n

$$n! = n(n-1)(n-2)(n-3)\ldots 3 \cdot 2 \cdot 1$$

$n!$ is read as "n factorial."
0! is defined as 1. (Note that this will be explained later.)

Example 11-32

Find 8!

Solution

$$8! = 8 \cdot 7 \cdot 6 \cdot 5 \cdot 4 \cdot 3 \cdot 2 \cdot 1 = 40,320$$

The formulas used later in this chapter require you to divide factorials. First notice that $\frac{n!}{n!} = 1$. For example, $\frac{5!}{5!} = \frac{5 \cdot 4 \cdot 3 \cdot 2 \cdot 1}{5 \cdot 4 \cdot 3 \cdot 2 \cdot 1} = \frac{120}{120} = 1$. Next notice that one need not write a factorial as a product ending in 1. For example, $8! = 8 \cdot 7! = 8 \cdot 7 \cdot 6! = 8 \cdot 7 \cdot 6 \cdot 5!$, etc. Now the quotient of $\frac{8!}{5!}$ is $\frac{8!}{5!} = \frac{8 \cdot 7 \cdot 6 \cdot 5 \cdot 4 \cdot 3 \cdot 2 \cdot 1}{5 \cdot 4 \cdot 3 \cdot 2 \cdot 1} = \frac{40,320}{120} = 336$. This can be shortened to $\frac{8!}{5!} = \frac{8 \cdot 7 \cdot 6 \cdot \cancel{5!}}{\cancel{5!}} = 8 \cdot 7 \cdot 6 = 336$ since $\frac{5!}{5!} = 1$.

Example 11-33

Evaluate each.

(a) $\dfrac{9!}{4!}$

(b) $\dfrac{12!}{10!}$

—*Continued*

Example **11-33** *Continued—*

Solution

(a) $\dfrac{9!}{4!} = \dfrac{9 \cdot 8 \cdot 7 \cdot 6 \cdot 5 \cdot \cancel{4!}}{\cancel{4!}} = 9 \cdot 8 \cdot 7 \cdot 6 \cdot 5 = 15{,}120$

(b) $\dfrac{12!}{10!} = \dfrac{12 \cdot 11 \cdot \cancel{10!}}{\cancel{10!}} = 12 \cdot 11 = 132$

Try This One

11-S Evaluate each.

(a) $\dfrac{11!}{6!}$

(b) $\dfrac{7!}{3!}$

For the answer, see page 594.

Permutations

The second rule that can be used to determine the total number of outcomes of a sequence of events is the *permutation rule*.

> An arrangement of n distinct objects in a specific order is called a **permutation** of the objects.

For example, if a photographer wanted to arrange three people (Sue, Mary, Bill) in a row for a photograph, he could do this in six different ways.

Sue	Mary	Bill	Mary	Sue	Bill
Mary	Bill	Sue	Sue	Bill	Mary
Bill	Mary	Sue	Bill	Sue	Mary

Since there were to be three people in the photograph, using the fundamental counting rule, there are $3! = 3 \cdot 2 \cdot 1 = 6$ arrangements for the photograph.

The number of permutations of n distinct objects using all of the objects is $n!$ Example 11-34 illustrates this principle.

Example **11-34**

Suppose a business owner has a choice of five locations in which to establish her business. She decides to rank each location according to certain criteria, such as the price of the store and parking facilities. How many different ways can she rank the five locations?

—Continued

Example 11-34 *Continued—*

Solution

Since there are five locations, she has five choices for the first location, four choices for the second location, three choices for the third location, etc.

$$5! = 5 \cdot 4 \cdot 3 \cdot 2 \cdot 1 = 120$$

She has 120 different possible rankings.

Try This One

11-T How many different ways can nine different books be displayed in a row on a shelf in a bookstore?

For the answer, see page 594.

In the previous examples, all of the objects were used in finding the number of permutations for *n* objects. When only some of the objects are selected, the **permutation rule** given next applies.

Calculator Explorations

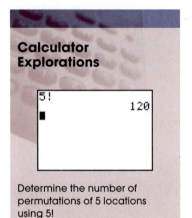

Determine the number of permutations of 5 locations using 5!

Permutation Rule

The arrangement of *n* objects in a specific order using *r* objects at a time is called a *permutation of* n *objects taking* r *objects at a time.* It is written as $_nP_r$ and the formula is

$$_nP_r = \frac{n!}{(n-r)!}$$

Math Note

The problem can also be solved using the multiplication rule. There are seven choices for the first letter, six choices for the second letter, and five choices for the third. Hence, the total number of arrangements is $7 \cdot 6 \cdot 5 = 210$.

Example 11-35

How many different arrangements of three letters can be made using the letters a, c, e, g, i, k, and m? (Repetitions are not permitted.)

Solution

Since we are selecting three letters from seven letters in a specific order, $n = 7$ and $r = 3$. Hence there are

$$_7P_3 = \frac{7!}{(7-3)!} = \frac{7!}{4!} = \frac{7 \cdot 6 \cdot 5 \cdot \cancel{4!}}{\cancel{4!}} = 7 \cdot 6 \cdot 5 = 210$$

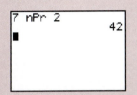
Example 11-36

How many ways can a manager and assistant manager be selected from a department consisting of 10 employees?

Solution

In this case, $n = 10$ and $r = 2$. Hence there are $_{10}P_2$ ways to select two people to fill the two positions.

$$_{10}P_2 = \frac{10!}{(10-2)!} = \frac{10!}{8!} = \frac{10 \cdot 9 \cdot 8!}{8!} = 10 \cdot 9 = 90$$

Try This One

11-U How many different ways can four books be arranged on a shelf if they can be selected from nine different books?

For the answer, see page 594.

A brief illustration is necessary for why the formula $_nP_r$ works. Given the letters a, b, c, d, and e, how many permutations of two letters can be made? Since there are two positions to fill, the multiplication rule states that the first one can be filled in five ways, and the second position can be filled in four ways, as shown

$$5 \cdot 4 = 20$$

In the formula $_5P_2$, the numerator is 5! and the denominator is $(5 - 2)!$ or 3!. That is,

$$_5P_2 = \frac{5!}{3!} = \frac{5 \cdot 4 \cdot 3 \cdot 2 \cdot 1}{3 \cdot 2 \cdot 1} = \frac{5 \cdot 4 \cdot \cancel{3} \cdot \cancel{2} \cdot \cancel{1}}{\cancel{3} \cdot \cancel{2} \cdot \cancel{1}}$$

Since the last three products in the numerator and denominator cancel, what is left is

$$5 \cdot 4 = 20$$

which is the same result as that given by the fundamental counting rule.

Recall that 0! is defined to be 1. The reason for this can be explained by looking at the photograph example. In this example, you were asked to find the number of different ways that three people can be arranged in a row for a photograph. The answer was found to be $3! = 3 \cdot 2 \cdot 1 = 6$. This problem can also be solved using the permutation formula by asking you to find how many permutations of three people can be made by selecting three people. Here $n = 3$ and $r = 3$. Using the permutation formula:

$$_3P_3 = \frac{3!}{(3-3)!} = \frac{3!}{0!} = \frac{3 \cdot 2 \cdot 1}{1} = 6$$

The denominator becomes 0! and has to be equal to 1 so that the answer obtained from the permutation formula is the same as the answer obtained from the fundamental counting rule.

In summary then, the formula for permutations is used when the order or arrangement of the objects is important and repetition of the objects is not allowed. That is, once an object is selected, it cannot be used again.

Sidelight

Win a Million or Be Struck by Lightning?

Do you think you would be more likely to win a large lottery and become a millionaire or more likely to be struck by lightning? The answer is that you would be more likely to be struck by lightning.

An article in the Associated Press noted that researchers have found that the chance of winning one million or more dollars is about 1 in 1.9 million. The chances of winning a million dollars in a recent Pennsylvania lottery were 1 in 9.6 million. The chances of winning a $10 million prize in Publisher's Clearinghouse Sweepstakes were 1 in 200 million. In contrast, the chances of being struck by lightning are about 1 in 600,000. Hence a person is at least three times more likely to be struck by lightning than to win one million dollars.

One way to guarantee winning a lottery is to buy all possible combinations of the winning numbers. In 1992, an Australian investment group purchased five million of the possible seven million possible combinations of the lottery numbers in a Virginia State Lottery. Because of the time, they could not purchase the other two million tickets. However, they were able to purchase the winning number and won $27 million. Their profit was about $22 million. Not bad!

States have written laws to prevent this from happening today, and they are devising lottery games with many more possibilities so that it would be impossible to purchase all the possible tickets to win.

Exercise Set 11-6

Computational Exercises

Evaluate each.

1. $10!$
2. $5!$
3. $9!$
4. $1!$
5. $0!$
6. $_8P_2$
7. $_7P_5$
8. $_{12}P_{12}$
9. $_5P_3$
10. $_6P_6$
11. $_6P_0$
12. $_8P_0$
13. $_8P_8$
14. $_{11}P_3$
15. $_6P_2$

Real World Applications

16. How many different four-letter permutations can be formed from the letters in the word *decagon*?

17. In a board of directors composed of eight people, how many ways can a chief executive officer, a director, and a treasurer be selected?

18. How many different ID cards can be made if there are six digits on a card and no digit can be used more than once?

19. How many different ways can seven different types of soaps be displayed on a shelf in a grocery store?

20. How many different ways can four tickets be selected from 50 tickets if each ticket wins a different prize?

21. How many different ways can a researcher select five rats from 20 rats and assign each to a different test?

22. How many different signals can be made by using at least three distinct flags if there are five different flags from which to select?

23. An investigative agency has seven cases and five agents. How many different ways can the cases be assigned if only one case is assigned to each agent?

24. An inspector must select three tests to perform in a certain order on a manufactured part. He has a choice of seven tests. How many ways can he perform three different tests?

25. A mother has five different chores and wishes to assign one to each of her five children. In how many different ways could she make the assignments?

26. How many different ways can a visiting nurse see six patients if she sees them all in one day?

27. A store owner has 50 items to advertise, and she can select one different item each week for the next 6 weeks to put on special. How many different ways can the selection be made?

28. In a club consisting of 17 people, how many different ways can a president, vice president, secretary, and treasurer be selected?

29. How many different ways can you visit four different stores in a shopping mall?

30. How many ways can a research company select three geographic areas from a list of six geographic areas to test market its product? One area will be selected and tested in September, a different area in October, and a third area will be used in November.

Writing Exercises

31. Explain the fundamental counting rule.

32. What is the meaning of a permutation of n distinct objects?

Critical Thinking

33. The number of permutations of n objects in which k_1 are alike, k_2 alike, etc. is

$$\frac{n!}{k_1!k_2 \ldots k_p!}$$

where $k_1 + k_2 + \cdots + k_p = n$

Use this formula for parts a and b.

(a) How many different permutations can be made from the letters in the word *Mississippi*?

(b) How many different code words can be made using the letters A, A, A, B, B, C, D, D, D, D if the word must contain 10 letters?

11-7 Combinations

Suppose a dress designer wishes to select two different colors of material to design a new dress, and he has on hand four different colors. How many different possibilities can there be in this situation?

This type of problem differs from the previous ones in that the order of selection is not important. That is, if the designer selects yellow and red, this selection is the same as the selection red and yellow. This type of arrangement is called a combination. The difference between a permutation and a combination is that in a combination, the order or arrangement of the objects is not important, but order *is* important in a permutation. Example 11-37 illustrates this difference.

> A selection of objects without regard to order is called a **combination.**

Example 11-37

Given the letters A, B, C, and D, list the permutations and combinations of selecting two letters.

Solution

The listings follow.

Permutations				Combinations	
AB	BA	CA	DA	AB	BC
AC	BC	CB	DB	AC	BD
AD	BD	CD	DC	AD	CD

Note that in permutations AB is different from BA. But in combinations, AB is the same as BA, and so only one of the two is listed.

Combinations are used when selecting a subset of objects from a given set and the order or arrangement is not important in the selection process. Suppose a committee of five students is to be selected from 25 students. The students represent a combination, since it does not matter who is selected first, second, etc.

Combination Rule

The number of combinations of *r* objects selected from *n* objects is denoted by $_nC_r$ and is given by the formula

$$_nC_r = \frac{n!}{(n-r)!r!}$$

Sidelight

Probability and Your Fears

All of us at one time or another have thought about dying. Some people have fears of dying in a plane crash or dying from a heart attack. In Mathematics in Our World, it was explained that it is safer to fly across the United States than to drive. Statisticians who work for insurance companies (called actuaries) also calculate probabilities for dying from other causes. For example, based on deaths in the United States, the risks of dying from various other causes are shown.

Motor vehicle accident	1 in 7000
Shot by a gun	1 in 10,000
Walking across the street	1 in 60,000
Lightning strike	1 in 3 million
Shark attack	1 in 100 million

The death risk for various diseases is much higher as shown.

Heart attack	1 in 400
Cancer	1 in 600
Stroke	1 in 2000

As you can see, the chances of dying from diseases are much higher than dying from accidents.

Example 11-38

How many combinations of four objects are there taken two at a time?

Solution

Since this is a combination problem, the answer is

$$_4C_2 = \frac{4!}{(4-2)!2!} = \frac{4!}{2!2!} = \frac{4 \cdot 3 \cdot 2!}{2 \cdot 1 \cdot 2!} = 6$$

This is the same result shown in Example 11-37.

Notice that the formula for $_nC_r$ is

$$\frac{n!}{(n-r)!r!}$$

which is the formula for permutations,

$$\frac{n!}{(n-r)!}$$

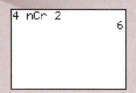
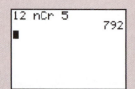
with an $r!$ in the denominator. This $r!$ divides out the duplicates from the number of permutations, as shown in Example 11-38. For each two letters there are two permutations, but only one combination. Hence, dividing the number of permutations by $r!$ eliminates the duplicates. This result can be verified for other values of n and r.

Example 11-39

Harry has to visit 10 cities. He can visit any three in one day. How many different ways can he select three cities? Assume distance is not a factor.

Solution

$$_{10}C_3 = \frac{10!}{(10-3)!3!} = \frac{10!}{7!3!} = 120$$

Example 11-40

In a club there are seven women and five men. A committee of three women and two men are to be chosen. How many different possibilities are there?

Solution

Here, one must select three women from seven women, which can be done in $_7C_3$, or 35 ways. Next, two men must be selected from five men, which can be done in $_5C_2$, or 10 ways. Finally, by the counting rule, the total number of different ways is $35 \cdot 10 = 350$.

Example 11-41

A committee of five people must be selected from five men and eight women. How many ways can selection be done if there are to be at least three women on the committee?

Solution

A committee of at least three women means that the committee can consist of three women and two men, or four women and one man, or five women. To find the different possibilities, find each separately, using combinations and the fundamental counting principal, and then add them.

$$_8C_3 \cdot _5C_2 + _8C_4 \cdot _5C_1 + _8C_5 = \frac{8!}{5!3!} \cdot \frac{5!}{3!2!} + \frac{8!}{4!4!} \cdot \frac{5!}{4!1!} + \frac{8!}{3!5!}$$

$$= 56 \cdot 10 + 70 \cdot 5 + 56 = 966$$

Table 11-1 summarizes the counting rules.

Table **11-1**

Summary of Counting Rules

Rule	Definition	Formula
Fundamental counting rule	The number of ways a sequence of n events can occur if the first event can occur in k_1 ways, the second event can occur in k_2 ways, etc. (Counts are independent.)	$k_1 \cdot k_2 \cdot k_3 \cdot \cdots \cdot k_n$
Permutation rule	The number or permutations of n objects taking r objects at a time. (Order is important.)	$\dfrac{n!}{(n-r)!}$
Combination rule	The number of combinations of r objects taken from n objects. (Order is not important.)	$\dfrac{n!}{(n-r)!r!}$

Try This One

11-V On an exam, a student must select two essay questions from six essay questions and 10 multiple choice questions from 20 multiple choice questions to answer. How many different ways can the student select questions to answer?
For the answer, see page 594.

Exercise Set **11-7**

Computational Exercises

For Exercises 1–10, evaluate each expression.

1. $_5C_2$ 2. $_8C_3$ 3. $_7C_4$

4. $_6C_2$ 5. $_6C_4$ 6. $_3C_0$

7. $_3C_3$ 8. $_9C_7$ 9. $_{12}C_2$

10. $_4C_3$

Real World Applications

11. How many different ways can five cards be selected from a standard deck of 52 cards?

12. How many ways are there to select three coins from a box containing a penny, a nickel, a dime, a quarter, a half-dollar, and a silver dollar?

13. How many ways can a student select five questions from an exam containing nine questions? How many ways are there if he must answer the first question and the last question?

14. How many ways can a committee of four people be selected from a group of 10 people?

15. If a person can select three presents from 10 presents under a Christmas tree, how many different combinations are there?

16. How many different possible tests can be made from a test bank of 20 questions if the test consists of five questions?

17. The general manager of a fast-food restaurant chain must select six restaurants from 11 for a promotional program. How many different possible ways can this selection be done?

18. How many ways can three cars and four trucks be selected from eight cars and 11 trucks to be tested for a safety inspection?

19. In a train yard there are four tank cars, 12 boxcars, and seven flatcars. How many ways can a train be made up consisting of two tank cars, five boxcars, and three flatcars?

20. There are seven women and five men in a department. How many ways can a committee of four people be selected if there must be two men and two women on the committee? How many ways can this committee be selected if there must be at least two women on the committee?

21. Wake Up cereal comes in two types: crispy and crunchy. If a researcher has 10 boxes of each, how many ways can she select three boxes of each for a quality control test?

22. How many ways can a student select three statistics books and two algebra books from the library if there are eight statistics books and five algebra books on the shelf?

23. How many ways can a person select two tapes, three records, and three compact disks from eight tapes, six records, and 10 compact disks?

24. How many ways can a foursome of two men and two women be selected from 10 men and 12 women in a golf club?

25. The state narcotics bureau must form a five-member investigative team. If it has 25 agents to choose from, how many different possible teams can be formed?

26. How many different ways can a computer programmer select three jobs from a possible 15?

27. The Environmental Protection Agency must investigate nine mills for complaints of air pollution. How many different ways can a representative select five of these to investigate this week?

28. How many ways can a person select eight videotapes from 10 tapes?

29. A buyer decides to stock 20 different coffee mugs. How many different ways can she select these 20 if there are 30 to choose from?

30. An advertising manager decides to have an ad campaign in which eight special items will be hidden at various locations in a shopping mall. If he has 17 locations to pick from, how many different possible combinations can he choose?

Writing Exercises

31. Define a combination.

32. Explain the difference between a permutation and combination.

Critical Thinking

33. A mathematician named Pascal wrote a treatise showing, among other things, how combinations can be derived from a triangular array of numbers. The triangle became known as Pascal's Triangle. Part of the triangle is shown next.

$n \backslash r$	0	1	2	3	4
0	1				
1	1	1			
2	1	2	1		
3	1	3	3	1	
4	1	4	6	4	1	$_4C_2$

The row value corresponds to the values of n and the column values correspond to the values for r. For example, the value of $_4C_2$ can be found in the intersection of row 4 and column 2. It is 6. A specific value in the table, other than the first or last values in each row, can also be found by adding the value immediately above it and the one next (going left) to the value immediately above it. For example, the value 6, found in the intersection of row 4 and column 2 can be found by adding $3 + 3$.

Complete Pascal's triangle for rows five and six, and verify the answers by using combinations.

11-8 Probability Using Permutations and Combinations

The permutation and combination rules can be combined with the probability rules to solve a variety of problems. First use the permutation rule or the combination rule to find the number of ways that the given event can occur, and then use the permutation rule or the combination rule to find the total number of outcomes in the sample space.

Finally, divide the first number by the second number. Example 11-42 uses combinations.

Example 11-42

A student has the option of selecting three books to read for a humanities course. The suggested book list consists of 10 biographies and five current events books. The student decides to select the three books at random. Find the probability that all three books selected will be current events books.

—Continued

Example 11-42 *Continued—*

Solution

Since there are five current events books and the student will need to select three of them, then there are $_5C_3$ or 10 ways of doing this.

$$_5C_3 = \frac{5!}{(5-3)!3!} = \frac{5 \cdot 4 \cdot \cancel{3!}}{2 \cdot 1 \cdot \cancel{3!}} = 10$$

The total number of outcomes in the sample space is $_{15}C_3$ or 455 since the student has to select three books from 15 books.

$$_{15}C_3 = \frac{15!}{(15-3)!3!} = \frac{15!}{12!3!} = \frac{15 \cdot 14 \cdot 13 \cdot \cancel{12!}}{\cancel{12!} \cdot 3 \cdot 2 \cdot 1} = 455$$

Hence, the probability of selecting three current events books is

$$\frac{10}{455} = \frac{2}{91} \approx 0.022$$

Example 11-43 uses the permutation rule.

Example 11-43

A combination lock shows 26 letters of the alphabet. Find the probability that if the combination to unlock the lock consists of three letters, it will contain the letters A, B, and C. (*Note:* A combination lock is really a permutation lock since the order of the letters is important when unlocking the lock.) Repetitions are not permitted.

Solution

The number of ways A, B, and C can be used is $_3P_3$.

$$_3P_3 = \frac{3!}{(3-3)!} = \frac{3!}{0!} = \frac{3 \cdot 2 \cdot 1}{1} = 6$$

The number of ways to select three letters from 26 letters in order is $_{26}P_3$.

$$_{26}P_3 = \frac{26!}{(26-3)!} = \frac{26!}{23!} = \frac{26 \cdot 25 \cdot 24 \cdot \cancel{23!}}{\cancel{23!}} = 15{,}600$$

Hence, the probability of selecting A, B, and C in any order is

$$\frac{6}{15{,}600} = \frac{1}{2600} \approx 0.00038$$

Permutations determine the number of combinations that can open a combination lock.

Example 11-44 uses combinations and the formula for classical probability.

Example   11-44

A store has six different fitness magazines and three different news magazines. If a customer buys three magazines at random, find the probability that the customer will select two fitness magazines and one news magazine.

Solution

There are $_6C_2$ or 15 ways to select two fitness magazines from six fitness magazines as shown.

$$_6C_2 = \frac{6!}{(6-2)!2!} = \frac{6!}{4!2!} = \frac{6 \cdot 5 \cdot 4!}{4! \cdot 2 \cdot 1} = 15$$

There are $_3C_1$ or three ways to select one magazine from three news magazines as shown.

$$_3C_1 = \frac{3!}{(3-1)!1!} = \frac{3!}{2! \cdot 1!} = \frac{3 \cdot 2!}{2! \cdot 1} = 3$$

Hence, there are $15 \cdot 3$ or 45 ways to select two fitness magazines *and* one news magazine. Note that "and" in these topics generally means to multiply.

Thus there are $_9C_3$ or 84 ways to select three magazines from nine magazines as shown.

$$_9C_3 = \frac{9!}{(9-3)!3!} = \frac{9!}{6!3!} = \frac{9 \cdot 8 \cdot 7 \cdot 6!}{6! \cdot 3 \cdot 2 \cdot 1} = 84$$

Now the probability of selecting two fitness magazines and one news magazine is

$$\frac{45}{84} \approx 0.536$$

Try This One

11-W A box contains 24 transistors and four of them are defective. If three transistors are selected at random, find the possibility that

(a) Exactly two are defective

(b) None are defective

(c) All three are defective

For the answer, see page 594.

As stated at the beginning of this section, a large variety of probability problems can be solved using permutations and combinations in conjunction with the probability rules.

Sidelight

The Classical Birthday Problem

What do you think the chances are that in a classroom with 23 students, two students have the same birthday (day and month)? Most people think that the probability would be very low since there are 365 days in a year. You may be surprised to find out that it is greater than 0.5 or 50%! Furthermore, as the number of people increases, the probability becomes even greater than 0.5 very rapidly. In a room of 30 students, there is a greater than 70% chance that two students have the same birthday. If you have 50 students in the room, the probability jumps to 97%!

The problem can be solved by using probability and permutation rules. It must be assumed that all birthdays are equally likely, but this assumption will have little effect on the answers. The way to solve the problem is to find the probability that nobody has the same birthday, and then subtract this probability from one. In other words, P(two students have the same birthday) = $1 - P$(all students have different birthdays).

For example, suppose that there were only three students in a room. Then the probability that each would have a different birthday is

$$\left(\frac{365}{365}\right) \cdot \left(\frac{364}{365}\right) \cdot \left(\frac{363}{365}\right) = \frac{_{365}P_3}{365^3} = 0.992$$

Hence, the probability that at least two of the three students have the same birthday is

$$1 - 0.992 = 0.008$$

In general, in a room with k people, the probability that at least two people have the same birthday is

$$1 - \frac{_{365}P_k}{365^k}$$

In a room with 23 students, then, the probability that at least two students will have the same birthday is

$$1 - \frac{_{365}P_k}{365^k} = 0.507 \quad \text{or} \quad 50.7\%$$

It is interesting to note that two presidents, James K. Polk and Warren G. Harding, were both born on November 2. Also, John Adams and Thomas Jefferson both died on July 4. The unusual thing about this is that they died on the same day of the same year, July 4, 1826.

Exercise Set   11-8

Real World Applications

1. A parent-teacher committee consisting of four people is to be formed from 20 parents and five teachers. Find the probability that the committee will consist of the following. (Assume that the selection will be random.)

   (a) All teachers

   (b) Two teachers and two parents

   (c) All parents

   (d) One teacher and three parents

2. In a company there are seven executives: four women and three men. Three are selected to attend a management seminar. Find these probabilities.

   (a) All three selected will be women.

   (b) All three selected will be men.

   (c) Two men and one woman will be selected.

   (d) One man and two women will be selected.

3. A city council consists of 10 members. Four are Republicans, three are Democrats, and three are Independents. If a committee of three is to be selected, find the probability of selecting

   (a) All Republicans

   (b) All Democrats

   (c) One of each party

   (d) Two Democrats and one Independent

   (e) One Independent and two Republicans

4. In a class of 18 students, there are 11 men and seven women. Four students are selected to present a demonstration on the use of the calculator. Find the probability that the group consists of

   (a) All men   (b) All women

   (c) Three men and one woman   (d) One man and three women

   (e) Two men and two women

5. A package contains 12 resistors, three of which are defective. If four are selected, find the probability of getting

   (a) No defective resistors   (b) One defective resistor

   (c) Three defective resistors

6. If 50 tickets are sold and two prizes are to be awarded, find the probability that one person will win two prizes if that person buys two tickets.

7. Find the probability of getting a full house (three cards of one denomination and two of another) when five cards are dealt from an ordinary deck.

8. A committee of four people is to be formed from six doctors and eight dentists. Find the probability that the committee will consist of

   (a) All dentists   (b) Two dentists and two doctors

   (c) All doctors   (d) Three doctors and one dentist

   (e) One doctor and three dentists

9. An insurance sales representative selects three policies to review. The group of policies he can select from contains eight life policies, five automobile policies, and two homeowner's policies. Find the probability of selecting

   (a) All life policies

   (b) Both home owner's policies

   (c) All automobile policies

   (d) One of each policy

   (e) Two life and one automobile policies

10. Find the probability of getting any triple-digit number, where all the digits are the same, on a lotto that consists of selecting a three-digit number.

11. Find the probability of selecting three science books and four math books from eight science books and nine math books. The books are selected at random.

12. Find the probability of randomly selecting two mathematics books and three physics books from four mathematics books and eight physics books in a box.

13. To win a state lottery, a person must select five numbers from 40 numbers. Find the probability of winning if a person buys one ticket. (*Note:* The numbers can be selected in any order.)

14. A five-digit identification card is made. Find the probability that the card will contain the digits 0, 1, 2, 3, and 4 in any order.

Writing Exercises

15. If three women and two men line up at random in a row for a picture, explain in writing how you would find the probability that the row would consist of a woman, a man, a woman, a man, and a woman.

16. Explain why the probability when selecting a three-digit number for the state lottery is not $\dfrac{1}{_{10}C_3}$.

Critical Thinking

Probabilities can be computed for poker hands. A poker hand consists of five cards dealt at random. For Exercises 17–20, find the probability of each poker hand.

17. Four aces   18. Four of a kind

19. A royal flush   20. A straight flush

Eleven

Summary

Section	Important Terms	Important Ideas
11-1	probability experiment outcome sample space event classical probability empirical probability	**Flipping** coins, drawing cards from a deck, and rolling a die are examples of probability experiments. The set of all possible outcomes of a probability experiment is called a sample space. The two types of probability are classical and empirical. Classical probability uses sample spaces. Empirical probability uses frequency distributions and is based on observation. The range of probability is from zero to one inclusive. When the probability of an event is close to zero, the event is highly unlikely to occur. When the probability of an event is near one, the event is almost certain to occur.
11-2	tree diagram table	**A** tree diagram is a device that can be used to determine the outcomes in the sample space. Sample spaces can also be represented by tables.
11-3	odds expectation (expected value)	**In** order to determine payoffs, gambling establishments give odds. There are two ways to compute odds for a game of chance. They are "odds in favor of an event" and "odds against the event." Another concept related to probability and odds is the concept of expectation or expected value. Expected value is used to determine what happens over the long run. It is found by multiplying the outcomes of a probability experiment by their corresponding probabilities and then finding the sum of the products.
11-4	mutually exclusive events addition rules	**Two** events are said to be mutually exclusive if they cannot occur at the same time. When one wants to find the probability of one event or another event occurring, one of two addition rules can be used, depending on whether or not the events are mutually exclusive.
11-5	multiplication rules independent events dependent events conditional probability	**Events** can be classified as independent and dependent. Events are said to be independent if the occurrence of the first event does not affect the probability of the occurrence of the next event. If the probability of the second event occurring is changed by the occurrence of the first event, then the events are dependent. When one wants to find the probability of one event and another event occurring, one of the two multiplication rules can be used, depending on whether the two events are independent or dependent. If the probability of an event B occurring is affected by an event A occurring, then we say that a condition has been imposed on the event and the conditional probability formula can be used to solve the problem.
11-6	fundamental counting rule factorial notation permutation permutation rule	**In** order to determine the total number of outcomes for a sequence of events, the fundamental counting rule or the permutation rule can be used. When the order or arrangement of the objects in a sequence of events is important, then the result is called a permutation of the objects.
11-7	combination combination rule	**When** the order of the objects is not important, then the result is called a combination.
11-8		**Probabilities** of events can be found by using the fundamental counting rule, the permutation rule, or the combination rule, depending on the situation.

Review Exercises

1. When a die is rolled, find the probability of getting

   (a) A 5

   (b) A 6

   (c) A number less than 5

2. When a card is drawn from a deck, find the probability of getting

   (a) A heart

   (b) A 7 and a club

   (c) A 7 or a club

   (d) A jack

   (e) A black card

3. In a survey conducted at a local restaurant during breakfast hours, 20 people preferred orange juice, 16 preferred grapefruit juice, and nine preferred apple juice with breakfast. If a person is selected at random, find the probability that he or she prefers grapefruit juice.

4. If a die is rolled one time, find these probabilities:

   (a) Getting a 5

   (b) Getting an odd number

   (c) Getting a number less than 3

5. A recent survey indicated that in a town of 1500 households, 850 have cordless telephones. If a household is randomly selected, find the probability that it has a cordless phone.

6. During a sale at a men's store, 16 white sweaters, three red sweaters, nine blue sweaters, and seven yellow sweaters were purchased. If a customer is selected at random, find the probability that he bought

   (a) A blue sweater

   (b) A yellow or white sweater

   (c) A red, blue, or a yellow sweater

   (d) A sweater that was not white

7. At a swimwear store, the managers found that 16 women bought white bathing suits, four bought red suits, three bought blue suits, and seven bought yellow suits. If a customer is selected at random and she buys one suit, find the probability that she bought

   (a) A blue suit

   (b) A yellow or red suit

   (c) A white or a yellow or a blue suit

   (d) A suit that was not red

8. When two dice are rolled, find the probability of getting

   (a) A sum of 5 or 6

   (b) A sum greater than 9

   (c) A sum less than 4 or greater than 9

   (d) A sum that is divisible by 4

   (e) A sum of 14

   (f) A sum less than 13

9. Two dice are rolled. Find the probability of getting a sum of 8 if the number on one die is a 5.

10. In a family of three children, find the probability that all the children will be girls if it is known that at least one of the children is a girl.

11. A Gallup Poll found that 78% of Americans worry about the quality and healthfulness of their diet. If five people are selected at random, find the probability that all five worry about the quality and healthfulness of their diet.

    Source: The Book of Odds, Michael D. Shook and Robert C. Shook (New York: Penguin Putnam, Inc., 1991), p. 33.

12. Twenty-five percent of the engineering graduates of a university received a starting salary of $25,000 or more. If three of the graduates are selected at random, find the probability that all have a starting salary of $25,000 or more.

13. Three cards are drawn from an ordinary deck *without* replacement. Find the probability of getting

    (a) All black cards

    (b) All spades

    (c) All queens

14. A coin is tossed and a card is drawn from a deck. Find the probability of getting

    (a) A head and a 6

    (b) A tail and a red card

    (c) A head and a club

15. A box of candy contains six chocolate-covered cherries, three peppermint patties, two caramels, and two strawberry creams. If a piece of candy is selected at random, find the probability of getting a caramel or a peppermint patty.

16. Find the odds for an event when $P(E) = \frac{1}{4}$.

17. Find the odds against an event when $P(E) = \frac{5}{6}$.

18. Find the probability of an event when the odds for the event are 6:4.

19. There are five envelopes in a box. One envelope contains a penny, one a nickel, one a dime, one a quarter, and one a half-dollar. A person selects an envelope. Find the expected value of the draw.

20. A person selects a card from a deck. If it is a red card, he wins \$1. If it is a black card between and including 2 and 10, he wins \$5. If it is a black face card, he wins \$10, and if it is a black ace, he wins \$100. Find the expectation of the game.

21. An automobile license plate consists of three letters followed by four digits. How many different plates can be made if repetitions are allowed? If repetitions are allowed in the letters but not in the digits?

22. How many different arrangements of the letters in the word *bread* are there?

23. How many different three-digit combinations can be made by using the numbers 1, 3, 5, 7, and 9 without repetitions if the "right" combination can open a safe? Does a combination lock really use combinations?

24. How many two-card pairs (i.e., the same rank) are there in a standard deck?

25. A person rolls an eight-sided die and then flips a coin. Draw a tree diagram and find the sample space.

26. A student can select one of three courses at 8:00 A.M. They are English, mathematics, and chemistry. The student can select either psychology or sociology at 11:00 A.M. Finally, the student can either select world history or economics at 1:00 P.M. Draw a tree diagram and find all the different ways the student can make a schedule.

27. How many ways can five different television programs be selected from 12 programs?

28. A quiz consists of six multiple-choice questions. Each question has three possible answer choices. How many different answer keys can be made?

29. How many different ways can a buyer select four television models from a possible choice of six models?

30. A card is selected from a deck. Find the probability that it is a diamond given that it is a red card.

31. A person has six bond accounts, three stock accounts, and two mutual fund accounts. If three investments are selected at random, find the probability that one of each type of account is selected.

32. A newspaper advertises five different movies, three plays, and two baseball games. If a couple selects three activities at random, find the probability they will attend two plays and one movie.

Chapter Test

1. When a card is drawn from an ordinary deck, find the probability of getting

   (a) A jack

   (b) A 4

   (c) A card less than 6 (an ace is considered above 6)

2. When a card is drawn from a deck, find the probability of getting

   (a) A diamond

   (b) A 5 or a heart

   (c) A 5 and a heart

   (d) A king

   (e) A red card

3. At a men's clothing store, 12 men purchased blue golf sweaters, eight purchased green sweaters, four purchased gray sweaters, and seven bought black sweaters. If a customer is selected at random, find the probability that he purchased

   (a) A blue sweater

   (b) A green or gray sweater

   (c) A green or black or blue sweater

   (d) A sweater that was not black

4. When two dice are rolled, find the probability of getting

   (a) A sum of 6 or 7

   (b) A sum greater than 3 or greater than 8

   (c) A sum less than 3 or greater than 8

   (d) A sum that is divisible by 3

   (e) A sum of 16

   (f) A sum less than 11

5. There are six cards numbered 1, 2, 3, 4, 5, and 6. A person flips a coin. If it lands heads up, he will select a

card with an odd number. If it lands tails up, he will select a card with an even number. Draw a tree diagram and find the sample space.

6. Of the physics graduates of a university, 30% received a starting salary of $30,000 or more. If five of the graduates are selected at random, find the probability that all had a starting salary of $30,000 or more.

7. Five cards are drawn from an ordinary deck *without* replacement. Find the probability of getting

   (a) All red cards

   (b) All diamonds

   (c) All aces

8. Four coins are tossed. Find the probability of getting four heads if it is known that two of the four coins landed heads up.

9. A card is drawn from a deck. Find the probability of getting a diamond if it is known that the card selected was a red card.

10. A die is rolled. Find the probability of getting a four if it is known that the result of the roll was an even number.

11. A coin is tossed and a die is rolled. Find the probability of getting a head on the coin if it is known that the number on the die is even.

12. One company's ID cards consist of five letters followed by two digits. How many cards can be made if repetitions are allowed? If repetitions are not allowed?

13. A physics test consists of 25 true–false questions. How many different possible answer keys can be made?

14. How many different ways can four radios be selected from a total of seven radios?

15. The National Bridge Association can select one of four cities for its playoff tournament next year. The cities are Pasadena, Wilmington, Chicago, and Charleston. The following year, it can hold the tournament in Hyattsville or Green Springs. How many different possibilities are there for the next 2 years? Draw a tree diagram and show all possibilities.

16. How many ways can five sopranos and four altos be selected from seven sopranos and nine altos?

17. How many different ways can eight kindergarten children be seated in a row?

18. Employees can be classified according to gender (male, female), income (low, medium, high), and rank (staff

nurse, charge nurse, head nurse). Draw a tree diagram and show all possible outcomes.

19. A soda machine servicer must restock and collect money from 15 machines, each one at a different location. How many ways can she select four machines to service in one day?

20. How many different ways can three cubes be drawn from a bag containing four differently colored cubes, if

   (a) Each individual cube is replaced after being drawn?

   (b) There is no replacement?

21. If a man can wear a shirt or a sweater and a pair of dress slacks or a pair of jeans, how many different outfits can he wear?

22. Find the odds in favor of an event when $P(E) = \frac{3}{8}$.

23. Find the odds against an event when $P(E) = \frac{4}{9}$.

24. Find the probability of an event when the odds against the event are 3:7.

25. There are six cards placed face down in a box. Each card has a number written on it. One is a 4. One is a 5. One is a 2. One is a 10. One is a 3, and one is a 7. A person selects a card. Find the expected value of the draw.

26. A person selects a card from an ordinary deck of cards. If it is a black card, she wins $2. If it is a red card between or including 3 and 7, she wins $10. If it is a red face card, she wins $25, and if it is a black jack, she wins $100. Find the expectation of the game.

27. On a lunch counter, there are five oranges, four apples, and two bananas. If a person selects three pieces of fruit at random, find the probability that the selection will include one orange, one apple, and one banana.

28. At a campus club meeting, there are six seniors, four juniors, and three sophomores. If a committee of four students is selected at random, find the probability that it will consist of two seniors, one junior, and one sophomore.

Projects

1. Make a set of three cards—one with a red star on both sides, one with a black star on both sides, and one with a black star on one side and a red star on the other side. With a partner, play the game described in Exercise 22 of Section 11-3 (pages 550–551) 100 times and record

how many times your partner wins. (*Note:* Do not change options during the 100 trials.)

(a) Do you think the game is fair (i.e., does one person win approximately 50% of the time)?

(b) If you think the game is unfair, explain what the probabilities might be and why.

2. Take a coin and tape a small weight (e.g., part of a paper clip) to one side. Flip the coin 100 times and record the results. Do you think you have changed the probabilities of the results of flipping the coin?

3. This game is called "Diet Fractions." Roll two dice and use the numbers to make a fraction less than or equal to one. Player A wins if the fraction cannot be reduced; otherwise, player B wins.

(a) Play the game 100 times and record the results.

(b) Decide if the game is fair or not. Explain why or why not.

(c) Using the sample space for two dice, compute the probabilities of player A winning and player B winning. Do these agree with the results obtained in part a?

Source: George W. Bright, John G. Harvey, and Margariete Montaque Wheeler, "Fair Games, Unfair Games." Chapter 8, *Teaching Statistics and Probability.* *NCTM 1981 Yearbook.* Reston, Virginia: The National Council of Teachers of Mathematics, Inc., 1981, p. 49. Used with permission.

4. Often when playing gambling games or collecting items in cereal boxes, one wonders how long will it be before one achieves a success. For example, suppose there are six different types of toys with one toy packaged at random in a cereal box. If a person wanted a certain toy, about how many boxes would that person have to buy before obtaining that particular toy? Of course, there is a possibility that the particular toy would be in the first box opened or that the person might never obtain the particular toy. These are the extremes.

(a) To find out, simulate the experiment using dice. Start rolling dice until a particular number, say 3, is obtained and keep track of how many rolls are necessary. Repeat 100 times. Then find the average.

(b) You may decide to use another number, such as 10 different items. In this case, use 10 playing cards (ace through 10 of diamonds), select a particular card (say an ace), shuffle the deck each time, deal the cards, and count how many cards are turned over before the ace is obtained. Repeat 100 times; then find the average.

(c) Summarize the findings for both experiments.

Mathematics in Our World
▶Revisited

Would You Bet Your Life?

In his book *Probabilities in Everyday Life*, John D. McGervey states that the chance of being killed on any given commercial airline flight is almost 1 in 1 million and that the chance of being killed during a transcontinental auto trip is about 1 in 8000. The corresponding probabilities are $\frac{1}{1,000,000} = 0.000001$ as compared to $\frac{1}{8000} = 0.000125$. Since the second number is 125 times greater than the first number, you have a much higher risk driving than flying across the United States.

Answers to **Try This One**

11-A. (a) $\frac{1}{9}$ (b) $\frac{5}{9}$ (c) $\frac{1}{3}$ (d) $\frac{2}{9}$

11-B. (a) $\frac{9}{44}$ (b) $\frac{17}{44}$ (c) $\frac{13}{22}$

11-C.

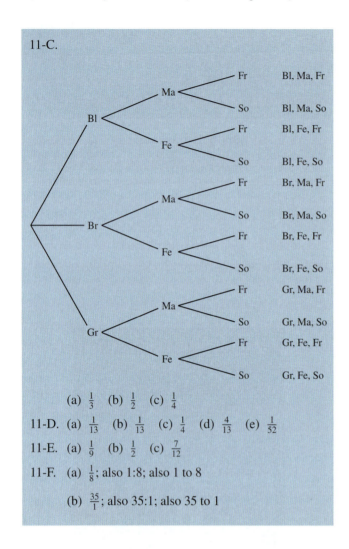

    (a) $\frac{1}{3}$ (b) $\frac{1}{2}$ (c) $\frac{1}{4}$

11-D. (a) $\frac{1}{13}$ (b) $\frac{1}{13}$ (c) $\frac{1}{4}$ (d) $\frac{4}{13}$ (e) $\frac{1}{52}$

11-E. (a) $\frac{1}{9}$ (b) $\frac{1}{2}$ (c) $\frac{7}{12}$

11-F. (a) $\frac{1}{8}$; also 1:8; also 1 to 8

      (b) $\frac{35}{1}$; also 35:1; also 35 to 1

11-G. $\frac{8}{9}$

11-H. \$58,000

11-I. $\frac{19}{39}$

11-J. $\frac{11}{26}$

11-K. $\frac{16}{25}$

11-L. $\frac{1}{169}$

11-M. (a) $\frac{1}{1296}$ (b) $\frac{1}{16}$

11-N. Approximately 0.01

11-O. (a) $\frac{1}{15}$ (b) $\frac{1}{5}$

11-P. $\frac{1}{13}$

11-Q. 84

11-R. (a) 16 (b) 12

11-S. (a) 55,440 (b) 840

11-T. 362,880

11-U. 3024

11-V. 2,771,340

11-W. (a) $\frac{15}{253} \approx 0.059$

      (b) $\frac{285}{506} \approx 0.56$

      (c) $\frac{1}{506} \approx 0.002$

Chapter
Twelve

Statistics

Outline

Objectives

After completing this chapter, you should be able to

1 Define statistics (12-1)

2 Explain the difference between a population and a sample (12-1)

3 Explain the four basic sampling methods (12-1)

4 Construct a frequency distribution for a data set (12-1)

5 Construct a stem and leaf plot for a set of data (12-1)

6 Draw a bar graph and a pie graph for the data in a categorical frequency distribution (12-2)

7 Draw a histogram and a frequency polygon for the data in a grouped frequency distribution (12-2)

8 Draw a time series graph (12-2)

9 Find the mean, median, mode, and midrange for a set of data (12-3)

10 Find the range, variance, and standard deviation for a set of data (12-4)

11 Find the percentile rank for a data value (12-5)

12 Find the data corresponding to a percentile rank (12-5)

13 State the properties of a normal distribution (12-6)

14 Find the z value for a specific data value (12-6)

15 Find the area under the standard normal distribution corresponding to various z values (12-6)

16 Answer questions about a normally distributed variable by finding areas under the normal distribution (12-7)

17 Draw and analyze a scatter plot (12-8)

18 Find the value for a correlation coefficient (12-8)

19 Determine whether or not the correlation coefficient is significant (12-8)

20 Find the equation of a regression line (12-8)

21 Given a value for the independent variable, find the corresponding value for the dependent variable using the equation of the regression line (12-8)

Introduction

Even though people may not be aware of it, statistics is a branch of mathematics that almost everyone knows something about. For example, one may read or hear about these types of statements:

"The median price of homes sold in February was $129,200." (*USA Today*)

"According to a Roper poll, 75% of American homes have a VCR, but only 58% of owners say they ever tape a show." (*TV Guide*)

"Among older men, the mortality rate for smokers is twice the rate of those who have never smoked." (*AARP Bulletin*)

Since statistics are used in almost all fields of human endeavor, it is important that you become familiar with the basic concepts of this branch of mathematics. This chapter presents the basic concepts of statistics, such as collecting and organizing data; drawing graphs; finding measures of average, variation, and position; and solving problems using the standard normal distribution. ■

Medical research makes extensive use of statistics, both in describing problems and in explaining how effective new treatments are compared to old treatments.

12-1 The Nature of Statistics and Organizing Data

In order to gain information about seemingly haphazard events, statisticians collect *data* for these events.

> **Data** are measurements or observations that are gathered for an event under study.

In sports, a statistician keeps records of the number of yards a running back gains during a football game or the number of hits a baseball player obtains in a season. In public health, an administrator keeps track of the number of residents of a municipality who

Mathematics in Our World

Who Is a Typical First-Time Home Buyer?

Real estate agents, bankers, and insurance company executives can use statistics to obtain a profile of the typical first-time home buyer. With this knowledge, they can tailor their advertising to target certain groups and provide their best services to their customers. In order to devise this profile, statistics such as the means, medians, modes, ranges, variances, and standard deviations are used.

This chapter will show you how to obtain these statistics and explain how each can be used to create a profile of a typical first-time buyer.

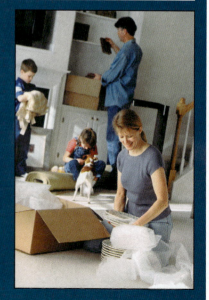

contract a new strain of flu. In education, a researcher may want to know if a new method of teaching is better than an older method. The media report the results of Nielson, Harris, and Gallop polls. All of these endeavors involve collecting data.

Once data are collected, it is necessary to organize, summarize, and present data in order to draw general conclusions. Hence, statistics can be defined as follows:

> **Statistics** is the branch of mathematics that involves collecting, organizing, summarizing, and presenting data and drawing general conclusions from data.

Populations and Samples

When statistical studies are performed, the statisticians must first define the *population* for the study.

> A **population** consists of all subjects under study.

For example, suppose a researcher wishes to conduct a study on the effect of drinking sugared soda on a person's teeth. The first step would be to define the population for the study. In this case, the population might be all people in the United States who drink three or more sugared soda beverages per day. In this case, the population is very large, and the researcher could not have everybody's teeth checked by a dentist. Hence, the researcher must select a smaller representative group of the population. This group is called a *sample*.

> A **sample** is a representative subgroup or subset of the population.

Wildlife biologists capture and tag animals to study how they live in the wild. The population is all the animals of the species in this region. The tagged animals constitute the sample. If the sample is not representative of the species (i.e., the wrong animal is tagged), the data will be skewed.

For example, in this study, the researcher might select a group of 100 individuals from the defined population. In order to make generalizations from a sample to a population, this sample must be *representative* of the population. That is, the characteristics in terms of ages, gender, occupations, etc. of the sample subjects must be similar to those of the subjects in the population. In order to obtain a representative sample, researchers use a variety of methods. Four of these sampling methods are explained next.

Sampling Methods

The four basic sampling methods that can be used to obtain a representative sample are random, systematic, stratified, and cluster.

In order to obtain a **random sample,** each subject of the population must have an equal chance of being selected. The best way to obtain a random sample is to use a list of random numbers. Random numbers can be obtained from a table or from a computer or calculator. Subjects in the population are numbered, and then they are selected by using the corresponding random numbers.

Using a random number generator such as a calculator, computer, or table of random numbers is like selecting numbers out of a hat. The difference is that when random numbers are generated by a calculator, computer, or table, there is a better chance that every number has an equally likely chance of being selected. When numbers are placed in a hat and mixed, one can never be sure that they are thoroughly mixed and selected so that each number has an equal chance of being selected.

A **systematic sample** is taken by numbering each member of the population and then selecting every *k*th member, where *k* is a natural number. For example, one might select every tenth person. The starting number must be selected at random though.

When the population is divided into groups where the members of each group have similar characteristics (such as female freshmen, male freshmen, female sophomores, and male

sophomores) and members of each group are selected at random, the result is called a **stratified sample.** For example, one might select five students at random from each group—female freshmen, male freshmen, female sophomores, and male sophomores—making a sample of 20 students. Twenty students is a fairly small sample. If we had selected our sample using random numbers alone (i.e., a random sample), we may have ended up with a sample that did not include any members from one of the four groups; for example, there may be no female freshmen. This could have jeopardized our study if female freshmen were likely to answer the questions of a survey differently than the other groups. In other words, the purpose of a stratified sample is to ensure that all groups will be represented in our study.

When an intact group of subjects that represent the population is used for a sample, it is called a **cluster sample.** For example, an inspector may select at random a carton of calculators and examine each one to determine how many are defective. The group in this carton represents a cluster. In this case, the researcher assumes that the calculators in the carton represent the population of all calculators manufactured by the company.

Samples are used in the majority of statistical studies, and if they are selected properly, the results of a study can be generalized to the population.

Descriptive and Inferential Statistics

There are two main branches of statistics, descriptive and inferential. Statistical techniques that are used to *describe* data are called **descriptive statistics.** For example, a researcher may wish to determine the average age of the full-time students enrolled in your college. Furthermore, the researcher may wish to know the percentage of the students who own automobiles. These and other summary measures will be explained in this chapter.

Statistical techniques used to make *inferences* are called **inferential statistics.** For example, every month the Bureau of Labor and Statistics estimates the number of people in the United States who are unemployed. Since it would be impossible to survey every adult resident of the United States, the Bureau selects a sample of adult individuals in the United States to see what percent are unemployed. In this case, the information obtained from a sample is used to estimate a population measure.

Another area of inferential statistics is called *hypothesis testing.* A researcher tries to test a hypothesis to see if there is enough evidence to support it. Here are some research questions that lend themselves to hypothesis testing:

Is one brand of aspirin better than another brand?

Does taking vitamin C prevent colds?

Are children more susceptible to ear infections than adults are?

A third aspect of inferential statistics is determining whether or not a relationship exists between two or more variables. This area of statistics is called *correlation and regression;* for example:

Is caffeine related to heart trouble?

Is there a relationship between a person's age and his or her blood pressure?

Is the birth weight of a certain species of animal related to the life span of the animal?

These are only a few examples of descriptive and inferential statistics. If you look around, you will see many other examples of statistics.

Frequency Distributions

The data collected for a statistical study are called **raw data.** In order to describe situations and draw conclusions, the researcher must organize the data in a meaningful way.

Statistics can reveal unsuspected variables in an experiment. For example, a drug trial found very different results between patients taking a drug to lower cholesterol with grapefruit juice and with other juices.

Sidelight

Mathematics and the Law

Many times mathematicians are called to testify in court. A couple was convicted of a robbery in Los Angeles on probability theory only. Based on the physical descriptions (i.e., hair color, beard, etc.) and the color of their automobile, a mathematician calculated that there was only one chance in 12 million that there could be another couple with the exact characteristics that the witnesses described for the robbers. However, their conviction was overturned based on faulty calculations and the lack of other evidence.

In the presidential election of 2000, the Democrats contested the election in Florida. To support their theory of voter irregularities, they brought in a statistician to testify. The Republicans then brought in their own statistician to refute the Democratic position. The case finally reached the Supreme Court, and the Republicans won.

Data can be organized by constructing a *frequency distribution* or a *stem and leaf plot.* There are two types of frequency distributions, the categorical frequency distribution and the grouped frequency distribution.

A categorical frequency distribution is used when the data are categorical rather than numerical. Example 12-1 shows how to construct a **categorical frequency distribution.**

Example 12-1

Twenty-five army inductees were given a blood test to obtain their blood types. The data follow.

A	B	B	AB	O
O	O	B	AB	B
B	B	O	A	O
AB	A	O	B	A
A	O	O	O	AB

Construct a frequency distribution for the data.

Solution

Step 1 Make a table as shown.

Type	Tally	Frequency
A		
B		
O		
AB		

—Continued

12-5

Thermometer

Calculator Explorations

A graphing calculator will assist you in sorting data. Sample screens: L1 is original data; L2 is sorted (ascending) data.

L1	L2	L3	2
112	100	------	
100	104		
127	105		
120	105		
134	105		
105	106		
110	107		

L2(1)=100

L1	L2	L3	2
117	120		
105	120		
118	121		
112	122		
114	122		
114	127		
110	134		

L2(50)=134

Example 12-1 Continued—

Step 2 Tally the data using the second column.

Step 3 Count the tallies and place the numbers in the third column. The completed frequency distribution is shown.

Type	Tally	Frequency
A	////	5
B	//// //	7
O	//// ////	9
AB	////	4

Try This One

12-A   A health-food store recorded the type of vitamin pills 35 customers purchased during a 1-day sale. Construct a categorical frequency distribution for the data.

C	C	C	A	D	E	C
E	E	A	B	D	C	E
C	E	C	C	C	D	A
B	B	C	C	A	A	E
E	E	E	A	B	C	B

(*Note:* Vitamin B represents vitamin B complex.)

For the answer, see page 677.

Another type of frequency distribution that can be constructed uses numerical data and is called a **grouped frequency distribution.** Example 12-2 shows this procedure.

Example 12-2

These data represent the record high temperatures for each of the 50 states in degrees Fahrenheit. Construct a grouped frequency distribution for the data.

112	100	127	120	134	105	110	109	112	118
110	118	117	116	118	114	114	105	109	122
107	112	114	115	118	118	122	106	110	117
116	108	110	121	113	119	111	104	111	120
120	113	120	117	105	118	112	114	114	110

Source: The World Almanac Book of Facts.

—Continued

Example   12-2   *Continued—*

Solution

Step 1   Subtract the lowest value from the highest value: $134 - 100 = 34$.

Step 2   Decide how many classes you want (in this case, seven), and divide 34 by 7 to get 4.8. Round up to 5. Note that rounding up is not the same as rounding off. When you round up, you always go to the next whole number when there is a remainder.

Step 3   Start with the lowest value and add 5 to get the lower class limits: 100, 105, 110, 115, 120, 125, 130, 135.

Step 4   Set up the classes by subtracting one from each lower class limit except the first lower class limit.

Step 5   Tally the data and record the frequencies as shown.

Class	Tally	Frequency
100–104	//	2
105–109	//// ///	8
110–114	//// //// //// ///	18
115–119	//// //// ///	13
120–124	//// //	7
125–129	/	1
130–134	/	1

Try This One

12-B   The data shown represent the number of miles employees travel to work. Construct a frequency distribution for the data. Use six classes.

1	2	6	7	12	13	2	6	9	5
18	7	3	15	15	4	17	1	14	5
4	16	4	5	8	6	5	18	5	2
9	11	12	1	9	2	10	11	4	10
9	18	8	8	4	14	7	3	2	6

For the answer, see page 677.

Stem and Leaf Plots

Another way to organize data is to use a **stem and leaf plot** (sometimes called a *stem plot*). Each data value or number is separated into two parts. For a two-digit number such as 53, the tens digit, 5, is called the *stem,* and the ones digit, 3, is called its *leaf.* For the number 72, the stem is 7, and the leaf is 2. For a three-digit number, say 138, the first two digits, 13, are used as the stem, and the third digit, 8, is used as the leaf. Example 12-3 shows how to construct a stem and leaf plot. Stem and leaf plots are analyzed by looking at the shape, peaks, and any gaps in the figure.

Example 12-3

At an outpatient testing center, a sample of 20 days showed the number of cardiograms done each day as presented next. Construct a stem and leaf plot for the data.

25	31	20	32	13
14	43	2	57	23
36	32	33	32	44
32	52	44	51	45

Solution

Step 1   Arrange the data in order. (Note that the data value 2 is written as 02.)

02 13 14 20 23 25 31 32 32 32 32 33 36 43 44 44 45 51 52 57

Step 2   Separate the data according to the first digit, as shown:

02   13 14   20 23 25   31 32 32 32 32 33 36   43 44 44 45   51 52 57

Step 3   Using the first digits as the stems and the second digits as leaves, write the stems in a vertical row and the leaves horizontally, as shown:

Stem and Leaf Plot for the Number of Cardiograms

Stems	Leaves
0	2
1	3 4
2	0 3 5
3	1 2 2 2 2 3 6
4	3 4 4 5
5	1 2 7

There is a peak in the plot in the 30–39 group.

Try This One

12-C   These data are the high temperatures in degrees Fahrenheit on a November day for a sample of cities in the United States. Construct a stem and leaf plot for the data.

62	17	62	59	64	56	62	46	67	55
58	65	48	56	46	50	74	70	56	60
47	50	47	58	62	86	75	56	68	
65	70	73	80	43	39	63	64	66	
62	65	63	70	72	62	78	57	63	
62	56	61	56	61	58	48	52	63	

Source: Pittsburgh Tribune Review

For the answer, see page 677.

For the answer, see page 677.

Math Note

When the numbers have three digits, such as 325, the stem is 32 and the leaf is 5.

Exercise Set 12-1

Real World Applications

1. At a college financial aid office, students who applied for a scholarship were classified according to their class rank: Fr = freshman, So = sophomore, Jr = junior, Se = senior. Construct a frequency distribution for the data.

Fr	Fr	Fr	Fr	Fr
Jr	Fr	Fr	So	Fr
Fr	So	Jr	So	Fr
So	Fr	Fr	Fr	So
Se	Jr	Jr	So	Fr
Fr	Fr	Fr	Fr	So
Se	Se	Jr	Jr	Se
So	So	So	So	So

2. A questionnaire about how people primarily get news resulted in the following information from 25 respondents. Construct a frequency distribution for the data (N = newspaper, T = television, R = radio, M = magazine).

N	N	R	T	T
R	N	T	M	R
M	M	N	R	M
T	R	M	N	M
T	R	R	N	N

3. The number of games won by the pitchers who were inducted into the Baseball Hall of Fame through 1992 are shown here. Construct a frequency distribution for the data using six classes.

373	254	237	243	308
210	266	253	201	266
239	114	224	373	286
329	236	284	247	273
198	361	416	207	243
326	251	169	360	311
215	189	344	268	363
21	270	165	240	48
150	300	207	314	197
209	210	260	327	

Source: The Universal Almanac

4. The ages of the signers of the Declaration of Independence are shown here. (Age is approximate since only the birth year appeared in the source, and one has been

omitted since his birth year is unknown.) Construct a frequency distribution for the data using seven classes.

41	54	47	40	39	35	50	37	49	42	70	32
44	52	39	50	40	30	34	69	39	45	33	52
44	62	60	27	42	34	50	42	52	38	36	45
35	43	48	46	31	27	55	63	46	33	60	62
35	46	45	34	53	50	50					

Source: The Universal Almanac

5. The number of automobile fatalities in 27 states where the speed limits were raised in 1996 are shown here. Construct a frequency distribution using seven classes.

1100	460	85
970	480	1430
4040	405	70
620	690	180
125	1160	3630
2805	205	325
1555	300	875
260	350	705
1430	485	145

Source: USA Today

6. The data (in cents) are the cigarette taxes per pack imposed by each state. Construct a frequency distribution. Use 0–19, 20–39, 40–59, etc.

16.5	12.0	76.0	56.0	41.0
100.0	80.0	75.0	5.0	51.1
58.0	28.0	48.0	44.0	44.0
32.5	58.0	18.0	24.0	2.5
37.0	15.5	17.0	23.0	82.5
20.0	36.0	18.0	68.0	17.0
50.0	24.0	34.0	31.0	59.0
24.0	3.0	35.0	71.0	12.0
20.0	37.0	7.0	33.0	36.0
33.9	74.0	80.0	21.0	13.0

Source: USA Today

7. The acreage (in thousands of acres) of the 39 U.S. National Parks is shown here. Construct a frequency distribution for the data using eight classes.

41	66	233	775	169
36	338	233	236	64
183	61	13	308	77
520	77	27	217	5

650	462	106	52	52
505	94	75	265	402
196	70	132	28	220
760	143	46	539	

Source: The Universal Almanac

8. The heights in feet above sea level of the major active volcanoes in Alaska are given here. Construct a frequency distribution for the data using 10 classes.

4,265	3,545	4,025	7,050	11,413
3,490	5,370	4,885	5,030	6,830
4,450	5,775	3,945	7,545	8,450
3,995	10,140	6,050	10,265	6,965
150	8,185	7,295	2,015	5,055
5,315	2,945	6,720	3,465	1,980
2,560	4,450	2,759	9,430	
7,985	7,540	3,540	11,070	
5,710	885	8,960	7,015	

Source: The Universal Almanac

9. During the 1998 baseball season, Mark McGwire and Sammy Sosa both broke Roger Maris's home run record of 61. The distances in feet for each home run follow. Construct a frequency distribution for each player using the same eight classes.

McGwire

306	370	370	430
420	340	460	410
440	410	380	360
350	527	380	550
478	420	390	420
425	370	480	390
430	388	423	410
360	410	450	350
450	430	461	430
470	440	400	390
510	430	450	452
420	380	470	398
409	385	369	460
390	510	500	450
470	430	458	380
430	341	385	410
420	380	400	440
377	370		

Sosa

371	350	430	420
430	434	370	420
440	410	420	460
400	430	410	370
370	410	380	340
350	420	410	415
430	380	380	366
500	380	390	400
364	430	450	440
365	420	350	420
400	380	380	400
370	420	360	368
430	433	388	440
414	482	364	370
400	405	433	390
480	480	434	344
410	420		

Source: USA Today

10. The data (in millions of dollars) are the values of the 30 National Football League franchises. Construct a frequency distribution for the data using seven classes.

170	191	171	235	173	187	181	191
200	218	243	200	182	320	184	239
186	199	186	210	209	240	204	193
211	186	197	204	188	242		

Source: The Pittsburgh Press

11. Twenty-nine executives reported the number of telephone calls made during a randomly selected week as shown here. Construct a stem and leaf plot for the data and analyze the results.

22	14	12	9	54	12
16	12	14	49	10	14
8	21	37	28	36	22
9	33	58	31	41	19
3	18	25	28	52	

12. The National Insurance Crime Bureau reported that these data represent the number of registered vehicles per car stolen for 35 selected cities in the United States. For example, in Miami, one automobile is stolen for every 38 registered vehicles in the city. Construct a stem and leaf plot for the data and analyze the distribution. (The data have been rounded to the nearest whole number.)

38	53	53	56	69	89	94
41	58	68	66	69	89	52
50	70	83	81	80	90	74
50	70	83	59	75	78	73
92	84	87	84	85	84	89

Source: USA Today

13. The growth (in centimeters) of a plant after 20 days is shown below. Construct a stem and leaf plot for the data.

20	12	39	38
41	43	51	52
59	55	53	59
50	58	35	38
23	32	43	53

14. The data shown represent the percentage of unemployed males for a sample of countries of the world. Using whole numbers as stems and the decimals as leaves, construct a stem and leaf plot.

8.8	1.9	5.6	4.6	1.5
2.2	5.6	3.1	5.9	6.6
9.8	8.7	6.0	5.2	5.6
4.4	9.6	6.6	6.0	0.3
4.6	3.1	4.1	7.7	

Source: The Time Almanac

Writing Exercises

15. What are *data*?

16. Define *statistics*.

17. Explain the difference between a population and a sample.

18. How is a random sample selected?

19. How is a systematic sample selected?

20. How is a stratified sample selected?

21. How is a cluster sample selected?

22. Name two ways to organize data.

23. Classify each sample as random, systematic, stratified, or cluster.

    (a) In a large school district, all teachers from two buildings are interviewed to determine whether they believe the students have less homework to do now than in previous years.

    (b) Every seventh customer entering a shopping mall is asked to select his or her favorite store.

    (c) Nursing supervisors are selected using random numbers in order to determine annual salaries.

    (d) Every hundredth hamburger manufactured is checked to determine its fat content.

    (e) Mail carriers of a large city are divided into four groups according to gender (male or female) and according to whether they walk or ride on their routes. Then 10 are selected from each group and interviewed to determine whether they have been bitten by a dog in the last year.

Critical Thinking

24. For each statement, decide whether descriptive or inferential statistics is used.

    (a) A recent study showed that eating garlic can lower blood pressure.

    (b) The average number of students in a class at White Oak University is 22.6.

    (c) It is predicted that the average number of automobiles each household owns will increase next year.

    (d) Last year's total attendance at Long Run High School's football game was 8325.

    (e) The chance that a person will be robbed in a certain city is 15%.

25. In addition to the four basic sampling methods, other methods are also used. Some of these methods are *sequence sampling, double sampling,* and *multiple sampling.* Investigate these methods and explain the advantages and disadvantages of each method.

12-2 Picturing Data

After the data have been collected and organized into frequency distributions, it is necessary to present that data in a meaningful and easily understood way. The most frequently used way to present data is by using statistical graphs. There are many types of statistical graphs. The most common ones are presented in this section.

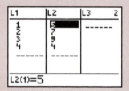

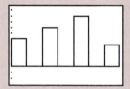
The Bar Graph and Pie Graph

Horizontal or vertical bars can be used to present data that have been organized in a categorical frequency distribution. Example 12-4 shows this method of using a **bar graph.**

Example 12-4

Using the data for blood type in Example 12-1, draw a vertical bar graph. The frequency distribution is presented here.

Type	Frequency
A	5
B	7
O	9
AB	4

Solution

Step 1   Draw the axes.

Step 2   Label the axes.

Step 3   Draw the bars with heights corresponding to the frequencies.

The completed graph is shown in Figure 12-1.

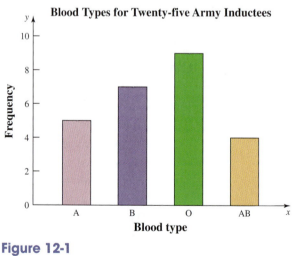

Figure 12-1

Another type of graph that can be drawn for the data that are categorical in nature is the **pie graph.** A pie graph is a circle that is divided into sections in proportion to the frequencies corresponding to the categories. The purpose of a pie graph is to show the relationship of the parts to the whole by visually comparing the size of the sections. Example 12-5 shows the procedure for constructing a pie graph.

Example 12-5

Construct a pie graph for the frequency distribution used in the previous example. The distribution is shown here.

Type	Frequency
A	5
B	7
O	9
AB	4
	$n = 25$

Solution

Step 1 The frequency for each class is converted into degrees by using the formula

$$\text{degrees} = \frac{f}{n} \cdot 360°$$

where

f = frequency for each class

n = sum of the frequencies

(*Note:* There are 360° in a circle.)

Hence, these conversions are obtained:

$$\text{A} \quad \frac{5}{25} \cdot 360° = 72°$$

$$\text{B} \quad \frac{7}{25} \cdot 360° = 100.8°$$

$$\text{O} \quad \frac{9}{25} \cdot 360° = 129.6°$$

$$\text{AB} \quad \frac{4}{25} \cdot 360° = 57.6°$$

Step 2 Using a protractor, graph each section on the circle, as shown in Figure 12-2.

Blood Types for Twenty-five Army Inductees

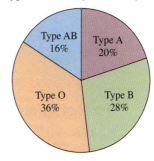

Figure 12-2

—Continued

Example   12-5   Continued—

Step 3   Calculate the percents for each section using the formula

$$\text{percent} = \frac{f}{n} \cdot 100\%$$

A $\frac{5}{25} \cdot 100\% = 20\%$

B $\frac{7}{25} \cdot 100\% = 28\%$

O $\frac{9}{25} \cdot 100\% = 36\%$

AB $\frac{4}{25} \cdot 100\% = 16\%$

Label each section with its percent value.

Try This One

12-D   The distribution shown here represents the number of sales for each type of automobile sold by a car dealer for the month of July. Draw a bar graph and a pie graph for the distribution.

Type	Frequency
Convertibles	3
Station Wagons	2
Compacts	5
Coupes	30
Sedans	20
	$n = 60$

For the answer, see page 677.

The Histogram and Frequency Polygon

When the data are organized into a grouped frequency distribution, two types of graphs can be drawn. They are the *histogram* and the *frequency polygon*.

The **histogram** is similar to a vertical bar graph in that the height of the bars corresponds to the frequencies; however, the class limits are placed on the horizontal axis. The procedure for constructing a histogram is shown in the next example.

Calculator Explorations

By using the minimum and the maximum of the data, the highest frequency in any class, and the range of each class assists you in setting up the window to view the frequency distribution histogram using a graphing calculator.

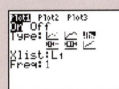

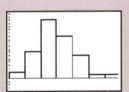

What information can be accessed using the trace function?

Example 12-6

Construct a histogram for the data in the frequency distribution used in Example 12-2. The data are the record high temperatures in degrees Fahrenheit for each of the 50 states. The distribution is shown here.

Class	Frequency
100–104	2
105–109	8
110–114	18
115–119	13
120–124	7
125–129	1
130–134	1

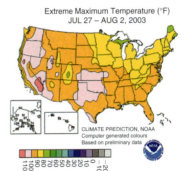

Extreme Maximum Temperature (°F)
JUL 27 – AUG 2, 2003

CLIMATE PREDICTION, NOAA
Computer generated colours
Based on preliminary data

Solution

Step 1 Represent the scale for the frequencies on the vertical axis and the class limits on the horizontal axis.

Step 2 Draw vertical bars corresponding to the frequencies for each class. See Figure 12-3.

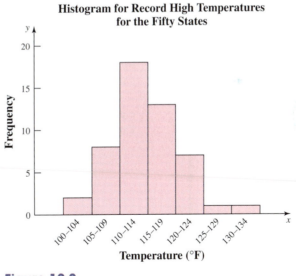

Figure 12-3

Math Note

When drawing a histogram, make sure the bars touch each other unless there is a class whose frequency is zero. When drawing a bar graph, make sure that there is a gap between bars.

The **frequency polygon** is similar to the histogram except that a broken line is drawn through the midpoints of the classes. The midpoint of a class is found by finding the sum of the lower class limit and the upper class limit and dividing the sum by two. For a class whose limits are 5–9, the midpoint is $\dfrac{5+9}{2} = 7$.

Example 12-7

Construct a frequency polygon for the data in the frequency distribution used in Example 12-6. The distribution is shown here.

Class	Frequency	Class	Frequency
100–104	2	120–124	7
105–109	8	125–129	1
110–114	18	130–134	1
115–119	13		

Solution

Step 1 Find the midpoints for each class.

$$\frac{100 + 104}{2} = 102$$

$$\frac{105 + 109}{2} = 107$$

$$\frac{110 + 114}{2} = 112$$

etc.

The midpoints are 102, 107, 112, 117, 122, 127, and 132.

Step 2 Label the vertical axis with an appropriate scale for the frequencies and the horizontal axes with the midpoints for the classes.

Step 3 Connect adjacent midpoints with straight lines.

Step 4 Finish the graph by drawing a line back to the horizontal at the beginning and end of the graph.

The horizontal length of the segments should be the same as the others; e.g., if the distance between the midpoints is 5, the line should touch the axis 5 units after the first and last midpoints. See Figure 12-4.

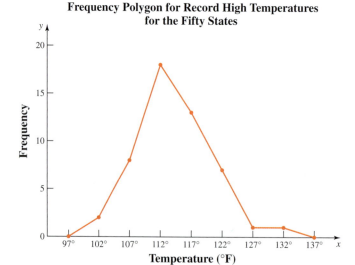

Frequency Polygon for Record High Temperatures for the Fifty States

Figure 12-4

> **Math Note**
>
> The frequency polygon must touch the horizontal axis at both ends.

Distribution Shapes

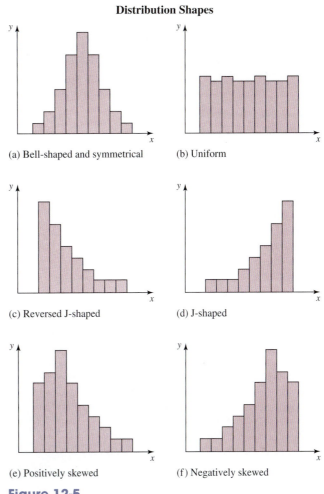

(a) Bell-shaped and symmetrical (b) Uniform

(c) Reversed J-shaped (d) J-shaped

(e) Positively skewed (f) Negatively skewed

Figure 12-5

In order to analyze a histogram to determine the nature of a frequency distribution, statisticians must look at the shape of the distribution. Figure 12-5 shows some common distribution shapes.

Distributions rarely conform to exact shapes; therefore, it is necessary to look at the general shape of the histogram in order to determine the nature of the frequency distribution.

A *bell-shaped distribution* has a single peak and is approximately the same on both sides of the peak (i.e., symmetrical). See Figure 12-5(a). When the data are evenly distributed in the classes, the bars will be approximately equal in height, and the distribution is said to be *uniform*. See Figure 12-5(b). When the frequencies of the classes are decreasing from the lowest class to the highest class, the distribution is said to be a *reverse J-shaped* distribution. See Figure 12-5(c). When the frequencies of the classes are increasing from the lowest class to the highest class, the distribution is said to be *J-shaped*. See Figure 12-5(d). When the distribution has a single peak to the left of center, we say the distribution is *positively skewed*. See Figure 12-5(e). This means that the data values "cluster" around a value located in the lower end of the distribution, and the data values in the higher classes are more spread out. Finally, when the distribution has a single peak to the right of center, we say that the distribution is *negatively skewed*. See Figure 12-5(f). This means that the data values "cluster" around a value located in the higher end of the distribution, and the data values in the lower classes are more spread out.

Florence Nightingale (1820-1910)

Florence Nightingale was a British nurse who worked in hospitals during the Crimean War. She is considered to be the founder of modern medical nursing, and she pioneered the cause of improving sanitary conditions in hospitals to save the lives of the soldiers she treated.

She is reported to have received private instruction in mathematics and statistics from James Joseph Sylvester, who at one time worked as an actuary for an insurance company. With her knowledge of statistics, she used charts and graphs to lobby the British government and the general public to improve sanitary conditions in hospitals. She is credited with developing several original types of graphs such as the one shown called a polar area diagram. This graph shows that more soldiers died from preventable diseases and war wounds than other causes during the Crimean War. The outermost regions show the number of deaths from preventable diseases. The middle regions show the number of deaths from war wounds, and the innermost regions show the number of deaths from all other causes. She also used statistical techniques in other areas such as the social sciences.

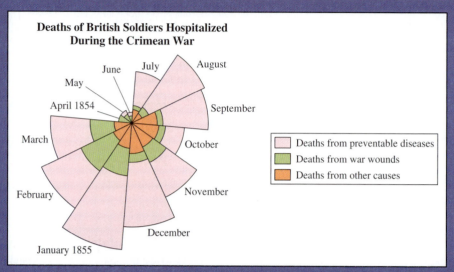

Deaths of British Soldiers Hospitalized During the Crimean War

Legend:
- Deaths from preventable diseases
- Deaths from war wounds
- Deaths from other causes

The histogram shown for the record high temperatures (Example 12-6) has a single peak with the class 110–114 containing the largest number of temperatures. It is positively skewed, meaning that the peak is to the left of the distribution and the values in the higher classes are more spread out than those in the lower classes.

Try This One

12-E The following data represent the net sales (in millions of dollars) for a chain of 25 retail clothing stores nationwide. Draw a histogram and frequency polygon for the data and analyze the graphs.

Class Limits	Frequency	Class Limits	Frequency
10–19	7	50–59	2
20–29	5	60–69	2
30–39	4	70–79	1
40–49	4		

For the answer, see page 678.

Time Series Graphs

A **time series graph** can be drawn for data collected over a period of time. This type of graph is used primarily to show various trends, such as prices rising or falling, for the time period. There are three types of trends. Secular trends are viewed over a long period of time, such as yearly. Cyclical trends show oscillating patterns. Seasonal trends show the values of a commodity for shorter periods of the year, such as fall, winter, spring, and summer. Example 12-8 shows how to draw a time series graph.

Example 12-8

Shown here is the average price in dollars of 1000 feet of lumber for the month of August for the years 1995–2000. Draw a time series graph for the data.

Year	1995	1996	1997	1998	1999	2000
Cost	327	443	413	355	404	287

Source: Pittsburgh Tribune Review

Solution

Represent the years on the x axis and the cost on the y axis, and then draw straight lines through the points. See Figure 12-6.

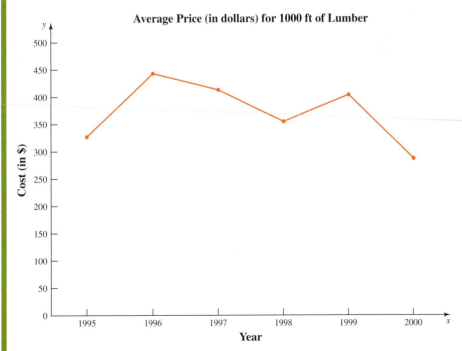

Figure 12-6

Calculator Explorations

Try This One

12-F   The following data represent the number of forest fires in the United States. Draw a time series graph for the data.

Year	Number of Fires
1994	57,176
1995	62,368
1996	84,460
1997	47,924
1998	58,160
1999	69,449
2000	72,968

Note: Round the data to the nearest thousand.

Source: National Interagency Fire Center

For the answer, see page 678.

In addition to the graphs shown here, there are many other graphs that are used in statistics. Some of these include Pareto charts, pictographs, box plots, and quality control charts.

Exercise Set   12-2

Real World Applications

1. Construct a bar graph for the number of transplants of various types performed for a specific year.

Type	Number
Kidney	11,390
Liver	3,653
Pancreas	844
Heart	2,340
Lung	737

Source: USA Today

2. Construct a bar graph for the number of health conditions per 100 reported by the elderly in a survey.

Condition	Number	Condition	Number
Arthritis	48	Cataracts	17
Hypertension	36	Sinusitis	16
Heart disease	32	Diabetes	11
Hearing impairments	32	Visual impairments	9
Orthopedic impairments	19		

Source: The Senior Profile

3. Construct a bar graph for the number of registered taxicabs in the selected cities.

City	Number
New York	11,787
Washington, D.C.	8,348
Chicago	5,300
Philadelphia	1,480
Baltimore	1,151

Source: Yellow Cab Service at a Glance

4. Construct a bar graph for the number of unemployed people in the selected states for June of 1995.

State	Number
Texas	605,000
New York	494,000
Pennsylvania	364,000
Florida	363,000
Ohio	269,000

Source: Employment in the USA

5. The frequency distribution here shows the number of freshmen, sophomores, juniors, and seniors who have part-time jobs after school. Construct a pie graph for the data.

Rank	Frequency
Freshmen	12
Sophomores	25
Juniors	36
Seniors	17

6. In an insurance company study of the causes of 1000 deaths, these data were obtained. Construct a pie graph to represent the data.

Cause of death	Number of deaths
Heart disease	432
Cancer	227
Stroke	93
Accidents	24
Other	224
	1000

7. In a study of 100 women, the numbers shown here indicate the major reason why each woman surveyed worked outside the home. Construct a pie graph for the data and analyze the results.

Reason	Number
To support self/family	62
For extra money	18
For something different to do	12
Other	8

8. A survey of the students in the school of education of a large university obtained the following data for students enrolled in specific fields. Construct a pie graph for the data and analyze the results.

Major field	Number
Preschool	893
Elementary	605
Middle	245
Secondary	1096

9. For 108 randomly selected college applicants, the frequency distribution shown here for entrance exam scores was obtained. Construct a histogram and frequency polygon for the data.

Class limits	Frequency
90–98	6
99–107	22
108–116	43
117–125	28
126–134	9

10. For 75 employees of a large department store, the distribution shown here for years of service was obtained. Construct a histogram and frequency polygon for the data.

Class limits	Frequency
1–5	21
6–10	25
11–15	15
16–20	0
21–25	8
26–30	6

11. Thirty automobiles were tested for fuel efficiency in miles per gallon (mpg). The frequency distribution shown here was obtained. Construct a histogram and frequency polygon for the data.

Class limits	Frequency
8–12	3
13–17	5
18–22	15
23–27	5
28–32	2

12. In a study of reaction times of dogs to a specific stimulus, an animal trainer obtained these data, given in seconds. Construct a histogram and frequency polygon for the data and analyze the results.

Class limits	Frequency
2.3–2.9	10
3.0–3.6	12
3.7–4.3	6
4.4–5.0	8
5.1–5.7	4
5.8–6.4	2

13. These data represent the number of babies born in a local hospital's maternity ward for the years listed. Draw a time series graph for the data.

Year	1980	1985	1990	1995	2000
Number	1981	2895	3027	1651	1432

14. These data represent the yards gained by Pittsburgh Steeler Lynn Swann. Draw a time series graph for the data.

Year	'74	'75	'76	'77	'78	'79	'80	'81	'82
Yards	208	781	516	789	880	808	710	505	265

Source: Pittsburgh Tribune Review

15. These data represent the number of books borrowed from a traveling bookmobile. Draw a time series graph for the data.

Year	1992	1994	1996	1998	2000
Books	12,413	15,160	18,201	20,206	19,143

Writing Exercises

16. Explain the difference between a bar graph and a pie graph.

17. How are the histogram and frequency polygon related? How are they different?

18. Explain the purpose of a time series graph.

Critical Thinking

19. State which type of graph (bar graph, pie graph, or time series graph) would most appropriately represent the given data.

    (a) The number of students enrolled at a local college each year for the last 5 years.

    (b) The budget for the student activities department at your college.

    (c) The number of students who get to school by automobile, bus, train, or by walking.

(d) The record high temperatures of a city for the last 30 years.

(e) The areas of the five lakes in the Great Lakes.

(f) The amount of each dollar spent for wages, advertising, overhead, and profit for a corporation.

20. Find out what a pictograph is and explain its purpose.

12-3   Measures of Average

What is an *average* student at your school like?

In the book *American Averages* by Mike Feinsilber and William B. Meed, the authors state:

> *"Average" when you stop to think of it is a funny concept. Although it describes all of us it describes none of us. . . . While none of us wants to be the average American, we all want to know about him or her.*

The authors go on to give examples of averages:

> *The average American man is five feet, nine inches tall; the average woman is five feet, 3.6 inches.*
> *The average American is sick in bed seven days a year—missing five days of work.*
> *On the average day, 24 million people receive animal bites.*
> *By his or her 70th birthday, the average American will have eaten 14 steers, 1050 chickens, 3.5 lambs, and 25.2 hogs.*[1]

In these examples, the word *average* is ambiguous, since several different methods can be used to obtain an average. Loosely stated, the average means the center of the distribution or the most typical case. Measures of average are also called *measures of central tendency* and include the *mean, median, mode,* and *midrange*.

The Mean

The *mean,* also known as the arithmetic average, is found by adding the values of the data and dividing by the total number of values. For example, the mean of 3, 2, 6, 5, and 4 is found by adding $3 + 2 + 6 + 5 + 4 = 20$ and dividing by 5; hence, the mean of the data is $20 \div 5 = 4$. The values of the data are represented by X's. In this data set, $X_1 = 3$, $X_2 = 2$, $X_3 = 6$, $X_4 = 5$, and $X_5 = 4$. To show a sum of the total X values, the symbol $\sum$ (the Greek capital letter sigma) is used, and $\sum X$ means to find the sum of the X values in the data set.

> The **mean** is the sum of the values divided by the total number of values. The symbol $\overline{X}$ represents the mean.
>
> $$\overline{X} = \frac{X_1 + X_2 + X_3 + \cdots + X_n}{n} = \frac{\sum X}{n}$$

Example   12-9

The number of home subscribers (in millions) for four satellite television companies are 5.5, 2.8, 1.9, and 1.7. Find the mean.

Source: Skyreport

—Continued

[1] Mike Feinsilber and William B. Meed: *American Averages* (New York, Bantam Doubleday Dell, 1980).

Example 12-9 Continued—

Solution

$$\overline{X} = \frac{\sum X}{n} = \frac{5.5 + 2.8 + 1.9 + 1.7}{4} = \frac{11.9}{4} = 2.975$$

The mean of the number of home subscribers is 2,975,000 (i.e., 2.975 × 1,000,000).

Try This One

12-G The number of seconds it takes ten websites to appear on the screen is shown here. Find the mean.

20.68   16.22   16.76   16.90   20.98   18.84   14.18   40.04   24.92   22.96

Source: Keystone Systems

For the answer, see page 678.

The procedure for finding the mean for grouped data uses the midpoints and the frequencies of the classes as shown in Example 12-10. This procedure will give only an approximate value for the mean, and it is used when the data set is very large or when the original raw data are unavailable but have been grouped by someone else.

Formula for finding the mean for grouped data:

$$\overline{X} = \frac{\sum f \cdot X_m}{n}$$

where

f = frequency

X_m = midpoint

$n = \sum f$ or sum of the frequencies

Example 12-10

Find the mean for the record high temperatures using the frequency distribution found in Example 12-2.

Class	Frequency
100–104	2
105–109	8
110–114	18
115–119	13
120–124	7
125–129	1
130–134	1
	$n = 50$

—Continued

Example 12-10 Continued—

Solution

Step 1 Find the midpoint of each class by adding the limits of the class and dividing by 2.

$$\frac{100 + 104}{2} = 102, \quad \frac{105 + 109}{2} = 107, \quad \frac{110 + 114}{2} = 112, \text{ etc.}$$

Place the midpoints in column C.

Step 2 Multiply the midpoint by the frequency for the class and place the product in column D.

$$102 \cdot 2 = 204, \quad 107 \cdot 8 = 856, \quad 112 \cdot 18 = 2016, \text{ etc.}$$

Step 3 Find the sum of the numbers in column D and divide by n (the sum of the frequencies in column B).

$$\overline{X} = \frac{5710}{50} = 114.2$$

The completed frequency distribution is shown next.

A Class	B Frequency	C Midpoint	D Frequency × Midpoint
100–104	2	102	204
105–109	8	107	856
110–114	18	112	2016
115–119	13	117	1521
120–124	7	122	854
125–129	1	127	127
130–134	1	132	132
	50		5710

Hence the mean for a grouped frequency distribution is found by finding the sum of the products of the frequency and the midpoint for each class, then dividing by n, which is the sum of the frequencies.

Try This One

12-H Find the mean for the frequency distribution.

Class	f
3–5	2
6–8	3
9–11	5
12–14	4
15–17	1

For the answer, see page 678.

The Median

An article recently reported that the median income for college professors was $43,250.00. This measure of average means that half of all the professors surveyed earned more than $43,250.00, and half earned less than $43,250.00.

The median is the halfway point in a data set. Before one can find this point, the data must be arranged in order. When the data set is ordered, it is called a **data array.** The median either will be a specific value in the data set or will fall between two values, as shown in the following examples.

> The **median** is the midpoint of the data array. The symbol for the median is MD.

The mean and median age for this group are different, because the grandmother's age pulls up the mean.

Steps in Computing the Median of a Data Array

Step 1   Arrange the data in order from the smallest value to the largest value.

Step 2   Select the middle point.

Example   12-11

The weights (in pounds) of seven army recruits are 180, 201, 220, 191, 219, 209, and 186. Find the median.

Solution

Step 1   Arrange the data in order.

180   186   191   201   209   219   220

Step 2   Select the middle value.

180   186   191   201   209   219   220
$\uparrow$
Median

Example   12-12

Find the median for the ages of seven preschool children. The ages are 1, 3, 4, 2, 3, 5, and 1.

Solution

1   1   2   3   3   4   5
$\uparrow$
Median

Examples 12-11 and 12-12 had an odd number of values in the data set; hence, the median was an actual data value. When there is an even number of values in the data set, the median will fall between two given values, as illustrated in Examples 12-13, 12-14, and 12-15.

Example 12-13

The number of tornadoes that have occurred in the United States in the last 8 years follows. Find the median.

684   764   656   702   856   1133   1132   1303

Source: The Universal Almanac

Solution

656   684   702   764   856   1132   1133   1303
↑
Median

Since the middle point falls halfway between 764 and 856, find the median by adding the two values and dividing the sum by 2.

$$\text{MD} = \frac{764 + 856}{2} = \frac{1620}{2} = 810$$

The median number of tornadoes is 810.

> **Math Note**
>
> When finding the median, make sure the data values are arranged in order.

Example 12-14

The ages of 10 college students are given here. Find the median.

18   24   20   35   19   23   26   23   19   20

Solution

18   19   19   20   20   23   23   24   26   35
↑
Median

$$\text{MD} = \frac{20 + 23}{2} = 21.5$$

Hence, the median age is 21.5 years.

Example 12-15

Six customers purchased the following number of magazines: 1, 7, 3, 2, 3, 4. Find the median.

Solution

$$1 \quad 2 \quad 3 \quad 3 \quad 4 \quad 7$$
$$\uparrow$$
$$\text{Median}$$

$$MD = \frac{3 + 3}{2} = 3$$

Hence, the median number of magazines purchased is 3.

Try This One

12-I   Find the median for each data set.

(a) The scores for a bowling team consisting of five members are 213, 197, 240, 184, and 160.

(b) The number of accidents for a seven-day period on an expressway is 3, 1, 0, 0, 3, 4, and 6.

(c) The retail prices of 1 gallon of milk at six randomly selected stores are $2.60, $2.75, $2.10, $2.75, $2.65, and $2.95.

For the answer, see page 678.

The Mode

The third measure of average is called the *mode*. The mode is the value that occurs most often in the data set. It is sometimes said to be the most typical case.

> The value that occurs most often in a data set is called the **mode.**

A data set can have more than one mode or no mode at all. These situations will be shown in some of the examples that follow.

Example 12-16

These data represent the duration (in days) of U.S. space shuttle voyages for the years 1992–1994. Find the mode.

8   9   9   14   8   8   10   7   6   9   7   8   10   14   11   8   14   11

Source: The Universal Almanac

—Continued

Example 12-16 Continued—

Solution

It is helpful to arrange the data in order, although it is not necessary.

6 7 7 8 8 8 8 8 9 9 9 10 10 11 11 14 14 14

Since 8-day voyages occurred five times—a frequency larger than any other number—the mode for the data set is 8.

Math Note

When there is no mode, do not say that the mode is zero. That would be incorrect because in some data, such as temperature, zero can be an actual value.

Example 12-17

Six strains of bacteria were tested to see how long they could remain alive outside their normal environment. The time, in minutes, is recorded below. Find the mode.

2 3 5 7 8 10

Solution

Since each value occurs only once, there is no mode.

When a data set has two modes, it is called *bimodal*, as shown in Example 12-18.

Example 12-18

Eleven different automobiles were tested at a speed of 15 miles per hour for stopping distances. The data, in feet, are shown below. Find the mode.

15 18 18 18 20 22 24 24 24 26 26

Solution

Since 18 and 24 both occur three times and no other value occurs more than three times, the modes are 18 and 24 feet. This data set is said to be bimodal.

Try This One

12-J Find the mode for each data set.

(a) The number of hunting accidents for a season in 10 randomly selected counties in Pennsylvania is 4, 2, 8, 3, 14, 11, 5, 14, 4, and 9.

Source: Pennsylvania Game Commission

(b) The number of problems (per 100 vehicles) reported during the first 90 days of ownership for 8 randomly selected cars is 93, 110, 93, 78, 93, 73, 111, and 110.

Source: J.D. Power and Associates

(c) The number of NFL football games won by a field goal for six seasons is 21, 29, 18, 28, 19, and 42.

Source: Elias Sports Bureau

For the answer, see page 678.

The mode is the only measure of central tendency that can be used in finding the most typical case when the data are classified by groups or categories, such as those shown in Example 12-19.

Example   12-19

A survey showed the distribution shown for the number of students enrolled in each field. Find the mode.

Business	1425
Liberal arts	878
Computer science	632
Education	471
General studies	95

Solution

Since the category with the highest frequency is business, the most typical case is a business major.

To find out which movie was most popular this month, use the mode.

Calculator Explorations

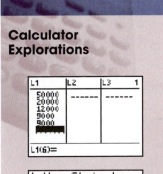

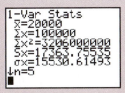

What is the shift in the mean and the median if the technicians make $11,000 each?

Comparing the Mean, Median, and Mode

For a data set, the mean, median, and mode can be quite different. Consider Example 12-20.

Example   12-20

A small company consists of the owner, the manager, the salesperson, and two technicians, all of whose annual salaries are listed here.

Staff	Salary
Owner	$50,000
Manager	20,000
Salesperson	12,000
Technician	9,000
Technician	9,000

Find the mean, median, and mode.

Solution

$$\overline{X} = \frac{\sum X}{n} = \frac{50,000 + 20,000 + 12,000 + 9,000 + 9,000}{5} = \$20,000$$

Hence, the mean is $20,000, the median is $12,000, and the mode is $9,000.

In Example 12-20, the mean is much higher than the median or the mode. This is because the extremely high salary of the owner tends to raise the value of the mean. In this and similar situations, the researcher may want to use the median as the measure of central tendency.

The Midrange

The **midrange** (MR) is a rough estimate of the middle. It is found by adding the lowest and the highest data values and dividing by 2.

$$MR = \frac{L + H}{2}$$

The midrange can be affected by one extremely high or low data value.

Example 12-21

The number of public companies filing for bankruptcy for the last 10 years is shown. Find the midrange.

115   123   91   86   70   85   86   83   122   145

Source: Bankruptcy.com

Solution

$$MR = \frac{70 + 145}{2} = 107.5$$

Try This One

12-K The data represent the number of cloudy days per year for 10 selected cities in the northwestern part of the United States. Find the midrange.

240   223   211   240   213   229   212   209   211   227

Source: USA Today

For the answer, see page 678.

Statisticians must know which measure of central tendency to use in a given situation, and some of the advantages and disadvantages of each measure. The mean is the most commonly used measure of average. It has the advantage over the median and mode since it is calculated using all the data values. It is unique. That is, there is only one mean for any given data set. The mean is used in calculating other statistics, such as the standard deviation (see Section 12-4). The biggest disadvantage in using the mean as a measure of central tendency is that its value is affected by extremely high or extremely low values in the data set, and may not be the most appropriate measure of average in these situations.

The median is used to find a value that is at the middle or center of a distribution. It can also be used when it is necessary to divide a distribution into two equal or approximately equal groups where the larger data values are placed in one group, and the smaller data values are placed in the other group. (*Note:* When the median is a specific data value as opposed to falling between two data values, it is up to the statistician to decide in which group the median value is to be placed. In this case, one group will have one more data value than the other

Sidelight

History of Statistics

The origin of descriptive statistics can be traced to the census taken by the Babylonians and Egyptians in 4500–3000 B.C.E. In addition, Roman Emperor Augustus (27 B.C.E.–17 C.E.) conducted surveys on births and deaths of the citizens of the empire as well as the amount of livestock each owned and the crops each harvested. In order to use this information, the Romans had to develop methods of collecting, organizing, and summarizing data.

During the 14th century, people began keeping records of births, deaths, and accidents in order to determine insurance rates. John Graunt (1620–1674) studied the number of males and females born and discovered that slightly more males than females died during the first year of life. Gregor Mendel (1822–1884) used probability and statistics in his studies of heredity at a monastery in Brunn; Sir Francis Galton (1822–1911) did correlation studies of heredity using peas, moths, dogs, and humans. Sir Ronald Fisher (1890–1962) developed statistical methods in inferential techniques, experimental design, estimation, and analysis of variance. Many new statistical methods have been developed since 1900, and new techniques in research and statistics are being studied and perfected each year.

With the advent of the computer, the statistician can process large amounts of data and do many complex calculations that were impossible a few years ago.

Gregor Mendel, 1866

group.) Finally, the median is less affected by an extremely large or extremely small data value than the mean and may be a more appropriate measure of average in this case.

The mode is used when the most typical case is desired. It is easy to find, but it may not be unique. Also, the data set may not have a mode. Finally, the mode is the only "average" that can be used with categorical data such as religious preference, political affiliation, etc.

The midrange is also easy to compute and gives the midpoint of the data set. It is affected by extremely high or low values in the data set.

Exercise Set 12-3

Real World Applications

For Exercises 1–10, find the mean, median, mode, and midrange.

1. These data are the number of burglaries reported in 1996 for nine Western Pennsylvania universities.

   61   11   1   3   2   30   18   3   7

   Source: Pittsburgh Post Gazette

2. The number of attorneys in 10 law firms in Pittsburgh is 87, 109, 57, 221, 175, 123, 170, 80, 66, and 80.

   Source: Pittsburgh Tribune Review

3. The number of dairy cows (in thousands) for 10 selected states is 700, 298, 638, 260, 1380, 280, 270, 1350, 380, and 570.

   Source: U.S. Department of Agriculture

4. The number of wins per season for selected college football teams that played in a post season bowl is 8, 9, 6, 6, 6, 8, 7, 9, 7, 9, 7, 8, 8, 8, 8, 7, 7, 9, and 11.

   Source: USA Today

5. The number of cigarettes (in billions) sold for eight randomly selected years is 516.5, 540.3, 584.7, 615.3, 603.6, 560.7, 416.7, and 478.6.

   Source: Federal Trade Commission

6. The tax millage for 23 selected communities near Pittsburgh is as shown.

42	24.5	26.75	21.5	26.5	36
28.9	27.7	12.25	41	24.5	19
21.5	12.1	48	34.8	33	38.33
23	20.5	20.5	46.5	28	

   Source: Pittsburgh Post Gazette

7. The number of hospitals for the five largest hospital systems is shown here.

   340   75   123   259   151

   Source: USA Today

8. During 1993, the major earthquakes had Richter magnitudes as shown here.

   7.0   6.2   7.7   8.0   6.4   6.2

   7.2   5.4   6.4   6.5   7.2   5.4

   Source: The Universal Almanac

9. Twelve members of the high school cross-country team were asked how many minutes each ran during practice sessions. Their answers are recorded here.

   32   28   35   37   43   51   61   39   48   51   53   49

10. The exam scores of 18 English composition students were recorded as shown.

   78   62   98   90   88   73   79   86   81   84   93   97   63   59   78   82   87   93

For Exercises 11–18, find the mean.

11. For 50 antique car owners, the distribution of the cars' ages was obtained as shown.

Class	Frequency	Class	Frequency
16–18	20	22–24	8
19–21	18	25–27	4

12. Thirty automobiles were tested for fuel efficiency (in miles per gallon). This frequency distribution was obtained:

Class	Frequency	Class	Frequency
8–12	3	23–27	5
13–17	5	28–32	2
18–22	15		

13. In a study of the time it takes an untrained mouse to run a maze, a researcher recorded these data in seconds.

Class	Frequency
2.1–2.7	5
2.8–3.4	7
3.5–4.1	12
4.2–4.8	14
4.9–5.5	16
5.6–6.2	8

14. Eighty randomly selected lightbulbs were tested to determine their lifetimes (in hours). The frequency distribution was obtained as shown.

Class boundaries	Frequency
53–63	6
64–74	12
75–85	25
86–96	18
97–107	14
108–118	5

15. These data represent the net worth (in millions of dollars) of 45 national corporations.

Class limits	Frequency
10–20	2
21–31	8
32–42	15
43–53	7
54–64	10
65–75	3

16. The cost per load (in cents) of 35 laundry detergents tested by a consumer organization is shown here.

Class limits	Frequency
13–19	2
20–26	7
27–33	12
34–40	5
41–47	6
48–54	1
55–61	0
62–68	2

17. The frequency distribution shown represents the commission earned (in dollars) by 100 salespeople employed at several branches of a large chain store.

Class limits	Frequency	Class limits	Frequency
150–158	5	186–194	20
159–167	16	195–203	15
168–176	20	204–212	3
177–185	21		

18. This frequency distribution represents the data obtained from a sample of 75 copying machine service technicians. The values represent the days between service calls for various copying machines.

Class boundaries	Frequency	Class boundaries	Frequency
16–18	14	25–27	10
19–21	12	28–30	15
22–24	18	31–33	6

Writing Exercises

19. Explain how to find the mean.
20. Explain how to find the median.
21. Explain how to find the mode.
22. Explain how to find the midrange.

Critical Thinking

23. For these situations, state which measure of central tendency—mean, median, or mode—should be used.

    (a) The most typical case is desired.

    (b) The data are categorical.

    (c) The values are to be divided into two approximately equal groups, one group containing the larger and one containing the smaller values.

24. Describe which measure of central tendency—mean, median, or mode—was probably used in each situation.

    (a) Half of the factory workers make more than $5.37 per hour and half make less than $5.37 per hour.

    (b) The average number of children per family in the Plaza Heights is 1.8.

    (c) Most people prefer red convertibles to any other color.

    (d) The average person cuts the lawn once a week.

    (e) The most common fear today is fear of speaking in public.

    (f) The average age of college professors is 42.3 years.

12-4 Measures of Variation

In addition to knowing the measures of central tendency (average), statisticians are also interested in *measures of variation,* which is how the data vary in the data set. The three most commonly used measures of variation are the *range,* the *variance,* and the *standard deviation.*

Range

The range is the simplest of the three measures and is defined next.

> The **range** is the highest value minus the lowest value. The symbol R is used for the range.
>
> $$R = \text{highest value} - \text{lowest value}$$

The two groups of dogs have about the same mean size, but the range of sizes is quite different.

Example 12-22

The data represent the number of larceny crimes for a sample of 11 colleges in Western Pennsylvania. Find the range.

24 359 6 16 64 135 10 472 25 35 518

Source: Pennsylvania State Police Uniform Reports

Solution

$$R = 518 - 6 = 512$$

The range for the data set is 512.

Try This One

12-L The data represent the number of vandalism crimes for a sample of 11 colleges in Western Pennsylvania. Find the range.

21 87 3 6 15 71 21 222 61 20 171

Source: Pennsylvania State Police Uniform Reports

For the answer, see page 678.

Variance and Standard Deviation

The range is a somewhat limited measure of variation since it uses only two values in the data set, namely, the largest value and the smallest value. In addition, one extremely large or small data value can make the range very large and make the data set appear more variable than it might actually be. For this reason, statisticians use two other measures of variation, the *variance* and the *standard deviation.*

The steps for finding the variance and standard deviation are summarized in the following table.

Procedure for Finding the Variance and Standard Deviation

Step 1 Find the mean.

Step 2 Subtract the mean from each data value in the data set.

Step 3 Square the differences.

Step 4 Find the sum of the squares.

Step 5 Divide the sum by $n - 1$ to get the variance, where n is the number of data values.

Step 6 Take the square root of the variance to get the standard deviation.

Example 12-23 shows how to compute the variance and the standard deviation.

Example 12-23

A testing lab wishes to test an experimental brand of outdoor paint to see how long it will last before fading. The testing lab makes 6 gallons of paint to test. The results (in months) follow. Find the variance and standard deviation for the data for the paint.

Paint

10
60
50
30
40
20

Solution

Step 1 Find the mean for the data.

$$\overline{X} = \frac{\sum X}{n} = \frac{10 + 60 + 50 + 30 + 40 + 20}{6} = \frac{210}{6} = 35$$

Step 2 Subtract the mean from each data value.

$$10 - 35 = -25 \qquad 50 - 35 = 15 \qquad 40 - 35 = 5$$
$$60 - 35 = 25 \qquad 30 - 35 = -5 \qquad 20 - 35 = -15$$

Step 3 Square each result.

$$(-25)^2 = 625 \qquad (15)^2 = 225 \qquad (5)^2 = 25$$
$$(25)^2 = 625 \qquad (-5)^2 = 25 \qquad (-15)^2 = 225$$

Step 4 Find the sum of the squares.

$$625 + 625 + 225 + 25 + 25 + 225 = 1750$$

Step 5 Divide the sum by $n - 1$ to get the variance (n is the sample size).

$$\text{variance} = 1750 \div 5 = 350$$

Step 6 Take the square root of the variance to get the standard deviation.

—Continued

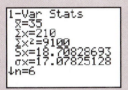
Example **12-23** *Continued—*

Hence, the standard deviation equals $\sqrt{350} \approx 18.7$.

It is helpful to make a table as follows:

A	B	C
X	$X - \overline{X}$	$(X - \overline{X})^2$
10	−25	625
60	25	625
50	15	225
30	−5	25
40	5	25
20	−15	225
		1750

The values in column A are the data values, X. The values in column B are the differences, $X - \overline{X}$. The values in column C are the squares of the differences, $(X - \overline{X})^2$.

The preceding computational procedure reveals several things. First, the square root of the variance gives the standard deviation, and vice versa, squaring the standard deviation gives the variance. Second, the variance is roughly the average of the squared distances that each value is from the mean. Therefore, if the values are near the mean, the variance will be small. In contrast, if the values are farther away from the mean, the variance will be large.

One might wonder why the squared distances are used instead of the actual distances. The reason is that the sum of the distances from the mean will always be zero. To verify this result for a specific case, add the values in column B of the table in Example 12-23. When each value is squared, the negative signs are eliminated.

Finally, why is it necessary to take the square root? The reason is that since the distances were squared, the units of the resultant numbers are the squares of the units of the original raw data. Taking the square root of the variance puts the standard deviation in the same units as the raw data.

When taking the square root, always use its positive or principal value, since the variance and standard deviation of a data set can never be negative.

The variance and standard deviation can now be formally defined.

The **variance** is an approximate average of the squares of the distance each value is from the mean. The symbol for the variance is s^2. The formula for the variance is

$$s^2 = \frac{\sum(X - \overline{X})^2}{n - 1}$$

where

X = individual value

$\overline{X}$ = mean

n = sample size

The **standard deviation** is the square root of the variance. The symbol for the standard deviation is s. The corresponding formula for the standard deviation is

$$s = \sqrt{s^2} = \sqrt{\frac{\sum(X - \overline{X})^2}{n - 1}}$$

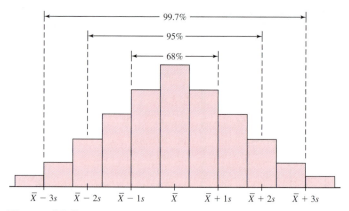

Figure 12-7

Try This One

12-M Find the variance and standard deviation for the data which represent the number of larceny crimes for a sample of 11 colleges in Western Pennsylvania. The data are repeated here.

24   359   6   16   64   135   10   472   25   35   518

Source: Western Pennsylvania State Police Uniform Reports

For the answer, see page 678.

Interpretation of the Standard Deviation

If two data sets have the same mean but one data set has a larger standard deviation, it can be said that the data are more variable in the set with the larger standard deviation. For example, suppose that the test scores for Section 01 of Statistics 101 have a mean of 70 and a standard deviation of 3, and suppose the test scores for Section 02 have a mean of 70 and a standard deviation of 8. It can be concluded that the test scores in Section 02 are more variable since the standard deviation of the scores is larger.

Another important aspect of the standard deviation is called the *empirical* or *normal* rule. The rule states that if the data are bell-shaped or normally distributed (see Section 12-6), approximately 68% of the data values fall within one standard deviation of the mean, approximately 95% of the data values fall within two standard deviations of the mean, and approximately 99.7% of the data values will fall within three standard deviations of the mean. See Figure 12-7.

Exercise Set 12-4

Real World Applications

For Exercises 1–12, find the range, variance, and standard deviation.

1. These data are the number of burglaries reported in 1996 for nine Western Pennsylvania universities. Which measure of variation might be the best in this case? Explain your answer.

   61   1   1   3   2   30   18   3   7

   Source: The Pittsburgh Post Gazette

2. The number of hospitals for the five largest hospital systems is shown here.

   340   75   123   259   151

   Source: USA Today

3. Ten used trail bikes are randomly selected, and the odometer reading of each is recorded as follows.

   1902   103   653   1901   788   361   216   363   223   656

4. Fifteen students were selected and asked how many hours each studied for the final exam in statistics. Their answers are recorded here.

   8   6   3   0   0   5   9   2   1   3   7   10   0   3   6

5. The weights of nine football players are recorded as follows.

   206   215   305   297   265   282   301   255   261

6. Shown here are the numbers of stories in the 11 tallest buildings in St. Paul, Minnesota.

   32   36   46   20   32   18   16   34   26   27   26

   Source: The World Almanac and Book of Facts

7. The heights (in inches) of nine male army recruits are shown here.

   78   72   68   73   75   69   74   73   72

8. The number of calories in 12 randomly selected microwave dinners is shown here.

   560   832   780   650   470   920   1090   970   495   550   605   735

9. The following data are the prices of a gallon of premium gasoline in U.S. dollars in seven foreign countries.

   $3.80 $3.80 $3.20 $3.57 $3.62 $3.74 $3.69

   Source: Pittsburgh Post Gazette

10. The number of attorneys in 10 law firms in Pittsburgh is 87, 109, 57, 221, 175, 123, 170, 80, 66, and 80.

    Source: Pittsburgh Tribune Review

11. The number of dairy cows (in thousands) for 10 selected states is 700, 298, 638, 260, 1380, 280, 270, 1350, 380, and 570.

    Source: U.S. Department of Agriculture

12. The number of wins per season for selected college football teams that played in a postseason bowl is 8, 9, 6, 6, 6, 8, 7, 9, 7, 9, 7, 8, 8, 8, 8, 7, 7, 9, and 11.

    Source: USA Today

Writing Exercises

13. Name three measures of variation.

14. What is the range?

15. Why is the range not usually the best measure of variation?

16. What is the relationship between the variance and standard deviation?

17. Explain the procedure for finding the standard deviation for data.

18. Explain how the variation of two data sets can be compared by using the standard deviations.

Critical Thinking

19. The three data sets have the same mean and range, but is the variation the same? Explain your answer.

   (a) 5   7   9   11   13   15   17
   (b) 5   6   7   11   15   16   17
   (c) 5   5   5   11   17   17   17

20. Using this set—10, 20, 30, 40, and 50,

   (a) Find the standard deviation.

   (b) Add 5 to each value and then find the standard deviation.

   (c) Subtract 5 from each value and then find the standard deviation.

   (d) Multiply each value by 5 and then find the standard deviation.

   (e) Divide each value by 5 and then find the standard deviation.

   (f) Generalize the results of (a)–(e).

12-5   Measures of Position

A *percentile* is used in statistics to measure the position of a data value in a data set.

> A **percentile** or *percentile rank, P,* of a data value indicates the percent of data values that are below the given data value.

Percentiles divide the distribution into 100 parts. The bottom percentile is 0, and the top percentile is 99. Percentiles are used in education to rank students with their peers. For example, if a student scored at the 82nd percentile on a test, this means that the student's test score is higher than 82% of the other students who took the test. (*Note:* It does not mean, however, that the student got 82% of the questions correct.) Scores on standardized tests like the SAT are usually given in percentiles.

Percentiles are also used in health care. For example, in order to determine if a child is developing properly, a physician can compare the child's height and weight at a certain age to that which is normal for the other children of the same age. By using percentiles, the physician can make a judgment as to whether or not the child is developing properly.

Example   12-24

For the 10 test scores given here, find the percentile rank of 77.

   93   82   64   75   98   52   77   88   90   71

Solution

Step 1   Arrange the scores in order.

   52   64   71   75   77   82   88   90   93   98

—Continued

> **Math Note**
>
> In this section, it will be assumed that all data values are distinct. That is, no two values are equal.

> **Math Note**
>
> Percentile measures are most useful for large data sets, such as the SAT test scores of students nationwide. There are several different methods that can be used to calculate percentiles. The method shown here is used for small data sets, and the percentile values found are only approximate values.

Physician's check that a baby's height, weight, and head size are not too different from the percentile at the last visit. A sudden drop to a much lower percentile suggests there may be a problem with the baby's health.

Example 12-24 *Continued—*

Step 2 Find the number of data values below 77. There are 4 values below 77.

Step 3 Divide the number below the score by the total number of data values and change the answer to a percent.

$$\frac{4}{10} = 0.40 = 40\%$$

Hence a test score of 77 is equivalent to the 40th percentile.

Try This One

12-N For the weights of 12 students, find the percentile rank of a student who weighs 97 pounds.

101   120   88   72   75   80   98   91   105   97   78   85

For the answer, see page 678.

Example 12-24 showed how to find the percentile rank of a data value in a data set. Example 12-25 shows how to find a data value in a data set corresponding to a percentile.

Example 12-25

These values are the lengths in minutes for 10 exercise tapes. Find the data value which corresponds to the 30th percentile.

108   120   80   92   56   87   101   32   48   63

Solution

Step 1 To find the score such that 30% of the data values fall below, take 30% of 10. This is equal to 3.

Step 2 Arrange the data in order from small to large and find the data value such that there are 3 values below it. It is the 4th value.

32   48   56 | 63   80   87   92   101   108   120

Hence, a score of 63 corresponds to the 30th percentile.

Percentile ranks can be used to compare data values that come from two different data sets, as shown in Example 12-26.

Example 12-26

Mike's class rank was 27 in a class of 50 students. Harry's class rank was 110 in a class of 200 students. Which student's position is higher?

—Continued

Example   **12-26**   *Continued—*

Solution

Find the percentile rank of each student. Since Mike's rank was 27 in a class of 50 students, there are $50 - 27 = 23$ students below him. Hence the percentile rank is

$$\frac{23}{50} = 0.46 \text{ or } 46\%$$

Since Harry's rank is 110 in a class of 200 students, there are $200 - 110 = 90$ students below Harry. Hence, his percentile rank is

$$\frac{90}{200} = 0.45 \text{ or } 45\%$$

Hence, Mike's position is higher since he has a percentile rank of 46 as compared to Harry's percentile rank of 45.

Try This One

12-O   Kimberly's rank in a class of 80 students was 20. Find her percentile rank.
For the answer, see page 678.

Another statistical measure is called a quartile. A **quartile** divides the distribution into quarters. The first quarter is symbolized as Q_1, the second quartile as Q_2, and the third quartile as Q_3. To find Q_1, Q_2, and Q_3, arrange the data in order from the smallest value to the largest value, and then find the median. This is Q_2. To find Q_1, find the median of the data values less than Q_2; and to find Q_3, find the median of the data values larger than Q_2.

Example   **12-27**

Find Q_1, Q_2, and Q_3 for the number of aircraft stolen during the last 8 years.

   14   11   20   21   42   24   36   35
Source: USA Today

Solution

Step 1   Arrange the data in order.

   11   14   20   21   24   35   36   42

Step 2   Find the median. This is Q_2.

   11   14   20   21   24   35   36   42
                      ↑
               $Q_2 = 22.5$

Step 3   Find the median of the data values less than Q_2. This is Q_1.

   11   14   20   21
         ↑
   $Q_1 = 17$

—Continued

Example **12-27** *Continued—*

Step 4 Find the median of the data values above Q_2. This is Q_3.

24   35   36   42

↑

$Q_3 = 35.5$

Hence, $Q_1 = 17$, $Q_2 = 22.5$, and $Q_3 = 35.5$.

Try This One

12-P Find the values for Q_1, Q_2, and Q_3 for the data shown.

18   32   54   36   27   42   31   15   60   25

For the answer, see page 678.

Another measure of position is called a z score. This measure is presented in Section 12-6.

Exercise Set 12-5

Real World Applications

1. The scores for 20 students on a 50-point mathematics exam are 42, 48, 50, 36, 35, 27, 47, 38, 32, 43, 24, 33, 38, 49, 44, 40, 29, 30, 41, and 37.

   (a) Find the percentile rank for a score of 32.

   (b) Find the percentile rank for a score of 44.

   (c) Find the percentile rank for a score of 36.

   (d) Find the percentile rank for a score of 27.

   (e) Find the percentile rank for a score of 49.

2. The heights (in inches) of 12 students are 73, 68, 64, 63, 71, 70, 65, 67, 72, 66, 60, and 61.

   (a) Find the percentile rank for a height of 67 in.

   (b) Find the percentile rank for a height of 70 in.

   (c) Find the percentile rank for a height of 63 in.

   (d) Find the percentile rank for a height of 68 in.

   (e) Find the percentile rank for a height of 66 in.

3. In a class of 500 students, Carveta's rank was 125. Find her percentile rank.

4. In a class of 400 students, John's rank was 80. Find his percentile rank.

5. In a bicycle race, there were 200 participants. Jessica finished 43rd. Find her percentile rank.

6. In a speed boat race, there were 25 participants. Nate came in 4th place. What is his percentile rank?

7. On an exam, Angela scored in the 20th percentile. If there were 50 students in the class, how many students scored lower than she?

8. On an achievement test for a large school district, Mike scored in the 25th percentile. How many students scored below him if there were 600 students who took the test?

9. Lea's percentile rank on an exam in a class of 600 students is 60. Bill's class rank is 220. Who is ranked higher?

10. In an English class of 30 students, Audrelia's percentile rank is 20. Maranda's class rank is 20. Whose rank is higher?

11. In an evening statistics class, the ages of 20 students are as follows.

18 24 19 20 33
42 43 27 32 37
21 43 27 32 37
34 23 35 28 24

(a) What is the percentile rank of 33?

(b) What is the rank (from the top) of 33?

(c) What age corresponds to the 20th percentile?

12. The test scores of 20 students were as follows.

87 98 62 75 96
77 89 65 79 68
67 91 95 74 84
92 80 72 88 73

(a) What is the percentile rank of 79?

(b) What is the rank (from the top) of 79?

(c) What score corresponds to the 25th percentile?

For Exercises 13–16, find the values for Q_1, Q_2, and Q_3.

13. Number of drive-in theaters in nine selected states:

59 20 21 34 52 48 24 29 55

Source: National Association of Theater Owners

14. Average cost in cents per kilowatt hour of producing electricity using nuclear energy for the past 8 years:

1.83 2.18 2.36 2.04 2.10 2.25 2.48 2.57

Source: Nuclear Energy Institute

15. Number of speeding tickets issued by 10 local municipalities in Southwestern Pennsylvania:

952 407 583 352 883 310 278 539 390 327

Source: Pittsburgh Tribune Review

16. Number of passengers (in millions) at 10 selected airports in the United States:

32.6 28.9 15.5 16.6 19.4 15.9 31.6 15.4 29.1 15.7

Source: Aviation Statistics

Writing Exercises

17. What is a percentile?

18. What is a class rank?

19. Is a percentile rank the same as a class rank?

20. What assumption is made in this section?

21. On an exam, would your score be higher if you scored at the 10th percentile or the 90th percentile?

22. What is a quartile?

Critical Thinking

23. If you were "average," what would your percentile rank be?

24. Explain why in any set of test scores where no score is repeated, half of the class scores below average.

| 12-6 |

The Normal Distribution

In a normal distribution, values cluster around a central value and fall off in either direction. Waiting times can be normally distributed.

Many real-life attributes, such as intelligence test scores, weights of sixth-grade male students, life spans of batteries, SAT scores, etc., have a distribution that is bell-shaped. This type of distribution is called a *normal distribution.*

Suppose a researcher selects a random sample of 100 adult women, measures their heights, and constructs a histogram. The researcher gets a graph similar to the one shown in Figure 12-8(a). Now if the researcher increases the sample size and decreases the width of the classes, the histograms will look like the ones shown in Figures 12-8(b) and 12-8(c). Finally, if it were possible to measure exactly the heights of all adult females in the United States and plot them, the histogram would approach what is called the *normal distribution,* shown in Figure 12-8(d). This distribution is also known as a *bell curve* or a *Gaussian distribution,* named for the German mathematician Carl Friedrich Gauss (1777–1855) who derived its equation.

A normal distribution is defined formally as follows.

> A **normal distribution** is a continuous, symmetric, bell-shaped distribution.

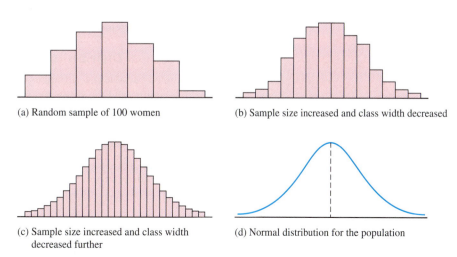

(a) Random sample of 100 women

(b) Sample size increased and class width decreased

(c) Sample size increased and class width decreased further

(d) Normal distribution for the population

Figure 12-8

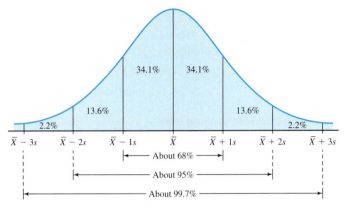

Figure 12-9

The properties of a normal distribution, including those mentioned in the definition, are explained next.

Summary of the Properties of a Normal Distribution

1. It is bell-shaped.
2. The mean, median, and mode are equal and located at the center of the distribution.
3. It is unimodal (i.e., it has only one mode).
4. It is symmetrical about the mean, which is equivalent to saying that its shape is the same on both sides of a vertical line passing through the center.
5. It is continuous—i.e., there are no gaps or holes.
6. It never touches the x axis. Though the curve gets increasingly close to the x axis, the two will never touch.
7. The total area under a normal distribution curve is equal to 1.00, or 100%. This feature may seem unusual, since the curve never touches the x axis, but this fact can be proven mathematically by using calculus. (The proof is beyond the scope of this book.)
8. The area under a normal curve that lies within one standard deviation of the mean is approximately 0.68, or 68%; within two standard deviations, about 0.95 or 95%; and within three standard deviations, about 0.997, or 99.7%. See Figure 12-9, which also shows the area in each region.

The Standard Normal Distribution

Each normally distributed variable (i.e., lifetimes of batteries, SAT scores, etc.) has its own mean and standard deviation, and the shape and location of these curves will vary accordingly. In practical applications, then, one would have to have a table of areas under each curve to answer questions about the variable. To simplify this situation, statisticians use what is called the *standard normal distribution.*

> The **standard normal distribution** is a normal distribution with a mean of 0 and a standard deviation of 1.

The standard normal distribution is shown in Figure 12-10. The values under the curve shown in Figure 12-10 indicate the proportion of area in each section. For example, the area between the mean and one standard deviation above or below the mean is about 0.341, or 34.1%.

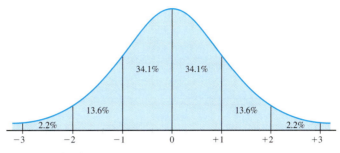

Figure 12-10

All normally distributed variables can be transformed into the standard normally distributed variable by using the formula

$$z = \frac{\text{value} - \text{mean}}{\text{standard deviation}}$$

The values obtained from the formula are called z values, or standard scores. The **z value** is actually the number of standard deviations that a particular X value is from the mean. The use of this formula and its applications will be explained in Section 12-7.

As stated previously, in order to answer real-world questions about normally distributed variables, you must be able to find areas under the standard normal distribution corresponding to various z values. Hence the major emphasis of this section will be to show you how to find these areas. The real-world applications will come later in Section 12-7. The table in Appendix C gives the area (to three decimal places) under the standard normal curve for any z value from 0 to 3.25.

Finding Areas Under the Standard Normal Distribution

When finding areas under the standard normal distribution, three situations can occur. Hence, there are three rules. It is recommended that you draw a picture, label the axis, and shade the desired area for each exercise. *Remember that the area under the normal distribution in each half is 0.500, and the area given in the table found in Appendix C is the area between $z = 0$ and any positive z value to the right of $z = 0$.*

For example, the area under the standard normal distribution between $z = 0$ and $z = 1.71$ is 0.456 or 45.6%. This area will be referred to as the area corresponding to $z = 1.71$.

Math Note

Remember that the mean for the standard normal distribution is zero, and the standard deviation is one.

z	A
1.51	.435
1.52	.436
1.53	.437
1.54	.438
⋮	⋮
1.69	.455
1.70	.455
1.71	.456
1.72	.457
1.73	.458
1.74	.459

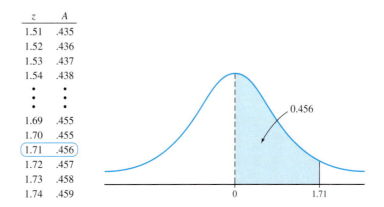

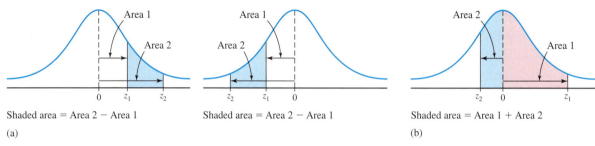

Shaded area = Area 2 − Area 1

(a)

Shaded area = Area 2 − Area 1

Shaded area = Area 1 + Area 2

(b)

Figure 12-11

Situation 1

Finding the area under the standard normal distribution between any two given z values when

(a) The z values are on the same side of the mean. Find the areas in the table in Appendix C corresponding to the z values and subtract the areas. See Figure 12-11(a).

(b) The z values are on opposite sides of the mean. Find the areas in the table in Appendix C corresponding to the z values and add them. See Figure 12-11(b).

Example   12-28

Find the area under the standard normal distribution.

(a) Between $z = 1.56$ and $z = 2.22$.
(b) Between $z = -0.62$ and $z = -1.33$.
(c) Between $z = 1.48$ and $z = -1.75$.

Solution

(a) Draw the picture, show z coordinates, and shade the desired area. See Figure 12-12.

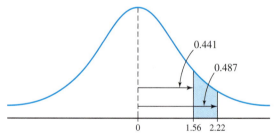

Shaded area = 0.487 − 0.441 = 0.046

Figure 12-12

Using the table, the area corresponding to $z = 2.22$ is 0.487, and the area corresponding to $z = 1.56$ is 0.441. Since the z values are on the same side of the mean, the area is found by subtracting:

$$0.487 - 0.441 = 0.046$$

Hence, the area between $z = 1.56$ and $z = 2.22$ is 0.046 or 4.6%.

—Continued

Example 12-28 *Continued—*

(b) Draw the picture, show z coordinates, and shade the desired area. See Figure 12-13.

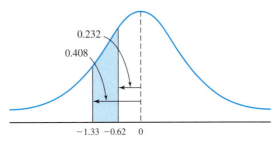

Shaded area $= 0.408 - 0.232 = 0.176$

Figure 12-13

In this case, both z values are negative, but the area under the distribution between these values is the same as if the values were positive. The reason is that the distribution is symmetric about the mean. Hence, find the areas corresponding to the z values and subtract the areas. The area corresponding to $z = -0.62$ is 0.232, and the area corresponding to $z = -1.33$ is 0.408.

$$0.408 - 0.232 = 0.176 \text{ or } 17.6\%$$

Hence, the area under the normal distribution curve between $z = -0.62$ and $z = -1.33$ is 0.176 or 17.6%.

(c) Draw the picture, show the z coordinates, and shade the desired area. See Figure 12-14.

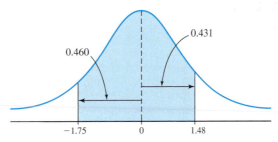

Shaded area $= 0.431 + 0.460 = 0.891$

Figure 12-14

Since the z values are on opposite sides of the mean, the areas corresponding to the z values are added. The area corresponding to $z = -1.75$ is 0.460, and the area corresponding to $z = 1.48$ is 0.431.

$$0.460 + 0.431 = 0.891 \text{ or } 89.1\%$$

Situation 2

Finding the area under the standard normal distribution that is to the right of a z value when

(a) The z value is to the right of the mean. Find the area corresponding to the z value in the table and subtract the area from 0.500. See Figure 12-15(a).

(b) The z value is to the left of the mean. Find the area corresponding to the z value in the table and add 0.500 to the area. See Figure 12-15(b).

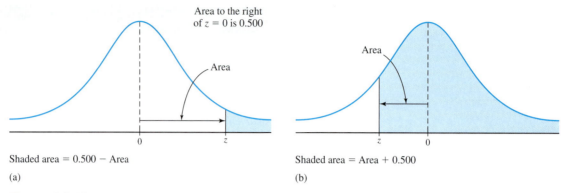

Shaded area $= 0.500 -$ Area

(a)

Shaded area $=$ Area $+ 0.500$

(b)

Figure 12-15

Example 12-29

Find the area under the standard normal distribution

(a) To the right of $z = 1.71$.

(b) To the right of $z = -0.96$.

Solution

(a) Draw the picture, show the z coordinates, and shade the area. See Figure 12-16.

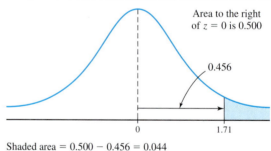

Shaded area $= 0.500 - 0.456 = 0.044$

Figure 12-16

Since the z value is to the right of the mean, find the area corresponding to $z = 1.71$ and subtract it from 0.500. The area corresponding to $z = 1.71$ is 0.456.

$$0.500 - 0.456 = 0.044$$

Hence, the area to the right of $z = 1.71$ is 0.044 or 4.4%.

(b) Draw the picture, show the z coordinates, and shade the area. See Figure 12-17.

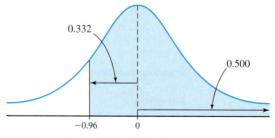

Shaded area $= 0.500 + 0.332 = 0.832$

Figure 12-17

—Continued

Example 12-29 *Continued—*

Since the z value is to the left of the mean, the area corresponding to $z = -0.96$ is found and then added to 0.500.

The area corresponding to $z = -0.96$ is 0.332.

$$0.500 + 0.332 = 0.832$$

Hence, the area to the right of $z = -0.96$ is 0.832 or 83.2%.

Situation 3

Finding the area under the standard normal distribution to the left of a z value when

(a) The z value is to the left of the mean. Find the area corresponding to the z value in the table and subtract the area from 0.500. See Figure 12-18(a).
(b) The z value is to the right of the mean. Find the area corresponding to the z value in the table and add 0.500 to the area. See Figure 12-18(b).

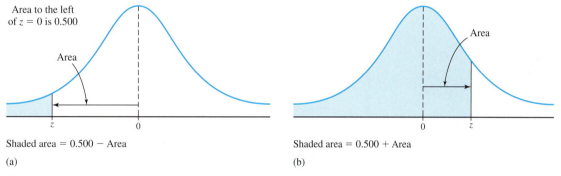

Area to the left of $z = 0$ is 0.500

Area

Shaded area = 0.500 − Area

(a)

Area

Shaded area = 0.500 + Area

(b)

Figure 12-18

Example 12-30

Find the area under the standard normal distribution.

(a) To the left of $z = -2.22$.
(b) To the left of $z = 1.96$.

Solution

(a) Draw the picture, show the z coordinates, and shade the area. See Figure 12-19.

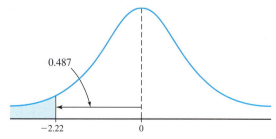

0.487

−2.22   0

Shaded area = 0.500 − 0.487 = 0.013

Figure 12-19

—Continued

Example 12-30 *Continued—*

The area corresponding to $z = -2.22$ is 0.487. Since z is on the left side of the mean, the area to the left of $z = -2.22$ is found by subtracting from 0.500.

$$0.500 - 0.487 = 0.013$$

Hence, the area to the left of $z = -2.22$ is 0.013 or 1.3%.

(b) Draw the picture, show the z coordinates, and shade the area. See Figure 12-20.

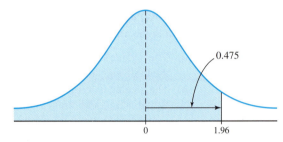

Shaded area = 0.500 + 0.475 = 0.975

Figure 12-20

Since the z value is on the right side of the mean, the area corresponding to $z = 1.96$ is added to 0.500 to get the area to the left side of $z = 1.96$. The area corresponding to $z = 1.96$ is 0.475.

$$0.500 + 0.475 = 0.975$$

Hence, the area to the left of $z = 1.96$ is 0.975 or 97.5%.

Math Note

Area is always positive, whereas the z values can be positive, negative, or zero.

Once you understand how the standard normal distribution table works, it should not be necessary to consult the rules for the different situations. By drawing the picture, showing the coordinates, and shading the area, you should be able to use the table and do the mathematics to find the areas.

Try This One

12-Q Find the area under the standard normal distribution for each.

(a) Between $z = 2.03$ and 2.41

(b) Between $z = 1.68$ and -1.33

(c) To the left of $z = -1.06$

(d) To the right of $z = -0.4$

(e) Between $z = -2.37$ and -1.47

For the answer, see page 678.

It should be noted that no variable fits the normal distribution perfectly, since the normal distribution is a theoretical distribution. However, the normal distribution can be used to describe many variables because the deviations from the normal distribution are very small. This concept will be explained in Section 12-7.

Exercise Set 12-6

Computational Exercises

For Exercises 1–18, find the area under the normal distribution curve,

1. Between $z = 0$ and $z = 1.97$.
2. Between $z = 0$ and $z = 0.56$.
3. Between $z = 0$ and $z = -0.48$.
4. Between $z = 0$ and $z = -2.07$.
5. To the right of $z = 1.02$.
6. To the right of $z = 0.23$.
7. To the left of $z = -0.42$.
8. To the left of $z = -1.43$.
9. Between $z = 1.23$ and $z = 1.90$.
10. Between $z = 0.79$ and $z = 1.28$.
11. Between $z = -0.87$ and $z = -0.21$.
12. Between $z = -1.56$ and $z = -1.83$.
13. Between $z = 0.24$ and $z = -1.12$.
14. Between $z = 2.47$ and $z = -1.03$.
15. To the left of $z = 1.22$.
16. To the left of $z = 2.16$.
17. To the right of $z = -1.92$.
18. To the right of $z = -0.18$.

Writing Exercises

19. What are the characteristics of the normal distribution?
20. Why is the normal distribution important in statistical analysis?
21. What is the total area under any normal distribution curve?
22. What percentage of the area falls below the mean? Above the mean?

Critical Thinking

23. Find a z value to the right of the mean so that 53.98% of the distribution lies to the left of it.
24. Find a z value to the left of the mean so that 98.6% of the area lies to the right of it.
25. Find two z values, one positive and one negative but having the same absolute value, so that the areas in the two tails (ends) total these values.
    (a) 5%
    (b) 10%
    (c) 1%

12-7 Applications of the Normal Distribution

The standard normal distribution curve can be used to solve a wide variety of practical problems. The only requirement is that the variable be normally or approximately normally distributed. There are several mathematical tests to determine whether a variable is normally distributed. However, those tests are not included here; for all the problems presented in this chapter, one can assume that the variable is normally or approximately normally distributed.

To solve problems by using the standard normal distribution, transform the original variable into a standard normal distribution variable by using the formula

$$z = \frac{\text{value} - \text{mean}}{\text{standard deviation}}$$

This is the same formula presented in Section 12-6. This formula transforms the values of the variable into standard scores or z values. Once the variable is transformed, the table in Appendix C can be used to solve problems.

For example, suppose that the scores for a standardized test are normally distributed, have a mean of 100, and have a standard deviation of 15. When the scores are transformed into z values, the two distributions coincide, as shown in Figure 12-21. (Recall that the z distribution has a mean of 0 and a standard deviation of 1.)

To solve the application problems in this section, transform the values of the variable into z values and then use the table, as shown in the next examples.

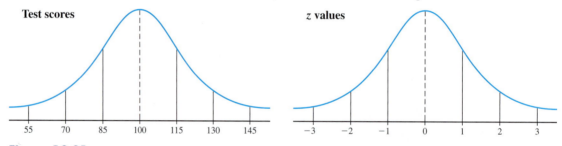

Figure 12-21

Example 12-31

If the scores for a test are normally distributed and have a mean of 100 and a standard deviation of 15, find the percentage of scores that will fall below 112.

Solution

Step 1   Draw the figure and represent the area, as shown in Figure 12-22.

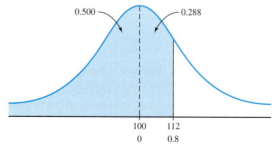

Figure 12-22

—Continued

Example 12-31 *Continued—*

Step 2 Find the z value corresponding to a score of 112. The mean is 100 and the standard deviation is 15.

$$z = \frac{\text{value} - \text{mean}}{\text{standard deviation}} = \frac{112 - 100}{15} = \frac{12}{15} = 0.8$$

Hence, 112 is 0.8 standard deviation above the mean of 100, as shown for the z distribution in Figure 12-22.

Step 3 Find the area using the table in Appendix C. The area between $z = 0$ and $z = 0.8$ is 0.288. Since the area under the curve to the left of $z = 0.8$ is desired, add 0.500 to 0.288 ($0.500 + 0.288 = 0.788$). Therefore, 78.8% of the scores fall below 112.

The area under the standard normal distribution curve can also be used to determine probabilities. For example, since the area under the standard normal distribution curve between $z = 0$ and $z = 1$ is 0.314, we can say that the *probability* of randomly selecting a z between $z = 0$ and $z = 1$ is 0.314 or 31.4%. An application of this concept is shown in Example 12-32.

Example 12-32

Each month, an American household generates an average of 28 pounds of newspaper for garbage or recycling. Assume the standard deviation is two pounds. If a household is selected at random, find the probability of its generating

(a) Between 27 and 31 pounds per month.
(b) More than 30.2 pounds per month.

Assume the variable is approximately normally distributed.

Source: Michael D. Shook and Robert L. Shook, *The Book of Odds* (New York: Plume, 1991)

Solution

(a) **Step 1** Draw the figure and represent the area. See Figure 12-23.

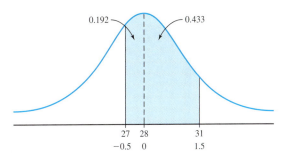

Figure 12-23

—Continued

Example **12-32** *Continued—*

Step 2 Find the two z values when the mean is 28 and the standard deviation is 2.

$$z_1 = \frac{\text{value} - \text{mean}}{\text{standard deviation}} = \frac{27 - 28}{2} = -\frac{1}{2} = -0.5$$

$$z_2 = \frac{\text{value} - \text{mean}}{\text{standard deviation}} = \frac{31 - 28}{2} = \frac{3}{2} = 1.5$$

Step 3 Find the appropriate area, using the table in Appendix C. The area between $z = 0$ and $z = -0.5$ is 0.192. The area between $z = 0$ and $z = 1.5$ is 0.433. Add 0.192 and 0.433 ($0.192 + 0.433 = 0.625$). Thus, the total area is 62.5%. See Figure 12-23.

Hence, the probability that a randomly selected household generates between 27 and 31 pounds of newspapers per month is 62.5%.

Solution

(b) **Step 1** Draw the figure and represent the area, as shown in Figure 12-24.

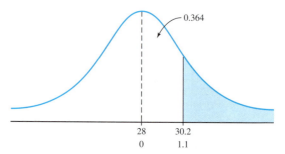

0.364

| 28 | 30.2 |
| 0 | 1.1 |

Figure 12-24

Step 2 Find the z value for 30.2. The mean is 28 and the standard deviation is two.

$$z = \frac{\text{value} - \text{mean}}{\text{standard deviation}} = \frac{30.2 - 28}{2} = \frac{2.2}{2} = 1.1$$

Step 3 Find the appropriate area. The area between $z = 0$ and $z = 1.1$ obtained from the table is 0.364. Since the desired area is in the right tail, subtract 0.364 from 0.500.

$$0.500 - 0.364 = 0.136$$

Hence, the probability that a randomly selected household will accumulate more than 30.2 pounds of newspapers is 0.136, or 13.6%.

Try This One

12-R Assume that for a certain group of people, the mean for systolic blood pressure is 120 and the standard deviation is 8 and the variable is normally distributed. Find each.

(a) The percent of individuals who have a blood pressure between 120 and 130

(b) The percent of individuals who have a blood pressure above 126

(c) The percent of individuals who have a blood pressure between 114 and 118

(d) The percent of individuals who have a blood pressure between 124 and 114

(e) The percent of individuals who have a blood pressure lower than 110

For the answer, see page 678.

The normal distribution can also be used to answer questions of "How many?" An example of this is shown next.

Example 12-33

The American Automobile Association reports that the average time it takes to respond to an emergency call is 25 minutes. Assume the variable is approximately normally distributed and the standard deviation is 4.5 minutes. If 80 calls are randomly selected, approximately how many will be responded to in less than 15 minutes?

Source: Michael D. Shook and Robert L. Shook, *The Book of Odds* (New York: Penguin Putnam, Inc., 1991)

Solution

To solve the problem, find the area under the normal distribution curve to the left of 15.

Step 1 Draw a figure and represent the area as shown in Figure 12-25.

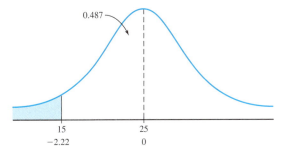

Figure 12-25

Step 2 Find the z value for 15. The mean is 25 and the standard deviation is 4.5.

$$z = \frac{\text{value} - \text{mean}}{\text{standard deviation}} = \frac{15 - 25}{4.5} = -2.22$$

—Continued

Math Note

For problems that use percents, always change the percents to decimals before multiplying. When necessary, round the answer to the nearest whole number.

Example   **12-33**   *Continued—*

Step 3   Find the appropriate area. The area obtained from the table in Appendix C is 0.487, which corresponds to the area between $z = 0$ and $z = -2.22$. (Use $+2.22$.)

Step 4   Subtract 0.487 from 0.500 to get 0.013.

Step 5   To find how many calls will be made in less than 15 minutes, multiply the sample size (80) by the area (0.013) to get 1.04. Hence 1.04, or approximately one, call will be responded to in under 15 minutes.

Try This One

12-S   The mean for a reading test given nationwide is 80, and the standard deviation is 8. The variable is normally distributed. If 10,000 students take the test, find each.

(a)   The number of students who will score above 90

(b)   The number of students who will score between 78 and 88

(c)   The number of students who will score above 74

(d)   The number of students who will score between 70 and 74

(e)   The number of students who will score below 76

For the answer, see page 679.

Exercise Set   12-7

Real World Applications

1. The average hourly wage of production workers in manufacturing is $11.76. Assume the variable is normally distributed. If the standard deviation of earning is $2.72, find these probabilities for a randomly selected production worker.

    (a)   The production worker earns more than $12.55.

    (b)   The production worker earns less than $8.00.

    Source: Statistical Abstract of the United States

2. The Speedmaster IV automobile gets an average of 22.0 miles per gallon in the city. The standard deviation is 3 miles per gallon. Assume the variable is normally distributed. Find the probability that on any given day, the car will get more than 26 miles per gallon when driven in the city.

3. If the mean salary of high school teachers in the United States is $29,835 and the standard deviation is $3000, find these probabilities for a randomly selected teacher. Assume the variable is normally distributed.

   (a) The teacher earns more than $35,000.

   (b) The teacher earns less than $25,000.

4. For a specific year, Americans spent an average of $71.12 for books. Assume the variable is normally distributed. If the standard deviation of the amount spent on books is $8.42, find these probabilities for a randomly selected American.

   (a) He or she spent more than $60 per year on books.

   (b) He or she spent less than $80 per year on books.

   Source: Statistical Abstract of the United States 1994

5. A survey found that people keep their television sets an average of 4.8 years. The standard deviation is 0.89 year. If a person decides to buy a new TV set, find the probability that he or she has owned the old set for the given amount of time. Assume the variable is normally distributed.

   (a) Less than 2.5 years

   (b) Between 3 and 4 years

   (c) More than 4.2 years

6. The average age of CEOs is 56 years. Assume the variable is normally distributed. If the standard deviation is 4 years, find the probability that the age of a randomly selected CEO will be in the given range.

   (a) Between 53 and 59 years old

   (b) Between 58 and 63 years old

   (c) Between 50 and 55 years old

7. The average life of a brand of automobile tires is 30,000 miles, with a standard deviation of 2000 miles. If a tire is selected and tested, find the probability that it will have the given lifetime. Assume the variable is normally distributed.

   (a) Between 25,000 and 28,000 miles

   (b) Between 27,000 and 32,000 miles

   (c) Between 31,500 and 33,500 miles

8. The average time a person spends at the West Newton Zoo is 62 minutes. The standard deviation is 12 minutes. If a visitor is selected at random, find the probability that he or she will spend the time shown at the zoo. Assume the variable is normally distributed.

   (a) At least 180 minutes

   (b) At most 50 minutes

9. The average amount of snow per season in Trafford is 44 inches. The standard deviation is 6 inches. Find the probability that next year Trafford will receive the given amount of snowfall. Assume the variable is normally distributed.

   (a) At most 50 inches of snow

   (b) At least 53 inches of snow

10. The average waiting time for a drive-in window at a local bank is 9.2 minutes, with a standard deviation of 2.6 minutes. When a customer arrives at the bank, find the

probability that the customer will have to wait the given time. Assume the variable is normally distributed.

(a) Between 5 and 10 minutes

(b) Less than 6 minutes or more than 9 minutes

11. The average time it takes college freshmen to complete the Mason Basic Reasoning Test is 24.6 minutes. The standard deviation is 5.8 minutes. Find these probabilities. Assume the variable is normally distributed.

(a) It will take a student between 15 and 30 minutes to complete the test.

(b) It will take a student less than 18 minutes or more than 28 minutes to complete the test.

12. A brisk walk at 4 miles per hour burns an average of 300 calories per hour. If the standard deviation of the distribution is 8 calories, find the probability that a person who walks 1 hour at the rate of 4 miles per hour will burn the given calories. Assume the variable is normally distributed.

(a) More than 280 calories

(b) Less than 293 calories

(c) Between 285 and 320 calories

13. During September, the average temperature of Laurel Lake is 64.2° and the standard deviation is 3.2°. Assume the variable is normally distributed. For a randomly selected day, find the probability that the temperature will be

(a) Above 62°.

(b) Below 67°.

(c) Between 65° and 68°.

14. If the systolic blood pressure for a certain group of people has a mean of 132 and a standard deviation of 8, find the probability that a randomly selected person in the group will have the given blood pressure. Assume the variable is normally distributed.

(a) Above 130

(b) Below 140

(c) Between 131 and 136

15. A national test has a mean of 100 and a standard deviation of 15. The test scores are normally distributed. If 2000 people take the test, find the number of people who will score

(a) Below 93.

(b) Above 120.

(c) Between 80 and 105.

(d) Between 75 and 82.

16. The average size (in square feet) of homes built in the United States is 1810. Assume the variable is normally distributed and the standard deviation is 92 square feet. In a sample of 500 recently built homes, find the number of homes that will

(a) Have between 1900 and 2000 square feet.

(b) Have more than 1780 square feet.

(c) Have less than 1850 square feet.

(d) Have more than 1720 square feet.

Source: Congressional Research Service

17. For certain types of homes, the average price is $145,500. Assume the variable is normally distributed and the standard deviation is $1500. If 800 homes are built, find the number of homes that will cost

    (a) More than $150,000.

    (b) Between $141,000 and $151,000.

    (c) Less than $147,500.

    (d) More than $139,000.

18. The average price of tropical fish sold is $9.52, and the standard deviation is $1.02. Assume the variable is normally distributed. If a national pet shop sells 1000 tropical fish during August, find the number of fish that were sold

    (a) For less than $8.00.

    (b) For more than $10.00.

    (c) Between $9.50 and $10.50.

    (d) Between $9.80 and $10.05.

Writing Exercises

19. Explain why the normal distribution can be used to solve many real-life problems.

20. Suppose that the mathematics SAT scores for high school seniors for a specific year have a mean of 456 and a standard deviation of 100 and are approximately normally distributed. If a subgroup of these high school seniors, those who are in the National Honor Society, is selected, would you expect the distribution of scores to have the same mean and standard deviation? Explain your answer.

21. Given a data set, how could you decide if the distribution of the data was approximately normal?

Critical Thinking

22. If a distribution of raw scores were plotted and then the scores were transformed into z scores, would the shape of the distribution change?

23. An instructor gives a 100-point examination in which the grades are normally distributed. The mean is 60 and the standard deviation is 10. If there are 5% A's and 5% F's, 15% B's and 15% D's, and 60% C's, find the scores that divide the distribution into those categories.

24. A researcher who is in charge of an educational study wishes subjects to perform some special skill. Fearing that people who are unusually talented or unusually untalented could distort the results, he decides to use people who scored in the middle 50% on a certain test. If the mean for the population is 100 and the standard deviation is 15, find the two limits (upper and lower) for the scores that would enable a volunteer to participate in the study. Assume the variable is normally distributed.

25. An athletic association wants to sponsor a footrace. The average time it takes to run the course is 58.6 minutes, with a standard deviation of 4.3 minutes. If the association decides to include only the top 20% of the racers, what should the cutoff time be in the tryout run? Assume the variable is normally distributed.

12-8 Correlation and Regression Analysis

Statisticians are also interested in determining if a relationship exists between two variables. For example, is there a relationship between the horsepower of an automobile engine and the number of miles per gallon the automobile gets?

In order to determine whether or not a relationship exists between two variables, the statistician must collect data for both variables in such a way that each data value for one variable can be paired with one data value for the other variable. In the previously mentioned example, the statistician would select a sample of automobiles and determine the horsepower of each engine and calculate the number of miles per gallon each obtained. A graph can then be plotted designating one variable, such as horsepower, as the *x variable* or **independent variable;** and the other variable, miles per gallon, as the *y variable* or **dependent variable.** This graph is called a *scatter plot.*

> A **scatter plot** is a graph of the ordered pairs (x, y) of the data values for two variables.

After the scatter plot is drawn, the statistician would analyze the graph to see if there is a pattern. If there is a noticeable pattern, such as the points falling in an approximately straight line, then a possible relationship between the two variables may exist. The next step, then, would be to determine the strength of the relationship. In order to do this, the statistician would calculate what is called the *correlation coefficient.*

> The **correlation coefficient** is a value that is computed from the paired data values in order to determine the strength of a relationship between two variables. The symbol for the correlation coefficient computed from data obtained from a sample is r.

If a strong relationship between the two variables exists, the researcher would then continue the statistical analysis by using what is called *regression analysis.* This will be explained later. It is best, however, to start with an explanation of the scatter plot.

Scatter Plots

Example 12-34 explains how to construct a scatter plot.

Example 12-34

A researcher selects a sample of small hospitals in a specific state and wishes to see if there is a relationship between the number of beds and the number of personnel employed by the hospital. Construct a scatter plot for the data shown.

No. of beds (x)	28	56	34	42	45	78	84	36	74	95
Personnel (y)	72	195	74	211	145	139	184	131	233	366

—Continued

Example 12-34 *Continued—*

Solution

Step 1 Draw and label the *x* and *y* axes, as shown.

Step 2 Plot the data pairs. See Figure 12-26.

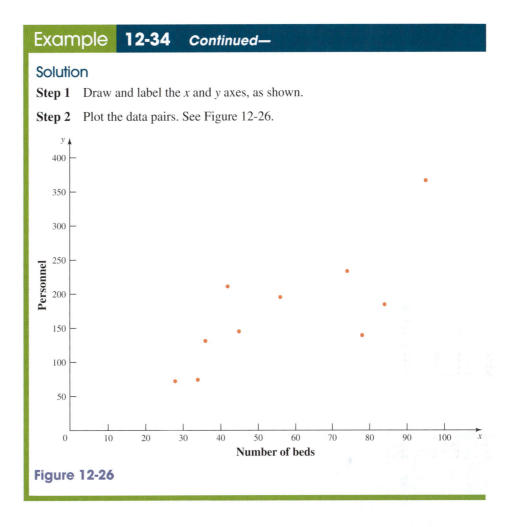

Figure 12-26

Analyzing the Scatter Plot

There are several types of relationships that can exist between the *x* values and the *y* values. These relationships can be identified by looking at the pattern of the points on the graphs. The types of patterns and corresponding relationships are given next.

1. *A positive linear relationship* exists when the points fall approximately in an ascending straight line from left to right, and both the *x* and *y* values increase at the same time. See Figure 12-27(a).
2. *A negative linear relationship* exists when the points fall approximately in a descending straight line from left to right. See Figure 12-27(b). The relationship then is as the *x* values are increasing, the *y* values are decreasing.
3. *A nonlinear relationship* exists when the points fall in a curved line. See Figure 12-27(c). The relationship is described by the nature of the curve.
4. *No relationship* exists when there is no discernable pattern to the points. See Figure 12-27(d).

The relationship between the variables in Example 12-34 might be a positive linear relationship. In other words, as the size of the hospital based on the number of beds increases, the number of personnel is also increasing.

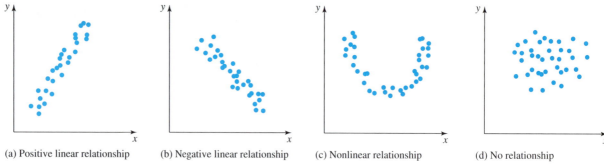

(a) Positive linear relationship   (b) Negative linear relationship   (c) Nonlinear relationship   (d) No relationship

Figure 12-27

Try This One

12-T   The data represent the heights in feet and the number of stories of the tallest buildings in Pittsburgh. Draw a scatter plot for the data and describe the relationship.

Height (x)	485	511	520	535	582	615	616	635	728	841
No. of stories (y)	40	37	41	42	38	45	31	40	54	64

Source: The World Almanac and Book of Facts

For the answer, see page 679.

The Correlation Coefficient

Since the conclusions based on the observation of a scatter plot are somewhat subjective in nature, the next step would be to find the value of the correlation coefficient.

The **correlation coefficient** is calculated in such a way that when there is a strong positive linear relationship, its value will be close to positive one. When there exists a strong negative linear relationship, the value of the correlation coefficient will be close to negative one. When the value of the correlation coefficient is near zero, the relationship between the x variable and the y variable is weak or nonexistent. Hence, the value of r will always range from -1 to $+1$; i.e., $-1 \le r \le +1$. See Figure 12-28. Furthermore, the value of r does not depend on the units used, and if x and y are interchanged, the value of r remains unchanged although the position of the data values on the graph will change.

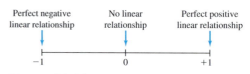

Perfect negative linear relationship   No linear relationship   Perfect positive linear relationship

-1 0 $+1$

Figure 12-28

Calculating the Value of the Correlation Coefficient

In order to find the value of the correlation coefficient, you can use the next formula.

Math Note

The sign of r will indicate the nature of the relationship if one exists. That is, if r is positive, the linear relationship will be positive. If r is negative, the linear relationship will be negative. For example, if the scatter plot was similar to the one shown in Figure 12-27(a), the value of r would be positive. If the scatter plot was similar to the one shown in Figure 12-27(b), the value of r would be negative. If the scatter plot was similar to the one shown in Figure 12-27(c) or 12-27(d), the value of r would be close to zero.

Formula for Finding the Value of r:

$$r = \frac{n(\sum xy) - (\sum x)(\sum y)}{\sqrt{\left[n(\sum x^2) - (\sum x)^2\right]\left[n(\sum y^2) - (\sum y)^2\right]}}$$

where

$n =$ the number of data pairs

$\sum x =$ the sum of the x values

$\sum y =$ the sum of the y values

$\sum xy =$ the sum of the products of the x and y values for each pair

$\sum x^2 =$ the sum of the squares of the x values

$\sum y^2 =$ the sum of the squares of the y values

When calculating the value of r, it is helpful to make a table, as shown in Example 12-35.

Example 12-35

Find the value for r using the data given in the previous example.

Solution

Step 1 Make a table as shown with the following headings: x, y, xy, x^2, and y^2.

Step 2 Find the values for the product xy, the values for x^2, the values for y^2, and place them in the columns as shown. Then find the sums of the columns.

x	y	xy	x^2	y^2
28	72	2,016	784	5,184
56	195	10,920	3,136	38,025
34	74	2,516	1,156	5,476
42	211	8,862	1,764	44,521
45	145	6,525	2,025	21,025
78	139	10,842	6,084	19,321
84	184	15,456	7,056	33,856
36	131	4,716	1,296	17,161
74	233	17,242	5,476	54,289
95	366	34,770	9,025	133,956
$\sum x = 572$	$\sum y = 1,750$	$\sum xy = 113,865$	$\sum x^2 = 37,802$	$\sum y^2 = 372,814$

—Continued

Example   12-35   *Continued—*

Step 3 Substitute in the formula and evaluate:

$$r = \frac{n(\sum xy) - (\sum x)(\sum y)}{\sqrt{\left[n(\sum x^2) - (\sum x)^2\right]\left[n(\sum y^2) - (\sum y)^2\right]}}$$

$$= \frac{10(113,865) - (572)(1750)}{\sqrt{[10(37,802) - (572)^2][10(372,814) - (1750)^2]}}$$

$$= \frac{137,650}{\sqrt{(50,836)(665,640)}}$$

$$\approx 0.748$$

Hence, the value of the correlation coefficient is 0.748.

For convenience, computer programs and graphic calculators are usually used to find r.

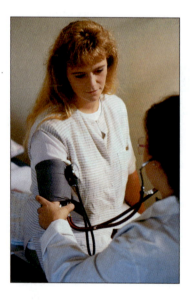

Try This One

12-U   Find the value for the correlation coefficient for these data, which represent the age and systolic blood pressure of six randomly selected subjects.

Age, x	43	48	56	61	67	70
Pressure, y	128	120	135	143	141	152

For the answer, see page 679.

The value of r in Example 12-35 is 0.748 and since this value was calculated from data obtained from a sample, statisticians need to know whether or not a relationship exists between the variables in the population. (The population in this case was all hospitals in the specific state that have a number of beds less than 100.) When information obtained from samples is used to make an inference about the population, one always has a chance of being wrong since only a part of the population (i.e., a sample) is used. Hence, we can never be 100% sure that our inference is correct. Statisticians have generally agreed that when we assume there is a relationship between variables in the population based on information obtained from a sample, we can be satisfied with having either a 5% chance or a 1% chance of being wrong. These are called **significance levels.**

In order to make an inference about correlations, the table in Appendix D can be used. The values in this table were computed mathematically, and the computations are beyond the scope of this book. What is important here is to know how to use the table.

Notice that the two significance levels, the 5% significance level and the 1% significance level, are shown. Before making an inference, the statistician needs to decide which level to use. Deciding which level to use depends on the seriousness of making an incorrect inference (that is, on inferring there is a relationship between the variables in the population when there is none). Generally, the 5% significance level is used more often than the 1% level. More will be said about this later.

The rule of thumb here is that if |r| is greater than or equal to the value given in the table in Appendix D for a specific significance level (either 5% or 1%), then we can infer that the variables in the population are related. Remember that when making such an inference, there is either a 5% or a 1% chance of being wrong, which means that even though r is significant, it may have occurred by chance.

Example 12-36

Determine if the correlation coefficient $r = 0.748$ found in the previous example is significant at the 5% level.

Solution

Since the sample size is $n = 10$ and the 5% significance level is used, the value of |r| must be greater than or equal to the number found in the table in Appendix D to be significant. In this case, $|r| = 0.748$, which is greater than or equal to 0.632 obtained from the table. Hence, it can be concluded that there is a significant relationship between the variables.

Researchers could use correlation and regression analysis to find out if there is a correlation between the type of a car a family owns and the number of children in the family.

In Example 12-36, it was concluded that r is significant at the 5% level. Remember that when making this conclusion, there is a 5% chance of being wrong, which means that even though r was greater than or equal to 0.632, there is really no relationship between the variables in the population, and the value $r = 0.748$ was due to the fact that the sample did not accurately represent the population.

Also, we cannot conclude that there is a relationship between the variables at the 1% significance level since r would need to be greater than or equal to 0.765. In this example, $r = 0.748$, which is less than 0.765.

Try This One

12-V   Test the significance of the correlation coefficient obtained from the previous "Try This One." Use 5%, and then 1%.
For the answer, see page 679.

Multi-symptom cold and cough relief without drowsiness

INDICATIONS: For the temporary relief of nasal congestion, minor aches, pains, headache, muscular aches, sore throat, and fever associated with the common cold. Temporarily relieves cough occurring with a cold. Helps loosen phlegm (mucus) and thin bronchial secretions to drain bronchial tubes and make coughs more productive.

DIRECTIONS: Adults and children 12 years of age and over, 2 liquid caps every 4 hours, while symptoms persist, not to exceed 8 liquid caps in 24 hours, or as directed by a doctor. Not recommended for children under 12 years of age.

WARNINGS: Do not exceed recommended dosage. If nervousness, dizziness, or sleeplessness occur, discontinue use and consult a doctor. Do not take this product for more than 10 days. A persistent cough may be a sign of a serious condition. If symptoms do not improve or if cough persists for more than 7 days, tends to recur, or is accompanied by rash, persistent headache, fever that lasts for more than 3 days, or if new symptoms occur, consult a doctor. Do not take this product for persistent or chronic cough such as occurs with smoking, asthma, chronic bronchitis, or emphysema, or where cough is accompanied by excessive phlegm (mucus) unless directed by a doctor. If sore throat is severe, persists for more than 2 days, is accompanied or followed by fever, headache, rash, nausea, or vomiting, consult a doctor promptly. Do not take this product if you have heart disease, high blood pressure, thyroid disease, diabetes, or difficulty in urination due to enlargement of the prostate gland unless directed by a doctor. As with any drug, if you are pregnant or nursing

Medication labels help users to become aware of the effects of taking medications.

Recall that the selection of the 5% or 1% significance level depends on the seriousness of making an inference that the variables are related in the population when they really are not. So how is one to determine which significance level to use? Here is an example. Suppose medical researchers think that a certain medication helps patients with a specific illness to improve or recover better or faster than the medication that is currently being used to treat the ailment, but suppose that a few patients experienced serious side effects such as heart attacks or strokes. Then one would use the 1% significance level in evaluating the relationships between the medication and the recovery of the patient since there may be serious consequences of giving the medication to some patients if it is not really effective. On the other hand, suppose the side effects were somewhat milder, such as headache, nausea, or dizziness. Then one could use the 5% significance level in determining whether or not there is a relationship between the medication and recovery, since the consequences of being wrong are not so serious as in the first situation. Remember that medications do not work in the same way for all individuals.

Regression

Once we have concluded that there is a significant relationship between the two variables, the next step is to find the *equation* of the **regression line** through the data points.

The regression line is a line that best fits the data. Broadly speaking, we say it is the line that passes through the points in such a way that the overall distance each point is from the line is at a minimum. Thus the regression line is also called the *line of best fit*.

Recall from algebra that the equation of a line in the slope-intercept form is $y = mx + b$, where m is the slope and b is the y intercept. (See Section 8-1.) In statistics, the equation of the regression line is written as $y = a + bx$, where a is the y intercept and b is the slope. This is the equation that will be used here. In order to find the values for a and b, two formulas are used.

Formulas for Finding the Values of a and b for the Equation of the Regression Line

$$b = \frac{n(\sum xy) - (\sum x)(\sum y)}{n(\sum x^2) - (\sum x)^2} \qquad \text{slope}$$

$$a = \frac{\sum y - b(\sum x)}{n} \qquad y \text{ intercept}$$

Example 12-37

Find the equation for the regression line for the data in Example 12-34.

Solution

Using the values at the bottom of the table found in Example 12-35, substitute in the first formula to find the value for the slope, b.

$$b = \frac{n(\sum xy) - (\sum x)(\sum y)}{n(\sum x^2) - (\sum x)^2}$$

$$= \frac{10(113,865) - (572)(1750)}{10(37,802) - (572)^2}$$

$$= \frac{137,650}{50,836}$$

$$= 2.71 \text{ (rounded)}$$

Substitute in the second formula to find the value for the y intercept, a, when $b = 2.71$.

$$a = \frac{\sum y - b(\sum x)}{n}$$

$$= \frac{(1750) - 2.71(572)}{10}$$

$$= \frac{199.88}{10}$$

$$= 19.988$$

The equation of the regression line is $y = 19.988 + 2.71x$.

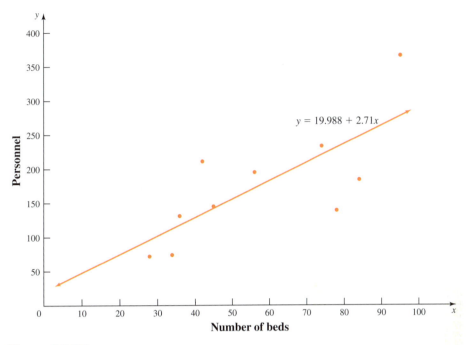

Figure 12-29

Computer programs or graphing calculators can be used to find the equation of the regression line.

After the equation of the regression line is found, it can be drawn on the scatter plot by one of the methods shown in Chapter 7. For example, you can find the coordinates of two points on the line by selecting two values for x, then substituting each in the regression line equation, and find the corresponding values for y.

In the first case, let $x = 30$; then

$$y = 19.988 + 2.71x$$
$$= 19.988 + 2.71(30)$$
$$= 101.288$$

In the second case, let $x = 70$; then

$$y = 19.988 + 2.71x$$
$$= 19.988 + 2.71(70)$$
$$= 209.688$$

Round to 101 and 210. Hence, the coordinates of the first point are (30, 101), and the coordinates of the second point are (70, 210). Next, plot the two points and draw a line through each point. See Figure 12-29.

The Relationship between r and the Regression Line

Two things should be noted concerning the relationship between the value for r and the regression line. First, the value of r and the value of the slope, b, will have the same sign. That is, when r is positive, b will be positive, and when r is negative, b will be negative. Second, the closer the value of r is to $+1$ or -1, the better the points will fit the line. In other words, the stronger the relationship, the better the fit. When $r = +1$ or -1, all the data values will fall exactly on the line. Perfect relationships (i.e., $r = +1$ or $r = -1$) occur in the theoretical studies. For example, the value of r will be one in the corresponding values for the Celsius and Fahrenheit temperatures. Figure 12-30 shows the relationship between the correlation coefficient and the regression line.

Once the equation of the regression line is found, one can predict in general a value for y given a value x as shown in the next example. Note that if r is not significant, a regression line/equation is meaningless, and should not be used.

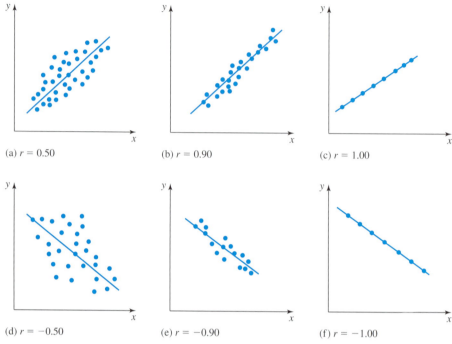

(a) $r = 0.50$ (b) $r = 0.90$ (c) $r = 1.00$

(d) $r = -0.50$ (e) $r = -0.90$ (f) $r = -1.00$

Figure 12-30

Math Note

It is not necessary to round all values for x and y to whole numbers. Rounding is based on the nature of the data. Since the number of beds and the number of personnel are counted using whole numbers, the value of y for a specific x is rounded to the nearest whole number.

Example 12-38

Using the equation for the regression line found in Example 12-37 for the data in Exercise 12-34, predict the approximate number of personnel for a hospital with 65 beds.

Solution

Substitute 65 for x in the equation $y = 19.988 + 2.71x$ and find the value for y.

$$y = 19.988 + 2.71(65)$$
$$= 196.138 \text{ (which can be rounded to 196)}$$

Hence, a hospital with 65 beds will have approximately 196 personnel.

Try This One

12-W   A medical researcher wishes to determine how the dosage (in milligrams) of a drug affects the heart rate. The data for seven patients are shown. Draw a scatter plot and find the equation for the regression line. Draw the line on the graph and predict the heart rate for a patient who takes 0.28 mg of the drug.

Dosage, x	0.125	0.20	0.25	0.30	0.35	0.40	0.50
Heart rate, y	95	90	93	92	88	80	82

For the answer, see page 679.

Correlation and Causation

Researchers must understand the nature of the relationship between the independent variable x and the dependent variable y. When a significant relationship exists between the variables, researchers must consider the possibilities outlined next.

Possible Relationships between Variables

When a significant relationship exists for the variable at a specific significance level, any of these five possibilities can exist.

1. *There is a direct cause-and-effect relationship between the variables.* That is, x causes y. For example, water causes plants to grow, poison causes death, and heat causes ice to melt.
2. *There is a reverse cause-and-effect relationship between the variables.* That is, y causes x. For example, suppose a researcher believes excessive coffee consumption causes nervousness, but the researcher fails to consider that the reverse situation may occur. That is, it may be that an extremely nervous person craves coffee to calm his or her nerves.
3. *The relationship between the variables may be caused by a third variable.* For example, if a statistician correlated the number of deaths due to drowning and the number of cans of soft drinks consumed during the summer, he or she would probably find a significant relationship. However, the soft drink is not necessarily responsible for the deaths, since both variables may be related to heat and humidity.

4. *There may be a complexity of interrelationships among many variables.* For example, a researcher may find a significant relationship between students' high school grades and college grades. But there probably are many other variables involved, such as IQ, hours of study, influence of parents, motivation, age, and instructors.
5. *The relationship may be coincidental.* For example, a researcher may be able to find a significant relationship between the increase in the number of people who are exercising and the increase in the number of people who are committing crimes. But common sense dictates that any relationship between these two values must be due to coincidence.

Exercise Set   12-8

Computational Exercises

For Exercises 1–6,

(a) Draw a scatter plot.
(b) Find the value for r.
(c) Test the significance of r at the 5% level and at the 1% level.
(d) Find the equation of the regression line and draw the line on the scatter plot, but only if r is significant.
(e) Describe in words the nature of the relationship if one exists.

1.

x	1	4	6	2	3	5	7
y	8	15	20	10	11	16	25

2.

x	21	25	24	30	36	40
y	12	8	9	5	3	2

3.

x	75	80	85	90
y	10	5	11	4

4.

x	9	12	15	11	10	13
y	50	60	71	55	53	60

5.

x	27	35	48	43	32
y	19	13	8	10	15

6.

x	31	34	37	40	46
y	3	15	2	13	5

Real World Applications

For Exercises 7–12,

(a) Draw a scatter plot.

(b) Find the value for r.

(c) Test the significance of r at the 5% level and at the 1% level.

(d) Find the equation of the regression line and draw the line on the scatter plot, but only if r is significant.

(e) Describe the nature of the relationship if one exists.

7. The data represent the heights in feet and the number of stories of the tallest buildings in Pittsburgh.

Height, x	485	511	520	535	582	615	616	635	728	841
No. of stories, y	40	37	41	42	38	45	31	40	54	64

Source: The World Almanac and Book of Facts

Predict y when $x = 500$.

8. A researcher wishes to determine whether the number of hours a person jogs per week is related to the person's age.

Age, x	34	22	48	56	62
Hours, y	3.5	7	3.5	3	1

Predict y when $x = 35$.

9. A study was conducted to determine if the amount a person spends per month on recreation is related to the person's income.

Monthly income, x	$800	$1200	$1000	$900	$850	$907	$1100
Amount, y	$ 60	$ 200	$ 160	$135	$ 45	$ 90	$ 150

Predict y when $x = \$925$.

10. A researcher wishes to determine if there is a relationship between the number of days an employee missed a year and the person's age.

Age, x	22	30	25	35	65	50	27	53	42	58
Days missed, y	0	4	1	2	14	7	3	8	6	4

Predict y when $x = 56$.

11. A statistics instructor wishes to determine if a relationship exists between the final exam score in Statistics 102 and the final exam scores of the same students who took Statistics 101.

Stat 101, x	87	92	68	72	95	78	83	98
Stat 102, y	83	88	70	74	90	74	83	99

Predict y when $x = 90$.

12. The data shown indicate the number of wins and the number of points scored for teams in the National Hockey League.

No. of wins, x	10	9	6	5	4	12	11	8	7	5	9	8	6	6	4
No. of points, y	23	22	15	15	10	26	26	26	21	16	12	19	16	16	11

Source: USA Today

Predict y when $x = 8$.

Writing Exercises

13. What is the name of the graph that is used to investigate whether or not two variables are related?

14. In relationship studies, what are the names of the two variables used?

15. Explain what is meant when two variables are positively linearly related. What would the scatter plot look like?

16. Explain what is meant when two variables are negatively linearly related. What would the scatter plot look like?

17. Explain how the values of r are related to the nature of the relationship.

18. What is meant by the regression line?

19. When all the points fall on the regression line, what will the value of r be equal to?

Critical Thinking

20. Explain why when r is significant, one cannot absolutely say that x causes y.

21. Find the value for r, then interchange the values for x and y and find the value of r. Explain the results.

x	1	2	3	4	5
y	3	5	7	9	11

22. Draw a scatter plot and determine if the variables are related. Explain the results. Then find the value for r for the data shown.

x	−3	−2	−1	0	1	2	3
y	9	4	1	0	1	4	9

Summary

Section	Important Terms	Important Ideas
12-1	data statistics population sample random sample systematic sample stratified sample cluster sample descriptive statistics inferential statistics raw data categorical frequency distribution grouped frequency distribution stem and leaf plot	**Statistics** is the branch of mathematics that involves the collection, organization, summarization, and presentation of data. In addition, researchers use statistics to make general conclusions from the data. When a study is conducted, the researcher defines a population, which consists of all subjects under study. Since populations are usually large, the researcher will select a representative subgroup of the population, called a sample, to study. There are four basic sampling methods. They are random, systematic, stratified, and cluster. Once the data are collected, they are organized into a frequency distribution. There are two types of frequency distributions. They are categorical and grouped. When the data set consists of a small number of values, a stem and leaf plot can be constructed. This plot shows the nature of the data while retaining the original data values.
12-2	bar graph pie graph histogram frequency polygon time series graph	**In** order to represent the data pictorially, graphs can be drawn. From a categorical frequency distribution, a bar graph and a pie graph can be drawn. From a grouped frequency distribution, a histogram and a frequency polygon can be drawn to represent the data. To show how data vary over time, a time series graph can be drawn.
12-3	mean data array median mode midrange	**In** addition to collecting data, organizing data, and representing data using graphs, various summary statistics can be found to describe the data. There are four measures of average. They are the mean, median, mode, and midrange. The mean is found by adding all of the data values and dividing the sum by the number of data values. The median is the middle point of the data set. The mode is the most frequent data value. The midrange is found by adding the lowest and the highest data values and dividing by two.
12-4	range variance standard deviation	**There** are three commonly used measures of variation. They are the range, variance, and standard deviation. The range is found by subtracting the lowest data value from the highest data value. The standard deviation measures the spread of the data values. When the standard deviation is small, the data values are close to the mean. When the standard deviation is large, the data values are farther away from the mean.
12-5	percentile quartile	**The** position of a data value in a data set can be determined by its percentile rank. The percentile rank of a specific data value gives the percent of data values that fall below the specific value. Quartiles divide a distribution into quarters.

—Continued

Summary   Continued—

Section	Important Terms	Important Ideas
12-6	normal distribution standard normal distribution z value (standard score)	**Many** variables have a distribution which is bell-shaped. These variables are said to be approximately normally distributed. Statisticians use the standard normal distribution to describe these variables. The standard normal distribution has a mean of zero and a standard deviation of one. A variable can be transformed into a standard normal variable by finding its corresponding z value. A table showing the area under the standard normal distribution can be used to find areas for various z values.
12-7		**Since** many real-world variables are approximately normally distributed, the standard normal distribution can be used to solve many real-world applications.
12-8	independent variable dependent variable scatter plot correlation coefficient significance levels regression line	**Statisticians** are also interested in determining whether or not two variables are related. In order to determine this, they draw and analyze a scatter plot. After the scatter plot is drawn, the value of the correlation coefficient is computed. If the correlation coefficient is significant, the equation of the regression line is found. Then prediction for y can be obtained given a specific value of x.

Review Exercises

1. A sporting goods store kept a record of sales of five items for one randomly selected hour during a recent sale. Construct a frequency distribution for the data (B = baseballs, G = golf balls, T = tennis balls, S = soccer balls, F = footballs).

F	B	B	B	G	T	F
G	G	F	S	G	T	
F	T	T	T	S	T	
F	S	S	G	S	B	

2. Construct a bar graph for the number of homicides reported for these cities.

City	Number
New Orleans	363
Washington, D.C.	352
Chicago	824
Baltimore	323
Atlanta	184

Source: USA Today

3. A student whose part time job income is $16,000 a year decides to spend the following amount for each:

Food	$4000
Clothing	$1920
Savings	$1600
Rent	$4800
Other	$3680

Construct a pie graph for the data.

4. The data set shown below represents the number of hours 25 part-time employees worked at the Sea Side Amusement Park during a randomly selected week in

June. Construct a stem and leaf plot for the data and analyze the results.

16	25	18	39	25	17	29	14	37
22	18	12	23	32	35	24	26	
20	19	25	26	38	38	33	29	

5. During June, a local theater company recorded the given number of patrons per day. Construct a grouped frequency distribution for the data. Use six classes.

102	116	113	132	128	117
156	182	183	171	168	179
170	160	163	187	185	158
163	167	168	186	117	108
171	173	161	163	168	182

6. Draw a histogram and frequency polygon for the frequency distribution obtained from the data in Exercise 8.

7. These data show the expenses of Chemistry Lab, Inc., that were used for research and development for the years 1995 to 1999. Each number represents thousands of dollars. Draw a time series graph for the data.

Year	Amount
1995	$ 8,937
1996	9,388
1997	11,271
1998	13,877
1999	19,203

8. These data represent the number of deer killed by motor vehicles for eight counties in Southwestern Pennsylvania.

2343	1240	1088	600	497	1925	1480	458

Source: Pittsburgh Post-Gazette

Find each of these.

(a) Mean   (b) Median

(c) Mode   (d) Midrange

(e) Range   (f) Variance

(g) Standard deviation

9. Twelve batteries were tested to see how many hours they would last. The frequency distribution is shown here.

Hours	Frequency
1–3	1
4–6	4
7–9	5
10–12	1
13–15	1

Find the mean.

10. The number of previous jobs held by each of six applicants is shown here.

2 4 5 6 8 9

(a) Find the percentile for each value.

(b) What value corresponds to the 30th percentile?

11. The data shown represent the number of days' inventory eight high-tech firms have on hand. Find the values for Q_1, Q_2, and Q_3.

158	151	91	45	74	118	285	29

Source: USA Today

12. Find the area under the standard normal distribution curve.

(a) Between $z = 0$ and $z = 1.95$

(b) Between $z = 0$ and $z = 0.37$

(c) Between $z = 1.32$ and $z = 1.82$

(d) Between $z = -1.05$ and $z = 2.05$

(e) Between $z = -0.03$ and $z = 0.53$

(f) Between $z = 1.10$ and $z = -1.80$

(g) To the right of $z = 1.99$

(h) To the right of $z = -1.36$

(i) To the left of $z = -2.09$

(j) To the left of $z = 1.68$

13. The average number of years a person takes to complete a graduate degree program is 3. The standard deviation is 4 months or $\frac{1}{3}$ of a year. Assume the variable is normally distributed. If an individual enrolls in the program, find the probability that it will take

(a) More than 4 years to complete the program.

(b) Less than 3 years to complete the program.

(c) Between 3.8 and 4.5 years to complete the program.

(d) Between 2.5 and 3.1 years to complete the program.

14. On the daily run of an express bus, the average number of passengers is 48. The standard deviation is 3. Assume the variable is normally distributed. Find the probability that the bus will have

    (a) Between 36 and 40 passengers.

    (b) Fewer than 42 passengers.

    (c) More than 48 passengers.

    (d) Between 43 and 47 passengers.

15. The average weight of an airline passenger's suitcase is 45 pounds. The standard deviation is 2 pounds. Assume the variable is normally distributed. If an airline handles 2000 suitcases in one day, find the number that will weigh less than 43.5 pounds.

16. The average cost of XYZ brand running shoes is $83.00 per pair, with a standard deviation of $8.00. If 90 pairs of shoes are sold, how many will cost between $80.00 and $85.00? Assume the variable is normally distributed.

17. A study is done to see whether there is a relationship between a student's grade point average and the number of hours the student watches television each week. The data are shown here.

Hours, x	6	10	8	15	5	6	12
GPA, y	2.4	4	3.2	1.6	3.7	3	3.5

Draw a scatter plot for the data and describe the relationship. Find the value for r and determine whether or not it is significant at the 5% significance level. If yes, find the equation of the regression line, and predict y when $x = 9$.

Chapter Test

1. A questionnaire about how people get news resulted in the information shown from 25 respondents. Construct a frequency distribution for the data (N = newspaper, T = television, R = radio, M = magazine).

N	N	R	T	T
R	N	T	M	R
M	M	N	R	M
T	R	M	N	M
T	R	R	N	N

2. Draw a bar graph for the frequency distribution obtained in Exercise 1.

3. Draw a pie graph for the data found in Exercise 2.

4. The data (in millions of dollars) are the values of the 30 National Football League franchises. Construct a frequency distribution for the data using eight classes.

170	191	171	235	173	187	181	191
200	218	243	200	182	320	184	239
186	199	186	210	209	240	204	193
211	186	197	204	188	242		

   Source: *Pittsburgh Post Gazette*

5. Draw a histogram and frequency polygon using the frequency distribution for the data in Exercise 4.

6. A special aptitude test is given to job applicants. The data shown here represent the scores of 30 applicants. Construct a stem and leaf plot for the data and summarize the results.

204	210	227	218	254
256	238	242	253	227
251	243	233	251	241
237	247	211	222	231
218	212	217	227	209
260	230	228	242	200

7. The given data represent the federal minimum hourly wage in the years shown. Draw a time series graph to represent the data and analyze the results.

Year	Wage
1960	$1.00
1965	1.25
1970	1.60
1975	2.10
1980	3.10
1985	3.35
1990	3.80
1995	4.25
2000	5.25

8. These temperatures were recorded in Pasadena for a week in April.

87	85	80	78	83	86	90

Find each of these.

(a) Mean   (b) Median

(c) Mode   (d) Midrange

(e) Range   (f) Variance

(g) Standard deviation

9. The distribution of the number of errors 10 students made on a typing test is shown.

Errors	Frequency
0–2	1
3–5	3
6–8	4
9–11	1
12–14	1

Find the mean.

10. The number of credits in business courses eight job applicants had is shown here.

9 12 15 27 33 45 63 72

(a) Find the percentile for each value.

(b) What value corresponds to the 40th percentile?

11. Find the area under the standard normal distribution for each.

(a) Between 0 and 1.50

(b) Between 0 and -1.25

(c) Between 1.56 and 1.96

(d) Between -1.20 and -2.25

(e) Between -0.06 and 0.73

(f) Between 1.10 and -1.80

(g) To the right of $z = 1.75$

(h) To the right of $z = -1.28$

(i) To the left of $z = -2.12$

(j) To the left of $z = 1.36$

12. The mean time it takes for a certain pain reliever to begin to reduce symptoms is 30 minutes, with a standard deviation of 4 minutes. Assuming the variable is normally distributed, find the probability that it will take the medication

(a) Between 34 and 35 minutes to begin to work

(b) More than 35 minutes to begin to work

(c) Less than 25 minutes to begin to work

(d) Between 35 and 40 minutes to begin to work

13. The average height of a certain age group of people is 53 inches. The standard deviation is 4 inches. If the variable is normally distributed, find the probability that a selected individual's height will be

(a) Greater than 59 inches.

(b) Less than 45 inches.

(c) Between 50 and 55 inches.

(d) Between 58 and 62 inches.

14. The average repair cost of a microwave oven is $55.00, with a standard deviation of $8.00. The costs are normally distributed. If 200 ovens are repaired, how many of them cost more than $60.00 to repair?

15. A study is conducted to determine the relationship between a driver's age and the number of accidents he or she has over a 1-year period. The data are shown here. Draw a scatter plot for the data and explain the nature of the relationship. Find the value for r and determine whether or not it is significant at the 5% significance level. If r is significant, find the equation of the regression line, and predict y when x is 61.

Driver's age, x	63	65	60	62	66	67	59
No. of accidents, y	2	3	1	0	3	1	4

Projects

1. Collect data on a single variable such as the number of miles a sample of students travels to school. Construct a frequency distribution, find various summary statistics, and draw a frequency polygon or histogram for the data. Write a brief report on your findings.

2. Collect data on two variables such as the number of credits a sample of students has and the number of hours per week the students spend doing schoolwork. Draw a scatter plot and determine the nature of the relationship.

3. Research and write a paper on the A. C. Nielson Company. Explain such things as how they select their samples, etc.

4. Write a short paper on the statistical work of Gregor Mendel.

Mathematics in Our World
►Revisited

Who Is a Typical First-Time Home Buyer?

The *USA Today* Snapshot shown here uses several measures of average and two percentages (63% and 82%). Although it is not possible to know which measures were used, one could make an educated guess. For example, the mean was probably used for family size (2.7 people) since it is in decimal form. The age of the buyer (32) is probably a median. The time (5 months) it took to find a home is probably a mode.

The information presented in this Snapshot is very vague. After mastering the concepts in this chapter, you should be able to present a more accurate profile if given the raw data.

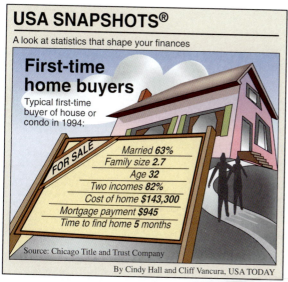

USA SNAPSHOTS®

A look at statistics that shape your finances

First-time home buyers

Typical first-time buyer of house or condo in 1994:

FOR SALE
Married **63%**
Family size **2.7**
Age **32**
Two incomes **82%**
Cost of home **$143,300**
Mortgage payment **$945**
Time to find home **5 months**

Source: Chicago Title and Trust Company

By Cindy Hall and Cliff Vancura, USA TODAY

Source: USA Today, February 22, 1995. Copyright 1995, USA TODAY. Reprinted with permission.

Answers to **Try This One**

12-A.

Type	Tally	Frequency
A	卌 |	6
B	卌	5
C	卌 卌 ||	12
D	|||	3
E	卌 ||||	9

12-B.

Class	Tally	Frequency
1–3	卌 卌	10
4–6	卌 卌 ||||	14
7–9	卌 卌	10
10–12	卌 |	6
13–15	卌	5
16–18	卌	5

12-C.

```
1 | 7
2 |
3 | 9
4 | 3 6 6 7 7 8 8
5 | 0 0 2 5 6 6 6 6 6 7 8 8 8 9
6 | 0 1 1 2 2 2 2 2 2 3 3 3 3 4 4 5 5 5 6 7 8
7 | 0 0 0 2 3 4 5 8
8 | 0 6
```

12-D.

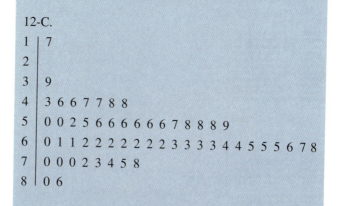

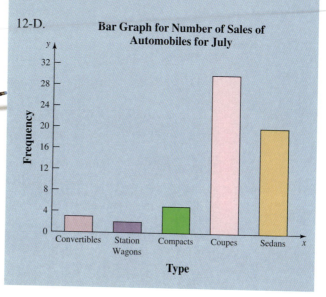

Bar Graph for Number of Sales of Automobiles for July

**Pie Graph for Number of Sales
of Automobiles for July**

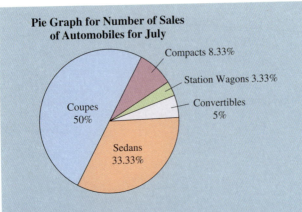

12-E. **Histogram for Net Sales for 25 Retail
Clothing Stores**

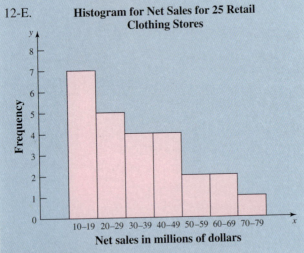

**Frequency Polygon for Net Sales
for 25 Retail Clothing Stores**

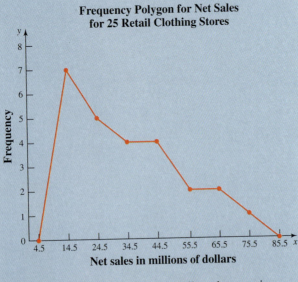

Analysis of both graphs: Generally, a decrease in
frequency occurs from the lower classes to the higher ones;
also reverse J shaped.

12-F. **Time Series Graph for Number of Forest Fires
in the United States**

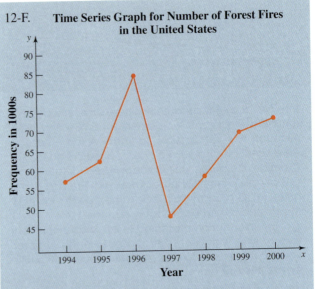

12-G. 21.25 seconds

12-H. 9.8

12-I. (a) 197
(b) 3
(c) $2.70

12-J. (a) 4 and 14
(b) 93
(c) None

12-K. 224.5

12-L. 219

12-M. $s^2 = 39{,}341.02$
$s = 198.35$

12-N. 58th percentile

12-O. 75th percentile

12-P. $Q_1 = 25$
$Q_2 = 31.5$
$Q_3 = 42$

12-Q. (a) 0.013
(b) 0.862
(c) 0.145
(d) 0.655
(e) 0.62

12-R. (a) 39.4%
(b) 22.7%
(c) 17.1%
(d) 46.2%
(e) 10.6%

12-S. (a) 1060
     (b) 4400
     (c) 7730
     (d) 1210
     (e) 3080

12-T. Relationship: There is an approximate, or rough, positive linear relationship.

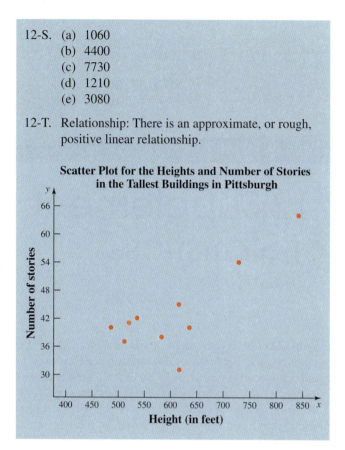

Scatter Plot for the Heights and Number of Stories in the Tallest Buildings in Pittsburgh

12-U. 0.897

12-V. There exists a significant relationship at the 5% level, but not at the 1% level.

12-W. Prediction for 0.28 mg is 89.5, by letting $x = 0.28$ in regression equation.

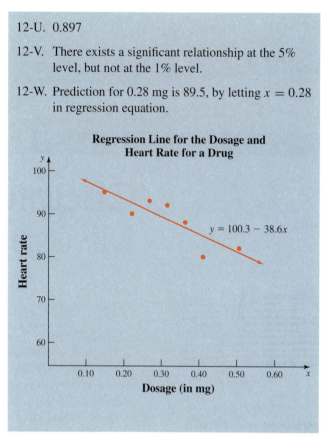

Regression Line for the Dosage and Heart Rate for a Drug

$y = 100.3 - 38.6x$

Chapter 12 Supplement: Misuses of Statistics

Every day we hear or read the results obtained from research studies, surveys, and opinion polls regarding our health, our opinions, and our behavior. Statistical research is used to provide knowledge and information in order to enable us to make intelligent decisions about our health and welfare. However, there is another aspect of statistics that needs to be addressed, and that is the fact that there are people who will misuse statistics to sell us products that don't work, to attempt to prove something that is not true, or to get our attention by using such techniques as fear, shock, and outrage supported by bad statistics.

There are some sayings that have been around for a long time that illustrate this point—for example,

"There are three types of lies—lies, damn lies, and statistics."

"Figures don't lie, but liars figure."

Just because we read or hear the results of a research study or an opinion poll in the media, this does not mean that these results are reliable or that they can be applied to any and all situations. For example, reporters sometimes leave out critical details such as the size of the sample used or how the research subjects were selected. Without this information, one cannot properly evaluate the research and properly interpret the conclusions of the study or survey.

It is the purpose of this section to present some ways that statistics can be misused. One should not infer that all research studies and surveys are suspect, but that there are many factors to consider when making decisions based on the results of research studies and surveys. Here are some ways that statistics can be misrepresented.

Suspect Samples

The first thing to consider is the sample that was used in the research study. Sometimes researchers use very small samples to obtain information. Several years ago, advertisements contained statements such as, "Three out of four doctors surveyed recommend brand such and such." If only four doctors were surveyed, the results could have been obtained by chance alone; however, if 100 doctors were surveyed, the results might be quite different.

People who are annoyed by unsolicited phone calls won't appear in a sample.

Not only is it important to have a sample size that is large enough, but it is also necessary to see how the subjects in the sample were selected. Studies using volunteers sometimes have a built-in bias. Volunteers generally do not represent the population at large. Sometimes they are recruited from a particular socioeconomic background, and sometimes unemployed people volunteer for research studies in order to get a stipend. Studies that require the subjects to spend several days or weeks in an environment other than their home or workplace automatically exclude people who are employed and cannot take time off from work. Sometimes college students or retirees are used. In the past, many studies have used only men but have attempted to generalize the results to both men and women. Opinion polls that require a person to phone or mail in a response most often are not representative of the population in general since only those with strong feelings for or against the issue usually call or respond by mail.

Another type of sample that may not be representative is the convenience sample. Educational studies sometimes use students in intact classrooms since it is convenient. Quite often, the students in these classrooms do not represent the student population of the entire school district.

When interpreting results from studies using small samples, convenience samples, or volunteer samples, care should be used when generalizing the results to the entire population.

Ambiguous Averages

In a study to determine how long it takes to fall asleep, would college students be a good sample of the overall population?

In this chapter, four statistical measures are loosely called "averages." They are the mean, the median, the mode, and the midrange, and for a given set of data, the values of these averages may sometimes differ markedly. People who know that there are several types of average can, without lying, select the one for the data that most lends evidence to support their position. For example, suppose a storeowner employs five salespeople. The number of years each has been employed by the store is 22, 10, 2, and 2. The mean number of years' service is 9. The median is 6, and the mode is 2. Now if the storeowner wanted to advertise the fact that his employees have many years of experience, which average do you think he would use? Obviously, he would use 9. However, 9 is not very high, so since the owner sometimes doubles as a salesman and since he has owned the store for 42 years, he includes his years of service in computing the average. Hence, the mean is now 15.6 years. Quite impressive, isn't it? (Actually the midrange, 22 years, is much more impressive.) Whenever the word "average" is used instead of mean, median, mode, or midrange, ask yourself, "What average is being used?"

Another type of statistical distortion can occur when different values are used to represent the same data. For example, one political candidate who is running for reelection might say, "During my administration, expenditures increased a mere 3%." His opponent, who is trying to unseat him, might say, "During my opponent's administration, expenditures have increased a whopping $6,000,000." Here both figures are correct; however, expressing $6,000,000 as a mere 3% makes it seem like a very small increase whereas expressing a 3% increase as $6,000,000 makes it sound like a very large increase. Here again, ask yourself, "Which measure best represents the data?"

Detached Statistics

A claim that uses a detached statistic is one in which no comparison is made. For example, you may hear a claim such as, "Our brand of crackers has $\frac{1}{3}$ fewer calories." Here, no comparison is made. One-third fewer calories than what? Another example is a claim that uses a

People on one side of an issue sometimes frame statistics to support their point.

detached statistic such as, "Brand A aspirin works four times faster." Four times faster than what? When you see statements such as this, always ask yourself, "As compared to what?"

Implied Connections

Many claims attempt to imply connections between variables that may not actually exist. For example, consider this statement: "Eating fish may help to reduce your cholesterol." Notice the words "may help." There is no guarantee that eating fish will definitely help you reduce your cholesterol.

"Studies suggest that using our exercise machine will reduce your weight." Here the word *suggest* is used, and again, there is no guarantee that you will lose weight using the exercise machine advertised.

Another claim might say, "Taking calcium will lower blood pressure in some people." Notice the word *some* is used. You may not be included in the group of "some" people. Be careful when drawing conclusions from claims that use words such as "may," "in some people," "might help," etc.

Misleading Graphs

Graphs give a visual representation that enables readers to analyze and interpret data more easily than they could simply by looking at numbers; however, inappropriately drawn graphs can misrepresent the data and lead the reader to false conclusions. For example, a car manufacturer's ad stated that 98% of the vehicles it had sold in the past 10 years were still on the road. The ad then showed a graph similar to the one in Figure 12-31(a). The graph shows the percentage of the manufacturer's automobiles still on the road and the percentage of its competitors' automobiles still on the road. Is there a large difference? Not necessarily. Notice the scale on the vertical axis in Figure 12-31(a). It has been cut off (or

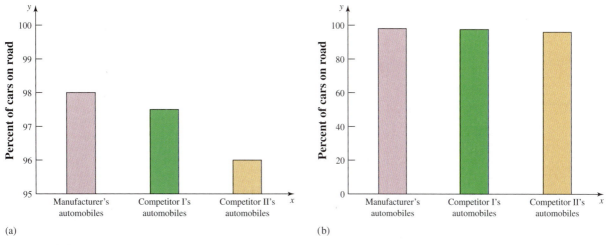

(a)   (b)

Figure 12-31

truncated), and it starts at 95%. When the graph is redrawn using a scale that goes from 0% to 100%, as in Figure 12-31(b), there is hardly a noticeable difference in the percentages. Thus, changing the units at the starting point on the axis can convey a very different visual representation of the data.

It is not wrong to truncate an axis of the graph; many times it is necessary to do so; however, the reader should be aware of this fact and interpret the graph accordingly. Do not be misled if an inappropriate explanation is given.

Let's consider another example. The percentage of the world's total motor vehicles produced by manufacturers in the United States declined from 25% in 1986 to 18% in 1991, as shown by these data.

Year	1986	1987	1988	1989	1990	1991
Percent produced in the U.S.	25	23.8	23.3	22.1	20.3	18.0

When one draws the graph, as shown in Figure 12-32(a), a scale ranging from 0% to 100% shows a slight decrease. However, this decrease can be emphasized by using a scale that ranges from 15% to 25%, as shown in Figure 12-32(b). Again, by changing the units or the starting point on the *y* axis, one can change the visual message.

Another misleading graphing technique sometimes used is exaggerating a one-dimensional increase by showing it in two dimensions. For example, the average cost of a 30-second Super Bowl commercial has increased from $40,000 in 1967 to $1 million in 1995 (Source: Nielsen Media Research). The increase shown by the bar graph in Figure 12-33(a) represents the change by a comparison of the heights of two bars in one dimension. The same data are shown two-dimensionally with circles in Figure 12-33(b). Notice that the difference seems much larger because the eye is comparing the areas of the circles rather than the lengths of the diameters.

Faulty Survey Questions

Surveys and opinion polls obtain information by using questionnaires. There are two types of studies: interviews, and self-administered questionnaires. The interview survey requires a person to ask the questions either in person or by telephone. Self-administered questionnaire surveys require the participant to answer the questions by mail, computer, or in a group setting, such as a classroom. When analyzing the results of a survey using questionnaires, you should be sure that the questions are properly written, since the way questions are

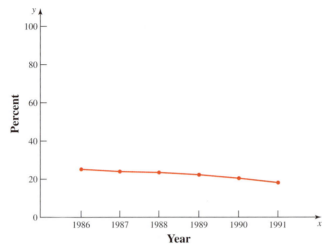

(a) Using a scale from 0% to 100%

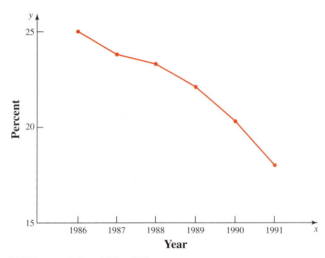

(b) Using a scale from 15% to 25%

Figure 12-32

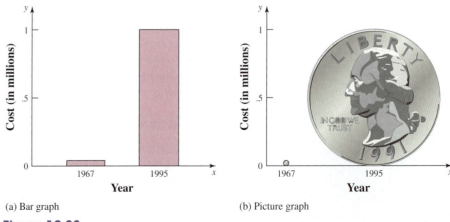

(a) Bar graph

(b) Picture graph

Figure 12-33

phrased can often influence the way people answer them. For example, when a group of people was asked, "Do you favor a waiting period before guns are sold?" 91% of the respondents said "Yes" and 6% said "No." However, when the question was rephrased as, "Do you favor a national gun-registration program costing about 20% of all dollars spent on crime control?" 37% responded "Yes" and 61% responded "No." As you can see, although the questions pertain to some form of gun control, each asks something a little different, and the responses are radically different. When reading and interpreting the results obtained from questionnaire surveys, watch out for some of these common mistakes made in the writing of the survey questions.

Asking Biased Questions

By asking a question in a certain way, the researcher can lead the respondents to answer the question the way the researcher wants them to. For example, the question, "Are you going to vote for Candidate Jones, even though the latest survey shows he will lose the election?" may lead the respondent to say, "No" since many people do not want to vote for a loser or admit that they have voted for a loser.

Using Confusing Words

Using words in a survey question that are not well defined or understood can invalidate the responses. For example, a question such as, "Do you think people would live longer if they were on a diet?" would mean many different things to people since there are many types of diets, such as low-salt diets, high-protein diets, etc.

Asking Double-Barreled Questions

Sometimes two ideas are contained in one question, and the respondent may answer one or the other in his or her response. For example, consider the question, "Are you in favor of a national health program and do you think it should be subsidized by a special tax as opposed to other ways to finance it, such as a national lottery?" Here the respondent is really answering two questions.

Using Double Negatives

Survey questions containing double negatives often confuse the respondent. For example, what is the question "Do you feel that it is not appropriate to have areas where people cannot smoke?" really asking? Other factors that could bias a survey would include anonymity of the participant, the time and place of the survey, and whether the questions were open-ended or closed-ended.

Participants will, in some cases, respond differently to questions based on whether or not their identity is known. This is especially true if the questions concern sensitive issues such as income, sexuality, abortion, etc. Researchers try to ensure confidentiality rather than anonymity; however, many people will be suspicious in either case.

The time and place where a survey is taken can influence the results. For example, if a survey on airline safety is taken right after a major airline crash, the results may differ from those obtained in a year with no major airline disasters.

To restate the premise of this section, statistics, when used properly, can be beneficial in obtaining much information, but when used improperly, they can lead to much misinformation. Therefore, it is important to understand the concepts of statistics and use them correctly.

Exercise Set | Chapter 12 Supplement

1. According to a pilot study of 20 people conducted at the University of Minnesota, daily doses of a compound called arabinogalactan over a period of 6 months resulted in a significant increase in the beneficial *lactobacillus* species of bacteria. Why can't it be concluded that the compound is beneficial for the majority of people?

2. Comment on this statement taken from a magazine advertisement: "In a recent clinical study, Brand ABC* was proved to be 1950 percent better than creatine!"
   *Actual brand will not be named.

3. In an ad for women, the following statement was made: "For every hundred women, 91 have taken the road less traveled." Comment on this statistic.

4. In many ads for weight loss products, under the product claims and in small print, the following statement is made: "These results are not typical." What does this say about the product being advertised?

5. An article in a leading magazine stated that, "When 18 people with chronic, daily, whip-lash related headaches received steroid injections in a specific neck joint, 11% had no more headaches." Think of a possible reason why the figure 11% was used.

6. In an ad for moisturizing lotion, the following claim is made: ". . . it's the #1 dermatologist recommended brand." What is misleading about the claim?

7. An ad for an exercise product stated that, "Using this product will burn 74% more calories." What is misleading about that statement?

8. "Vitamin E is a proven antioxidant and may help in fighting cancer and heart disease." Is there anything ambiguous about this claim?

9. "Just one capsule of Brand X* can provide 24 hours of acid control." What needs to be more clearly defined in this statement?
   *Actual brand will not be named.

10. ". . . male children born to women who smoke during pregnancy run a risk of violent and criminal behavior that lasts well into adulthood." Can we infer that smoking during pregnancy is responsible for criminal behavior in people?

For Exercises 11–13, explain why each pair of graphs is misleading.

11. A company advertises that their brand of energy pills gets into the user's blood faster than a competitor's brand and show these two graphs to prove their claim.

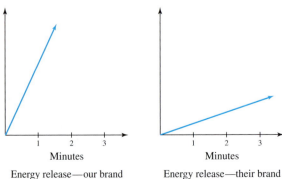

Minutes   Minutes
Energy release—our brand   Energy release—their brand

12. The graph shows the difference in sales of pumpkins during October for the years 1990 and 2000.

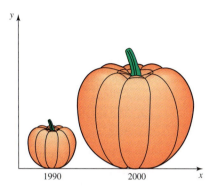

13. Explain this contradiction: The two graphs were drawn using the same data, yet the first graph shows sales remaining stable, and the second graph shows sales increasing dramatically.

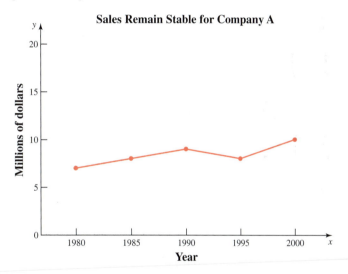

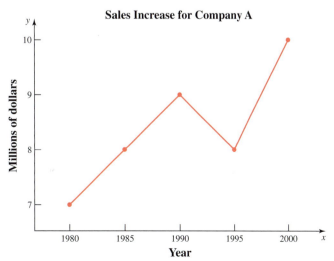

For Exercises 14–16, explain why each survey question might lead to an erroneous conclusion.

14. "How often do you run red lights?"

15. "Do you think gun manufacturers should put safety locks on all guns sold even though it would increase the cost of the gun by 20%?"

16. "Do you think that it is not important to give extra tutoring to students who are not failing?"

17. The results of a survey reported in *USA Weekend* stated that:

    "9% would drive through a toll booth without paying."

    "13% would steal cable television or inflate their resumes."

    Explain why these figures may not be representative of the population in general.

18. In an article in *USA Weekend,* this statement was made: "More serious seems to be coffee's potential to raise blood pressure levels of homocysteine, a protein that promotes artery clogging. A recent Norwegian study found 20% higher homocysteine in heavy coffee drinkers (more than 9 cups a day) than in non-coffee drinkers." Based on this statement, should we give up our daily cup of coffee?

19. An article in a newspaper with the headline, "Lead: The Silent Killer" listed the number of confirmed childhood lead poisoning cases in Allegheny County (PA) in 1985 as 15 and in 1997 as 124. Can you conclude that the incidence of childhood lead poisoning cases is increasing in Allegheny County? Suggest a factor that might cause an increase in reported lead poisoning in children.

    Source: Pittsburgh Tribune Review

20. In a recent article, the author states that 71% of adults do not use sunscreen. Although 71% is a large percentage, explain why it could be misleading.

21. In a book on probabilities, the author states that in the United States every 20 minutes on average, someone is murdered. Based on this statement, can we conclude that crime is rampant in the United States? Note: The population of the United States is about 250 million people.

22. For a specific year, there were 6067 male fatalities in the workplace and 521 female deaths. A government official made this statement: "Over 90 percent of the fatal injuries the past year were men, although men accounted for only 54 percent of the nation's employment." Can we conclude that women are more careful on the job?

Chapter Thirteen

Voting Methods

Outline

Objectives

After completing this chapter, you should be able to

1 Construct a preference table given a number of preference ballots (13-1)

2 Answer questions about an election when the results are summarized in a preference table (13-1)

3 Determine the winner of an election using the plurality method (13-1)

4 Determine whether an election using the plurality method violates the head-to-head criterion (13-1)

5 Determine the winner of an election using the Borda count method (13-2)

6 Determine whether an election using the Borda count method violates the majority criterion (13-2)

7 Determine the winner of an election using the plurality-with-elimination method (13-2)

8 Determine whether an election using the plurality-with-elimination method violates the monotonicity criterion method (13-2)

9 Determine the winner of an election using the pairwise comparison method (13-3)

10 Determine whether an election using the pairwise comparison method violates the irrelevant alternatives criterion (13-3)

11 Determine the winner of an election using the approval voting method (13-3)

Introduction

Voting is one of the basic concepts of a democratic society. It is used to elect leaders, pass amendments, or choose a course of action. One often hears the statement, "The majority rules." However, that is not always true, as evidenced by the U.S. presidential elections of 1992 and 2000. During these years, presidents were elected with a minority (i.e., less than 50%) of the popular votes.

The simplest voting method is choosing between two alternatives (i.e., candidates, amendments, or courses of action). Here all votes are treated equally and the alternatives are considered equal. A voter then can vote for either alternative. The alternative with the most votes is the winner. One problem with this situation is that if the number of voters is even, then the possibility of a tie exists.

When there are three or more alternatives, as in the Oscars, voting procedures become more complex. Throughout the years, people have proposed different methods of voting to select the winner in these elections. Four of the most common voting methods will be explained in this chapter. They include

1. The plurality method

2. The Borda count method

3. The plurality-with-elimination method

4. The pairwise comparison method

You will learn that each method has at least one inherent flaw. So which method is the best? In the 1950s, Kenneth Arrow proved that a voting method for choosing one of three or more alternatives that is always fair is an impossibility. His theorem is called *Arrow's impossibility theorem*. This is explained in Section 13-3. ■

Mathematics in Our World

Academy Awards

Each year the Academy of Motion Picture Arts and Sciences awards "Oscars" to recognize achievements in the motion picture industry. Some of these awards include best picture, best director, best actor, best actress, best supporting actor, and best supporting actress. These awards were started in 1928 when the best picture was *Wings*. The best actor was Emil Jannings and the best actress was Janet Gaynor. The "Oscar" gets its name from an executive director, Margaret Herrick, who remarked that the award looked like her uncle Oscar.

In the early years, 36 studio executives selected the winners. However, in recent years, the number of voting members has increased to well over 3000. There are 13 branches of the organization that nominate the top five candidates for each category. Then each member votes for his or her favorite candidate.

Just how does the voting take place? This chapter explains some of the various types of voting methods used today in areas where voting is required to determine a winner. To find out what voting methods are used to select the winners for the Oscars, see Mathematics in Our World—Revisited.

13-1 Preference Tables and the Plurality Method

Preference Tables

Suppose there are three candidates running for club president. We'll call them A, B, and C. Now, instead of voting for the single candidate of your choice, you are asked to rank each candidate in order of preference. This type of ballot is called a *preference ballot*.

In this case, there are six possible ways to rank the candidates, as shown.

First choice	A	A	B	B	C	C
Second choice	B	C	A	C	A	B
Third choice	C	B	C	A	B	A

Now, suppose that the 20 club members voted as follows.

A	B	A	A	A	A	A	B	A	A	B	A	A	A
B	C	B	C	B	C	B	C	B	B	C	B	C	C
C	A	C	B	C	B	C	A	C	C	A	C	B	B

A	C	C	A	A	B
C	B	B	B	B	C
B	A	A	C	C	A

In this election, nine people voted for the candidates in order of preference ABC, five people voted ACB, four people voted BCA, and two people voted CBA.

A **preference table** can be made showing the results.

Number of voters	9	5	4	2
First choice	A	A	B	C
Second choice	B	C	C	B
Third choice	C	B	A	A

The sum of the numbers in the top row indicates the total number of voters. Also note that $9 + 5$ or 14 voters selected candidate A as their first choice, four selected candidate B as their first choice, and two voters selected candidate C as their first choice.

Because no voters cast ballots ranking candidates as BAC or CAB, those possible rankings are not listed as columns in the table.

Example 13-1

Four candidates, W, X, Y, and Z, are running for class president. The students were asked to rank all candidates in order of preference. The results of the election are shown in the preference table.

Number of voters	86	42	19	13	40
First choice	X	W	Y	X	Y
Second choice	W	Z	Z	Z	X
Third choice	Y	X	X	W	Z
Fourth choice	Z	Y	W	Y	W

—Continued

Example 13-1 Continued—

(a) How many students voted?
(b) How many people voted for candidates in the order Y, Z, X, W?
(c) How many students selected candidate Y as their first choice?
(d) How many students selected candidate W as their first choice?

Solution

(a) To find the total number of voters, find the sum of the numbers in the top row.
   $86 + 42 + 19 + 13 + 40 = 200$
(b) To find the number of voters who selected the slate Y, Z, X, W, look at the number above column number 3. It is 19. Hence, 19 students voted for Y, Z, X, W in that order of preference.
(c) To find the number of voters who selected candidate Y as their first choice, find the sum of the numbers in columns 3 and 5. That is, $19 + 40 = 59$.
(d) To find the number of voters who selected candidate W as their first choice, find the column with W as the first choice. In this case, it is column 2. Hence, 42 voters selected candidate W as their first choice.

Try This One

13-A A committee of 10 people needs to decide in which of four restaurants to have a dinner meeting. The choices are Airport Restaurant (A), Bobs' Lounge (B), Crab Shack (C), and Diner's House (D). The preference ballots are as shown. Construct a preference chart and determine how many people voted for the Crab Shack as their first choice.

A	C	C	C	C	C	C	C	C	A
B	D	A	B	D	A	B	D	A	B
C	B	B	D	B	B	D	B	B	C
D	A	D	A	A	D	A	A	D	D

For the answer, see page 722.

The next sections will explain the four common voting methods.

The Plurality Method

The easiest method of determining a winner in an election with three or more candidates is called the *plurality method.*

> *Math Note*
>
> Plurality does not necessarily mean majority; it simply means more votes than any other candidate receives, whereas majority means more than 50% of the votes cast.

In an election with three or more candidates that uses the **plurality method** to determine a winner, the candidate with the most first-place votes is the winner.

Example 13-2

The preference table for a club presidential election consisting of three candidates is shown. Using the plurality method, determine the winner.

Number of votes	4	3	6	3
First choice	B	A	C	B
Second choice	C	C	A	A
Third choice	A	B	B	C

Solution

In this situation, only the first-place votes for each candidate are considered. Candidate A received 3 first-place votes (column 2). Candidate B received $4 + 3$ or 7 first-place votes (columns 1 and 4). Candidate C received 6 first-place votes (column 3). Hence, Candidate B is the winner since that candidate received the most first-place votes.

In Example 13-2, the top row consists of the number of voters who ranked the candidate in the order shown in the column. Instead of numbers in the top row, percents can also be used. In order to change the counts to percents, divide the count by the total number of ballots cast and multiply by 100. Using the preference table shown in the preceding example, the percent of voters ranking the candidates as BCA is $\frac{4}{16} \times 100\% = 25\%$. The percent of voters ranking the candidates as ACB is $\frac{3}{16} \times 100\% = 18.75\%$. The percent of voters ranking the candidates as CAB is $\frac{6}{16} \times 100\% = 37.5\%$. The percent of voters ranking the candidates as BAC is $\frac{3}{16} \times 100\% = 18.75\%$. Hence, the preference table could also be shown as

Percent of votes	25%	18.75%	37.5%	18.75%
First choice	B	A	C	B
Second choice	C	C	A	A
Third choice	A	B	B	C

Percentage of First Place Votes

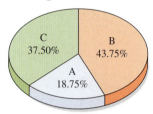

Although the plurality method is an easy way to determine a winner, it does have some flaws. First, notice that a candidate can win an election without receiving the majority of votes (i.e., over 50%) as shown in the previous example. The winner, candidate B, received $25\% + 18.75\% = 43.75\%$ of the votes cast. Second, there is always a possibility of a tie. Third, this method ignores any other information about voters' preferences except the first-place votes. Fourth, this method sometimes, but not always, violates what is called the *head-to-head comparison criterion*.

A *criterion* is a way of measuring or evaluating a situation. In this chapter, we will discuss various criteria for assessing the fairness of voting systems. The first of these is the head-to-head comparison criterion.

> The **head-to-head comparison criterion** states that if a particular candidate wins all head-to-head comparisons with all other candidates, then that candidate should win the election.

Sidelight

Does History Repeat Itself?

It seems that the United States' presidents who were elected in a year with "0" at the end and in increments of 20 years have shared some sad coincidences.

    1840: William Henry Harrison died in office.

    1860: Abraham Lincoln was assassinated.

    1880: James A. Garfield was assassinated.

    1900: William McKinley died in office.

    1920: Warren G. Harding died in office.

    1940: Franklin D. Roosevelt died in office.

    1960: John F. Kennedy was assassinated.

    1980: Ronald Reagan survived an assassination attempt.

It can be shown that the election held in Example 13-2 violates the head-to-head comparison criterion. Using the preference table in the example, one needs to compare each candidate with each of the other candidates taking two candidates at a time and ignoring the rest. The candidate with the higher ranking is the preferred one for each column. Here the preference table for the club president's election is reprinted again.

Number of votes	4	3	6	3
First choice	B	A	C	B
Second choice	C	C	A	A
Third choice	A	B	B	C

First, compare A with B:

In column 1, B was preferred over A by 4 voters.

In column 2, A was preferred over B by 3 voters.

In column 3, A was preferred over B by 6 voters.

In column 4, B was preferred over A by 3 voters.

Hence, A was preferred over B by $3 + 6 = 9$ voters; B was preferred over A by $4 + 3 = 7$ voters, meaning in a head-to-head comparison, A wins over B.

Next, compare A with C:

In column 1, C was preferred over A by 4 voters.

In column 2, A was preferred over C by 3 voters.

In column 3, C was preferred over A by 6 voters.

In column 4, A was preferred over C by 3 voters.

Hence, A was preferred over C by $3 + 3 = 6$ voters; C was preferred over A by $6 + 4 = 10$ voters, meaning in a head-to-head comparison, C wins over A.

Finally, compare B with C:

In column 1, B was preferred over C by 4 voters.

In column 2, C was preferred over B by 3 voters.

In column 3, C was preferred over B by 6 voters.

In column 4, B was preferred over C by 3 voters.

Hence, B was preferred over C by $4 + 3 = 7$ voters; C was preferred over B by $3 + 3 = 6$ voters, meaning in a head-to-head comparison, B wins over C.

Now, the head-to-head comparison criterion is violated because no one single candidate won in all head-to-head comparisons. By the plurality method of voting, candidate B won the election. In order to fulfill the head-to-head comparison criterion, because candidate B won the election, B should have been preferred over A, and B should have been preferred over C in a head-to-head comparison. In this case, it is not true.

The head-to-head criterion is called a *fairness criterion.* It is one of four fairness criteria that political scientists have agreed on that a fair voting system should meet.

> **Math Note**
>
> It is not necessary to perform all head-to-head comparisons. It is only necessary to compare pairwise the winner in a plurality election with each other candidate, in order to determine whether the election satisfies the criterion.

> **Math Note**
>
> You should not get the impression that in *all* elections which use the plurality method, the head-to-head criterion is violated. This is not the case. However, if it is possible to violate the head-to-head criterion in at least one election using the plurality method, then we say the voting method does not satisfy the head-to-head criterion.

Try This One

13-B An election was held for the chairperson of the Psychology Department. There were three candidates: Professor Jones (J), Professor Kline (K), and Professor Lane (L). The preference table for the ballot is shown.

Number of votes	2	4	1	3
First choice	L	J	K	L
Second choice	J	K	L	K
Third choice	K	L	J	J

(a) Who won the election if the plurality method of voting was used?

(b) In this case, does the voting method violate the head-to-head comparison criterion?

For the answer, see page 722.

Exercise Set 13-1

Real World Applications

1. The preference ballots for the election of a CEO by the board of directors are shown. Make a preference table for the results of the election and answer each question.

   (a) How many people voted?

   (b) How many people voted for the candidates in the order of preference XZY?

   (c) How many people voted for candidate Y as their first choice?

   (d) Using the plurality method, determine the winner of the election.

X	X	Y	Z	X	Y	Z	Z	X	Y	X	Y	X	Y
Y	Z	Z	Y	Y	Z	Y	Y	Y	Z	Z	Z	Y	Z
Z	Y	X	X	Z	X	X	X	Z	X	Y	X	Z	X

X	Y	Z	Z	X	X	Y	Y
Z	Z	Y	Y	Y	Z	Z	Z
Y	X	X	X	Z	Y	X	X

2. The Tube City Talkers Club held its annual speech contest. The preference ballots for the best speaker are shown. The candidates were Cortez (C), Lee (L), and Smith (S). Make a preference table for the results of the election and answer each question.

   (a) How many people voted?

   (b) How many people voted for the candidates in the order of preference CLS?

   (c) How many people voted for Lee for first place?

   (d) Using the plurality method, determine the winner of the election.

C	S	S	L	L	C	S	S	L	L	C	C	S	S
L	C	C	S	S	L	C	C	S	S	L	L	C	C
S	L	L	C	C	S	L	L	C	C	S	S	L	L

C	L	S	L	L
L	S	C	S	S
S	C	L	C	C

3. The preference ballots of the board of directors for the selection of a city in which to hold the next National Mathematics Instructors' Association Conference are shown. The three cities under consideration are Chicago (C), Philadelphia (P), and Miami (M). Make a preference table for the results of the election and answer each question.

   (a) How many people voted?

   (b) How many people voted for the candidates in the order of preference PMC?

   (c) How many people voted for Chicago as their first choice?

   (d) Using the plurality method, determine the winner of the election.

P	C	M	P	P	P	P	C	M	M	P	P	C
M	P	P	M	M	M	M	P	P	P	M	M	P
C	M	C	C	C	C	C	M	C	C	C	C	M

P	M	P	C	M
M	P	M	P	P
C	C	C	M	C

4. A group of club members decide to vote to select the color of their meeting room. The color choices are white (W), light blue (B), and light yellow (Y). Make a preference table for the election and answer each question.

   (a) How many club members voted?

   (b) How many club members voted for white as their first choice?

   (c) How many club members voted for the colors in the order BYW?

   (d) Using the plurality method, determine the winner.

W	B	B	Y	B	Y	W	B	Y	Y	W	B	Y
B	Y	Y	W	Y	W	B	Y	W	W	W	Y	W
Y	W	W	B	W	B	Y	W	B	B	B	W	B

5. Students at a college were asked to rank three improvements that they would like to see at their college. The choices were build a new gymnasium (G), build a swimming pool (S), or build a baseball/football field (B). The votes are summarized in the preference table.

Number of votes	83	56	42	27
First choice	G	S	S	B
Second choice	S	G	B	S
Third choice	B	B	G	G

   (a) How many students voted?

   (b) What option won if the plurality method was used to determine the winner?

6. A college fraternity wanted to provide a free drink stand for people during the Homecoming Parade. They decided to vote on the choice of beverage. The choices

were milk (M), water (W), tea (T), or soda (S). The results of the election are shown in the preference table.

Number of votes	8	6	5	3	2
First choice	S	M	W	T	W
Second choice	W	W	T	S	S
Third choice	T	T	S	W	M
Fourth choice	M	S	M	M	T

(a) How many people voted?

(b) What drink was selected as the winner if the plurality method was used to determine the winner?

7. Students from a sorority voted to select a floral bouquet for their representative in a homecoming parade. The choices were roses (R), gardenias (G), carnations (C), and daisies (D). The preference table is shown.

Number of votes	3	5	2	6	4
First choice	R	D	C	C	R
Second choice	G	C	R	G	C
Third choice	C	G	G	D	D
Fourth choice	D	R	D	R	G

(a) How many students voted?

(b) What flower bouquet won if the plurality method was used to determine the winner?

8. The students in Dr. Lee's mathematics class were asked to vote on the starting time for their final examination. Their choices were 8:00 A.M., 10:00 A.M., 12:00 P.M., or 2:00 P.M. The results of the election are shown in the preference table.

Number of votes	8	12	5	3	2	2
First choice	8	10	12	2	10	8
Second choice	10	8	2	12	12	2
Third choice	12	2	10	8	8	10
Fourth choice	2	12	8	10	2	12

(a) How many students voted?

(b) What time was the final exam if the plurality method was used to determine the winner?

9. Using the election results given in Exercise 5, has the head-to-head comparison criterion been violated? Explain your answer.

10. Using the election results given in Exercise 6, has the head-to-head comparison criterion been violated? Explain your answer.

11. Using the election results given in Exercise 7, has the head-to-head criterion been violated? Explain your answer.

12. Using the election results given in Exercise 8, has the head-to-head criterion been violated? Explain your answer.

Writing Exercises

13. What is a preference table?
14. Explain how to determine a winner when the plurality voting method is used.
15. What is meant by the head-to-head comparison criterion?
16. Is it possible for the plurality voting method to violate the head-to-head comparison criterion?

Critical Thinking

Use the following information for Exercises 17 through 19.

Suppose 100 votes are cast in an election involving three candidates, A, B, and C, and 80 votes are counted so far. The results are

A	36
B	32
C	12

(*Note:* This is not a preference list. Voters are submitting a single name on a ballot.)

17. What is the minimum number of remaining votes candidate A needs to guarantee that he or she wins the election using the plurality method? Explain your answer.
18. What is the minimum number of remaining votes candidate B needs to guarantee that he or she wins the election using the plurality voting method?
19. Can candidate C win the election using the plurality voting method? Explain your answer.
20. If there are 408 votes cast in an election with four candidates, what is the smallest number of votes a candidate needs to win if the plurality method is used?
21. Is it possible to have a tie if the head-to-head comparison method is used? If so, give an example.
22. If an election is held with four candidates and 204 votes are cast, what is the smallest number of votes needed for one candidate to win if the plurality method is used?

13-2 The Borda Count Method and the Plurality-with-Elimination Method

The Borda Count Method

A second method of voting when there are three or more alternatives is called the Borda count method. This method was developed by a French naval captain and mathematician, Jean-Charles de Borda.

> The **Borda count method** of voting requires the voter to rank each candidate from most favorable to least favorable then assigns 1 point to the last-place candidate, 2 points to the next-to-the-last-place candidate, 3 points to the third-from-the-last-place candidate, etc. The points for each candidate are totaled separately, and then the candidate with the most points wins the election.

Example 13-3 shows how to use the Borda count method.

Example   13-3

The preference table for a club presidential election consisting of three candidates is shown. Use the Borda count method to determine the winner.

Number of votes	15	8	3	2
First choice	B	A	C	A
Second choice	C	B	A	C
Third choice	A	C	B	B

Solution

Assign 1 point for the third choice, 2 points for the second choice, and 3 points for the first choice. Then multiply the number of votes by the number of the choice for each candidate to get the total points.

For candidate A:

Because A is ranked third in column 1, A gets 1 point from each of the 15 voters at the top of the column. Since A is ranked first in column 2, A gets 3 points from each of the 8 voters. Since A is ranked second in column 3, A gets 2 points from each of the 3 voters. Finally, since A is ranked first in the last column, A gets 3 points from each of the 2 voters at the top. Hence, $15 \cdot 1 + 8 \cdot 3 + 3 \cdot 2 + 2 \cdot 3 = 51$.

For candidate B:

Because candidate B is ranked first in column 1, B gets 3 points from each of the 15 voters at the top of the column. Since candidate B is ranked second in column 2, B gets 2 points from each of the 8 voters at the top of the column. Candidate B is ranked third in column 3 and gets 1 point from each of the 3 voters at the top of the column. Candidate B is ranked third in column 4 and gets 1 point from each of the voters at the top of the column. Hence, $15 \cdot 3 + 8 \cdot 2 + 3 \cdot 1 + 2 \cdot 1 = 66$.

For candidate C:

Because candidate C was ranked second in column 1, C gets 2 points from each of the 15 voters at the top of the column. Since candidate C is ranked third in column 2, C gets 1 point from each of the 8 voters at the top of the column. Candidate C is ranked first in column 3 and gets 3 points from each of the 3 voters at the top of the column. Candidate C is ranked second in column 4 and gets 2 points from each of the 2 voters at the top of the column. Hence, $15 \cdot 2 + 8 \cdot 1 + 3 \cdot 3 + 2 \cdot 2 = 51$.

In this case, candidate B received the most points; hence, candidate B is declared the winner.

Notice that candidate B also received a majority of first-place votes; hence, candidate B would have also won using the plurality voting method.

The Borda count method, like the plurality method, also has its shortcomings. This method sometimes violates the fairness criterion called the *majority criterion*.

Jean-Charles de Borda (1733–1799)

Jean-Charles de Borda was born in France. As a young man, he served in both the army and the navy. During the Revolutionary War, he commanded the ship *Solitaire* and served with "great distinction."

Besides being a teacher of mathematics, Borda made many contributions to science and architecture.

He wrote papers on the mathematics of projectiles, the construction of naval vessels, the physics of hydraulics, and navigation. He founded a school of naval architecture in France.

In 1770, as a member of the Academy of Science, he devised the voting method that bears his name in order to elect officers of the academy.

The **majority criterion** states that if a candidate receives a majority of first-place votes, then that candidate should be the winner of the election.

Example 13-4 illustrates this.

Example 13-4

Consider the election summarized in the preference table shown here. Show that the winner violates the majority criterion.

Number of votes	15	8	3	2
First choice	B	A	C	A
Second choice	A	C	A	B
Third choice	C	B	B	C

Solution

Find the winner using the Borda count method.

Candidate A:

$$15 \cdot 2 + 8 \cdot 3 + 3 \cdot 2 + 2 \cdot 3 = 66$$

Candidate B:

$$15 \cdot 3 + 8 \cdot 1 + 3 \cdot 1 + 2 \cdot 2 = 60$$

Candidate C:

$$15 \cdot 1 + 8 \cdot 2 + 3 \cdot 3 + 2 \cdot 1 = 42$$

According to the Borda count method, candidate A is the winner, but candidate B received 15 first-place votes, which is a majority. Hence, the majority criterion rule is violated.

Sidelight

Heisman Trophy Winners

The Heisman Memorial Trophy is an award presented to the best college football player of the year. It was originally awarded to the best college football players east of the Mississippi River by the Downtown Athletic Club of New York City in 1935.

In 1936 the rule was changed to include any college player in the United States. The award was called the Heisman Memorial Trophy named after an outstanding college coach and Director of Athletics, John W. Heisman, who died in 1936. The first Heisman Trophy winner was Jay Berwanger.

Voters who receive a ballot for the Heisman Trophy are asked to select and rank their choices for the top three candidates. A Borda count method is used to determine the winner. The top choice gets 3 points, the second choice gets 2 points, and the third choice gets

1 point. Any candidate not listed gets 0 points. The winner is the candidate who gets the highest Borda count.

Try This One

13-C   Four couples, denoted by K, L, M, and N, are running for homecoming parade couple. The preference table for the election is shown.

Number of votes	232	186	95	306
First choice	M	K	M	L
Second choice	K	L	L	K
Third choice	L	N	K	N
Fourth choice	N	M	N	M

(a) Using the Borda count method, determine who won the election.

(b) Does this particular election violate the majority criterion?

For the answers, see page 722.

The Borda count method is used to select the Heisman Trophy winners, the American and National Baseball Leagues' Most Valuable Player awards, and the Country Music Vocalists of the year, among other elections.

The Plurality-with-Elimination Method

The plurality-with-elimination method of voting has been dubbed the "survival of the fittest" method. In this method of voting, if no candidate receives a majority of votes, a series of "rounds" is used. Formally defined,

> When using the **plurality-with-elimination method,** the candidate with the majority of first-place votes is declared the winner. If no candidate has a majority of first-place votes, the candidate (or candidates) with the least number of first-place votes is eliminated, then the candidates who were below the eliminated candidate move up on the ballot, and the number of first-place votes is counted again. If a candidate receives the majority of first-place votes, that candidate is declared the winner. If no candidate receives a majority of first-place votes, the one with the least number of first-place votes is eliminated, and the process continues.

Example 13-5 shows how to use the plurality-with-elimination method.

Example 13-5

Use plurality-with-elimination method to determine the winner of the election shown in the preference table.

Number of votes	6	27	17	9
First choice	A	B	C	D
Second choice	D	A	D	B
Third choice	C	C	B	C
Fourth choice	B	D	A	A

Solution

Round 1: There were 59 votes cast, and no one received a majority of votes (30 or more), so candidate A is eliminated since he has the fewest first-place votes, 6. The other candidates in the first column move up.

Number of votes	6	27	17	9
First choice	A̶	B	C	D
Second choice	D	A̶	D	B
Third choice	C	C	B	C
Fourth choice	B	D	A̶	A̶

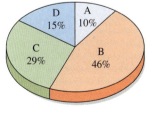

Round 1 results

Round 2: In this round, the 6 first-place votes candidate A received go to candidate D because she moved up in the first column when candidate A was eliminated. But

—Continued

Example 13-5 *Continued—*

candidate D is eliminated because she has the fewest first-place votes, 15, in this round.

Number of votes	6	27	17	9
First choice	D̶	B	C	D̶
Second choice	C	C	D̶	B
Third choice	B	D̶	B	C

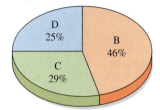

Round 2 results

Round 3: The first-place votes candidate D received in column 1 to go to candidate C while the 9 first-place votes candidate D received in the fourth column go to candidate B. With A and D eliminated, the preference table looks like this.

Number of votes	6	27	17	9
First choice	C	B	C	B
Second choice	B	C	B	C

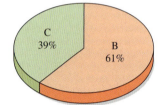

Round 3 results

Now candidate B has 36 $(27 + 9)$ first-place votes and candidate C has 23 $(6 + 17)$; hence, candidate B is declared the winner.

The plurality-with-elimination method, like the other two methods, also has some shortcomings. One shortcoming is that it sometimes fails the fairness criterion known as the *monotonicity criterion*.

> The **monotonicity criterion** states that if a candidate wins an election, and a reelection is held in which the only changes in voting favor the original winning candidate, then that candidate should win the reelection.

Consider this election:

Number of votes	7	13	11	10
First choice	X	Z	Y	X
Second choice	Z	X	Z	Y
Third choice	Y	Y	X	Z

Using the plurality-with-elimination method, candidate Y is eliminated in round 1. With Y eliminated, the preference table looks like this.

Number of votes	7	13	11	10
First choice	X	Z	Z	X
Second choice	Z	X	X	Z

Z wins with 24 first-place votes.

Now suppose the first election was declared invalid for some reason, and on a second election, the voters in column 1 change their ballots in favor of candidate Z and vote ZXY.

The new preference table will be

Number of votes	7	13	11	10
First choice	Z	Z	Y	X
Second choice	X	X	Z	Y
Third choice	Y	Y	X	Z

Here X is eliminated on the first round and the preference table becomes

Number of votes	7	13	11	10
First choice	Z	Z	Y	Y
Second choice	Y	Y	Z	Z

Here Y is the winner with 21 votes compared to 20 votes for Z.

In this case, the plurality-with-elimination method fails the monotonicity criterion. That is, on the second election, even though candidate Z had received seven more first-choice votes, he lost the election! By doing better the second time, the candidate did worse!

Try This One

13-D The planning committee for a company's annual picnic votes for an afternoon activity. The choices are a softball game (S), a touch football game (F), a *bocce* game (B), and a volleyball game (V). The preference table is shown.

Number of votes	3	5	2	1	1
First choice	B	S	V	F	B
Second choice	S	F	S	S	V
Third choice	F	B	B	V	F
Fourth choice	V	V	F	B	S

Determine the winner using the plurality-with-elimination method. Explain your answer. *For the answer, see page 722.*

The plurality-with-elimination voting method was used by the Olympic Committee to select a city for the 2000 Summer Olympics. This method is also used by the Academy of Motion Pictures to determine its academy awards.

Exercise Set 13-2

Real World Applications

1. A book discussion club wishes to decide which type of book to read and discuss at their next meeting. The choices are a novel (N), a biography (B), or a science fiction book (S). The preference table for the results is shown.

Number of votes	8	5	4	2
First choice	S	N	N	B
Second choice	B	S	B	N
Third choice	N	B	S	S

Using the Borda count method of voting, determine the winner.

2. The McKees' Point Yacht Club Board of Directors wants to decide where to hold their fall business meeting. The choices are the Country Club (C), Frankie's Fine Foods (F), West Oak Golf Club (W), and Rosa's Restaurant (R). The results of the election are shown in the preference table.

Number of votes	8	6	5	2
First choice	R	W	C	F
Second choice	W	R	F	R
Third choice	C	F	R	C
Fourth choice	F	C	W	W

Determine the winner using the Borda count method of voting.

3. A local movie theater asks its patrons which movies they would like to view during next month's "Oldies but Goodies" week. The choices are *Gone with the Wind* (G), *Casablanca* (C), *Anatomy of a Murder* (A), and *Back to the Future* (B). The preference table is shown.

Number of votes	331	317	206	98
First choice	G	A	C	B
Second choice	A	C	B	G
Third choice	C	B	G	A
Fourth choice	B	G	A	C

Use the Borda count voting method to determine the winner.

4. Parents of a kindergarten class are given the option of choosing a location for the annual spring field trip. The choices are the zoo (Z), the local amusement park (A), or the local museum (M). The preference table is shown.

Number of votes	15	9	5	4
First choice	Z	M	A	Z
Second choice	A	Z	M	M
Third choice	M	A	Z	A

Using the Borda count voting method, determine the winner.

5. Students at a college were asked to rank three improvements that they would like to see at their college. Their choices were build a new gymnasium (G), build a baseball/football field (B), or build a swimming pool (S). The votes are summarized in the preference table.

Number of votes	83	56	42	27
First choice	G	S	S	B
Second choice	S	G	B	S
Third choice	B	B	G	G

(a) Using the Borda count method of voting, determine the winner.

(b) Refer to Exercise 5 in Section 13-1. Is the winner the same as the one determined by the plurality method?

6. A college fraternity wanted to provide a free drink stand for people during the homecoming parade. They decided to vote on the choice of beverage. The choices

were milk (M), water (W), iced tea (T), or soda (S). The results of the election are shown in the preference table.

Number of votes	8	6	5	3	2
First choice	S	M	W	T	W
Second choice	W	W	T	S	S
Third choice	T	T	S	W	M
Fourth choice	M	S	M	M	T

(a) Using the Borda count method of voting, determine the winner.

(b) Refer to Exercise 6 in Section 13-1. Is the winner the same as the one determined by the plurality method?

7. Does the election in Exercise 1 violate the majority criterion?

8. Does the election in Exercise 2 violate the majority criterion?

9. Does the election in Exercise 3 violate the majority criterion?

10. Does the election in Exercise 4 violate the majority criterion?

11. Does the election in Exercise 5 violate the majority criterion?

12. Does the election in Exercise 6 violate the majority criterion?

13. The English department is voting for a new department chairperson. The three candidates are Professor Greene (G), Professor Williams (W), and Professor Donovan (D). The results of the election are shown in the preference table.

Number of votes	4	3	2
First choice	D	W	G
Second choice	W	G	D
Third choice	G	D	W

Using the plurality-with-elimination method of voting, determine the winner.

14. The Association of Self-Employed Working Persons must select a speaker for their next meeting. The choices for a topic are health care (H), investments (I), or advertising (A). The results of the election are shown in the preference table.

Number of votes	6	4	9	2
First choice	H	H	I	I
Second choice	I	A	H	A
Third choice	A	I	A	H

Using the plurality-with-elimination method of voting, determine the winner.

15. Students in a sorority voted to select a floral bouquet for their representative to wear in the homecoming parade. The choices included roses (R), gardenias (G), carnations (C), and daisies (D). The preference table is shown.

Number of votes	3	5	2	6	4
First choice	R	D	C	C	R
Second choice	G	C	R	G	C
Third choice	C	G	G	D	D
Fourth choice	D	R	D	R	G

(a) Using the plurality-with-elimination method, determine the winner.

(b) Is the winner the same as the one determined by the plurality method used in Exercise 7 of Section 13-1?

16. The students in Dr. Lee's mathematics class were asked to vote on the starting time for the final examination. Their choices were 8:00 A.M., 10:00 A.M., 12:00 P.M., or 2:00 P.M. The results of the election are shown in the preference table.

Number of votes	8	12	5	3	2	2
First choice	8	10	12	2	10	8
Second choice	10	8	2	12	12	2
Third choice	12	2	10	8	8	10
Fourth choice	2	12	8	10	2	12

(a) Using the plurality-with-elimination method, determine the winner.

(b) Is the winner the same as the one determined by using the plurality method used in Exercise 8 of Section 13-1?

17. Determine whether the election shown in Exercise 13 violates the monotonicity criterion if Professor Greene cannot serve.

18. Determine whether the election shown in Exercise 14 violates the monotonicity criterion if the health care speaker cannot be there.

19. Determine whether the election shown in Exercise 15 violates the monotonicity criterion if the florist does not have any daisies.

20. Determine whether the election shown in Exercise 16 violates the monotonicity criterion if no room is available at 8:00 A.M.

Writing Exercises

21. Explain how to determine the winner of an election using the Borda count method.

22. Explain what is meant by the majority criterion.

23. Explain how to determine the winner of an election using the plurality-with-elimination method.

24. What is meant by the monotonicity criterion?

Critical Thinking

25. Construct a preference table for an election involving three candidates such that candidate A wins the election using the Borda count method, but the majority criterion is violated.

26. Construct a preference table for an election involving three candidates such that candidate B wins using the plurality-with-elimination method, but the monotonicity criterion is violated.

27. Construct a preference table for an election such that one candidate wins the election using the Borda count method and a different candidate wins the same election using the plurality-with-elimination method.

28. If the candidates on a preference ballot are ranked such that the lowest candidate gets 0 points, the next to the lowest candidate gets 1 point and so on, will the winner be the same using the Borda count method as the winner when the candidates are ranked the way they are explained in this section? Explain your answer using an illustration.

29. If the candidates on a preference ballot are ranked by giving the top candidate a score of 1 (first place), the next highest candidate a score of 2 (second place) and so on, can the Borda count method be used to determine a winner? Explain why or why not.

13-3 The Pairwise Comparison Method and Approval Voting

The Pairwise Comparison Method

The pairwise comparison method uses the preference table to compare each pair of candidates. For example, if there are four candidates, A, B, C, and D, running in an election, then there would be six comparisons, as shown.

A vs. B	B vs. C
A vs. C	B vs. D
A vs. D	C vs. D

In general, the number of pairwise comparisons when n candidates are running is equal to

$$\frac{n(n-1)}{2}$$

If there were five candidates running, you would need to make

$$_5C_2 = \frac{5(5-1)}{2} = \frac{5 \cdot 4}{2} = \frac{20}{2} = 10 \text{ pairwise comparisons}$$

Math Note

The formula for the number of possible pairwise comparisons of n candidates in an election can be found using the combination formula in Chapter 11 where $r = 2$:

$$_nC_2 = \frac{n!}{(n-2)!2!}$$

$$_nC_2 = \frac{n!}{(n-2)!2!}$$

$$= \frac{n \cdot (n-1)(n-2)!}{(n-2)!2!}$$

$$= \frac{n(n-1)}{2}$$

Example 13-6

Find the number of pairwise comparisons needed if six candidates are running in an election.

Solution

Here,

$$n = 6$$

So

$$\frac{6(6-1)}{2} = \frac{6(5)}{2} = 15$$

Hence, you would need to make 15 comparisons.

Try This One

13-E Find the number of pairwise comparisons needed if eight candidates are running in an election.
For the answer, see page 722.

When making pairwise comparisons between candidates, say A and B, on a preference table, count the number of ballots where A is ranked higher than B, and the number of ballots where B is ranked higher than A. The one with the most votes is given 1 point. In case of a tie, give each candidate $\frac{1}{2}$ point. After all pairwise comparisons are made, tally the points. The winner is the candidate with the most first-place points. Formally defined

> The **pairwise comparison method** of voting requires that all candidates be ranked by the voters. Then each candidate is paired with every other candidate in a one-to-one contest. For each one-to-one comparison, the candidate who wins on more ballots gets 1 point. In case of a tie, each candidate gets $\frac{1}{2}$ point. After all possible two-candidate comparisons are made, the points for each candidate are tallied, and the candidate with the most points wins the election.

Example 13-7 shows how to use the pairwise comparison method.

Example 13-7

Use the pairwise comparison method to find the winner of the elections whose results are shown in the following preference table.

Number of votes	14	13	16	15
First choice	B	A	C	B
Second choice	C	C	A	A
Third choice	A	B	B	C

Solution

You will need to make three pairwise comparisons A vs. B, A vs. C, and B vs. C.

First, A vs. B:

Number of votes	14	13	16	15
First choice	(B)	(A)	C	(B)
Second choice	C	C	(A)	(A)
Third choice	(A)	(B)	(B)	C

A is ranked higher than B in columns 2 and 3; therefore, the sum of the numbers at the top of columns 2 and 3 gives A $13 + 16 = 29$ votes. B is ranked higher than A in columns 1 and 3, so B gets $14 + 15 = 29$ votes. In this case, it is a tie, so assign $\frac{1}{2}$ point to A and $\frac{1}{2}$ point to B.

Next, compare A to C:

Number of votes	14	13	16	15
First choice	B	(A)	(C)	B
Second choice	(C)	(C)	(A)	(A)
Third choice	(A)	B	B	(C)

—Continued

Example 13-7 Continued—

A is ranked higher than C in columns 2 and 4, so A gets $13 + 15 = 28$ votes. C is ranked higher than A in columns 1 and 3, so C gets $14 + 16 = 30$ votes. Since C has more votes, assign 1 point to C.

Finally, compare B to C:

Number of votes	14	13	16	15
First choice	B	A	C	B
Second choice	C	C	A	A
Third choice	A	B	B	C

B is ranked higher than C in columns 1 and 4, so B gets $14 + 15 = 29$ votes. C is ranked higher than B in columns 2 and 3, so C gets $13 + 16 = 29$ votes. Since this is a tie, assign $\frac{1}{2}$ point to B and $\frac{1}{2}$ point to C.

Now find the totals.

		Total
Candidate A	$\frac{1}{2}$	$\frac{1}{2}$
Candidate B	$\frac{1}{2} + \frac{1}{2}$	1
Candidate C	$1 + \frac{1}{2}$	$1\frac{1}{2}$

Hence, candidate C is the winner.

It can be shown that the pairwise comparison voting method satisfies the majority criterion, the head-to-head criterion, and the monotonicity criterion, so this method seems at first glance to be an ideal voting method. However, there is a flaw in this voting method. It sometimes fails to satisfy the fairness criterion called the *irrelevant alternatives criterion*.

The **irrelevant alternatives criterion** requires that if a certain candidate X wins an election and one of the other candidates is removed from the ballot and the ballots are recounted, candidate X still wins the election.

In other words, if X wins, it should not matter what happens to the other candidates. X should win regardless of the removal of a defeated competing candidate.

Consider the election in the previous example. Suppose that for some reason after the fact, candidate A is declared ineligible. What happens now?

Consider the preference table with candidate A removed.

Number of votes	14	13	16	15
First place	B	C	C	B
Second place	C	B	B	C

There are $14 + 15 = 29$ voters who preferred B to C, and there are $13 + 16 = 29$ voters who preferred C to B. Hence, a tie has occurred. This violates the irrelevant alternatives criterion because C should still win no matter which other candidate drops out.

Another problem with the pairwise comparison method is that it is possible for all candidates to tie. Consider the next preference table.

Number of votes	8	5	7
First choice	A	C	B
Second choice	B	A	C
Third choice	C	B	A

Here A wins over B (13 to 7), B wins over C (15 to 5), and C wins over A (12 to 8); hence, there is a three-way tie for first place. Using the pairwise comparison, it is possible that in some cases, no winner can be declared. In all voting methods, the possibility of a tie should be considered before the votes are counted and some way of breaking a tie should be agreed on in advance of the election.

Try This One

13-F   The members of the Music Appreciation Club must decide whether they will attend an opera (O), a symphony (S), or a ballet (B). The results of the election are shown in the preference table.

Number of votes	13	9	6	3
First choice	O	B	S	B
Second choice	S	O	B	S
Third choice	B	S	O	O

Use the pairwise comparison voting method to determine the winning selection. Does this election violate the irrelevant alternatives criterion if the symphony is sold out and the members cannot attend it?
For the answer, see page 722.

The next table summarizes the four voting methods and the marks show which criterions are always satisfied by the voting methods.

	Head-to-head criterion	Majority criterion	Monotonicity criterion	Irrelevant alternatives criterion
Plurality		✓	✓	
Borda count			✓	
Plurality-with-elimination		✓		
Pairwise comparison	✓	✓	✓	

Arrow's Impossibility Theorem

As shown in the previous sections, none of the four voting methods is perfectly fair. Each violates one or more of the four fairness criteria: the majority criterion, the head-to-head criterion, the monotonicity criterion, or the irrelevant alternatives criterion.

Is there a voting method that satisfies all the criterion? The answer is *no!* In 1951, a famous economist, Kenneth Arrow, proved that there does not exist and never will exist a democratic voting method for three or more alternatives that will satisfy all of the fairness criteria discussed in this chapter. His theorem is called **Arrow's impossibility theorem.** (The proof is beyond the scope of this book.)

Sidelight

Kenneth J. Arrow (1921–   )

Kenneth J. Arrow received a B.S. in Social Science from the City College of New York and an M.A. and a Ph.D. from Columbia University. His research interests include the economics of information and organization, collective decision making, and general equilibrium theory.

In 1951, he proved the theorem that states that no voting system will satisfy all of the four fairness criteria. The theorem became known as Arrow's impossibility theorem.

In 1972, he received the Nobel Peace Prize in Economics for his work in social-choice theory. In 1986, he won the Von Neuman Theory Prize for his contributions to decision sciences.

He is a professor of Economics (Emeritus) at Stanford University. He has many professional affiliations including being past president of the Institute of Management Sciences.

Approval Voting

In the late 1970s, several political scientists and economists introduced a voting method called *approval voting*.

> With **approval voting,** each voter gives one vote to as many candidates on the ballot as he or she finds acceptable. The votes are counted, and the winner is the candidate who receives the most votes.

In this case, voters can select from no candidates to all candidates. Example 13-8 shows how approval voting works.

Example 13-8

Five candidates are nominated for the teacher of the year award, and there are 20 teachers who will vote. The results are shown in the table. Which candidate wins?

Number of votes	9	3	2	5	1
Candidate A	/		/		/
Candidate B		/	/	/	
Candidate C		/	/	/	
Candidate D	/	/		/	
Candidate E				/	

—Continued

Example 13-8 Continued—

Solution

From the table, count the number of votes for each candidate:

Candidate	Votes
A	12
B	10
C	10
D	17
E	5

In this election, candidate D received 17 votes and is declared the winner.

Try This One

13-G An election was held for employee of the month award using the approval voting method. The results are shown in the table. Which candidate won the election?

Number of votes	20	18	12	4
Candidate F		/		/
Candidate G	/		/	/
Candidate H		/	/	/
Candidate I	/			/
Candidate J			/	/

For the answer, see page 722.

Several political scientists and analysts independently developed approval voting in the late 1970s. This method is now used to elect the Secretary General of the United Nations, and it is also used to elect the leaders of some academic and professional societies such as the National Academy of Sciences. The advantages of approval voting are that it is simple to use and easy to understand. The ballot is also uncomplicated. There are some disadvantages, though. The major one is that there is no ranking or preference of the candidates. Most voters have favorite choices, but there is no way to indicate these preferences on an approval ballot. All votes for candidates are considered equal. The approval voting method then is another method that is being used today.

Tie Breaking

Regardless of the voting method selected, the possibility of a tie between two or more candidates always exists. There are many ways to break a tie and a fair tie-breaking method should always be decided upon in advance of an election.

In some cases, the chairperson of a committee does not vote on motions unless there is a tie. In this case, the chairperson would cast the tie-breaking vote. The most obvious method of breaking a tie is, of course, the age-old method of "flipping a coin." In other cases, drawing a name from a hat could be used. Using a third-party judge could be considered. For example, if a tie occurs in an election of a department chairperson, the dean could decide the winner. Another possibility might be to consider some other criteria such as seniority, education, or experience of the candidates.

Finally, in case of a tie, another voting method could be used. For example, if a tie occurred using the Borda count method, then perhaps the pairwise comparison method could be used to determine a winner.

Exercise Set 13-3

Real World Applications

1. If there are four candidates in an election, how many pairwise comparisons need to be made in order to determine a winner?

2. If there are seven candidates in an election, how many pairwise comparisons need to be made in order to determine a winner?

3. If there are 10 candidates in an election, how many pairwise comparisons need to be made in order to determine a winner?

4. If there are nine candidates in an election, how many pairwise comparisons need to be made in order to determine a winner?

5. A college band was invited to perform at three different shows on the same day. They were Real Town (R), Steel Center (S), and Temple Village (V). Since they could only perform at one show, they voted on which one they would do. The results are shown in the preference table.

Number of votes	26	19	15	6
First choice	R	T	S	R
Second choice	S	S	R	T
Third choice	T	R	T	S

Using the pairwise comparison voting method, determine the winner.

6. The senior class at a small high school must vote on the senior class trip. Their selections are Disneyland (D), Epcot Center (E), Sea World (S), and Six Flags Resort (F). The preference table is shown.

Number of voters	19	12	8	6
First choice	D	E	S	D
Second choice	E	S	D	F
Third choice	S	D	F	S
Fourth choice	F	F	E	E

Using the pairwise comparison voting method, determine the winner.

7. The English department is voting for a new department chairperson. The three candidates are Professor Greene (G), Professor Williams (W), and Professor Donovan (D). The results of the election are shown in the preference table.

Number of votes	4	3	2
First choice	D	W	G
Second choice	W	G	D
Third choice	G	D	W

(a) Determine the winner using the pairwise comparison voting method.

(b) Compare the winner with the one determined by the plurality-with-elimination method. See Exercise 13 in Exercise Section 13-2.

8. A book discussion club wishes to decide which type of book to read and discuss at their next meeting. The choices are a novel (N), a biography (B), or a science fiction book (S). The preference table is shown.

Number of votes	8	5	4	2
First choice	S	N	N	B
Second choice	B	S	B	N
Third choice	N	B	S	S

(a) Determine the winner using the pairwise comparison method.

(b) Compare the winner with the one determined by the Borda counting method used in Exercise 1 of Exercise Section 13-2.

9. The McKees' Point Yacht Club Board of Directors wants to decide where to hold their fall business meeting. The choices are the Country Club (C), Frankie's Fine Foods (F), West Oak Golf Club (W), and Rosa's Restaurant (R). The results of the election are shown in the preference table.

Number of votes	8	6	5	2
First choice	R	W	C	F
Second choice	W	R	F	R
Third choice	C	F	R	C
Fourth choice	F	C	W	W

(a) Determine the winner using the pairwise comparison method.

(b) Is this the same winner as the one using the Borda count method in Exercise 2 in Exercise Section 13-2?

10. The students in Dr. Lee's mathematics class are asked to vote on the starting time for their final examination. Their choices are 8:00 A.M., 10:00 A.M., 12:00 P.M., or 2:00 P.M. The results of the election are shown in the preference table.

Number of votes	8	12	5	3	2	2
First choice	8	10	12	2	10	8
Second choice	10	8	2	12	12	2
Third choice	12	2	10	8	8	10
Fourth choice	2	12	8	10	2	12

(a) Determine the starting time using the pairwise comparison method.

(b) Is this the same starting time as determined by the plurality-with-elimination method used in Exercise 16 of Exercise Section 13-2?

11. If Professor Donovan was unable to serve as Department Chairperson in the election shown in Exercise 7 and the votes are recounted, does this election violate the irrelevant alternative criterion?

12. If the club members decide to eliminate the science fiction book and the votes were recounted in the election shown in Exercise 8, is the irrelevant alternative criterion violated?

13. If the White Oak Golf Club is unavailable and the votes were recounted in the election shown in Exercise 9, is the irrelevant alternatives criterion violated?

14. If a room for Dr. Lee's final examination was not available at 2:00 P.M. and the votes were recounted in the election shown in Exercise 10, is the irrelevant alternatives criterion violated?

15. A sports committee of students must select a team physician. The result of the voting is shown. The approval method will be used.

Number of votes	15	18	12	10	5
Dr. Michaels	/		/	/	/
Dr. Jones		/	/	/	/
Dr. Philip	/	/		/	
Dr. Smith	/	/		/	

Which physician was selected?

16. The students voted for a mascot for their soccer team. The results of the election are shown. Using the approval voting method determine the winner.

Number of votes	235	531	436	374
Ravens	/		/	/
Panthers		/	/	/
Killer Bees	/	/	/	/
Termites		/		/

17. A nursing school committee decides to purchase a van for the school. They vote on a color using the approval voting method. The results are shown.

Number of votes	1	2	1	3	2	1
White	/		/	/	/	
Blue	/	/	/			
Green		/	/	/	/	
Silver			/	/		/

Which color was selected?

18. A research committee decides to test market a new flavor for a children's drink. The results of a survey at a local mall are shown.

Number of votes	38	32	16	5	3
Strawberry	/		/		/
Lime	/		/	/	/
Grape		/	/		
Orange			/	/	/
Bubble gum	/	/		/	

Which flavor was selected?

19. A parole board must release one prisoner on good behavior. After hearing the case, they decide to use the approval voting method. The result is shown here.

Number of votes	1	1	1	1	1	1
Inmate W	/		/		/	/
Inmate X		/	/		/	
Inmate Y	/	/	/			
Inmate Z		/	/	/	/	/

Which inmate was related?

20. The park association committee decided to make one improvement for the local park this spring. The result of an election using approval voting is shown.

Number of votes	2	1	3	4	2
Paint benches	/		/		
Trim bushes			/	/	/
Repair snack bar		/	/	/	
Patch cement walks	/	/	/	/	/

Which repair was made?

Writing Exercises

21. Explain how the winner is determined using the pairwise comparison method.
22. Explain the irrelevant alternatives criterion.
23. Explain how the winner is determined using the approval voting method.
24. Explain Arrow's impossibility theorem.

Critical Thinking

25. Construct a preference table with three candidates, X, Y, and Z, such that a candidate wins the election using all four voting methods.
26. Construct a preference table in which a candidate wins using the pairwise comparison method but that violates the irrelevant alternatives criterion.
27. Under the approval voting method, if a voter likes (or dislikes) all candidates equally, should the voter give every candidate one vote or every candidate no vote?

Summary

Section	Important Terms	Important Ideas
13-1	preference table plurality method head-to-head comparison criterion	**This** chapter presented four voting methods people have used to determine the results of an election where there are three or more choices (i.e., candidates, courses of action, etc.). These methods involve ranking the choices in order of preference on a preference ballot. A preference table can be used to summarize the results of the election. Using the plurality method, the candidate with the most first-place votes is the winner. Each method has at least one inherent flaw. That is, it fails to satisfy one of the fairness criteria for a voting method. The head-to-head criterion states that if a candidate wins all head-to-head comparisons with all other candidates, that candidate should win the election.
13-2	Borda count method majority criterion plurality-with-elimination method monotonicity criterion	**The** Borda count method states that the candidates are ranked on the ballot by voters. The candidate in last place on the ballot gets 1 point. The candidate in the next to the last place gets 2 points, etc. The points are tallied and the candidate with the most points is the winner. The majority criterion says that if a candidate receives a majority of first-place votes, then that candidate should be the winner. Plurality-with-elimination states that the candidate with a majority of first-place votes is the winner. If no candidate has a majority of first-place votes, the candidate with the least number of votes is eliminated, and all candidates who were below the eliminated candidate move up. The first-place votes are counted again. The process then is repeated until a candidate receives a majority of first-place votes. The monotonicity criterion states that if after an election a reelection is held with the same candidates and the only changes in voting favor the winner of the original election, then he or she should win the reelection.
13-3	pairwise comparison method irrelevant alternatives criterion Arrow's impossibility theorem approval voting	**The** pairwise comparison method states that each candidate is ranked by the voters. Then each candidate is paired with every other candidate in a head-to-head or one-to-one contest. The winner of each contest gets 1 point. In case of a tie, each candidate gets $\frac{1}{2}$ of a point. The candidate with the most points wins the election. The irrelevant alternatives criterion states that if candidate A wins a certain election and if one of the other candidates is removed from the ballot and the ballots are recounted, candidate A should still win the election. The search for a perfect voting method continued until the early 1950s when an economist, Kenneth Arrow, proved a theorem that states that in elections with three or more candidates, a voting method that is always fair and democratic is an impossibility. This fact is known as Arrow's impossibility theorem. In the late 1970s several political scientists devised a voting method that is called approval voting. Here each voter votes for as many candidates as he or she finds acceptable. The votes are tabulated and the candidate who receives the most votes is declared the winner.

Review Exercises

Use this information for Exercises 1–4: The preference ballots for an election for the best speaker in a contest are shown. There are three candidates: Peterson (P), Quincy (Q), and Ross (R).

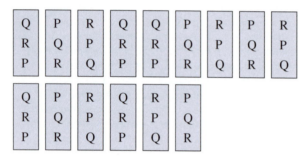

1. Construct a preference table for the results of the election.

2. How many people voted in the election?

3. How many people voted for Ross as the best speaker?

4. How many people voted for the contestants in the order Quincy, Ross, Peterson?

Use this information for Exercises 5–8: A class of students decided to have a pizza party on the last day of school. They voted to select a pizza eatery to get pizza. The candidates were Pizza Palace (P), Pizza Heaven (H), and Pizza City (C). The preference ballots are shown.

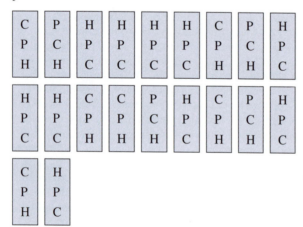

5. Construct a preference table for the results of the election.

6. How many people voted?

7. How many people voted for Pizza City as their first choice?

8. How many people voted for the Pizza Palace, Pizza City, and Pizza Heaven in that order?

Use this information for Exercises 9–17: A large city police department has the option to choose among three styles of bulletproof vests. They are A, B, and C. The preference table is shown.

Number of votes	26	15	10	7
First choice	A	B	C	B
Second choice	B	C	A	A
Third choice	C	A	B	C

9. How many police officers voted?

10. How many votes did the preference CAB receive?

11. Using the plurality method, what style won?

12. Using the Borda count method, what style won?

13. Using the plurality-with-elimination method, what style won?

14. Using the pairwise comparison method, what style won the election?

15. When the plurality method was used, was the head-to-head comparison criterion violated?

16. When the plurality-with-elimination method was used, was the majority criterion violated?

17. If style C was unavailable, and the votes were recounted, was the irrelevant alternative criterion violated if the pairwise comparison voting method is used?

Use this information for Exercises 18–27: A college theater class must decide which musical to perform as part of their course requirements. Their options are *Chorus Line* (C), *Guys and Dolls* (G), *Bye Bye Birdie* (B), and *Music Man* (M). They decide to vote on the musical, and the preference table is shown.

Number of votes	18	17	9	3
First choice	M	B	G	C
Second choice	B	G	C	G
Third choice	G	C	M	B
Fourth choice	C	M	B	M

18. How many votes did the preference GCMB receive?

19. How many students voted in the election?

20. Using the Borda count method which musical was selected?

21. Using the plurality method which musical was selected?

22. Using the pairwise comparison method which musical was selected?

23. Using the plurality-with-elimination method which musical was selected?

24. When the Borda count method was used, was the majority criterion violated?

25. When the plurality method was used, was the head-to-head comparison criterion violated?

26. If *Chorus Line* was too expensive to produce and had to be removed and the votes were recounted, was the irrelevant alternatives criterion violated if the pairwise comparison voting method is used?

27. When the plurality-with-elimination method was used, was the monotonicity criterion violated?

28. The play committee decided to vote on a musical for their next production. The results of the approval voting are shown.

Number of voters	3	1	1	2	4	1	1
Sound of Music	/		/		/		/
Cats	/	/	/				
Oklahoma			/	/		/	/	/
South Pacific					/	/	
Music Man		/				/	/

Which musical won the election?

29. The members of the student government must decide on a program for the students. The results of the approval voting are shown.

Number of voters	1	5	3	1	1	2
Magician	/			/	/	/
Speaker		/	/		/	
Rock band	/	/	/			
Comedian				/	/	/

What will be on the program?

Chapter Test

Use this information for Exercises 1–4: Customers in a coffee shop are asked to rank three brands of coffee. They are brand A, B, and C. The preference ballots are shown.

C	B	B	C	B	A	C	B	B
B	A	A	B	A	B	B	A	A
A	C	C	A	C	C	A	C	C

C	B	B
B	A	A
A	C	C

1. Construct a preference table for the ballots.

2. How many people voted?

3. How many people voted for brand B as their first choice?

4. How many people selected CBA as their ranking preference?

Use this information for Exercises 5–14: A small contracting company has been invited to relocate to one of three cities: Pittsburgh (P), Baltimore (B), and Richmond (R). The employees are asked to vote on the city. The preference table is shown.

Number of votes	43	27	18	12
First choice	P	R	B	P
Second choice	R	P	R	B
Third choice	B	B	P	R

5. How many employees voted?

6. How many selected the preference PBR?

7. Using the plurality method which city won?

8. Using the Borda count method which city won?

9. Using the plurality-with-elimination method which city won?

10. Using the pairwise comparison method which city won?

11. When the plurality method was used, was the head-to-head criterion violated?

12. When the Borda count method was used, was the majority criterion violated?

13. When the plurality-with-elimination method was used, was the head-to-head criterion violated?

14. If Baltimore was ruled out and the votes were recounted, was the irrelevant alternatives criterion violated if the Borda count method was used?

15. A sports committee of students must select a team physician. The result of the voting is shown. The approval method will be used.

Number of votes	15	18	12	10	5
Dr. Michaels	/		/	/	/
Dr. Jones		/	/	/	/
Dr. Philip	/	/		/	
Dr. Spoz	/	/		/	

Which physician was selected?

16. The students voted for a mascot for their soccer team. The result of the election is shown. Using the approval voting method, determine the winner.

Number of votes	235	531	436	374
Tigers	/		/	/
Ravens	/	/		
Panthers		/	/	/
Killer Bees	/	/		/
Termites		/		/

Projects

1. Research and write a paper explaining the Electoral College and how the electors elect our presidents. Explain.

2. How is it possible for a candidate to win the popular vote yet lose the presidential election in the Electoral College? Write a research paper explaining your answer.

3. Explain how the number of Representatives in the House of Representatives is determined by "apportionment."

Mathematics in Our World
▶Revisited
Academy Awards

As stated previously, each branch of the Academy of Motion Picture Arts and Sciences nominates five candidates for the award for a specific category. They use the plurality-with-elimination voting method to come up with the candidates. In each round, the candidate with the fewest first-place votes is eliminated. After the top five nominees are selected, then each member of the Academy votes for one candidate. The candidate with the most votes (i.e., the plurality method) is determined to be the winner.

Answers to **Try This One**

13-A.

Number of voters	2	3	3	2
First choice	A	C	C	C
Second choice	B	D	A	B
Third choice	C	B	B	D
Fourth choice	D	A	D	A

8 people voted for the Crab Shack as their first choice.

13-B. (a) Professor Lane won with five first-place votes.

(b) Yes. There is no head-to-head winner.

13-C. (a) The winner is K with 2548 points.

(b) No. No candidate received a majority of votes.

13-D. First remove F. Then remove V. S wins with nine first-choice votes.

13-E. 28

13-F. There is a three-way tie. Hence, there is no winner.

13-G. Candidate G wins with 36 votes.

Appendix A

Measurement

Introduction

Throughout history, people have developed methods of measuring things that they use in their daily lives. The first units of measurement were based on human body parts. For example, a **cubit,** used by the Egyptians in 3000 B.C.E., was equal to the length of a person's arm from his or her elbow to the outstretched middle finger. Another unit of measurement was called the **palm.** This was the distance across the base of the four fingers of your hands. A **digit** was the thickness of a person's middle finger. As you can readily see, these units are not very precise.

King Henry I of England decreed that the distance from the tip of his nose to the end of his thumb should be the official length of 1 **yard.** In the 13th century, King Edward I of England proclaimed that one third of a yard should be called a **foot.** Queen Elizabeth changed the measure of a mile from 5000 feet to 5280 feet because 1 furlong equaled 660 feet and a mile would become exactly 8 furlongs long. Because England was a world power at that time and had many colonies throughout the world, the **English system of measurement** became established as the recognized measurement system of the world.

The English system uses units such as feet, yards, pounds, quarts, gallons, etc. It is a very complicated system for a person to learn; however, most of us feel comfortable with it since we have been using it all of our lives.

Around 1670, a Vicar of St. Paul in Lyons, France, named Gabriel Mouton proposed a standard system of measurement based on measurements taken from the Earth's surface. The standard unit of length was based on the length of one minute ($\frac{1}{60}$ of a degree) of the arc of a great circle on Earth. A great circle is any circumference measure of the Earth.

Not much was done with the new system until 1790 when the French Academy of Sciences devised a new measurement system based on Mouton's ideas, and it became what is called the **metric** system. This system uses units based on the physical universe. The basic units of length, volume, and mass are interrelated, and the units are based on the decimal or base ten system. The metric system uses units such as meter, gram, and liter.

The metric system is used in almost all the countries of the world as the standard system of measure. It is even used somewhat in the United States, and there are ongoing efforts to make it the standard system of measurement in the United States.

Converting from One Unit of Measure to Another Unit of Measure

The units of length that are used in the United States are the inch, foot, yard, and mile. The basic conversion factors are shown in the next table.

Units of Length in the English System

12 inches (in.) = 1 foot (ft)

3 feet = 1 yard (yd)

36 inches = 1 yard

5280 feet = 1 mile (mi)

1760 yards = 1 mile

There are several ways to convert from one unit of measure to another. Many people use the "common sense" method. For example, to convert 6 yards to feet, multiply 6×3 feet = 18 feet, since there are 3 feet in 1 yard. To convert 564 inches to feet, divide 564 inches by 12 = 47 feet, since there are 12 inches in 1 foot.

A second way to convert one unit of measure to another is to use what is called **dimensional analysis.** Dimensional analysis uses what are called **unit fractions.** *A unit fraction is a fraction consisting of a conversion factor (from the table) having a numerator or denominator which is equal to 1.* For example, since 1 yard is equal to 3 feet, two unit fractions can be written.

$$\frac{1 \text{ yd}}{3 \text{ ft}} \quad \text{and} \quad \frac{3 \text{ ft}}{1 \text{ yd}}$$

Since 1 mile is equal to 5280 feet, two unit fractions can be written.

$$\frac{1 \text{ mi}}{5280 \text{ ft}} \quad \text{and} \quad \frac{5280 \text{ ft}}{1 \text{ mi}}$$

The value of a unit fraction is always equal to one.

To convert from one unit of measure to another unit of measure using dimensional analysis: multiply by one or more unit fractions so that the first measurement unit cancels out and the second measurement unit remains in the numerator, which will be retained in the final answer after multiplication.

Although this may sound complicated, it is really quite simple once you get the hang of it. Examples 1 through 4 will explain how to use dimensional analysis.

Example 1

Convert 6 yards to feet.

Solution

We need to start with 6 yards and end up with feet. The unit fraction to use in multiplication is

$$\frac{3 \text{ ft}}{1 \text{ yd}}$$

—Continued

Example 1   *Continued—*

This unit fraction is used because feet will appear in the numerator and the yards will cancel out in the multiplication as shown.

$$6 \text{ yd} = \frac{6 \cancel{\text{yd}}}{1} \cdot \frac{3 \text{ ft}}{1 \cancel{\text{yd}}} = 6 \cdot 3 \text{ ft} = 18 \text{ ft}$$

(Notice the yards cancel out and feet is left in the numerator.)

Example 2

Convert 546 inches to feet.

Solution

Start with 564 inches and multiply by the unit fraction $\frac{1 \text{ ft}}{12 \text{ in.}}$ as shown.

$$564 \text{ in.} = \frac{564 \cancel{\text{in.}}}{1} \cdot \frac{1 \text{ ft}}{12 \cancel{\text{in.}}} = \frac{564 \text{ ft}}{12} = 47 \text{ ft}$$

Example 3

Convert 23,760 feet to miles.

Solution

Start with 23,760 feet and multiply by the unit fraction $\frac{1 \text{ mi}}{5280 \text{ ft}}$ as shown.

$$23,760 \text{ ft} = \frac{23,760 \cancel{\text{ft}}}{1} \cdot \frac{1 \text{ mi}}{5280 \cancel{\text{ft}}} = \frac{23,760 \text{ mi}}{5280} \approx 4.5 \text{ mi}$$

Sometimes it is necessary to multiply by two or more unit fractions as shown in the next example.

Example 4

Convert 7 miles to inches.

Solution

Looking at the table, we can find no conversion factor for the number of inches in 1 mile, so we have to go from miles to feet $\left(\frac{5280 \text{ ft}}{1 \text{ mi}} \right)$ then from feet to inches $\left(\frac{12 \text{ in.}}{1 \text{ ft}} \right)$. The reason these two unit fractions are used is that we can start with miles, which will cancel out, move to feet, which will cancel out, and end up with inches as shown.

$$7 \text{ miles} = \frac{7 \cancel{\text{mi}}}{1} \cdot \frac{5280 \cancel{\text{ft}}}{1 \cancel{\text{mi}}} \cdot \frac{12 \text{ in.}}{1 \cancel{\text{ft}}} = 7 \cdot 5280 \cdot 12 \text{ in.} = 443,520 \text{ in.}$$

Dimensional analysis can be used in all types of unit conversions.

The Metric System

The metric system uses three basic units of measure. Length is measured by using **meters.** One meter is slightly larger than 1 yard. One meter is approximately 39.37 inches. The symbol for meter is "m."

The basic unit in the metric system that is used to measure capacity is the **liter.** One liter is slightly larger than a quart. One liter is approximately equal to 1.06 quarts. The symbol for liter is "L."

The basic unit to measure **mass** (weight) in the metric system is the **gram.** The gram is a very small unit of measure. A nickel weighs about 5 grams. The symbol for gram is "g."

Just as there are different units of measure for length, capacity, and weight in the English system (i.e., 12 in. = 1 ft, 16 oz = 1 lb, etc.) there are different units of measure for length, capacity, and weight in the metric system. The major difference, though, is regardless of whether we are measuring length, capacity, or weight in the metric system, all of the units use the same prefixes, and they are all based on the powers of 10. These prefixes are shown in Table A.

For example, 1 kilometer is equal to 1000 meters. One centigram is equal to $\frac{1}{100}$ gram or 0.01 g. The symbols for metric units consist of the prefix symbols followed by the unit symbol. For example, the symbol for kilogram is "kg." The symbol for dekaliter is "daL." The symbol for centimeter is "cm."

The relationship between units in the metric system is based on powers of 10 and is similar to the units used in our monetary system. Table B shows this relationship for meters.

Table A

Metric Prefixes

Prefix	Symbol	Meaning
kilo	k	1000 units
hecto	h	100 units
deka	da	10 units
	m, L, g	1 unit
deci	d	$\frac{1}{10}$ of a unit
centi	c	$\frac{1}{100}$ of a unit
milli	m	$\frac{1}{1000}$ of a unit

Table B

Relationship between Metric Units and the U.S. Currency

Metric Unit	Meaning	Money
kilometer	1000 meters	$1000
hectometer	100 meters	100
dekameter	10 meters	10
meter	1 meter	1
decimeter	0.1 meter	0.10
centimeter	0.01 meter	0.01
millimeter	0.001 meter	0.001

There are several ways to do conversions in the metric system. The easiest way is to multiply or divide by powers of 10. The basic rules are given next.

Conversion When Using the Metric System

To change a larger unit to a smaller unit in the metric system, multiply by 10^n, where n is the number of steps that you move down in Table C.

To change a smaller unit to a larger unit in the metric system, divide by 10^n, where n is the number of steps that you move up in Table C.

Table C

$$
\text{divide}
\begin{cases}
\text{kilo} \\
\text{hecto} \\
\text{deka} \\
\text{unit} \\
\text{deci} \\
\text{centi} \\
\text{milli}
\end{cases}
\text{multiply}
$$

Math Note

You can use a shortcut when multiplying by powers of 10. Move the decimal point to the right the same number of places as the value of the exponent. For 42.5×10^5, you can move the decimal point 5 places to the right.

$$42.5 \times 10^5 = 4\,2\,5\,0\,0\,0\,0$$
$$\underset{1\,2\,3\,4\,5}{}$$

$$= 4{,}250{,}000$$

When dividing by powers of 10, you can move the decimal point to the left the same number of places as the value of the exponent. For $1253.7 \div 10^3$, you can move the decimal point to the left 3 places.

$$1253.7 \div 10^3 = 1\,2\,5\,3\,7$$
$$\underset{3\,2\,1}{}$$

$$= 1.2537$$

Example 5

Change 42.5 kilometers to centimeters.

Solution

Since we are changing a larger unit to a smaller unit, we must multiply. The number of steps from kilometers to centimeters in Table C is 5; hence, we multiply 42.5 by 10^5.

$$42.5 \text{ km} \times 10^5 = 42.5 \times 100{,}000 = 4{,}250{,}000 \text{ cm}$$

Hence, 42.5 km is the same as 4,250,000 cm.

Example 6

Change 1253.7 milligrams to grams.

Solution

Since we are changing a smaller unit to a larger unit, we divide. The number of steps from milligrams to grams is 3; hence, we divide by 10^3.

$$1253.7 \text{ mg} \div 10^3 = 1253.7 \div 1000 = 1.2537 \text{ grams}$$

Hence, 1253.7 mg $= 1.2537$ g.

Conversion in the metric system can also be done using dimensional analysis as shown in Example 7.

Example 7

Convert 532 grams to kilograms.

Solution

$$532 \text{ g} = \frac{532 \text{ g}}{1} \cdot \frac{1 \text{ dag}}{10 \text{ g}} \cdot \frac{1 \text{ hg}}{10 \text{ dag}} \cdot \frac{1 \text{ kg}}{10 \text{ hg}} = 0.532 \text{ kg}$$

Hence, 532 g = 0.532 kg.

Exercise Set A-1

Complete each.

Measures of Length

1. 8 m = _____ cm

2. 0.25 m = _____ dm

3. 12 dam = _____ m

4. 24 hm = _____ dam

5. 0.6 km = _____ hm

6. 30 cm = _____ dm

7. 90 m = _____ dm

8. 18,426 mm = _____ m

9. 375.6 cm = _____ m

10. 63 m = _____ km

11. 405.3 m = _____ km

12. 0.6 dam = _____ hm

13. 12 km = _____ cm

14. 50,000 cm = _____ km

15. 1.85 km = _____ mm

16. 650,000 mm = _____ km

17. 12.62 km = _____ dm

18. 39 m = _____ cm

19. 8 hm = _____ km

20. 540 dm = _____ cm

Measures of Capacity

21. 500 mL = _____ L

22. 80 hl = _____ L

23. 92 L = _____ dL

24. 48 daL = _____ hL

25. 8 dL = _____ cL

26. 42 L = _____ mL

27. 6.7 hL = _____ daL

28. 92 L = _____ hL

29. 64 kL = _____ hL

30. 26.3 kL = _____ mL

31. 81 L = _____ kL

32. 4256 L = _____ kL

33. 7 L = _____ daL

34. 673 L = _____ hL

35. 117 L = _____ daL

36. 53 L = _____ mL

37. 142 cL = _____ L

38. 134 mL = _____ dL

39. 32,546 mL = _____ kL

40. 128,306,501 mL = _____ kL

Measures of Weight

41. 9 dg = _____ cg

42. 6 mg = _____ cg

43. 44 kg = _____ hg

44. 16.34 hg = _____ g

45. 18 dg = _____ cg

46. 27 cg = _____ g

47. 71 cg = _____ dg

48. 215 g = _____ dag

49. 5 g = _____ dg

50. 32 kg = _____ hg

51. 0.325 g = _____ mg

52. 3217 cg = _____ g

53. 4325 kg = _____ g

54. 5 dag = _____ kg

55. 86 mg = _____ g

56. 24 hg = _____ g

57. 400 g = _____ kg

58. 6.6 kg = _____ g

59. 5632 g = _____ hg

60. 150 mg = _____ cg

Conversion for Length

Recall that the basic unit for measuring length in the metric system is the meter. One meter is approximately 39.37 inches. The basic unit for measuring length in the English system is called the foot. The next table shows the basic equivalents for measuring length in the English system.

Linear Measure Equivalents in the English System
12 inches (in.) = 1 foot (ft)
36 inches = 3 feet = 1 yard (yd)
5280 feet = 1 mile (mi)

Example 8

Use dimensional analysis to convert each.

(a) 288 inches to yards
(b) 51.5 feet to inches
(c) 24,640 yards to miles

Solution

(a) We must start out with inches in the numerator of the first fraction and yards in the numerator of the second fraction; hence, we use the unit fraction $\frac{1 \text{ yd}}{36 \text{ in.}}$. Then

$$288 \text{ in.} = \frac{288 \text{ in.}}{1} \cdot \frac{1 \text{ yd}}{36 \text{ in.}} = \frac{288 \text{ yd}}{36} = 8 \text{ yd}$$

(b) Multiply by the unit fraction $\frac{12 \text{ in.}}{1 \text{ ft}}$ as shown.

$$51.5 \text{ ft} \cdot \frac{12 \text{ in.}}{1 \text{ ft}} = 51.5 \cdot 12 \text{ in.} = 618 \text{ in.}$$

(c) In this case, two unit fractions are needed as multipliers since we start with yards and end up with miles, and the direct conversion factor between yards and miles is not given in the table.

$$24{,}640 \text{ yd} \cdot \frac{3 \text{ ft}}{1 \text{ yd}} \cdot \frac{1 \text{ mi}}{5280 \text{ ft}} = 14 \text{ mi}$$

Sometimes it will be necessary to convert from the English system to the metric system or vice versa. The next table of conversion factors for lengths and dimensional analysis can be used. The conversion factors are approximate.

English and Metric Equivalents for Length

1 inch = 2.54 centimeters

1 foot = 30.48 centimeters

1 mile = 1.61 kilometers

Example 9

Convert each.

(a) 135 feet to centimeters
(b) 87 centimeters to inches
(c) 213.36 millimeters to feet

Solution

(a) To convert feet to centimeters use the unit fraction $\frac{30.48 \text{ cm}}{1 \text{ ft}}$.

$$135 \text{ ft} \cdot \frac{30.48 \text{ cm}}{1 \text{ ft}} = 135 \cdot 30.48 \text{ cm} = 4114.8 \text{ cm}$$

—Continued

Example 9 Continued—

(b) To convert centimeters to inches use the unit fraction $\frac{1 \text{ in.}}{2.54 \text{ cm}}$.

$$87 \text{ cm} \cdot \frac{1 \text{ in.}}{2.54 \text{ cm}} = \frac{87 \text{ in.}}{2.54} = 34.25 \text{ in. (rounded)}$$

(c) To convert from millimeters to feet, it is necessary to convert millimeters to centimeters using the unit fraction $\frac{1 \text{ cm}}{10 \text{ mm}}$ then convert centimeters to feet using the unit fraction $\frac{1 \text{ ft}}{30.48 \text{ cm}}$.

$$213.36 \text{ mm} \cdot \frac{1 \text{ cm}}{10 \text{ mm}} \cdot \frac{1 \text{ ft}}{30.48 \text{ cm}} = \frac{213.36 \text{ ft}}{(10 \cdot 30.48)} = 0.7 \text{ ft}$$

Exercise Set A-2

For Exercises 1–20, convert each unit to the specified equivalent unit.

1. 5 meters = _____ inches
2. 14 yards = _____ meters
3. 16 inches = _____ millimeters
4. 50 meters = _____ yards
5. 235 feet = _____ dekameter
6. 563 decimeters = _____ inches
7. 1350 meters = _____ feet
8. 4375 dekameters = _____ feet
9. 0.6 inch = _____ millimeters
10. 256 kilometers = _____ miles
11. 0.06 hectometers = _____ feet
12. 54 inches = _____ centimeters
13. 1345 feet = _____ decimeters
14. 44,000 millimeters = _____ yards
15. 2.35 kilometers = _____ miles
16. 837 miles = _____ kilometers
17. 42 decimeters = _____ yards
18. 75 centimeters = _____ inches
19. 333 inches = _____ meters
20. 1256 kilometers = _____ inches

Conversion for Area

The English and metric conversions for area units are shown in the next table.

English and Metric Conversions for Area

1 square inch (in.^{2}) = 6.5 square centimeters (cm^2)
1 square foot (ft^2) = 0.09 square meter (m^2)
1 square yard (yd^2) = 0.8 square meter
1 square mile (mi^2) = 640 acres = 2.6 square kilometers (km^2)

Example 10

Convert 42 square inches to square centimeters.

Solution

Use dimensional analysis as shown.

$$42 \text{ in.}^2 = \frac{42 \text{ in.}^2}{1} \cdot \frac{6.5 \text{ cm}^2}{1 \text{ in.}^2} = 273 \text{ cm}^2$$

Example 11

Convert 143.5 square kilometers to square miles.

Solution

Use dimensional analysis as shown.

$$143.5 \text{ km}^2 = \frac{143.5 \text{ km}^2}{1} \cdot \frac{1 \text{ mi}^2}{2.6 \text{ km}^2} = 55.192 \text{ mi}^2$$

Exercise Set A-3

For Exercises 1–20, convert each unit to the specified equivalent unit.

1. $18 \text{ in.}^2 = $ _____ cm^2

2. $52 \text{ ft}^2 = $ _____ m^2

3. $40 \text{ m}^2 = $ _____ yd^2

4. $72 \text{ mi}^2 = $ _____ km^2

5. $32 \text{ acres} = $ _____ km^2

6. $43 \text{ in.}^2 = $ _____ cm^2

7. $18 \text{ ft}^2 = $ _____ dm^2

8. $93 \text{ cm}^2 = $ _____ in.^2

9. $3 \text{ yd}^2 = $ _____ dm^2

10. $15.6 \text{ m}^2 = $ _____ ft^2

11. $103 \text{ km}^2 = $ _____ acres

12. $80,000 \text{ cm}^2 = $ _____ in.^2

13. $42 \text{ dm}^2 = $ _____ in.^2

14. $19.4 \text{ cm}^2 = $ _____ ft^2

15. $1875 \text{ in.}^2 = $ _____ dm^2

16. $35 \text{ yd}^2 = $ _____ m^2

17. $5326 \text{ mm}^2 = $ _____ in.^2

18. $152 \text{ acres} = $ _____ km^2

19. $777 \text{ dm}^2 = $ _____ ft^2

20. $1,000,000 \text{ m}^2 = $ _____ yd^2

Conversion for Weight

As stated previously, weight and mass are not exactly the same. Your weight varies with the force of gravity, but your mass always remains the same. On Earth, as your weight increases, so does your mass; hence, for the purposes of this section, weight and mass will be considered the same.

In the English system, weight is most often measured in ounces, pounds, and tons.

16 ounces (oz) = 1 pound (lb)

2000 pounds = 1 ton (T)

As stated previously, the basic unit of weight in the metric system is the gram. A gram is a very small unit of measure. A paper clip weighs almost 1 gram. Recall that 1000 grams equals 1 kilogram. A kilogram weighs about 2.2 pounds. A metric ton (t) is equal to 1000 kilograms. The basic conversion units for weight are shown in the next table.

English and Metric Equivalents for Weight

1 kilogram = 2.2 pounds

1 gram = 0.04 ounce

1 ounce = 28 grams

Example 12

Change 7 pounds to grams.

Solution
Use dimensional analysis as shown.

$$7 \text{ lb} = \frac{7 \text{ lb}}{1} \cdot \frac{16 \text{ oz}}{1 \text{ lb}} \cdot \frac{28 \text{ g}}{1 \text{ oz}} = 3136 \text{ g}$$

Example 13

Change 862 grams to pounds.

Solution
Use dimensional analysis as shown.

$$862 \text{ g} = \frac{862 \text{ g}}{1} \cdot \frac{0.04 \text{ oz}}{1 \text{ g}} \cdot \frac{1 \text{ lb}}{16 \text{ oz}} = 2.155 \text{ pounds}$$

Exercise Set A-4

For Exercises 1–20, convert each unit to the specified equivalent unit.

1. 120 grams = _____ ounces

2. $15\frac{3}{4}$ ounces = _____ grams

3. 4823 centigrams = _____ pounds

4. 27 metric tons = _____ pounds

5. 3 tons = _____ hectograms

6. 14 decigrams = _____ ounces

7. 357,201 pounds = _____ metric tons

8. 13 pounds = _____ decigrams

9. 5.75 tons = _____ metric tons

10. 16.3 metric tons = _____ tons

11. 213 ounces = _____ decigrams

12. 64 pounds = _____ dekagrams

13. 815 dekagrams = _____ ounces

14. 37 tons = _____ metric tons

15. 183 ounces = _____ decigrams

16. 550 hectograms = _____ ounces

17. 27 pounds = _____ decigrams

18. 14,625 milligrams = _____ pounds

19. 41 pounds = _____ grams

20. 42 grams = _____ ounces

Conversion for Volume and Capacity

Volume is measured in cubic units. In the English system, volume is measured in cubic inches (in.^{3}), cubic feet (ft^3), and cubic yards (yd^3). For example, 1 cubic foot consists of a cube whose measure is 12 inches on each side. Recall that to find the volume of a cube, we multiply the length times the width times the height. Since all these measures are the same for a cube, 1 cubic foot = 12 inches × 12 inches × 12 inches or 1728 cubic inches. Also, 1 cubic yard = 3 feet × 3 feet × 3 feet or 27 cubic feet. See Figure A-1.

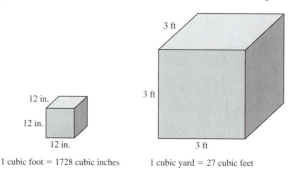

1 cubic foot = 1728 cubic inches   1 cubic yard = 27 cubic feet

Figure A-1

Likewise, in the metric system, a cubic centimeter consists of a cube whose measure is 1 centimeter on each side. Since there are 100 centimeters in 1 meter, 1 cubic meter is equal to 100 cm × 100 cm × 100 cm or 1,000,000 cubic centimeters. See Figure A-2.

Measures of volume also include measures of capacity. The capacity of a container is equal to the amount of fluid the container can hold. In the English system, capacity is measured in fluid ounces, pints, quarts, and gallons. The conversion factors for capacity are shown in the next table. Also one cubic centimeter is equal to one milliliter.

Units of Capacity in the English System

1 pint (pt) = 16 fluid ounces (oz)

1 quart (qt) = 2 pints (pt)

1 gallon (gal) = 4 quarts (qt)

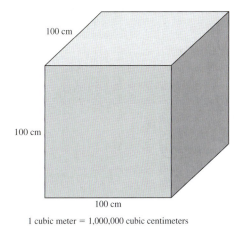

1 cubic centimeter   1 cubic meter = 1,000,000 cubic centimeters

Figure A-2

Now a cubic foot (cu. ft) of water is about 7.48 gallons and a cubic yard of water is about 202 gallons of water. If a gallon of water is poured into a container, it would take up about 231 cubic inches. Finally, a cubic foot of fresh water weighs about 62.5 pounds and a cubic foot of seawater weighs about 64 pounds. These measures are summarized in the next table.

Conversion Factors for Capacity in the English System

1 cubic foot = 7.48 gal
1 cubic yard = 202 gal
1 gal = 231 cubic inches
1 cubic foot fresh water = 62.5 lb
1 cubic foot sea water = 64 lb

Example 14

If a water tank has a volume of 3000 cubic feet, how many gallons of water will the tank hold?

Solution

Use dimensional analysis as shown.

$$3000 \text{ cubic feet} = \frac{3000 \text{ cubic feet}}{1} \cdot \frac{7.48 \text{ gal}}{1 \text{ cubic feet}} = 22{,}440 \text{ gallons}$$

Example 15

If the tank in the previous example contains fresh water, find its weight.

Solution

1 cubic foot of fresh water weighs 62.5 pounds; hence,

$$3000 \text{ cubic feet} = \frac{3000 \text{ cubic feet}}{1} \cdot \frac{62.5 \text{ pounds}}{1 \text{ cubic feet}} = 187{,}500 \text{ pounds}$$

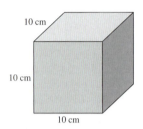

10 cm

10 cm

10 cm

1 liter = 1000 cubic centimeters

Figure A-3

Example 16

Find the weight of a quart of water.

Solution

Use dimensional analysis as shown.

$$1 \text{ qt} = \frac{1 \text{ qt}}{1} \cdot \frac{1 \text{ gal}}{4 \text{ qt}} \cdot \frac{1 \text{ cubic feet}}{7.48 \text{ gal}} \cdot \frac{62.5 \text{ lb}}{1 \text{ cubic feet}} = 2.1 \text{ lb}$$

The basic unit of capacity in the metric system is the liter. One liter of liquid is equivalent to a cube whose side is 10 centimeters in length. In other words, 1 liter = 1000 cubic centimeters. See Figure A-3. One liter is a little more than a quart. One liter = 1.06 quarts. One liter of fresh water weighs 1 kilogram at 4°C (Celsius).

Example 17

Convert 800 cubic meters to milliliters.

Solution

Use dimensional analysis as shown.

$$800 \text{ m}^3 = \frac{800 \text{ m}^3}{1} \cdot \frac{1000 \text{ cm}^3}{1 \text{ m}^3} \cdot \frac{1 \text{ L}}{1000 \text{ cm}^3} \cdot \frac{1000 \text{ mL}}{1 \text{ L}} = 800{,}000 \text{ mL}$$

Example 18

Find the number of liters in a gallon of milk.

Solution

Use dimensional analysis as shown.

$$1 \text{ gal} = \frac{1 \text{ gal}}{1} \cdot \frac{4 \text{ qt}}{1 \text{ gal}} \cdot \frac{1 \text{ L}}{1.06 \text{ qt}} = 3.77 \text{ L}$$

Exercise Set A-5

For Exercises 1–20, convert each unit to the specific equivalent unit.

1. 3 cubic feet = _____ fluid ounces

2. 6 cubic yards = _____ gallons

3. 400 gallons = _____ cubic feet

4. 3,724 gallons = _____ cubic yards

5. 12,561 fluid ounces = _____ gallons

6. 12,000 cubic inches = _____ cubic feet

7. 22,000 cubic feet = _____ gallons

8. 32 pints = _____ cubic inches

9. 8 cubic yards = ____ gallons

10. 4532 cubic inches = ____ gallons

11. 4.5 liters = ____ cubic centimeters

12. 97 milliliters = ____ cubic meters

13. 28.5 cubic centimeters = ____ milliliters

14. 140 cubic decimeters = ____ deciliters

15. 433 milliliters = ____ cubic centimeters

16. 87,250 cubic centimeters = ____ liters

17. 32 liters = ____ cubic centimeters

18. 437 centiliters = ____ cubic meters

19. 32 centiliters = ____ cubic meters

20. 1.6 liters = ____ cubic centimeters

21. Find the weight of 500 cubic feet of fresh water.

22. Find the weight of 23 cubic feet of salt water.

23. Find the capacity in cubic feet of 200 pounds of fresh water.

24. Find the capacity in cubic feet of 755 pounds of salt water.

Conversion for Temperature

Temperature is measured in the English system by the **Fahrenheit** scale. On this scale, water freezes at 32° and boils at 212°. The average temperature of the human body is 98.6°.

Temperature is measured in the metric system using the **Celsius** scale, sometimes called the Centigrade temperature. Here water freezes at 0° and boils at 100°. The temperature of the human body is 37°.

In order to convert from one scale to the other, these two formulas are used. First, to convert Celsius to Fahrenheit:

$$F = \tfrac{9}{5}C + 32$$

Second, to convert Fahrenheit to Celsius:

$$C = \tfrac{5}{9}(F - 32)$$

Example 19

Find the Fahrenheit temperature when the outside temperature is given as 28°C (meaning 28° Celsius).

Solution

Substitute in the first formula and solve.

$$F = \tfrac{9}{5}C + 32$$
$$= \tfrac{9}{5}(28) + 32$$
$$= 82.4°$$

Hence, when the Celsius thermometer reads 28°C outside, the Fahrenheit temperature will be 82.4°.

Example 20

Find the Celsius temperature when the Fahrenheit temperature is $-15°$.

Solution

Substitute in the second formula and solve.

$$C = \tfrac{5}{9}(F - 32)$$

$$= \tfrac{5}{9}(-15 - 32)$$

$$= -26.1 \text{ (rounded)}$$

Exercise Set A-6

For Exercises 1–10, convert each Celsius temperature to an equivalent Fahrenheit temperature.

1. $14°C$

2. $27°C$

3. $55°C$

4. $100°C$

5. $150°C$

6. $-5°C$

7. $-18°C$

8. $-20°C$

9. $-33°C$

10. $-50°C$

For Exercises 11–20, convert each Fahrenheit temperature to an equivalent Celsius temperature.

11. $5°F$

12. $27°F$

13. $32°F$

14. $158°F$

15. $100°F$

16. $-3°F$

17. $-10°F$

18. $-22°F$

19. $-14°F$

20. $212°F$

Appendix B

Trigonometric Ratios

Angle	Sine	Cosine	Tangent	Angle	Sine	Cosine	Tangent
0°	.0000	1.0000	.0000	28°	.4695	.8829	.5317
1°	.0175	.9998	.0175	29°	.4848	.8746	.5543
2°	.0349	.9994	.0349	30°	.5000	.8660	.5774
3°	.0523	.9986	.0524	31°	.5150	.8572	.6009
4°	.0698	.9976	.0699	32°	.5299	.8480	.6249
5°	.0872	.9962	.0875	33°	.5446	.8387	.6494
6°	.1045	.9945	.1051	34°	.5592	.8290	.6745
7°	.1219	.9925	.1228	35°	.5736	.8192	.7002
8°	.1392	.9903	.1405	36°	.5878	.8090	.7265
9°	.1564	.9877	.1584	37°	.6018	.7986	.7536
10°	.1736	.9848	.1763	38°	.6157	.7880	.7813
11°	.1908	.9816	.1944	39°	.6293	.7771	.8098
12°	.2079	.9781	.2126	40°	.6428	.7660	.8391
13°	.2250	.9744	.2309	41°	.6561	.7547	.8693
14°	.2419	.9703	.2493	42°	.6691	.7431	.9004
15°	.2588	.9659	.2679	43°	.6820	.7314	.9325
16°	.2756	.9613	.2867	44°	.6947	.7193	.9657
17°	.2924	.9563	.3057	45°	.7071	.7071	1.0000
18°	.3090	.9511	.3249	46°	.7193	.6947	1.0355
19°	.3256	.9455	.3443	47°	.7314	.6820	1.0724
20°	.3420	.9397	.3640	48°	.7431	.6691	1.1106
21°	.3584	.9336	.3839	49°	.7547	.6561	1.1504
22°	.3746	.9272	.4040	50°	.7660	.6428	1.1918
23°	.3907	.9205	.4245	51°	.7771	.6293	1.2349
24°	.4067	.9135	.4452	52°	.7880	.6157	1.2799
25°	.4226	.9063	.4663	53°	.7986	.6018	1.3270
26°	.4384	.8988	.4877	54°	.8090	.5878	1.3764
27°	.4540	.8910	.5095	55°	.8192	.5736	1.4281

—Continued

Continued—

Angle	Sine	Cosine	Tangent	Angle	Sine	Cosine	Tangent
56°	.8290	.5592	1.4826	74°	.9613	.2756	3.4874
57°	.8387	.5446	1.5399	75°	.9659	.2588	3.7321
58°	.8480	.5299	1.6003	76°	.9703	.2419	4.0108
59°	.8572	.5150	1.6643	77°	.9744	.2250	4.3315
60°	.8660	.5000	1.7321	78°	.9781	.2079	4.7046
61°	.8746	.4848	1.8040	79°	.9816	.1908	5.1446
62°	.8829	.4695	1.8807	80°	.9848	.1736	5.6713
63°	.8910	.4540	1.9626	81°	.9877	.1564	6.3138
64°	.8988	.4384	2.0503	82°	.9903	.1392	7.1154
65°	.9063	.4226	2.1445	83°	.9925	.1219	8.1443
66°	.9135	.4067	2.2460	84°	.9945	.1045	9.5144
67°	.9205	.3907	2.3559	85°	.9962	.0872	11.4301
68°	.9272	.3746	2.4751	86°	.9976	.0698	14.3007
69°	.9336	.3584	2.6051	87°	.9986	.0523	19.0811
70°	.9397	.3420	2.7475	88°	.9994	.0349	28.6363
71°	.9455	.3256	2.9042	89°	.9998	.0175	57.2900
72°	.9511	.3090	3.0777	90°	1.0000	.0000	
73°	.9563	.2924	3.2709				

Appendix C

Area Under the Standard Normal Distribution

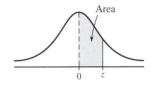

Area

The area under A is the area under the normal distribution between $z = 0$ and the positive value of z found under z.

z	A	z	A	z	A	z	A	z	A	z	A	z	A
.00	.000	.20	.079	.40	.155	.60	.226	.80	.288	1.00	.341	1.20	.385
.01	.004	.21	.083	.41	.159	.61	.229	.81	.291	1.01	.344	1.21	.387
.02	.008	.22	.087	.42	.163	.62	.232	.82	.294	1.02	.346	1.22	.389
.03	.012	.23	.091	.43	.166	.63	.236	.83	.297	1.03	.349	1.23	.391
.04	.016	.24	.095	.44	.170	.64	.239	.84	.300	1.04	.351	1.24	.393
.05	.020	.25	.099	.45	.174	.65	.242	.85	.302	1.05	.353	1.25	.394
.06	.024	.26	.103	.46	.177	.66	.245	.86	.305	1.06	.355	1.26	.396
.07	.028	.27	.106	.47	.181	.67	.249	.87	.308	1.07	.358	1.27	.398
.08	.032	.28	.110	.48	.184	.68	.252	.88	.311	1.08	.360	1.28	.400
.09	.036	.29	.114	.49	.188	.69	.255	.89	.313	1.09	.362	1.29	.402
.10	.040	.30	.118	.50	.192	.70	.258	.90	.316	1.10	.364	1.30	.403
.11	.044	.31	.122	.51	.195	.71	.261	.91	.319	1.11	.367	1.31	.405
.12	.048	.32	.126	.52	.199	.72	.264	.92	.321	1.12	.369	1.32	.407
.13	.052	.33	.129	.53	.202	.73	.267	.93	.324	1.13	.371	1.33	.408
.14	.056	.34	.133	.54	.205	.74	.270	.94	.326	1.14	.373	1.34	.410
.15	.060	.35	.137	.55	.209	.75	.273	.95	.329	1.15	.375	1.35	.412
.16	.064	.36	.141	.56	.212	.76	.276	.96	.332	1.16	.377	1.36	.413
.17	.068	.37	.144	.57	.216	.77	.279	.97	.334	1.17	.379	1.37	.415
.18	.071	.38	.148	.58	.219	.78	.282	.98	.337	1.18	.381	1.38	.416
.19	.075	.39	.152	.59	.222	.79	.285	.99	.339	1.19	.383	1.39	.418

—Continued

Continued—

z	A	z	A	z	A	z	A	z	A	z	A	z	A
1.40	.419	1.67	.453	1.94	.474	2.21	.487	2.48	.493	2.75	.497	3.02	.499
1.41	.421	1.68	.454	1.95	.474	2.22	.487	2.49	.494	2.76	.497	3.03	.499
1.42	.422	1.69	.455	1.96	.475	2.23	.487	2.50	.494	2.77	.497	3.04	.499
1.43	.424	1.70	.455	1.97	.476	2.24	.488	2.51	.494	2.78	.497	3.05	.499
1.44	.425	1.71	.456	1.98	.476	2.25	.488	2.52	.494	2.79	.497	3.06	.499
1.45	.427	1.72	.457	1.99	.477	2.26	.488	2.53	.494	2.80	.497	3.07	.499
1.46	.428	1.73	.458	2.00	.477	2.27	.488	2.54	.495	2.81	.498	3.08	.499
1.47	.429	1.74	.459	2.01	.478	2.28	.489	2.55	.495	2.82	.498	3.09	.499
1.48	.431	1.75	.460	2.02	.478	2.29	.489	2.56	.495	2.83	.498	3.10	.499
1.49	.432	1.76	.461	2.03	.479	2.30	.489	2.57	.495	2.84	.498	3.11	.499
1.50	.433	1.77	.462	2.04	.479	2.31	.490	2.58	.495	2.85	.498	3.12	.499
1.51	.435	1.78	.463	2.05	.480	2.32	.490	2.59	.495	2.86	.498	3.13	.499
1.52	.436	1.79	.463	2.06	.480	2.33	.490	2.60	.495	2.87	.498	3.14	.499
1.53	.437	1.80	.464	2.07	.481	2.34	.490	2.61	.496	2.88	.498	3.15	.499
1.54	.438	1.81	.465	2.08	.481	2.35	.491	2.62	.496	2.89	.498	3.16	.499
1.55	.439	1.82	.466	2.09	.482	2.36	.491	2.63	.496	2.90	.498	3.17	.499
1.56	.441	1.83	.466	2.10	.482	2.37	.491	2.64	.496	2.91	.498	3.18	.499
1.57	.442	1.84	.467	2.11	.483	2.38	.491	2.65	.496	2.92	.498	3.19	.499
1.58	.443	1.85	.468	2.12	.483	2.39	.492	2.66	.496	2.93	.498	3.20	.499
1.59	.444	1.86	.469	2.13	.483	2.40	.492	2.67	.496	2.94	.498	3.21	.499
1.60	.445	1.87	.469	2.14	.484	2.41	.492	2.68	.496	2.95	.498	3.22	.499
1.61	.446	1.88	.470	2.15	.484	2.42	.492	2.69	.496	2.96	.499	3.23	.499
1.62	.447	1.89	.471	2.16	.485	2.43	.493	2.70	.497	2.97	.499	3.24	.499
1.63	.449	1.90	.471	2.17	.485	2.44	.493	2.71	.497	2.98	.499	3.25	.499
1.64	.450	1.91	.472	2.18	.485	2.45	.493	2.72	.497	2.99	.499		*
1.65	.451	1.92	.473	2.19	.486	2.46	.493	2.73	.497	3.00	.499		
1.66	.452	1.93	.473	2.20	.486	2.47	.493	2.74	.497	3.01	.499		

*For *z* values beyond 3.25 use $A = 0.500$.

Significant Values for the Correlation Coefficient

Sample Size	5%	1%
4	.950	.990
5	.878	.959
6	.811	.917
7	.754	.875
8	.707	.834
9	.666	.798
10	.632	.765
11	.602	.735
12	.576	.708
13	.553	.684
14	.532	.661
15	.514	.641
16	.497	.623
17	.482	.606
18	.468	.590
19	.456	.575
20	.444	.561
21	.433	.549
22	.423	.537
23	.412	.526
24	.403	.515
25	.396	.505
30	.361	.463
40	.312	.402
60	.254	.330
120	.179	.234

Appendix E

Using the TI-83 Plus Graphing Calculator

This appendix is intended to give you brief instructions and tips for some useful features that can help you explore some of the concepts in this book.

Animator

We can use the Animator feature to draw and compare the right side and the left side of an equation.

To graph the equation $2(4x - 5) = 8x - 10$, press [Y=] and enter $2(4x - 5)$ in Y_1. Press either [▼] or [ENTER] to move to Y_2 and enter $8x - 10$. To access the Animated Line feature, use the left arrow key to move to the left of Y_2 and press [ENTER] four times.

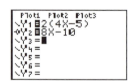

Using the Zoom features, we can quickly access a viewing window that shows the graph of the left and right sides (with animated line) of the equation.

Press [ZOOM] [4] .

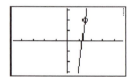

To show the animator, press the [ENTER] key to pause the graphing process. The graphing process is restarted by pressing [ENTER] a second time.

Contrast

We can contrast different graphs by choosing different types of lines for our graphs.

To contrast the left and right sides of $3x + 2 = -3x + 2$, press Y= and enter $3x + 2$ in Y_1 and $-3x + 2$ in Y_2. Use the left arrow key to move to the left of Y_2 and press ENTER to access the bold line feature. This will help us differentiate the two lines in our graph.

Using the Zoom features we can quickly access a viewing window that shows the graph of the left and right sides (with bold line) of the equation.

Press ZOOM 4 .

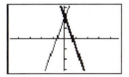

Edit

You can use previously entered expressions to create and edit new expressions.

If you had entered $1 * 5^{\wedge}1 + 0 * 5^{\wedge}0$ and want to edit it to $1 * 5^{\wedge}1 + 3 * 5^{\wedge}0$, begin by pressing 2nd ENTER to recall the previously entered expression.

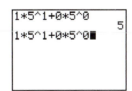

There are two types of editing modes: type-over mode and insert mode. Type-over mode is the default mode. To replace a number, move the cursor over the number to be replaced; then type the new number(s) or symbol(s) and press ENTER .

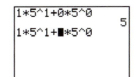
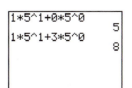

If you want to change $1*5^\wedge 1+0*5^\wedge 0$ to $2*5^\wedge 2+4*5^\wedge 1+1*5^\wedge 0$, begin by pressing [2nd] [ENTER] twice.

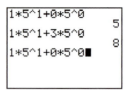

Enter insert mode by moving the cursor to the insertion point and press [2nd] [DEL] ; then type the new number(s) or symbol(s).

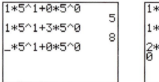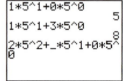

To get out of insert mode, move the cursor to the left or the right.

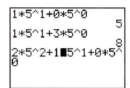

Finish editing the expression and press [ENTER] .

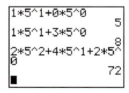

You can easily tell which editing mode you are in by the type of cursor on the screen. Type-over mode's cursor is a flashing block, while insert mode has a flashing underline.

Horizontal and Vertical Lines

To graph a horizontal line, press [Y=] and enter a numerical value.

We can quickly access a viewing window that shows the graph of the horizontal line by pressing (GRAPH) .

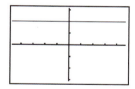

To graph a vertical line, first clear or turn off equations in the (Y=) screen. While this is not necessary it makes it clear where the vertical line is when graphed. From the home screen (accessed by pressing (2nd) (MODE)), press (2nd) (PRGM) to access the DRAW menu.

Press (4) to access the draw Vertical command. The command Vertical −2 will draw the vertical line $x = -2$.

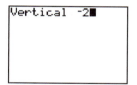

Press (ENTER) to execute the command.

To clear this drawing from the graph screen, press (2nd) (PRGM) (1) (ENTER) from the home screen or the graph screen.

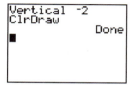

Intersect

Using the Intersect feature will assist you in viewing the graphical representation of the solution to an algebraic equation.

The first step is to enter the expression from each side of the equation $5x + 9 = 29$ in the  screen. Enter $5x + 9$ in Y_1 and 29 in Y_2.

Using the Zoom features we can quickly access a viewing window that shows the intersection of the left and right sides of the equation.

Press 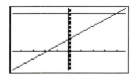 to access a window that fits the equation, showing the intersection.

We can use the Calculate menu to calculate the intersection of the left and right sides of the equation, the solution.

Press to access the calculate menu.

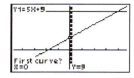

Press to access the intersect command.

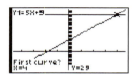

Press to move the cursor to (or close to) the intersection. It may be necessary to press more than once. In this example, we need to press it eight times.

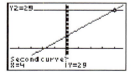

Press to continue the calculation of the intersection.

Press ENTER to continue the calculation of the intersection.

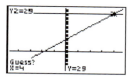

Press ENTER to complete the calculation of the intersection.

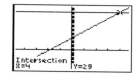

Parentheses

On most calculators, parentheses replace brackets and braces used in expressions. The resulting expression uses nested parentheses.

Example:

$$84 \div 4 - \{3 \times [10 + (15 - 2)]\}$$

becomes

$$84/4 - (3 * (10 + (15 - 2)))$$

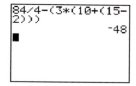

Probability

Graphing calculators offer probability functions. The probability functions include: combination notation, factorial notation, and permutation notation.

Combination notation example:

How many combinations of four objects are there taken two at a time?

To determine the answer to this exercise use the calculator function nCr, where n is the number of objects and r is the number taken at a time.

To enter the information for this example using the combination function, press [4], [MATH] ▶ ▶ ▶ [3] (to access nCr), [2].

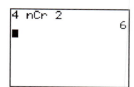

Factorial notation example:

How many different ways can five cities be ranked?

To determine the answer to this exercise use the factorial function.

To enter the information for this example using the factorial function, enter ,
MATH ▶ ▶ ▶ 4 (to access !).

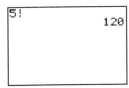

Permutation notation example:

How many different ways can a chairperson and an assistant chairperson be selected for a research project if there are seven scientists available?

To determine the answer to this exercise use the permutation function nPr, where n is the number of objects and r is the number of objects taken at a time.

To enter the information for this example using the combination function, enter 4 **MATH**
▶ ▶ ▶ 2 (to access nPr), 2 .

Scientific Notation

The calculator can be set to scientific notation mode by pressing **MODE** ▶ **ENTER** .

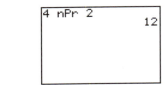

Press **2nd** **MODE** to return to the HOMESCREEN.

Scientific notation expressions can be entered into the calculator using the EE feature,
2nd **,** or using powers of 10, as shown in the next screen.

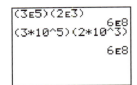

Shade

The Shade feature in  screen can be used to show the graphic solution set when graphing inequalities.

To graph $x - y \geq 6$, first solve for y. The inequality becomes $y \leq x - 6$.

Press $\boxed{\text{Y=}}$ $\boxed{\triangleleft}$ $\boxed{\triangleleft}$ to move to the left side of Y_1. Press $\boxed{\text{ENTER}}$ three times to have the graph shaded below the line (for $<$ or $\leq$). Move the cursor to the right side of the equal sign to enter the expression $x - 6$.

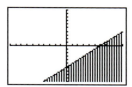

Using the Zoom features we can quickly access a viewing window that shows the shaded graph of the equation.

Press $\boxed{\text{ZOOM}}$ $\boxed{6}$ $\boxed{\text{ENTER}}$.

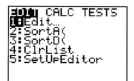

Statistics

Graphing calculators offer statistics functions. The statistics functions included in this appendix are edit/entering data, clearing lists, copy list to list, sort (ascending or descending), and graphing vertical bar graphs (histograms) and frequency polygons.

Entering Data

To enter data, press $\boxed{\text{STAT}}$,

then 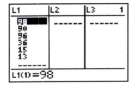 (for Edit).

Previously entered data may exist in the list(s). If the list column (L1, L2, etc.) you wish to use to enter the data contains data, press ▲ to move to the column title (L1 in this case).

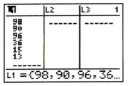

Press CLEAR , then ENTER to clear the column of data.

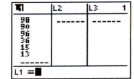 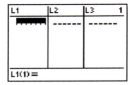

Next, enter data one item at a time in the column, pressing ENTER after each number.

Sample data:

53, 75, 27, 32, 15, 18

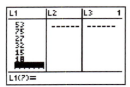

Alternatively, you can enter the data in the list title edit line, using braces to enclose the data and separating each data point with a comma.

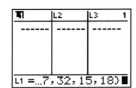 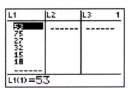

Ninety-nine data entries are possible.

For some of the functions discussed next, it is necessary to access your saved lists. This is easily done by pressing 2nd LIST and selecting the needed list.

Copy List

To copy a list, begin in the title edit line of the list where you want the data to appear.

Then, type in the number of the list to be copied and press ENTER . In this case, we begin in the title line of L2 and type L1 to copy the list. As stated previously, you can also press 2nd LIST and select L1.

 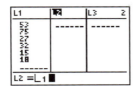

Sorting

To sort a list into ascending order, press STAT 2 then the list title (in this case, L2) and ENTER .

 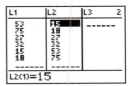

To sort a list into descending order, press STAT 3 then the list title (in this case, L2) and ENTER .

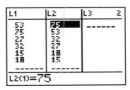

Vertical Bar Graph

To graph a vertical bar graph, begin by entering data in lists.

Sample:

Type	Frequency
1	5
2	7
3	9
4	4

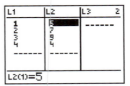

Press Y= and clear any entries. Press 2nd Y= 1 to access a statistical plot.

Turn on the plot, and select the vertical bar graph. Make sure the Xlist is L1, and the Freq is L2.

Press [ZOOM] [9].

Histogram

To graph a histogram, use the same procedure as for the vertical bar graph. Then press [WINDOW] and edit the options. The minimum and maximum of the data are Xmin and Xmax, respectively. The range of each class is the Xscl. The highest frequency in any class is the Ymax. The next window shows the settings using our sample data.

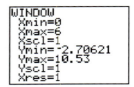

Next, press [GRAPH].

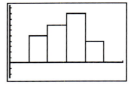

Frequency Polygon

To graph a frequency polygon, use the same procedure as for the vertical bar graph, except the Type should be frequency polygon (line graph).

Press (WINDOW) and edit the settings as was done for the histogram.

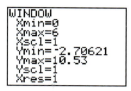

Next, press (GRAPH).

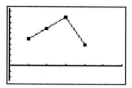

Scatter Plot

To graph a scatter plot, you will need to enter your data into lists, set up a plot, use the Zoom Stat feature, and adjust the window as needed.

Suppose we have these sample data:

L1 = {376, 650, 844, 1162, 1513, 1650, 2236, 3002, 4028, 4010}

L2 = {5, 20, 20, 28, 26, 34, 35, 56, 68, 55}

Press (STAT) (1) and enter the data.

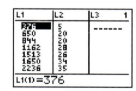
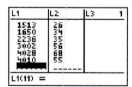

Press (2nd) (Y=) (1) to set up the scatter plot.

Press (ZOOM) (9).

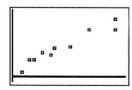

Press 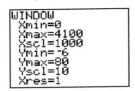 and edit the options for better viewing. A sample is shown.

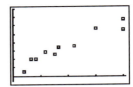

Store

The Store feature, STO▸, can be used to store values as letters or words.

To store 1 for I, 5 for V, and 10 for X press:

1 STO▸ ALPHA x^2 ENTER ;

5 STO▸ ALPHA 6 ENTER ; and

10 STO▸ X,T,θ,n ENTER .

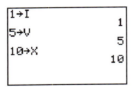

Alternatively, you could store the numbers as follows:

1 STO▸ ALPHA x^2 2nd · 5 STO▸ ALPHA 6 2nd · 10 STO▸ X,T,θ,n ENTER

Table Setup

Using tables will assist you in viewing numerical values of an algebraic expression.

The first step in viewing the numerical representation is to enter the expression in the Y= screen. Press Y= and enter $((2X + 6)/2) - X$.

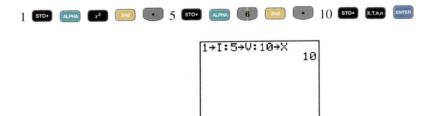

Next, we set up the table to view a set of integers in place of the x variable and value of the expression at each of the integer variables.

Press ⬚ ⬚ to access the TABLE SETUP menu. Start the table at an appropriate value and select an appropriate increment for the *x* values. In our example, we will start the table with −3 and choose 1 for our increment. The starting value and increment can vary depending on the expression we wish to evaluate.

To view the TABLE press ⬚ ⬚ .

In this case, we see that no matter what variable we use for *x*, the value of the expression is 3.

Another expression may show different values for *y*.

In ⬚ , enter $2x + 3$.

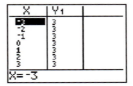

To view the TABLE press ⬚ ⬚ . Note that our table settings haven't changed from the previous example. We are still starting with −3 and our *x* increments are still 1.

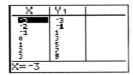

We see that for each value of *x* the value of *y* changes.

If we want to see the value of *y* when $x = 100$, we could use the down arrow to scroll to 100. We could also use the Ask feature from the Table Setup menu.

Press ⬚ ⬚ to access TBLSET. Press ⬚ twice, then ⬚ ⬚ to turn on the Ask feature for the independent variable (X in this case).

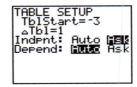

To view the TABLE press [2nd] [GRAPH] .

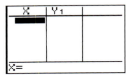

Begin with 100 [ENTER] .

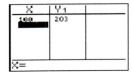

We see t when $x = 100$ and $y = 203$. Next try 1000 [ENTER] ; then 10000 [ENTER] . Next try 2345 [ENTER] ; then 67890 [ENTER] .

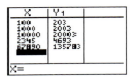

Any value of x can be entered to determine the corresponding value of y.

Test

The TEST feature includes symbols for both equalities and inequalities:

Keystrokes	Symbol
[2nd] [MATH] [1]	$=$
[2nd] [MATH] [2]	$\neq$
[2nd] [MATH] [3]	$>$
[2nd] [MATH] [4]	$\geq$
[2nd] [MATH] [5]	$<$
[2nd] [MATH] [6]	$\leq$

The calculator can be used to verify that an inequality is true. When using the TEST feature, the calculator displays a result of 1 for true or 0 for false.

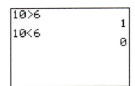

Zoom

The first step in viewing the graphical representation of the equation $5x + 9 = 29$ is to enter the expression from each side of the equation in the [Y=] screen. Press [Y=] and enter $5x + 9$ for Y_1 and 29 for Y_2.

Using the Zoom features we can quickly access a viewing window that shows the graph of the left and right sides of the equation.

Press ZOOM 4 to begin with a centered window.

Then press ZOOM 0 to access a window that fits the equation.

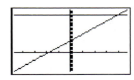

When Zoom Fit does not show the intersection, Zoom Out can assist you.

Enter each side of the equation $6x - 10 = 4x + 10$ in the Y= screen. Enter $6x - 10$ for Y_1 and $4x + 10$ for Y_2.

Using ZOOM 4 or ZOOM 0 for this equation does not show the intersection in the viewing window.

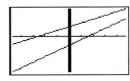

Press ZOOM 3 ENTER to Zoom Out.

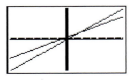

We can now see the intersection of the two lines.

Appendix F

Credits

Photos

Chapter 1

Page 2: © The British Museum
Page 3: © The McGraw-Hill Companies, Inc./Photo by C.P. Hammond
Page 4(top): © Mark Burnett/Stock, Boston/PictureQuest
Page 4(bottom): © PhotoDisc Website
Page 5: © Bettmann/Corbis Images
Page 10: © Alinari Archives/Corbis Images
Page 11: © Chuck Painter/Stanford News Service
Page 13: © The Granger Collection, New York
Page 14: © Jim Cummins/Corbis Images
Page 17: Courtesy of Constance Reid
Page 18: © The McGraw-Hill Companies, Inc./Photo by C.P. Hammond
Page 20: © Adam Smith/Getty Images

Chapter 2

Page 34(top): Library of Congress
Page 34(bottom): © Susan Van Etten/PhotoEdit
Page 35: © Frank Siteman/PhotoEdit
Page 36: © AP/Wide World Photos
Page 41: © Michael J. Doolittle/Image Works
Page 45: © Bob Krist/Corbis Images
Page 49: © Matthias Kulka/Corbis Images

Chapter 3

Page 74: © Bettmann/Corbis Images
Page 76: National Library of Medicine, #81192
Page 78: © PhotoDisc/Vol.#46
Page 79: © PhotoDisc/Vol.#OS50
Page 85: © Brian Sytnyk/Masterfile
Page 86: © Kevin Dodge/Corbis Images
Page 88: © Robert Landau/Corbis Images
Page 89: © Mark Richards/PhotoEdit
Page 93: © Corbis/R-F Website
Page 96: Lynne Siler/Focus Group/PictureQuest
Page 100: © Martyn Goddard/Corbis Images
Page 103: © PhotoDisc Website

Chapter 4

Page 106: © Jeff Vanuga/Corbis Images
Page 115: © Bettmann/Corbis Images

Chapter 4

Page 123: © Scala/Art Resource, NY. Iraq Museum, Baghdad, Iraq
Page 125: © Cathy Melloan/Photoedit
Page 132: National Radio Astronomy Observatory
Page 137: © AP/Wide World Photos
Page 140: Courtesy, Naval Historical Center
Page 156: © Jeff Smith/Getty Images

Chapter 5

Page 160(top): © PhotoDisc/Vol.#76
Page 160(bottom): © The McGraw-Hill Companies, Inc./Photo by C.P. Hammond
Page 164: © Granger Collection
Page 166: © James Zipp/Photo Researchers
Page 167: © Granger Collection
Page 171: © PhotoDisc Website
Page 176: © PhotoDisc Website
Page 183: © James Leynse/Corbis Images
Page 186: © Corbis/Vol.#149
Page 187: © Reuters/Larry Downing/Corbis Images
Pages 195, 205: © Granger Collection
Page 214: © Phillip Hayson/Photo Researchers
Page 223: © Granger Collection
Page 224: © Corbis Images
Page 236: © The McGraw-Hill Companies, Inc./Photo by C.P. Hammond

Chapter 6

Page 240: © Matthias Kulka/Corbis Images
Page 244: © Granger Collection
Page 246: © Bettmann/Corbis Images
Page 250: © PhotoDisc Website
Page 251: © Granger Collection
Page 252: © PhotoDisc Website
Page 259: © David Young-Wolff/PhotoEdit
Page 260: © Matthias Kulka/Corbis Images

Chapter 7

Page 268: © PhotoDisc Website
Page 272: © The British Museum
Page 274: © PhotoDisc/Vol.#21
Page 285(top): © AP/IBM/Wide World Photos
Page 285(bottom): © Coco McCoy/Rainbow/PictureQuest
Page 289: © Kathy Ferguson-Johnson/PhotoEdit
Page 292(top): © Image State-Pictor/PictureQuest
Page 292(bottom): © PhotoDisc/Vol.#1
Page 295: © Tony Freeman/PhotoEdit
Page 298: © PhotoDisc Website
Page 300: © Michael Newman/PhotoEdit
Page 304: © Steve Bein/Corbis Images
Page 306(top): © PhotoDisc/Vol.#31
Page 306(bottom): © AP/Wide World Photos
Page 307: © PhotoDisc/Vol.#7
Page 309: © PhotoDisc Website
Page 316: © Corbis/R-F Website

Chapter 8

Page 334: © Corbis/Vol.#62
Page 335: © Granger Collection
Page 336(top): © Lowell Georgia/Corbis Images
Page 336(bottom): © PhotoDisc/Vol.#OS55
Page 341: © PhotoDisc/Vol.#DT8
Page 344: © PhotoDisc/Vol.#86
Page 345(left): © PhotoDisc/Vol.#51
Page 345(right): © PhotoDisc/Vol.#36
Page 346(left): © PhotoDisc/Vol.#16
Page 346(right): © PhotoDisc/Vol.#74
Page 353: © Stuart Cohen/Image Works
Page 360: © Granger Collection
Page 361: © Susan Van Etten/PhotoEdit
Page 367: © Granger Collection
Page 371: © Susan Van Etten/PhotoEdit
Page 373: © PhotoDisc/Vol.#21
Page 378: © PhotoDisc/Vol.#41
Page 382: © Joel Gordon
Page 385: © PhotoDisc/Vol.#13
Page 390(top): © PhotoDisc/Vol.#77
Page 390(bottom): © PhotoDisc/Vol.#103
Page 391: © James L. Amos/Photo Researchers

Chapter 9

Page 402(top): © Robert Brenner/PhotoEdit
Page 402(bottom): © Corbis/Vol.#165
Page 406: © Ulrike Schanz/Animals Animals
Page 408: © Setboun Michel/Corbis Sygma
Page 409: © Corbis/Vol.#144
Page 413: © PhotoDisc/Vol.#21
Page 414: © PhotoDisc/Vol.#21
Page 418: © Jeff Greenberg/PhotoEdit
Page 426: © David Young-Wolff/PhotoEdit
Page 437: © PhotoDisc/Vol.#17
Page 439: © Bob Daemmrich/Image Works
Page 442: © PhotoDisc/Vol.#OS49
Page 446: © Amy Etra/PhotoEdit
Page 447: © Tony Freeman/PhotoEdit

Chapter 10

Page 458: © AP/Wide World Photos
Page 459: © Lawrence Manning/Corbis Images
Page 460: © Corbis Images
Page 466: © Steve Dunwell/Getty Images
Page 471: © Jeremy Horner/Corbis Images
Page 472: © Journal-Courier/Steve Warmowski/Image Works
Page 480(top): © David Reed/Corbis Images
Page 480(center): © Scott Camazine/Photo Researchers
Page 480(bottom): © Andrew Syred/Photo Researchers
Page 481: © Bettmann/Corbis Images
Page 485: © Joseph Sohm/ChromoSohm/Corbis Images
Page 486: © PhotoDisc/Vol.#77
Page 489: © Scott Camazine/Photo Researchers
Page 490: NASA
Page 495(top): © Tony Freeman/PhotoEdit/PictureQuest
Page 495(bottom): © Nancy Richmond/Image Works
Page 501(top): © Bob Daemmrich/Image Works
Page 501(bottom): © George Shelley/Corbis Images
Page 506: © Patrick Ward/Corbis Images
Page 513: © Corbis/R-F Website
Page 521: © AP/Wide World Photos

Chapter 11

Page 524(top): © Tom Stewart/Corbis Images
Page 524(bottom): © PhotoDisc/Vol.#88
Page 525: © PhotoDisc/Vol.#EO5
Page 526: © PhotoDisc/Vol.#8
Page 528: © Bohemian Nomad Picturemakers/Corbis Images
Page 529: © The Granger Collection
Page 530: © Syracus Newspapers/Image Works
Page 531: © The McGraw-Hill Companies, Inc./Photo by C.P. Hammond
Page 535: © Anthony Redpath/Corbis Images
Page 537: © Bettmann/Corbis Images
Page 538: © Corbis/R-F Website
Page 540: © The McGraw-Hill Companies, Inc./Photo by C.P. Hammond
Page 544: © Michael Philip Manheim/Photo Network/PictureQuest
Page 546: © Al Bello/Getty Images News Services
Page 548(top): © Jan Miele/Corbis Images
Page 548(bottom): © AFP/Corbis Images
Page 551: © Reuters Newsmedia Inc./Corbis Images
Page 555(top): © Peter Saloutos/Corbis Images
Page 555(bottom): © The Granger Collection
Page 560: © Ariel Skelley/Corbis Images
Page 564: © Corbis/Vol.#126
Page 566: © AP/Wide World Photos
Page 570: © PhotoDisc/Vol.#OS28
Page 573: © John Henley/Corbis Images
Page 574: © Robert Brenner/PhotoEdit
Page 576: © A & J Verkaik/Corbis Images
Page 579: © Mike Parry/Minden Pictures
Page 583: © Jose Luis Pelaez/Corbis Images
Page 584: © PhotoDisc/Vol.#8
Page 586: © Anna Clopet/Corbis Images
Page 593: © Giry Daniel/Corbis Sygma

Chapter 12

Page 596(top): © PhotoDisc/Vol.#40
Page 596(bottom): © Ariel Skelley/Corbis Images

Page 597(top): © Dan Guravich/Corbis Images

Page 597(bottom): © Richard Lord/Image Works

Page 598(top): © Fritz Hoffmann/Image Works

Page 598(bottom): © The McGraw-Hill Companies, Inc./Photo by
C.P. Hammond

Page 599: © M. Craig/Commercial/Corbis Sygma

Page 601: © Image Works

Page 611: Climate Prediction Center/NOAA

Page 615: © PhotoDisc/Vol.#39

Page 616: © AP/The Daily Interlake/Wide World Photos

Page 620: © PhotoDisc/Vol.#118

Page 621: © David Young-Wolff/PhotoEdit

Page 623: © Luc Beziat/Getty Images

Page 624: © AP/The Reporter/Wide World Photos

Page 625(top): © A. Ramey/PhotoEdit

Page 625(bottom): NASA

Page 627: © Tom Prettyman/PhotoEdit

Page 629: © Corbis Images

Page 633(top): © Tim Davis/Corbis Images

Page 633(bottom): © Chris Collins/Corbis Images

Page 634: © PhotoDisc/Vol.#7

Page 639: © Barros and Barros/Getty Images

Page 640: © Corbis/Vol.#79

Page 643: © PhotoDisc/Vol.#48

Page 653: © PhotoDisc/Vol.#31

Page 655(top): © PhotoDisc/Vol.#18

Page 655(bottom): © PhotoDisc/Vol.#18

Page 660(top): © Derek Trask/Corbis Images

Page 660(bottom): © Ryan McVay/Getty Images

Page 662: © Corbis/Vol.#54

Page 664: © PhotoDisc/Vol.#40

Page 665(top): © Steve Dunwell/Getty Images

Page 665(bottom): © The McGraw-Hill Companies, Inc./Photo by
C.P. Hammond

Page 669(top): © David Young-Wolff/PhotoEdit

Page 669(bottom): © PhotoDisc/Vol.#51

Page 681(top): © Dwayne Newton/PhotoEdit

Page 681(bottom): © Esbin-Anderson/Photo Network/PictureQuest

Page 682: © The McGraw-Hill Companies, Inc./Photo by C.P. Hammond

Page 683: © Mary Kate Denny/PhotoEdit

Chapter 13

Page 690(top): © AP/Wide World Photos

Page 690(bottom): © Reuters NewMedia/Corbis Images

Page 694: © Brooks Kraft/Corbis Images

Page 702: © AP/Wide World Photos

Page 705(top): © David Madison/Getty Images

Page 705(bottom): © Reuters NewMedia, Inc./Corbis Images

Page 712: © Kevin Fleming/Corbis Images, L.A.

Page 713: © AP/Wide World Photos

Page 722: © AP/Wide World Photos

Appendix G

Answers

NOTE: Answers to Critical Thinking exercises are not always given in detail. Sometimes only suggestions are given, and different answers to some exercises are also possible.

Chapter 1

Exercise Set 1-1

1. 37

3. 10

5. 72

7.

9. $5 + 13 + 17 = 35 =$ odd

11. $5^2 \div 2 = 12.5$

13. Final answer is always -10.

15. Final answer is always 20.

17. Answers will vary.

19. It may be impossible or impractical to verify the conclusion for *all* applicable cases.

21. It cannot be done.

23. 6, 15, 20, 15, 6

Exercise Set 1-2

1. 7 quarters; 5 half dollars

3. 20

5. 5 chickens; 7 cows

7. 43

9. 8 and 15

11. 7

13. Pete's age: 18; Lashanna's age: 9

15. Sam receives $45; Pete receives $15.

17. 84 pounds

19. $1093

21. 42,900

G-1

23. $206.35

25. $395

27. $228.96

29. $465.50

31. 1. Understand the problem.
2. Devise a plan to solve the problem.
3. Carry out the plan to solve the problem.
4. Check the answer.

33. Make a cut at 1 inch and again at 3 inches; then you will have a 1 inch piece, a 2 inch piece, and a 3 inch piece. Then give the knight the 1 inch piece at the end of the first day. At the end of the second day, take the 1 inch piece back and give the knight the 2 inch piece, etc.

35. $22,500

37. $2400

Exercise Set 1-3*

1. 2900

3. 3,260,000

5. 63

7. 200,000

9. 3.67

11. 327.1

13. 5,460,000

15. 300,000

17. 264.9735

19. 563.27

21. $480

23. 6 hours

25. $72

27. $4

*When using estimation, other correct answers are possible.

29. Three possible answers, in hours: 9, 10, 11

31. $18,000

33. $540

35. $64

37. 23,000 square miles

39. 25,000 square miles

41. Approximately 305

43. 350 billion

45. 560 miles

47. (a) Buying groceries

(b) "Having an idea" of the time needed for a trip to the post office

(c) Figuring the number of gallons of milk produced per cow, per month

49. If the estimate falls within a "reasonable distance" of the exact answer, then the latter has a better (greater) probability of being correct.

51. Answers will vary.

Review Exercises*

1. 18, 19, 21

3. {q, 1024, n}

5.

7. $5(7)(11) = 385 =$ odd

9. Final answer $= 13$ more than $(\frac{1}{2})$ the even number

11. 9

13. 110 lb

15. 30

17. Approximately $6.67

19. 20

21. 3

23. Harry has 5; Bill has 7.

25. 13

27. 12.5 and 7.5

29. $800 at 8%; $200 at 6%

31. 187

33. 0.6

35. $1320

37. About 1172 million pounds

39. About $350

41. 60 miles

Chapter Test*

1. 13, 12, 18

3.

10

*When using estimation, other correct answers are possible.

5. Final answer $=$ (original number) $+ 13$

7. On the eighth day

9. Use Roman numerals:

$$|\ V\ -\ |\ | \ =\ |\ | \quad \text{or} \quad 4 - 2 = 2$$

11. 84

13. 12

15. $36 and $24

17. 93%

19. 65,400 ft

21. Mark is 17; Mother is 49.

23. 1.38

25. 35.5 hours

27. 310

Chapter 2

Exercise Set 2-1

1. {s, t, r, e}

3. {51, 52, 53, 54, 55, 56, 57, 58, 59}

5. {1, 3, 5, 7, 9, 11, 13}

7. {11, 12, 13, 14, . . .}

9. {2001, 2002, 2003, . . . , 2999}

11. {Monday, Tuesday, Wednesday, Thursday, Friday, Saturday, Sunday}

13. {diamond, club, spade, heart}

15. {even natural numbers}

17. {the first four multiples of 9}

19. {letters in Mary}

21. {natural numbers from 100 to 199}

23. $\{x \mid x$ is a multiple of 10$\}$

25. $\{x \mid x \in N$ and $x > 20\}$

27. $\{x \mid x$ is an odd natural number less than 10$\}$

29. There are none.

31. {7, 14, 21, 28, 35, 42, 49, 56, 63}

33. {102, 104, 106, . . .}

35. Well defined

37. Well defined

39. Not well defined

41. Not well defined

43. True

45. True

47. True

49. Infinite

51. Finite

53. Infinite

55. Finite

57. Equal and equivalent

59. Neither

61. Equivalent

63. Neither

65. {10 20 30 40}

$\updownarrow$ $\updownarrow$ $\updownarrow$ $\updownarrow$

{40 10 20 30}

67. {1 2 3 ..., 25 26}

$\updownarrow$ $\updownarrow$ $\updownarrow$ $\updownarrow$ $\updownarrow$

{a b c ..., y z}

69. {Carnival, Royal Caribbean, Princess, Holland America, Norwegian}

71. {Seaborn, Windstar}

73. {'94, '95}

75. ∅

77. A *set* is a well-defined collection of objects.

79. Equal sets have exactly the *same elements*. Equivalent sets have exactly the *same number* of elements.

81. Each element of one set can be associated (paired) with exactly one element of the other set, and no element in either set is left alone.

83. Yes

85. Answers will vary.

87. Yes

Exercise Set 2-2

1. ∅; {r}; {s}; {t}; {r, s}; {r, t}; {s, t}; {r, s, t}

3. ∅; {1}; {3}; {1, 3}

5. {} or ∅

7. ∅; {5}; {12}; {13}; {14}; {5, 12}; {5, 13}; {5, 14}; {12, 13}; {12, 14}; {13, 14}; {5, 12, 13}; {5, 12, 14}; {5, 13, 14}; {12, 13, 14}; {5, 12, 13, 14}

9. ∅; {1}; {10}; {20}; {1, 10}; {1, 20}; {10, 20}

11. ∅

13. None

15. True

17. False

19. False

21. False

23. True

25. 8

27. 1

29. 4

31. {10, 30, 40, 50, 60, 70, 90}

33. {20, 40, 60, 80, 100}

35. {20, 40, 60, 80, 100}

37. ∅

39. {20, 40, 60, 80, 100}

41. {b, d}

43. {a, c, e, h}

45. {a, c, h}

47. {e}

49. {b, d, f, g}

51. {2, 4, 6}

53. Universal set

55. ∅

57. {2, 5, 6, 11}

59. ∅

61. Set B

63. {x | x is an odd multiple of 3 or an even multiple of 9}
{3, 9, 15, 18, 21, 27, 33, 36, 39, ... }

65. (a) None

(b) Phone

(c) Television

(d) Internet

(e) Phone, television

(f) Phone, Internet

(g) Television, Internet

(h) Phone, television, Internet

67. 128

69. 16

71. If every element of set A is also in set B, then A is a *subset* of B.

73. 2^n, which includes ∅

75. A subset is a set in its own right, hence a *collection* of well-defined objects. An element of a set is just an *individual member* of the set.

77. The union of sets A and B consists of all elements that are not in *at least one* of A and B. The intersection of A and B consists of all elements that are *in both* A and B.

79. The set of all elements used in a particular problem or situation is called a universal set.

81. Answers will vary.

83. Answers will vary.

Exercise Set 2-3

1.

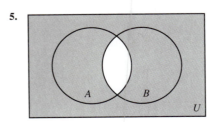

3.

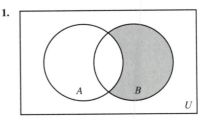

5.

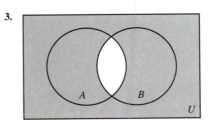

7.

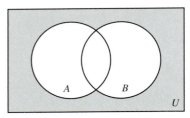

9.

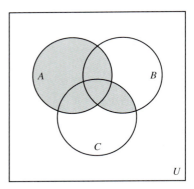

11.

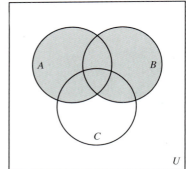

13.

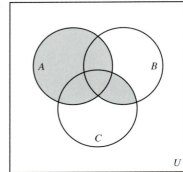

15.

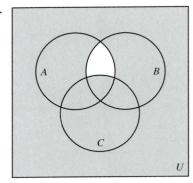

17.

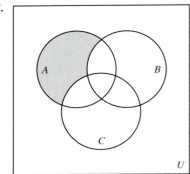

19.

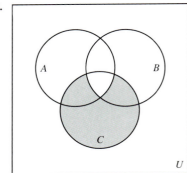

21.

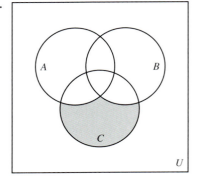

23.

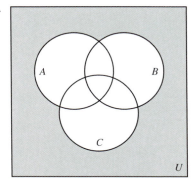

25.

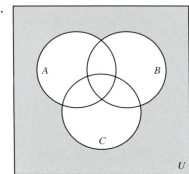

27. Yes

29. Yes

31. No

33. No

35. In the figure are shown all those elements (represented by points) that belong: (1) strictly to A, (2) strictly to B, and (3) to both A and B. These mentioned points form precisely the union of A and B.

37. Disjoint sets have no elements in common; accordingly, the circles representing these sets cannot touch each other anywhere.

39. $\{2, 3\}$

41. $\{1, 2\}$

43.

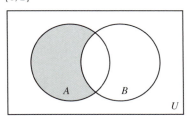

Exercise Set 2-4

1. (a) 13   (b) 18   (c) 2

3. (a) 5   (b) 7   (c) 6

5. (a) 2   (b) 9   (c) 35

7. (a) 18   (b) 7   (c) 14   (d) 17

9. (a) 37   (b) 2   (c) 37

11. (a) 22   (b) 21   (c) 15

13. (a) 32   (b) 37   (c) 58

15. 6

Exercise Set 2-5

1. $7n$

3. 4^n

5. $-3n$

7. $\dfrac{1}{n+1}$

9. $4n - 2$

11. $\{3, \ 6, \ 9, \ 12, \ 15, \ldots, \ 3n, \ldots\}$
$\updownarrow \ \updownarrow \ \updownarrow \ \updownarrow \ \updownarrow \qquad \updownarrow$
$\{6, \ 12, \ 18, \ 24, \ 30, \ldots, \ 6n, \ldots\}$

13. $\{9, \ 18, \ 27, \ 36, \ 45, \ldots, \ 9n, \ldots\}$
$\updownarrow \ \updownarrow \ \updownarrow \ \updownarrow \ \updownarrow \qquad \updownarrow$
$\{18, \ 36, \ 108, \ 144, \ 180, \ldots, \ 18n, \ldots\}$

15. $\{2, \ 5, \ 8, \ 11, \ldots, \ 3n - 1, \ldots\}$
$\updownarrow \ \updownarrow \ \updownarrow \ \updownarrow \qquad \updownarrow$
$\{5, \ 11, \ 17, \ 23, \ldots, \ 6n - 1, \ldots\}$

17. $\{10, \quad 100, \ldots, \quad 10^n, \ldots\}$
$\updownarrow \qquad \updownarrow \qquad\quad \updownarrow$
$\{100, \ 10{,}000, \ldots, \quad 10^{2n}, \ldots\}$

19. $\left\{\dfrac{5}{1}, \dfrac{5}{2}, \dfrac{5}{3}, \ldots, \quad \dfrac{5}{n}, \ldots\right\}$
$\updownarrow \ \updownarrow \ \updownarrow \qquad\quad \updownarrow$
$\left\{\dfrac{5}{2}, \dfrac{5}{3}, \dfrac{5}{4}, \ldots, \quad \dfrac{5}{n+1}, \ldots\right\}$

21. An infinite set is a set that can be placed in a one-to-one correspondence with a proper subset of itself.

23. A set is countable if it is finite or if there is a one-to-one correspondence between the members of the set and the natural numbers.

25. $\aleph_0$

Review Exercises

1. $\{52, 54, 56, 58\}$

3. $\{l, e, t, r\}$

5. $\{501, 502, 503, \ldots\}$

7. $\varnothing$

9. $\{x \mid x \in E \text{ and } 16 < x < 26\}$

11. $\{x \mid x \text{ is an odd natural number greater than } 100\}$

13. Infinite

15. Finite

17. Finite

19. $\varnothing$; $\{r\}$; $\{s\}$; $\{t\}$; $\{r, s\}$; $\{r, t\}$; $\{s, t\}$; $\{r, s, t\}$

21. 32 subsets; 31 proper subsets

23. $\{t, u, v\}$

25. $\varnothing$

27. {s, w, z}

29. {s, w, x, y, z}

31. {p, q, r, s, w, x, y, z}

33.

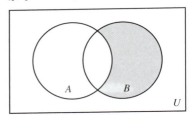

35.

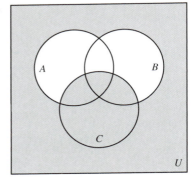

37. (a) 12   (b) 3

39. (a) 3   (b) 5   (c) 6

41. False

43. False

45. True

47. False

49. True

51. False

53. $-3 - 2n$

Chapter Test

1. {92, 94, 96, 98}

3. {e, n, v, l, o, p}

5. {1, 2, 3, 4, . . . , 79}

7. {January, June, July}

9. $\{x \mid x \in E \text{ and } 10 < x < 20\}$

11. $\{x \mid x$ is an odd natural number greater than 200$\}$

13. Infinite

15. Finite

17. Finite

19. {} or ∅; {p}; {q}; {r}; {p, q}; {p, r}; {q, r}; {p, q, r}

21. {a}

23. {b, c, d, e, f, h}

25. {a, b, c, d, e, f, g, i, k}

27.

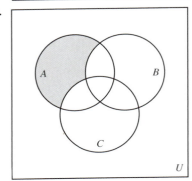

29.

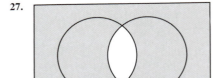

31. $15n$

Chapter 3

Exercise Set 3-1

1. No

3. Yes

5. No

7. Yes

9. No

11. Compound

13. Compound

15. Simple

17. Compound

19. Simple

21. The sky is not blue.

23. The blanket is red.

25. It is true that Harry failed statistics.

27. Conjunction.

29. Biconditional

31. Disjunction

33. Biconditional

35. $p \wedge q$

37. $(\sim q) \to p$

39. $\sim(\sim q)$

41. $q \vee \sim p$

43. $q \leftrightarrow p$

45. $\sim q$

47. $q \to p$

49. $(\sim q) \vee p$

51. $q \leftrightarrow p$

53. $(\sim q) \to p$

55. The plane is on time, and the sky is clear.

57. If the sky is clear, then the plane is on time.

59. The plane is not on time, and the sky is not clear.

61. The plane is on time, or the sky is not clear.

63. If the sky is clear, then the plane is or is not on time.

65. Trudy is not attractive.

67. Mark is handsome, or Trudy is not attractive.

69. If Mark is not handsome, then Trudy is not attractive.

71. Mark is handsome, or Trudy is attractive.

73. Trudy is attractive, or Mark is handsome.

75. A statement is a declarative sentence that can be classified as true or false, but not both.

77. ∧; ∨; →; ↔

79. It cannot be classified as true or false.

Exercise Set 3-2

1.

p	q	~	(p ∨ q)
T	T	F	T
T	F	F	T
F	T	F	T
F	F	T	F
		②	①

3.

p	q	~p	∧	q
T	T	F	F	T
T	F	F	F	F
F	T	T	T	T
F	F	T	F	F
		①	③	②

5.

p	q	~p	↔	q
T	T	F	F	T
T	F	F	T	F
F	T	T	T	T
F	F	T	F	F
		①	③	②

7.

p	q	~	(p ∧ q)	→	p
T	T	F	T	T	T
T	F	T	F	T	T
F	T	T	F	F	F
F	F	T	F	F	F
		②	①	④	③

9.

p	q	~q	∧	p	→	~p
T	T	F	F	T	T	F
T	F	T	T	T	F	F
F	T	F	F	F	T	T
F	F	T	F	F	T	T
		①	③	②	⑤	④

11.

p	q	(p ∧ q)	↔	q	∨	~p
T	T	T	T	T	T	F
T	F	F	T	F	F	F
F	T	F	F	T	T	T
F	F	F	F	F	T	T
		①	⑤	③	④	②

13.

p	q	(p ∧ q)	∨	p
T	T	T	T	T
T	F	F	T	T
F	T	F	F	F
F	F	F	F	F
		①	③	②

15.

p	q	r	(r ∧ q)	∨	(p ∧ q)
T	T	T	T	T	T
T	T	F	F	T	T
T	F	T	F	F	F
T	F	F	F	F	F
F	T	T	T	T	F
F	T	F	F	F	F
F	F	T	F	F	F
F	F	F	F	F	F
			①	③	②

17.

p	q	r	~	(p ∨ q)	→	~	(p ∧ r)
T	T	T	F	T	T	F	T
T	T	F	F	T	T	T	F
T	F	T	F	T	T	F	T
T	F	F	F	T	T	T	F
F	T	T	F	T	T	T	F
F	T	F	F	T	T	T	F
F	F	T	T	F	T	T	F
F	F	F	T	F	T	T	F
			②	①	⑤	④	③

19.

p	q	r	(~p	∨	q	∧	r
T	T	T	F	T	T	T	T
T	T	F	F	T	T	F	F
T	F	T	F	F	F	F	T
T	F	F	F	F	F	F	F
F	T	T	T	T	T	T	T
F	T	F	T	T	T	F	F
F	F	T	T	T	F	T	T
F	F	F	T	T	F	F	F
			①	③	②	⑤	⑥

21.

p	q	r	(p ∧ q)	↔	~r	∨	q
T	T	T	T	T	F	T	T
T	T	F	T	T	T	T	T
T	F	T	F	T	F	F	F
T	F	F	F	F	T	T	F
F	T	T	F	F	F	T	T
F	T	F	F	F	T	T	T
F	F	T	F	T	F	F	F
F	F	F	F	F	T	T	F
			①	⑤	②	④	③

23.

p	q	r	r	→	~	(p ∨ q)
T	T	T		F	F	T
T	T	F		T	F	T
T	F	T		F	F	T
T	F	F		T	F	T
F	T	T		F	F	T
F	T	F		T	F	T
F	F	T		T	T	F
F	F	F		T	T	F
				③	②	①

Note: The values for *p*, *q*, and *r* have sometimes been omitted in the truth tables in order to make the answers easier to check.

25.

p	q	r	p	→	(~q	∧	~r)
T	T	T	T	F	F	F	F
T	T	F	T	F	F	F	T
T	F	T	T	F	T	F	F
T	F	F	T	T	T	T	T
F	T	T	F	T	F	F	F
F	T	F	F	T	F	F	T
F	F	T	F	T	T	F	F
F	F	F	F	T	T	T	T
			①	⑤	②	④	③

27.

p	q	r	~	(q	→	p)	∧	r
T	T	T	F	T			F	T
T	T	F	F	T			F	F
T	F	T	F	T			F	T
T	F	F	F	T			F	F
F	T	T	T	F			T	T
F	T	F	T	F			F	F
F	F	T	F	T			F	T
F	F	F	F	T			F	F
			②	①			④	③

29.

p	q	r	(r ∨ q)	∧	(r ∧ p)
T	T	T	T	T	T
T	T	F	T	F	F
T	F	T	T	T	T
T	F	F	F	F	F
F	T	T	T	F	F
F	T	F	T	F	F
F	F	T	T	F	F
F	F	F	F	F	F
			①	③	②

31. A truth table enables us to decide on the truth value of a statement (usually a compound one) by examining the truth value of the components of the statement.

33. The *inclusive* disjunction is false only when both of its parts are false. The *exclusive* disjunction is false not only when both parts are false but also when both parts are true. At other times, it's true.

35. ↔; →; ∧ or ∨; ~

Exercise Set 3-3

1. Tautology
3. Self-contradiction
5. Tautology
7. Neither
9. Neither
11. Logically equivalent
13. Neither
15. Neither
17. Negation
19. Neither
21. $q → p$; $~p → ~q$; $~q → ~p$
23. $p → ~q$; $q → ~p$; $~p → q$
25. $~q → p$; $~p → q$; $q → ~p$
27. *Converse:* If he will get a job, then he graduated.
    Inverse: If he did not graduate, then he will not get a job.
    Contrapositive: If he will not get a job, then he did not graduate.

29. *Converse:* If I have a party, then it is my birthday.
   Inverse: If it is not my birthday, then I will not have a party.
   Contrapositive: If I will not have a party, then it is not my birthday.

31. When, in their respective truth tables, they have the same value (T or F) in the same relative position.

33. The converse, the inverse, and the contrapositive can be made from the conditional statement.

35. $p ∧ ~q$

37. Answers will vary.

Exercise Set 3-4

1. Valid
3. Valid
5. Valid
7. Valid
9. Valid
11. Invalid
13. Let p = "It rains." Let q = "I (will) do my homework."

    $p → q$
    $~p$
    —————
    $∴ ~q$
    Invalid

15. Let p = "Sam gains 15 pounds." Let q = "He makes the football team."

    $p → q$
    q
    —————
    $∴ p$
    Invalid

17. Let p = "I cut the grass." Let q = "It does not rain."

    $p ↔ q$
    q
    —————
    $∴ p$
    Valid

19. Let p = "I did not study." Let q = "I passed the exam."

    $p ∨ q$
    p
    —————
    $∴ ~q$
    Invalid

21. Let p = "You do 20 pushups." Let q = "You pass physical education." Let r = "You graduate."

    $p → q$
    $q → r$
    —————
    $∴ p → r$
    Valid

23. An argument consists of two parts: (a) Two or more statements, called *premises,* which are joined logically by and(s). (b) A last statement, called a *conclusion.* The argument is valid if part (b) follows logically from part (a), or else it's invalid.

25. Yes. An argument is invalid if it has at least one F in the → column. Whether or not the conclusion is actually true in real life does not matter, as far as the truth table is concerned.

27. Valid

Exercise Set 3-5

1.

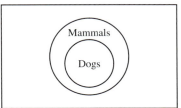

3.

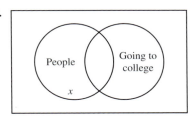

5.

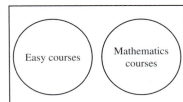

7.

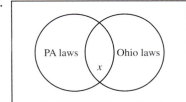

9.

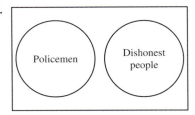

11. Invalid

13. Valid

15. Invalid

17. Valid

19. Invalid

21. Invalid

23. Invalid

25. Valid

27. Invalid

29. Invalid

31. (a) Universal Affirmative: All computers have a memory.

(b) Universal Negative: No chickens have horns.

(c) Particular Affirmative: Some mountains always have snow.

(d) Particular Negative: Some trees do not bear fruit.

33. All A is C.

35. No birds have teeth.

Review Exercises

1. $p \wedge q$.

3. $q \leftrightarrow p$

5. $\sim p \rightarrow \sim q$

7. $\sim(p \rightarrow q)$

9. $\sim(\sim q)$

11. It is cool or not cloudy.

13. It is cool if and only if it is cloudy.

15. It's not the case that it is not cool, or it is cloudy.

17.

p	q	r	$(p$	$\rightarrow$	$\sim q)$	$\vee$	r
T	T	T	T	F	F	T	T
T	T	F	T	F	F	F	F
T	F	T	T	T	T	T	T
T	F	F	T	T	T	T	F
F	T	T	F	T	F	T	T
F	T	F	F	T	F	T	F
F	F	T	F	T	T	T	T
F	F	F	F	T	T	T	F
			①	③	②	⑤	④

19.

p	q	$\sim p$	$\rightarrow$	$(\sim q$	$\vee$	$p)$
T	T	F	T	F	T	T
T	F	F	T	T	T	T
F	T	T	F	F	F	F
F	F	T	T	T	T	F
		①	⑤	②	④	③

21. Tautology

23. Neither

25. Neither

27. No

29. Invalid

31. Invalid

33. *Converse:* If it says "Meow," then it is a cat.

Inverse: If it is not a cat, then it does not say, "Meow."

Contrapositive: If it does not say, "Meow," then it is not a cat.

35. *Converse:* If the car is not very fast, then it is red.

Inverse: If it is not red, then it is very fast.

Contrapositive: If the car is very fast, then it is not red.

37. Invalid

Chapter Test

1. $p \wedge q$

3. $p \leftrightarrow q$

5. $\sim(\sim p \wedge q)$

7. It is sunny or not warm.

9. It is sunny if and only if it is warm.

11. It is not the case that it is not sunny or it is warm.

13.

p	q	r	$(p$	$\rightarrow$	$\sim q)$	$\wedge$	r
T	T	T	T	F	F	F	T
T	T	F	T	F	F	F	F
T	F	T	T	T	T	T	T
T	F	F	T	T	T	F	F
F	T	T	F	T	F	T	T
F	T	F	F	T	F	F	F
F	F	T	F	T	T	T	T
F	F	F	F	T	T	F	F
			①	③	②	⑤	④

15.

p	q	$(\sim q$	$\vee$	$p)$	$\wedge$	p
T	T	F	T	T	T	T
T	F	T	T	T	T	T
F	T	F	F	F	F	F
F	F	T	T	F	F	F
		①	③	②	⑤	④

17. Neither

19. Neither

21. Neither

23. Invalid

25. Invalid

27. Yes

29. Invalid

Chapter 4

Exercise Set 4-1

1. 35

3. 20,225

5. 30,163

7. 20,314

9. 1,112,010

11. ||||||

13. ∩∩∩||||||

15. ⊙∩∩∩∩∩∩|||||||

17. ⊙⊙⊙⊙⊙⊙⊙⊙|

19. 𝑓⊙⊙∩∩∩∩|||||

21. ⊙∩|||||||

23. ◁◁◁◁⌒⌒⌒⌒⌒∩∩|||||

25. 𝑓𝑓𝑓𝑓𝑓𝑓𝑓𝑓𝑓𝑓⊙⊙⊙⊙⊙⊙⊙
∩∩∩∩∩∩∩∩|||||||||||

27. 12

29. 51

31. 40,871

33. 109,284

35. 792

37. ⟨⟨⟨▼▼

39. ▼ ⟨▼▼▼▼▼▼▼▼

41. ▼▼▼▼ ⟨⟨⟨⟨⟨⟨▼▼

43. ⟨▼▼▼▼▼▼▼ ▼▼▼

45. ▼ ⟨⟨▼▼▼▼▼▼ ⟨⟨⟨⟨⟨▼▼▼▼▼▼

47. 17

49. 43

51. 86

53. 418

55. 490

57. XXXIX

59. DLXVII

61. MCCLVIII

63. MCDLXII

65. MMM

67. hundreds

69. ten thousands

71. ones

73. $8 \times 10^1 + 6$

75. $1 \times 10^3 + 8 \times 10^2 + 1 \times 10^1 + 2$

77. $6 \times 10^3 + 2$

79. $1 \times 10^5 + 6 \times 10^4 + 2 \times 10^3 + 8 \times 10^2 + 7 \times 10^1 + 3$

81. $1 \times 10^7 + 7 \times 10^6 + 5 \times 10^5 + 3 \times 10^4 + 1 \times 10^3 + 8 \times 10^2 + 1$

83. 1939

85. 1997

87. Each symbol has a certain value (= number); we add these values to obtain the total value.

89. It's primarily an additive system, in the sense that we add values to obtain the total; however, the letters must be in the correct relative position. Example: CCX = 100 + 100 + 10 = 210, but CXC = 100 + 90 = 190.

91. Answers will vary.

93. Answers will vary.

Exercise Set 4-2

1. 11

3. 33

5. 108

7. 106

9. 311

11. 69

13. 359

15. 184

17. 37,406

19. 1166

21. 11111_{two}

23. 1333_{six}

25. 22_{seven}

27. 1017_{nine}

29. 10110_{two}

31. 33_{five}

33. $939_{sixteen}$

35. 1042212_{five}

37. 44_{seven}

39. 100000000_{two}

41. 58_{ten}; 111010_{two}

43. 97_{ten}; 81_{twelve}

45. 52_{ten}; 310_{four}

47. 282_{ten}; $1B6_{twelve}$

49. 1370_{ten}; 3665_{seven}

51. 5 lb, 7 oz

53. 34 yd, 2 ft, 8 in.

55. STOP

57. CLASSISOVER

59. ITISRAINING

61. 13256

63. 67138

65. 44501

67.

69.

71.

73. Let b = the base = some integer (exceeding 1). In the number to base b, the rightmost digit has a place value of $b^0 = 1$; the next digit has a place value of $b^1 = b$; the next digit has a place value of $b^2 = b \times b$; etc.

75. Let y = the given number in base ten, and let b = the "other base." Divide y by the biggest power of b (say b^m) that goes into y; let q_m be the quotient, exclusive of the remainder r_m. Divide r_m, regardless of its value, by b^{m-1}; let q_{m-1} be the quotient, exclusive of the remainder r_{m-1}. Continue the process until you have divided by b^1. The desired answer = $q_m q_{m-1} q_{m-2} \cdots q_1 r_1$.

77. 6 manufacturer's

79. 0 product

Exercise Set 4-3

1. 32_{five}

3. 11212_{four}

5. 2020_{six}

7. 6657_{nine}

9. 31_{five}

11. 101_{seven}

13. 40040_{nine}

15. 571_{twelve}

17. 332_{six}

19. 56067_{nine}

21. $3976A_{twelve}$

23. 482_{nine}, with remainder = 2_{nine}

25. 230_{five}, with remainder = 3_{five}

27. 10000_{two}

29. $403_{sixteen}$

31. 1001_{two}

33. $4DCA_{sixteen}$

35. 110010_{two}

37. $2894_{sixteen}$

39. 11_{two} with remainder = 10_{two}

41. $B23_{sixteen}$ with remainder = $2_{sixteen}$

43. Consider $a - b$ in base eight. Find b at extreme left column of the table, and go horizontally to a. From here, go up to the top row of the table; the entry you meet is the answer. The idea is to "go backward" in the addition table because subtraction is the opposite of addition.

45. L

47. G

49. 0101 0011 0101 0100 0100 1111 0101 0000

51. 0100 1000 0100 0101 0100 1100 0100 1100 0100 1111

Review Exercises

1. 1,000,221

3. 681

5. 419

7. DCCCXCVI

9.

11. 119

13. 17,327

15. 1246

17. 2058

19. 215

21. 52_{six}

23. 2663_{nine}

25. 101011_{two}

27. 103_{four}

29. 1000_{seven}

31. 251_{nine}

33. 111011_{two}

35. $15AB6_{twelve}$

37. 101100_{two}

39. 331_{four}

41. 21331_{nine}

43. 1011011_{two}

45. 230_{five} with remainder = 2_{five}

47. 342_{eight} with remainder = 6_{eight}

Chapter Test

1. 2,000,312

3. 1271

5. 426

7. DLXVII

9.

11. 96

13. 17,375

15. 1231

17. 500

19. 241

21. 133_{five}

23. 6362_{nine}

25. 1001_{two}

27. 1123_{four}

29. 1160_{seven}

31. 282_{nine}

33. 1000111_{two}

35. $14B25_{\text{twelve}}$

37. 10010111_{two}

39. 1030_{four}

41. 1250_{six}

43. 151_{eight} with remainder $= 3_{\text{eight}}$

Chapter 5

Exercise Set 5-1

1. 1; 2; 4; 8; 16

3. 1; 2; 3; 6; 7; 9; 14; 18; 21; 42; 63; 126

5. 1; 2; 4; 8; 16; 32

7. 1; 3; 9

9. 1; 2; 3; 4; 6; 8; 12; 16; 24; 32; 48; 96

11. 1; 17

13. 1; 2; 4; 8; 16; 32; 64

15. 1; 3; 5; 7; 15; 21; 35; 105

17. 1; 2; 7; 14; 49; 98

19. 1; 71

21. 15; 18; 21; 24; 27

23. 20; 30; 40; 50; 60

25. 30; 45; 60; 75; 90

27. 34; 51; 68; 85; 102

29. 2; 3; 4; 5; 6

31. $2 \times 2 \times 2 \times 2$ or 2^4

33. $2 \times 2 \times 2 \times 2 \times 3 \times 3 \times 3 \times 3$, or $2^4 \times 3^4$

35. None exists

37. $2 \times 5 \times 5$, or 2×5^2

39. Seven factors, each $= 2$; or 2^7

41. $2 \times 2 \times 3 \times 5 \times 5$, or $2^2 \times 3 \times 5^2$

43. $5 \times 5 \times 19$, or $5^2 \times 19$

45. $7 \times 8 \times 8$, or 7×8^2

47. 13×19

49. $2 \times 3 \times 5 \times 5 \times 5$, or $2 \times 3 \times 5^3$

51. 3

53. 1

55. 12

57. 25

59. 25

61. 10

63. 150

65. 630

67. 308

69. 36

71. At noon

73. Pencils: four groups of six each; pictures: three groups of six each

75. 90 days

77. All the numbers in the infinite list 1, 2, 3, 4, . . .

79. It has exactly one factor $=$ itself $= 1$.

81. The factorization into primes is unique, if a certain order (say, ascending) is agreed on. That is, for a given composite number, there is exactly one set of prime factors.

83. Yes, but only for the trivial case when factor $=$ the number (call it n) $=$ multiple. At other times, No, because a factor would be smaller than n, whereas a multiple would be bigger than n.

85. One. It is 2.

87.

$3 + 1 = 4$	$7 + 7 = 14$
$3 + 3 = 6$	$3 + 13 = 16$
$3 + 5 = 8$	$5 + 13 = 18$
$5 + 5 = 10$	$3 + 17 = 20$
$5 + 7 = 12$	

Exercise Set 5-2

1. -1

3. 9

5. -11

7. -12

9. -13

11. 14

13. -5

15. 1

17. -5

19. -70

21. 45

23. -24

25. -36

27. 42

29. 0

31. 8

33. -5

35. -4

37. 7

39. 1

41. 0

43. 0

45. 111

47. 51

49. 36

51. 6

53. −758

55. −56

57. 59

59. −492

61. <

63. >

65. >

67. <

69. <

71. 8

73. 10

75. 8

77. −10

79. 0

81. $1180

83. 1399

85. (1) 2291; (2) 10,277; (3) 15,411; (4) 11,770; (5) 8518

87. 8 inches

89. A negative number is always less than zero and always to the left of zero on the number line. The opposite of a number may or may not be negative. Example: The respective opposites of 22 and −51 are −22 and 51.

91. (a) To indicate the operation of subtraction

(b) To indicate position to the left of zero on the number line

93. The set of integers includes all the whole numbers, as well as the integers −1, −2, −3, −4, ...

95. (a) 3   (b) −5   (c) −9   (d) 10   (e) 0

Exercise Set 5-3

1. $\frac{1}{6}$

3. $\frac{7}{10}$

5. $\frac{5}{6}$

7. $\frac{7}{8}$

9. $\frac{5}{9}$

11. $\frac{15}{48}$

13. $\frac{38}{48}$

15. $\frac{35}{45}$

17. $\frac{55}{80}$

19. $\frac{6}{30}$

21. $-\frac{1}{6}$

23. $-\frac{37}{24}$ or $-\left(1\frac{13}{24}\right)$

25. $\frac{7}{24}$

27. $\frac{7}{6}$ or $1\frac{1}{6}$

29. $\frac{7}{10}$

31. $-\frac{1}{16}$

33. $\frac{11}{12}$

35. $\frac{6}{5}$ or $1\frac{1}{6}$

37. $\frac{3}{4}$

39. $\frac{7}{36}$

41. 0.2

43. $0.\overline{6}$

45. 2.25

47. $0.30\overline{5}$

49. $0.\overline{75}$

51. $0.\overline{9411764705882352}$

53. $\frac{7}{8}$

55. $\frac{6}{11}$

57. $\frac{3}{8}$

59. 475 miles

61. $\frac{1}{7}$

63. 190 miles

65. $\frac{3}{40}$ meter or 0.075 meter or 7.5 cm

67. $7733.\overline{3}$ tons

69. Answers will vary.

Exercise Set 5-4

1. Rational

3. Irrational

5. Irrational

7. $2\sqrt{6}$

9. $\frac{\sqrt{2}}{6}$

11. $5\sqrt{10}$

13. $\frac{\sqrt{5}}{5}$

15. $\frac{\sqrt{6}}{2}$

17. $\frac{\sqrt{21}}{14}$

19. $\frac{\sqrt{6}}{3}$

21. $-2\sqrt{3}$

23. $4\sqrt{5}$

25. $-6\sqrt{5}$

27. $18\sqrt{2}$

29. $2\sqrt{5}$

31. $3\sqrt{30}$

33. $24\sqrt{3}$

35. $\sqrt{30}$

37. $2\sqrt{2}$

39. $7\sqrt{15}$

41. 4 seconds

43. 20 volts

45. 4π seconds ≈ 12.6 seconds

47. A rational number $= \frac{a}{b}$, where a and b are integers and $b \neq 0$. An irrational number cannot be written in such form.

49. The number π, being irrational, has an infinite number of digits after the decimal point.

51. The cube roots can be either rational or irrational; i.e., $\sqrt[3]{27} = 3$; $\sqrt[3]{9} = 2.080083823\ldots$

53. Yes; $\sqrt{2} \cdot \sqrt{2} = 2$

Exercise Set 5-5

1. Integer, rational, real
3. Rational, real
5. Rational, real
7. Irrational, real
9. Irrational, real
11. Rational, real
13. Natural, whole, integer, rational, real
15. Natural, whole, integer, rational, real
17. Closure property of addition
19. Commutative property of addition
21. Commutative property of multiplication
23. Distributive
25. Inverse property for multiplication
27. Commutative property of addition
29. Identity property for addition
31. Commutative property of multiplication
33. Addition; multiplication
35. Addition; subtraction; multiplication
37. None
39. No
41. They are either rational, or else irrational; and in both cases, the numbers are real.
43. The set of natural numbers = {1, 2, 3, 4, ...}. The set of integers is {..., −3, −2, −1, 0, 1, 2, 3, ...}. Thus every natural number is also an integer. Any natural number n is automatically rational, since $n = \frac{n}{1}$. Real numbers consist of all rationals and of all irrationals; therefore, "natural" implies (from above) "rational" whence, "real."
45. $2 − 3 = −1$. $−1$ is not a natural number. Yes. Yes. Yes.

Exercise Set 5-6

1. 243
3. 1
5. 1
7. $\frac{1}{243}$
9. $\frac{1}{64}$
11. $3^6 = 729$
13. $4^7 = 16,384$
15. $3^2 = 9$
17. 2
19. $5^6 = 15,625$
21. $\frac{1}{3^2} = \frac{1}{9}$
23. $\frac{1}{5^5} = \frac{1}{3125}$
25. $\frac{1}{2^2} = \frac{1}{4}$
27. $\frac{1}{4^3} = \frac{1}{64}$
29. $\frac{1}{7}$
31. 6.25×10^8
33. 7.3×10^{-3}

35. 5.28×10^{11}
37. 6.18×10^{-6}
39. 4.32×10^4
41. 8.14×10^{-2}
43. 3.2×10^{13}
45. 59,000
47. 0.0000375
49. 2400
51. 0.000003
53. 1000
55. 8,020,000,000
57. 7,000,000,000,000
59. 6×10^{10}
61. 2.67×10^{-7}
63. 8.8×10^{-3}
65. 1.5×10^{-9}
67. 2×10^2
69. 6, or 6×10^0
71. 6×10^{-2}
73. 2.58×10^{15}
75. 2.4×10^1
77. 1×10^{-12}
79. 186,000
81. 0.0000000000000000000000017
83. 5,880,000,000,000
85. About 1.65×10^{15} miles
87. About 421 million miles
89. About 43 minutes
91. Move the decimal point left or right until exactly one nonzero digit lies to the left of the point; let $n =$ number of slots moved. Multiply the new decimal number by
    (a) 10^n if the moving was to the left
    (b) 10^{-n} if the moving was to the right
93. If written in decimal notation:
    (a) The first number is readily seen to exceed 1 and usually very large
    (b) The second number is readily seen to be less than 1 and usually very small
95. Answers will vary.

Exercise Set 5-7

1. (a) 5; (b) 8; (c) 93; (d) 588
3. (a) 50; (b) −2; (c) 28; (d) 468
5. (a) $\frac{1}{8}$; (b) $\frac{2}{3}$; (c) $\frac{179}{24}$; (d) $\frac{91}{2}$
7. (a) 0.6; (b) 1; (c) 11.6; (d) 73.2
9. (a) 4; (b) 3; (c) 708,588; (d) 1,062,880
11. (a) $\frac{1}{2}$; (b) $\frac{1}{2}$; (c) $\frac{1}{4096} \approx 0.000244$; (d) $\frac{4095}{4096} \approx 0.998$
13. (a) −3; (b) −5; (c) 146,484,375; (d) 122,070,312
15. (a) 1; (b) 3; (c) 177,147; (d) 265,720
17. 1; 7; 13; 19; 25
19. −9; −12; −15; −18; −21
21. $\frac{1}{4}$; $\frac{5}{8}$; 1; $\frac{11}{8}$; $\frac{7}{4}$

23. 12; 24; 48; 96; 192

25. $-5; -\frac{5}{4}; -\frac{5}{16}; -\frac{5}{64}; -\frac{5}{256}$

27. $\frac{1}{6}; -1; 6; -36; 216$

29. Geometric

31. Arithmetic

33. (a) $1600; (b) $41,500

35. $1800

37. $814.45

39. A list of numbers related to each other by a specific rule is called a sequence.

41. A list of numbers such that any member, except the first, equals the preceding member multiplied by a fixed quantity is called a geometric sequence.

43. He added $1 + 100$ to get 101, $2 + 99$ to get 101, and then multiplied $50 \times 101 = 5050$.

45. $a_1 = \frac{3}{10}; r = \frac{1}{10}; S_n = \frac{1}{3}$

Review Exercises

1. 1; 2; 3; 6; 13; 26; 39; 78

3. 1; 3; 5; 9; 15; 45

5. 1; 2; 4; 5; 7; 10; 14; 20; 28; 35; 70; 140

7. 8; 12; 16; 20; 24

9. 18; 27; 36; 45; 54

11. $2 \times 2 \times 2 \times 2 \times 2 \times 3$

13. $2 \times 5 \times 5 \times 5$

15. $2 \times 2 \times 2 \times 3 \times 5 \times 5$

17. 2; 30

19. 5; 280

21. 20; 1200

23. 18

25. -45

27. 7

29. -6

31. 2709

33. $\frac{15}{19}$

35. $\frac{4}{5}$

37. $\frac{23}{24}$

39. $\frac{5}{21}$

41. $\frac{6}{17}$

43. $\frac{13}{18}$

45. $-\frac{17}{7}$

47. $\frac{41}{40}$

49. 0.9

51. $0.\overline{857142}$

53. $\frac{11}{16}$

55. $\frac{23}{90}$

57. $4\sqrt{3}$

59. $\frac{7\sqrt{5}}{5}$

61. $\frac{\sqrt{6}}{4}$

63. $-\sqrt{5} + 10\sqrt{3}$

65. $9\sqrt{21}$

67. 2

69. $2\sqrt{3} + \sqrt{30}$

71. Rational; real

73. Rational; real

75. Whole; integer; rational; real

77. Inverse for multiplication

79. Closure for addition

81. 1024

83. 1

85. $\frac{1}{7776}$

87. 625

89. $\frac{1}{4}$

91. 3.83×10^3

93. 3.27×10^{-6}

95. 580,000,000,000

97. 0.000627

99. 9.2×10^{-2}

101. 2×10^{10}

103. 8; 18; 28; 38; 48; 58; ninth term $= 88$; sum $= 432$

105. $-13; -18; -23; -28; -33; -38$; ninth term $= -53$; sum $= -297$

107. 7.5; 15; 30; 60; 120; 240; ninth term $= 1920$; sum $= 3832.5$

109. $\frac{1}{9}; \frac{1}{36}; \frac{1}{144}; \frac{1}{576}; \frac{1}{2304}; \frac{1}{9216}$; ninth term $= \frac{1}{589,824}$; sum ≈ 0.148

111. 44 million

Chapter Test

1. Integer; rational; real

3. Rational; real

5. Irrational; real

7. Whole; integer; rational; real

9. Integer; rational; real

11. 14; 168

13. 25; 4200

15. $\frac{3}{7}$

17. $\frac{16}{25}$

19. $\frac{7}{10}$

21. 36

23. $\frac{15}{16}$

25. $\frac{27}{8}$

27. $2(\sqrt{3} + \sqrt{6})$

29. 3

31. $\frac{7}{8}$

33. $\frac{2}{9}$

35. Commutative for addition

37. Identity for addition

39. Associative for multiplication

41. 4096

43. 1

45. $\frac{1}{3125}$

47. 2.36×10^{-3}

49. -0.00006

51. 3×10^3

53. $\frac{3}{4}; -\frac{1}{8}; \frac{1}{48}; -\frac{1}{288}; \frac{1}{1728}; -\frac{1}{10,368}; \frac{1}{62,208};$
15th term = about 9.57×10^{-12}; sum = about $\frac{9}{14}$

55. $320; $620

Chapter 6

Exercise Set 6-1

1. 3

3. 11

5. 6

7. 7

9. 7

11. 9

13. 2

15. 10

17. 3

19. 6

21. 11

23. 10

25. 2

27. 10

29. 6

31. 3

33. 11

35. 4

37. 6

39. 12

41. 10

43. 9

45. 6

47. 12

49. 12

51. 7

53. 10

55. 5

57. 5

59. None

61. None

63. 1

65. $2 + (5 + 8) = (2 + 5) + 8$. Both sides reduce to 3.

67. $4 + 12 = 4$

69. $7 \times (5 + 11) = 7 \times 5 + 7 \times 11$. Both sides reduce to 4.

71. 10

73. 1

75. 8

77. 6

79. 6

81. 5:00 P.M.

83. 3:11 A.M.

85. 3:42 P.M.

87. 10:18 P.M.

89. 0656

91. 0400

93. 1727

95. 2342

97. 10:09 A.M.

99. 2:36 P.M.

101. It's a nonempty set of elements together with (a) operation(s) for the elements, (b) properties of said operations, and (c) definitions.

103. Given m on the clock, find that positive clock integer r that obeys $m + r = 12$; this r = the sought inverse.

105. Answers will vary.

107. They are all equal to 4 on the clock.

Exercise Set 6-2

1. 2

3. 2

5. 4

7. 1

9. 4

11. 3

13. 3

15. 2

17. 0

19. 6

21. 6

23. 6

25. 4

27. 3

29. 11

31. 2

33. 2

35. 7

37. 2

39. 3

41. 4

43. 3

45. 2

47. 2

49. 5

51. Tuesday

53. Friday

55. Tuesday

57. Not valid

59. Not valid

61. Let r = any number in the system. The identity, usually denoted by 0, is a number (also in the system) with the property that $r + 0 = r = 0 + r$. In modulo m, the identity = $0 = m$.

63. The numbers 0, 1, 2, 3, ... $m - 1$, for any mod m are equivalent to the numbers in base ten. Hence, performing operations in base ten will result in an answer that can be converted to mod m by counting on a clock in mod m.

65.

+	0	1	2	3	4	5	6
0	0	1	2	3	4	5	6
1	1	2	3	4	5	6	0
2	2	3	4	5	6	0	1
3	3	4	5	6	0	1	2
4	4	5	6	0	1	2	3
5	5	6	0	1	2	3	4
6	6	0	1	2	3	4	5

×	0	1	2	3	4	5	6
0	0	0	0	0	0	0	0
1	0	1	2	3	4	5	6
2	0	2	4	6	1	3	5
3	0	3	6	2	5	1	4
4	0	4	1	5	2	6	3
5	0	5	3	1	6	4	2
6	0	6	5	4	3	2	1

67. Closure, commutative, identity, inverse, associative, distributive

Exercise Set 6-3

1. C
3. E
5. E
7. D
9. D
11. Yes
13. Yes
15. D
17. ☆
19. □
21. △
23. ☆
25. □
27. Yes
29. Yes

31.

∪	U	A	∅
U	U	U	U
A	U	A	A
∅	U	A	∅

33. Yes

35.

∩	U	A	∅
U	U	A	∅
A	A	A	∅
∅	∅	∅	∅

37. Yes
39. A nonempty set of elements (symbols); one or more operations on these elements (symbols); definitions; properties
41. Draw the main diagonal, which is a segment drawn from top left to bottom right. If all entries are symmetrical with respect to this diagonal, then the operation is commutative.

43.

∨	T	F
T	T	T
F	T	F

45.

↔	T	F
T	T	F
F	F	T

Closure, commutative, associative, identity, inverse, distributive

Review Exercises

1. 2
3. 4
5. 2
7. 7
9. 2
11. 2
13. 0
15. 0
17. 1
19. 4
21. 3
23. 1
25. 4
27. 1
29. 2
31. 4
33. 2
35. 5
37. 2
39. 6
41. 4
43. 3
45. 1
47. 11
49. 1
51. -1
53. i
55. -1
57. -1
59. -1
61. $-i$
63. Yes
65. Yes
67. 1
69. -1

Chapter Test

1. 1
3. 1
5. 2
7. 4
9. 5
11. 1
13. 2
15. 2
17. 4; also 8; also 0
19. z
21. s
23. z
25. Yes

Chapter 7

Exercise Set 7-1

1. $11x$
3. $-18y$
5. $9p - q - 17$
7. $11x^2 - x + 5$
9. $30x - 35$
11. $-48x + 40$
13. $x + 27$
15. $-26x - 50$
17. $8x - 23$
19. $-x + 13$
21. 83
23. 16
25. 79
27. 466
29. 104
31. 136
33. 997
35. 205
37. 17
39. $127\frac{7}{12}$
41. 376.8 sq in.
43. $6050
45. about 267.95 mm^3
47. about $31,876.96
49. 678.24
51. $100
53. 3.48
55. $10°$C
57. 600,000 ergs
59. about 0.89
61. $14,281.87
63. Variables are used in expressions and formulas.
65. Like terms are terms having the same variables, as well as the same exponents for each variable.
67. For any three real numbers a, b, and c, we have $a(b + c) = ab + ac$ and also $(a + b)c = ac + bc$.
69. Formulas are a short/compact "message" for indicating a relationship among certain variables, using algebra rather than words.

Exercise Set 7-2

1. $\{26\}$
3. $\{59\}$
5. $\{-3\}$
7. $\{3\}$
9. $\{-12\}$
11. $\{6\}$
13. $\{16\}$

15. $\{20\}$
17. $\{11\}$
19. $\{5\}$
21. $\{4\}$
23. $\{12\}$
25. $\{\frac{20}{3}\}$ or $\{6\frac{2}{3}\}$
27. $\{3\}$
29. $\{5\}$
31. $\{54\}$
33. $\{3\}$
35. $\{36\}$
37. $\{64\}$
39. $\{\frac{76}{3}\}$ or $\{25\frac{1}{3}\}$
41. $\{\frac{180}{7}\}$ or $\{25\frac{5}{7}\}$
43. $\{\frac{84}{13}\}$ or $\{6\frac{6}{13}\}$
45. $\{\frac{30}{11}\}$ or $\{2\frac{8}{11}\}$
47. $\{\frac{4}{9}\}$
49. $\{-\frac{9}{53}\}$
51. $\dfrac{3x + 4}{2}$ or $\dfrac{3x}{2} + 2$
53. $\dfrac{7y + 16}{5}$
55. $\dfrac{9 - 7x}{2}$
57. $\{x \mid x$ is a real number$\}$
59. $\varnothing$
61. $L = \dfrac{Rd^2}{k}$
63. $h = \dfrac{V}{\pi r^2}$
65. $h = \dfrac{V}{lw}$
67. $m = \dfrac{E}{c^2}$
69. $h = \dfrac{2A}{b}$
71. $r = \dfrac{mv^2}{F}$
73. A closed equation contains no variables. An open equation contains at least one variable.
75. Solution set = that set whose individual members are all the solutions of the equation.
77. Multiply both sides by the lowest common denominator of all fractions present. This eliminates all such fractions.
79. Answers will vary.

Exercise Set 7-3

1. $x - 3$
3. $x + 9$
5. $11 - x$

7. $x - 9$

9. $7 - x$

11. $8 \cdot x$

13. $3x + 5$

15. $5x + 3$

17. $2x$

19. $\frac{x}{14}$ or $x \div 14$

21. 8

23. 16

25. 11 and 17

27. 12

29. 28

31. $18 billion for Coca-Cola; $29 billion for PepsiCo

33. Bill's age is 12 and Pete's age is 36.

35. $3000

37. 8755 in 1990; 8229 in 1998

39. 18 in.; 24 in.; 30 in.

41. 160 acres

43. $68,333.33

45. Let some letter = the unknown item, and translate appropriate information into an equation. Solve the equation and check the solution.

47. Decreased by; less; less than; subtracted from; diminished

49. Divide by; quotient; out of; ratio

29. $\left\{x \mid x \le 8\frac{2}{5}\right\}$

31. $\left\{x \mid x < -\frac{6}{5}\right\}$

33. $\{x \mid x < 14\}$

35. $\left\{x \mid x \le -10\frac{3}{5}\right\}$

37. $\left\{x \mid x \ge -46\frac{1}{2}\right\}$

39. $\{x \mid x \ge -20\}$

41. $7364.48

43. 64

45. Texas, Florida, Kansas, Colorado

47. Missouri, West Virginia

49. Maryland, California, Pennsylvania, Missouri, West Virginia

51. The open dot indicates the solution set does *not* include that point. The closed dot indicates inclusion.

53. In the original inequality, replace all occurrences of the unknown by any one member of the proposed solution set. If the inequality now holds true, then the proposed solution set could be correct.

55. 78% or greater

Exercise Set 7-4

1.

3.

5.

7.

9.

11. $\{x \mid x < 5\}$

13. $\{x \mid x \le 30\}$

15. $\{x \mid x \ge 6\}$

17. $\{x \mid x < -35\}$

19. $\{x \mid x < 27\}$

21. $\{x \mid x > -3\}$

23. $\{x \mid x \ge 12\}$

25. $\{x \mid x \ge -8\}$

27. $\{x \mid x \le 21\}$

Exercise Set 7-5

1. $\frac{9}{14}$

3. $\frac{7}{16}$

5. $\frac{4}{5}$

7. $\frac{3}{8}$

9. $\frac{2}{1}$

11. $\frac{135}{14} \approx 9.64$

13. 35

15. 10

17. $10\frac{1}{3}$ or $\frac{31}{3}$

19. $\frac{21}{4}$ or $5\frac{1}{4}$

21. 15

23. $\frac{14}{3}$ in. $= 4\frac{2}{3}$ in.

25. 4

27. 99

29. 40

31. 8

33. 20

35. 4 cans

37. $1200

39. 10.19 inches

41. A ratio is a comparison of two quantities using division.

43. A proportion is a statement of equality of two equal ratios.

45. 25 pounds for $7.49

47. 7 ounces for $1.99

49. 34.5 ounces for $7.49

Exercise Set 7-6

1. $x^2 + 16x + 63$

3. $x^2 - 17x + 70$

5. $x^2 - 7x - 120$

7. $14x^2 - 67x + 63$

9. $15x^2 - 19x - 56$

11. $\{-2, -3\}$

13. $\{-4, 3\}$

15. $\{-3, 17\}$

17. $\{-27, 3\}$

19. $\{3, 5\}$

21. $\left\{-3, \frac{7}{2}\right\}$

23. $\left\{-\frac{4}{3}, \frac{3}{2}\right\}$

25. $\left\{-\frac{4}{3}, \frac{3}{2}\right\}$

27. $\left\{-2, \frac{3}{5}\right\}$

29. $\left\{\frac{4}{3}, -\frac{3}{2}\right\}$

31. $\left\{\dfrac{-1+\sqrt{13}}{6}, \dfrac{-1-\sqrt{13}}{6}\right\}$

33. $\left\{-\frac{3}{2}, 4\right\}$

35. $\left\{\dfrac{-5+\sqrt{13}}{6}, \dfrac{-5-\sqrt{13}}{6}\right\}$

37. $\{-1, 9\}$

39. $\left\{\dfrac{-5+\sqrt{37}}{2}, \dfrac{-5-\sqrt{37}}{2}\right\}$

41. 9 seconds

43. 12 and 13

45. 4 ft and 10 ft

47. The term $(x - 1) = 0$. Division by 0 is undefined.

Review Exercises

1. $-2x + 5y - 7$

3. $7x - 36$

5. $13x - 21$

7. 99

9. -70

11. 120

13. $\{-10\}$

15. $\{20\}$

17. $\{-13\}$

19. $\{67\}$

21. $\{-4\}$

23. $8n - 4$

25. $4n + 3$

27. $\frac{40}{9}$ hours to get there; $\frac{32}{9}$ hours to return

29. 158 at $8 each; 79 at $10 each; 89 at $12 each

31. $x > 10$

33. $x > 7$

35. $x \le -\frac{93}{4}$

37. 28

39. $\dfrac{82 \text{ miles}}{15 \text{ gallons}}$

41. $\frac{1}{6}$

43. 9

45. 72

47. 750

49. $10,800 for wife; $7,200 for son

51. $4320

53. 4.44 amps

55. $-2; 3$

57. $-3; 7$

59. $-1; \frac{2}{3}$

61. $\dfrac{5+\sqrt{53}}{2}; \dfrac{5-\sqrt{53}}{2}$

63. $\dfrac{-7+\sqrt{17}}{8}; \dfrac{-7-\sqrt{17}}{8}$

65. $\dfrac{-7+\sqrt{69}}{4}; \dfrac{-7-\sqrt{69}}{4}$

67. 11 and 12

Chapter Test

1. $5x - 10y + 5$

3. 91

5. $\left\{\frac{9}{7}\right\}$

7. $\dfrac{mv^2}{F}$

9. $\left\{x \mid x \le -\frac{3}{2}\right\}$

11. 4

13. $2x^2 - 13x - 24$

15. $\{-3, 17\}$

17. $\left\{-\frac{3}{2}, \frac{4}{3}\right\}$

19. $\left\{-1, \frac{3}{5}\right\}$

21. $2500 at 6%; $500 at 8%

23. About 21.9 hours

25. 12

27. width $= 8$ inches; length $= 12$ inches

Chapter 8

Exercise Set 8-1

1.

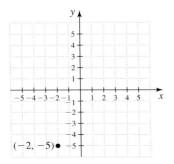

3.

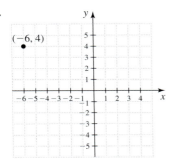

5.

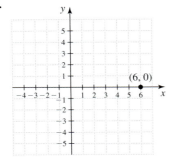

7.

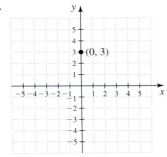

9.

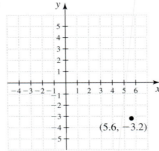

11.

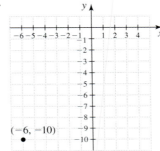

13.

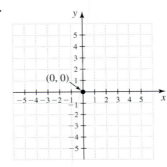

15.

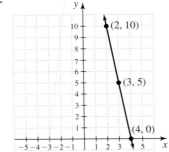

17.

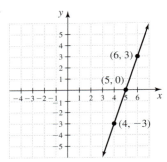

19.

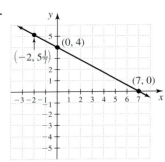

21.

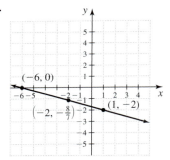

23.

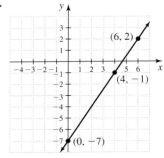

25. 1

27. −0.5

29. $-\frac{3}{5}$

31. About 2.4

33. x intercept $= (8, 0)$; y intercept $= (0, 6)$

35. x intercept $= (-6, 0)$; y intercept $= (0, -5)$

37. x intercept $= (9, 0)$; y intercept $= (0, -18)$

39. x intercept $= (3, 0)$; y intercept $= (0, -7.5)$

41. $y = -\frac{7}{5}x + 7$; slope $= -\frac{7}{5}$; y intercept $= (0, 7)$

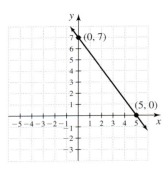

43. $y = \frac{1}{4}x - 4$; slope $= \frac{1}{4}$; y intercept $= (0, -4)$

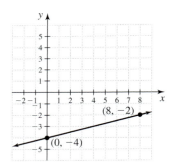

45. $y = \frac{8}{3}x - 8$; slope $= \frac{8}{3}$; y intercept $= (0, -8)$

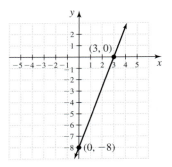

47. $y = 2x - 19$; slope $= 2$; y intercept $= (0, -19)$

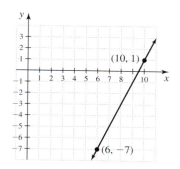

49.

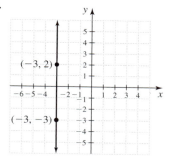

51.

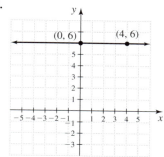

53. (a) $69.50; (b) $82.50; (c) $115

55. (a) $109.30; (b) $86.20; (c) $179.70

57. 31,500

59. 37.8 million

61. Quadrants I, II, III, and IV, respectively, have points whose coordinates are $(+, +), (-, +), (-, -), (+, -)$.

63. Let $(x_0, 0) =$ the x intercept. To find x_0, let $y = 0$ in the equation of the line and solve for x. Let $(0, y_0) =$ the y intercept. To find y_0, let $x = 0$ in the equation of the line and solve for y.

65. If the equation reduces to the form $y =$ constant, then the line is horizontal.

67. The denominator of the fraction for the slope will always be zero.

69. $d = \sqrt{(y_2 - y_1)^2 + (x_2 - x_1)^2}$

Exercise Set 8-2

1. $\{(7, 0)\}$

3. $\{(-2, 0)\}$

5. $\{(2, -2)\}$

7. $\{(6, -2)\}$

9. $\{(-3, 3)\}$

11. $\{(4, -1)\}$

13. $\{(\frac{54}{11}, \frac{16}{11})\}$

15. $\{(5, 3)\}$

17. $\{(\frac{76}{21}, \frac{3}{7})\}$

19. $\{(2, -1)\}$

21. $\{(x, y) \mid 5x - 2y = 11\}$

23. $\{(4, 2)\}$

25. Consistent; $\{(1, -2)\}$

27. Consistent; $\{(7, -2)\}$

29. Consistent; $\{(-\frac{15}{2}, -\frac{11}{2})\}$

31. Consistent; $\{(\frac{41}{19}, -\frac{45}{19})\}$

33. 15 and 17

35. 25 adults and 15 children

37. 19 and 23

39. $2.19 for one sandwich; $1.15 for one order of French fries

41. 322 students; 178 general admission tickets

43. Algebraically: The system has exactly one solution, or else an infinite number of solutions. Geometrically: The lines intersect at exactly one point, or else the lines coincide.

45. Algebraically: Any solution for either line is also a solution for the other; hence, an infinite number of solutions exist. Geometrically: The two lines coincide; that is, they are the same line.

47. Say the two variables are y and z. In either equation solve for one variable, say y, and substitute into the other equation to solve for z. Substitute this actual value of z into either original equation, to solve for the (actual, numerical) value of y.

49. System is inconsistent if the solution procedure leads to some constant $=$ some different constant (e.g., $20 = 14$).

51. Answers will vary.

Exercise Set 8-3

1.

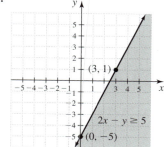

3.

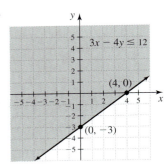

5.

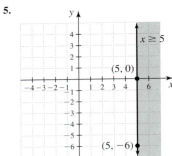

7.

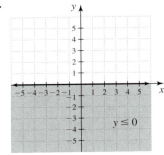

9.

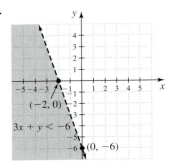

11.

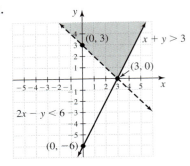

13.

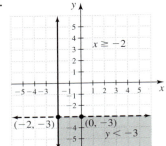

15.

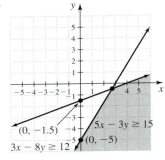

17.

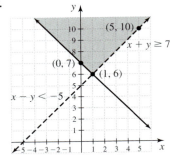

19.

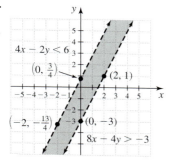

$4x - 2y < 6$
$\left(0, \frac{3}{4}\right)$
$(2, 1)$
$\left(-2, -\frac{13}{4}\right)$ $(0, -3)$
$8x - 4y > -3$

21. A solid line indicates all points of it are part of (belong to) the graph. A dashed line indicates no points of it are part of (belong to) the graph.

23. When the signs of both inequalities are a combination of $\geq$ or $\leq$.

25. $\varnothing$

Exercise Set 8-4

1. $x \leq 50$

3. $x = 2y$, where x = number of pens and y = number of pencils

5. $P = 85d + 130l$

7. $P = 4t + 3h$, where t = number of turkeys and h = number of hams

9. $2v + 3c \leq 6$, where v = number of VCRs assembled and c = number of CD players assembled

11. P at $(0, 0) = \$0$; P at $(5, 6) = \$390$; P at $(10, 0) = \$300$; P at $(0, 8) = \$320$. Hence, vertex $(5, 6)$ gives maximum profit.

13. P at $(0, 0) = \$0$; P at $(4, 8) = \$100$; P at $(0, 6) = \$30$; P at $(7, 0) = \$105$. Hence, vertex $(7, 0)$ gives maximum profit.

15. P at $(0, 0) = \$0$; P at $(20, 50) = \$19,400$; P at $(0, 62) = \$21,080$; P at $(43, 0) = \$5160$. Hence, vertex $(0, 62)$ gives maximum profit.

17. 20 television sets; 10 VCRs.

19. 50 rabbits; 0 chicks

21. 50 acres of beans; 0 acres of corn

23. Linear programming is used in making decisions and in finding cost effective solutions. A common problem is to find "the most . . ." or "the least. . . ."

25. It's a function (such as an equation) that relates the variables in linear programming.

27. Evaluate the objective function at the coordinates of each vertex of the region. From the various results, choose the maximum or minimum, as desired.

Exercise Set 8-5

1. Domain = $\{5, 6, 7, 8\}$; range = $\{8, 9, 10, 11\}$; relation is a function.

3. Domain = $\{6, 7, 8, 9\}$; range = $\{11\}$; relation is a function.

5. Domain = $\{0\}$; range = $\{0\}$; relation is a function. (The requirement for function is trivially or automatically satisfied.)

7. Domain = $\{-10, -40, -60\}$; range = $\{20, 40, 60\}$; relation is not a function.

9. 17

11. -48

13. 0

15. 5.56

17.

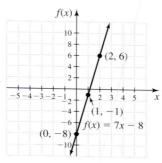

$(2, 6)$
$(1, -1)$
$f(x) = 7x - 8$
$(0, -8)$

19.

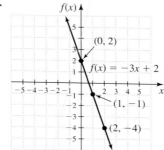

$(0, 2)$
$f(x) = -3x + 2$
$(1, -1)$
$(2, -4)$

21.

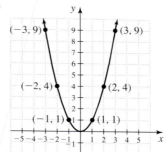

$(-3, 9)$ $(3, 9)$
$(-2, 4)$ $(2, 4)$
$(-1, 1)$ $(1, 1)$

23.

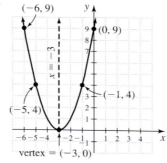

$(-6, 9)$
$(0, 9)$
$x = -3$
$(-1, 4)$
$(-5, 4)$
vertex = $(-3, 0)$

25.

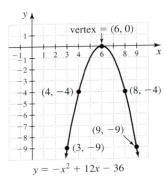

$y = -x^2 + 12x - 36$

27.

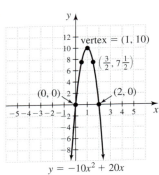

$y = -10x^2 + 20x$

29.

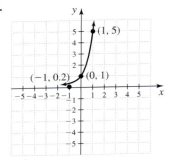

31.

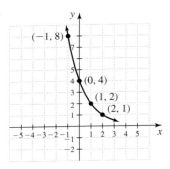

33. Function

35. Not a function

37. Function

39. 62.25 ft = maximum height; 3.85 seconds = duration of entire flight

41. 8 ft for the base; 4 ft for each of the folded sides.

43. 4,732,864

45. 0.206

47. A function is a special relation; more specifically: a relation in which there exists one and only one second element for each first element in the ordered pairs.

49. The set of second elements of the ordered pairs is called the range.

51. A parabola is a **U**-shaped curve, which is the graph of a quadratic function.

53. Have the equation in the form $f(x)$[or y] $= ax^2 + bx + c$, where a, b, and c, are constants. If $a > 0$, then the parabola opens upward.

55. (a) Compound interest. This occurs when interest on an account is not collected but kept, and thus contributes to the earnings.

   (b) Radioactive decay. The total amount (of a substance) lost depends on the temporary amount present "along the way."

57. The graph opens to the left.

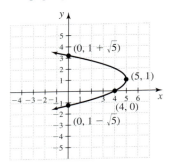

Review Exercises

1.

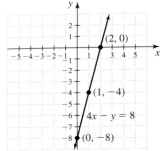

3.

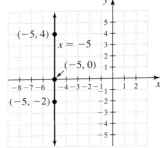

5.

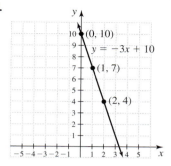

7. $-\frac{1}{4}$

9. 7

11. 0

13. $y = -3x + 12$; slope $= -3$; x intercept $= (4, 0)$; y intercept $= (0, 12)$

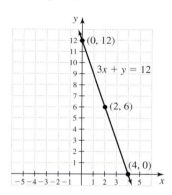

15. $y = \frac{4}{7}x - 4$; slope $= \frac{4}{7}$; x intercept $= (7, 0)$; y intercept $= (0, -4)$

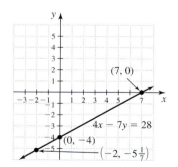

17. Betty has $7.64; Mary has $24.36.

19. $x = 10$; $y = 2$

21. $x = 5$; $y = 2$

23. $x = 10$; $y = -1$

25. $x = 6$; $y = 3$

27. $x = 4$; $y = 0$

29. $x = \frac{3}{2}$; $y = -\frac{5}{2}$

31. $2.50 for 1 lb of coffee; $1.75 for 1 lb of tea

33.

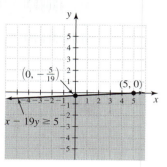

35.

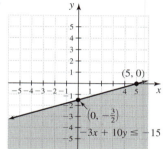

37.

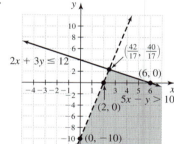

39.

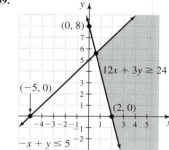

41. 25 of model A

43. Domain $= \{2, 5, 6\}$; range $= \{5, -7, -10\}$; function

45. Domain $= \{4, 5, 6\}$; range $= \{10\}$; function

47. 13

49. 40

51.

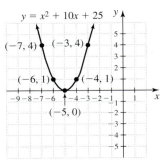

53.

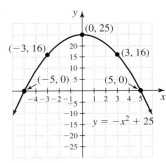

55.

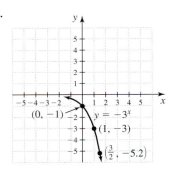

57. From point of launch: 100 ft; from ground level: 104 ft

Chapter Test

1.

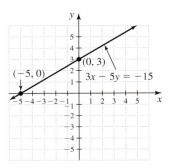

3.

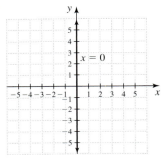

5.

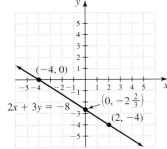

7. $\frac{17}{7}$

9. $y = -\frac{x}{5} + 4$; slope $= -\frac{1}{5}$; x intercept $= (20, 0)$; y intercept $= (0, 4)$

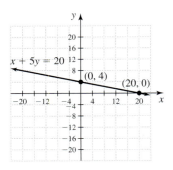

11. $x = 3$; $y = -1$

13. Dependent system. Solution $= \{(x, y) \mid 3x + 4y = 19\}$

15.

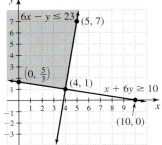

17. Domain = {−4, −10, 12}; range = {6, 18, 5}; function

19. 55

21.

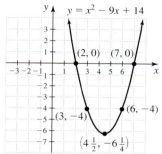

23.

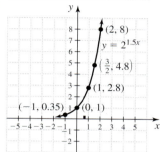

25. 26 and 31

27. $5.50 for servers; $7.50 for cooks

29. 6 in. for each of the 2 heights; 12 in. for width of base

Chapter 9

Exercise Set 9-1

1. 63%

3. 2.5%

5. 156%

7. 20%

9. $66.\overline{6}\%$ or 66.7%

11. 125%

13. 0.18

15. 0.06

17. 0.625

19. 3.2

21. $\frac{6}{25}$

23. $\frac{9}{100}$

25. $2\frac{9}{25}$

27. $\frac{1}{200}$

29. $\frac{1}{6}$

31. Sales tax = $15; total cost = $314.99

33. Sales tax = $9.00; total cost = $158.99

35. 60%

37. 24%

39. $95.99

41. $149.99

43. $1256

45. 2446

47. Hundredths, or part of a hundred

49. Change the fraction to a decimal; change this decimal to a percent.

51. Use the formula Part = (Rate = Percentage) × Base: $P = R \times B$.

    I. Finding a percent of a number: Have R as a decimal and solve for P in $P = R \times B$.

    II. Finding what percent one number is of another: Solve for R in the formula $P = R \times B$; change the answer to a percent.

    III. Finding a number when a percent of it is known: Solve for B in the formula $P = R \times B$, but first change R into a decimal.

53. No, it will be worth $91.

Exercise Set 9-2

1. $1440

3. 2 years

5. 5%

7. $985

9. 3.2%

11. $128.25

13. 6.5 years

15. 10%

17. 6%

19. 3 years

21. Interest = $396.20; MV = $1221.20

23. Interest = $14.67; MV = $89.67

25. Interest = $991.92; MV = $1616.92

27. Interest = $688.05; MV = $2683.05

29. 8.43%

31. $2500

33. 3 years

35. 4.25 years = 51 months

37. $2096.25

39. $12,175.94

41. $335,095.61

43. 6.14%

45. 6.66%

47. $20,548.25

49. $27,223.76

51. $24,664.64

53. $59,556.16

55. Payment for the use of money is called interest.

57. The *term* of a loan is the time (duration) that the loan is in effect.

59. The effective rate is the simple interest rate that would yield the same maturity value over 1 year as the compound interest rate.

61. $10,000 at 9% for 6 years

63. $18,000 at $12\frac{1}{2}$% for 8 years

Exercise Set 9-3

1. Finance charge = $16.65; new balance = $1124.15
3. Finance charge = $39.49; new balance = $3368.60
5. Finance charge = $13.32; new balance = $411.35
7. (a) $602.14
   (b) $7.23
   (c) $669.15
9. (a) $370.55
   (b) 5.19
   (c) $350.76
11. (a) $370.92
    (b) $4.08
    (c) $533.30
13. 15.67%
15. 22.26%
17. 18.33%
19. $180.13
21. $8.22
23. $5.18
25. It's a method of borrowing money, whereby portions of the debt (including charges) are paid back from time to time. In the case of credit-card companies, these payments need not be equal or equally spaced. In the case of lending institutions, the payments typically are constant and once a month.
27. The *unpaid balance method* entails interest (for 1 month) charged only on the balance from the previous month. With the *average daily balance method*, the balances for each day of the month are added, and the sum is divided by the number of days involved. Then interest (for 1 month) is charged on the quotient. (That is, on the average daily balance.)
29. The rule of 78s is one method of computing the money saved by a borrower for paying off early (usually a short-term one), and thus reducing the otherwise total interest. It's a fact that $\frac{12}{78} + \frac{11}{78} + \frac{10}{78} + \cdots + \frac{1}{78} = \frac{78}{78} = 1$. For a 12-month loan, this fact is useful to a lending institution that wants to collect during the first month: $\frac{12}{78}$ of the total interest; second month: $\frac{11}{78}$ of the total interest; etc.
31. $3636.36
33. $32; $12.31

Exercise Set 9-4

1. (a) $21,750
   (b) $123,250
   (c) $949.03
   (d) $161,459
3. (a) $80,000
   (b) $120,000
   (c) $1142.40
   (d) $291,264
5. (a) $2625
   (b) $49,875
   (c) $523.19
   (d) $201,256.20

7. (a) $60,000
   (b) $140,000
   (c) $1492.40
   (d) $218,176

9.

Payment Number	Interest	Payment on Principal	Balance of Loan
1	$718.96	$230.07	$123,019.93
2	$717.62	$231.41	$122,788.52
3	$716.27	$232.76	$122,555.76

11.

Payment Number	Interest	Payment on Principal	Balance of Loan
1	$1341.67	$150.73	$139,849.27
2	$1340.22	$152.18	$139,697.09
3	$1338.76	$153.64	$139,543.45

13. $3000
15. $6843.75
17. $660
19. $23,100
21. In a *fixed-rate* mortgage, the rate of interest stays constant throughout the entire loan. In an *adjustable-rate mortgage,* the rate of interest may decrease and/or increase.
23. Multiply the (constant) monthly payment by the number of such payments. Decrease this sum by the amount borrowed. In short: total interest = (sum of all payments) minus (loan).
25. The *market value* is the present monetary worth of the home. The *assessed value* is a certain percentage of the market value.
27. 20 years at 9% with a 25% down payment is better.

Exercise Set 9-5

1. $S = $3487.50; M = 1237.50
3. $C = $72; M = 8
5. $R = 41.7\%; M = 25
7. $R = 35\%; M = 70
9. $C = $44; R = 47.6\%$
11. $C = $200; S = 290
13. $S = $540; M = 360
15. $C = $33.33; S = 83.33
17. $M = $6; S = 36
19. $C = $66.67; S = 86.67
21. $M = $20; R = 40\%$
23. $M = $260; C = 260
25. $S = $187.50; C = 37.50
27. 46.7%
29. 35%
31. 40%
33. Markdown = $39.20; reduced price = $58.80
35. Markdown = $9; reduced price = $6
37. $42.12
39. 35.5%

41. 186%

43. The markup on cost is a certain percentage of the cost the merchant paid for some item. The markup on selling price is a certain percentage of the intended selling price. In both cases, the markup is an amount added to the cost of an item, to give the selling price.

45. Selling price is 100% when it is the base.

47. 25%

Review Exercises

1. 0.875; 87.5%

3. $\frac{4}{5}$; 0.8

5. $\frac{37}{20}$; 1.85

7. 5.75; 575%

9. $\frac{91}{200}$; 0.455

11. 69.12

13. 1100

15. $60

17. $2322

19. 2.5 years

21. 7%

23. $1375

25. Interest = $603.67; MV = $2378.67

27. Interest = $12.07; MV = $57.07

29. Interest = $1800; MV = $7800

31. 1.4%

33. 12.55%

35. $8129.93

37. Finance charge = $124.13; new balance = $7424.53

39. $15.24

41. $2835

43. $C = \$20.50$; $M = \$4.50$

45. $R = 25\%$; $M = \$225$

47. 72.4%

Chapter Test

1. 31.25%

3. $\frac{7}{25}$

5. 80%

7. 300

9. $2568

11. $486

13. Interest = $8.16; MV = $443.16; monthly payment = $73.86

15. Interest = $216; MV = $2016; monthly payment = $168

17. Interest = $7885.08; MV = $17,635.08

19. 8.16%

21. Finance charge = $20; new balance = $1030

23. 11.17%

25. (a) $4000

(b) $76,000

(c) $611.80

(d)

Payment Number	Interest	Payment on Principal	Balance on Loan
1	$570	$41.80	$75,958.20
2	$569.69	$42.11	$75,916.09

27. 66.7% = rate on cost; 40% rate on selling price

29. 25%

Chapter 10

Exercise Set 10-1

1. Open segment; $\overset{\circ-\circ}{PQ}$

3. Line; $\overleftrightarrow{RS}$

5. Half open segment; $\overset{\bullet-\circ}{CD}$

7. Segment; $\overline{TU}$

9. $\measuredangle RST$; $\measuredangle TSR$; $\measuredangle S$; $\measuredangle 3$

11. Straight

13. Obtuse

15. Vertical

17. Corresponding

19. Corresponding

21. Alternate exterior

23. 82°

25. 58°

27. 12°

29. 24°

31. 118°

33. 60°

35. $m\measuredangle 1 = 37° = m\measuredangle 3$; $m\measuredangle 2 = 143°$

37. 90° for each of angles 1, 2, and 3

39. 165° for each of angles 2, 4, and 6; 15° for each of angles 1, 3, 5, and 7

41. 90°

43. 60°

45. A *point* can be considered as a dot, but theoretically smaller than any conceivable size. A *line* is a set of connected points forming a "taut string" of infinite length, but theoretically with no thickness, or widths, or gaps. A *plane* is a flat surface (theoretically without thickness), extending infinitely in every direction.

47. A half line, plus its end point, constitutes a ray.

49. The sum of the measures equals 90° for two complementary angles; 180° for two supplementary angles.

51. $m\measuredangle 3 = 55°$; $m\measuredangle 4 = 125°$

Exercise Set 10-2

1. Isosceles; also acute

3. Obtuse

5. Scalene; also acute

7. 70°

9. 15°

11. 34 ft

13. 560 cm

15. 810 yd

17. 30 in.

19. 17 m

21. 11 ft

23. 84.85 ft

25. 104 ft

27. 270 yd

29. 12.8 ft

31. 212.2 mi

33. *Equilateral:* all sides equal; *isosceles:* two sides equal; *scalene:* no two sides are equal.

35. Draw a triangle on a piece of paper and cut off the angles. Place the angles next to each other on a flat surface in such a way that their vertices coincide and two pairs of angles have a common side. The set of angles forms a 180° angle.

37. Similar triangles have the same shape, but not necessarily the same size. More specifically, the corresponding angles of such triangles have equal measure.

39. $\dfrac{\pi a^2}{8}, \dfrac{\pi b^2}{8}, \dfrac{\pi c^2}{8}$

Exercise Set 10-3

1. Octagon; 1080°

3. Triangle; 180°

5. Hexagon; 720°

7. Rectangle

9. Trapezoid

11. 76 yd

13. 28 ft

15. 44 in.

17. 42 ft

19. 42 mi

21. 360 ft

23. 241 ft

25. 66 in.

27. 5.08

29. Similarities:

   (a) In both, opposite sides are = and ∥.

   (b) In both, sum of measures of angles is 360°.

   (c) In both, opposite angles have equal measure.

   Differences:

   (a) In a rhombus, all four sides are = in length; in a rectangle, this need not hold.

   (b) A rectangle always has four right angles; this may, or may not, be true in a rhombus.

31. A square is a rectangle because in both

   (a) Opposite sides are equal in length.

   (b) All four angles are right.

   A square is also a rhombus since in both

(a) All four sides are equal in length.

(b) Opposite angles have equal measure.

33. Sum $= (n - 2)180°$, where $n =$ number of sides the polygon has.

35. $\angle A + \angle B = \angle BCD$

Exercise Set 10-4

1. 289 in.^{2}

3. 450 yd^2

5. 400 m^2

7. 105 miles^2

9. 467.5 in.^{2}

11. 72 in.^{2}

13. $C = 50.24$ in.; $A = 200.96$ in.^{2}

15. $C = 50.24$ m; $A = 200.96$ m^2

17. $C = 131.88$ km; $A = 1384.74$ km^2

19. 11.11

21. 60

23. $1620

25. 7.5 ft^2

27. $188.40

29. 11,304 square miles

31. 16.28 ft^2

33. The *perimeter* is the distance around the border of the polygon. The *area* is the number of unit squares needed to cover completely the space enclosed by the polygon, but no other space. (A "unit square" means a square, each of whose sides measures 1 cm, or 1 inch, or 1 ft, etc.)

35. Take any circle; the circumference divided by the diameter is always the same answer, which we denote by π.

37. 30 in.

Exercise Set 10-5

1. SA $= 150$ square inches; $V = 125$ cubic inches

3. SA $= 214$ square meters; $V = 210$ cubic meters

5. SA $= 96$ square meters; $V = 48$ cubic meters

7. SA $= 4044.32$ square centimeters; $V = 19,694.08$ cubic centimeters

9. SA $= 628$ square feet; $V = 1004.8$ cubic feet

11. SA $= 5024$ square inches; $V = 33,493.33$ cubic inches

13. 461.58 cm^3

15. 648 cubic feet

17. 190,228,656 ft^3

19. 13,677.84 in.^{3}

21. 126.86 in.^{2} (exclude base)

23. 3.88×10^{10} mi^3

25. Pretend we can detach a very thin skin off the outermost surface of the solid. Lay this skin flat on a flat surface and measure the area covered.

27. Approximately 154 in.^{3}

Exercise Set 10-6

1. 92.7 cm

3. 76.89 in.

5. 1.392 mm

7. 122.98 in.

9. 1857.29 ft

11. 77.46 miles

13. 48°

15. 60°

17. 29°

19. 45°

21. 23,648.41 ft

23. 4980.78 meters

25. 1218.91 ft

27. 68°

29. 359.47 ft

31. 3930.34 ft

33. Triangle measurement

35. All right triangles having a 30° angle are similar. Hence, given any one of these triangles, the ratio of any two sides equals the corresponding ratio from any other of said triangles.

37. It's the angle made by a horizontal line and the downward line of sight to an object.

39. 737.6 ft, 673.8 ft

Exercise Set 10-7

1. 5 evens, 4 odds; not traversable

3. 2 evens, 2 odds; traversable

5. 3 evens, 2 odds; traversable

7. 3 evens, 2 odds; traversable

9. 3 evens, 4 odds; not traversable

11.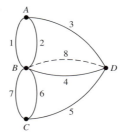

Solution I: Add bridge 8. Then $A =$ odd, $D =$ even; $B =$ even; $C =$ odd; hence, network is traversable.

Solution II: Do not have bridge 8 or bridge 2. Then $A =$ even, $B =$ even, $C =$ odd, $D =$ odd; hence, network is traversable.

Note: Other solutions do exist.

13. A network is traversable if it's possible to pass through, or trace, each path exactly once, without lifting your pencil. However, a vertex can be crossed more than once.

15.

One possibility: The vertex at the center is even; all six other vertices are odd.

17. Answers will vary.

Review Exercises

1. Line

3. Open segment

5. Trapezoid

7. Triangle (equilateral)

9. Hexagon (regular)

11. Obtuse

13. Vertical

15. Alternate exterior

17. (a) 63°

   (b) 2°

19. 43°

21. 72 ft

23. 42° for angles 2, 3, and 6; 138° for angles 1, 4, 5, and 7

25. 16 ft

27. 60.8 m^2

29. 116 ft^2

31. 24,416.64 cm^3

33. 84.78 yd^3

35. 339.62 in.^{2}

37. 776.47 or 776.09 (depending on rounding)

39. 38.77 cm^3

41. 39°

43. Not traversable

45. Not traversable

47. Traversable

Chapter Test

1. 17°

3. Respectively: 152°; 28°; 152°

5. 48°

7. 2.5 ft

9. 1080°

11. 18 in.^{2}

13. 432 yd^2

15. 345.96 mi^2

17. 884.74 in.^{3}

19. 100 ft^3

21. 150 ft^3

23. 1331 ft^2

25. 346.2 in.^{2}

27. 24.49 ft

29. 0.21 lb

31. 307.97 ft

33. Traversable

35. Not traversable

Chapter 11

Exercise Set 11-1

1. (a) $\dfrac{1}{6}$ (b) $\dfrac{1}{2}$ (c) $\dfrac{1}{3}$ (d) 1 (e) 1 (f) $\dfrac{5}{6}$ (g) $\dfrac{1}{6}$

3. (a) $\dfrac{1}{7}$ (b) $\dfrac{3}{7}$ (c) $\dfrac{3}{7}$ (d) 1 (e) 0

5. $\dfrac{5}{9}$

7. $\dfrac{9}{16}$

9. $\dfrac{12,166}{12,249} \approx 0.99$

11. $84\% = 0.84$

13. (a) $\dfrac{8}{15}$ (b) $\dfrac{1}{3}$ (c) $\dfrac{1}{5}$ (d) $\dfrac{1}{3}$ (e) $\dfrac{1}{5}$

15. (a) $\dfrac{1}{5}$ (b) $\dfrac{3}{5}$ (c) $\dfrac{2}{5}$ (d) $\dfrac{2}{5}$

17. A probability experiment is a process that leads to well-defined results called outcomes.

19. An event is a subset whose elements are outcomes.

21. From zero (inclusive) to one (inclusive)

23. 0

25. The sample space consists of two outcomes: rain and no rain. Since all the probabilities add to 1, we have $P(\text{rain}) + P(\text{no rain}) = 1$, whence $0.45 + P(\text{no rain}) = 1$, $P(\text{no rain}) = 0.55$.

27. b, d, f, i

29. Answers will vary.

31. $\dfrac{1}{36}$

Exercise Set 11-2

1. (a) $\dfrac{3}{8}$ (b) $\dfrac{1}{8}$ (c) $\dfrac{7}{8}$

3. (a) $\dfrac{5}{12}$ (b) $\dfrac{1}{4}$ (c) $\dfrac{1}{6}$

5.

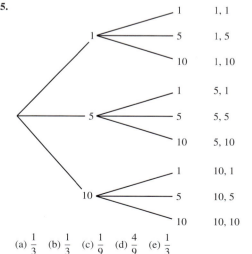

(a) $\dfrac{1}{3}$ (b) $\dfrac{1}{3}$ (c) $\dfrac{1}{9}$ (d) $\dfrac{4}{9}$ (e) $\dfrac{1}{3}$

7.

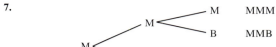
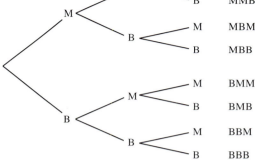

(a) $\dfrac{1}{4}$ (b) $\dfrac{3}{4}$ (c) $\dfrac{1}{4}$ (d) $\dfrac{1}{8}$

9.

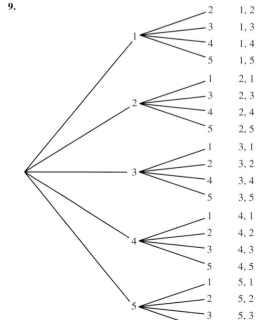

(a) $\dfrac{3}{5}$ (b) $\dfrac{1}{2}$ (c) $\dfrac{4}{5}$

11.

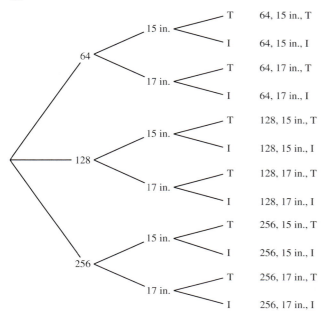

T   64, 15 in., T

15 in.

I   64, 15 in., I

64

T   64, 17 in., T

17 in.

I   64, 17 in., I

T   128, 15 in., T

15 in.

I   128, 15 in., I

128

T   128, 17 in., T

17 in.

I   128, 17 in., I

T   256, 15 in., T

15 in.

I   256, 15 in., I

256

T   256, 17 in., T

17 in.

I   256, 17 in., I

(a) $\dfrac{1}{2}$ (b) $\dfrac{1}{6}$ (c) $\dfrac{1}{4}$

13. (a) $\dfrac{1}{13}$ (b) $\dfrac{1}{4}$ (c) $\dfrac{1}{52}$ (d) $\dfrac{2}{13}$ (e) $\dfrac{4}{13}$ (f) $\dfrac{4}{13}$ (g) $\dfrac{1}{2}$ (h) $\dfrac{1}{26}$

(i) $\dfrac{15}{26}$ (j) $\dfrac{1}{26}$

15. (a) $\dfrac{1}{9}$ (b) $\dfrac{2}{9}$ (c) $\dfrac{1}{6}$ (d) $\dfrac{5}{18}$ (e) $\dfrac{11}{36}$ (f) $\dfrac{1}{2}$ (g) $\dfrac{3}{4}$ (h) 1

17. Use branches (segments) emanating from one point to show possibilities for first stage (i.e., for first experiment); then use segments from the tip of each of these segments to show possibilities for second stage; and so on.

19. 216

21. $\dfrac{5}{108}$

Exercise Set 11-3

1. (a) $\dfrac{1}{11}$ or 1:11;   (b) $\dfrac{1}{35}$ or 1:35;   (c) $\dfrac{5}{1}$ or 5:1;

(d) $\dfrac{17}{1}$ or 17:1;   (e) $\dfrac{1}{5}$ or 1:5

3. (a) $\dfrac{1}{12}$ or 1:12;   (b) $\dfrac{3}{10}$ or 3:10;   (c) $\dfrac{3}{1}$ or 3:1;

(d) $\dfrac{1}{12}$ or 1:12;   (e) $\dfrac{1}{1}$ or 1:1

5. (a) $\dfrac{7}{11}$;   (b) $\dfrac{5}{7}$;   (c) $\dfrac{3}{4}$;   (d) $\dfrac{4}{5}$

7. $\dfrac{5}{14}$

9. −$3.00

11. \$ $\left(\dfrac{5}{6}\right) \approx$ \$0.83

13. −$1

15. −$0.50; −$0.52

17. Odds in favor of event E give an indication of the frequency for E to occur—called a "win." Example: If such odds are 4:11, then there would be, on average, a win four times out of 15 tries. Odds against entail a similar concept, but have to do with E not occurring—called a "loss." Example: if the odds against E are 8:5, then there would be, on average, eight losses out of 13 tries.

19. Odds are used to make a game mathematically fair, or to give one party a mathematical advantage over the other party. If, on average, a player has as much chance of winning an amount as of losing the *same* amount, then the game is considered fair.

21. It's a mathematically expected result or average, such as over the long run. More specifically: If the "parts" or outcomes of event E are $X_1, X_2, \ldots, X_n$, and the corresponding probabilities are $P(X_1)$ etc., then the expected value $= X_1 P(X_1) + X_2 P(X_2) + \cdots + X_n P(X_n)$.

23. 0.518; 0.491

Exercise Set 11-4

1. $\dfrac{1}{6}$

3. $\dfrac{11}{19}$

5. (a) $\dfrac{17}{20}$;   (b) $\dfrac{11}{20}$;   (c) $\dfrac{3}{5}$

7. (a) $\dfrac{8}{17}$;   (b) $\dfrac{6}{17}$;   (c) $\dfrac{9}{17}$;   (d) $\dfrac{12}{17}$

9. (a) $\dfrac{6}{7}$;   (b) $\dfrac{4}{7}$;   (c) 1

11. (a) $\dfrac{67}{118}$;   (b) $\dfrac{81}{118}$;   (c) $\dfrac{44}{59}$

13. (a) $\dfrac{38}{45}$;   (b) $\dfrac{22}{45}$;   (c) $\dfrac{2}{3}$

15. (a) $\dfrac{14}{31}$;   (b) $\dfrac{23}{31}$;   (c) $\dfrac{19}{31}$

17. (a) $\dfrac{3}{13}$;   (b) $\dfrac{3}{4}$;   (c) $\dfrac{19}{52}$;   (d) $\dfrac{7}{13}$;   (e) $\dfrac{15}{26}$

19. $\dfrac{7}{10}$

21. They are events that cannot occur at the same time.

23. 0.06

25. 0.30

Exercise Set 11-5

1. 0.0058

3. 0.073

5. 0.0625

7. $\dfrac{1}{144}$

9. $\dfrac{1}{8}$

11. $\dfrac{1}{15}$

13. 0.000216

15. (a) 0.00018;   (b) 0.013;   (c) 0.12

17. $\dfrac{1}{56} \approx 0.018$

19. $\dfrac{1}{6}$

21. $\dfrac{2}{11}$

23. $\dfrac{1}{4}$

25. $\dfrac{2}{3}$

27. 1

29. 0.61

31. $\dfrac{1}{2}$

33. Two events A and B are independent if the fact that A occurs has no effect on the probability of B occurring. They are dependent if such an effect does exist. Examples of two events are

Independent: cutting your tree in your backyard, and choosing Brand B for your calculator.

Dependent: Being on time for the bus, and being on time for the meeting.

35. $\dfrac{25}{216}$

37. $\dfrac{1}{133,225}$

Exercise Set 11-6

1. 3,628,800

3. 362,880

5. 1

7. 2520

9. 60

11. 1

13. 40,320

15. 30

17. 336

19. 5040

21. 1,860,480

23. 2520

25. 120

27. 11,441,304,000

29. 24

31. Suppose, in a sequence of n events, are available:

k_1 possibilities for the first event;

k_2 possibilities for the second event;

k_3 possibilities for the third; and so forth.

Then the total number of ways in which the sequence of n events can occur is $k_1 \cdot k_2 \cdot k_3 \cdot \, \cdots \, \cdot k_n$.

33. (a) 34,650;   (b) 12,600

Exercise Set 11-7

1. 10

3. 35

5. 15

7. 1

9. 66

11. 2,598,960

13. 126; 21

15. 120

17. 462

19. 166,320

21. 14,400

23. 67,200

25. 53,130

27. 126

29. 30,045,015

31. A combination is a selection of objects without regard to order or arrangement.

33. 1   5   10   10   5   1
　　 1   6   15   20   15   6   1

Exercise Set 11-8

1. (a) $\dfrac{1}{2530} \approx 0.0004$ (b) $\dfrac{38}{253} \approx 0.15$

(c) $\dfrac{969}{2530} \approx 0.383$ (d) $\dfrac{114}{253} \approx 0.45$

3. (a) $\dfrac{1}{30} \approx 0.033$ (b) $\dfrac{1}{120} \approx 0.0083$ (c) $\dfrac{3}{10}$ (d) $\dfrac{3}{40}$ (e) $\dfrac{3}{20}$

5. (a) $\dfrac{14}{55} \approx 0.25$ (b) $\dfrac{28}{55} \approx 0.51$ (c) $\dfrac{1}{55} \approx 0.018$

7. 0.00144

9. (a) $\dfrac{8}{65} \approx 0.123$ (b) $\dfrac{1}{35} \approx 0.029$ (c) $\dfrac{2}{91} \approx 0.022$

(d) $\dfrac{16}{91} \approx 0.176$ (e) $\dfrac{4}{13} \approx 0.308$

11. $\dfrac{882}{2431} \approx 0.363$

13. $\dfrac{1}{658,008} \approx 1.5 \times 10^{-6}$

15. Method A: From Section 11-5, use multiplication rule 2, extended to more than two events: P(woman, man, woman, man, woman) = $\frac{3}{5} \cdot \frac{2}{4} \cdot \frac{2}{3} \cdot \frac{1}{2} \cdot \frac{1}{1} = \frac{12}{120} = \frac{1}{10}$. Method B: From Exercise Set 11-6, use the formula given in Exercise 33 for the number of permutations of n objects, where k_1 are alike, k_2 are alike, etc. In the present case, $k_1 = 3$, $k_2 = 2$, and $k_1 + k_2 = n = 5$. Then the number of permutations is $\dfrac{n!}{k_1!k_2!} = \dfrac{5!}{3!2!} = \dfrac{120}{12} = 10$. Since only one of these 10 meets our order, the desired probability $= \dfrac{1}{10}$.

17. $\dfrac{1}{54,145}$

19. $\dfrac{1}{649,740}$

Review Exercises

1. (a) $\frac{1}{6}$ (b) $\frac{1}{6}$ (c) $\frac{2}{3}$

3. $\frac{16}{45}$

5. $\frac{17}{30}$

7. (a) 0.1   (b) $\frac{11}{30}$ (c) $\frac{13}{15}$ (d) $\frac{13}{15}$

9. $\frac{2}{11}$

11. 0.289

13. (a) $\frac{2}{17} \approx 0.118$ (b) $\frac{11}{850} \approx 0.013$ (c) $\frac{1}{5525} \approx 1.81 \times 10^{-4}$

15. $\frac{5}{13}$

17. 1:5

19. 18.2 cents

21. 175,760,000; 88,583,040

23. 60; No.

25.

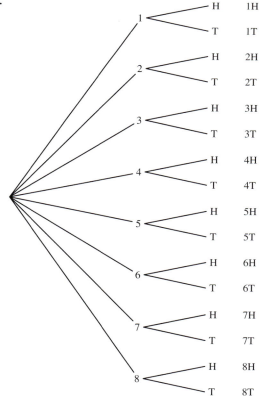

27. 792

29. 15

31. $\frac{12}{55}$

Chapter Test

1. (a) $\frac{1}{13}$ (b) $\frac{1}{13}$ (c) $\frac{4}{13}$

3. (a) $\frac{12}{31}$ (b) $\frac{12}{31}$ (c) $\frac{27}{31}$ (d) $\frac{24}{31}$

5.

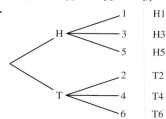

7. (a) 0.025   (b) 0.000495   (c) 0

9. $\frac{1}{2}$

11. $\frac{1}{2}$

13. 33,554,432

15.

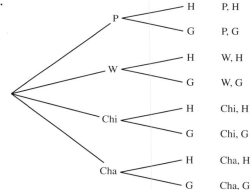

8 = total number of possibilities

17. 40,320

19. 1365

21. 4

23. 5:4

25. $5\frac{1}{6}$

27. $\frac{8}{33}$

Chapter 12

Exercise Set 12-1

1.

Rank	Tally	Frequency
Fr	卌 卌 卌 ///	18
So	卌 卌 //	12
Jr	卌 /	6
Se	////	4

3.

Class	Tally	Frequency
21–86	//	2
87–152	//	2
153–218	ʇʜʜ ʇʜʜ //	12
219–284	ʇʜʜ ʇʜʜ ʇʜʜ ///	18
285–350	ʇʜʜ ////	9
351–416	ʇʜʜ /	6

5.

Class	Tally	Frequency
70–637	ʇʜʜ ʇʜʜ ʇʜʜ	15
638–1205	ʇʜʜ /	6
1206–1773	///	3
1774–2341		0
2342–2909	/	1
2910–3477		0
3478–4045	//	2

7.

Class	Tally	Frequency
5–101	ʇʜʜ ʇʜʜ ʇʜʜ //	17
102–198	ʇʜʜ /	6
199–295	ʇʜʜ /	6
296–392	//	2
393–489	//	2
490–586	///	3
587–683	/	1
684–780	//	2

9. Note: Combined highest − combined lowest = 550 − 306.

Class	McGwire: Tally & Frequency		Sosa: Tally & Frequency	
306–336	/	1		0
337–367	ʇʜʜ /	6	ʇʜʜ ʇʜʜ	10
368–398	ʇʜʜ ʇʜʜ ʇʜʜ ////	19	ʇʜʜ ʇʜʜ ʇʜʜ /	16
399–429	ʇʜʜ ʇʜʜ ʇʜʜ	15	ʇʜʜ ʇʜʜ ʇʜʜ ʇʜʜ /	21
430–460	ʇʜʜ ʇʜʜ ʇʜʜ ///	18	ʇʜʜ ʇʜʜ ʇʜʜ	15
461–491	ʇʜʜ /	6	///	3
492–522	///	3	/	1
523–553	//	2		0

11.

Stems	Leaves
0	3 8 9 9
1	0 2 2 2 4 4 4 6 8 9
2	1 2 2 5 8 8
3	1 3 6 7
4	1 9
5	2 4 8

Analysis: Ten executives made 10–19 calls, while six made 21–28 calls.

13.

1	2
2	0 3
3	2 5 8 8 9
4	1 3 3
5	0 1 2 3 3 5 8 9 9

15. Measurements or observations that are gathered for an event under study are called data.

17. A population consists of all subjects under study; a sample is a representative subgroup, or subset, of the population.

19. Number each member of the population, and then select every kth member. The starting number, though, must be selected at random.

21. Take an intact group of subjects that represent the population.

23. (a) Cluster
(b) Systematic
(c) Random
(d) Systematic
(e) Stratified

25. Answers will vary.

Exercise Set 12-2

1.

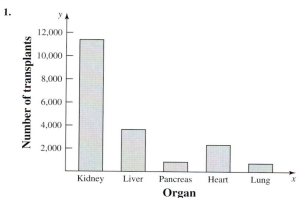

3.

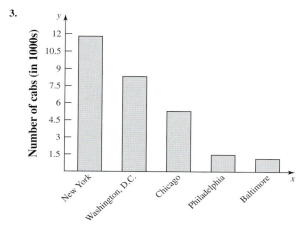

5.

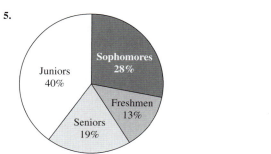

7.

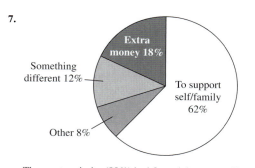

The great majority (80%) had financial reasons. About 1 out of 10 wanted to do "something different."

9.

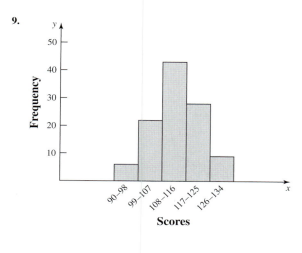

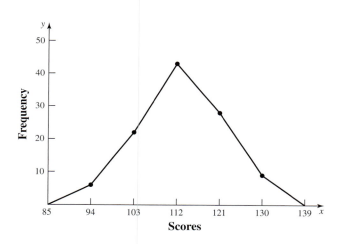

11.

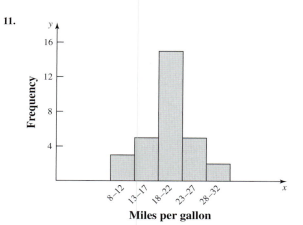

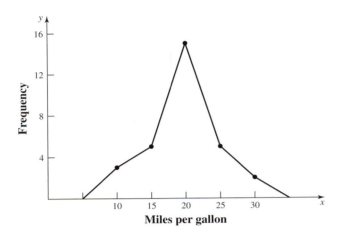

Miles per gallon

13.

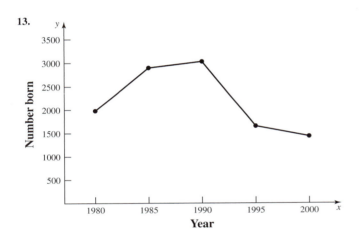

Year

15.

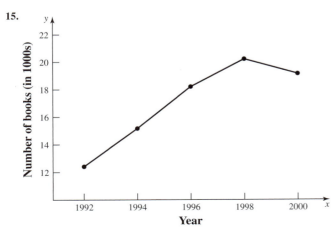

Year

17. *Relationship:* Both indicate the frequency for each class.

Difference: The histogram represents each entire class by means of rectangles/bars touching their neighbors. In the frequency polygon, we use the midpoints of each class, and connect the "frequency points" with straight segments.

19. (a) Time series graph

(b) Pie graph

(c) Bar graph

(d) Time series graph

(e) Piegraph

(f) Bar graph

Exercise Set 12-3

1. Mean = 15.11; MD = 7; mode = 3; MR = 31

3. Mean = 612,600; MD = 475,000; mode = none; MR = 820,000

5. (All answers are in billions.) mean = 539.55; MD = 550.5; mode = none; MR = 516

7. Mean = 189.6; MD = 151; mode = none; MR = 207.5

9. Mean = 43.92; MD = 45.5; mode = 51; MR = 44.5

11. 19.76

13. 4.4 seconds

15. $42.87 million

17. $180.28

19. Add all the values. Divide this sum by the total number of values. In the case of grouped data, use the formula $\overline{X} = \dfrac{\sum f \cdot X_m}{n}$, where f = frequency, X_m = midpoint of class, and n = sum of frequencies.

21. Arrange data set in order (for easier spotting of certain entries), although this is not essential. The mode = the values or value, if any, that occur(s) most often. If no such values exists, we say there is no mode.

23. (a) Mode   (b) Mode   (c) Median

Exercise Set 12-4

1. $R = 60$; $s^2 = 406.75$; $s = 20.17$; since there is one extremely high value (61), and since s takes into account all values, s is the best measure in this case.

3. $R = 1799$; $s^2 = 438{,}113.6$; $s = 661.90$

5. $R = 99$; $s^2 = 1288.19$; $s = 35.89$

7. $R = 10$; $s^2 = 9$; $s = 3$

9. $R = \$0.60$; $s^2 = 0.04$; $s = 0.2$ (or 0.21) depending on rounding

11. (Reminder: Each of the given numbers was in thousands.) $R = 1{,}120{,}000$; $s^2 = 183{,}128{,}933.3$; $s = 427{,}935.66$

13. Range, variance, and standard deviation

15. Because (a) the range uses only two of the values in the data set, and (b) an extremely large, and/or an extremely low, value can make the range very large—thus giving the impression of more variability than is actually the case.

17. Find the mean and subtract it from each value in the data set. Square each difference, and find the sum of all such differences. Divide this sum by $(n - 1)$, where n = number of values in data set. Take the square root of the quotient to obtain the standard deviation.

19. *Hint:* Find the standard deviation for each set.

Exercise Set 12-5

1. (a) 20th percentile
   (b) 75th percentile
   (c) 35th percentile
   (d) 10th percentile
   (e) 90th percentile
3. 75th percentile
5. (78.5 or 79)th percentile
7. 10
9. Bill, because he is at the 63rd percentile
11. (a) 60th percentile   (b) 8   (c) 23
13. $Q_1 = 22.5$; $Q_2 = 34$; $Q_3 = 53.5$
15. $Q_1 = 327$; $Q_2 = 398.5$; $Q_3 = 583$
17. A percentile is a percentage that equals the percent of data values lying below a certain given point.
19. No, they are not the same. The *numerical class rank* tells how far down the list, starting from the top, a person or object is (Thus, "first" gives rank 1, "second" gives rank 2, etc.). The *percentile rank* is the percentage of values lying below a certain point—that is, below a certain value—in the data set.
21. The 90th percentile would be higher.
23. About 50th percentile

Exercise Set 12-6

1. 0.476
3. 0.184
5. 0.154
7. 0.337
9. 0.080
11. 0.225
13. 0.464
15. 0.889
17. 0.973
19. The normal distribution is bell-shaped, unimodal, symmetrical about the mean, and continuous. Furthermore,
   (a) It never touches the x axis.
   (b) The area under the curve is 1, and is divided approximately as
      (i) 0.68 within 1 standard deviation of the mean
      (ii) 0.95 within 2 standard deviations of the mean
      (iii) 0.997 within 3 standard deviations of the mean
21. 1
23. +0.100
25. (a) ±1.96   (b) ±1.65   (c) ±2.58

Exercise Set 12-7

1. (a) $0.386 = 38.6\%$ (b) $0.084 = 8.4\%$
3. (a) $0.043 = 4.3\%$ (b) $0.054 = 5.4\%$
5. (a) $0.005 = 0.5\%$ (b) $0.165 = 16.5\%$ (c) $0.749 = 74.9\%$
7. (a) $0.153 = 15.3\%$ (b) $0.774 = 77.4\%$ (c) $0.187 = 18.7\%$
9. (a) $0.841 = 84.1\%$ (b) $0.067 = 6.7\%$
11. (a) $0.776 = 77.6\%$ (b) $0.405 = 40.5\%$
13. (a) $0.755 = 75.5\%$ (b) $0.811 = 81.1\%$ (c) $0.284 = 28.4\%$

15. (a) 638   (b) 184   (c) 1074   (d) 136
17. (a) 1   (b) 799   (c) 726   (d) 800
19. Many real-life situations, with a large and random population, closely resemble the normal distribution (that is, the theoretical one). Now, the mathematics-statistics of this distribution are well known; hence, certain conclusions or probabilities can be drawn from an appropriate real-life situation.
21. Plot a graph, and see if the graph has properties quite similar to those of the normal distribution. (In more advanced treatments, certain tests do exist for deciding.)
23. (Answers are approximate.) A—above 76, B—76–68, C—67–53, D—52–44, F—below 44
25. Approximately 55 minutes

Exercise Set 12-8

1. (a)

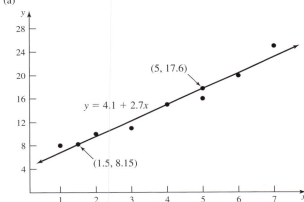

(b) $r = 0.977$
(c) r is significant at the 5% and the 1% level.
(d) See graph.
(e) There is a positive linear relationship.

3. (a)

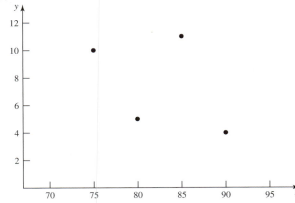

(b) $r = -0.441$
(c) r is not significant at 5% nor at 1%.
(d) Since r is not significant, the computing and drawing of a regression line would be meaningless.
(e) No relationship exists. The dots go "up and down."

5. (a)

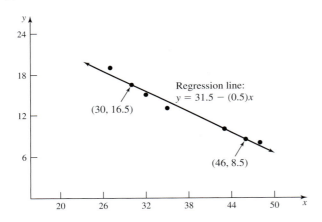

(b) $r = -0.983$

(c) r is significant at both the 5% level and 1% level.

(d) See graph.

(e) There exists a negative linear relationship.

7. (a)

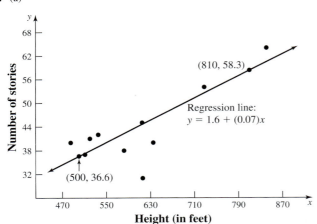

(b) $r = 0.798$

(c) r is significant at both the 5% level and the 1% level.

(d) See graph.

(e) There is a positive linear relationship, except at the lower-left portion of the plot.

When $x = 500$, y is predicted to be about 37.

9. (a)

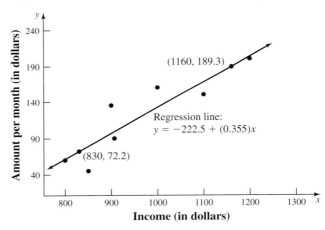

(b) $r = 0.896$

(c) r is significant at both the 5% level and the 1% level.

(d) See graph.

(e) There is a positive linear relationship.

When $x = \$925$, y is predicted to be $\$105.88$.

11. (a)

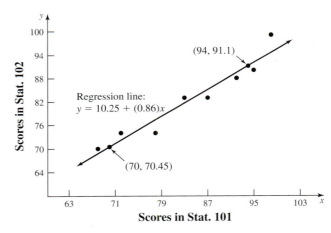

(b) $r = 0.963$

(c) r is significant at both the 5% level and 1% level.

(d) See graph.

(e) There is a positive linear relationship.

When $x = 90$, y is predicted to be 87.65.

13. Scatter plot

15. Generally, as x increases, so does y. The points would form a straight, or roughly straight, "stream" from lower left to upper right.

17. The values of r range from -1 to $+1$. At -1, a perfect negative linear relationship exists; at 0, no relationship exists; at $+1$, a perfect positive linear relationship exists.

19. Either $+1$ or -1

21. The value of r is the same.

Review Exercises

1.

Item	Tally	Frequency
B	////	4
F	⊬⊬⊬	5
G	⊬⊬⊬	5
S	⊬⊬⊬	5
T	⊬⊬⊬ /	6

3.

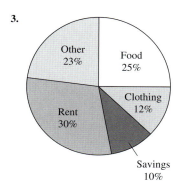

The greatest amount of money is spent for rent; then next is for food.

5.

Class	Tally	Frequency
102–116	////	4
117–131	///	3
132–146	/	1
147–161	////	4
162–176	⊬⊬⊬ ⊬⊬⊬ /	11
177–191	⊬⊬⊬ //	7

7.

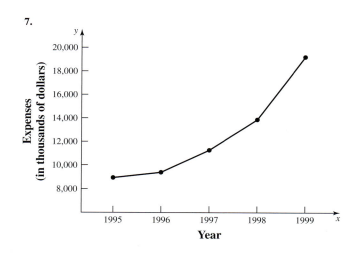

9. 7.25

11. $Q_1 = 59.5$; $Q_2 = 104.5$; $Q_3 = 154.5$

13. (a) $0.001 = 0.1\%$
    (b) $0.5 = 50\%$
    (c) $0.008 = 0.8\%$
    (d) $0.551 = 55.1\%$

15. 454

17.

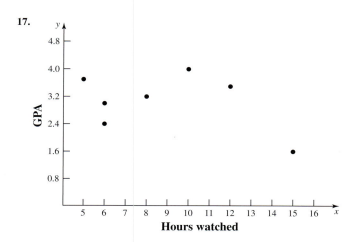

No discernable relationship; $r = -0.388$ is not significant at 5% level; no regression line is appropriate (since r is not significant), and no prediction for $x = 9$ is appropriate.

Chapter Test

1.

Source	Tally	Frequency
M	⊬⊬⊬ /	6
N	⊬⊬⊬ //	7
R	⊬⊬⊬ //	7
T	⊬⊬⊬	5

3.

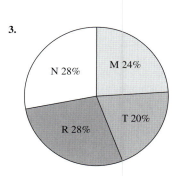

5.

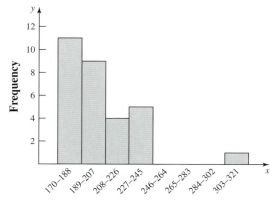

(d) 0.103

(e) 0.291

(f) 0.828

(g) 0.040

(h) 0.900

(i) 0.017

(j) 0.913

13. (a) $0.067 = 6.7\%$

(b) $0.023 = 2.3\%$

(c) $0.465 = 46.5\%$

(d) $0.094 = 9.4\%$

15.

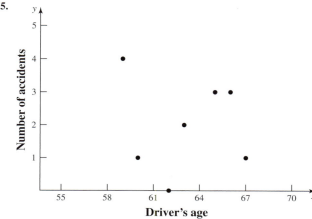

No relationship can be discerned from the scatter plot. $r = -0.078$ and is not significant at the 5% level; hence, no regression equation and no prediction are appropriate.

7.

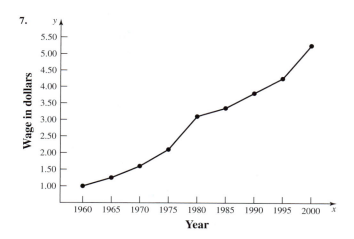

The graph shows an increase in the wage for all periods. The two steepest increases were during the jumps from 1975 to 1980 and from 1995 to 2000.

9. 6.4

11. (a) 0.433

(b) 0.394

(c) 0.034

Chapter Supplement

1. (a) The number of people (20) is too small of a sample when compared to the general population of, say, the state of Minnesota.

(b) How were the 20 subjects chosen? Did they represent the general population, or were they chosen from a specific group such as college students or others?

3. Various questions need to be answered:

(a) "Less traveled" in what respect?

(b) How many groups of 100 women were there?

(c) Were the women selected from the general population or from a specific group?

5. Perhaps the article's originator wanted to deceive readers into associating 11 with 18 (thus exceeding more than half), whereas 11% of 18 is only $1.98 \approx 2$. Moreover, the article should also address the presence or absence of side effects.

7. The ad does not state what the "74% more" is compared with, nor how the conclusion was reached.

9. The words "acid control" are vague. Is the control, say, 5%, 50%, or what? Does the word "control" mean neutralization of acid, and into what? Or, does it mean nonproduction of acid? If Brand X is indeed effective, are there side effects?

11. The two graphs do not have any labeling or any scales on the vertical axes. Hence, the apparent faster speed in the first graph is not necessarily appreciable, as the graph "suggests"; in fact, the two speeds might be very close.

13. In the second graph, the vertical distance to represent $1 is much bigger than in the first graph. Accordingly, "steepness" and "jumps" are much stronger in the second graph.

15. Disadvantages to the safety locks should be pointed out, such as: additional cost, additional time to operate the gun, etc.

17. Responders may not be telling the truth.

19. Better methods of detecting lead poisoning have been developed.

21. There are no factors given to make comparisons.

Chapter 13

Exercise Set 13-1

1.

Number of votes	5	4	8	5
First choice	X	X	Y	Z
Second choice	Y	Z	Z	Y
Third choice	Z	Y	X	X

(a) 22 (b) 4 (c) 8 (d) X

3.

Number of votes	9	4	5
First choice	P	C	M
Second choice	M	P	P
Third choice	C	M	C

(a) 18 (b) 9 (c) 4 (d) Philadelphia (P)

5. (a) 208 (b) Swimming pool (S)

7. (a) 20 (b) Carnations (C)

9. No. In a head-to-head comparison, S won over G and B.

11. No. In a head-to-head comparison, C won over R, G, and D.

13. A preference table is a summary of an election where candidates are ranked by voters as to first choice, second choice, etc.

15. The head-to-head comparison criterion states that if a particular candidate wins all head-to-head comparisons with all other candidates, then that candidate should win the election.

17. To assure winning the election, candidate A needs a total of 51 votes. Hence, to be assured of winning the election, candidate A needs 51 − 36 or 15 votes. However, candidate A can also win if all the remaining 20 votes go to candidate C since C would have a total of 32 votes. Other possibilities also exist.

19. No. If candidate C received all of the 20 remaining votes, he or she would only have a total of 32 votes and candidate A would win with 36 votes.

21. Yes

Exercise Set 13-2

1. The winner is science fiction (S).

3. The winner is *Anatomy of a Murder* (A).

5. (a) The winner is build a swimming pool (S).

   (b) Yes. The winner is the same as the one determined by the plurality method.

7. Yes, since no book type received the majority of first-place votes.

9. Yes, since no movie received the majority of first-place votes.

11. Yes, since no choice received the majority of first-place votes.

13. The winner is Professor Donovan (D).

15. (a) The winner is carnations (C).

   (b) The winner is the same when using the plurality method.

17. No

19. No

21. The Borda count method of voting requires that each candidate be ranked from most favorable to least favorable, and then 1 point is assigned to the last-place candidate, 2 points to the next-to-last place candidate, etc. The candidate with the most points is declared the winner.

23. The plurality-with-elimination method states that if a candidate has a majority of first-place votes, that candidate is declared the winner. If no candidate has a majority of first-place votes, the candidate with the least number of first-place votes is eliminated, and then another count is taken with each of the other candidates moving up. This continues until a candidate receives a majority of first-place votes.

25. Answers will vary.

27. Answers will vary.

29. Yes

Exercise Set 13-3

1. 6

3. 45

5. Steel center (S)

7. (a) There is a three-way tie.

   (b) The results are different since Professor Donovan won when using the plurality-with-elimination method.

9. (a) Rosa's Restaurant

   (b) Rosa's Restaurant also won when the Borda count method was used.

11. Yes, since Professor Williams (W) wins.

13. No, the winner is still Rosa's Restaurant (R).

15. Dr. Jones

17. Green

19. Inmate Z

21. Each candidate is ranked by the voters. Then each candidate is paired with every other candidate in a head-to-head contest. The winner of each contest gets 1 point. In case of a tie, each candidate gets a $\frac{1}{2}$ point. The candidate with the most points wins the election.

23. With approval voting, each voter gives one vote to as many candidates on the ballot as he or she finds acceptable. The votes are counted, and the winner is the candidate who receives the most votes.

25. Answers will vary.

27. Answers will vary.

Review Exercises

1.

Number of votes	5	5	5
First choice	Q	P	R
Second choice	R	Q	P
Third choice	P	R	Q

3. 5

5.

Number of votes	6	4	10
First choice	C	P	H
Second choice	P	C	P
Third choice	H	H	C

7. 6

9. 58

11. A

13. A

15. No

17. No

19. 47

21. *Music Man* (M)

23. *Music Man* (M)

25. Yes

27. No

29. Tie between a speaker and a rock band

Chapter Test

1.

Number of votes	1	7	4
First choice	A	B	C
Second choice	B	A	B
Third choice	C	C	A

3. 7

5. 100

7. Pittsburgh (P)

9. Pittsburgh (P)

11. No

13. No

15. Dr. Jones

Appendix A

Exercise Set A-1

1. 800 cm

3. 120 m

5. 6 hm

7. 900 dm

9. 3.756 m

11. 0.4053 km

13. 1,200,000 cm

15. 1,850,000 mm

17. 126,200 dm

19. 0.8 km

21. 0.5 L

23. 920 dL

25. 80 cL

27. 67 daL

29. 640 hL

31. 0.081 kL

33. 0.7 daL

35. 11.7 daL

37. 1.42 L

39. 0.032546 kL

41. 90 cg

43. 440 hg

45. 180 cg

47. 7.1 dg

49. 50 dg

51. 325 mg

53. 4,325,000 g

55. 0.086 g

57. 0.4 kg

59. 56.32 hg

Exercise Set A-2

1. 196.85 in.

3. 406.4 mm

5. 7.16 dam

7. 4429.13 ft

9. 15.24 mm

11. 19.69 ft

13. 4099.56 dm

15. 1.46 mi

17. 4.59 yd

19. 8.46 m

Exercise Set A-3

1. 117 cm^2

3. 50 yd^2

5. 0.13 km^2

7. 162 dm^2

9. 240 dm^2

11. 25,353.85 acres

13. 672 in.^2

15. 121.88 dm^2

17. 8.19 in.^2

19. 86.33 ft^2

Exercise Set A-4

1. 4.8 ounces

3. 0.106 pounds

5. 27,272.73 hectograms

7. 162.364 metric tons

9. 5.227 metric tons

11. 59,640 decigrams

13. 326 ounces

15. 51,240 decigrams

17. 122,727.27 decigrams

19. 18,636.36 grams

Exercise Set A-5

Note: Answers can vary, depending on rounding and on the conversion factors used.

1. 2872.32 fluid ounces
3. 53.48 cubic feet
5. 98.13 gallons
7. 164,560 gallons
9. 1616 gallons
11. 4500 cubic centimeters
13. 28.5 millimeters
15. 433 cubic centimeters
17. 32,000 cubic centimeters
19. 0.00032 cubic meters
21. 31,250 lb
23. 3.2 cubic feet

Exercise Set A-6

1. 57.2°F
3. 131°F
5. 302°F
7. −0.4°F
9. −27.4°F
11. −15°C
13. 0°C
15. 37.78°C
17. −23.33°C
19. −25.56°C

Index